The physiology of excitable cells

The physiology of excitable cells

THIRD EDITION

David J. Aidley

School of Biological Sciences
University of East Anglia, Norwich

CAMBRIDGE UNIVERSITY PRESS

Cambridge

New York Port Chester

Melbourne Sydney

Published by the Press Syndicate of the University of Cambridge
The Pitt Building, Trumpington Street, Cambridge CB2 1RP
40 West 20th Street, New York NY 10011, USA
10 Stamford Road, Oakleigh, Melbourne 3166, Australia

First published 1971
Reprinted 1973, 1974, 1975, 1976
Second edition 1979
Reprinted 1979, 1981, 1982, 1983, 1985, 1986, 1988
Third edition 1989

Printed in Great Britain by The Bath Press, Avon

British Library cataloguing in publication data

Aidley, D J (David John)
The physiology of excitable cells. – 3rd ed
1. Animals. Excitable cells. Physiology
I. Title
591.87′6

Library of Congress cataloguing in publication data

Aidley, D. J.
The physiology of excitable cells.
Bibliography: p.
Includes index.
1. Cell physiology. 2. Neurophysiology. I. Title.
QH631.A36 1989 596′.0188 88-34161

ISBN 0 521 30301 X (hardback)
ISBN 0 521 38863 5 (paperback)

BS

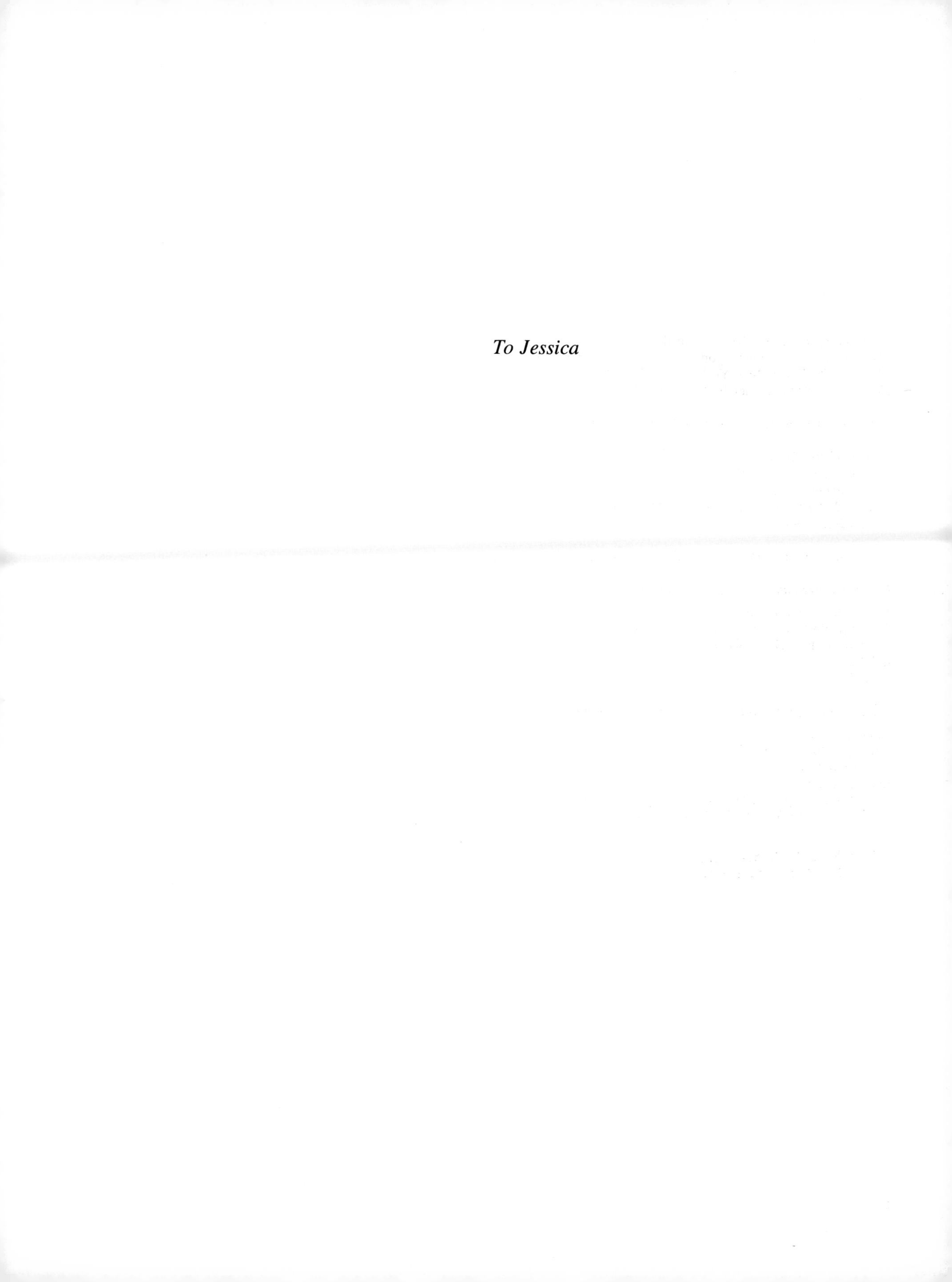

To Jessica

Contents

Preface

The physiology of nerve, muscle and sensory cells is a subject in which, as in many other fields in the biological sciences, there has been a great flowering of knowledge in recent decades. This book aims to help the reader to learn about the subject by giving an account of some of the experimental evidence on which our present knowledge is based. It is primarily intended for use by students taking courses in physiology, biophysics, neuroscience or zoology, but it may also prove useful to those beginning research and to scientists of related disciplines.

There have been many exciting advances in the subject matter of this book since its second edition was prepared. Two particular technical innovations have proved to be very fruitful: the patch clamp technique for examining the currents flowing through single membrane channels, and the application of recombinant DNA methods for determining the structures of membrane proteins. These and other advances are reflected in the content of this third edition, which has been extensively revised and restructured.

I am very grateful to those authors and publishers who have allowed me to reproduce here diagrams which originally appeared in their publications. The sources of each of these will be found in the reference list at the end of the book.

It is a pleasure to thank those of my colleagues who have read and criticized the text at various stages of its preparation. I would thus like to repeat my thanks to Drs Michael Brown and George Duncan, Professor Sir Alan Hodgkin, Dr Peter Miller, Professor Eduardo Rojas, and Drs Graham Shelton, Richard Tregear and David White for their help with the first and second editions, and to thank Drs Alan Coddington, Alan Dawson, George Duncan and Paul Taylor for their advice on some of the new material for this edition. Needless to say, the responsibility for such shortcomings as remain is entirely my own.

1
Introduction

Suppose a man has a tomato thrown at his head, and that he is able to take suitable evasive action. His reactions would involve changes in the activity of a very large number of cells in his body. First of all, the presence of a red object would be registered by the visual sensory cells in the eye, and these in turn would excite nerve cells leading into the brain via the optic nerve. A great deal of activity would then ensue in different varieties of nerve cell in the brain and, after a very short space of time, nerve impulses would pass from the brain to some of the muscles of the face and, indirectly, to muscles of the neck, legs and arms. The muscle cells there would themselves be excited by the nerve impulses reaching them, and would contract so as to move the body and so prevent the tomato having its desired effect. These movements would then result in excitation of numerous sensory endings in the muscles and joints of the body and in the organs of balance in the inner ear. The resulting impulses in sensory nerves would then cause further activity in the brain and spinal cord, possibly leading to further muscular activity.

A chain of events of this type involves the activity of a group of cell types which we can describe as 'excitable cells': a rather loose category which includes nerve cells, muscle cells, sensory cells and some others. An excitable cell, then, is a cell which readily and rapidly responds to suitable stimuli, and in which the response includes a fairly rapid electrical change at the cell membrane.

The study of excitable cells is, for a number of reasons, a fascinating one. These are the cells which are principally involved in the behavioural activities of animals: these are the cells with which we move and think. Yet just because their functioning must be examined at the cellular and subcellular levels of organization, the complexities that emerge from investigating them are not too great for adequate comprehension; it is frequently possible to pose specific questions as to their properties, and to elicit some of the answers to these questions by suitable experiments. It is perhaps for this reason that the subject has attracted some of the foremost physiologists of this century. As a consequence, the experimental evidence on which our knowledge of the physiology of excitable cells is based is often elegant, clearcut and intellectually exciting, and frequently provides an object lesson in the way a scientific investigation should be carried out. Nevertheless, there are very many investigations still to be done in this field, many questions which have yet to be answered, and undoubtedly very many which have not yet been asked.

Most readers of this book will possess a considerable amount of information on basic ideas in the biological and physical sciences. But it may be as well in this introductory chapter to remind them, in a rather dogmatic fashion, of some of the background which is necessary for a more detailed study: to formulate, in fact, a few axioms.

The biological material

Cells

All large organisms are divided into a number of units called cells, and every cell is the progeny of another cell. This statement constitutes the cell theory. Every cell is bounded by a cell membrane and contains a nucleus in which the genetic material is found. The main part of the living matter of the cell is a highly organized system called the cytoplasm, which is concerned with the day-to-day activity of the cell. The cell membrane separates this highly organized system inside from the relative chaos that exists outside the cell. In order to maintain and increase its high degree of organization and in order to respond to and alter its environment, the cell requires a continual supply of energy. This energy must be ultimately derived

Table 1.1. *The amino acids found in proteins. They have the general formula R—CH(NH$_2$)COOH, where R is the side chain or residue. Proline is actually an imino acid. Cystine is two cysteines linked by a disulphide bridge. Abbreviations are given in three- and one-letter codes*

Type	Amino acid	Side chain	Abbreviations		Hydropathy index[a]
Nonpolar	Isoleucine	$—CH(CH_3)CH_2.CH_3$	Ile	I	4.5
	Valine	$—CH(CH_3)_2$	Val	V	4.2
	Leucine	$—CH_2.CH(CH_3)_2$	Leu	L	3.8
	Phenylalanine	$—CH_2.C_6H_5$	Phe	F	2.5
	Methionine	$—CH_2.CH_2.SCH_3$	Met	M	1.9
	Alanine	$—CH_3$	Ala	A	1.8
	Tryptophan	$—CH_2$ —CH$_2$— (indole ring, N H)	Trp	W	—0.9
	Proline	$—CH_2.CH_2.CH_2—$	Pro	P	—1.6
Uncharged Polar	Cysteine/cystine	$—CH_2.SH$	Cys	C	2.5
	Glycine	—H	Gly	G	—0.4
	Threonine	$—CH(OH)CH_3$	Thr	T	—0.7
	Serine	$—CH_2OH$	Ser	S	—0.8
	Glutamine	$—CH_2.CH_2.CO.NH_2$	Gln	Q	—3.5
	Asparagine	$—CH_2.CO.NH_2$	Asn	N	—3.5
Acidic	Aspartic acid	$—CH_2.COO^-$	Asp	D	—3.5
	Glutamic acid	$—CH_2CH_2COO^-$	Glu	E	—3.5
Basic	Histidine	$—CH_2$ —CH$_2$— (imidazole ring, HN NH$^+$)	His	H	—3.2
	Lysine	$—(CH_2)_3.NH_3^+$	Lys	K	—3.9
	Arginine	$—(CH_2)_3NH.C(NH_2){=}NH_2^+$	Arg	R	—4.5

[a]The hydropathy index is from Kyte and Doolittle (1982), and referred to in chapter 3 and elsewhere.

from the environment, usually in the form of chemical energy such as can be extracted by the cell from glucose molecules. Thus, thermodynamically speaking, the cell is an open system maintained in a rather improbable steady state by the continual expenditure of energy. Its life is a continual battle against the second law of thermodynamics (which we may state without gross inaccuracy as 'things tend to get mixed up').

The cells of nervous systems are called *neurons*. Their primary function is the carriage of information in the form of changes in the electrical potential across the cell membrane, especially as unitary events known as nerve impulses or *action potentials*. The idea that the nervous system is composed of discrete cells is known as the neuron theory. This theory, which is merely a particular application of the cell theory, was developed during the nineteenth century, and is now generally accepted. The alternative proposal, that nervous systems are not divided into separate membrane-bounded entities (the reticular theory) was extremely difficult to reconcile with the observations of light microscopists, and seems to be conclusively refuted by the evidence of electron microscopy.

Proteins

So pervasive are the functions of proteins in cells that one way of defining living material is to say that it contains active proteins. Proteins are composed of chains made up from different combinations of twenty different amino acids, and their properties depend critically upon the sequence in which these amino acids are arranged.

The protein's amino acid sequence is specified by the nucleotide base sequence in the DNA molecules which form the genetic material of the cell. This means that proteins are the products of evolution, so that present-day proteins largely represent stable and successful sequences. Animals whose cells produced too many unstable or non-functional sequences would have died before producing viable offspring, so the genes specifying those sequences have been largely eliminated.

The twenty amino acids in protein chains have different properties. Three of them are basic, with positive charges, two are acidic, with negative charges, seven are polar but uncharged, and eight are non-polar (table 1.1). The amino acid sequence determines the way in which the protein chain is folded. Non-polar residues tend to occur in the middle of the molecule or in association with lipids in the cell membrane. Polar and charged residues are more likely to be found on the outside of the molecule in contact with the aqueous environment. Hydrogen bonds, electrostatic interactions and disulphide links (between pairs of cysteine residues) all serve to hold the chain in its folded conformation.

The shape of many protein molecules changes when they react with smaller molecules or other proteins. Changes of this type underly much protein activity, such as enzymic hydrolysis, opening of membrane channels, and muscular contraction. The day to day activity of the cell can thus be largely described in terms of the actions of proteins; the reader of modern accounts of cell biology (such as those by Alberts *et al.*, 1983, and Darnell *et al.*, 1986) will find ample illustration of this statement.

Animals

Every animal has a history: every animal owes its existence to the success of its ancestors in combating the rigours of life, that is to say, in surviving the rigours of natural selection. Hence every animal is adapted to its way of life, and its organs, its tissues and it cells are adapted to performing their functions efficiently. There is an enormous variety in the form and functioning of animals, connected with a similar variety in ways of living. Nevertheless, it seems that in many cases there is only a limited number of ways in which animals can solve particular physiological problems. For instance, there is only a limited number of respiratory pigments, only a limited number of designs for hearing organs, and so on. Hence it is possible for us to detect certain principles in, and make certain generalizations about, the ways in which particular physiological problems are solved in different animals.

An animal is a remarkably stable entity. It is able to survive the impact of a variety of different environments and situations, and its cells and tissues are able to survive a variety of different demands upon their capacities. The main reason for this seems to be that an animal is a complex of self-regulating (homeostatic) systems. These systems are themselves coordinated and regulated so that the physiology and behaviour of the animal form an integrated whole.

Nervous systems

A nervous system is that part of an animal which is concerned with the rapid transfer of information through the body in the form of electrical signals. The activity of a nervous system is initiated partially by the input elements – the sense organs – and partially by endogenous activity arising in certain cells of the system. The output of the system is ultimately expressed via effector organs – muscles, glands, chromatophores, etc.

Primitive nervous systems consist of scattered but usually interconnected nerve cells, forming a nerve net, as in coelenterates. Increase in the complexity of responses is associated with the aggregation of nerve cell bodies to form ganglia, and when the ganglia themselves are collected and connected together, we speak of a *central nervous system.* The *peripheral nervous system* is then mainly composed of nerve fibres originating from the central nervous system. Peripheral nerves contain afferent (sensory) neurons taking information inwards into the central nervous system and efferent (motor) neurons taking information outwards; neurons confined to the central nervous system are known as interneurons. Ganglia which remain or arise outside the central nervous system, and the nerve fibres which lead to and arise from them to innervate the animal's viscera, are fre-

Table 1.2. *Some electrical units*

Quantity	SI unit	Symbol	SI equivalent
Current	ampere	A	—
Charge	coulomb	C	A s
Potential difference	volt	V	$J\ C^{-1}$
Resistance	ohm	Ω	$V\ A^{-1}$
Conductance	siemens	S	Ω^{-1}
Capacitance	farad	F	$C\ V^{-1}$

Table 1.3. *Some prefixes for multiples of scientific units*

Multiple	Prefix	Symbol
10^{-2}	centi	c
10^{-3}	milli	m
10^{-6}	micro	μ
10^{-9}	nano	n
10^{-12}	pico	p
10^{-15}	femto	f
10^{3}	kilo	k
10^{6}	mega	M
10^{9}	giga	G

quently described as forming the *autonomic nervous system*.

One of the simplest, but possibly not one of the most primitive, modes of activity of a nervous system is the *reflex*, in which a relatively fixed output pattern is produced in response to a simple input. The stretch reflexes of mammalian limb muscles provide a well-known example (fig. 9.3). Stretching the muscle excites the endings of sensory nerve fibres attached to certain modified fibres (muscle spindles) of the muscle. Nerve impulses pass up the sensory fibres into the spinal cord where they meet motor nerve cells (the junctional regions are called *synapses*) and excite them. The nerve impulses so induced in the motor nerve fibres then pass out of the cord along peripheral nerves to the muscle, where their arrival causes the muscle to contract. Much more complicated interactions occur in the analysis of complex sensory inputs, the coordination of locomotion, the expression of the emotions and instinctive reactions, in learning and other 'higher functions'. These more complicated interactions are outside the scope of this book.

Electricity

Matter is composed of atoms, which consist of positively charged nuclei and negatively charged electrons. Static electricity is the accumulation of electric charge in some region, produced by the separation of electrons from their atoms. Current electricity is the flow of electric charge through a conductor. Current flows between two points connected by a conductor if there is a potential difference between them, just as heat will flow from a hot body to a cooler one placed in contact with it. The unit of potential difference is the *volt*. The current, i.e. the rate of flow of charge, is measured in *amperes*, and the quantity of charge transferred is measured in *coulombs*. Thus one coulomb is transferred by a current of one ampere flowing for one second.

In many cases it is found that the current (I) though a conductor is proportional to the potential difference (V) between its ends. This is *Ohm's law*. Thus if the constant of proportionality, the *resistance* (measure in *ohms*) is R, then

$$V = I \cdot R.$$

The specific resistance of a substance is the resistance of a 1 cm cube of the substance. The resistance of a wire of constant specific resistance is proportional to its length and inversely proportional to its cross-sectional area. The reciprocal of resistance is called *conductance* (G).

Let us apply Ohm's law to a simple calculation. In chapter 6 we shall see that under certain conditions small channels open to let sodium ions flow through. If we can measure this current flow and we know what the driving voltage is, we can calculate the conductance of the channel. Thus in one experiment the single channel current was 1.6 pA with a driving voltage of 90 mV. (Table 1.2 shows selected electrical units and table 1.3 gives prefix names for multiples and submultiples.) Applying Ohm's law, the conductance of the channel is given by

$$G = I/V$$

i.e.

$$\text{conductance (siemens)} = \frac{\text{current (amps)}}{\text{voltage (volts)}}$$

$$= \frac{1.6 \times 10^{-12}}{90 \times 10^{-3}}$$
$$= 17.8 \text{ pS.}$$

The total resistance of a number of resistive elements arranged in series is the sum of their individual resistances, whereas the total conductance of a number of elements in parallel is the sum of their conductances. A patch of membrane containing five channels each with a conductance of 17.8 pS, for example, will have a conductance of 89 pS if all the channels are open.

Two plates of conducting material separated by an insulator form a capacitor. If a potential difference V is applied across the capacitor, a quantity of charge Q, proportional to the potential difference, builds up on the plates of the capacitor. Thus

$$Q = V \cdot C$$

where C, the constant of proportionality, is the *capacitance* of the capacitor. When the voltage is changing, charge flows away from one plate and into the other, so that we can speak of current, I, through a capacitor, given by

$$I = C \cdot \mathrm{d}V/\mathrm{d}t$$

where $\mathrm{d}V/\mathrm{d}t$ is the rate of change of voltage with time. The capacitance of a capacitor is proportional to the area of the plates and the dielectric constant (a measure of the ease with which the molecules of a substance can be polarized) of the insulator between them, and inversely proportional to the distance between the plates. The total capacitance of capacitors in parallel is the sum of the individual capacitances, whereas the reciprocal of the total capacitance of capacitors in series is the sum of the reciprocals of their individual capacitances.

Scientific investigation

Science is concerned with the investigation and explanation of the phenomena of the natural world. Any particular investigation usually starts with an idea – a hypothesis – about the relations between some of the factors in the system to be studied. The hypothesis must then be tested by suitable observations or experiments. This business of testing the hypothesis is what distinguishes the scientific method from other attempts at the acquisition of knowledge, and hence it follows that a scientific hypothesis must be capable of being tested. We must therefore understand what is meant by 'testing' a hypothesis.

In mathematics and deductive logic it is frequently possible to prove, given a certain set of axioms, that a certain idea about a particular situation is true or not true. For instance, it is possible to prove absolutely conclusively that, in the system of Euclidean geometry, the angles of an equilateral triangle are all equal to one another. But this absolute proof of the truth of an idea is not possible in science. For example, consider the hypothesis 'No dinosaurs are alive today'. This statement would be generally accepted by biologists as being almost certainly true, but, of course, it is just possible that there are some dinosaurs alive which have never been seen. Some years ago the statement 'No coelacanths are alive today' would also have been accepted as almost certainly correct.

However, in many cases, it *is* possible to prove that a hypothesis is false. The hypothesis 'No coelacanths are alive today' has been proved, conclusively, to be false. If we were to find just one living dinosaur, the hypothesis 'No dinosaurs are alive today' would also have been shown to be false. It follows from this argument that in order to test a hypothesis it is necessary to attempt to disprove it. When a hypothesis has successfully survived a number of attempts at disproof, it seems more likely that it provides a correct description of the situation to which it applies (Popper, 1963).

If we can only test a hypothesis by attempting to disprove it, it follows that a scientific hypothesis must be formulated in such a way that it is open to disproof – so that we can think of an experiment or observation in which one of the possible results would disprove the hypothesis. Any idea which we cannot see how to disprove is not a scientific hypothesis.

But where do the ideas come from? Science is a progressive activity. Advances are usually made step by step. Ideas arise in a controlled imagination: the scientist usually starts from a generally accepted understanding of the situation and makes a small conjecture into the unknown. A high rate of progress follows two particular types of advance: ideas which provide a major reinterpretation of what we know, and new techniques. In 1954, for example, as we shall see in chapter 12, the study of muscular contraction entered a new and highly productive phase as the result of the formulation of of the sliding filament theory, which itself arose in the context of advances in X-ray diffraction methods and electron microscopy. More recently, the advent of the patch clamp technique (chapter 6) has led to a great flowering of work on the ionic channels of cell membranes.

What implications does the nature of science have for learning about science? To be worth his salt, any student must get to grips with the intellectual credentials of his subject. For the science student, this implies that simply comprehending a proposition that we believe to be true is not enough. It is also necessary to understand why we believe it to be true, what the evidence for the proposition is, and hence what sort of evidence might lead us to revise our opinion about it.

It is for this reason that this book is much concerned with experiments and observations, and not simply with the understanding that has arisen from them. The conclusions from some of these experiments will stand the tests of further investigations in the future, those of others will have to be revised. The science student cannot hope to know everything about his subject, but if he understands just why he believes what he does, then he can look the future in the face.

2
Electrophysiological methods

Excitable cells can be and are studied by the great variety of techniques that are available for the study of cells in general. These include light and electron microscopy, X-ray diffraction measurements, experiments involving radioactive tracers, cell fractionation techniques, biochemical methods, and so on. We shall not deal with these methods here. The techniques which are particular to the study of excitable cells are those involving the measurement of rapid electrical events. Consequently, in order to understand the subject, it is necessary to have some idea of how these measurements are made. In this chapter we shall look briefly at some of the more general methods used. The powerful techniques of voltage clamping and patch clamping are introduced in later chapters. Duncan (1989) gives more detail on a variety of electrical measurement techniques.

Recording electrodes

If we wish to record the potential difference between two points, it is necessary to position electrodes at those points and connect them to a suitable instrument for measuring voltage. It is desirable that these electrodes should not be affected by the passage of small currents through them, i.e. that they should be non-polarizable. For many purposes fine silver wires are quite adequate. Slightly better electrodes are made from platinum wire or from silver wire that has been coated electrolytically with silver chloride. For very accurate measurements of steady potentials, calomel half-cells (mercury/mercuric chloride electrodes) may have to be used.

If the site we wish to record from is very small in size (such as occurs in extracellular recording from cells in the central nervous system), the electrode must have a very fine tip, and be insulated except at the end. Successful electrodes of this type have been made from tungsten wire which is sharpened by dipping it into a solution of sodium nitrate while current is being passed from the electrode into the solution; insulation is produced by coating all except the tip of the electrode with a suitable varnish.

The manufacture of a suitable electrode is rather more difficult if we wish to record the potential inside a cell. Apart from a few large cells such as squid giant axons, this necessitates the use of an electrode which is fine enough to penetrate the cell membrane without causing it any appreciable damage. The problem was solved with the development of glass capillary microelectrodes by Ling and Gerard in 1949. These are made on a suitable device which will heat a small section of a hard glass tube and then very rapidly extend it as it cools. The heated section gets thinner and cooler as the pulling proceeds, so that finally the pulling force exceeds the cohesive forces in the glass, and the tube breaks to give two micropipettes. If the machine designed to do this has been correctly adjusted, the outside diameter of the micropipette at its tip will be about 0.5 μm.

The micropipette now has to be filled with a strong electrolyte solution such as 3 M potassium chloride solution. This is most readily done if the glass tube from which the electrode is made has a fine filament of glass fused to its inner wall; the angle between the filament and the wall leads to high capillary forces, so that the pipette can be filled to its tip by injecting the solution into the barrel. The connection to the recording apparatus is made via a non-polarizing electrode such as a silver wire coated with silver chloride.

An electrode of this type has a very high resistance, from five to several hundred megohms; in fact the suitability of an electrode is usually tested by measuring its resistance, since the tip is too small to be examined satisfactorily by light microscopy. Further details of microelectrode methods are given in Standen *et al.* (1987).

Electronic amplification

The potential differences which are measured in investigations on the activity of excitable cells vary in size from just over 0.1 V down to as little as 20 μV or so. Before being measured by a recording instrument, these voltages usually have to be amplified. This is done by means of suitable electronic circuits involving thermionic valves or transistors, the details of which need not concern us. However, there are three aspects of any amplifier used for electrophysiological recording purposes which we must consider: the frequency response, the noise level and the input resistance.

All amplifiers show a higher gain (the gain is the ratio of the output to the input voltages) at some frequencies than at others. A typical amplifier for use with extracellular electrodes might have a constant gain over the range 10 Hz–50 kHz, the gain falling at frequencies outside this range (fig. 2.1). An amplifier of this type is known as an a.c. amplifier, since it measures voltages produced by alternating current. A d.c. amplifier is one which can measure steady potential differences, i.e. its frequency response extends down to zero hertz. If we wish to measure steady potentials, or slow potential changes without distortion, then obviously we must use a d.c. amplifier. Amplifiers used with intracellular electrodes are usually d.c. amplifiers, so that the steady potential difference across the cell membrane can be measured, as well as the rapid (high-frequency) changes involved in its activity.

Any amplifier will produce small fluctuations in the output voltage even when there is no input signal. These fluctuations are known as *noise*, and are caused by random electrical activity in the amplifier. The existence of noise sets a lower limit to the signal voltage that can be measured, since it is difficult to distinguish very small signals from the noise. Consequently it is necessary to use an amplifier with a low noise level if we wish to measure signals of very small size.

If we connect a potential difference across the input terminals of an amplifier, a very small current flows between them, which is proportional to the potential difference. The proportionality factor is called the *input resistance* of the amplifier. It is determined by application of Ohm's law and is measured in ohms; for instance, the input resistance of a cathode ray oscilloscope amplifier is usually about 1 MΩ. Now suppose we connect an intracellular microelectrode whose resistance is, say, 10 MΩ to an amplifier with an input resistance of 1 MΩ. The equivalent circuit is shown in fig. 2.2*a*. The two resistances form a potential divider, so that the voltage input to the amplifier is only $10^6/10^6 + 10^7)$ i.e. one-eleventh of the signal voltage. Obviously this is of little use for measuring the signal voltage. Hence, when using high-resitance electrodes, it is necessary to use an input stage with a very high input resistance, such as is given by the use of a junction field effect transistor. Suppose the input resistance of such a device is 1000 GΩ (fig. 2.2*b*); then the voltage recorded is $10^{12}/(10^{12} + 10^7)$, which is effectively equal to the signal voltage.

Fig. 2.1. The frequency response of an a.c. amplifier.

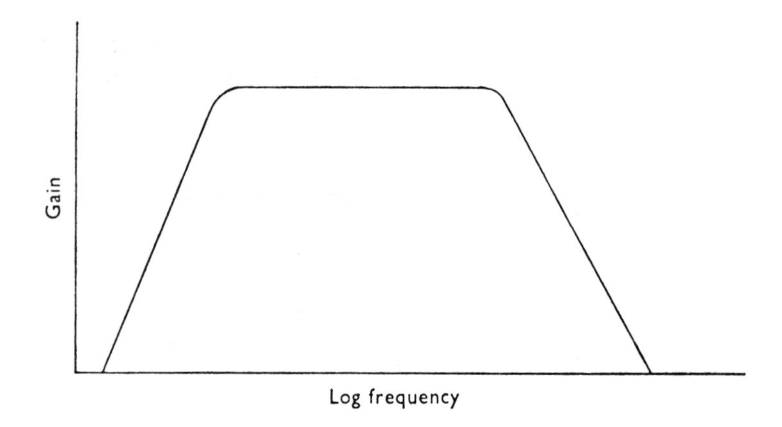

The cathode ray oscilloscope

We now have to consider how the voltage output from an amplifier is to be recorded and measured. If the voltage is steady, or only changing very slowly, we could use an ordinary voltmeter in which the current through a coil of wire placed between the poles of a magnet causes a pointer on which the coil is mounted to move. However, a device of this nature is no use for measuring rapid electrical changes since the inertia of the pointer is too great. What we need, in effect, is a voltmeter with a weightless pointer. This is provided by the beam of electrons in the cathode ray tube of a cathode ray oscilloscope.

The cathode ray tube is an evacuated glass tube containing a number of electrodes and a screen, which is coated with phosphorus compounds so that it luminesces when and where it is bombarded by electrons, at one end (fig. 2.3). The cathode is heated and made negative (by about 2000 V) to the anode; consequently electrons are emitted from the cathode and accelerate towards the anode. Since the anode has a hole in its centre, some of the electrons continue moving at a constant high velocity beyond it; these form the electron beam. The intensity and focusing of the beam can be controlled by other electrodes, not shown in fig. 2.3, placed in the vicinity of the anode and cathode. When the electron beam hits the screen, it produces a spot of light at the point of impact. Between the anode and the screen, the beam passes between two pairs of plate electrodes placed at right angles to each other, known as the *X* and *Y* plates. If a potential difference is connected across one of these pairs, the electrons in the beam will move towards the positive plate. Thus when the electrons pass out of the electric field of the pair of plates, their direction will have been changed, and so the light spot on the screen will move, by an amount proportional to the potential difference between the plates. The *Y* plates are connected to the output of an amplifier whose input is the signal voltage to be measured; hence the signal voltage appears as a vertical deflection of the spot on the screen. The *X* plates are usually connected to a waveform generating circuit (the 'time-base' generator) which produces a sawtooth waveform. This sawtooth waveform thus moves the spot horizontally across the screen at a constant velocity, flying back and starting again at the end of each 'tooth'. As a consequence of these arrangements, the spot on the screen traces out a graph with signal voltage on the *y*-axis and time on the *x*-axis. By making the rise-time of the time-base sawtooth sufficiently fast, it is possible to measure the form of very rapid voltage changes.

Many oscilloscopes have tubes with two beams, each with separate *Y* plates and amplifiers, so that one can measure two signals at once. The time-base unit can frequently be arranged so that a single sawtooth wave, leading to a single sweep of the beam, can be initiated (or 'triggered') by some suitable electrical signal; this facility is essential for much electrophysiological work. In some cases it is possible to connect the *X* plates to another input amplifier, instead of to the time-base generator, so that a signal related to some quantity other than time is measured on the *x*-axis.

On the screen of a standard oscilloscope the waveform traced out by the electron beam persists for only a fraction of a second. A permanent record of the trace can be obtained by photographing it; many of the diagrams in this book are photographs of oscilloscope traces.

Storage oscilloscopes are devised so that the trace can be held on the screen for some time.

Fig. 2.2. Equivalent circuits of a glass capillary microelectrode whose resistance is 10 MΩ connected to *a*: an 'ordinary' amplifier with an input resistance of 1 MΩ, and *b*: an input stage with an input resistance of 1000 GΩ.

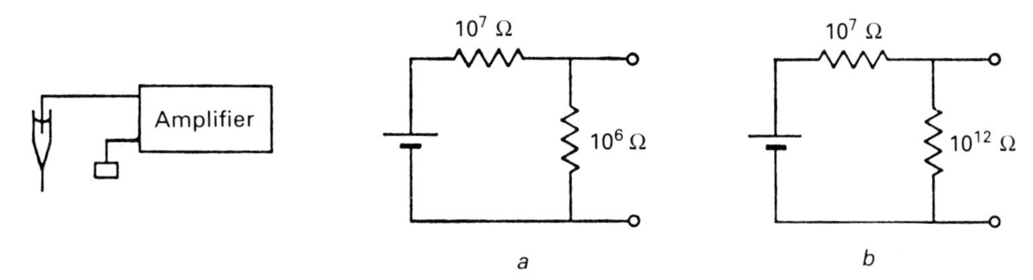

Analogue storage devices depend upon rather expensive modifications to the cathode ray tube. Digital storage oscilloscopes store the trace in memory, from which it is continually read out and displayed on the screen. An advantage of this type of instrument is that the memory can be read out into other devices, such as a chart recorder for hard-copy production.

Oscilloscopes are always easier to use if one can have a second or third look at what one wishes to display, so it is often useful and sometimes essential to be able to store the recorded signals in some permanent form. Such immediate data storage can be supplied by magnetic tape (using a frequency modulated tape recorder) or a computer memory.

Electrical stimulation

An electrical stimulus must be applied via a pair of electrodes. Stimulating electrodes may be of any of the types previously described for recording purposes. The simplest way of providing a stimulating pulse is to connect the electrodes in series with a battery and a switch, but this is not satisfactory if brief pulses are required. In the past, stimulating pulses have been produced by such means as discharge of condensors or by using an induction coil, but nowadays most investigators use electronic stimulators which produce 'square' pulses of constant voltage, beginning and ending abruptly. A good stimulator unit will be able to produce pulses which can be varied in strength, duration and frequency. We need not be concerned here with the design of these instruments (see Young, 1973).

Fig. 2.3. Simplified diagram to show the principal components of a cathode ray oscilloscope.

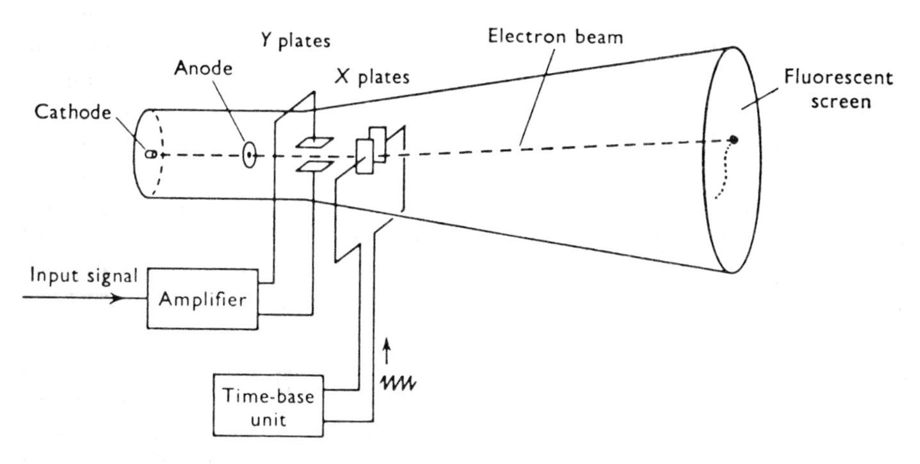

3
The resting cell membrane

If an intracellular microelectrode is inserted into a nerve or muscle cell, it is found that the inside of the cell is electrically negative to the outside by some tens of millivolts. This potential difference is known as the *resting potential*. If we slowly advance a microelectrode so that it penetrates the cell, the change in potential occurs suddenly and completely when the electrode tip is in the region of the cell membrane; thus the cell membrane is the site of the resting potential. In this chapter we shall consider some of the properties of the cell membrane that are associated with the production of the resting potential.

The structure of the cell membrane

Plasma membranes are usually composed of roughly equal amounts of protein and lipid, plus a small proportion of carbohydrate. Human red cell membranes, for example, contain about 49% protein, 44% lipid and 7% carbohydrate. Intracellular membranes tend to have a higher proportion of protein, whereas the protein content of myelin (p. 44) is only about 23%.

Fig. 3.1 shows the chemical structure of some membrane lipids. Phospholipid molecules are esters of glycerol with two long-chain fatty acids, with the glycerol moiety attached via a phosphate group to various small molecules. The fatty acid chains thus form non-polar tails attached to polar heads. The fatty acid chains are usually 14–20 carbon atoms long, and some of them are unsaturated, with one or more double bonds in the chain. Glycolipids have one or more sugar residues attached to hydrocarbon chains. Cholesterol is a sterol.

When lipids are spread on the surface of water, they form a monolayer in which the polar ends of the molecules are in contact with the water surface and the non-polar hydrocarbon chains are oriented more or less at right angles to it. This monolayer can be laterally compressed until the lipid molecules are in contact with each other; at this point the lateral pressure exerted by the monolayer has reached a maximum, as can be seen by the use of a suitable surface balance. Using such a balance, Gorter and Grendel (1925) measured the minimum area of the monolayer produced by the lipids extracted from red blood cells, and compared it with the surface area of the cells. They found that the monolayer area was almost double the surface area of the cells, and concluded that the lipids in the cell membrane are arranged in a layer two molecules thick. (Later work showed that this result was somewhat fortuitous in that an underestimate of the surface area seems to have been compensated for by incomplete extraction of the lipids; see Bar *et al.*, 1966). Davson and Danielli (1943) later suggested that this bimolecular layer was stabilized by a thin layer of protein molecules on each side of the lipid layer.

The idea that the essential barrier to movement of substances across the cell membrane is a layer of lipid molecules receives some corroboration from the observation that lipid-soluble substances appear to penetrate the cell membrane more readily than many non-lipid-soluble substances. The electrical capacitance of cell membranes is usually about 1 μF cm^{-2}; this is what one would expect if the membrane were a bimolecular layer of lipid 50 Å thick with a dielectric constant of 5.

High-resolution electron microscopy of sections of cells (usually fixed with potassium permanganate) shows that the cell membrane appears as two dense lines separated by a clear space, the whole unit being about 75 Å across (see Robertson, 1960). This accords well with Davson and Danielli's model, since one would expect electron-dense stains to be taken up by the polar groups of the lipid molecules and the proteins associated with them, but not by the non-polar lipid chains in the middle of the membrane. X-ray diffraction studies (Wilkins *et al.*, 1971) fit well with the bilayer hypothesis, giving a distance of 45 Å between the polar head groups in erythrocyte membranes. This is about what we would expect from the dimensions of phospholipid molecules (fig. 3.2).

About this time it became clear that the Davson–Danielli model, with its thin layer of protein on each side of the lipid bilayer, did not really fit the facts. Wilkins *et al.* calculated that in red cell membranes there was insufficient lipid to account for the observed thickness of the membrane, and suggested that part of its area was occupied by protein. Enzymes which would hydrolyse phospholipids were able to attack cell membranes, suggesting that the phospholipids were not protected by an overlying layer of protein. But perhaps the most graphic evidence comes from a special technique of electron microscopy known as freeze-fracture. A portion of tissue is frozen and then broken with a sharp knife. The cell membranes then cleave along the middle so as to separate the inner and outer lipid leaflets. A replica of the fractured face is made, 'shadowed' with some electron dense material and then examined in the electron

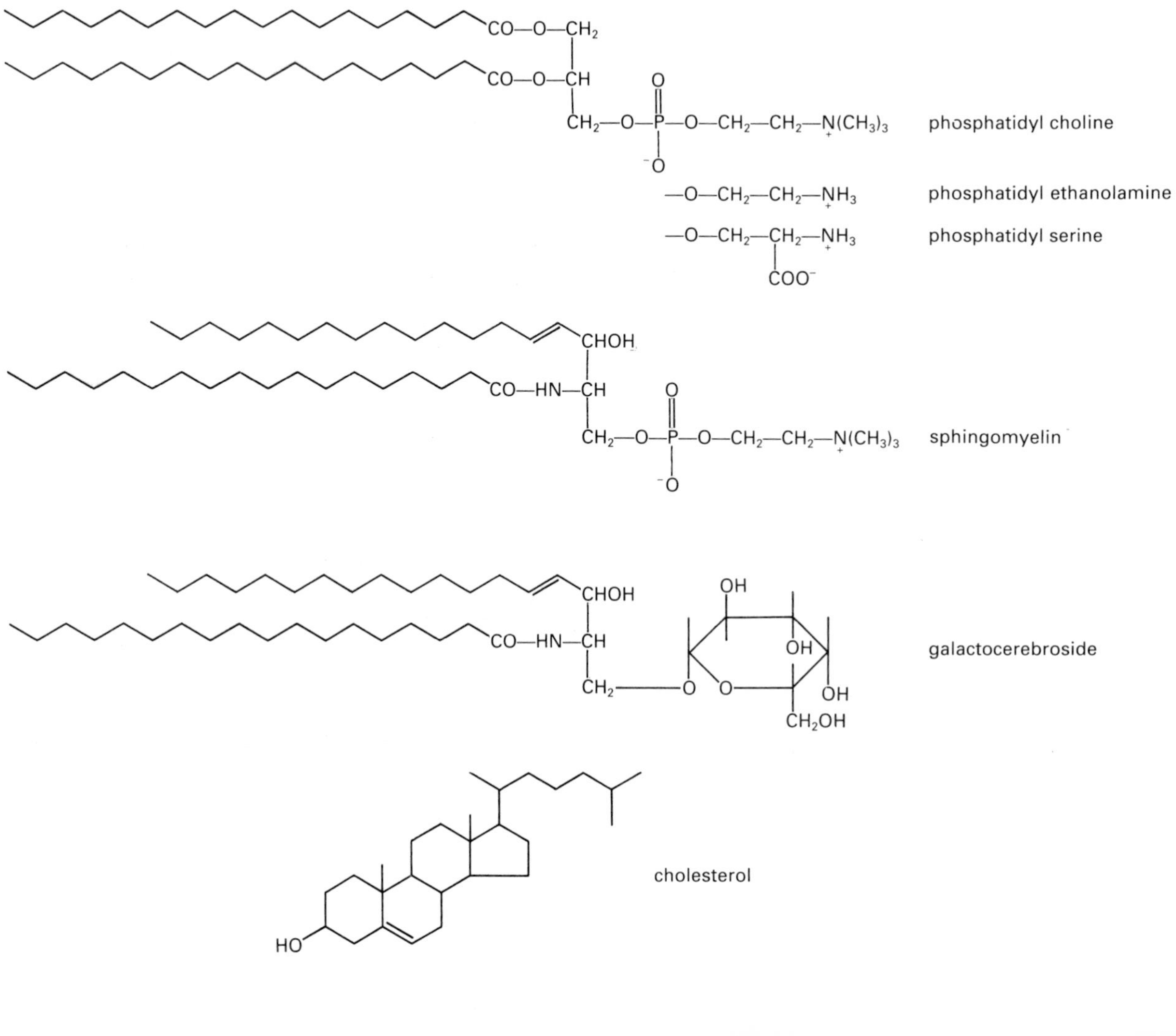

Fig. 3.1. Chemical formulae of some plasma membrane lipids. The fatty acid chains on the left of the formulae may be of variable length and may have one or more double bonds.

phosphatidyl choline
phosphatidyl ethanolamine
phosphatidyl serine
sphingomyelin
galactocerebroside
cholesterol

microscope. Small particles are seen projecting from both faces, but especially from the inner one; an example is shown in fig. 7.25*a*. If these are protein molecules (and it is difficult to see what else they could be) then they are clearly embedded within the lipid bilayer, rather than simply applied to its surface as in the Davson and Danielli model.

These results led to the conclusion that much of the protein of the plasma membrane penetrates the lipid bilayer, as is shown in fig. 3.3 (Singer and Nicolson, 1972; Bretscher, 1973). Singer and Nicolson suggested that the non-polar parts of these intrinsic membrane proteins are embedded in the non-polar environment formed by the hydrocarbon chains of the lipid molecules, whereas their polar sections project from the membrane into the

Fig. 3.2. The arrangement of phospholipid molecules in the cell membrane of the lipid bilayer. (Based on Griffith *et al.*, 1974, and Marsh, 1975).

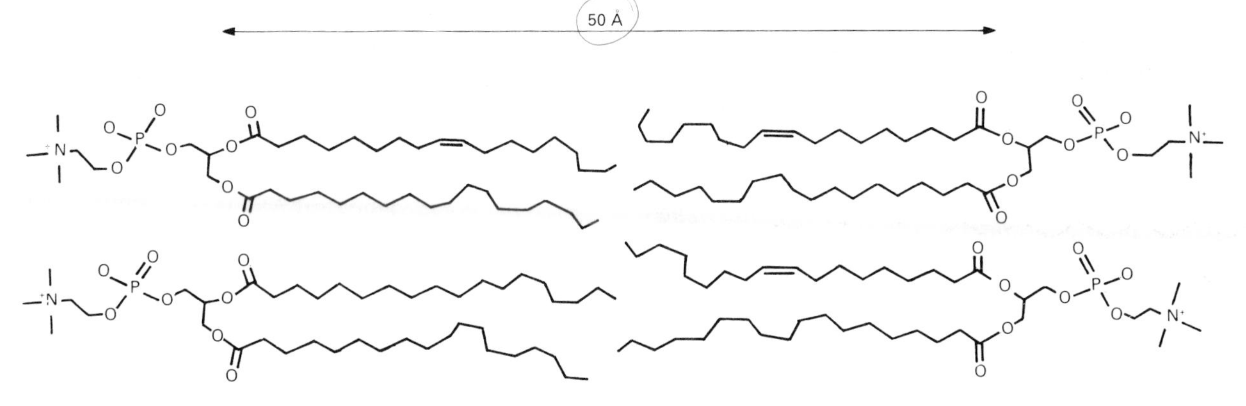

Fig. 3.3. Fluid mosaic model of the cell membrane. The phospholipid bilayer is shown split into two leaflets (as it might be by the freeze fracture technique) at the right. Integral proteins traverse the bilayer, peripheral proteins sit on its surface. Oligosaccharides occur on the outer surface of the membrane, attached mainly to proteins (forming glycoproteins) but also to lipids. (From Darnell *et al.*, 1986.)

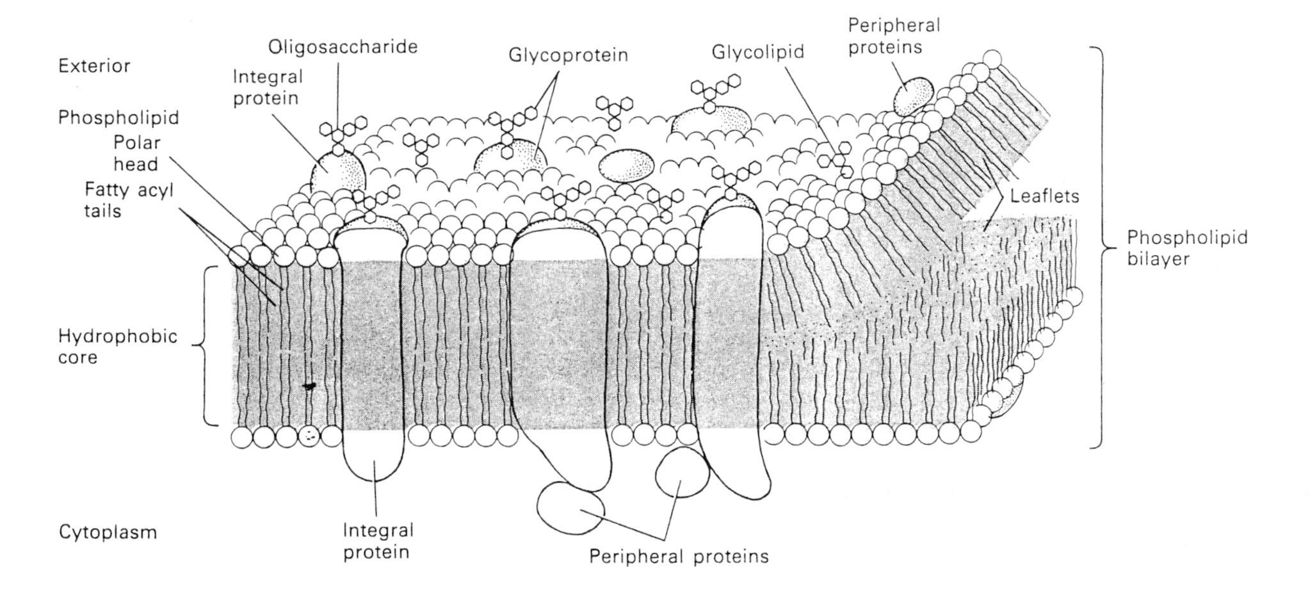

polar environment provided by the aqueous media on each side of the membrane.

Singer and Nicolson view the membrane as a mosaic in which a variety of protein molecules serve different functions. There is evidence that some of the protein molecules are able to move in the plane of the membrane (rather like icebergs in the surface waters of some arctic sea) and hence the structure illustrated in fig. 3.3 has become known as the fluid mosaic model. The model has been refined since its original formulation, to include peripheral proteins, which are attached to the membrane but not embedded in it, and to recognize that many of the proteins may have their movements restricted by attachment to each other, to peripheral proteins or to proteins of the cytoskeleton (Nicolson, 1976).

Plasma membranes are highly asymmetrical structures (Bretscher, 1973; Rothman and Lenard, 1977). Phospholipids are differentially distributed in the two leaflets of the bilayer, so that there is more sphingomyelin and phosphatidylcholine in the outer leaflet and more phosphatidylethanolamine and phosphatidylserine in the inner one (Verkleij *et al.*, 1973). Proteins show an absolute asymmetry: they are nearly all positioned in their own particular way with respect to the inner or outer surface of the membrane. Thus, for example, transporting enzymes always have their ATP-binding sites on the inner (cytoplasmic) surface, glycoproteins have their sugar residues on the outer surface, and the peripheral proteins on the outer surface have quite different functions from those on the cytoplasmic surface.

We shall see later that the cell membrane, largely because of the impermeability of its lipid bilayer, acts as a barrier separating the contents of the cell from the outside world. But the proteins embedded in the membrane may act as doors in the barrier through which particular information or specific substances can be transferred from one side to the other. We shall see that there are many different doors and many different keys to open them.

Concentration cells

Consider the system shown in fig. 3.4. The two compartments contain different concentrations of an electrolyte XY in aqueous solution, and are separated by a membrane which is permeable to the cation X^+ but impermeable to the anion Y^-. The concentration in compartment 1 is higher than that in compartment 2. Obviously X^+ will tend to move down its concentration gradient, so that a small number of cations will move from compartment 1 to compartment 2, carrying positive charge with them. This movement of charges causes a potential difference to be set up between the two compartments. The higher the potential gets, the harder it is for the X^+ ions to move against the electrical gradient. Hence an equilibrium position is reached at which the electrical gradient (tending to move X^+ from 2 to 1) just balances the chemical or concentration gradient (tending to move X^+ from 1 to 2). Since the potential difference at equilibrium (known as the *equilibrium potential* for X^+) arises from the difference in the concentration of X^+ in the two compartments, the system is known as a *concentration cell.*

What is the value of this potential difference? Suppose a small quantity, δn moles, of X is to be moved across the membrane, up the concentration gradient, from compartment 2 to compartment 1. Then, applying elementary thermodynamics, the work required to do this, δW_c, is given by

$$\delta W_c = \delta n \cdot RT \log_e \frac{[X]_1}{[X]_2}$$

where R is the gas constant (8.314 J deg^{-1} mole^{-1}), T is the absolute temperature, and $[X]_1$ and $[X]_2$ are the molar concentrations* of X in compartments 1 and 2 respectively. Now consider the electrical work, δW_e, required to move δn moles of X against the electrical gradient, i.e.

* Or, more strictly, the activities of X in the two compartments. The simplification given here is valid as long as the activity coefficient of X is the same in each compartment.

Fig. 3.4. Concentration cell. See text for details.

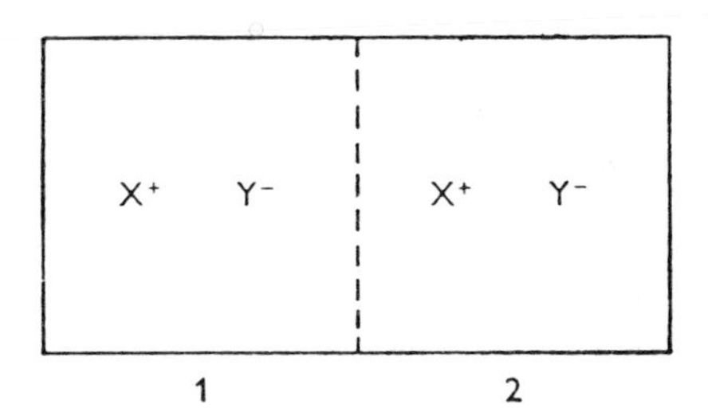

from compartment 1 to compartment 2. This is given by

$$\delta W_e = \delta n \cdot zFE$$

where z is the charge on the ion, F is Faraday's constant (96 500 coulombs mole^{-1}) and E is the potential difference in volts between the two compartments (measured as the potential of compartment 2 with respect to compartment 1). Now, at equilibrium, there is no net movement of X, and therefore

$$\delta W_e = \delta W_c$$

or $$\delta n \cdot zFE = \delta n \cdot RT \log_e \frac{[X]_1}{[X]_2}$$

i.e. $$E = \frac{RT}{zF} \log_e \frac{[X]_1}{[X]_2}. \qquad (3.1)$$

This equation, known as the *Nernst equation* after its formulator, is most important for our purposes since it helps us to understand the origins of electric potentials in excitable cells, which are in part dependent upon differences in the ionic composition of the cytoplasm and the external fluid.

Equation (3.1) can be simplified by enumerating the constants to become, at 18 °C,

$$E = \frac{25}{z} \log_e \frac{[X]_1}{[X]_2} \qquad (3.2)$$

or $$E = \frac{58}{z} \log_{10} \frac{[X]_1}{[X]_2} \qquad (3.3)$$

where E is now given in millivolts. For instance, in fig. 3.4, suppose $[X]_1$ were ten times greater than $[X]_2$, then E would be +58 mV if X were K^+ and +29 mV if X were Ca^{2+}; if the membrane were permeable to Y and not to X, then E would be −58 mV if Y were Cl^-, and −29 mV if Y were SO_4^{2-}.

How many X ions have to cross the membrane to set up the potential? The answer depends upon the valency of the ion, the value of the potential set up and the capacitance of the membrane. Let us consider a membrane with a capacitance (C) of 1 μF cm^{-2} and a potential of 70 mV across it. Then the charge on 1 cm^2 is given by

$$Q = CV,$$

where Q is measured in coulombs, C in farads and V in volts. The number of moles of X moved will be CV/zF, where z is the charge on the ion and F is Faraday's constant. In this case, assuming that X is monovalent,

$$\frac{CV}{zF} = \frac{10^{-6} \times 7 \times 10^{-2}}{96\,500}$$
$$= 6.8 \times 10^{-13} \text{ moles cm}^{-2}$$

which is a very small quantity. For instance (to anticipate a little), supposing we are dealing with a squid giant axon, diameter 1 mm, with a membrane capacitance of 1 μF cm^{-2}, and a membrane potential of 70 mV set up primarily by a potassium ion concentration cell: 1 cm^2 of membrane will enclose a volume containing about 3×10^{-5} moles of potassium ions, and thus a loss of 6.8×10^{-13} moles would be undetectable. In this case the difference between the numbers of positive and negative charges inside the axon would be about 0.000 002 %.

It is important to realize that the movements of ions from one compartment to another which we have been discussing are *net* movements, At the equilibrium position, although there is now no net ionic movement, X ions still move across the membrane, but the rate of movement (the *flux*) is equal in each direction. Before the equilibrium position is reached, the flux in one direction will be greater than that in the other. Fluxes can only be measured by using isotopic tracer methods. In describing fluxes across cell membranes, movement of ions into the cell is called the *influx*, and movement out, the *efflux*.

Ionic concentrations in the cytoplasm

Various microchemical methods are available for determining the quantities of ions present in a small mass of tissue. These include various microtitration techniques, flame photometry (in which the quantity of an element is determined from the intensity of light emission at a particular wavelength from a flame into which it is injected) and activation analysis (in which the element is converted into a radioactive isotope by prolonged irradiation in an atomic pile). All estimations made on a mass of tissue must include corrections made for the fluid present in the extracellular space. The size of the extracellular space is usually estimated by measuring the concentration in the tissue of a substance which is thought not to penetrate the cell membrane, such as inulin. Table 3.1 gives a simplified balance sheet of the ionic concentrations in

Table 3.1. *Ionic concentrations in frog muscle fibres and in plasma*

	Concentration in muscle fibres (mM)	Concentration in plasma (mM)
K	124	2.25
Na	10.4	109
Cl	1.5	77.5
Ca	4.9	2.1
Mg	14.0	1.25
HCO_3	12.4	26.6
Organic anions	*ca.* 74	*ca.* 13

Simplified after Conway, 1957.

Table 3.2. *Ionic concentrations in squid axoplasm and blood*

	Concentration in axoplasm (mM)	Concentration in blood (mM)
K	400	20
Na	50	440
Cl	40–150	560
Ca	0.4	10
Mg	10	54
Isethionate	250	—
Other organic anions	*ca.* 110	—

Simplified after Hodgkin, 1958.

frog muscle determined in this way. In the giant axons of squids, which may be up to 1 mm in diameter, the situation is much more favourable, since the axoplasm can be squeezed out of an axon like toothpaste out of a tube; table 3.2 shows the concentrations of the principal constituents.

The main features of the ionic distribution between the cytoplasm and the external medium are similar in both the squid axon and the frog muscle fibre. In each case the intracellular potassium concentration is much greater than that of the blood, whereas the reverse is true for sodium and chloride. Moreover, the cytoplasm contains an appreciable concentration of organic anions. These features seem to be common to all excitable cells. As we shall see, these inequalities in ionic distribution are essential to the electrical activity of nerve and muscle cells. They are dependent upon active transport processes which are driven by metabolic energy.

Active transport of ions

Sodium

The Nernst potential for any particular ion species shows what the cell membrane potential would be if the distribution of that ion across the membrane had reached equilibrium. If the actual membrane potential is not equal to the Nernst potential, then the ion is not in equilibrium and will tend to flow down its electrochemical gradient if the membrane is permeable to it.

The resting potential of frog muscle cells is usually around −90 to −100mV, and that for squid axon is typically about −60 mV. Table 3.3 shows the Nernst potentials for the major monovalent ions in these cells. Clearly the distribution of potassium and chloride ions is fairly close to equilibrium in both cases, since their Nernst potentials are not far from the resting potential. Sodium ions, however, have a Nernst potential which is more than 100 mV more positive than the resting potential, and hence there is a large electrochemical gradient for sodium ions, tending to drive them into the cell.

One explanation for this situation might assume that the cell membrane is impermeable to sodium ions. However, this is not so; resting nerve and muscle cells do take up radioactive sodium ions (for example, the resting influx of sodium into giant axons of the cuttlefish *Sepia* is about 35 pmoles cm^{-2} s^{-1}). If nothing were done about this influx, the system would in time run down, so as to produce equal concentrations of each ion on both sides of the membrane. In fact, the cell prevents this by means of a continuous extrusion of sodium ions (frequently known as the 'sodium pump'). Since such extrusion must occur against an electrochemical gradient, we would expect this process to be an active one, involving the consumption of metabolic energy.

Let us first consider some experiments by Hodgkin and Keynes (1955*a*) on the extrusion of sodium from *Sepia* giant axons. An axon was

Table 3.3. *Application of the Nernst equation to nerve and muscle cells*

	Ion	Ionic concentrations (mM) External	Internal	Nernst potential (mV)
Frog muscle	K	2.25	124	−101
	Na	109	10.4	+59
	Cl	77.5	1.5	−99
Squid axon	K	20	400	−75
	Na	440	50	+55
	Cl	560	108	−41

placed in sea water containing radioactive sodium ions and stimulated repetitively for some time, so that (as we shall see in chapter 5) the interior of the axon became loaded with radioactive sodium. It was then placed in a capillary tube through which nonradioactive sea water could be drawn (fig. 3.5), and the efflux of sodium determined by measuring the radioactivity of samples of this sea water at intervals. This particular arrangement has two advantages: the efflux from the cut ends of the axon is not measured, and the measured efflux occurs from a known length of axon into a known volume of sea water. It was found that the relation between the logarithm of the efflux and time is a straight line of negative slope (as is shown in the first section of the graph in fig. 3.6), indicating that the efflux of radioactive sodium falls exponentially with time. This is just what we would expect to see if both the internal sodium ion concentration and the rate of extrusion of sodium ions are constant, since under these conditions a constant proportion of the radioactive sodium inside the axon will be removed in successive equal time intervals.

When 2:4-dinitrophenol (DNP) was added to the sea water, the sodium efflux fell markedly, and then recovered when the DNP was washed away (fig. 3.6). Now DNP is an inhibitor of metabolic activity, and probably acts by uncoupling the formation of the energy-rich compound adenosine triphosphate (ATP) from the electron chain in aerobic respiration, hence this experiment implies that the extrusion of sodium is probably dependent on metabolic energy supplied directly or indirectly in the form of ATP. More conclusive evidence for this view is provided by a series of experiments by Caldwell *et al.* (1960), one of which is shown in

Fig. 3.5. Apparatus used to measure the efflux of radioactive sodium from *Sepia* giant axons. (From Hodgkin and Keynes, 1955*a*.)

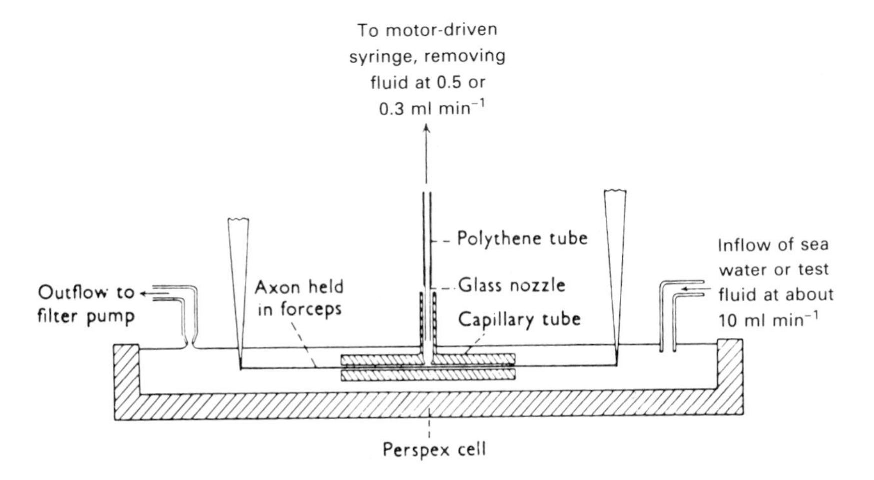

fig. 3.7. A solution containing radioactive sodium ions was injected into a giant axon of the squid *Loligo* by means of a very fine glass tube inserted longitudinally into the axon, and the efflux was then measured. Soon after the start of the experiment the axon was poisoned with cyanide. Cyanide ions interfere with the energy-producing processes in aerobic respiration (by inhibiting the action of cytochrome oxidase), and so the efflux of sodium fell to a low level. Injections of ATP, arginine phosphate or phosphoenol pyruvate into the axon produced transient increases in the rate of sodium extrusion, so confirming the view that the sodium pump is driven by the energy derived from the breakdown of energy-rich phosphate compounds.

Hodgkin and Keynes also showed that the

Fig. 3.6. The effect of the metabolic inhibitor 2:4-dinitrophenol (DNP) on the efflux of radioactive sodium from a *Sepia* giant axon. (From Hodgkin and Keynes, 1955*a*.)

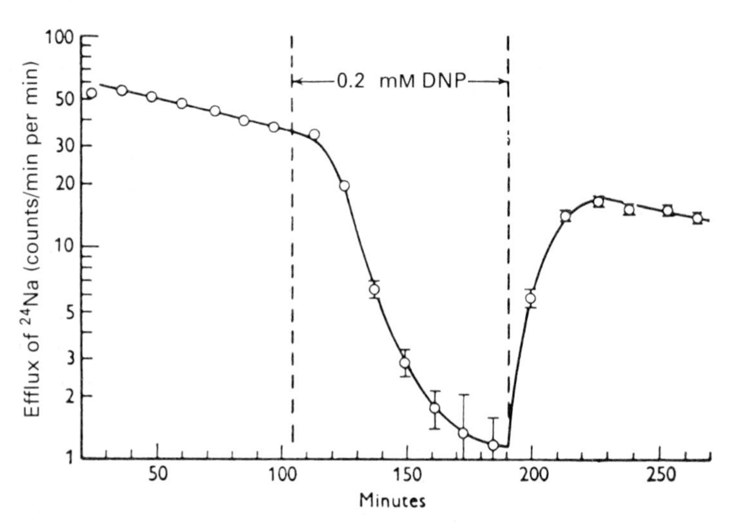

Fig. 3.7. The effect of ATP injection on the sodium efflux from a squid giant axon poisoned with cyanide. More ATP was introduced with the second injection than with the first. (From Caldwell *et al.*, 1960.)

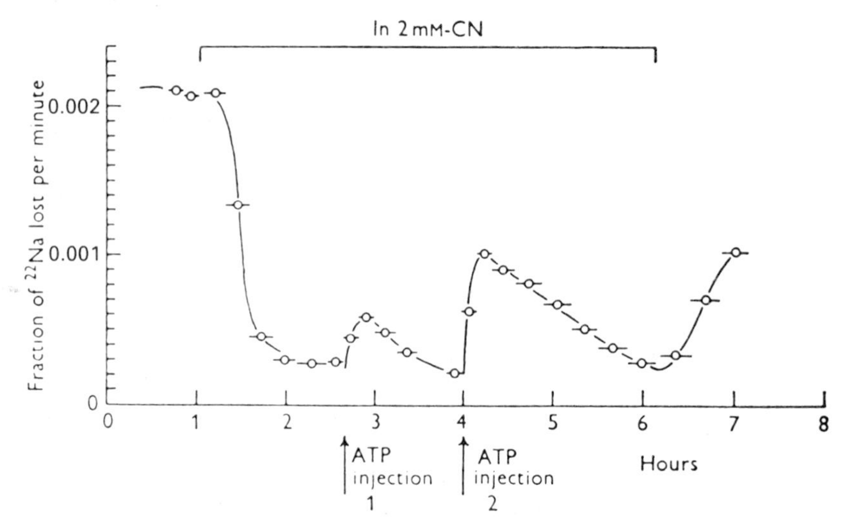

sodium efflux from *Sepia* axons is dependent upon the external potassium ion concentration; potassium-free sea water reduced the efflux to about a third of its normal value. This implies that the sodium extrusion process is coupled to the uptake of potassium.

In terms of the fluid mosaic model of the cell membrane (fig. 3.3), we may suppose that sodium ions are pumped at particular sites each consisting of a particular protein molecule or group of molecules. How many of these sites are there in any particular area of membrane? The problem has been approached by measuring the binding of the drug ouabain to nerve axons. Ouabain is a glycoside found in certain plants and used as an arrow poison in parts of Africa. It acts by inhibiting the sodium pump of cell membranes. By measuring the uptake of radioactive oubain it is possible to estimate the number of sodium-pumping sites in a system, assuming that each site binds one molecule of ouabain. In squid giant axons, Baker and Willis (1972) estimate that there are 1000–5000 sites per square micrometre of cell membrane, and Ritchie and Straub (1975) estimate that for garfish olfactory nerve there are 350 sites per square micrometre.

More information about the sodium pump came from experiments on human red blood cells. Thus Post and Jolly (1957) measured the ratio of sodium ions to potassium ions moved by the pump. They stored red cells for two days at 2 °C; this would stop the activity of the pump and allow sodium entry and potassium loss by passive fluxes. On raising the temperature to 37 °C the pump became active again and so after a few hours it was possible to measure the change in the amounts of sodium and potassium ions in the cells. The results showed that two potassium ions were taken up for every three sodium ions extruded. In similar experiments with resealed red cell ghosts (red cells which have had their haemoglobin removed), Garrahan and Glynn (1967) found that one ATP molecule was split for every three sodium ions extruded.

Our understanding of the sodium pump was much enhanced when Skou (1957) isolated an ATPase that was stimulated by sodium and potassium ions from crab nerves. An ATPase is an enzyme which splits ATP, usually with the object of utilizing the energy of the ATP molecule for some energy-demanding function of the cell. Later work showed that the enzyme is a dimeric protein consisting of an α subunit (molecular weight 100 kDa) which contains the ATPase activity and the ouabain-binding site, and a β subunit (molecular weight 55 kDa).

The structure of the α subunit has been determined using complementary DNA cloning techniques. The base sequence of the complementary DNA was determined, and from this the amino acid sequence of the protein could be deduced. Details of such methods are given by Watson *et al.* (1983), for example, and we shall look at their application to the acetylcholine receptor (one of the first large membrane proteins to be examined with these techniques) in chapter 8.

The sequences of the Na,K-ATPase from sheep kidney (Shull *et al.*, 1985) and from the electric organ of the electric ray Torpedo (Kawakami *et al.*, 1985) are very similar, with over 85% of the sequence identical in the two enzymes. The α chain in the sheep kidney enzyme consists of a sequence of 1016 amino acids. From this sequence it is possible to make some deductions about the shape of the molecule.

For membrane proteins it is particularly useful to know which parts of the molecule are embedded in the membrane and which parts are in contact with the cytoplasm or the extracellular space. Different amino acid side chains have different affinities for aqueous and lipid environments; thus valine and leucine, for example, are hydrophobic (they will tend to be found preferentially in a lipid environment) and glutamic and aspartic acids, lysine and arginine are hydrophilic and so more likely to be found in an aqueous environment.

Kyte and Doolittle (1982) arranged the twenty amino acids commonly found in proteins on a scale of hydropathicity (also known as hydropathy or hydrophobicity) which measures their affinity for nonpolar environments. The scale runs from a hydropathicity index from +4.5 for the hydrophobic isoleucine to −4.5 for the hydrophilic arginine; the values for the different amino acids are shown in table 1.1. The scale can be used to tell us something about the structure of a protein whose amino acid sequence is known. For an amino acid at position i, the average hydropathicity index of all the amino acids from position $i - n$ to position $i + n$ (where n is 7 or 9, for example) is plotted. The result is a hydropathicity profile whose peaks show hydrophobic sections of the protein chain and whose valleys show hydrophilic sections. If the hydrophobic sections are long enough, we may suspect that they are embedded in the lipid environment of the membrane and cross from one side of it to the other.

Application of the Kyte–Doolittle analysis to the Na,K-ATPase sequence (fig. 3.8) shows that there are at least six and perhaps eight major hydrophobic sequences, suggesting that the protein chain traverses the membrane six or eight times (fig. 3.9). ATP hydrolysis seems to involve binding to a lysine residue at position 501 and phosphorylation of an aspartic acid residue at 369; the chain is probably coiled so as to bring these two residues together. Binding of ouabain probably occurs at the tryptophan residue at position 310 in the short external portion between the third and fourth hydrophobic sequences.

Jorgensen (1975, 1982, 1985) found that the digestive enzyme trypsin cleaves the α chain at different sites according to whether sodium or potassium ions are present. This implies that the molecule changes shape when combined with these two different ions; it seems probable that this shape change is involved in the pumping activity.

Chloride

In squid giant axons, Keynes (1963) showed that, besides the potassium-linked sodium pump, there is another active transport system, concerned with the inward movement of chloride ions. If chloride were passively distributed across the axonal membrane, then the internal chloride concentration would be given by

$$E = \frac{RT}{-F} \log_e \frac{[\mathrm{Cl}]_o}{[\mathrm{Cl}]_i}$$

(where E is the resting potential), or, in isolated axons,

$$-60 = -58 \log_{10} \frac{560}{[\mathrm{Cl}]_i}$$

i.e. $[\mathrm{Cl}]_i = 55$ mM.

In fact, however, the average internal chloride concentration in Keynes's experiments was 108 mM, or about twice what one would expect. There are two possible explanations for this situation: either about half of the internal chloride was bound in non-ionic form, or there was an active

Fig. 3.9. A two-dimensional impression of the structure of the Na,K-ATPase molecule. In fact the molecule is probably folded so that the hydrophobic regions 1 to 8 are closer together.

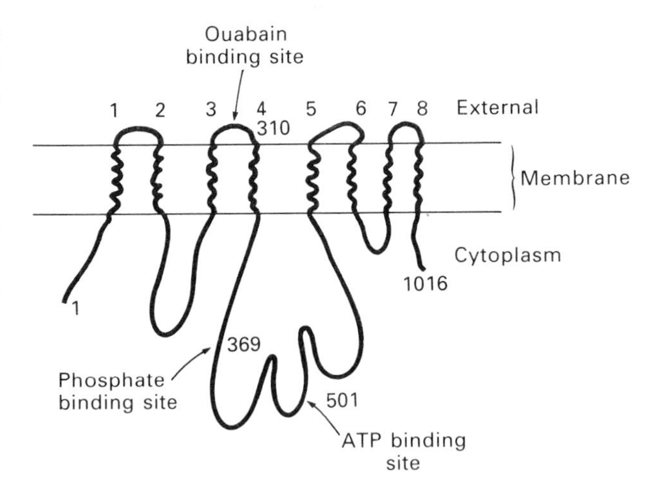

Fig. 3.8. Hydropathy plot of the amino acid sequence of sheep kidney Na,K-ATPase. Hydrophobic regions are above the x axis and hydrophilic ones below it. The major hydrophobic regions 1 to 8 are probably places where the chain crosses the membrane. (From Shull *et al.*, 1985).

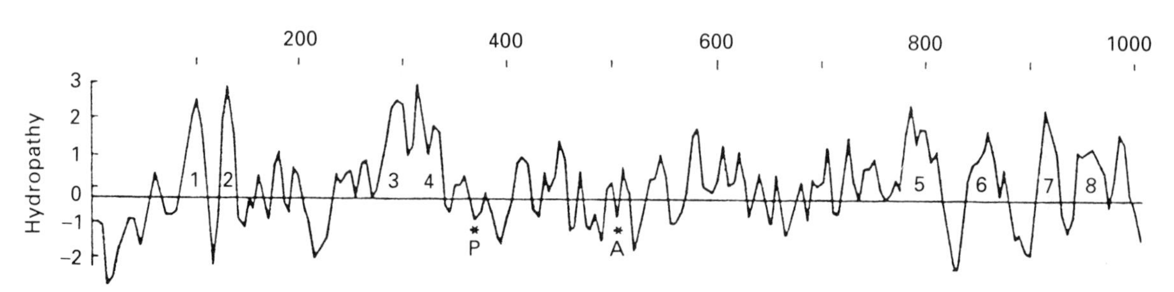

inward transport of chloride. Measurements on extruded axoplasm with a chloride-sensitive electrode (a silver wire coated with silver chloride) showed that the activity coefficient of chloride in the axoplasm was much the same as that in free solution, which indicates that very little of the chloride can be bound. The chloride influx was halved by treating the axon with DNP, thus confirming the conclusion that there is some active uptake of chloride ions. The functional importance of this 'chloride pump' is obscure; Russell (1984) suggests that it may be important in the regulation of intracellular pH.

Fig. 3.10 summarizes the conclusions reached in this and the previous sections.

Calcium

The total concentration of calcium in axoplasm is low, usually in the range 50–200 μM (Baker and Dipolo, 1984).

If a small quantity of radioactive calcium ions is injected into a squid axon, the resulting patch of radioactivity does not spread out by diffusion or move in a longitudinal electric field (in contrast to potassium, for example – see fig. 3.12). This suggests that nearly all the calcium is bound in some way, and that only a small proportion is in free solution in the ionic form (Hodgkin and Keynes, 1957; Baker and Crawford, 1972). By using the protein aequorin, which emits light in the presence of calcium ions, it is possible to show that the concentration of ionized calcium is as low as 0.1 μM. Similar results are obtained using a calcium-sensitive microelectrode (Dipolo *et al.*, 1983).

The calcium is bound in a number of different ways by the axoplasm (Baker and Schlaepfer, 1978; Brinley, 1978). The mitochondria split ATP to take up an appreciable quantity of calcium. There is also an energy-independent binding capacity, part of which is accounted for by the calcium-binding protein calmodulin.

With an internal concentration around 0.1 μM, there is thus clearly a very great concentration gradient of calcium ions across the membrane, and it is therefore not surprising that there is a continual influx of calcium ions. This influx is normally balanced by a corresponding efflux in which the calcium ions are being moved up their electrochemical gradient, a process which must consume energy.

There are two components to this efflux. Firstly, there is an ATP-dependent 'calcium pump' which is a membrane-bound Ca-ATPase, active at low internal calcium ion concentrations. The enzyme has a molecular weight of 138 kDa and is activated by calmodulin. One ATP molecule is split for each calcium ion transported, and (in mammalian cells at 30 °C) 25–100 calcium ions are transported per second at each site (Carafoli and Zurini, 1982; Carafoli *et al.*, 1986).

The second system is one on which internal calcium ions are exchanged for external sodium ions. It comes into action when the internal calcium ion concentration is somewhat raised, and it is driven by sodium ions moving down their concentration gradient (Blaustein and Hodgkin, 1969; Dipolo *et al.*, 1983; Baker, 1986). Under suitable conditions a similar exchange working in the oppo-

Fig. 3.10. A schematic representation of the movements of the major monovalent ions across the squid giant axon membrane in the resting condition. Passive fluxes are represented by straight arrows whose thickness suggests the magnitude of the flux. Curved arrows indicate active transport. Concentrations are suggested by the size of the chemical symbols. A^- represents indiffusible organic anions. In frog muscle cells the chloride fluxes are relatively larger, and chloride movement seems to be largely passive.

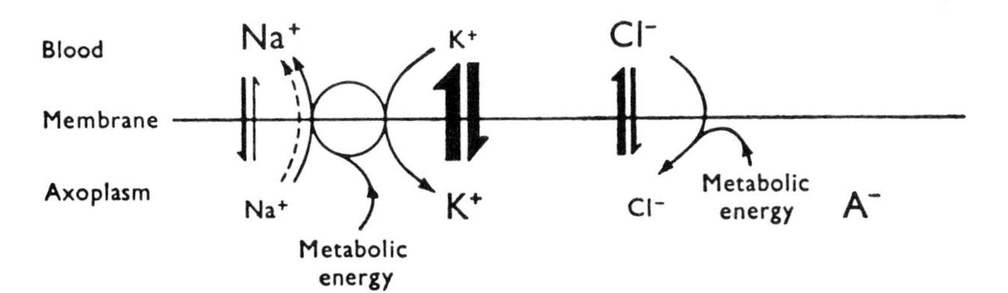

site direction can be detected; this requires ATP for its activity (Baker *et al.*, 1969; Baker, 1986).

Baker pointed out that the very low intracellular concentration of ionized calcium allows calcium ions to be used as a potent physiological trigger. Thus release of 1% of the bound calcium in squid axoplasm would increase the free ionic concentration forty-fold, from 0.1 to 4 μM. We shall see in later chapters that such trigger actions are fundamental to the release of synaptic transmitter substances and the initiation of muscular contraction.

The resting potential

The potassium electrode hypothesis

At the beginning of this century, Bernstein produced his 'membrane theory' of the resting potential. At that time there was some doubt as to whether there actually was any resting potential in the intact, uninjured cell. Bernstein held that there was such a potential, and that it arose as a result of the selective permeability of the cell membrane to potassium ions: he was the first, it seems, to apply the Nernst equation to the living cell.

If it is correct that the resting membrane is selectively permeable to potassium ions, then we would expect the membrane potential to be given by the Nernst equation for potassium ions, which my be written, at 18 °C, as

$$E = 58 \log_{10} \frac{[\mathrm{K}]_o}{[\mathrm{K}]_i} \qquad (3.4)$$

The best method of readily testing this hypothesis is to vary the external potassium concentration and observe the change in membrane potential; (3.4) predicts that the relation should be a straight line with a slope of 58 mV per unit increase in $\log_{10}[\mathrm{K}]_o$.

Fig. 3.11 shows the membrane potential of isolated frog muscle fibres in solutions containing different potassium ion concentrations (Hodgkin and Horowicz, 1959). It is evident that (3.4) fits the results very well at potassium concentrations

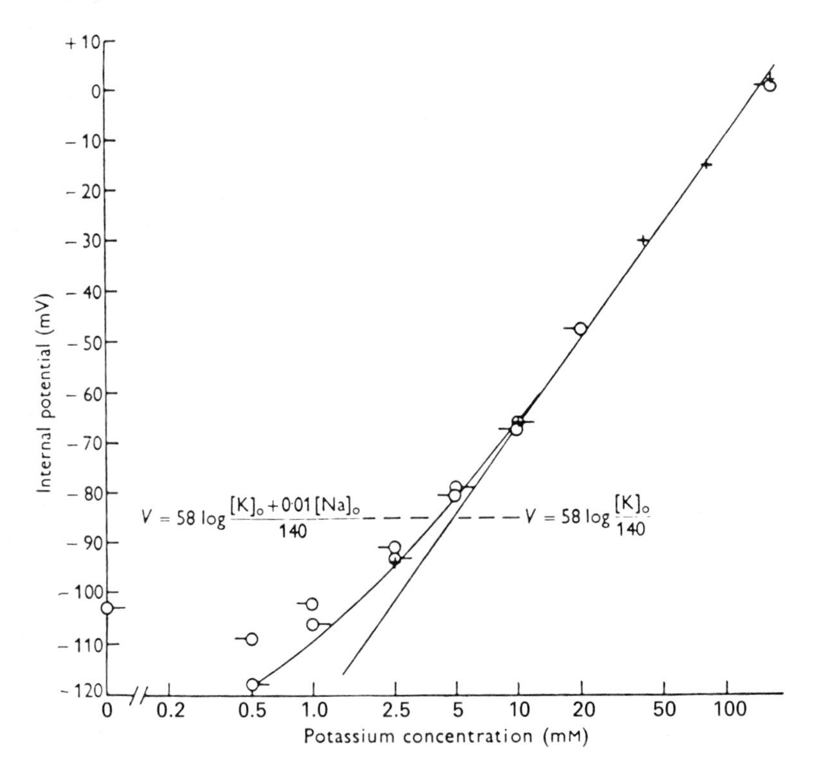

Fig. 3.11. The effect of the external potassium ion concentration on the membrane potential of isolated frog muscle fibres. The external solutions were chloride-free, the principle anion being sulphate. (From Hodgkin and Horowicz, 1959.)

above about 10 mM, but below this value the membrane potential is rather less than one would expect. Similar results have been obtained in squid axons (Curtis and Cole, 1942; Hodgkin and Keynes, 1955*b*) and many other cells (Hodgkin, 1951).

What is the reason for the departure from the simple potassium electrode hypothesis at low external potassium ion concentrations? We have seen that sodium ions are distributed across the cell membrane so as to produce the potentiality of a concentration cell with the inside positive, and since the membrane is not completely impermeable to sodium ions, we might expect this concentration cell to play some part in determining the membrane potential. Just how this effect will make itself felt will depend upon the properties of the membrane. A useful approach, known as the 'constant field theory' has been developed by Goldman (1943). This theory assumes that ions move in the membrane under the influence of electrical fields and concentration gradients just as they do in free solution, that the concentrations of ions in the membrane at its edges are proportional to those in the aqueous solutions in contact with it, and that the electrical potential gradient across the membrane is constant. From these assumptions it is possible to show (see Hodgkin and Katz, 1949) that, when there is no current flowing through the membrane, the membrane potential is given by

$$E = \frac{RT}{F} \log_e \frac{P_K[K]_o + P_{Na}[Na]_o + P_{Cl}[Cl]_i}{P_K[K]_i + P_{Na}[Na]_i + P_{Cl}[Cl]_o}, \quad (3.5)$$

where P_K, P_{Na} and P_{Cl} are permeability coefficients.

Equation (3.5) is often known as the Goldman–Hodgkin–Katz equation. The permeability coefficients are measured in centimetres per second and defined as $u\beta RT/aF$, where u is the mobility of the ion in the membrane, β is the partition coefficient between the membrane and aqueous solution, a is the thickness of the membrane, and R, T and F have their usual significance. Actual values for the permeability coefficients must be determined by measuring the fluxes of the different ions through the membrane, but their ratios (such as P_K/P_{Na}) can be estimated from membrane potential measurements.

If chloride ions are omitted from the system, or if they are assumed to be in equilibrium so that the equilibrium potential for chloride is equal to the membrane potential, (3.5) becomes

$$E = \frac{RT}{F} \log_e \frac{[K]_o + \alpha[Na]_o}{[K]_i + \alpha[Na]_i}, \quad (3.6)$$

where α is equal to P_{Na}/P_K. Equation (3.6) is plotted in fig. 3.11, assuming α to be 0.01; it is clear that it provides a good fit for the experimental results. Results similar to those shown in fig. 3.11 have also been obtained from observations on the membrane potentials of nerve axons (Hodgkin and Katz, 1949; Huxley and Stämpfli, 1951). We can therefore conclude that sodium ions play some small part in determining the membrane potential of nerve and muscle cells, their effect being greater when the external potassium ion concentration is low.

One further piece of information is required before we can accept the hypothesis that the resting potential is determined mainly by the Nernst equation for potassium ions. It is necessary to show that the internal concentration of free potassium ions (which is involved in (3.4)–(3.6)) is the same, or almost the same, as the total internal potassium concentration (which is what can be measured by chemical methods of analysis). For instance, if half the internal potassium were found in non-ionic form, the predicted values for membrane potentials would have to be reduced by nearly 20 mV. This question was investigated by Hodgkin and Keynes (1953) on *Sepia* giant axons. They placed an axion in oil, but with a short length of it passing through a drop of sea water containing radioactive potassium ions. Thus, after a time, this short length of axon contained radioactive potassium. It was then placed in a longitudinal electric field, so that the potassium ions would move along the axon towards the cathode if they were free to do so, and the longitudinal distribution of radioactivity was measured at intervals by means of a Geiger counter masked by a piece of brass with a thin slit in it.

The result of one of these experiments is shown in fig. 3.12, from which it is clear that the internal radioactive potassium ions are free to move in an electric field. From the rate of movement, it was possible to calculate the ionic mobility and diffusion coefficient of radioactive potassium ions in the axon: the values were found to be close to those in a 0.5 M solution of potassium chloride. Thus the radioactive potassium is present inside the axon in a free, ionic form. Since Keynes and Lewis (1951) had previously shown that radioactive potassium exchanges with at least 97% of the potassium present in crab axons, it seemed very reasonable to conclude that almost all the potassium in the axoplasm is effectively in free solution and so can contribute to the production of the resting potential.

Experiments on internally perfused giant axons

It has been possible to test further the potassium electrode hypothesis by means of the technique of intracellular perfusion (Baker *et al.*, 1962*a*; Tasaki and Shimamura, 1962). In the method used by Baker and his colleagues, the axoplasm is squeezed out of a squid axon, leaving a flattened tube consisting mainly of the axon membrane and the Schwann cells and endoneurial sheath. This tube is then reinflated by filling it with a perfusion fluid isotonic with sea water.

When the perfusion fluid is an isotonic potassium chloride solution, the resting potential is about −55 mV. Isotonic solutions of other potassium salts, such as sulphate or isethionate, produce resting potentials a few millivolts greater, indicating that the membrane is slightly permeable to chloride ions (Baker *et al.*, 1962*b*). If the internal solution is now replaced by an isotonic sodium chloride solution, the membrane potential falls to near zero.

By gradually increasing the internal potassium ion concentration, it is possible to see how far the membrane behaves as a potassium electrode; it was found that the membrane potential tended to reach a saturating value of −50 to −60 mV at an internal potassium ion concentration of about 300 mV. This relation is compatible with (3.5) if we assume that P_K decreases with increasing (more negative) membrane potentials. If the solutions inside and outside the membrane were identical in composition, the membrane potential was within 1 mV of zero. Finally, when the internal solution was isotonic sodium chloride and the external isotonic potassium chloride, the inside became positive to the outside by 40–60 mV.

The electrogenic nature of the sodium pump

Consider a membrane ion pump which is concerned solely with the active transport of one species of ion in one direction. There would then be a current flow across the membrane and a change in membrane potential caused by the depletion of charge on one side of the membrane. There would be similar consequences from a pump which coupled inflow of one ion to outflow of another but in which the numbers of ions transported in the two directions were not equal. A pump in which such a net transfer of charge does take place is described

Fig. 3.12. Movement of radioactive potassium in a longitudinal electric field in a *Sepia* giant axon. The two curves show the distribution of radioactivity immediately before (*a*) and 37 minutes after (*b*) application of the longitudinal current. Arrows at 11 and 23 minutes show the positions of peak radioactivity at these times. (From Hodgkin and Keynes, 1953.)

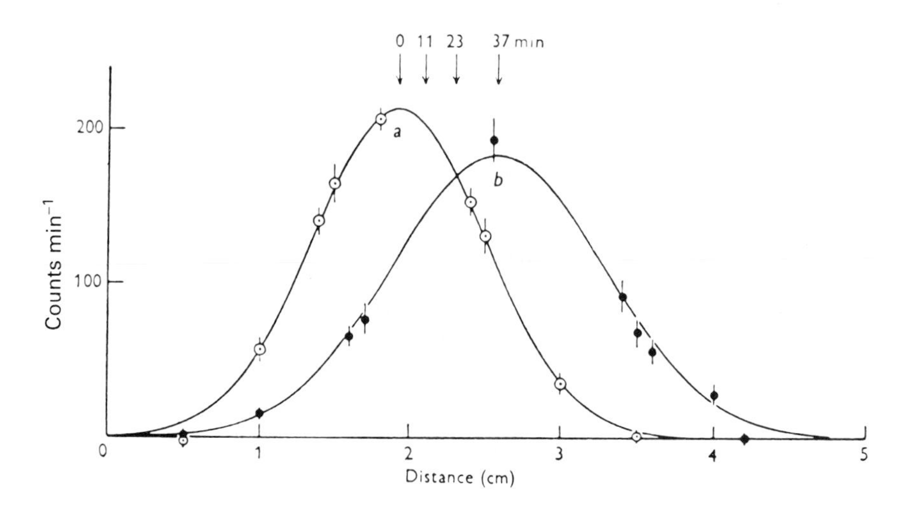

as being electrogenic. In mammalian red blood corpuscles, for example, there is very clear evidence that two potassium ions are taken up for every three sodium ions extruded (Post and Jolly, 1957); there must thus be a net flow of charge equal to one third of the sodium ion flow.

The sodium pump of nerve axons was once thought to be electrically neutral, because of a supposed one-for-one exchange of sodium for potassium ions. Later experiments on a variety of cells indicated that this is not so and that the pump is electrogenic (Ritchie, 1971; Glynn, 1984). Some of the best evidence for this conclusion comes from experiments by Thomas (1969, 1972) on snail neurons; let us examine them.

The essence of Thomas's method was to inject sodium ions into the cell body of a neuron and observe the subsequent changes in membrane potential. In order to perform the injection two microelectrodes filled with a solution of sodium acetate were inserted into the cell and current passed between them; sodium ions would thus be carried out of the cathodal electrode into the cytoplasm, so raising the internal sodium concentration. A third microelectrode was used to record the membrane potential.

Thomas found that injection of sodium ions into the cell is followed by a hyperpolarization of up to 15 mV, after which the membrane potential returned to normal over a period of about 10 min, as is shown in the first parts of the traces in fig. 3.13. Fig. 3.13*a* shows that ouabain greatly reduces this hyperpolarization, suggesting that it is connected with an active sodium extrusion. In fig. 3.13*b*, the second injection of sodium ions occurs while the cell is in a potassium-free environment, when there is no hyperpolarization, suggesting that sodium extrusion is coupled to potassium

Fig. 3.13. Responses of snail neurons to injections of sodium ions, demonstrating the electrogenic nature of the sodium pump. The records show pen recordings of the membrane potential; black bars indicate the duration of the injecting currents. The first response in each trace shows the hyperpolarization which normally follows sodium injection. After treatment with ouabain (*a*) this is greatly reduced. In a potassium-free external solution (*b*), the hyperpolarization does not occur until external potassium is replaced, indicating that sodium extrusion is coupled to potassium uptake. (The thickness of the trace at less negative membrane potentials is caused by spontaneous activity in the neuron, the individual action potentials being too rapid to be registered fully by the pen recorder.) (From Thomas, 1969.)

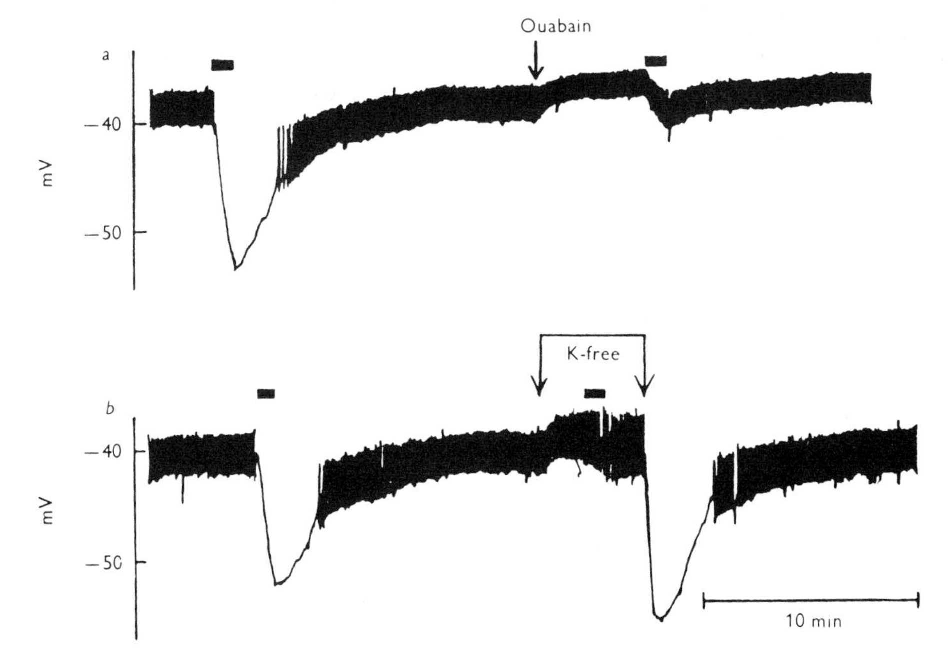

uptake. (Incidentally this experiment also eliminates the hypothesis that the hyperpolarization is caused by a local reduction in potassium ion concentration at the outer surface of the cell, caused by such a coupled uptake of potassium.) Injections of potassium or lithium ions produced no membrane hyperpolarization.

These observations suggest that, while sodium extrusion is coupled to potassium uptake, the number of sodium ions moved is greater than the number of potassium ions. There is thus a net flow of positive charge outward during the action of the pump, seen as a hyperpolarization of the membrane. Is it possible to estimate the ratio of sodium ion to potassium ion movements? Thomas attacked this problem by using an ingenious method known as the voltage-clamp technique. We shall examine the principle of this method in chapter 5; suffice it to say here that in involves a negative feedback control system whereby the membrane potential is maintained at a constant level ('clamped') by passing current through it via another electrode. Membrane currents can thus be measured at constant membrane potentials. Thomas found that sodium injection was followed by an outwardly directed current which took about 10 min to fall to zero.

By integrating this current with respect to time it was possible to estimate the amount of charge transferred. Thomas found that this was always much less than the quantity of charge injected as sodium ions. And yet he also found that all of the injected sodium was extruded during the period of membrane current flow; he did this by using an intracellular sodium-sensitive microelectrode (one which produces an electrical signal proportional to the sodium ion concentration in its environment). This means that the sodium outflow must be partly balanced by an inflow of some other cation, for which the obvious candidate is potassium. Thomas calculated that the net charge transfer in the pump was about 27% of the sodium ion flow. This figure is fairly near to the 33% that would be expected from a system like that in red blood cells, where three sodium ions move for every two potsssium ions.

The Donnan equilibrium system in muscle

In the system shown in fig. 3.14, two potassium chloride solutions are separated by a membrane which is permeable to both potassium and chloride ions. The system is at constant volume so that, although the membrane is permeable to water, no net flow of water can occur between compartments i and o. Let us assume that the two potassium chloride solutions are initially at the same concentration. A quantity of a potassium salt KA is now added to compartment i, and it dissolves to give K^+ and A^- ions; we shall assume that the anion A^- cannot pass through the membrane. The concentration of potassium ions in compartment i, $[K]_i$ is now greater than in compartment o, $[K]_o$ so that potassium ions move from i to o. Since the total number of positive and negative charges in i must be approximately* equal (the same applies to o, of course), chloride ions move from i to o also. Now since $[K]_i \neq [K]_o$ and $[Cl]_i \neq [Cl]_o$, a concentration cell potential must arise between the two compartments in each case, such that, applying the Nernst equation,

$$E_K = \frac{RT}{F} \log_e \frac{[K]_o}{[K]_i} \tag{3.7}$$

and

$$E_{Cl} = \frac{RT}{-F} \log_e \frac{[Cl]_o}{[Cl]_i} \tag{3.8}$$

where E_K is as the pctassium equilibrium potential and E_{Cl} is the chloride equilibrium potential. But there can only be a single potential difference across the membrane; hence potassium chloride will continue to move from i to o until $E_K = E_{Cl}$.

When this stage is reached, the system is in

* See p. 15 for what is meant by 'approximately' in this context.

Fig. 3.14. Donnan equilibrium system. See text for details.

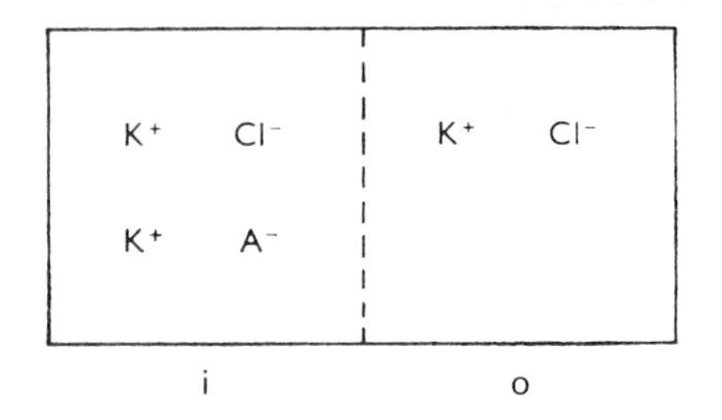

equilibrium and it follows that

$$\frac{RT}{F}\log_e\frac{[\mathrm{K}]_o}{[\mathrm{K}]_i}=\frac{RT}{-F}\log_e\frac{[\mathrm{Cl}]_o}{[\mathrm{Cl}]_i}.$$

Therefore

$$[\mathrm{K}]_o/[\mathrm{K}]_i=[\mathrm{Cl}]_i/[\mathrm{Cl}]_o$$

or

$$[\mathrm{K}]_i\times[\mathrm{Cl}]_i=[\mathrm{K}]_o\times[\mathrm{Cl}]_o \qquad (3.9)$$

Equation (3.9) is the Donnan rule, which states that the product of the concentrations of the diffusible ions in one compartment is equal to the product of the concentrations of the diffusible ions in the other compartment, at equilibrium.

Thus the presence of the indiffusible anions in compartment i results in an inequality of distribution of the diffusible ions, potassium and chloride. However, we cannot apply this simple system to an animal cell, because an animal cell is not a constant volume system, and therefore the osmotic concentration must be the same on each side of the membrane. In the simple system of fig 3.14 this is not so, as the following analysis shows. Applying the condition for approximate electrical neutrality in each compartment,

$$[\mathrm{K}]_o=[\mathrm{Cl}]_o$$

and

$$[\mathrm{K}]_i>[\mathrm{Cl}]_i\,.$$

Now

$$[\mathrm{K}]_i\times[\mathrm{Cl}]_i=[\mathrm{K}]_o\times[\mathrm{Cl}]_o\,,$$

therefore

$$[\mathrm{K}]_i+[\mathrm{Cl}]_i>[\mathrm{K}]_o+[\mathrm{Cl}]_o\,,$$

therefore

$$[\mathrm{K}]_i+[\mathrm{Cl}]_i+[\mathrm{A}]>[\mathrm{K}]_o+[\mathrm{Cl}]_o\,,$$

i.e. the osmotic concentration in i is greater than that in o. Thus if the constant-volume constraint were removed from the system, water would move from compartment o to compartment i. This would change the ionic concentrations in the two compartments and so disturb the Donnan equilibrium. Hence potassium chloride would move from o to i, again upsetting the osmotic equilibrium. These processes would continue until the concentration of A in i was infinitesimal and the concentrations of potassium and chloride were equal throughout the system.

Now consider a similar system (fig. 3.15) in which the membrane is effectively impermeable to sodium ions and some sodium chloride is added to compartment o. Potassium and chloride will move until the Donnan equilibrium is established, i.e. until

$$[\mathrm{K}]_i\times[\mathrm{Cl}]_i=[\mathrm{K}]_o\times[\mathrm{Cl}]_o$$

But now $[\mathrm{K}]_o$ is less than $[\mathrm{Cl}]_o$, so that if the constant volume constraint is removed, it *is* possible for the system to be in Donnan equilibrium and osmotic equilibrium at the same time. In effect, the osmotic effect of the indiffusible anion in i is balanced by that of the effectively indiffusible cation (sodium) in o.

We must conclude that a system similar to that shown in fig. 3.15 might account for the inequalities of ionic distribution seen in nerve and muscle cells. If this is so, then it should be possible to show that the [K][Cl] product is equal inside and outside after equilibrium has been reached in various conditions. This experiment was performed by Boyle and Conway (1941), who developed the theory outlined above. They soaked frog muscles in solutions containing various potassium chloride concentrations for 24 h, and then determined the intracellular potassium and chloride concentrations. The results conformed to the Donnan equilibrium hypothesis at all external potassium concentrations above about 10 mM.

Non-equilibrium conditions

We must now consider the contribution of chloride ions to the resting potential in muscle. This, together with some related problems, was extensively investigated by Hodgkin and Horowicz (1959) on isolated frog muscle fibres. The great advantage of using single fibres is that, with a suitable

Fig. 3.15. Double Donnan equilibrium system. See text for details.

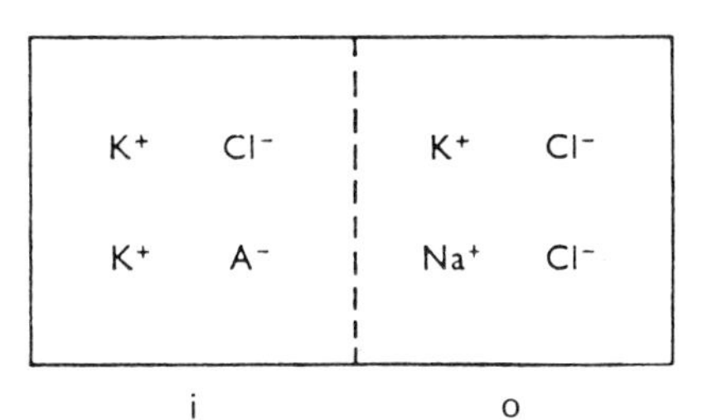

apparatus, it is possible to change the ionic concentrations at the cell surface within a fraction of a second, so eliminating the inevitable diffusion delays involved in work with whole muscles. When the external solution was changed for one with different potassium and chloride concentrations but with the same $[\mathrm{K}]_o[\mathrm{Cl}]_o$ product, the new membrane potential (given by (3.7) and (3.8)) was reached within two or three seconds, and thereafter remained constant. However, if the chloride concentration was changed without altering the potassium concentration, the membrane potential jumped rapidly to a new value, but this transient effect gradually decayed over the next few minutes and the potential returned to very nearly its original value (fig. 3.16). The explanation of this effect is based on the Donnan equilibrium hypothesis. The chloride equilibrium potential is given by (3.4), or, at 18 °C,

$$E_{\mathrm{Cl}} = -58 \log_{10} \frac{[\mathrm{Cl}]_o}{[\mathrm{Cl}]_i}. \quad (3.10)$$

At the start of the experiment shown in fig. 3.16 (when $[\mathrm{Cl}]_o$ is 120 mM) we assume that chloride is in equilibrium, i.e. that E_{Cl} is equal to the membrane potential of −98.5 mV; this gives a value of 2.4 mM for $[\mathrm{Cl}]_i$. When $[\mathrm{Cl}]_o$ is reduced to 30 mM, E_{Cl} will change by 35 mV to −63.5 mV. The membrane potential changes to −77 mV, which is intermediate between E_{K} and E_{Cl}, indicating that the membrane is permeable to both potassium and chloride ions. Then, in order to restore the Donnan equilibrium, potassium chloride moves out of the cell until E_{Cl} is equal to E_{K}. This point is reached when $[\mathrm{Cl}]_i$ has fallen to about 0.6 mM and, since there is also some movement of water out of the cell in order to maintain osmotic equilibrium, $[\mathrm{K}]_i$ is practically unchanged. Hence the new steady potential (reached after 10–15 min in fig. 3.16) is the same as it originally was. When $[\mathrm{Cl}]_o$ is returned to 120 mM, there is a similar transient

Fig. 3.16. The effect of a sudden reduction in the external chloride concentration on the membrane potential of an isolated frog muscle fibre. (From Hodgkin and Horowicz, 1959.)

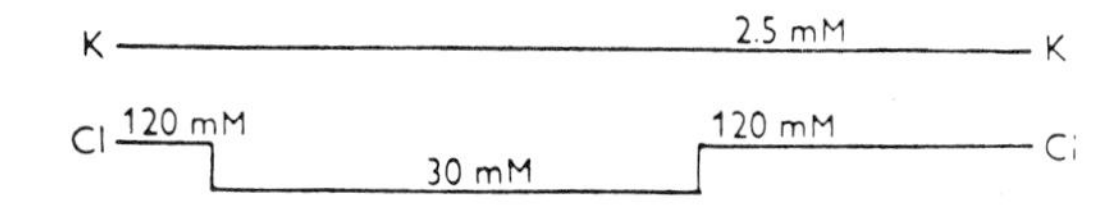

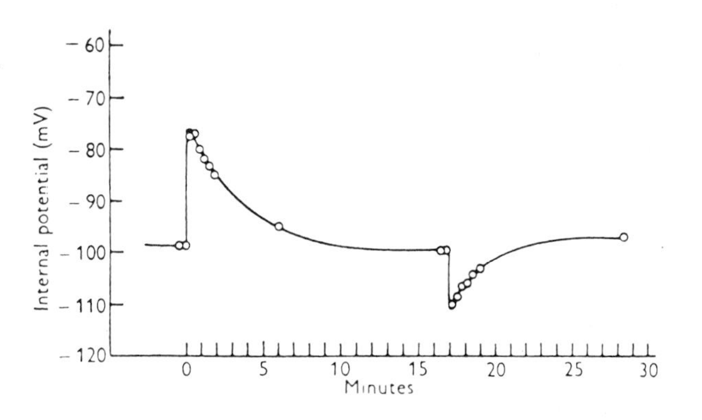

Fig. 3.17. The effect of changes in the external potassium ion concentration on the membrane potential of an isolated frog muscle fibre. (From Hodgkin and Horowicz, 1959.)

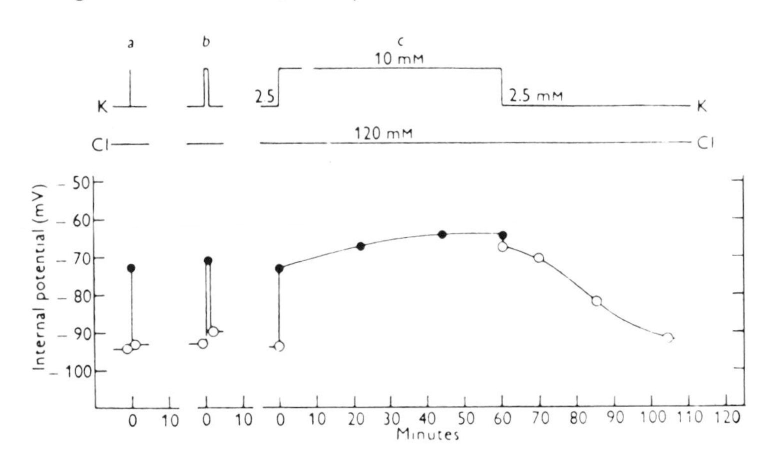

membrance potential change in the opposite direction, and equilibrium is then restored by movement of potassium chloride and water into the fibre. The fact that these experiments can be so readily interpreted in this way provides further evidence in favour of Boyle and Conway's Donnan equilibrium hypothesis.

Fig. 3.17*c* shows the results of a similar experiment in which $[K]_o$ is changed from 2.5 to 10 mM. This will change E_K by 35 mV, and so the membrane potential jumps to a new value intermediate between E_K and E_{Cl}. Equilibrium is then reached by entry of potassium chloride (and water) into the fibre, so that the membrane potential gradually moves from −73 to −65 mV. On returning $[K]_o$ to 2.5 mM, there is a small instantaneous repolarization, and then equilibrium (and the normal resting potential) is slowly restored by loss of potassium chloride from the fibre. Notice that, in this case, the fourfold increase in $[K]_o$ causes an instantaneous depolarization of 21 mV, whereas the later fourfold decrease causes an instantaneous repolarization of only 3 mV. This must indicate that there is some rectification process in the potassium pathway, so that potassium ions can move inwards very easily but outwards with much more difficulty. The larger repolarizations in figs. 3.17*a* and *b* are caused by the fact that the internal chloride concentration has not had time to change much from its normal level, so that E_{Cl} is still near the normal resting potential. Hodgkin and Horowicz calculated that P_K is about 8×10^{-6} cm s^{-1} for inward current but may be as little as 0.05×10^{-6} cm s^{-1} for outward current, whereas P_{Cl} remains at about 4×10^{-6} cm s^{-1} irrespective of the direction of the current flow.

4

Electrical properties of the nerve axon

The most striking anatomical feature of nerve cells is that part of the cell is produced into an enormously elongated cylindrical process, the axon. It is this part of the cell which we shall be concerned with in this chapter and the next. The essential function of the axon is the propagation of nerve impulses.

Action potentials in single axons

Let us consider a simple experiment on the giant fibres in the nerve cord of the earthworm. These fibres are anatomically not axons, because they are multicellular units divided by transverse septa in each segment, but physiologically each fibre acts as a single axon. These are three giant fibres, one median and two lateral, which run the length of the worm; the laterals are interconnected at intervals. The experimental arrangement for eliciting and recording impulses in the giant fibres is shown in fig. 4.1. The stimulator produces a square voltage pulse which is applied to the nerve cord at the stimulating electrodes. The recording electrodes pick up the electrical changes in the nerve cord and feed them into the amplifier. Here they are amplified about 1000 times, and then passed to the oscilloscope where they are displayed on the screen of the cathode ray tube. The output of the stimulator is also fed into the oscilloscope so that it is displayed on the second trace on the screen. The timing of the oscilloscope sweep is arranged so that both traces start at the moment that the pulse from the stimulator arrives.

We begin with a stimulating pulse of low intensity (fig. 4.2*a*). The lower trace shows the stimulating pulse, but nothing appears on the upper

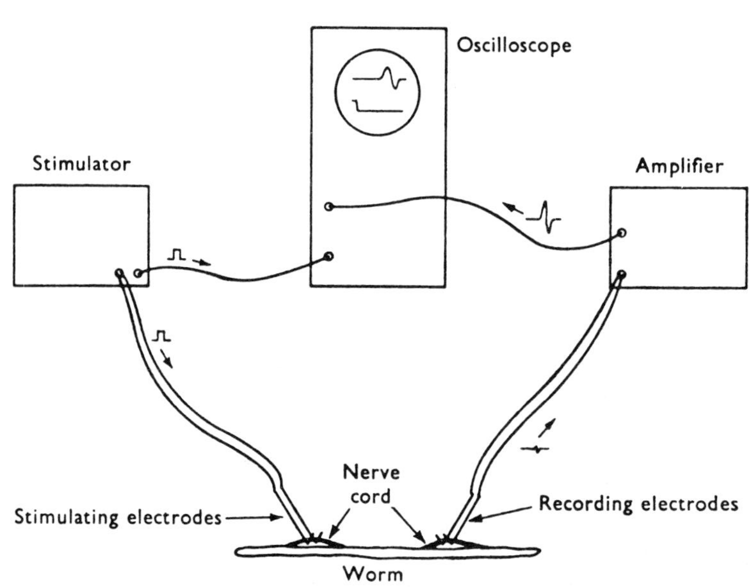

Fig. 4.1. Arrangement for recording action potentials from the giant fibres in the nerve cord of the earthworm.

trace except a deflection which is coincident with the stimulus and (as we can show by varying the stimulus intensity) proportional to its size. This phenomenon is known as the *stimulus artefact*; it is not a property of the worm, and is merely due to the recording electrodes picking up the electric field set by the stimulus pulse. As we increase the intensity of the stimulus, a new deflection appears on the upper trace (fig. 4.2*b*). This is the nerve impulse, or *action potential*, in the median fibre. Further increase in the stimulus intensity does not change the size of this deflection, and if we turn the stimulus intensity down again it suddenly disappears, without any preparatory decrease in size. In other words, the size of the action potential is independent of the size of the stimulus; it is either there or not there. This phenomenon is known as the *'all-or-nothing' law*. The stimulus intensity which is just sufficient to produce an action potential is called the *threshold stimulus intensity*. If we further raise the stimulus intensity, a second deflection may appear at a later time on the trace (fig. 4.2*c*); this is caused by the action potentials of the two lateral fibres.

Now let us perform another experiment. Keeping the stimulus intensity constant, above the threshold for the median fibre, the recording electrodes are moved nearer to the stimulating electrodes. We find that the action potential occurs earlier on the trace. If we measure the time between the stimulus and the action potential at a number of different distances between the two pairs of electrodes, we can plot a graph showing the position of the action potential at various times after the stimulus (fig. 4.3). The points lie very nearly on a straight line which passes near the origin. This means that the action potential arises at the stimulating electrodes and is then conducted

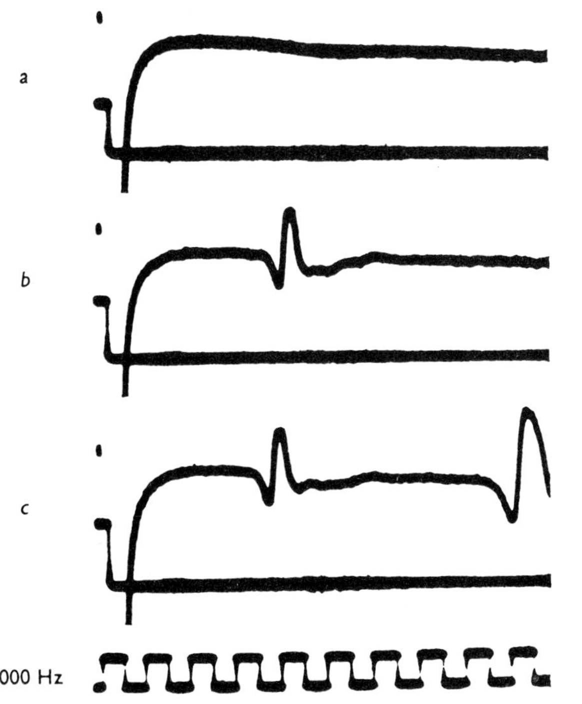

Fig. 4.2. Oscilloscope records from the experiment shown in fig. 4.1. In each case the upper trace is a record of the potential changes at the recording electrodes and the lower trace (at a much lower amplification) monitors the stimulus pulse. Further details in the text.

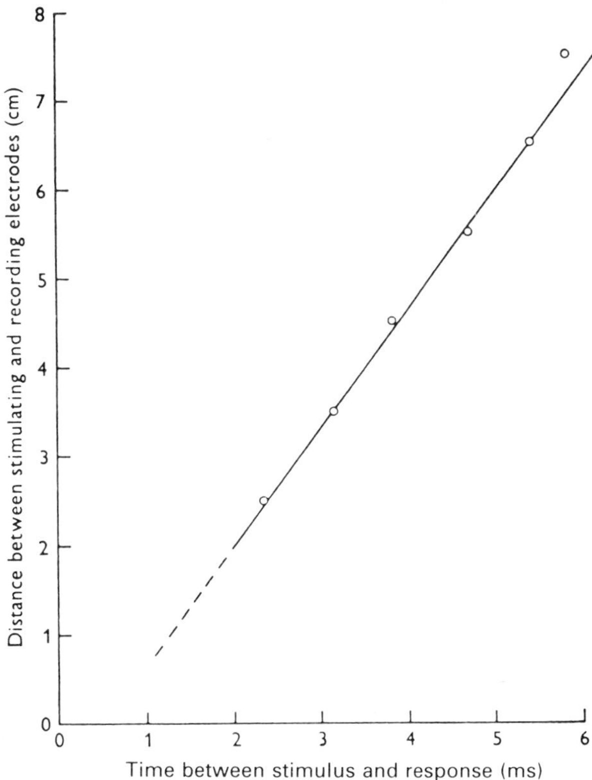

Fig. 4.3. Results of an experiment to measure the conduction velocity in the median giant fibre of an earthworm. In this case it was about 12.5 m s^{-1}.

along the fibre at a constant velocity given by the slope of this straight line.

In this experiment on conduction in earthworm giant fibres, we have been measuring potential differences between points at different places outside the axon. Obviously it would be informative to measure potential changes across the membrane, using an electrode inside the cell. This was first done in 1939 by Hodgkin and Huxley at Plymouth and by Curtis and Cole at Woods Hole, Massachusetts, using the giant axons innervating the mantle muscle of squids. As was mentioned in the previous chapter, these axons are of unusually large diameter (0.1–1 mm) and have been much used by nerve physiologists since their description by J. Z. Young in 1936.

The intracellular electrode used in these experiments was a glass capillary tube, about 100 μm in diameter, filled with sea water (Hodgkin and Huxley, 1945) or a potassium chloride solution isotonic with sea water (Curtis and Cole, 1942). It was inserted through the cut end of axon and pushed along so that its tip was level with an undamaged part of the axon. The potential difference between the electrode and the external sea water was then measured. The axon could be electrically stimulated via a pair of external electrodes. Fig. 4.4 shows a typical record of an action potential obtained by this method. The trace begins with

Fig. 4.4. An action potential recorded intracellularly from a squid giant axon. The vertical scale shows the potential (mV) of the internal electrode with respect to the external sea water. The action potential is preceded by a small stimulus artefact. The sine wave at the bottom is a time marker, frequency 500 Hz. (From Hodgkin and Huxley, 1945.)

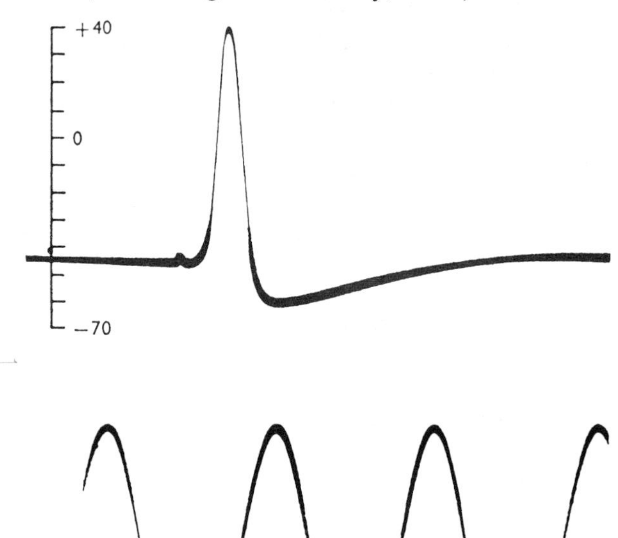

Fig. 4.5. Diagram to show the nomenclature applied to an action potential and the after-potentials which follow it.

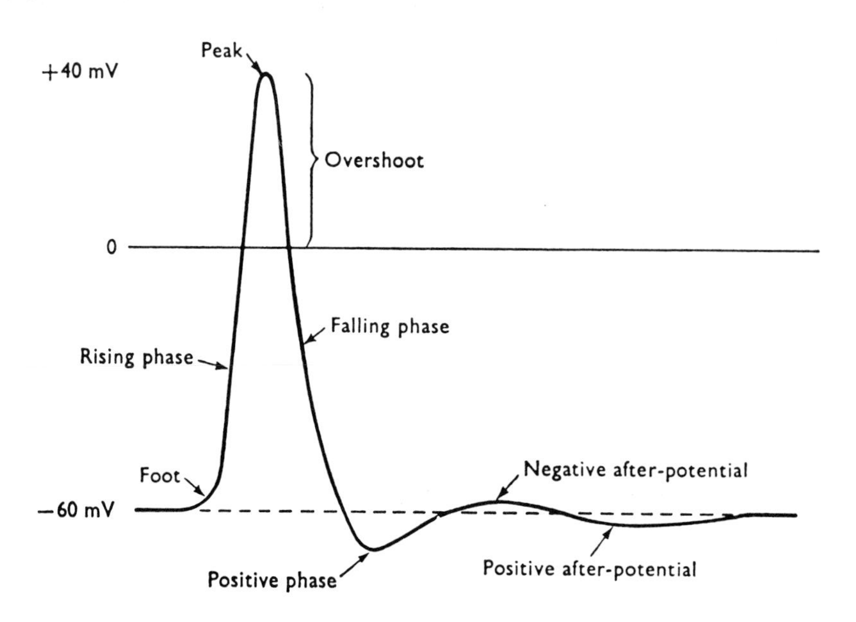

the inside of the axon negative to the outside (this is the resting potential), but during the action potential the membrane potential changes so that the inside becomes positive. The discovery of this 'overshoot' was unexpected at the time; its significance will be considered in the next chapter.

In many cases a careful examination of the time course of an action potential shows that the membrane potential does not immediately settle down at the resting level after its completion. As is shown in fig. 4.5, the action potential may be followed by (i) a *positive phase*, evident in fig. 4.4, in which the membrane potential 'underswings' towards a more negative value than the normal resting potential, (ii) a *negative after-potential*, in which the membrane potential is less negative than normal, and finally (iii) a *positive after-potential*, in which the membrane potential is again more negative than usual. The reason for the apparently paradoxical nomenclature of these after-potentials is that they were first observed using extracellular electrodes, which of course measure the potential on the outside of the axon membrane instead of the inside, so that the directions of all potential changes are reversed; this point is illustrated in fig. 4.6.

Fig. 4.6. Diagram to show the difference in sign of action potentials recorded by intracellular and by extracellular electrodes. Extracellular records are in fact frequently shown with negative potentials upwards.

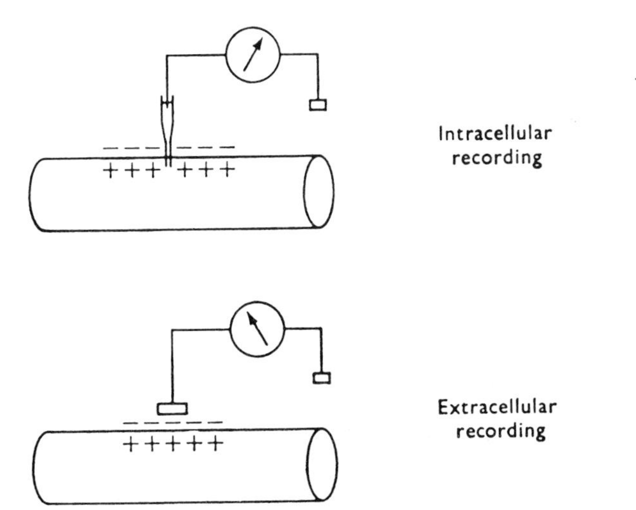

Subthreshold potentials

We will now consider some experiments performed by Hodgkin (1938) on isolated axons from the legs of the shore crab, *Carcinus*, with the object of examining the responses to subthreshold electrical stimuli. The axon was stimulated with brief (60 μs) electrical shocks, and the response recorded in the vicinity of the stimulating electrodes. The results of a typical experiment are shown in fig. 4.7.

The responses to inward current are shown as the deflections below the baseline in fig. 4.7. Notice that the voltage does not immediately return to the baseline after the end of the stimulating pulse, but decays gradually. The reason for this is that the nerve membrane behaves electrically as a resistance and capacitance in parallel. We can see how this system works by reference to a simple electrical model (fig. 4.8). A battery E is connected to a resistance R and capacitance C via a switch S, so that a voltage V appears across the resistance–capacitance network when the switch is closed. As a consequence there is a positive charge on one plate of the capacitance and a negative charge on the other. When the switch is opened, this charge flows through the resistance R as shown by the arrow, so that there is a potential difference across R which falls exponentially as the capacitance discharges. Thus, returning to fig. 4.7, the voltage decay following passage of inward current through the membrane (anodal stimulation) is caused by the discharge of the membrane capacitance through the membrane resistance. This type of electrical change (i.e. one that can be attributed to current flow through the resistance and capacitance of the resting membrane) is known as an *electrotonic potential.*

The responses above the baseline in fig. 4.7 were obtained with the cathode of the stimulating electrodes next to the recording electrode (i.e. with cathodal stimulation, producing a flow of current outwards through the membrane). When the stimulus intensity was very small, the responses were mirror images of the responses to anodal stimulation, and can similarly be described as electrotonic potentials. However, when the stimulus intensity was raised above half the threshold level, the resulting potential did not return to the baseline as rapidly as in the response to anodal stimulation of

Fig. 4.7. Subthreshold responses recorded extracellularly from a crab axon in the vicinity of the stimulating electrodes. The ordinate is a voltage scale on which the height of the action potential is taken as one unit. (From Hodgkin, 1938, by permission of the Royal Society.)

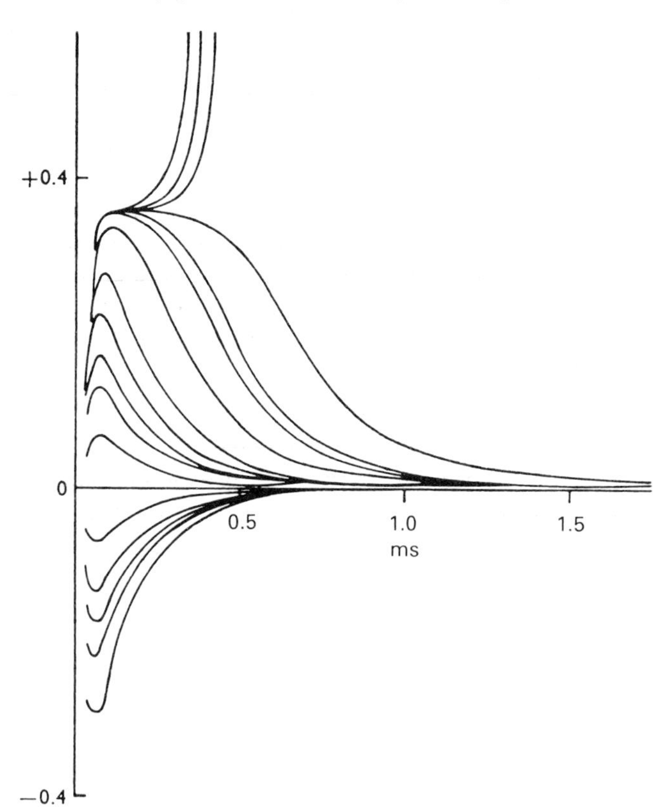

Fig. 4.8. Resistance and capacitance in parallel.

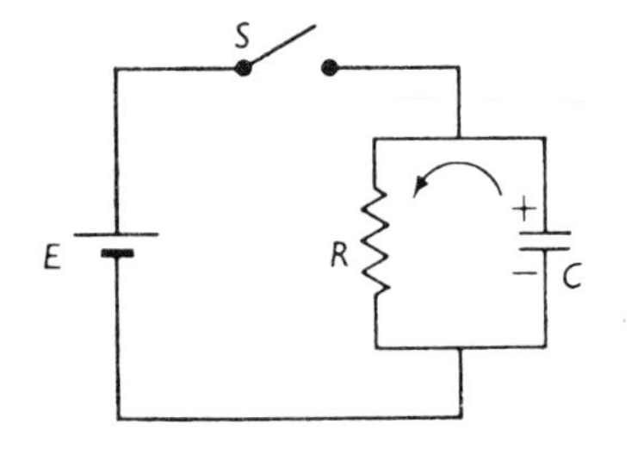

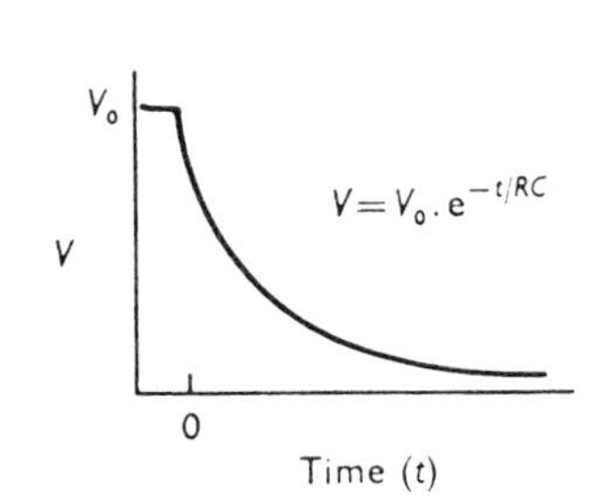

the same intensity. Hodgkin suggested that these responses were composed of a *local response* added to the electrotonic potential; the time course of the local response could therefore be obtained by subtracting the expected electrotonic response from the whole response. Finally, if the stimulus was large enough to produce a response which crossed a particular level of membrane potential, a propagated action potential was produced. This critical level of membrane potential is called the *threshold membrane potential.*

The results of this experiment are brought together in graphical form in fig. 4.9. For anodal stimuli, and for cathodal stimuli of intensities below about half threshold, the relation between stimulus and response is a straight line, so that the behaviour of the system (assuming the membrane capacitance to be constant) is in accordance with Ohm's law. Above this value this is not so, and thus we must assume that the resistance or capacitance of the membrane changes during local responses and action potentials.

Fig. 4.9. The relation between stimulus and response in a crab axon, derived from fig. 4.7. The abscissa shows the stimulus intensity, measured as a fraction of the threshold stimulus. The ordinate shows the recorded potential 0.29 ms after the stimulus, measured as a fraction of the action potential height. (From Hodgkin, 1938, by permission of the Royal Society.)

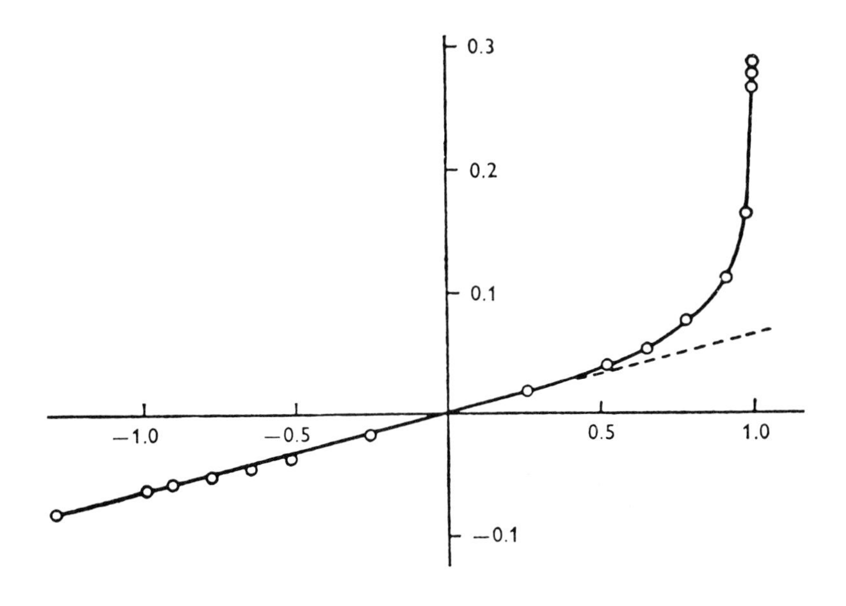

Impedance changes during activity

If we pass current (I) through a resistor, the voltage (V) across the resistor is proportional to the current (Ohm's law again), and the constant of proportionality is called the resistance (R), so that

$$V = IR.$$

Similarly, if we pass alternating current through a network containing a resistance and capacitance in parallel, the voltage appearing across the network is proportional to the current, and the constant of proportionality is called the *impedance (Z)*, so that

$$V = IZ.$$

The impedance is a complex quantity containing both resistive and capacitative elements, and dependent upon the frequency of the alternating current.

One method of determining the value of an unknown resistance R_x is by means of a simple Wheatstone bridge circuit (fig. 4.10*a*). The two fixed resistances R_1 and R_2 are known, and R_v is a known variable resistance. R_v is altered until there is no voltage between C and D. In this condition (when the bridge is said to be 'balanced') it is evident that

$$R_1/R_2 = R_x/R_v$$

from which R_x can be calculated.

A similar method can be used to measure impedance. In fig. 4.10*b* the unknown circuit contains a resistance R_x in parallel with a capacitance C_x. If the two fixed resistances are equal, then the bridge is balanced when the variable resistance R_v is equal to R_x and the variable capacitance C_v is equal to C_x. This is the principle of the method used by Cole and Curtis (1939) to measure the impedance changes in the axon membrane during the passage of an action potential. A squid giant axon was placed in a trough so that it passed between two plate electrodes which were connected to a bridge circuit similar to that shown in fig. 4.10*b*. A high-frequency alternating current (2–1000 kHz) was applied to the bridge, and the output was converted to a 175 kHz signal of the same amplitude and then displayed on an oscilloscope. Thus the oscilloscope trace was a thin line when the bridge was in balance, and became a broadened band (the 175 kHz a.c. signal) when it was out of balance. Knowing the values of the variable resistance and capacitance necessary to balance the bridge, and also the geometry of the axon–electrode system, it was possible to calculate the resistance and capacitance of the axon membrane.

Fig. 4.11*a* shows the effect of stimulation of the axon, with the bridge initially balanced in the resting condition. A short time after the start of the action potential, the oscilloscope trace broadens, indicating that the bridge is out of balance and therefore that the impedance of the system has changed. By starting with the bridge out of balance (as in fig. 4.11*b* and *c*), it is possible to find points at which the bridge is in balance at different times during the passage of the action potential and so to determine the resistance and capacitance of the membrane at these times. The results of these experiments showed that the membrane capacitance

Fig. 4.10. Wheatstone bridges. *a*: Simple d.c. bridge for the measurement of resistance. *b*: a.c. bridge used to measure resistance and capacitance in parallel.

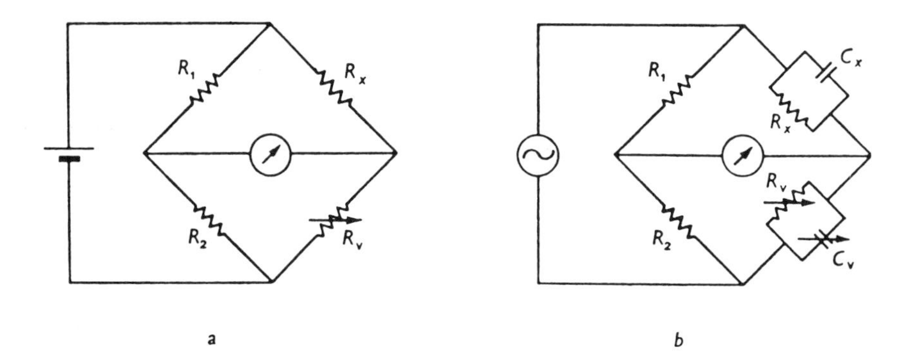

Fig. 4.11. Transverse impedance changes in a squid giant axon during the passage of an impulse. In *a*, the bridge was in balance during the resting state; in *b* and *c* it was out of balance during the resting state but was brought into balance by the impedance changes associated with the passage of the action potential. (After Cole and Curtis, 1939; redrawn.)

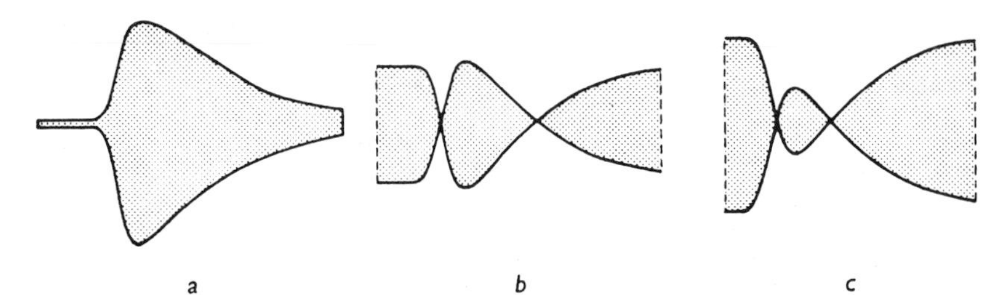

was 1.1 μF cm^{-2} in the resting condition, and fell by about 2% during activity. The resistance of the membrane, on the other hand, fell very markedly during activity, from its resting value of about 1000 Ω cm^2 to 25 Ω cm^2.*

The passive electrical properties of the axon membrane

From the results of the experiments so far described, it is evident that the axon membrane has a transverse resistance (r_m) and a transverse capacitance (c_m). If we wish to draw a circuit diagram to show the electrical properties of a length of axon, we must also include two other components, the longitudinal resistance of the external medium (r_o) and the longitudinal resistance of the axoplasm (r_i); the complete network is shown in the upper half of fig. 4.12. The suggestion that the axon can be represented in this way is known as the *core-conductor theory*, since it implies that the axon behaves as a poorly insulated cable. This representation is a simplification in some respects: (i) the membrane is a continuous structure, not a discontinuous series of elements; (ii) the resting potential is not represented – this could be incorporated by inserting a series of batteries in series with r_m, but since the object of the analysis is essentially to determine *changes* in membrane potential, this is an unnecessary refinement; and (iii) it omits the radial resistances of the external medium and axoplasm.

If we pass current through a part of the membrane so that a voltage (V_0) is set up across the membrane at that point, then the voltage across the membrane at some distant point $x(V_x)$ must be dependent in the distance x, and (because of the capacitances in the system) it must also vary with time, t. Hence it is necessary for us to know how V varies with x and t. We assume that the system is regular, i.e. that r_o, r_i, r_m and c_m do not change along the length of the axon, and that it is infinitely long. These quantities refer to unit lengths (e.g. 1 cm) of axon. Now the transverse current flowing through the membrane (i_m) will be the sum of the currents flowing through r_m and c_m, i.e.

$$i_m = \frac{V}{r_m} + c_m \frac{dV}{dt} \tag{4.1}$$

The current (i) flowing through the longitudinal resistances r_o and r_i must get progressively less as we move from its source, since a constant fraction in each unit length is diverted through the membrane; thus it follows that

$$i = -\frac{dV}{dx}\left(\frac{1}{r_o + r_i}\right) \tag{4.2}$$

and

$$i_m = -\frac{di}{dx} \tag{4.3}$$

Hence

$$i_m = \left(\frac{1}{r_o + r_i}\right)\frac{d^2V}{dx^2}. \tag{4.4}$$

Then, equating (4.4) with (4.1),

$$\frac{V}{r_m} + c_m \frac{\partial V}{\partial t} = \left(\frac{1}{r_o + r_i}\right)\frac{\partial^2 V}{\partial x^2},$$

or

$$V = \left(\frac{r_m}{r_o + r_i}\right)\frac{\partial^2 V}{\partial x^2} - r_m c_m \frac{\partial V}{\partial t}. \tag{4.5}$$

We now define two constants, a *space constant* λ, and a *time constant* τ, by

$$\lambda^2 = \frac{r_m}{r_o + r_i}$$

and $\tau = r_m c_m$.

Substituting these definitions in (4.5), we get

$$V = \lambda^2 \frac{\partial^2 V}{\partial x^2} - \tau \frac{\partial V}{\partial t}. \tag{4.6}$$

This is the relation between V, x and t that we have been looking for. The solution of (4.6) requires the use of various transform methods (see Hodgkin and Rushton, 1946; Taylor, 1963; and Jack *et al.*, 1975); we shall merely consider some relatively simple results that follow from it.

First, consider the situation when a constant current has been applied for a long (effectively infinite) time, to set up the voltage V_0 at that point and the voltage V_x at a distance x from it. Then (4.6) simplifies to become

$$V = \lambda_2 \frac{d^2V}{dx^2}.$$

* Notice that the membrane resistance must decrease with increase in area, hence it is measured in ohms *times* square centimetres, whereas the membrane capacitance increases with increase in area and is therefore measured in microfarads *per* square centimetre.

The relevant solution of this equation is

$$V_x = V_0 e^{-x/\lambda}. \tag{4.7}$$

This means that the voltage across the membrane falls off exponentially with the distance from the point at which the current is applied. This feature is shown in the lower half of fig. 4.12. If we make x equal to λ in (4.7), V_x becomes $V_0 e^{-1}$; λ can thus be defined as the length over which the voltage across the membrane falls to 1/e of its original value.

Now consider the total charge (Q) on the membrane. This is given by

$$Q = c_m \int_0^{\infty} V \mathrm{d}x, \tag{4.8}$$

so that Q is obtained by integrating (4.6) with respect to x. If the applied current is constant, beginning at $t = 0$, then the solution of (4.8) is

$$Q_t = Q_\infty (1 - e^{-t/\tau}), \tag{4.9}$$

where Q_t is the charge at time t, and Q_∞ is the charge at infinite time (or, effectively, when $t \gg \tau$). If a current which has produced a charge Q_0 is suddenly switched off at $t = 0$, then the charge on the membrane decays according to the equation

$$Q_t = Q_0 e^{-t/\tau}. \tag{4.10}$$

If we put t equal to τ in (4.9) or (4.10) it becomes evident that τ can be defined as the time taken for the change in total charge on the membrane to reach 1/e of completion.

The equations describing the time course of the voltage change at any particular point are rather complex, but their general effect is that the potential rises and falls less rapidly the farther that point is from the point at which the current is applied. The responses to the onset and cessation of constant currents are, of course, limited by (4.7), (4.9) and (4.10). Fig. 4.13 shows the results of Hodgkin and Rushton's calculations of the voltage changes.

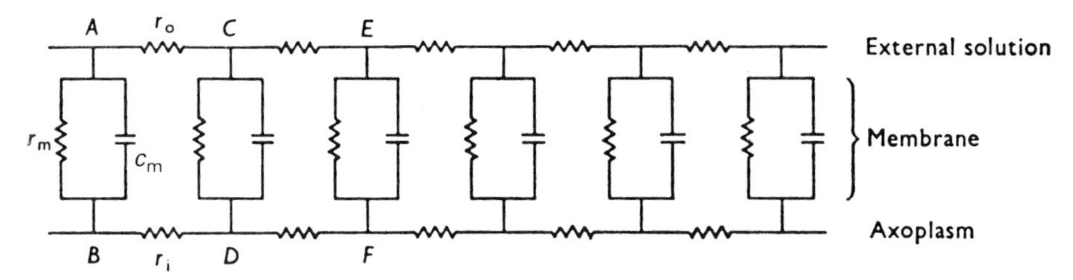

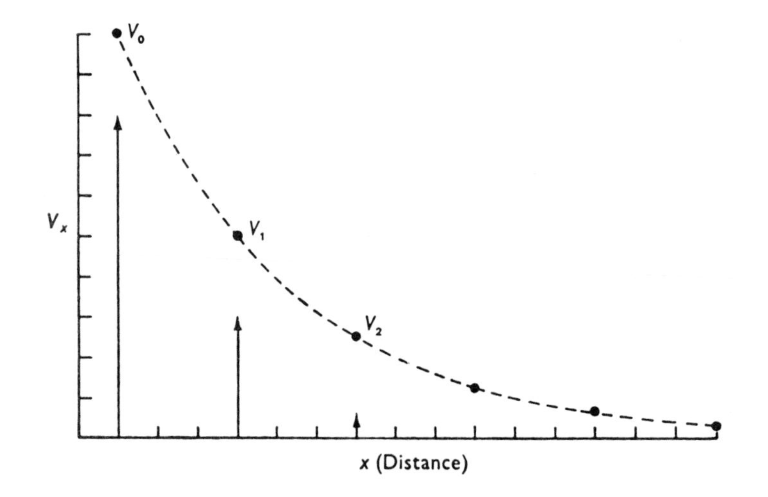

Fig. 4.12. Electrical model of the passive (electrotonic) properties of a length of axon. The graph shows the steady-state distribution of transmembrane potential along the model when points A and B on the model are connected to a constant current source. The vertical arrows on the graph are used to indicate that the potential at any point on the membrane rises gradually to its final value when the current is applied.

The core-conductor equations we have examined so far have been based on quantities referring to a unit length of axon. For many purposes, however, it is useful to possess measurements of membrane properties in terms of unit area; the conversion from one to the other can easily be made if we know the radius of the axon. The relation between the membrane resistance of a unit length of axon, r_m, and the resistance of a unit area of membrane, R_m, is given by

$$r_m = R_m / 2\pi a, \tag{4.11}$$

where a is the axon radius. Similarly, the membrane capacitance per unit length, c_m, is related to the membrane capacitance per unit area, C_m, by

$$c_m = 2\pi a C_m \,. \tag{4.12}$$

Fig. 4.13. Theoretical distribution of potential difference across a passive nerve membrane in response to onset (*a* and *c*) and cessation (*b* and *d*) of a constant current applied at the point $x = 0$. *a* and *b* show the spatial distribution of potential difference at different times, and *c* and *d* show the time courses of the potential change at different distances along the axon. Time (t) is expressed in units equal to the time constant, τ, and distance (x) is expressed in units to the space constant, λ. (From Hodgkin and Rushton, 1946, by permission of the Royal Society.)

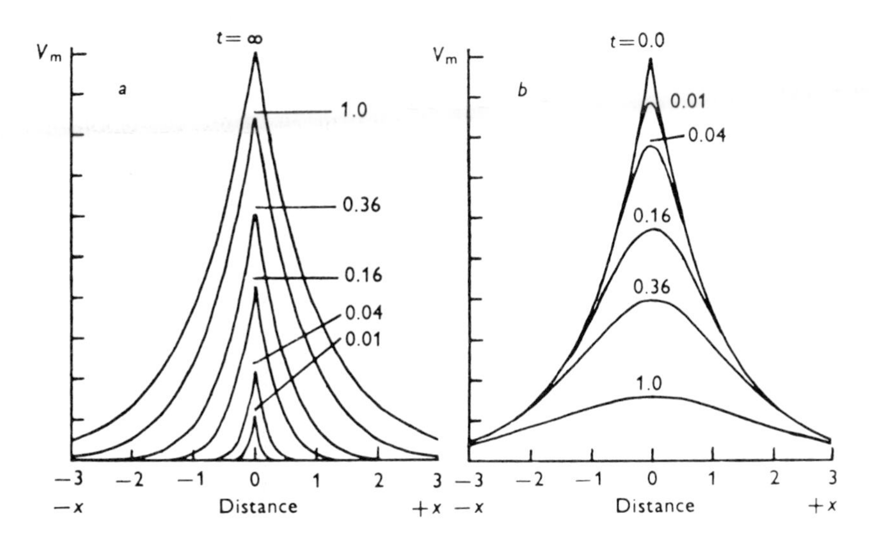

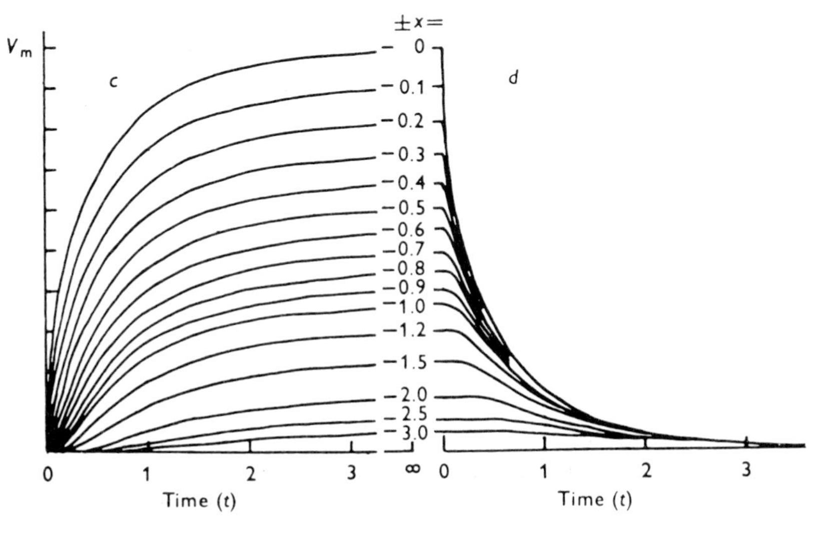

And the longitudinal resistance of the axoplasm in a unit length of axon, r_i, is related to the resistivity of the axoplasm, R_i, by

$$r_i = R_i / \pi a^2 . \tag{4.13}$$

To take a numerical example, consider an axon of radius 25 μm, membrane resistance (R_m) 2000 Ω cm^2, axoplasm resistivity (R_i) 60 Ω cm and membrane capacitance (C_m) 1 μF cm^{-2}. Then r_m (membrane resistance of a unit length) is 127 000 Ω cm, r_i (internal longitudinal resistance) is 3 060 000 Ω cm^{-1}, and c_m (membrane capacitance per unit length) is 0.0157 μF cm^{-1}.

These relations can usefully be inserted in some of the core-conductor equations. For simplicity we shall assume that the axon is in a large volume of external medium, so that r_o is very much less than r_i and can therefore be omitted from the equations. First, consider the space constant λ. If r_o is very small compared with r_i, then

$$\lambda = \sqrt{\frac{r_m}{r_i}} = \sqrt{\frac{R_m / 2\pi a}{R_i / \pi a^2}} = \sqrt{\frac{a \cdot R_m}{2R_i}} .$$

Hence λ increases with increasing axon radius; if R_m and R_i remain constant, λ is proportional to the square root of the radius.

We have seen that the time constant is defined by

$$\tau = r_m c_m .$$

Substituting (4.11) and (4.12) in this we get

$$\tau = (R_m / 2\pi a)(2\pi a C_m) = R_m C_m .$$

Thus τ is essentially independent of the axon radius.

A further useful relation follows from (4.4); substituting (4.13) in this we get

$$i_m = \frac{\pi a^2}{R_i} \cdot \frac{d^2 V}{dx^2} .$$

Now the membrane current density per unit area, I_m, will be related to the membrane current per unit length, i_m, by

$$i_m = I_m \cdot 2\pi a .$$

Hence

$$I_m = \frac{a}{2R_i} \cdot \frac{d^2 V}{dx^2} . \tag{4.14}$$

(This equation, as we shall see later, is of considerable importance in the further analysis of nervous conduction.)

The core-conductor theory was experimentally tested by Hodgkin and Rushton (1946) on large axons from the walking legs of the lobster *Homarus*. Since they used extracellular electrodes, the voltage change measured was that on the outside of the membrane; this will be proportional to the change in potential across the membrane (the proportionality constant will be $r_o/(r_o + r_i)$) so that the core-conductor equations can still be applied. It was found that the observed potential changes were in accordance with (4.7), (4.9) and (4.10) and with equations derived from (4.6) describing the change of potential with time at different distances from the point of application of the current. This correspondence between fact and theory is obviously good evidence for the core-conductor theory. The next step was to evaluate the quantities r_o, r_i, r_m and c_m. Then, knowing the diameter of the axon, the resistance and capacitance of a square centimetre of membrane could be calculated. Average values from experiments on ten axons were 2300 Ω cm^2 for R_m, the membrane resistance (range 600–7000 Ω cm^2 – this quantity was rather variable), and 1.3 μF cm^{-2} for C_m, the membrane capacitance. These values are in quite good agreement with those obtained for squid axons by Cole and Curtis using the a.c. bridge method, a fact which further increases our confidence in the applicability of the core-conductor model.

The local circuit theory

We are now in a position to consider how the action potential is propagated. Suppose a small length of axon is stimulated by a depolarizing pulse above threshold intensity, so that an action potential arises in the stimulated area. Regard this action potential as corresponding to the voltage V_0 in fig. 4.12 and assume that V_0 is about four times the threshold change in membrane potential. The potential across *CD* will now rise towards the point V_1, but at some point before reaching V_1 it will cross the threshold, so that an action potential appears across *CD*. This means that the potential across *EF*, originally moving towards V_2, now starts moving towards the potential level of V_1, and so it in turn crosses the threshold and an action potential now arises across *EF*. Thus an action potential has been propagated from *AB* through *CD* to *EF*, and will of course continue along the chain. Furthermore, unless there is any change along the length of the membrane in the threshold level or in the values of the component resistances and capacitances of the system, the conduction velocity will obviously be constant.

This hypothesis of the mechanism of conduction, proposed by Hermann at the turn of the century, is known as the *local circuit theory*, since it postulates that conduction is dependent on the electrotonic currents across the membrane set up in front of the action potential. The local circuits are shown in fig. 4.14*a*. Notice that, in order to set up passive (electrotonic) currents at the beginning of the action potential, there must be some inward current flow at the peak of the action potential. This inward current flow is from negative to positive and is therefore analogous to the flow *inside* a battery; in other words, the electrical energy needed to cause the electrotonic currents (which flow from positive to negative) is derived from this inward movement of positive charge at the peak of the action potential. This concept is illustrated in fig. 4.14*b*. The 'battery'* is brought into action at any particular point when the membrane potential crosses the threshold at that point. The quantitative distribution of membrane current can be deduced from the form of the action potential by applying (4.4).

If the local circuit theory is correct, then it should be possible to observe the local currents associated with the action potential. This was done by Hodgkin (1937) by cooling a short length of nerve so that the action potential could not pass this region. On the other side of the block, small *extrinsic potentials* appeared (fig. 4.15), which diminished in size along the length of the nerve in the same way that electrotonic potentials do.

A less direct method of testing the theory is by consideration of the factors which affect conduction velocity. Reverting to fig. 4.12 again, it is evident that the conduction velocity must be dependent upon the time taken for the potential at a point a given distance in front of the action potential to cross the threshold. This time, in turn, will be governed by the values of the resistances in the system. Hence changes in r_o or r_i should produce changes in conduction velocity. Hodgkin (1939) showed that an increase in r_o, produced by immersing an axon in paraffin oil instead of Ringer's solution, decreased the conduction velocity. Con-

* We shall see in the next chapter that this 'battery' is a sodium ion concentration cell and that the inward current it produces is an inward flow of sodium ions.

Fig. 4.14. *a*: Local circuit currents set up by a propagating action potential. *b*: Model system to show the local circuit currents (shown by arrows) set up by a battery inserted in the core-conductor model.

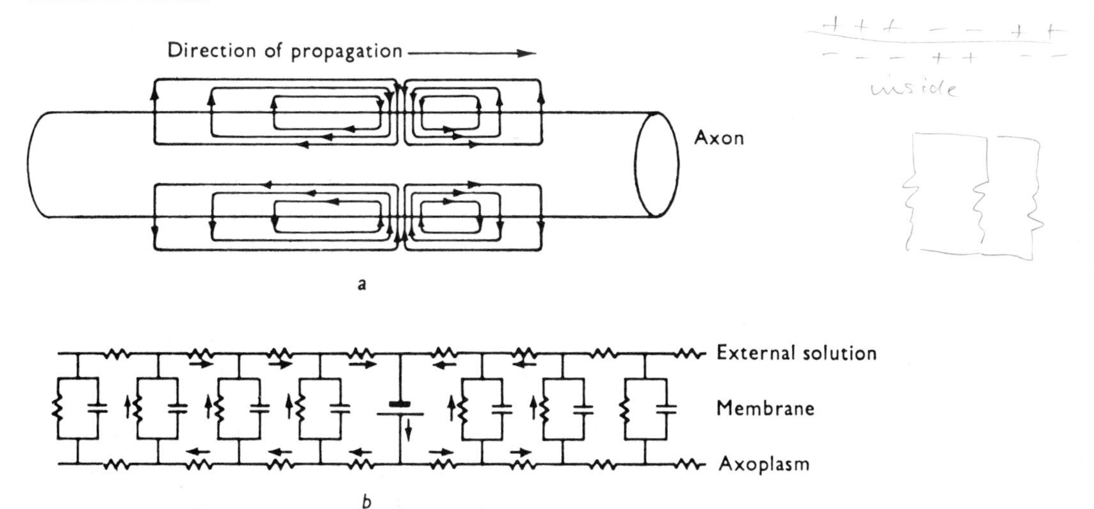

versely, a decrease in r_o, produced by laying the axon across a series of brass bars which could be connected by means of a mercury trough, resulted in an increase in conduction velocity. More recently, del Castillo and Moore (1959) have obtained large increases in conduction velocity by inserting a silver wire down the middle of an axon, so greatly reducing r_i.

Since r_i varies with axon radius, we would expect conduction velocity to vary with axon radius also. Theoretical arguments given by Hodgkin (1954) suggest that, other things being equal (a rather important proviso), conduction velocity is proportional to the square root of the axon radius. A simple argument why this should be so is as follows. For an action potential travelling at a constant velocity θ,

$$\frac{dV}{dx} = \frac{1}{\theta} \cdot \frac{dV}{dt}$$

and

$$\frac{d^2V}{dx^2} = \frac{1}{\theta^2} \cdot \frac{d^2V}{dt^2}.$$

Substituting this relation in (4.14), we get

$$I_m = \frac{a}{2R_i \cdot \theta^2} \frac{d^2V}{dt^2}. \tag{4.15}$$

If the membrane properties remain the same in axons of different radii, then I_m and d^2V/dt^2 will be constant. In this case (4.15) becomes

$$\frac{a}{\theta^2} = \text{constant}$$

or

$$\theta = \text{constant} \times \sqrt{a},$$

i.e. the conduction velocity is proportional to the square root of the fibre radius.

An experimental investigation of this relation was carried out by Pumphrey and Young (1938), using the giant axons of *Sepia* and *Loligo*; the best fit for their results was given by conduction velocity being proportional to $(\text{radius})^{0.6}$, which is very near to the expected value. Other investigations have not always provided such good agreement with the square root relation, but it is generally observed that, in a group of axons of the same type, increase in radius is associated with increase in conduction velocity.

The conclusion to be drawn from the experiments and observations described in this section is that the action potential propagates by means of local circuits, as shown in fig. 4.14*a*. This analysis enables us to explain the shapes of action potentials as recorded by external electrodes. In fig. 4.16*a* two electrodes *X* and *Y* are placed a short distance

Fig. 4.15. Extrinsic potentials in frog sciatic nerve. A 3 mm length of nerve was blocked by cooling, and the electrical response of the nerve was recorded 2 mm proximal to the block (*a*) and at various distances distal to the block (*b*–*f*). Notice that the gain for records *b* to *f* is five times that for *a*. The distal records were made at the following distances from the block (mm): *b*, 1.4; *c*, 2.5; *d*, 4.1; *e*, 5.5; and *f*, 8.3. (From Hodgkin, 1937).

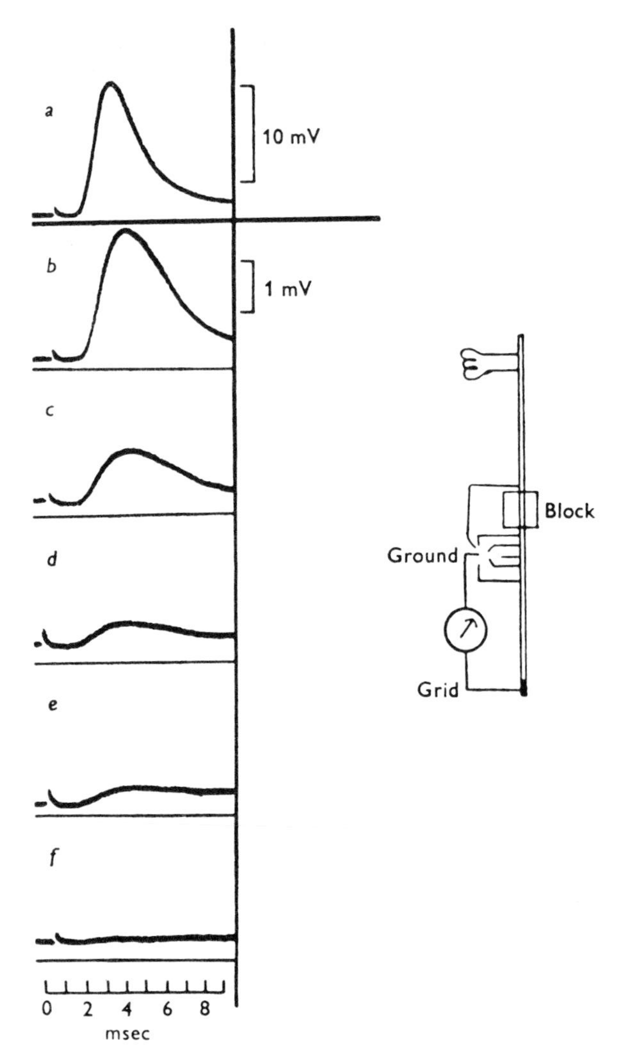

apart on an axon. We shall assume that the input resistance of the recording system used to measure the potential difference between X and Y is very high. As the active region moves under electrode X, X becomes negative to Y, so producing an upward peak in the record, However, when the active region reaches electrode Y, X becomes positive to Y and so the potential record shows a downward peak. This type of recording is known as diphasic. In fig. 4.16*b* the axon is crushed between the electrodes so that the action potential is unable to reach Y, and therefore Y never goes negative; this gives a monophasic record which is similar in shape to the potential across the membrane at X. In fig. 4.16*c* there is a shunt resistance between the electrodes (such as might be produced by recording from a nerve in Ringer's solution or *in situ*) so that currents can flow between them. In this case the electrode X records the local currents associated with the action potential and the record is therefore triphasic.

Fig. 4.16. The form of action potentials recorded from nerve axons with extracellular electrodes. The active region is stippled in each case. The records on the right show an upward deflection when electrode X is negative to electrode Y. Further explanation in the text.

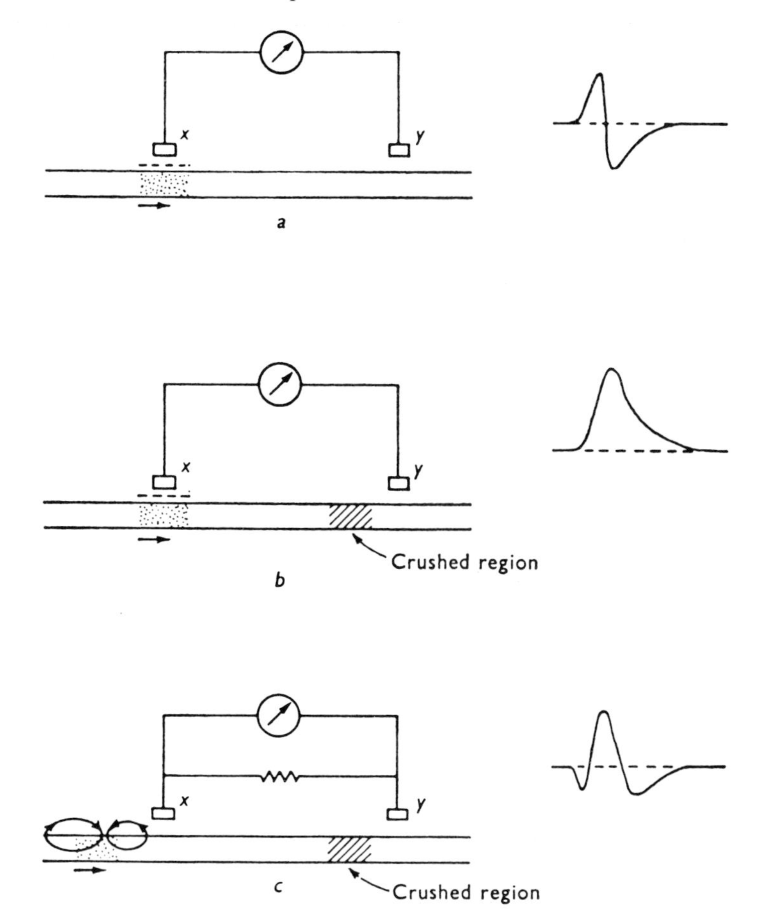

Saltatory conduction in myelinated nerves

All peripheral nerve axons are surrounded by accessory cells called *Schwann cells* (fig. 4.17). A connective tissue sheath, the *endoneurium*, which contains small endoneurial cells, surrounds the complex of Schwann cells and axon. Numbers of these Schwann cells and endoneurial cells are found distributed along the length of an axon. Transverse sections of large axons (such as squid giant axons, as shown in fig. 4.17*c*) shows that many Schwann cells are needed to surround them.

When the axons are small, the reverse situation may be seen (fig. 4.17*b*), in which a single Schwann cell surrounds a number of axons. In all these cases the axon is separated from the Schwann cell by a space about 150 Å wide, which is in communication with the extracellular fluid via channels (mesaxons) between the membranes of the Schwann cells. Examination of axons of this type by light microscopy does not reveal the presence of a fatty sheath surrounding the axon, and the nerve fibres (we shall regard a 'fibre' as comprising an axon plus its accessory cells) are therefore described as *unmyelinated.*

Many vertebrate and a few invertebrate axons (such as the large diameter axons of prawns; Holmes, 1942) are surrounded by a fatty sheath known as *myelin* and are described as being *myelinated.* The sheath is interrupted at intervals of the order of a millimetre or so, to form the *nodes of Ravier* (fig. 4.18). Electron microscope studies by Geren (1954) and Robertson (1960) have shown that myelin is formed from many closely packed layers of Schwann cell membrane, produced by the mesaxon being wrapped round and round the axon (fig. 4.19). As a consequence of this arrangement, there are no extracellular channels between the axon membrane and the external medium in the internodal regions. Such contact is available at the nodes, as is shown in fig. 4.20.

It is obvious that the presence of a myelin sheath will considerably affect the electrical properties of the nerve fibre. The overlapping Schwann cell membranes act as chains of resistances and capacitances in series. This means

Fig. 4.17. The form of Schwann cells surrounding unmyelinated axons. *a* shows the basic arrangement, *b* the situation with many small axons, and *c* the situation with large axons. (Based on Hodgkin, 1964.)

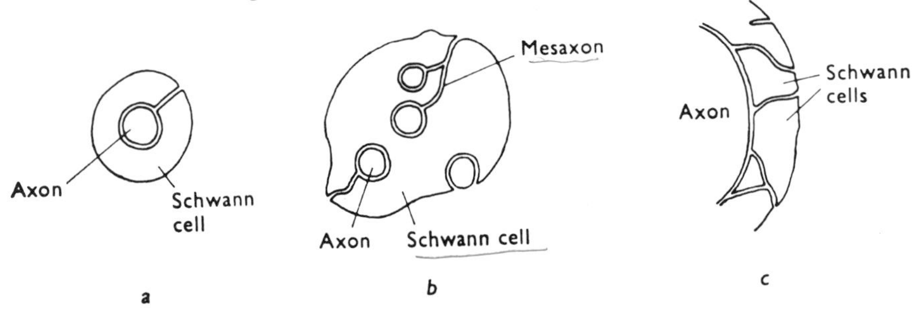

Fig. 4.18. Diagram to show the structure of a myelinated nerve fibre.

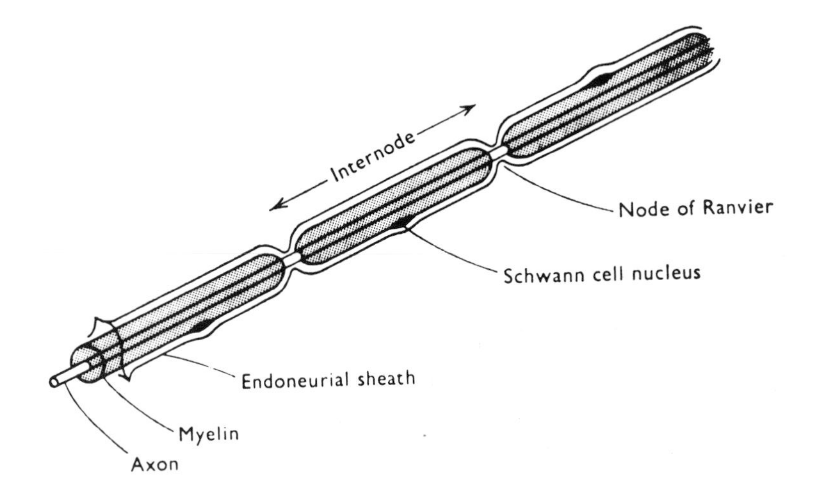

that the myelin sheath will have a much higher transverse resistance and a much lower transverse capacitance than a normal cell membrane; the actual figures for frog fibres are about 160 000 Ω cm² and 0.0025 μF cm⁻². The values for the resting membrane at the node are quite different, being of the order of 20 Ω cm² and 3 μF cm⁻². In view of these figures, we might expect that when current is passed across the membrane the current flow through the nodes would be greater than that at the internodes. If, in addition, the nodes are the regions at which excitation of the membrane occurs, then, applying the local circuit theory, we would expect conduction in a myelinated axon to be a discontinuous (or *saltatory*) process.

Verification of the saltatory conduction theory depends first of all on the ability of the experimenter to dissect out single fibres from a whole nerve trunk. This technique was first developed by Japanese workers in the nineteen-thirties. It was found that blocking agents such as cocaine or urethane were only effective when applied at the nodes, and that the threshold stimulus intensity was

Fig. 4.19. The development of the myelin sheath by vertebrate Schwann cells. (From Robertson, 1960, by permission of Pergamon Press Ltd.)

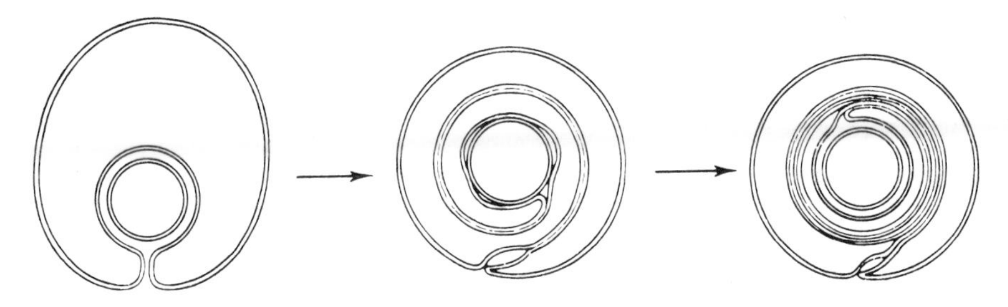

Fig. 4.20. Longitudinal section of a node of Ranvier, as seen by electron microscopy. (From Robertson, 1960, by permission of Pergamon Press Ltd.)

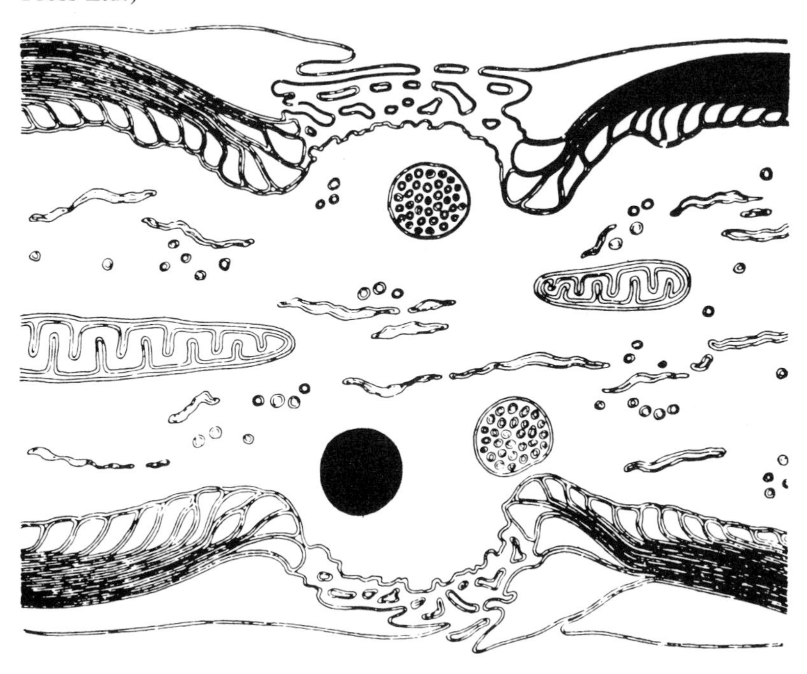

lowest at the nodes (fig. 4.21). These results accord well with the predictions of the saltatory theory, but it was possible to argue that the sensitivity of the nodal region was merely caused by the exposure of the axon membrane at this site. More conclusive evidence that conduction is saltatory would be provided if it could be shown that inward flow of current is restricted to the nodes. Experiments to test this idea have been performed by Tasaki and Takeuchi (1942) and by Huxley and Stämpfli (1949); we shall now consider some of the details of these experiments.

Tasaki and Takeuchi's technique is shown in fig. 4.22. The nerve fibre is placed in three pools of Ringer's solution insulated from each other by air gaps. The two outer pools are earthed, and the middle pool is connected to earth via a resistance. With this arrangement, all the radial current flowing across the axon membrane and myelin sheath of that part of the fibre which is in the

Fig. 4.21. The variation in threshold stimulus intensity along the length of a single myelinated fibre. N_1 and N_2 mark the positions of two nodes of Ranvier. (From Tasaki, 1953.)

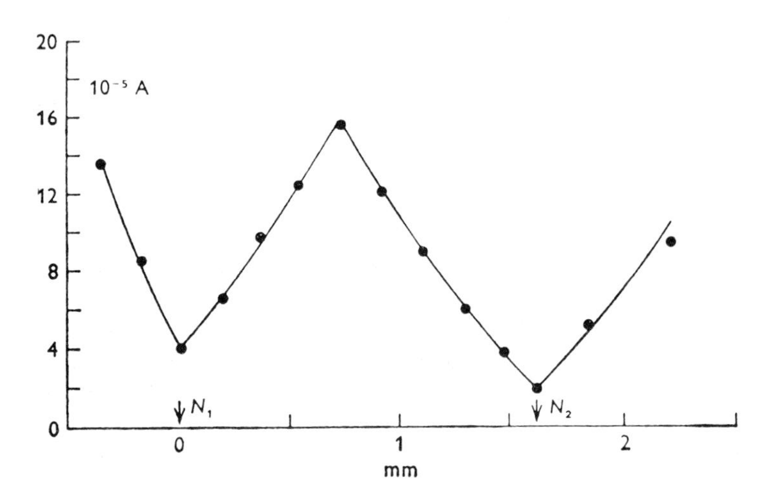

Fig. 4.22. The radial currents in a short length of a myelinated fibre during the passage of an action potential. *a*: The recording arrangement. *b*: The potential difference across the resistance *R* when the middle pool of Ringer does (right trace) or does not (left trace) contain a node. (From Tasaki and Takeuchi, 1942.)

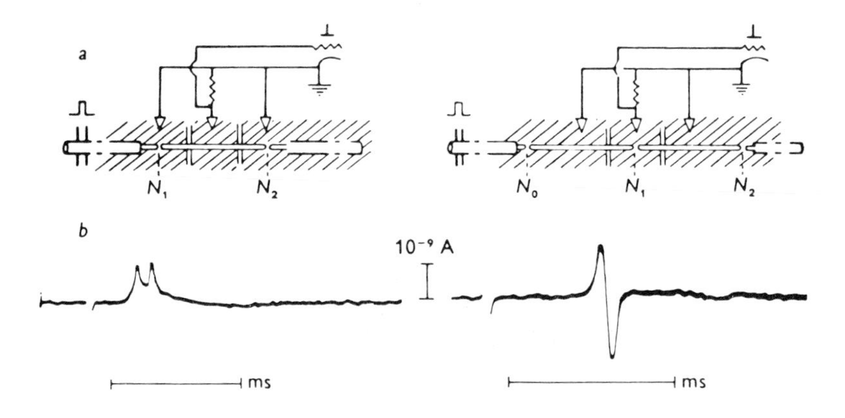

middle pool will flow through the resistance, and can therefore be measured by the potential of the middle pool with respect to earth. The results (fig. 4.22*b*) showed that inward currents occurred only when there was a node in the middle pool, and therefore inward current flow must be restricted to the nodal regions.

The technique used by Huxley and Stämpfli was rather different. The fibre was threaded through a fine hole in a glass slide, and the potential measured on each side of the hole during the passage of an impulse (fig. 4.23). If the resistance of the Ringer's solution surrounding the fibre in the hole is known, then the longitudinal current flow can readily be obtained from the measured potential by application of Ohm's law. By pulling the fibre through the hole, a series of such measurements can be made at intervals along its length, as shown in fig. 4.24*a*. The transverse current across the membrane is then given by (4.3), i.e. by the difference between the longitudinal currents at two adjacent recording sites. The calculated membrane currents are shown in fig. 4.24*b*, from which it is clear that here, again, inward current flow is restricted to the nodes. This implies that the transverse current in the internodal regions can be explained as a passive current through a resistance and capacitance in parallel, but there must be some 'active' component at the nodes of Ranvier.

The conduction velocity of myelinated fibres is normally greater than that of unmyelinated fibres of the same diameter. The reason for this is that the high resistance and low capacitance of the myelin sheath increases the lingitudinal spread of the local currents involved in propagation. Theoretical arguments given by Rushton (1951) suggest that conduction velocity should be proportional to the fibre diameter, and not, as in unmyelinated axons, to the square root of fibre diameter. This appears to be so (Hursh, 1939; Tasaki *et al.*, 1943). This conclusion implies that the conduction velocity of very small myelinated fibres should be less than that of unmyelinated fibres of the same diameter (fig. 4.25). Rushton calculated that the critical diameter below which myelination confers no advantage should be about 1 μm, which is near that of the largest unmyelinated fibres in mammalian peripheral nerves.

However, Waxman and Bennett (1972) pointed out that myelinated fibres less than 1 μm in diameter are found in the central nervous system, and suggested that the critical diameter is as low as 0.2 μm.

Why should there be such a difference between central and peripheral fibres? Ritchie (1982) concluded that the answer lies in the different mechanisms of myelination in the two cases. In peripheral nerve fibres each internode is produced by a separate Schwann cell and there is a minimum internodal distance, in the range 200–400 μm. This means that at small diameters (less than 4 μm) the internodal distance is longer than the optimum for conduction, and at very small diameters the electrotonic spread along the internode will be insufficient to cause excitation at the succeeding node. Ritchie calculates that the limiting diameter at which conduction will fail is about 0.8 μm. In central fibres, however, the myelin sheaths are produced by oligodendrocytes, each of which may myelinate up to 50 internodes (Hirano, 1981), and there seems to be no anatomical lower limit to the length of the internodes.

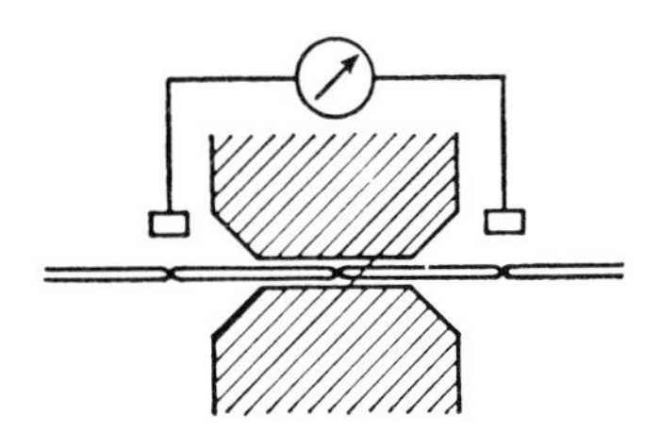

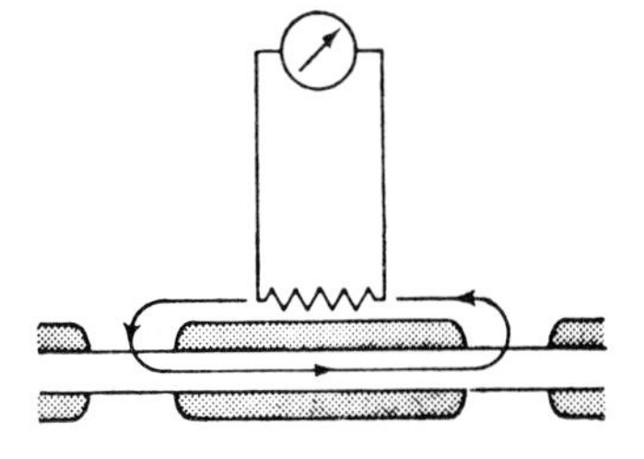

Fig. 4.23. The method used by Huxley and Stämpfli to measure longitudinal currents outside a short length of a myelinated fibre. *a* gives a schematic view of the recording arrangement, and *b* the equivalent circuit of the system. (After Huxley and Stämpfli, 1949.)

Electrical stimulation parameters

Strength–duration relation

If an axon is stimulated with square constant current pulses, it is found that the threshold stimulus intensity rises as the pulse length is lessened. This effect is known as the *strength–duration relation* (fig. 4.26). The curve in fig. 4.26 is quite well fitted by the empirical equation

$$\frac{I}{I_0} = \frac{1}{1 - e^{-t/k}} \tag{4.16}$$

where I is the intensity of the pulse, t is the length of the pulse, I_0 is the threshold stimulus intensity when t is large (the *rheobase*) and k is a constant (Lapique, 1907; see Hill, 1936). The pulse length when the threshold stimulus intensity is twice the rheobase is called the *chronaxie*.

Equation (4.16) is similar in form to that describing the change of voltage across a circuit consisting of a resistance and capacitance in parallel following the application of a constant current, i.e.

$$V = IR(1 - e^{-t/RC}) \tag{4.17}$$

where V is the voltage, I the current, t the time after application of the current, R the resistance and C the capacitance. If V is constant (for instance, if we regard V as the threshold membrane depolarization), this becomes

$$I \times \text{constant} = \frac{1}{1 - e^{-t/RC}}$$

which is equivalent to (4.16). But this equation involves a number of simplifications (see Noble, 1966). Firstly, the voltage across the passive membrane cannot be described by (4.17); the much

Fig. 4.24. Longitudinal and radial currents at different points along the length of a myelinated fibre during the passage of an action potential. *a*: Longitudinal currents, measured directly by the method shown in fig. 4.23.

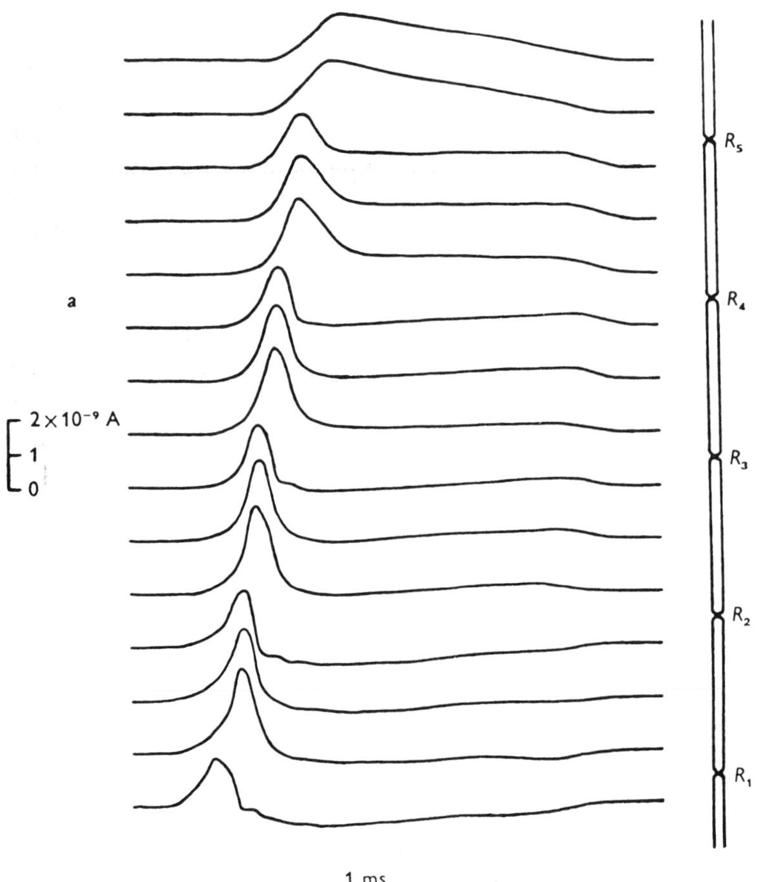

more complicated relations derived from (4.6) must be applied. Secondly, strong currents of short duration produce depolarizations which fall off rapidly with distance from the stimulating electrode, whereas the depolarizations produced by weaker currents of longer duration are more spread out, as will be evident from fig. 4.13. Since propagation is dependent on local circuits of sufficient intensity (involving movement of at least a minimal quantity of electric charge), the size of the active region is important, hence the threshold membrane depolarization is higher when smaller areas of membrane are depolarized, and therefore it is higher for brief constant current pulses than for longer ones. Finally, the existence of local responses (p. 35) means that (4.6) cannot be strictly applied for depolarizations greater than about half threshold. Thus, while it is easy to provide a qualitative explanation for the strength–duration relation (depolarization is more rapid with stronger currents; see fig. 4.27), a quantitative formulation of the various factors involved (Noble and Stein, 1966) must necessarily be very complex.

Fig. 4.24*b*: Radial (membrane) currents, obtained from the difference between the longitudinal currents at two points 0.75 mm apart. (From Huxley and Stämpfli, 1949.)

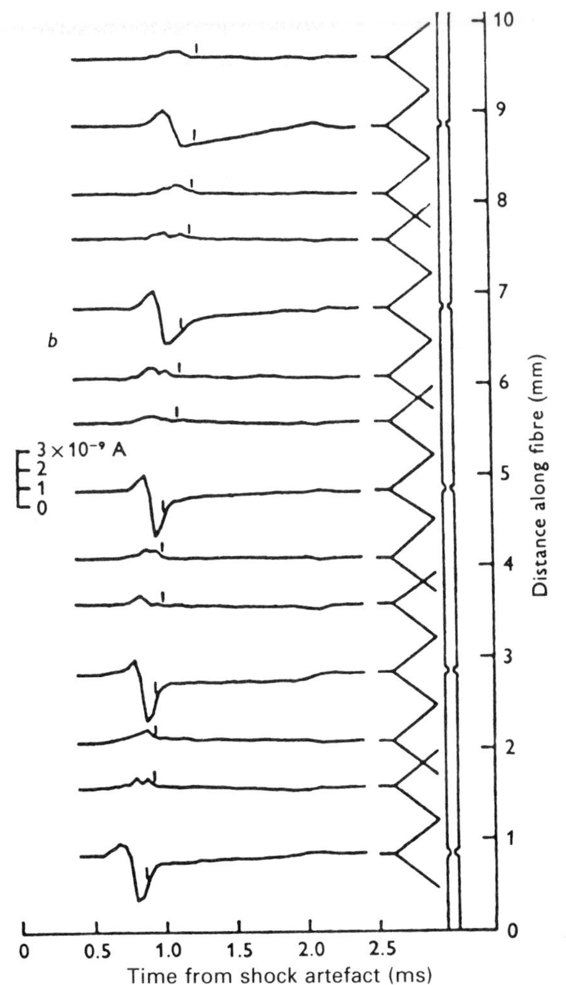

Latency

The time between the onset of a stimulus and the peak of the ensuing action potential is called the *latency* of the response. It is clear from fig. 4.27 that the latency decreases with increasing current strengths.

Latent addition

For a short time after a brief subthreshold cathodal stimulus, the membrane potential is nearer to the threshold membrane potential than usual. During this time, less current is required to depolarize the membrane by enough to cause excitation, so that stimuli which are normally of subthreshold intensity may be sufficient to elicit an

Fig. 4.25. Theoretical relations between fibre diameter and conduction velocity in myelinated and unmyelinated nerve fibres. (From Rushton, 1951.)

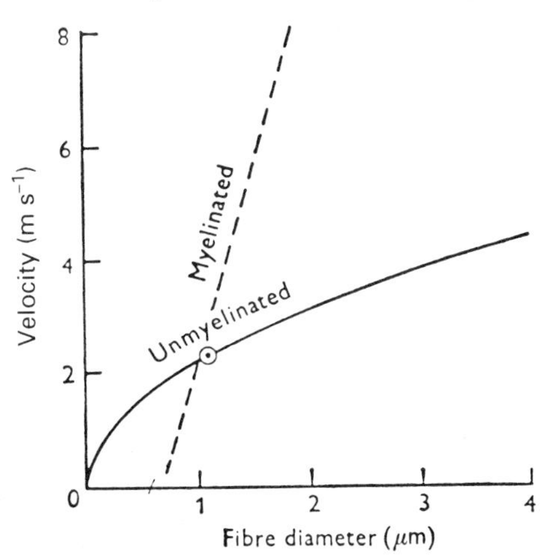

action potential (fig. 4.28). Conversely, if an anodal shock is applied, a cathodal stimulus soon afterwards has to be of a higher intensity than usual in order to cause excitation. These phenomena constitute *latent addition.* They can be used to investigate the time course of subthreshold responses. Katz (1937), for instance, demonstrated indirectly the presence of local responses in this way; his results were very similar to those obtained by Hodgkin (1938) from direct observations on crab nerves.

Accommodation

If a constant subthreshold depolarizing current is passed through an axon membrane, it is found that the threshold slowly rises during the passage of the current, and then falls again after its cessation. Conversely, when hyperpolarizing current is used, the threshold falls. This delayed dependence of the threshold on membrane poten-

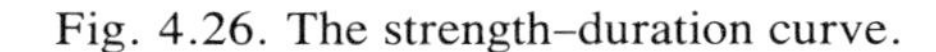
Fig. 4.26. The strength–duration curve.

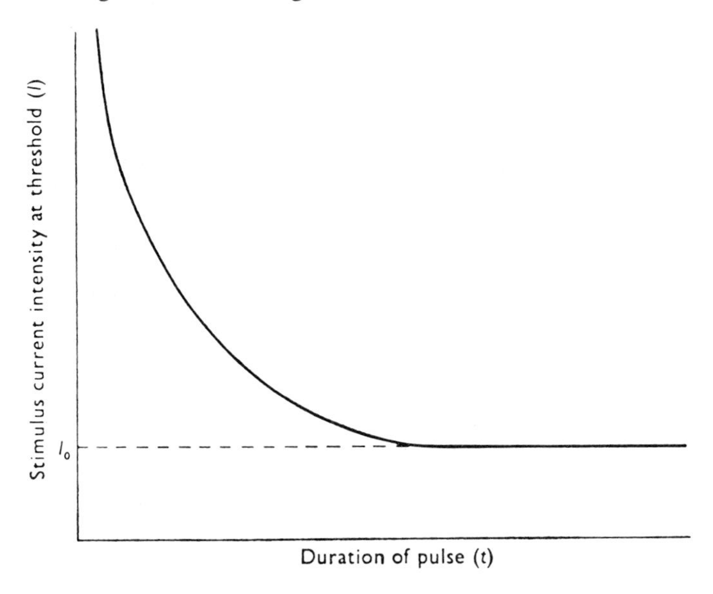

Fig. 4.27. Schematic diagram to show the effects of depolarizing currents on the membrane potential of a nerve axon. The current strengths increase in the order *a* to *f*. Responses *c* to *f* lead to excitation.

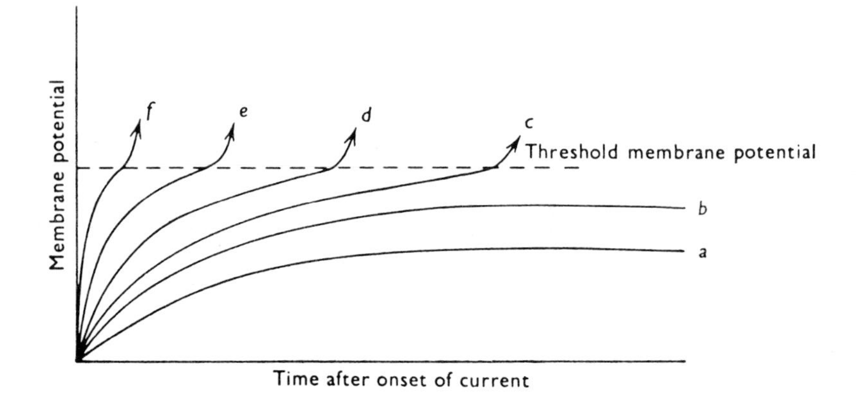

tial is known as accommodation (fig. 4.29). With hyperpolarizing currents of sufficient strength, the threshold may fall beyond the resting potential, so that when the current is switched off, the membrane potential is temporarily above threshold. This leads to the initiation of an action potential (fig. 4.29*c*), and the phenomenon is therefore known as *anode break excitation.*

If an axon is stimulated by a linearly rising cathodal current, the membrane potential change follows an approximately linear course, and, because of accommodation, the threshold membrane potential also changes slowly. Consequently, the threshold current intensity is lower for rapidly rising currents than it is for slowly rising ones. When the rate of rise of current is sufficiently low, the change in membrane potential is too slow to be able to gain on the change in threshold level, and therefore no excitation occurs.

The refractory period

For a short time after the passage of an action potential, it is not possible to elicit a second action potential, however high the stimulus intensity is. This period is known as the *absolute refractory period* (fig. 4.30). Following the end of the absolute refractory period, there is a period during which it is possible to elicit a second action potential, but the threshold stimulus intensity is higher than usual. During this *relative refractory period* the second action potential, if it is recorded near the stimulating electrodes, may be reduced in size. The existence of refractoriness places an upper limit on the frequencies at which axons can conduct nerve impulses.

Fig. 4.28. To illustrate latent addition. A subthreshold depolarizing current pulse is applied to an axon at the time shown by the arrow. The upper trace shows the consequent change in membrane potential. The lower trace, which is proportional to the difference between the upper trace and the threshold membrane potential, shows the changes in threshold stimulus intensity which this produces.

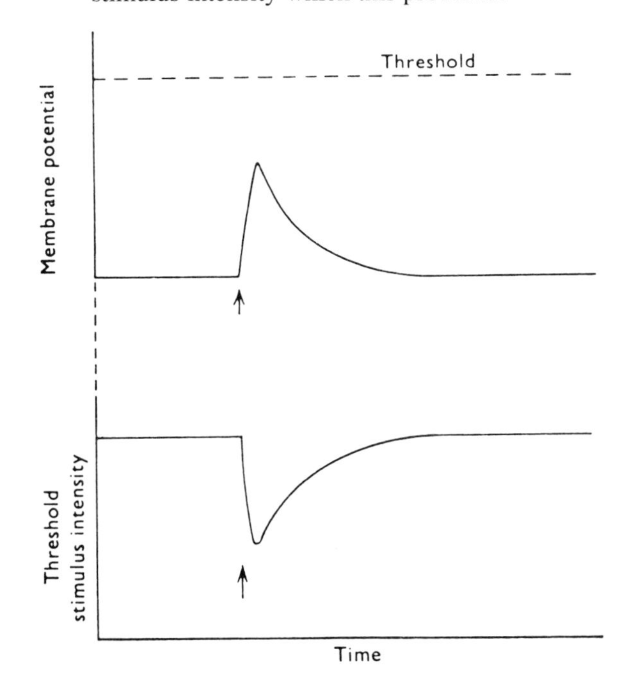

Fig. 4.29. Accommodation in response to constant currents. In each case the continuous line shows the membrane potential and the dotted line the threshold membrane potential. *a*: Depolarizing current; *b* and *c*: hyperpolarizing current. In *c* the current is strong enough to cause 'anode break excitation' after it is switched off, at the point marked by the arrow.

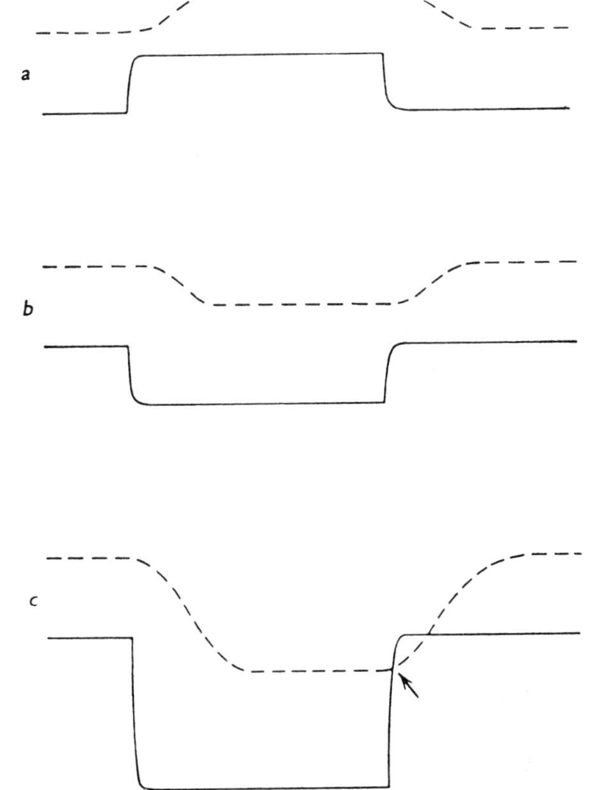

Fig. 4.30. Refractory periods.

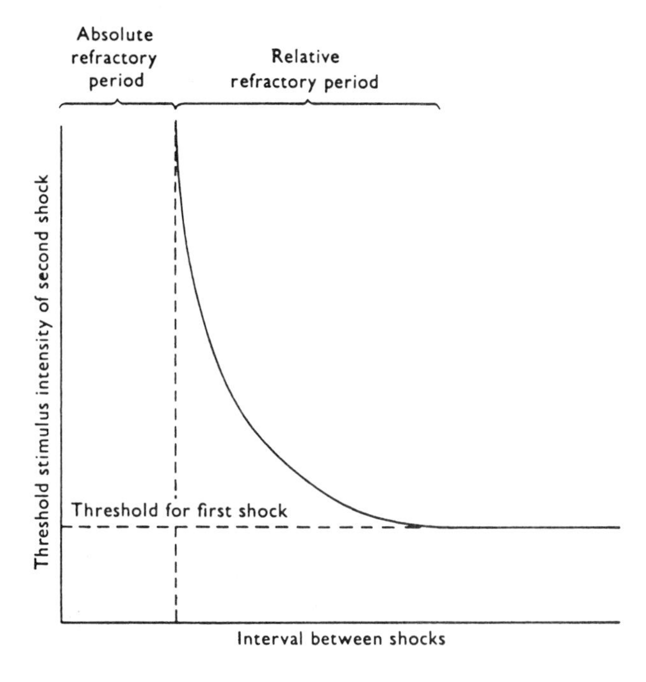

Fig. 4.31. The compound action potential in bullfrog sciatic nerve. The time scale is logarithmic; trace *b* is at a higher amplification than trace *a*. (From Keynes and Aidley, 1981, redrawn after Erlanger and Gasser, 1937.)

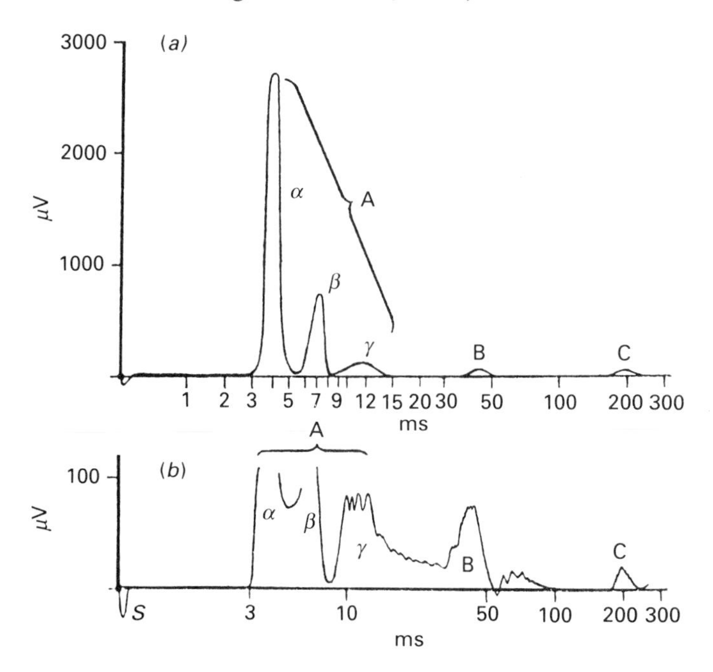

Compound action potentials

The isolated sciatic nerve of the frog has been much used in the study of nervous phenomena. If such a nerve is arranged so that it can be stimulated at one end and recorded from at the other, action potentials can be recorded in the usual manner. These action potentials differ from those of isolated single fibres in that their size is, over a restricted range, proportional to the size of the stimulus. The reason for this is that the recorded action potential is the result of simultaneous activity in a large number of axons (each of which itself obeys the all-or-nothing law) which have different thresholds.

When the distance between the stimulating and recording electrodes is large, it is found that the monophasic compound action potential (recorded as in fig. 4.16*b*) consists of a number of potential waves; this is because the nerve trunk contains fibres of different diameter and hence different conduction velocities. Separation of the different components of the whole response occurs because action potentials with higher conduction velocities reach the recording electrodes before those with lower ones (fig. 4.31).

Erlanger and Gasser (1937) classified vertebrate nerve fibres according to their conduction velocities into three groups, A, B and C, with A being further subdivided into α, β, γ and

Table 4.1. *Fibre groups in bullfrog sciatic nerve, with typical values for their conduction velocities at room temperature*

Fibre group		Conduction velocity ($m\ s^{-1}$)
	α	41
A	β	22
	γ	15
B		4
C		0.7

Based on Erlanger and Gasser, 1937.

δ (table 4.1). An alternative classification, due to Lloyd (1943), has been much used for the sensory fibres from mammalian muscles, with groups I to IV designated according to their fibre diameter (table 4.2).

Table 4.2. *The relation between function and diameter in the afferent fibres of mammalian muscle nerves*

Group	Diameter (μm)	Conduction velocity ($m\ s^{-1}$)	Sensory endings
Ia	12–20	72–120	Primary endings on muscle spindles
Ib	12–20	72–120	Golgi tendon organs
II	4–12	24–27	Secondary endings on muscle spindles
III	1–4	6–24	Pressure/pain receptors
IV	Non-myelinated fibres		Pain

Based mainly on Hunt, 1954.

5

The ionic theory of nervous conduction

In chapter 3, we saw that the resting potential is determined by the ionic concentration gradients across the cell membrane and the relative permeabilities of the membrane to the different ions in the system. The aim of this chapter is to show that the electrical excitability of the axon membrane, as described in the previous chapter, is dependent upon changes in its ionic permeability; in particular, upon changes in the permeability to sodium and potassium ions.

According to the ionic theory of nervous conduction, we can interpret the properties of the axon membrane in terms of a conceptual model, the electrical circuit shown in fig. 5.1. In this model, we assume that the concentration gradient of any particular ion in the system acts as a battery, whose electromotive force is given by the Nernst equation for that ion. For example, there is a 'potassium battery', E_K, whose electromotive force is given by

$$E_K = \frac{RT}{F} \log_e \frac{[K]_o}{[K]_i}. \tag{5.1}$$

In squid axons, E_K is usually about -75 mV (inside negative). Similarly, the electromotive force of the 'sodium battery', E_{Na}, is given by

$$E_{Na} = \frac{RT}{F} \log_e \frac{[Na]_o}{[Na]_i}. \tag{5.2}$$

Since $[Na]_o$ is usually greater than $[Na]_i$, E_{Na} is positive, typical values being in the region of $+55$ mV. The other ions in the system, of which chloride is usually the most important, contribute to a third battery, E_l, which may produce a 'leakage current'; the value of E_l is near the resting

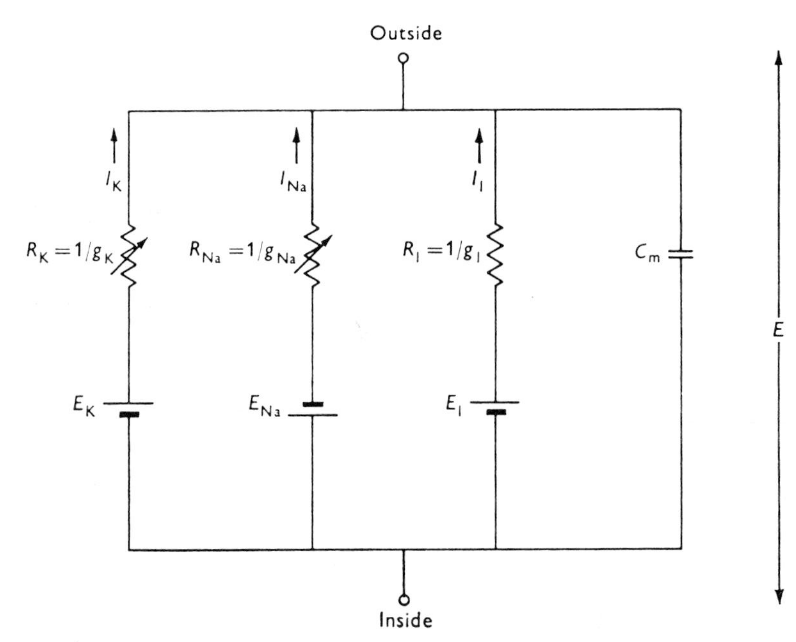

Fig. 5.1. Conceptual model of a patch of electrically excitable membrane. (After Hodgkin and Huxley, 1952*d*.)

potential, at about -55 mV. In series with each battery is a resistance (R_K, R_{Na} and R_l in fig. 5.1); but it is more convenient to talk about the conductances, g_K, g_{Na} and g_l (conductance is the reciprocal of resistance). These conductances represent the ease with which ions can pass through the membrane; they are related to the permeability coefficients mentioned in chapter 3, but of course the units of measurement are different. We assume that the sodium and potassium conductances are variable. These three ionic pathways are arranged in parallel, the current flowing through each pathway being represented by I_K, I_{Na} and I_l. Finally, the membrane capacitance is represented by a fourth unit C_m.

A moment's thought will show that the net potential across the membrane, E, will, in the absence of externally applied current, be determined by the relative values of the ionic conductances. If g_K is much higher than g_{Na}, or g_l, for instance, E will be near to E_K; if g_{Na} is then increased, E will move towards E_{Na}, and so on.

A crucial feature of the model shown in fig. 5.1 is the capability of the sodium and potassium conductances to adopt different values. For many years the physical basis of this was not at all clear. Much still remains to be discovered, but there is now good evidence that ionic flow through the membrane occurs through macromolecular structures called *channels*. The channels are usually permeable to only one or a few types of ion, so that we find sodium channels which open to allow the flow of sodium ions, potassium channels which open to allow potassium ions through, and so on. Each channel is normally closed, but can open for short periods of time. Hence the conductance of a single channel is either zero or a particular value such as 18 pS. So changes in the conductance of a whole pathway are brought about by changes in the number of membrane channels which are open.

Let us take an example. Suppose the conductance of a single sodium channel is 18 pS and suppose the sodium conductance of an area of membrane (represented by g_{Na} in fig. 5.1) rises to 36 mS. This increase in total membrane conductance is brought about by a corresponding increase in the number of sodium channels which are in the open state, and in order to get a total conductance of 36 mS we must have 2×10^9 open channels.

Work on the ionic theory of nervous conduction proceeded apace in the years after the second world war, reaching a peak with the remarkable analysis by Hodgkin and Huxley in 1952. Later work served to fill in the details and expand the boundaries of this analysis. From the nineteen-seventies onwards there has been an increasing emphasis on the nature and properties of the individual channels involved, and we shall examine some of that work in the following chapter. In this chapter we consider the evidence that the conceptual model shown in fig. 5.1 provides an adequate description of the properties of the axon membrane.

The 'sodium theory' of the action potential

The suggestion that the resting potential is due to the selective permeability of the membrane to potassium ions was first made by Bernstein in 1902. He also suggested that the action potential is brought about by a breakdown in this selectivity, so that the membrane potential falls to zero. However the discovery that the membrane potential 'overshoots' the zero level during the action potential, so that the inside becomes positive, implied that some other process must be involved. Hodgkin and Katz (1949) suggested that this process was a rapid and specific increase in the permeability of the membrane to sodium ions. In terms of the model shown in fig. 5.1, g_{Na} becomes temporarily very much larger than g_K, so that E moves towards E_{Na}.

In order to test the validity of the 'sodium theory' as this suggestion is known, Hodgkin and Katz measured the height of action potentials from squid giant axons placed in solutions containing different concentrations of sodium ions. No action potentials could be produced in the absence of sodium ions in the external medium. If the axon was placed in a solution with reduced sodium ion concentration (prepared by mixing sea water with an isotonic glucose solution), the action potential was reduced in height (fig. 5.2). Furthermore the slope of the curve relating the height of the action potential to $\log_{10}[Na]_o$ was close to 58 mV per unit, except at very low sodium ion concentrations where failure of conduction was imminent; this is just what one would expect from (5.2). It was also found that the height of the action potential was increased after addition of sodium chloride to the external solution.

Fig. 5.2. The effect of reducing the external sodium ion concentration on the size of the action potential in a squid giant axon. In each set of records, record 1 shows the response with the axon in sea water, record 2 in the experimental solution, and record 3 in sea water again. Experimental solutions were made by mixing sea water and isotonic glucose solutions, the proportions of sea water being *a*, 33%; *b*, 50%; and *c*, 70%. (From Hodgkin and Katz, 1949.)

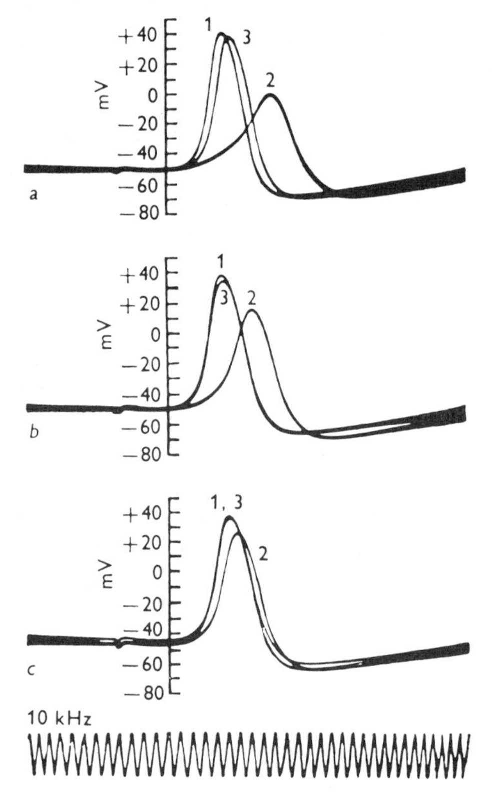

Fig. 5.3. The method used by Keynes to measure the radioactivity of *Sepia* giant axons. (From Keynes, 1951.)

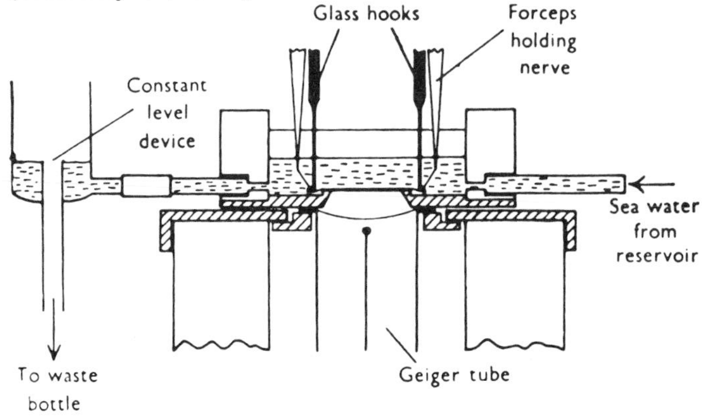

Similar results have been obtained from a variety of other electrically excitable cells, including frog myelinated axons (Huxley and Stämpfli, 1951), insect axons (Narahashi, 1963), frog 'fast' skeletal muscle fibres (Nastuk and Hodgkin, 1950) and mammalian heart muscle fibres (Draper and Weidmann, 1951). The 'sodium theory' thus seems to be of fairly general application, although certain arthropod muscle fibres provide an exception (see fig. 14.4).

Ionic movements during activity

Direct measurements of ionic movements, using radioactive isotopes as tracers and various accurate methods of chemical analysis, are obviously relevant to a study of the nature of the action potential. Let us consider some experiments by Keynes (1951) on the movements of radioactive sodium and potassium ions across the membrane of the giant axons of the cuttlefish *Sepia*. The principle of the technique is shown in fig. 5.3. The axon, held in two pairs of forceps, could be loaded with the radioactive isotope by immersion in a pot of sea water containing the isotope. In order to observe its radioactivity, it was transferred to a chamber containing flowing non-radioactive sea water and placed over a Geiger counter. Stimulating pulses could be applied via one pair of forceps, and action potentials could be recorded from the other pair so as to provide a check on the excitability of the axon.

Fig. 5.4 shows the results of one of Keynes's experiments on the movement of radioactive potassium ions. The axon was initially immersed in ^{42}K-sea water. After 15 min it was removed to the counting chamber and its radioactivity measured at intervals during the next 40 min; when plotted on a logarithmic scale, these values fell on a straight line, and so the radioactivity at 15 min could be estimated by producing this line back. From this value, knowing the potassium ion concentration of the sea water, and its specific activity, the rate of entry of potassium could be calculated. Then the axon was transferred to the radioactive sea water again and stimulated for 10 min; on returning to the measuring chamber its increase in radioactivity could be calculated as before, from which the extra influx on stimulation could be determined. The rate of loss of radioactivity in the resting condition could be obtained from the slope of the lines through the experimental points in fig. 5.4, and the rate of loss on stimulation could be found by stimulation in the non-radioactive sea water in the counting chamber. By making plausible assumptions as in the internal potassium concentration, these figures could be converted into absolute values of potassium efflux.

Similar experiments were performed with sodium ions (fig. 5.5). The main results of the investigation are shown in graphical form in fig. 5.6.

Fig. 5.4. Results of an experiment to determine potassium ion movements in a *Sepia* giant axon. (From Keynes, 1951.)

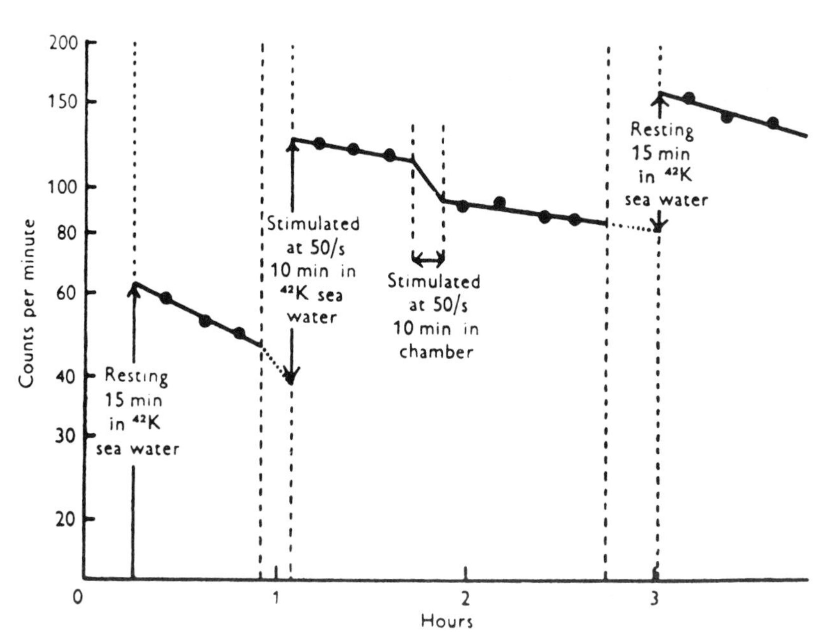

The net effect of each impulse was to produce a net entry of 3.7 pmoles cm^{-2} of sodium ions and a net exit of 4.3 pmoles cm^{-2} of potassium ions.

Entry of sodium ions will cause a build-up of positive charge on the inside of the axon membrane (making the membrane potential more positive), and exit of potassium ions will remove it. Hence it is reasonable to suggest, from Keynes's results, that the rising phase of the action potential is brought about by inward movement of sodium ions, and the falling phase by outward movement of potassium ions. If this suggestion is correct, we should be able to show that the quantity of sodium ions entering the axon is sufficient to cause the observed changes in membrane potential. The charge (Q, measured in coulombs) on a capacitance (C, measured in farads) is given by

$$Q = CV$$

where V is the voltage across the capacitance. If this charge is produced by a univalent ion, the number of moles of the ion moved from one side of the capacitance to the other is given by

$$n = CV/F$$

where F is Faraday's constant. In the case of an axon, C is 1 $\mu F\,cm^{-2}$ and V (the height of the action potential) is about 110 mV. Hence

Fig. 5.5. Results of an experiment to determine sodium ion movements in a *Sepia* giant axon. The values recorded immediately after a period of immersion in radioactive sea water probably included radioactivity carried over in the external medium; the straight lines are therefore drawn through the later points only. (From Keynes, 1951.)

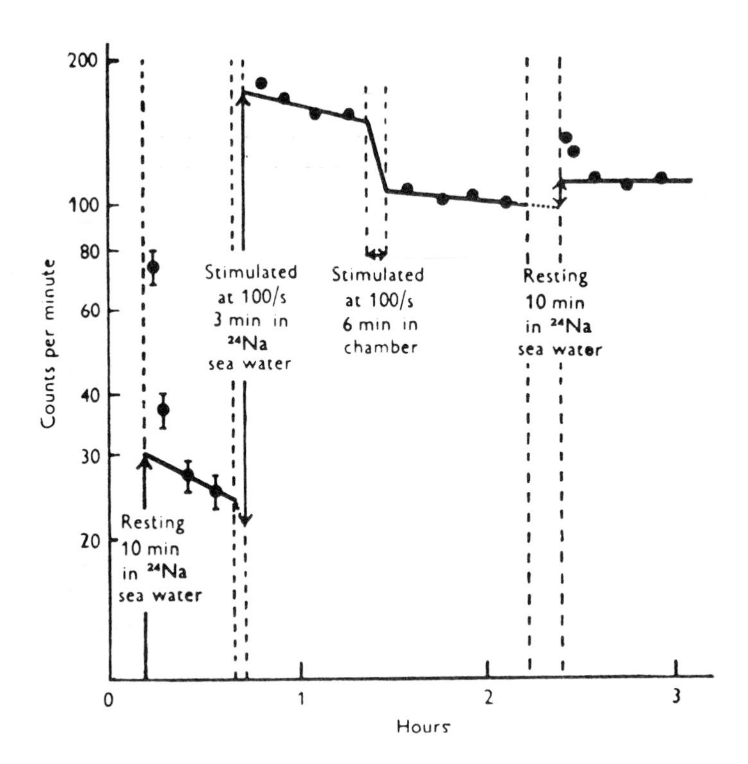

Fig. 5.6. Fluxes of sodium and potassium ions in *Sepia* giant axons. Shaded columns: resting. White columns: stimulated at 100 shocks per second. (Based on data of Keynes, 1951.)

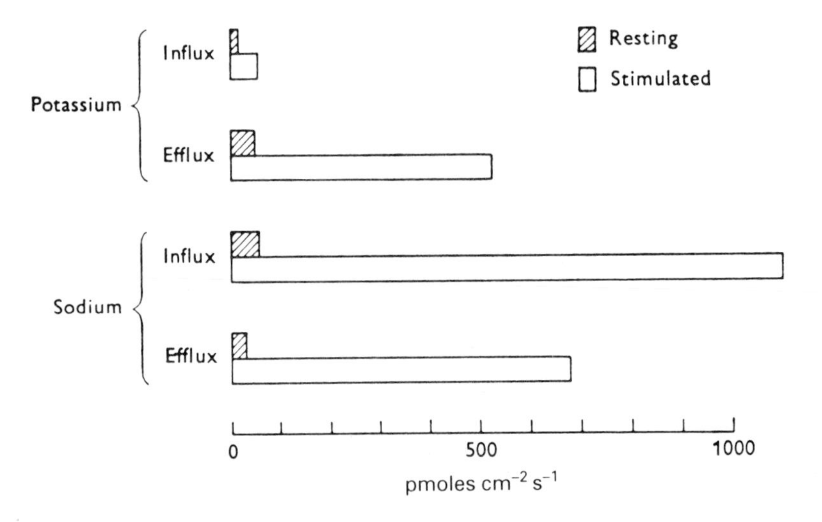

$$n = \frac{10^{-6} \times 0.11}{10^5} \text{ moles cm}^{-2}$$
$$= 1.1 \text{ pmoles cm}^{-2}.$$

It follows that the net flow of sodium ions is more than sufficient to account for the change in membrane potential during the action potential.

Isotopic experiments provide a very good measure of the gross ionic exchanges occurring on stimulation, but their time resolution is naturally extremely poor in relation to the duration of the action potential. A more detailed picture of ionic movements during activity can only be obtained by using electrical methods, which we must now consider.

Voltage-clamp experiments

If the 'sodium theory' is correct, then the initial depolarization which produces an action potential must result in an increase in sodium conductance, and this increase in sodium conductance will itself produce a further depolarization. Thus we are dealing with a positive feedback system, as shown in fig. 5.7*a*, and such systems are very difficult to analyse. These difficulties have been overcome by means of a special technique known as the voltage clamp method, which was initiated by Cole (1949) and Marmont (1949) in Chicago and further developed by Hodgkin, Huxley and Katz (1952) in Cambridge. The essence of the method is that the postive feedback effect of sodium conductance on membrane potential is eliminated by passing current through the nerve membrane so as to hold the membrane potential constant at any desired value. The use of this method by Hodgkin and Huxley (1952*a–d*) led to a remarkably complete analysis of the events occurring during the action potential.

A schematic diagram of the experimental arrangement is shown in fig. 5.8. Two thin silver wire electrodes *a* and *b* are passed down the middle of a squid giant axon. The axon is placed in a trough passing through pools of sea water separated by insulated partitions; on the outside of these

Fig. 5.7. Effect of depolarization on the sodium and potassium ion conductances of the axon membrane. *a*: Sodium: this is a positive feedback loop. *b*: Potassium: this is a negative feedback loop.

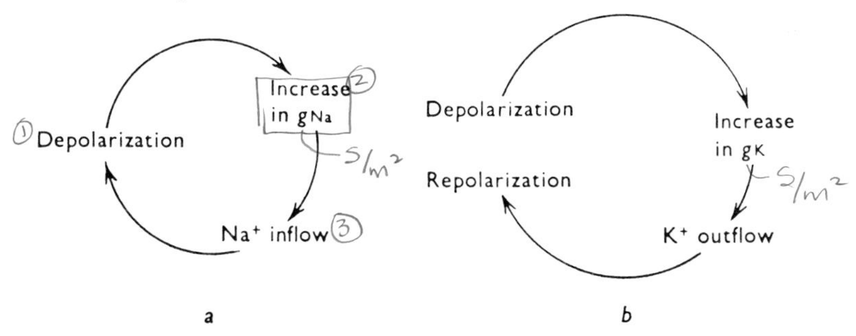

Fig. 5.8. Schematic diagram of the method used to determine membrane currents in a squid axon under voltage clamp. (Based on Hogkin *et al.*, 1952.)

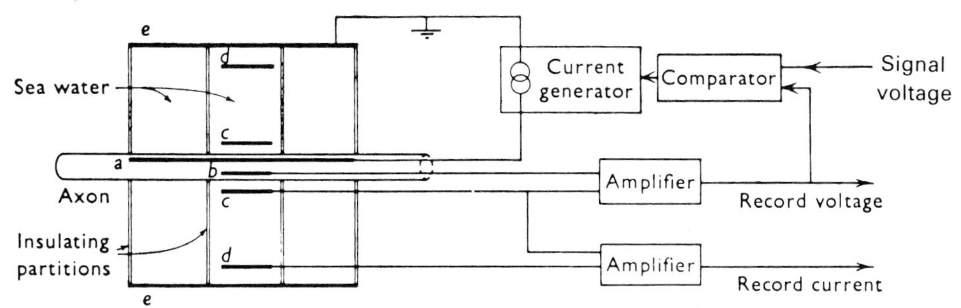

pools is an earthed electrode *e*. Current is passed through the nerve membrane by means of a generator connected to the internal electrode *a* and the earth electrode *e*. In the middle pool, this current must pass through the resistance provided by the sea water between electrodes *c* and *d*, and thus (applying Ohm's law) the voltage between *c* and *d* is proportional to the current flowing through the axon membrane in contact with the middle pool. The voltage across this part of the membrane is recorded by means of electrode *c* and the internal electrode *b*. This voltage, after amplification, is fed into a comparator which is also supplied by a signal voltage. The output of the comparator is passed to the current generator so as to increase or decrease the current flowing through the membrane, which makes the voltage across the membrane (after amplification) equal to the signal voltage. (In practice, a single high-gain differential d.c. amplifier acts both as the comparator and the current generator.) Hence this arrangement constitutes a negative feedback control system in which the voltage across the membrane is determined by the externally applied signal voltage. The presence of the outer pools, with part of the current between *a* and *e* passing through them, ensures that the membrane potential is constant over the length of axon in the middle pool.

The current flowing through the axon membrane is assumed to consist of two components, a capacity current (i.e. changes in charge density at the inner and outer surfaces of the membrane) and an ionic current (i.e. passage of ions through the membrane). Hence the total current, I, is given by

$$I = C_m \frac{dE}{dt} + I_i \qquad (5.3)$$

where C_m is the membrane capacitance, E is the membrane potential and I_i is the ionic current. Thus when the voltage is held constant ('clamped') $dE/dt = 0$, and so the record of current flow gives a direct measure of the total ionic flow.

Fig. 5.9*a* shows the type of record obtained by Hodgkin and Huxley. Notice that the voltage measured, V, corresponds to the difference between the resting potential (E_r) and the clamped membrance potential (E); the reason for this is that it is difficult to obtain accurate absolute values for the membrane potential with silver wire electrodes. The current record shows three components. First, there is a brief 'blip' of outward current; this is caused by discharge of the membrane capacitance. After this the current is inward for about 1 ms, and finally the current becomes outward and climbs to a steady level which is maintained while the clamp lasts.

Now, applying the 'sodium theory' of the action potential, we might expect that the initial inward current during the clamp is caused by sodium ions flowing inwards. If this is so, then it should disappear if the axon is placed in a solution containing no sodium ions. In fact, it was found (by using choline chloride as a substitute for sodium chloride) that the inward current is replaced by an outward current under these conditions (fig. 5.9*b*). This is understandable if we assume that depolarization causes a brief increase in sodium conductance, since the internal sodium concentration is now very much greater than that outside, and we

Fig. 5.9. Typical records of the membrane current during a voltage clamp experiment. *a* and *c*: In sea water; *b*: in a sodium-free choline chloride solution. (From Hodgkin, 1958, after Hodgkin and Huxley, 1952*a*.)

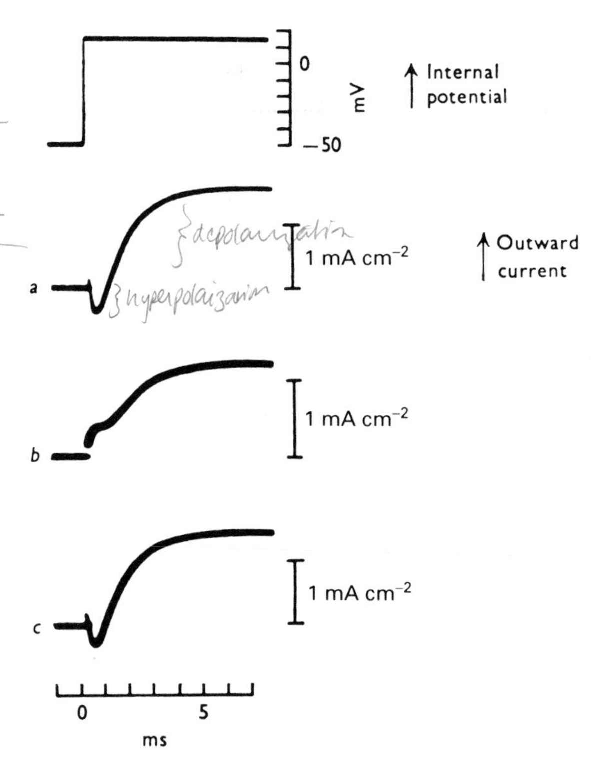

would therefore expect sodium ions to flow outwards.

The direction of sodium ion flow must be dependent upon the membrane potential as well as on the concentration gradient. When the membrane potential is equal to the sodium equilibrium potential E_{Na} (given by (5.2)), there will be no net flow of sodium ions, so that if the membrane potential is clamped at E_{Na} (produced by a depolarization of V_{Na}, where $V_{Na} = E_{Na} - E_r$) there should be no sodium current. If the depolarization does not reach V_{Na}, sodium current will be inwards, and if it is greater than V_{Na} it will be outwards. This effect is shown in fig. 5.10, in which it is evident that on this interpretation V_{Na} is approximately 117 mV. An alternative way of eliminating the sodium current is to alter the external sodium ion concentration, so changing E_{Na}; trace *b* in fig. 5.12 was obtained in this way.

These results imply very strongly that the initial current flow under voltage clamp conditions is caused by movement of sodium ions. In order to clinch the matter, it is necessary to show that the potential at which the initial current reverses really is the sodium equilibrium potential, E_{Na}. This could not be done directly in the original experiments since the internal sodium ion concentration was not known exactly, and also because of the uncertainties involved in translating the measured potentials (V) into membrane potentials (E). However, it was possible to test the hypothesis by measuring *changes* in E_{Na} (measured as changes in V_{Na} plus a small correction for any resting potential change) which occurred when the external sodium ion concentration was changed. The observed changes never differed from those expected by more than 3%.

The slowly rising, maintained outward current is only very slightly affected by changes in the external sodium ion concentration, and is therefore caused by movement of some other ion, probably (it was assumed) potassium. Direct evidence that this was so was provided by an experiment in which the efflux of ^{42}K from the region of an axon membrane under a cathode was measured (Hodgkin and Huxley, 1953). The increase in potassium ion efflux was linearly proportional to the current density with a slope equal to Faraday's constant (fig. 5.11), and hence the current was carried by potassium ions.

It was now possible to analyse the ionic current following depolarization into two components, due to the flow of sodium and potassium ions. In fig. 5.12, trace *b* represents the current produced by an axon in 10% sea water (90% isotonic choline chloride solution) when V was equal to V_{Na}. Trace *a* is the current produced by the same depolarization (56 mV in this case) with the same axon in sea water. Now since there is no sodium current (I_{Na}) when V is equal to V_{Na}, trace *b* must represent the potassium current I_K) plus a small constant component, the leakage current (I_l). Hence the difference between trace *a* and trace *b* must be equal to the sodium current; this is

Fig. 5.10. Membrane currents at large depolarizations. Values of V are shown at the right of each record. (From Hodgkin, 1958, after Hodgkin, Huxley and Katz, 1952.)

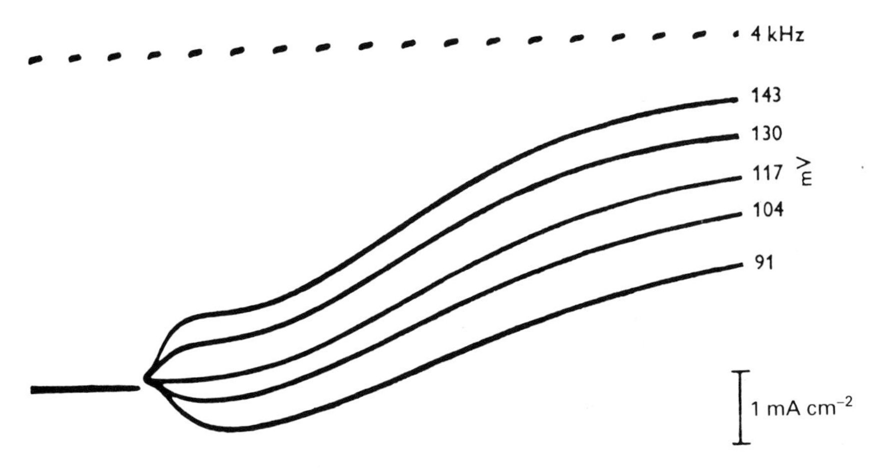

drawn as trace c in fig. 5.12. The problem of separating the sodium and potassium currents is a little more complicated when the clamped depolarization in the low-sodium solution was not equal to V_{Na}, but, by assuming that the sodium current in the low-sodium case was always a constant fraction of that in the high-sodium case, it is possible to calculate the sodium and potassium currents in each case.

The next step in the analysis was to determine the conductance of the membrane to sodium and potassium ions during a clamped depolarization. Applying Ohm's law again, the ionic current flow through the membrane is equal to the product of the conductance and the electromotive force. In the case of sodium ions, for instance, the electromotive force is given by the difference between the membrane potential and the sodium equilibrium potential, and so the sodium current is given by

$$I_{Na} = g_{Na}(E - E_{Na}) = g_{Na}(V - V_{Na}). \quad (5.4)$$

Thus, from trace c in fig. 5.12, and knowing V_{Na} it is a simple matter to calculate g_{Na} throughout the course of the clamp. By a similar process, the potassium conductance can be calculated from trace b and the relation

$$I_K = g_K(V - V_K). \quad (5.5)$$

The results of these calculations are shown in fig. 5.13. Fig. 5.14 shows the conductance changes resulting from depolarizations of different magnitudes.

The results of this investigation at this stage can be summarized as follows. Depolarization produces three effects: (1) a rapid increase in sodium conductance, followed by (2) a slow decrease in sodium conductance (known as the sodium inactivation process) and (3) a slow increase in potassium conductance. The extent of these conductance changes increases with increasing depolarization, reaching a maximum (saturating) level of about 30 mS cm^{-2} at depolarizations of 100 mV and above. If the membrane potential is suddenly returned to its resting level, these changes are reversed, as is shown by the broken curves in fig. 5.13.

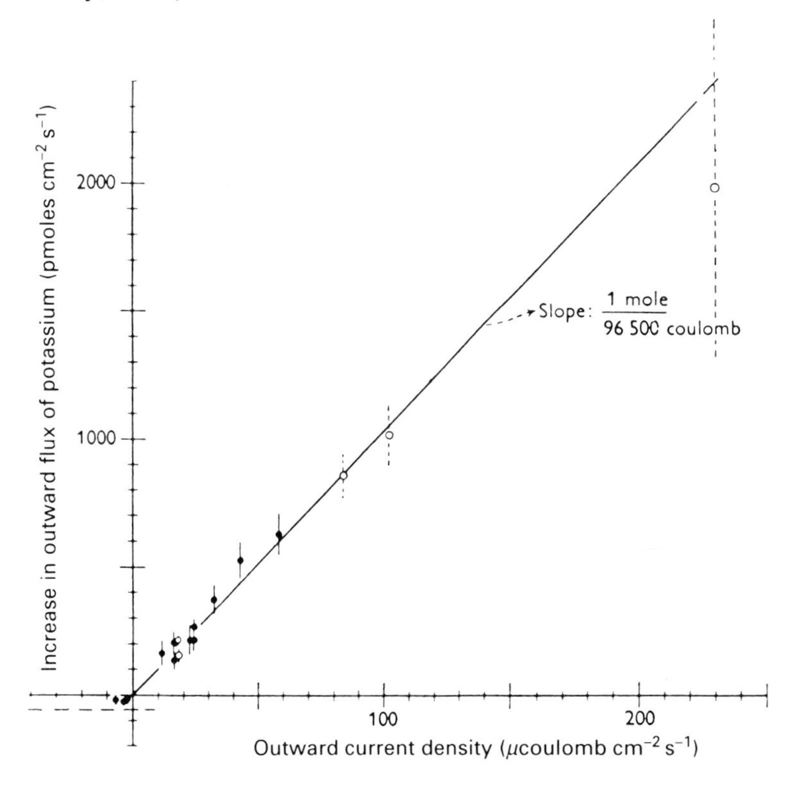

Fig. 5.11. Efflux of potassium under a cathode applied to a *Sepia* axon. (From Hodgkin and Huxley, 1953.)

Fig. 5.12. Analysis of the ionic current in a *Loligo* axon during a voltage clamp. Trace *a* shows the response to a depolarization of 56 mV with the axon in sea water. Trace *b* is the response with the axon in a solution comprising 10% sea water and 90% isotonic choline chloride solution. Trace *c* is the difference between traces *a* and *b*. Further explanation in the text. (From Hodgkin, 1958, after Hodgkin and Huxley, 1952*a*.)

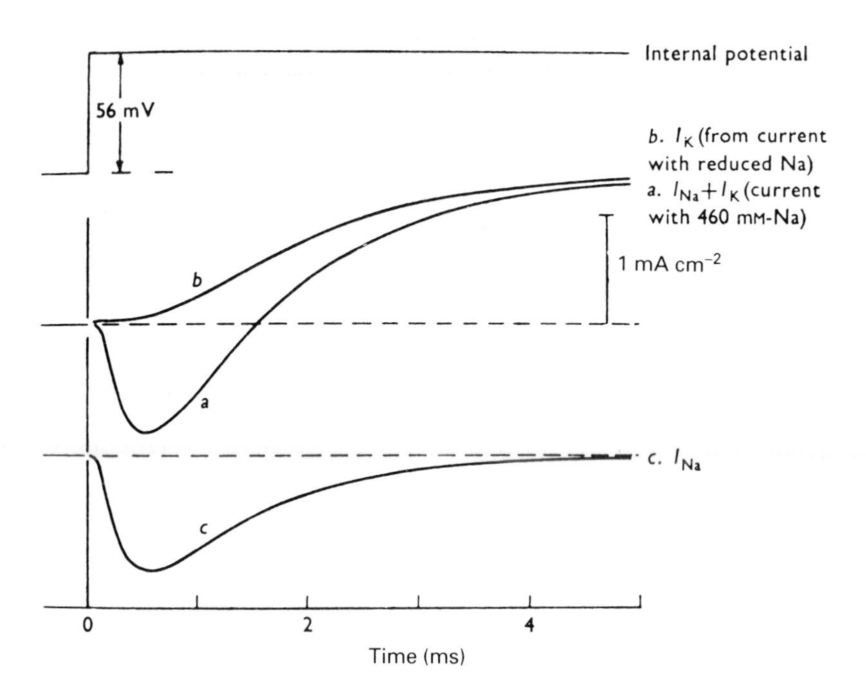

Fig. 5.13. Ionic conductance changes during a clamped depolarization, derived from the current curves shown in fig. 5.12. The broken curves show the effects of repolarization. (From Hodgkin, 1958, by permission of the Royal Society.)

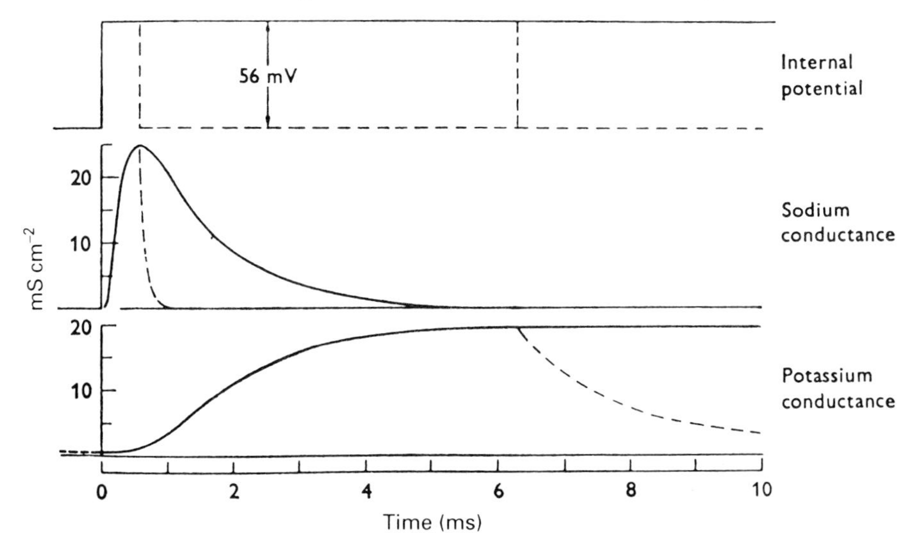

Calculation of the form of the action potential

If the conceptual model of the membrance shown in fig. 5.1 is correct, then it should be possible to calculate the form of the action potential if we know the values of the fixed components in the system (E_K, E_{Na}, E_l, R_l and C_m) and if equations describing the behaviour of the variables g_K and g_{Na} can be obtained. From their voltage clamp experiments, Hodgkin and Huxley (1952*d*) were able to provide a series of equations describing the changes of g_K and g_{Na} with depolarization and time (from which the continuous curves in fig. 5.14 were calculated), and so were able to undertake this rather laborious calculation. The results were dramatically accurate, with only slight differences between the predicted action potentials and those actually observed (fig. 5.15).

Fig. 5.16 shows the theoretical solution for a propagated action potential, and the associated changes in g_{Na} and g_K. It is instructive to follow the potential and conductance changes during its course. Initially, g_K is small but g_{Na} is much smaller, so that the resting potential is near E_K. Then the presence of an action potential approaching along the axon draws charge out of the membrane capacitance by local circuit action so causing a depolarization. When the membrane has been depolarized by about 10 mV, the sodium conductance begins to increase, so that a small number of sodium ions cross the membrane, flowing down their electrochemical gradient into the axon. This transfer of charge results in a further depolarization, which causes a further increase in g_{Na}, so

Fig. 5.14. Conductance changes brought about by clamped depolarizations of different extents. The circles represent values derived from the experimental measurements of ionic current, and the curves are drawn according to the equations used to describe the conductance changes. (From Hodgkin, 1958, after Hodgkin and Huxley, 1952*d*.)

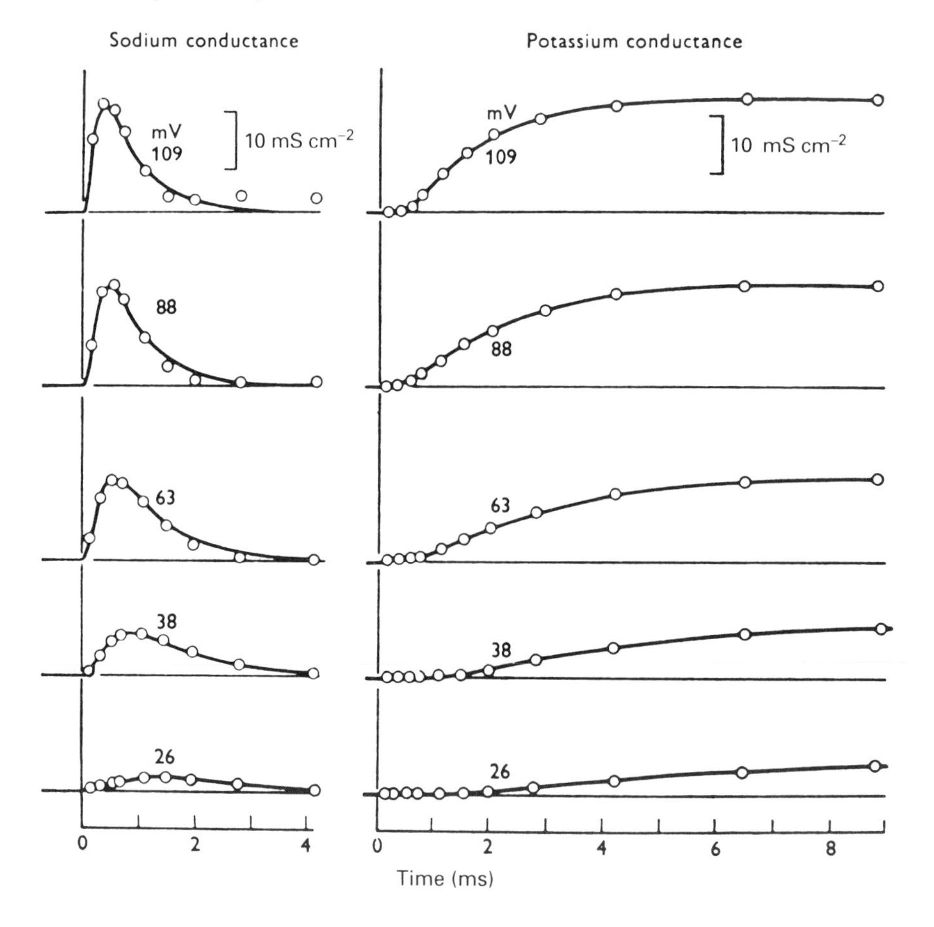

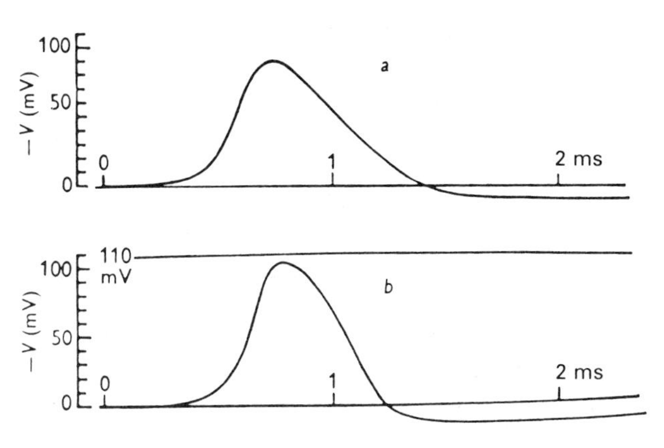

Fig. 5.15. Comparison of computed (*a*) and observed (*b*) propagated action potentials in squid axon at 18.5 °C. The calculated velocity of conduction was 18.8 m s^{-1}; the observed velocity was 21.2 m s^{-1}. (From Hodgkin and Huxley, 1952*d*.)

Fig. 5.16. Calculated changes in membrane potential (upper curve) and sodium and potassium conductances (lower curves) during a propagated action potential in a squid giant axon. The scale of the vertical axis is correct, but its position may be slightly inaccurate; it has been drawn here assuming a resting potential of −60 mV. The positions of E_{Na} and E_K are correct with respect to the resting potential. In the original calculations, voltages were measured from the resting potential, as in fig. 5.15. (After Hodgkin and Huxley, 1952*d*; redrawn.)

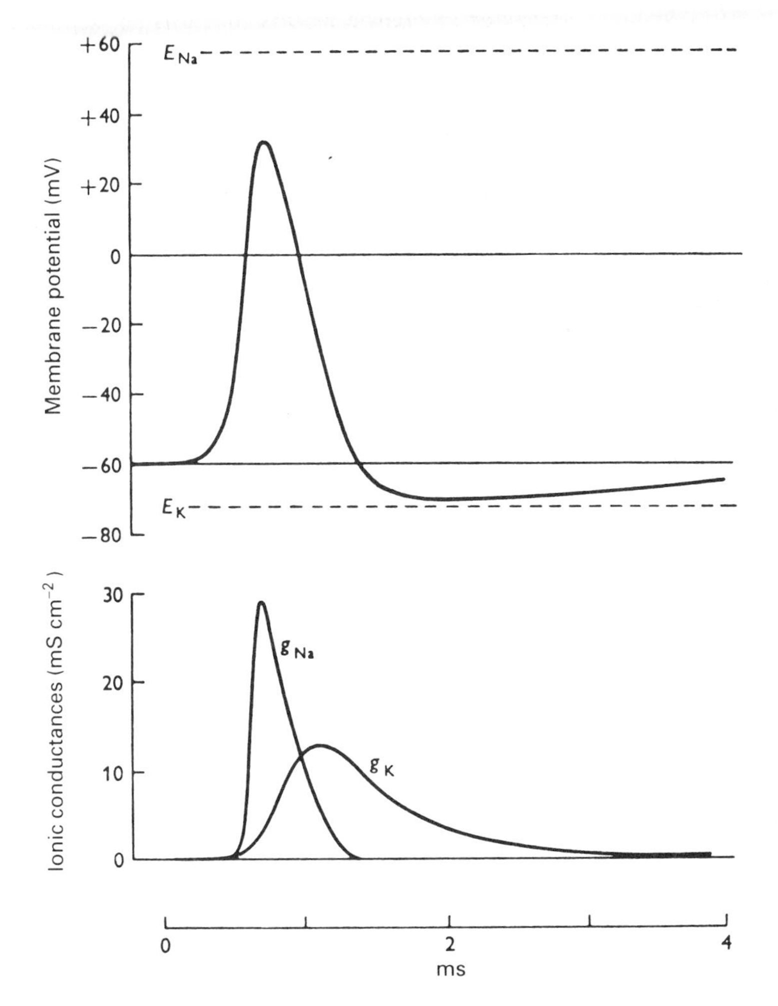

more sodium ions cross the membrane, and so on. The result of this regenerative action between depolarization and sodium conductance is that the sodium battery is relatively much more important than the potassium battery in determining the membrane potential, and so the membrane potential goes racing up towards the sodium equilibrium potential, E_{Na}. But now the two slower consequences of depolarization, sodium inactivation and the increase in potassium conductance, begin to take effect. This means that the potassium battery becomes more important and the sodium battery less important in determining the membrane potential (because sodium ion inflow declines and potassium ion outflow increases), and so the membrane potential begins to fall. This repolarization further reduces g_{Na} (it also reduces g_K, but more slowly) so that the membrane potential is brought rapidly back to its resting level. At this point, although g_{Na} is extremely low, g_K is still considerably higher than usual, so that the membrane potential passes the resting level and moves even nearer to the potassium equilibrium potential. Finally, as g_K declines to its normal low value, the membrane potential returns to its resting level.

The equations

Now let us have a brief look at the equations used by Hodgkin and Huxley (1952*d*) in the calculation of the form of the action potential. The first step was to find empirical equations to describe the conductance changes seen, for example, in fig. 5.14. The potassium conductance is given by

$$g_K = \bar{g}_K n^4$$

where $\bar{g}_K$ is a constant equal to the maximum value of g_K. The idea behind this equation was that potassium ions might be let through the membrane when four charged particles moved to a certain region of the membrane under the influence of the electric field, the quantity n being the probability that of these particles is in the right position. This idea may or may not be correct; the question is unimportant from the point of view of providing a mathematical description of the conductance changes. The variation of the quantity n with time is given by the equation

$$\frac{dn}{dt} = \alpha_n(1 - n) - \beta_n n, \tag{5.6}$$

in which α_n and β_n are rate constants which, at 6 °C, vary with voltage according to the empirical equations

$$\alpha_n = \frac{0.01(V + 10)}{\exp[V + 10)/10] - 1}$$

and

$$\beta_n = 0.125 \exp (V/80).$$

An alternative way of writing (5.6) is

$$\frac{dn}{dt} = (n_\infty - n)/\tau_n \tag{5.7}$$

where n_∞ (the steady state value of n at any particular voltage) is given by

$$n_\infty = \alpha_n/(\alpha_n + \beta_n) \tag{5.8}$$

and τ_n is a time constant given by

$$\tau_n = 1/(\alpha_n + \beta_n). \tag{5.9}$$

The sodium conductance is given by

$$g_{Na} = \bar{g}_{Na} m^3 h, \tag{5.10}$$

where $\bar{g}_{Na}$ is a constant equal to the maximum value of g_{Na}. This equation is based on the idea that the sodium channel can be opened by movement of three particles, each with a probability m of being in the right place, and inactivated by an event of probability $(1 - h)$. m is given by the equation

$$\frac{dm}{dt} = \alpha_m(1 - m) - \beta_m m, \tag{5.11}$$

where, at 6 °C,

$$\alpha_m = \frac{0.1(V + 2.5)}{\exp [(V + 25)/10] - 1}$$

and

$$\beta_m = 4 \exp (V/18).$$

h is given by the equation

$$\frac{dh}{dt} = \alpha_h(1 - h) - \beta_h h, \tag{5.12}$$

where, at 6 °C,

$$\alpha_h = 0.07 \exp (V/20)$$

and

$$\beta_h = \frac{1}{\exp [(V + 20)/10] + 1}.$$

Equations (5.11) and (5.12) can be written alternatively as

$$\frac{dm}{dt} = (m_\infty - m)/\tau_m$$

and

$$\frac{dh}{dt} = (h_\infty - h)/\tau_h$$

with m_∞, τ_m, h_∞ and τ_h being given by equations analogous to (5.8) and (5.9).

The voltage-dependence of the parameters n_∞, m_∞ and h_∞, and their corresponding time constants is shown in fig. 5.17.

The total membrane current, I, is given by (5.3), which is expanded to give

$$\begin{aligned} I &= C_m \frac{dV}{dt} + I_K + I_{Na} + I_l \\ &= C_m \frac{dV}{dt} + g_K(V - V_K) + g_{Na}(V - V_{Na}) + g_l(V - V_l) \\ &= C_m \frac{dV}{dt} + \bar{g}_K n^4 (V - V_K) + \bar{g}_{Na} m^3 h (V - V_{Na}) + g_l(V - V_l). \end{aligned}$$

If I is known, it is possible to work out V from this equation by numerical integration. The simplest case occurs when an appreciable stretch of the membrane is excited simultaneously by means of an internal silver wire electrode; there are then no local circuit currents and the net current flow through the membrane is zero, the ionic current being equal and opposite to the capacitance current. This is known as a 'membrane' action potential. In the case of the propagated action potential, the situation is a little more complicated; we have seen in (4.15) that

$$I = \frac{a}{2R\theta^2} \cdot \frac{d^2V}{dt^2},$$

where a is the radius of the axon, R is the resistivity of the axoplasm, and θ is the conduction velocity. The right value of θ has to be found by trial and error; wrong values lead to infinite voltage changes.

Other predictions

Besides predicting the form of the action potential, the equations derived from analysis of the voltage clamp experiments can account for a number of other features of the physiology of the axon.

By adding g_K and g_{Na}, the impedance change during the course of the action potential can be obtained; the result is very similar to that observed by Cole and Curtis (fig. 4.11). Knowing the time course of the ionic currents during the action potential it is possible, by integrating with respect to time, to calculate the flow of sodium and potassium ions across the membrane during an impulse. For a propagated action potential the calculated net sodium entry was 4.33 pmoles cm^{-2} and the calculated net potassium loss was 4.26 pmoles cm^{-2}. These values are very close to those obtained by

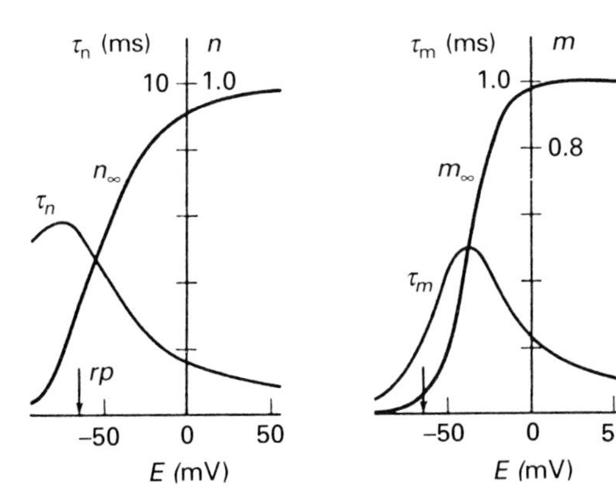

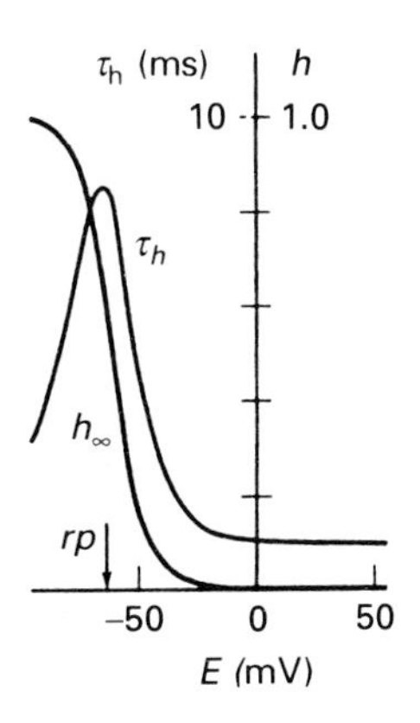

Fig. 5.17. How some of the Hodgkin–Huxley parameters vary with membrane potential. See equations (5.7)–(5.9) for n (the potassium activation parameter) and corresponding equations for m and h (the sodium activation and inactivation parameters). (From Cole, 1968.)

Keynes from radioisotopes measurements on *Sepia* axons.

By calculating the responses to instantaneous depolarizations of varying extent, curves very similar to those of fig. 4.7 were obtained. Thus the local responses produced by membrane potential displacements just less than threshold are caused by small increases in g_{Na} which are soon swamped by increase in g_K. The threshold for production of an action potential is the point at which the increase in g_{Na} is just large and rapid enough to avoid this effect.

At the end of an action potential, g_K is higher than usual, and the sodium inactivation process is well developed, so that g_{Na} cannot be much increased by depolarization. The membrane is thus inexcitable for a time; this corresponds to the duration of the absolute refractory period. A little later, g_K and the extent of the sodium inactivation process have fallen somewhat, so that a submaximal regenerative response can be initiated by sufficient depolarization. This period corresponds to the relative refractory period, where the threshold is higher than normal and the ensuing action potential may be reduced in size. Finally g_K and the sodium inactivation process return to their normal resting levels, and the membrane shows its normal excitability.

Accommodation is explained by the increase in the sodium inactivation process and potassium conductance, which is produced by a slowly rising membrane potential, so raising the threshold. Conversely, anode break excitation is caused by the lowering of the threshold by means of a decrease in sodium inactivation and potassium conductance.

It has long been known that the conduction velocity of an axon increases with increase in temperature. This is because the time course of the electrical changes constituting the action potential is much faster at higher temperatures, and these electrical changes are of course themselves dependent upon the time course of the conductance changes produced by depolarization. In terms of the Hodgkin–Huxley equations, the rate constants for the conductance changes (the alphas and betas in the equations) increase threefold for every 10 °C rise in temperature.

Direct measurement of the conductances during an action potential

Bezanilla, Rojas and Taylor (1970) provided a nice verification of the Hodgkin–Huxley analysis. Advances in electronic design enabled them to subject an axon membrane to a voltage clamp in the middle of an action potential and measure the immediate value of the resulting ionic current. Fig. 5.18 shows the results of one such experiment, in which the membrane was clamped at the potassium

Fig. 5.18. An experiment to determine the sodium conductance at different times during the 'membrane' action potential in a squid (*Dosidicus*) giant axon. Lower traces show the membrane potential, the voltage being clamped at progressively later times (*a* to *h*) during the action potential. Upper traces show membrane ionic current flow; the capacitance current is over within 30 μsec of the onset of the clamp. (From Bezanilla *et al.*, 1970.)

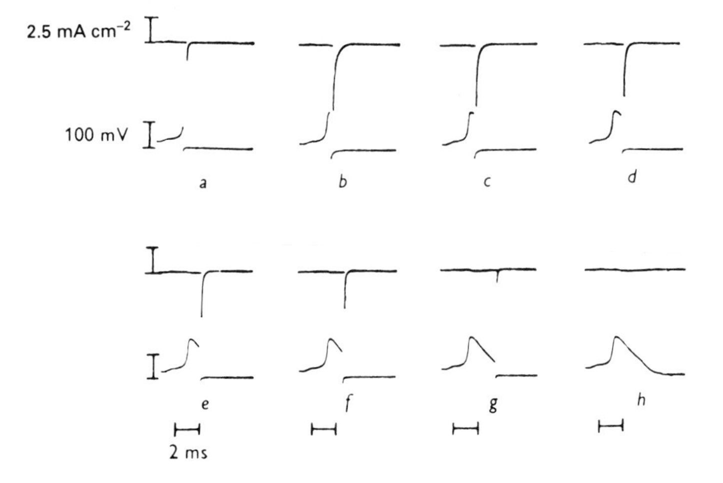

equilibrium potential at different times during a 'membrane' action potential. The current traces therefore measure the sodium current, from which the sodium conductance can be calculated as in (5.4). In each case it can be seen that the sodium conductance falls rapidly, as one would expect at this membrane potential, but its initial value will be the same as it was immediately before the application of the voltage clamp. (This conclusion arises from the kinetics of sodium conductance changes as determined by Hodgkin and Huxley, 1952*b*, and described by the Hodgkin–Huxley equations: the alphas and betas in these equations do change instantaneously with change in voltage, but the quantities m and h do not, as is evident in (5.11) and (5.12), where $\mathrm{d}m/\mathrm{d}t$ and $\mathrm{d}h/\mathrm{d}t$ always have finite values.) So by applying the clamp at different times it is possible to determine the time course of the sodium conductance during the action potential, with the result shown by the open circles in fig. 5.19.

Similarly, by clamping the membrane at the sodium equilibrium potential it is possible to measure the initial potassium currents at different times. The time course of the potassium conductance during the action potential is then readily calculated, and is shown by the black circles in fig. 5.19.

Comparison of fig. 5.19 with fig. 5.16 shows that the agreement between the experimental curves and those predicted by Hodgkin and Huxley is very good. A sharp eye will detect some differences: the fall of sodium conductance is a little more rapid in the experimental determination, and the time course of the potassium conductance change is rather slower. But in general the similarities are much more impressive than the differences.

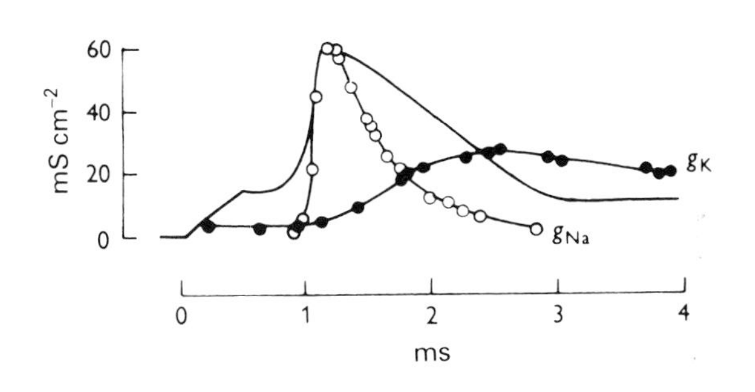

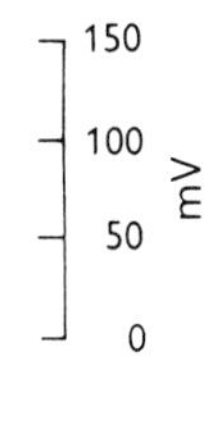

Fig. 5.19. Time course of the ionic conductance changes during a membrane action potential. The sodium conductance (open circles) was determined from an experiment such as that of fig. 5.18, the potassium conductance (black circles) was determined by clamping the membrane at the sodium equilibrium potential at different times. (From Bezanilla *et al.*, 1970.)

Myelinated nerves

The membrane potentials of myelinated nerve fibres, which are usually less than 15 μm in diameter, cannot be measured with the ease with which those of squid giant axons are. It is not normally possible to insert electrodes into the axon, and so indirect methods have to be used. These depend essentially upon using the axoplasm itself as the intracellular electrode by insulating the cut end of an internode from the external fluid surrounding an adjacent node (Dodge and Frankenhaeuser, 1958). The results (see Frankenhaeuser, 1965, and Hille, 1971*a*) show that the properties of the frog nodal membrane are remarkably similar to those of squid axon membrane. The currents produced during a voltage-clamped depolarization consist of an initial sodium current and a delayed potassium current (fig. 5.20), and can be described by relatively minor modifications of the Hodgkin–Huxley equations.

The situation in mammalian myelinated fibres is rather different. Voltage-clamp experiments show that the potassium current at the node is negligible (Chiu *et al.*, 1979). The rate of inactivation of the sodium conductance is more rapid than in frog nodes. Computer reconstructions of the action potential show that the initial depolarization is brought about by the transient inward sodium current (just as in frog nodes and squid axons), but that the repolarization phase can be accounted for by the leakage current alone.

It is interesting to find that there is some potassium conductance in the internodal region in both frogs and rabbits, so that potassium currents do occur in demyelinated axons (Chiu and Ritchie, 1981, 1982). Chiu and Ritchie (1984) suggest that the function of these channels is to maintain a resting potential under the myelin sheath so as to avoid depolarization of the nodes.

Separation of the ionic currents by drugs

Tetrodotoxin is a virulent nerve poison found in the tissues of the Japanese puffer fish *Spheroides rubripes*. Narahashi, Moore and Scott (1964), using voltage-clamped lobster axons, found that tetrodotoxin in the external solution blocks the increase in sodium conductance that occurs on depolarization but has no effect on the increase in potassium conductance (see fig. 5.20*b*). This provides further evidence in favour of the idea that sodium and potassium flow through separate channels. *Saxitoxin*, a poison produced by the dinoflagellate protozoan *Gonyaulax*, also blocks the sodium channels. Both these substances have no effect on nervous conduction if they are injected into the axoplasm. Hence their site of action must be at or near the external surface of the axon membrane.

Tetraethylammonium (TEA) ions produce prolongation of the action potential when injected into squid axons, but external application is ineffective. Voltage-clamp experiments have shown that this prolongation is caused by a blockage of the potassium channels, as in fig. 5.20*d*.

Various other drugs prevent or modify the opening of the ionic channels; we shall consider some of these actions in the following chapter. Less direct effects are produced by drugs which interfere with the maintenance processes of the cell, so that the ionic gradients across the membrane are altered. Cardiac glycosides, such as ouabain, prevent sodium extrusion directly; metabolic inhibitors do so by removing the energy supply for the process. As one would expect from the ionic theory of nervous conduction, poisoning of the extrusion system does not immediately affect the electrical properties of the axon. In an experiment by Hodgkin and Keynes (1955*a*), over 200 000 impulses were elicited from a *Sepia* axon which had been poisoned with dinitrophenol.

Fig. 5.20. Voltage-clamp records for frog nodes, showing separation of the sodium and potassium currents by the use of selective blocking agents. The membrane potential was clamped at −120 mV for 40 ms before the start of the records, and then depolarised to various levels ranging from −60 to +60 mV in 15 mV steps. Leakage and capacity currents were subtracted by computer. Records in *a* and *c* show the normal response. In *b* the node shown in *a* was treated with 300 nM tetrodotoxin; only the potasssium current remains. In *d* the node shown in *c* was treated with 6 mM TEA; only the sodium current remains. (From Hille, 1984, after Hille, 1966 and 1967.)

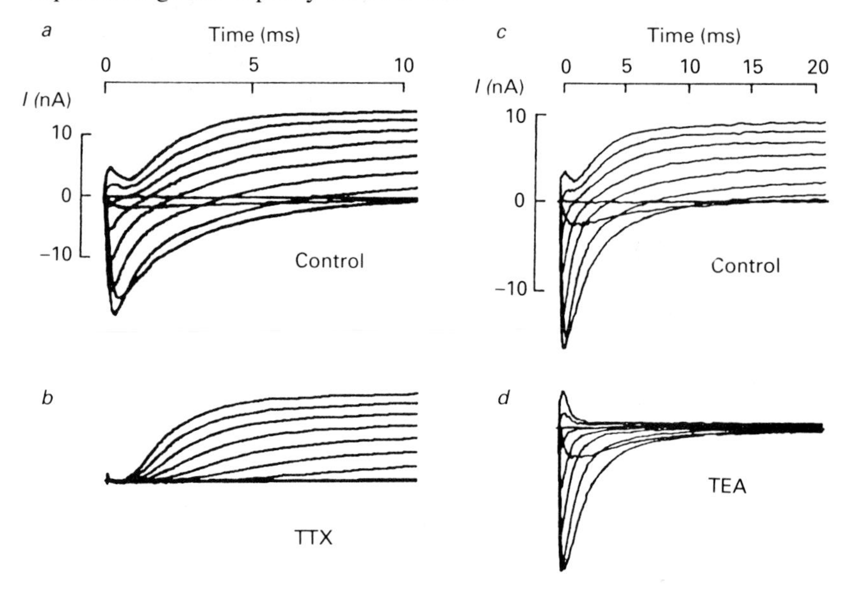

Experiments on perfused giant axons

It is possible to replace the axoplasm of squid giant axons by an internal perfusion solution whose ionic concentration can be altered at will, as is described in chapter 3. This technique has proved to be very useful in confirming and extending our knowledge of the membrane conductance changes involved in nervous conduction.

In the experiment of Hodgkin and Katz shown in fig. 5.2, the action potential was reduced in height by reducing the external sodium ion concentration, so reducing the sodium equilibrium potential, as given by (5.2). The intracellular perfusion method makes it possible to reduce the sodium equilibrium potential by the alternative method of increasing the internal sodium ion concentration. Baker, Hodgkin and Shaw (1962*b*) found that this reduced the overshoot of the action potential as expected (fig. 5.21).

The voltage-clamp technique coupled with internal perfusion has been used by Atwater, Bezanilla and Rojas (1969) to measure sodium fluxes during clamped depolarizations. They placed an axon in sea water containing ^{22}Na and perfused it with a potassium fluoride solution. By measuring the radioactivity of this perfusion solution after it had passed through the axon it was possible to calculate the sodium influx. First the sodium influx at rest was measured. Then the axon was subjected to a few thousand clamped depolarizations at a rate of 10 per second, each pulse lasting about 3 ms. The radioactivity of the emerging perfusion solution rose to reach a new steady level within a few minutes, from which the influx associated with each clamped depolarization could be calculated. The corresponding inward current was calculated from the difference between clamped currents measured in the presence and absence of tetrodotoxin, a substance which blocks the early inward current.

Fig. 5.22 summarizes the results obtained in these experiments. At a number of different membrane potentials the sodium influx almost entirely accounts for the inward current flow, in most convincing agreement with Hodgkin and Huxley's analysis.

Chandler and Meves (1965) carried out voltage clamp experiments on giant axons perfused internally with different solutions. When an axon was perfused with a 300 mM potassium chloride solution, they found that the initial inward current

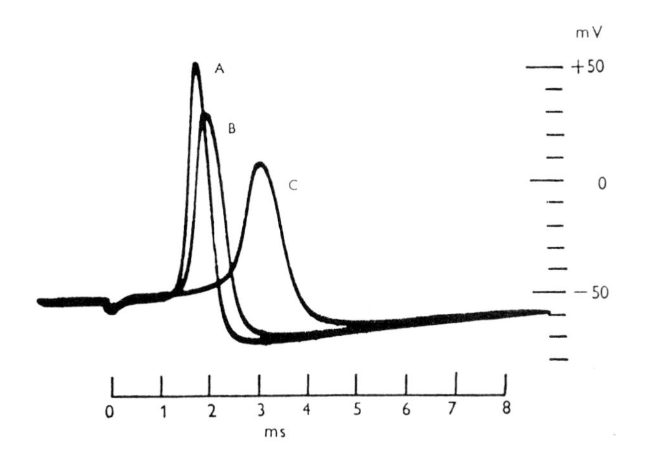

Fig. 5.21. Action potentials from an internally-perfused squid giant axon, showing the effect of increasing the internal sodium ion concentration. Record *A* shows the response with an isotonic potassium sulphate solution as the perfusion fluid, records *B* and *C* show responses with respectively a quarter and a half of the potassium replaced by sodium. (From Baker *et al.*, 1961.)

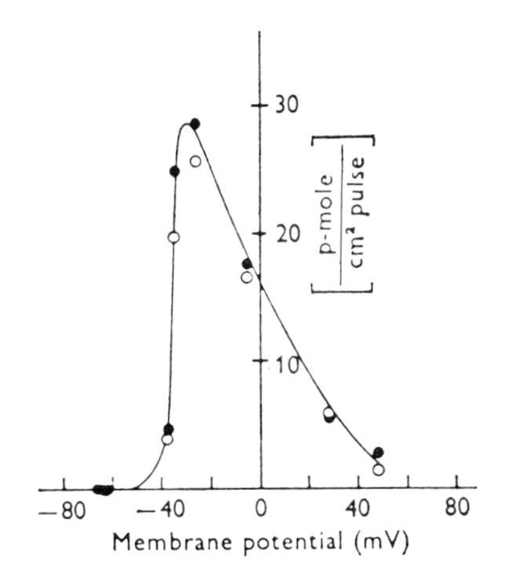

Fig. 5.22. The nature of the early current during voltage clamp in a squid giant axon. Open circles show the sodium ion influx as measured with a radioactive tracer method. Black circles show the charge movement (coulombs divided by Faraday's constant) attributable to the early inward current, measured from the difference between ionic current flow in the presence and absence of tetrodotoxin. (From Atwater *et al.*, 1969.)

(which is attributed entirely to sodium ions in the Hodgkin–Huxley analysis) became an outward current at a reversal potential in the region of +70 mV. But under these conditions, with no internal sodium ions, the sodium equilibrium potential is infinite and so a purely sodium ion current should be inward. Chandler and Meves concluded that potassium ions can flow through the sodium channels. By applying the constant field equation (3.5) it is possible to estimate the relative permeabilities of the channels to sodium and potassium ions. Thus with no internal sodium and no external potassium, and assuming that the channels are impermeable to anions, (3.5) becomes

$$E = \frac{RT}{F} \log_e \left(\frac{P_{Na}[Na]_o}{P_K [K]_i} \right) . \qquad (5.13)$$

At 0 °C, with 320 mM potassium ions inside and 472.5 mM sodium ions outside, the mean reversal potential for the early current was +67.8 mV. Substituting these values in (5.13) we get

$$67.8 = 54.2 \log_{10} \left(\frac{472.5 \, P_{Na}}{320 \, P_K} \right)$$

whence

$$P_{Na}/P_K = 12.0.$$

By repeating this type of experiment with other monovalent cations it was possible to estimate their permeabilities through the sodium channel. They fell in the order Li > Na > K > Rb > Cs, with relative values of 1.1, 1, 1/12, 1/40 and 1/61.

The potassium channel seems to be rather more selective, and indeed its permeability to potassium ions is reduced in the presence of other monovalent cations: sodium, rubidium and especially caesium cause reductions in the potassium current during a voltage clamp. Ammonium ions can pass through both the sodium and potassium channels, their permeability being about a third of that of the normal ion in each case (Binstock and Lecar, 1969).

The effect of internal anions on membrane function was investigated by Tasaki, Singer and Takenaka (1965). They found that some ions were much better than others in maintaining excitability, in the order F > HPO_4 > glutamate > aspartate > SO_4 > acetate > Cl > NO_3 > Br > I > SCN. Hence in recent years it has become quite common to use fluoride as the internal anion in perfusion experiments.

More about conductance

One way of representing the results of a typical voltage-clamp experiment is shown in fig. 5.23, for the giant axon of the tube-dwelling marine worm *Myxicola*. From records of current flow at different depolarizations, we plot the peak early current (curve *a*, largely sodium current) and the maintained outward current (curve *b*, potassium current) against membrane potential. Treatment with tetrodotoxin abolishes the sodium current, leaving a small outward current (curve *c*) which is composed of leakage current plus a small component of the potassium current. The normal sodium current is then the difference between curves *a* and *c* in fig. 5.23. Notice that curves *a* and *c* meet at the sodium equilibrium potential (E_{Na}), about +70 mV in this case.

The curves in fig. 5.23 are relations between current and voltage, and so they can tell us something about the membrane conductance. For any particular point on one of the curves, there are in fact two different measures of conductance (fig. 5.24). Firstly, we can draw a line connecting our point to the point where the driving force is zero (this will be at E_{Na} for the early current curve): the gradient of this gives the *chord conductance*. Secondly, we can measure the slope of the curve at our point by drawing the tangent at that point and measuring its gradient: this gives us the *slope conductance*. It is the chord conductance, given by g_{Na} and g_K in equations (5.4) and (5.5), which is utilized in the Hodgkin–Huxley analysis.

The Hodgkin–Huxley equations assume that chord conductances are linear, i.e. that Ohm's law applies strictly to the instantaneous relation between current and voltage. Hodgkin and Huxley (1952*b*) showed that this was so by measuring the ionic currents immediately after a second change in the clamped potential. In fig. 5.25, for example, the membrane was depolarised by 29 mV for 1.53 ms and then set at a new value and the ionic current (the 'tail' current) measured immediately after the change. It is clear from curve *A* in fig. 5.25 that the relation between this instantaneous current and voltage really is a straight line passing through E_{Na}. Curve *A* is in fact the chord conductance at the peak inward current at a depolarization of 29 mV. Instantaneous current–voltage curves for the potassium current in squid axons were also linear.

Now let us interpret the two curves in fig. 5.25 in terms of ionic channels. If the current–voltage relation of a single channel is linear then that of a number of open channels in a patch of membrane will also be linear. Hence we can conclude that the instantaneous current–voltage relation (curve A in fig. 5.25) represents the properties of a particular number of open channels. (Notice that this conclusion justifies the analysis used in the experiment by Bezanilla *et al.* shown in figs. 5.18 and 5.19.) The peak current curve α, however, in which the currents are measured some time after the change of voltage, it far from linear and so must indicate that there are different numbers of channels open at different points on the curve. The slope conductance is a reflection of how these numbers change with changes in membrance potential.

Rectification

When the resistance (or conductance) of a conductor is not independent of the voltage across it, the conductor is said to show rectification. The potassium current curve in fig. 5.23 shows relatively large currents for depolarization and negligible ones for hyperpolarization, hence the potassium pathway is sometimes described as showing 'delayed rectification'. Notice that this form of rectification arises from the increased number of channels which are opened when the membrane is depolarized: the individual channels are not (or not necessarily) rectifiers.

Fig. 5.23. Current–voltage relations in the axon membrane. Giant axons of the marine worm *Myxicola* were subjected to a series of clamped depolarizations to different membrane potentials. Curve *a* shows the peak early currents (largely I_{Na}), curve *b* shows the steady-state currents after 25 ms (largely I_K). Curve *c* shows currents corresponding to those in curve *a*, obtained in the presence of tetrodotoxin. (From Binstock and Goldman, 1969.)

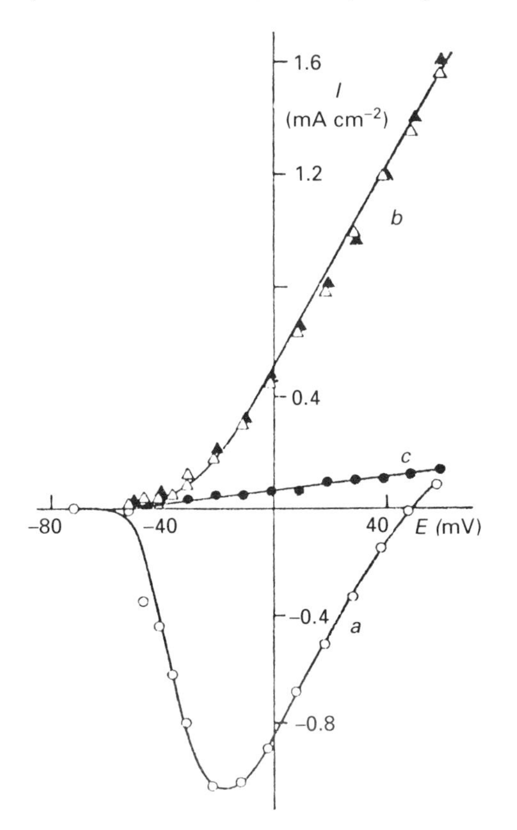

Fig. 5.24. Diagram to show the difference between chord and slope conductances. The curve shows a typical current–voltage relation for the peak inward current in an axon membrane. The slope conductance at point P is the slope of the tangent at that point. The chord conductance is the slope of the line connecting P to the point on the curve at which the voltage is equal to the sodium equilibrium potential.

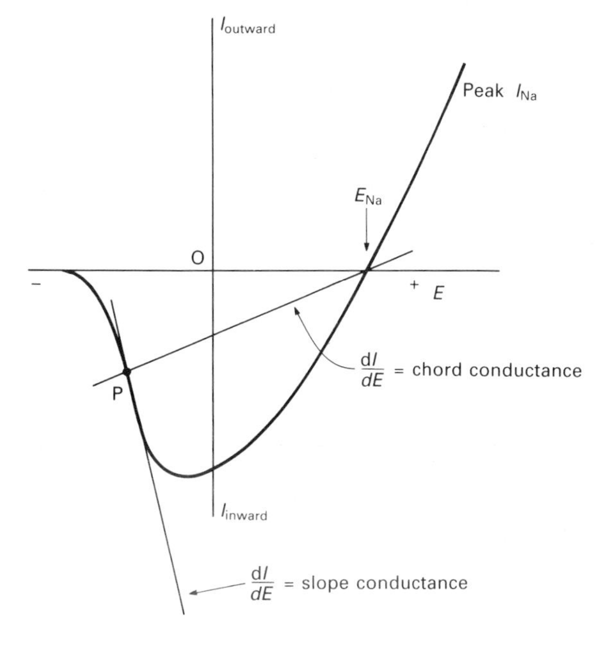

The constant-field theory (chapter 3) predicts a certain amount of rectification, known as 'constant-field rectification'. In a system in which this occurs, the instantaneous current–voltage relation will not be a straight line. Rectification of this type must be a property of the individual open channels. There is evidence that the individual ionic channels of frog nodes behave in this way, so that the constant-field permeability coefficients P_{Na} and P_K are used in the equations for ionic currents there, in place of the conductances g_{Na} and g_K (Dodge and Frankenhaeuser, 1959).

Fixed charges and the involvement of calcium ions

Several investigations have suggested that there are fixed negative charges on the outer and inner sides of the axon membrane (see for example Hille *et al.*, 1975, and Gilbert and Ehrenstein, 1984). We would expect such charges to occur at the polar ends of the membrane lipids and also on the protein molecules. Their existence would modify the electrical potential field in the membrane and in its immediate vicinity, perhaps in the manner shown in fig. 5.26*a*. We would also expect that the excitability mechanisms in the membrane (which must presumably be dependent upon the electric field in the membrane) would be affected by any alterations in the density of these fixed charges. Such alterations would be produced by changes in the divalent ion concentration, pH or ionic strength of the solutions on either side of the

Fig. 5.25. Sodium current–voltage relations in squid axon. Curve α shows the peak early currents in a voltage clamp (corresponding to curve *a* in fig. 5.23). Curve *A* shows the instantaneous current–voltage relation obtained by measuring the currents produced on changing the membrane potential again at 1.53 ms after the onset of a clamped depolarization of 29 mV. (After Hodgkin and Huxley, 1952*b*.)

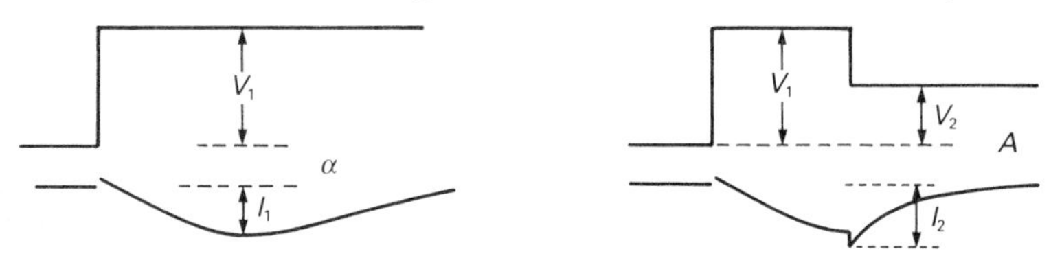

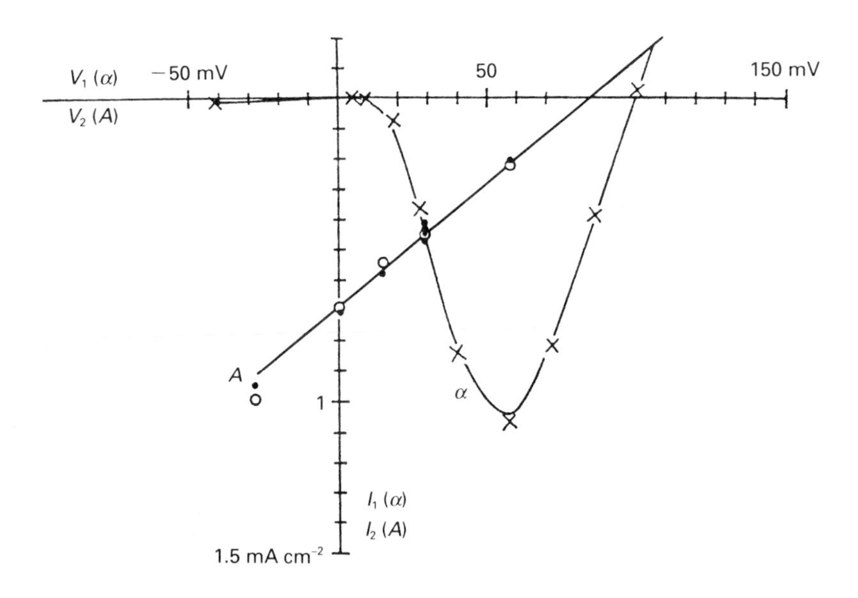

membrane. As an example of these effects, let us look at the action of calcium ions.

Reduction of the calcium ion concentration in the external medium produces an increase in the excitability of axons, seen as a reduction in threshold and a tendency towards spontaneous and repetitive activity. Frankenhaeuser and Hodgkin (1957), applying the voltage-clamp technique, concluded that these effects were caused by an increase in the sodium and potassium conductances of the membrane, such that a fivefold decrease in external calcium ion concentration produces effects similar to a depolarization of 10–15 mV.

Fig. 5.26 shows one way in which these results can be interpreted in terms of fixed charges. With low external calcium ion concentrations, there are many more fixed negative charges on the outer surface of the membrane than on its inner surface. These charges will not affect the total potential across the membrane as measured at short distances away from its surfaces, but they will have marked effects on the detailed form of the potential gradient within the membrane and in its immediate vicinity. In particular, the greater number of nega-

Fig. 5.26. Schematic diagrams to show how fixed charges might affect the potential distribution (heavy lines) in the membrane and in its immediate vicinity. The broken lines show the potential distribution expected in the absence of such fixed charges. The diagrams show low (*a* and *b*) and high (*c* and *d*) external calcium ion concentrations, with membrane potentials at the resting level (*a* and *c*) and partly depolarized (*b* and *d*). Remember that the fixed charges only influence the potential field in the immediate vicinity of the membrane: they cannot be measured with microelectrodes, which are much too large for the job.

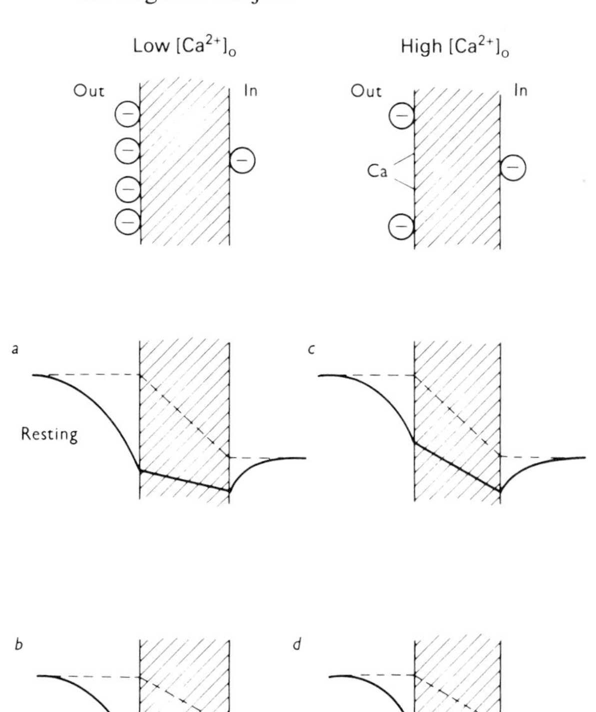

Fig. 5.27. Schematic diagrams to show how the gradient of potential across the membrane at low membrane potentials may be altered by internal perfusion with a solution of low ionic strength. Such a solution allows the fixed charges on the inner face of the membrane to exert a much greater effect. Compare with fig. 5.26.

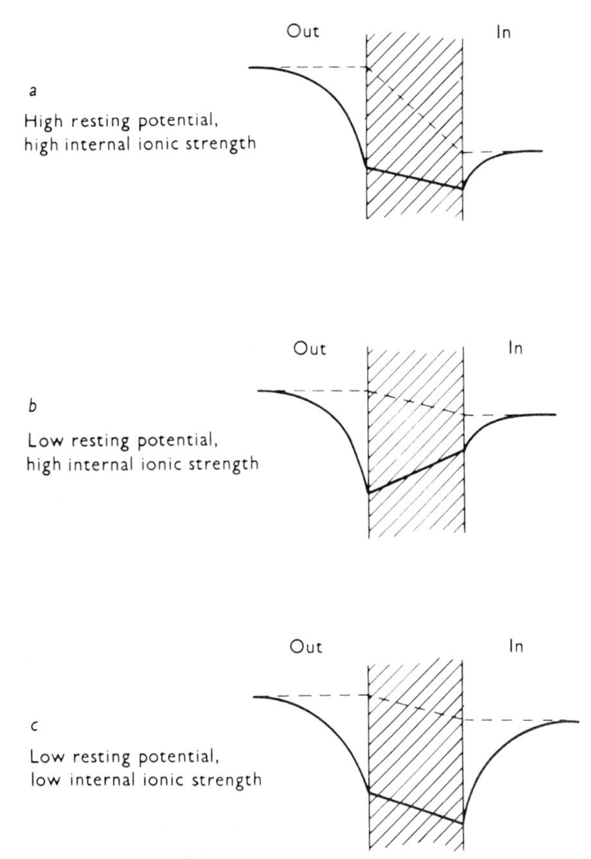

tive charges on the outer surface ensures that the gradient of the potential field within the membrane will be less than it would otherwise be (fig. 5.26*a*). We may assume that the voltage-dependent ionic conductances are determined by the gradient of the field within the membrane. An increase in external calcium concentration will reduce the negative charge density on the outer surface of the membrane, and so will increase the gradient of the potential field within the membrane and hence raise the membrane potential at which the channels become opened.

In an internally-perfused squid axon, dilution of the internal medium with a solution of sucrose or some other nonelectrolyte results in a reduced resting potential, as we would expect. But a similar reduction in excitability does not occur: large and often prolonged action potentials may be seen under these conditions. Using the voltage clamp technique it was possible to show that the relation between sodium conductance and membrane potential was moved to a less negative position; the sodium inactivation relation was moved similarly. In other words, the sodium channel behaved as though the membrane potential was more negative than it actually was. Chandler, Hodgkin and Meves (1965) explained this in terms of fixed charges on the inside of the membrane; when the ionic strength of the internal solution is low such charges will have a much larger effect on the potential field within the membrane. Fig. 5.27 gives an indication of how this system might work.

Gating currents

We have seen that the Hodgkin–Huxley equations were based on the idea that the ionic channels are opened by the movement of charged particles within the membrane. Indeed it seems almost inevitable that a voltage-dependent change should be initiated by some sort of movement of charge. There should therefore be some current flow in the membrane just prior to the increase in ionic permeability. The search for such 'gating currents' (so called because they would open the 'gates' for ionic flow) was eventually successful (Armstrong and Bezanilla, 1973, 1974; Keynes and Rojas, 1974; see also Armstrong, 1981, and Keynes, 1983).

The main difficulty in detecting these gating currents is that they are very small in comparison with the ionic and capacity currents. The ionic currents can be blocked pharmacologically (by using external tetrodotoxin to block the sodium current and internal caesium fluoride to block the potassium current, for example). The capacity current is symmetrical with respect to the direction of voltage change, so it can be set aside by subtracting the response to a hyperpolarizing voltage-clamp pulse from the response to a depolarizing pulse of the same magnitude; the difference is then the gating current.

The reason for the asymmetry of the gating current is as follows. Suppose the charges that are moved during the gating current are positively charged particles that are held near to the inside of the membrane by the resting potential (fig. 5.28). During a hyperpolarization they will not be able to move any further towards the inside of the membrane simply because they are at the limit of their possible movement. There will therefore be no gating current on hyperpolarization. On depolarization, however, the particles will move towards the outer side of the membrane and this movement will constitute the gating current. (It is possible, of course that the gating particles will reach the inner limits of their movement at membrane potentials more negative than the resting potential, but it would seem unlikely that the difference would be very great. It is for this reason that many measurements of gating currents have been made from 'holding potentials' of $-100\,\mathrm{mV}$ or so, rather more negative than the resting potential.) Notice that the arguments are similar in principle if the gating particles are negatively charged and move inwards or if they are rotating dipoles.

Fig. 5.29 shows gating currents of a squid axon for clamped depolarizations, and also the ionic currents obtained from the same axon. Notice that the 'on' gating current is outward and has a much faster time course than the ionic current. The total quantity of charge moved (obtained by integrating the current with respect to time) is about $30\,\mathrm{nC\,cm^{-2}}$ in a squid axon, which corresponds to about 2000 electronic charges per square micrometre of membrane (Keynes and Rojas, 1974).

On repolarization at the end of a voltage clamp pulse there is an inward gating current which has a time course similar to that of the ionic current produced in similar circumstances, as is shown in fig. 5.29. For short pulses (less than 1 ms), the total charge movement during the 'off' gating current is equal and opposite to that at the beginning of the pulse. For longer pulses there is some reduction in this 'off' charge movement in parallel with the sodium inactivation process (Armstrong and Bezanilla, 1977).

Fig. 5.28. Schematic diagram showing one way in which gating currents could arise. The capacity currents are much larger than the gating currents.

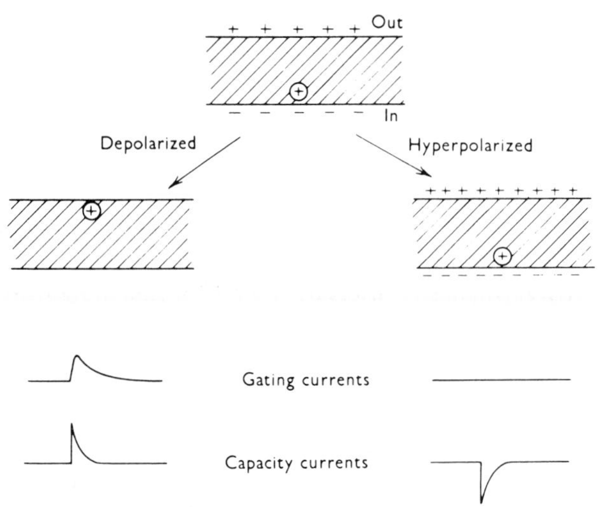

Fig. 5.29. Sodium ionic and gating currents in squid axon, produced during depolarizations under voltage clamp. The upper traces show currents recorded in an artificial sea water with only one fifth of the normal sodium concentration, for depolarizations from −70 mV to −20, 0 and +20 mV. The initial brief outward current is gating current, followed by the much larger inward sodium ionic current. The lower set shows the gating currents alone, after blockage of the sodium ionic currents with tetrodotoxin. Potassium currents were eliminated by using potassium-free solutions for both internal and external media. (From Bezanilla, 1986.)

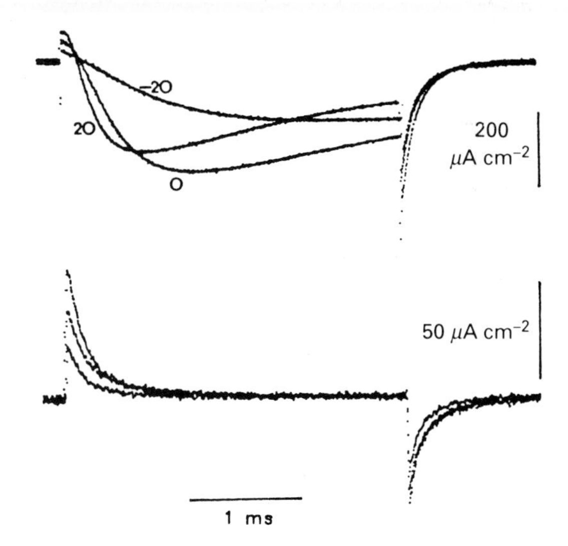

6
Voltage-gated channels

The Hodgkin–Huxley analysis of nervous conduction, which we examined in the previous chapter, showed that the sodium and potassium conductances of the axon membrane are switched on and off (or 'gated') by changes in membrane potential. Hence the channels through which the sodium and potassium ions flow are themselves gated by changes in membrane potential. Later we shall see that there are other channels which are gated by chemicals (ligands) produced in the cell or in other cells. There are thus two major groups of channels in excitable cells: voltage-gated and ligand-gated. In this chapter we examine some of the properties of voltage-gated channels.

The sodium channel

Voltage-clamp studies of the Hodgkin–Huxley type can provide information about the overall actions of large numbers of channels, but they cannot tell us how the individual channels behave. Two methods have been developed for this: fluctuation analysis and the patch-clamp technique. Fluctuation analysis was developed first, but the patch-clamp method allows more direct observation if the individual channel currents are large enough.

Patch clamping

Patch clamping is a technique whereby a very small area of membrane can be voltage-clamped, so allowing the current flow through individual channels to be measured directly. The first records of this type were published by Neher and Sakmann in 1976, from the acetylcholine channels of denervated muscle cells (see p. 142). Since then the technique has been used in a large number of investigations so that it now provides a major source of new information about membrane channels.

Fig. 6.1 illustrates the method. A glass microelectrode is polished to produce a smooth tip 1–2 μm in diameter, and is then coated with a resin to reduce the conductance and capacitance of the glass, and filled with an isotonic electrolyte solution. The electrode is pushed against the cell membrane so that the resistance between the inside of the electrode and the external solution rises to about 100 MΩ. Application of suction pulls the cell membrane into the tip of the electrode where it makes close contact with the glass wall and so forms a seal with a very high resistance, of the order of 50 GΩ (a 'gigaseal'). Such a seal greatly reduces the background noise from the system and so enables the very small currents flowing through the ionic channels to be measured. The gigaseal may not form if there is much extracellular material attached to the cell membrane, hence it is usual to use embryonic cells or to 'clean' the cell with an enzyme such as collagenase.

Once the gigaseal has been formed there is a choice of alternatives. The pipette electrode may be left as it is, in the 'cell-attached' position, or it may be pulled away, bringing the patch with it in the 'inside-out' arrangement. Application of more suction to the cell-attached arrangement breaks the patch so that the electrode now records from the whole cell, and pulling the electrode away from this may tear off a patch of membrane in the 'outside-out' configuration. Further details are provided by Hamill *et al.* (1981) and Sakmann and Neher (1983, 1984).

The first single-channel records of sodium channel currents were obtained by Sigworth and Naher (1980) by patch-clamping myoballs, which are spherical cells prepared by tissue culture of embryonic rat muscle in the presence of colchicine. They used pipettes containing TEA to block any potassium channels present and α-bungarotoxin to block acetylcholine channels (p. 120). The records in fig. 6.2 show successive responses to a clamped

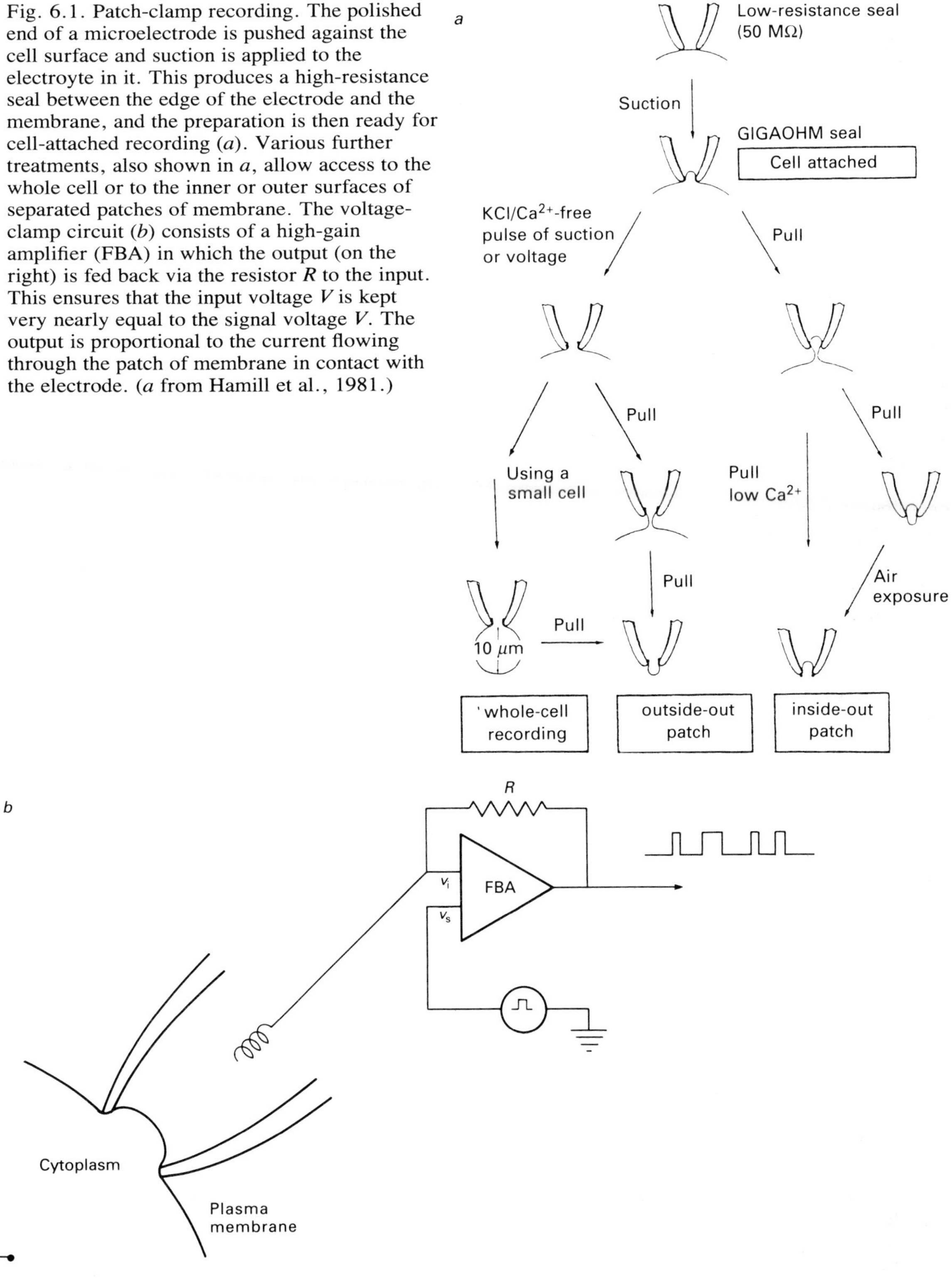

Fig. 6.1. Patch-clamp recording. The polished end of a microelectrode is pushed against the cell surface and suction is applied to the electroyte in it. This produces a high-resistance seal between the edge of the electrode and the membrane, and the preparation is then ready for cell-attached recording (*a*). Various further treatments, also shown in *a*, allow access to the whole cell or to the inner or outer surfaces of separated patches of membrane. The voltage-clamp circuit (*b*) consists of a high-gain amplifier (FBA) in which the output (on the right) is fed back via the resistor *R* to the input. This ensures that the input voltage *V* is kept very nearly equal to the signal voltage *V*. The output is proportional to the current flowing through the patch of membrane in contact with the electrode. (*a* from Hamill et al., 1981.)

depolarization by 40 mV. Most of the traces show square inward current pulses. These are always the same amplitude at the same membrane potential but they differ in duration.

How do we know that these unitary pulses of current are carried by sodium ions? Sigworth and Neher showed that (1) they were blocked by tetrodotoxin, (2) they decreased in size at less negative membrane potentials (and disappeared around +20 to +40 mV, which is presumably near E_{Na}) and (3) they were reduced when the sodium ion concentration in the pipette was reduced (fig. 6.2*d*). Trace *b* in fig. 6.2 shows the average of 300 patch-clamp records; it is equivalent to the current flow which we would expect to see in a single record from an area 300 times as large as the experimental one and so corresponds to the current flow produced in a conventional large-area voltage-clamp system.

The individual pulses in fig. 6.2 are much the same size but they vary considerably in duration and timing. This means that the opening and closing of the channels are stochastic processes: we cannot predict precisely when any particular channel will open or close, but we can in principle determine the probability that it will do either of these things in any one time interval.

We can now give a molecular interpretation of the sodium conductance changes during a voltage clamp of a large area of membrane. At any particular instant a channel is either open or closed: for practical purposes there is no halfway stage at which the conductance of the channel is at some intermediate value. This means that changes in the conductance of an area of axon membrane are produced by changes in the number of channels that are in the open state. When the membrane is depolarized the probability of any particular channel opening is increased so the total number of open channels rises and the conductance of the membrane as a whole rises. During inactivation or after repolarisation the number of open channels falls and so the overall conductance falls also.

Computer simulations of these effects have been performed by Clay and DeFelice (1983). In the Hodgkin–Huxley equations the rate constant α determines the increase of the parameters n, m and h, and the rate constant β determines their decrease. In stochastic terms α and β are also rate constants determining the opening and closing of channels, hence they can be used in equations describing the probabilities of these events at any particular time. So computer simulations can indicate the changes of state of individual channels,

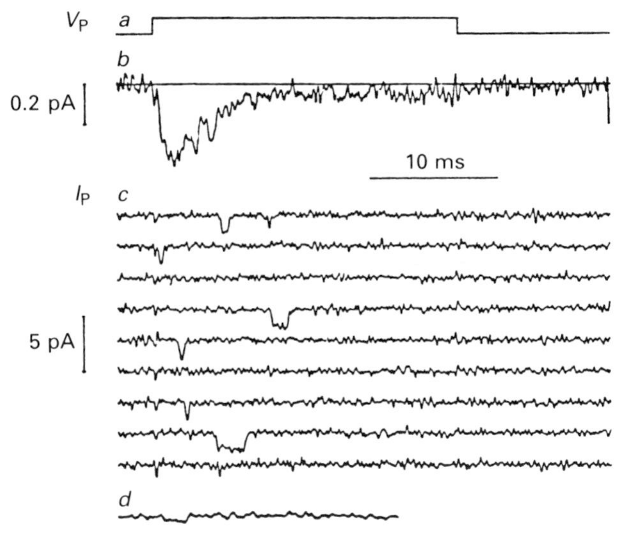

Fig. 6.2. Single sodium channel currents from cultured rat muscle cells, recorded with the cell-attached patch-clamp technique. Trace *a* shows the imposed membrane potential, held at $V = -30$ mV (where V = 0 is the resting potential) and depolarized by 40 mV to $V = +10$ mV for about 23 ms at 1 s intervals. Trace *b* shows the average of a set of 300 of current records elicited by these pulses. *c* shows nine successive individual records from this set. Square pulses of inward current (average size 1.6 pA) can be seen in most of the records; these correspond to the opening of individual channels. Trace *d* shows a record taken when two thirds of the sodium ions in the pipette had been replaced with tetramethylammonium ions; the single-channel current is reduced accordingly. (From Sigworth and Neher, 1980.)

and these can be added together to show the overall effect of large numbers of channels. Fig. 6.3 shows some of the results of these simulations. Although the current flow through any single channel is a series of discrete events of constant magnitude and variable duration, the total current through large numbers of channels follows the continuous and determinate changes described by the Hodgkin–Huxley equations.

The mean current during the single channel pulses in fig. 6.2 was 1.6 pA. We can calculate this in terms of sodium ions moved as follows:

number of ions per second

$$= \frac{\text{current in amps} \times \text{Avogadro's number}}{\text{Faraday's constant}}$$

$$= \frac{1.6 \times 10^{-12} \times 6 \times 10^{23}}{9.6 \times 10^{4}}$$

$$= 10^{7} \text{ ions s}^{-1}.$$

The mean channel lifetime was 0.7 ms so on average 7000 sodium ions pass through each channel before it closes. A transfer rate of 10^7 ions s^{-1} is much higher than the rates of carrier transport

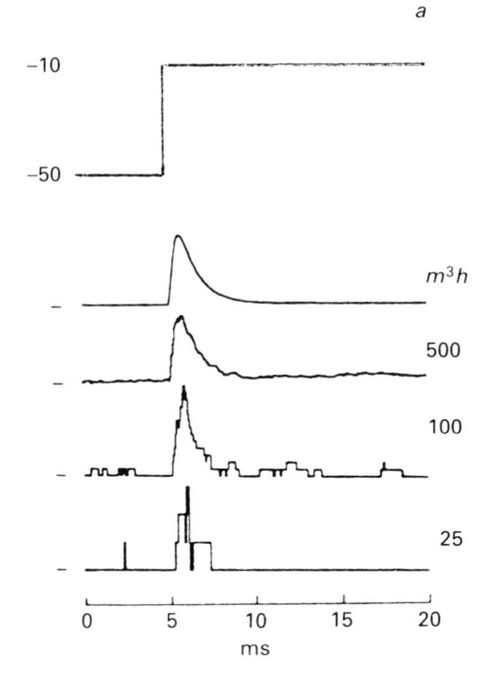

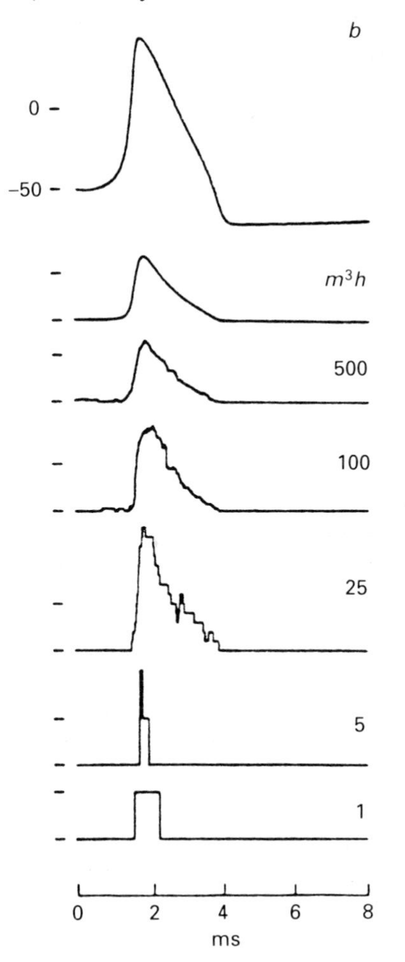

Fig. 6.3. Computer simulations of sodium channel gating in patches of axon membrane, based on the Hodgkin–Huxley equations. Responses are shown during (*a*) a voltage-clamped depolarization by 40 mV, and (*b*) an action potential. The upper trace in each case shows the membrane potential. The m^3h curve shows the probability of an individual channel being open and is proportional to the sodium current in a large area of membrane. The lower graphs show the fraction of open channels in patches of 1–500 channels. Compare the 100 channels trace in *a* with trace *b* in fig. 6.2. (From Clay and DeFelice, 1983.)

systems and enzyme actions and so provides excellent evidence that the open channel really is an aqueous pore through which sodium ions can diffuse.

We can calculate the conductance γ of the open channel from the equation

$$i = \gamma(V - V_{Na})$$

where i is the single channel current and $(V - V_{Na})$ is the difference between the clamped membrane potential and the sodium equilibrium potential. In this case

$$\gamma = 1.6\,\mathrm{pA}/0.09\,\mathrm{V}$$
$$= 18\,\mathrm{pS}.$$

Patch clamping provides us with perhaps the most direct evidence for the existence of discrete ionic channels in that we can actually observe and measure the currents flowing through them.

Fluctuation analysis

Fluctuation analysis is a technique whereby we can gain information about the nature of unit electrical events by examining the fluctuations (or 'noise') in the gross currents produced by the sum of a larger number of such events. It was developed before patch clamping and is useful where the membrane may not be accessible to a patch-clamp electrode (as in nerve axons surrounded by Schwann cells) or where channel densities are very high (as at the node of Ranvier or the neuromuscular junction). Details of the method are given by Stevens (1972), Neher and Stevens (1977) and DeFelice (1981). At the node of Ranvier fluctuation analysis was used to determine single sodium channel currents by Conti *et al.* (1976*a*), but here we shall consider a development of this method by Sigworth (1980*a*).

Sigworth's method depends essentially on a simple relationship between the variance and the mean in a binomial distribution. Suppose that under some specified situation (such as the time of peak sodium current during a clamped depolarization) the probability of any particular channel being open is p, and that there is a total of N channels of which r are open on any particular occasion when the specified situation occurs. The sodium current will be proportional to the proportion of channels open, r/N. The variance of r/N, determined from a large number of occasions, is given by

$$\mathrm{var}(r/N) = \mathrm{var}(r)/N = p(1 - p)/N.$$

Hence the variance of r/N enables us to estimate N; a high variance means that a small number of channels are involved.

Let us illustrate this by considering the finances of two analogical characters, Harry and Tom. Each day they are presented with a bag containing one dollar's worth of coins, always of the same denomination, and our problem is to know how many coins are in the bag, i.e. what denomination they are. Our characters share the coins between them by tossing: Harry takes those that fall head and Tom takes the tails. The amount each receives will fluctuate from day to day, and the variance of these amounts will be greater the higher the denomination of the coins. For example, if Harry tells us at the end of the year that he several times received the whole dollar's worth of coins we can be confident that the coins were 25 or 50 cent pieces, not 10 cents or less. More precisely, we can expect the mean daily income for each man to be near to 50 cents whatever the coinage used, but the standard deviations (the square roots of the variances) will be 25 cents for quarters ($N = 4$), 16 cents for 10-cent pieces ($N = 10$), 11 cents for 5-cent pieces ($N = 20$) and 5 cents for single cents ($N = 100$).

Sigworth's results are illustrated in fig. 6.4. Trace *a* shows six successive current records produced by depolarizations to -5 mV. By taking the mean of such a group of records and subtracting it from each individual record, we get the deviations of the individual records from the mean; this is shown in trace *b*. From a large number of such records we can calculate the variance of the current, as in *c*, which was computed from 65 groups of six records. (The use of local means to get traces as in *b* largely eliminates problems of long-term drift. Potassium currents were eliminated with TEA.) The variance shown in trace *c* contains a component caused by thermal noise in the system; this has to be subtracted to give var(I), the variance in the current due to fluctuations in conductance.

We assume that each of the sodium channels at the node can be either open or closed and that this is independent of what state other channels are in. N is the number of channels, p is the probability of any particular channel being open, i is the current flowing through an open channel and I is the mean total current. Then

$$I = Npi$$

and

$$\mathrm{var}(I) = Np(1-p)i^2. \quad (6.1)$$

Hence, substituting for p,

$$\mathrm{var}(I) = iI - I^2/N \quad (6.2)$$

Equation (6.2) is a parabola which hits the I axis at iN and has a slope of i where it hits the var(I) axis. Hence by plotting var(I) against I from records such as that in fig. 6.4, Sigworth was able to estimate i and N.

Fig. 6.5 shows one of Sigworth's variance-mean plots; let us see how the various parameters can be determined from it. The curve is a parabola fitted by eye; ideally it should go through the origin of the graph, but a small residual variance has been added. The slope of the curve is given by differentiating equation (6.1):

$$\frac{\mathrm{d}(\mathrm{var}(I))}{\mathrm{d}I} = i - 2I/N \quad (6.3)$$

so when $I = 0$ the slope of the curve is the single channel current i. At the maximum value of var(I), d(var(I))/d$I = 0$ and so $I = Ni/2$, which enables N to be calculated. In fig. 6.5 $i = -0.34$ pA and $N = 42\,000$. If all the channels were open at once the current would therefore be $42\,000 \times 0.34$ pA, i.e. 14.2 nA. The maximum current produced during the clamp was 8.2 nA and so at this time 58% of the sodium channels were open (the corresponding value at +125 mV was 93%). The clamp potential was −5 mV and E_{Na} was +54 mV, so the driving force was 59 mV and hence the single channel conductance γ was 0.34/0.059 i.e. 5.8 pS.

The method of fluctuation analysis described in this section is named non-stationary or ensemble analysis since it utilizes a large number of records in which the probability of channels being open changes with time. A simpler method, known as

Fig. 6.4. Sodium current fluctuations at the frog node of Ranvier, using the nonstationary (or ensemble) analysis method. Trace *a* shows six successive current records produced by clamped depolarizations to −5 mV. *b* shows the deviations of the individual currents in *a* from their mean. *c* shows the variance of 65 such groups of records. (From Sigworth, 1980*a*.)

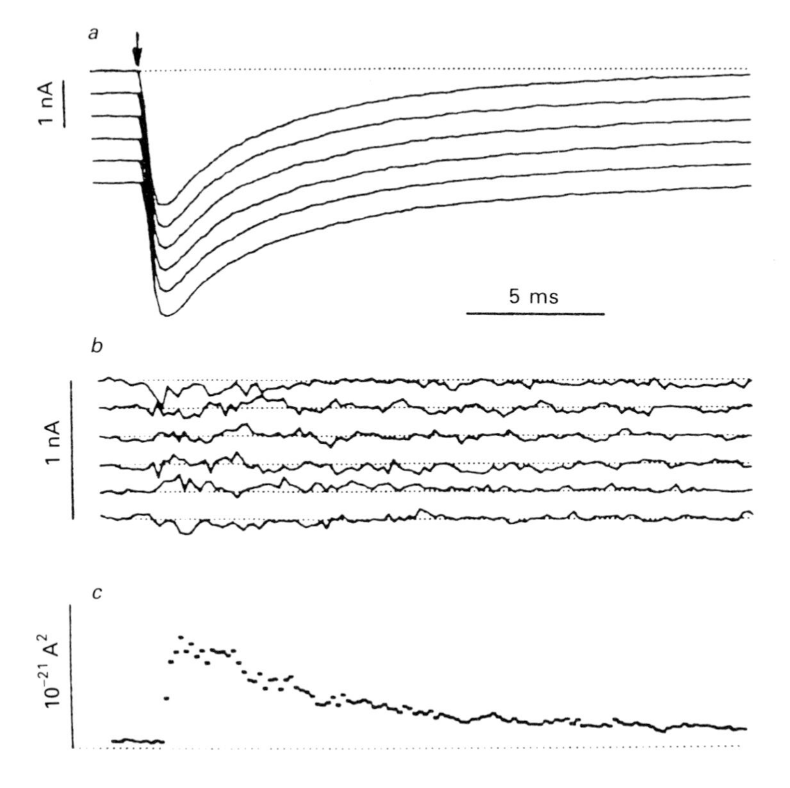

stationary fluctuation analysis, is used for situations where this is not so; we shall examine this form of the technique in chapter 8.

A physiologist's model for the sodium channel

Fig. 6.6 gives an impression of how the sodium channel might be constructed. This model is based on physiological evidence; we shall look at results of biochemical analyses later. The channel protein sits in the lipid membrane so that the aqueous pore which it contains connects the external medium to the internal cytoplasm. The pore is closed at rest by an activation gate. The gate is opened following movement of electrical charge within a sensor as a result of a change in membrane potential; this charge movement can be detected as the gating current.

Internal perfusion of a squid axon with pronase, a proteolytic enzyme, removes the sodium inactivation process (Armstrong *et al.*, 1973). So inactivation must be brought about by a part of the molecule which is readily accessible from the cytoplasmic side of the membrane. The model as drawn here suggests that an inactivating particle can swing into the pore to block it when the activation gate is open.

The sodium channel is highly selective of the ions that will pass through it (p. 72). Hille (1984) suggests that this implies that at some point the pore must be narrow enough to bring the permeant ion directly into contact with its walls, since otherwise it is difficult to see how the selectivity can arise if the ion is 'hidden' inside a shell of water molecules. This idea fits well with what we know about the permeability of sodium channels to organic cations (Hille, 1971*b*). Those which will pass through the channels will all fit snugly into a space 3.2 × 5.2 Å, as will a sodium ion plus one water molecule. Hence it is reasonable to suggest that this space gives the dimensions of the selectivity filter. It is notable that guanidinium ions will pass through the channel and that both tetrodotoxin and saxitoxin have guanidinium moieties in their molecules; it may be that these toxins cause block by getting stuck in the narrow part of the channel.

Selectivity may well not simply be a question of size alone. It seems very likely that the filter includes what is in effect an ion binding site, so that there is a sequence of very fast reactions such as

$$\mathrm{Na_o + S \rightarrow Na\text{–}S \rightarrow Na_i + S}$$

where S is the binding site in the selectivity filter and the subscripts o and i indicate ions on the outside and inside of the filter. There is some indication that two such binding sites may be involved (Begenisich and Cahalan, 1980.)

Blocking agents

Many natural neurotoxins have been produced by animals and plants for attack or defence. The actions of these and other substances on nervous conduction are interesting both in themselves and in relation to their possible uses in

Fig. 6.5. Variance-mean plot from non-stationary fluctuation analysis of the sodium currents at a frog node of Ranvier. The inserts show the mean sodium current I (lower trace, 1 nA per small division) and its variance var(I) (upper trace, 2×10^{-22} A^2 per small division) measured as in fig. 6.4, in response to a 20 ms clamped depolarization to −5 mV. The points on the main graph show these two quantities plotted against each other and the curve is equation 6.2 with $i = 0.34$ pA and $N = 42\,000$. (From Sigworth, 1980*a*.)

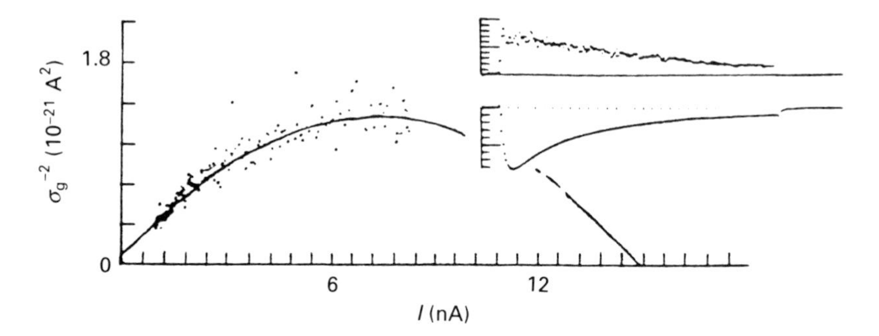

medicine or agriculture. They are also of much use as tools in the study of ionic channels (see Narahashi, 1974, and Catterall, 1980, 1986*a*, *b*).

Neurotoxins which act on voltage-gated ionic channels are of two types: those which block the channels and those which modify the gating process so as to keep them open for longer than usual.

The water-soluble substances tetrodotoxin and saxitoxin block the passage of sodium ions, probably by binding near the mouth of the channel. The dose–response relation of the effect of saxitoxin on sodium currents suggests that each saxitoxin molecule combines with one sodium channel so as to block it completely (Hille, 1968). A similar conclusion arises from fluctuation analysis experiments: in the presence of a low concentration of tetrodotoxin the number of open channels at a frog node fell but their single-channel conductance did not (Sigworth, 1980*a*).

A number of local anaesthetics such as procaine and lidocaine act by blocking sodium channels. These are all lipid-soluble compounds and it seems likely that they reach their binding site in the channel by first dissolving in the lipid phase of the membrane. A number of quaternary ammonium compounds of otherwise similar structure (such as QX-314, pancuronium and N-methylstrychnine) probably act at the same site. These quaternary compounds are charged and so they are not lipid-soluble. They are only effective when applied from the inside of the cell, which suggests that their site of action is on the inner side of the selectivity filter (Strichartz, 1973; Schwarz *et al.*, 1977).

Peak sodium currents in voltage-clamp experiments are reduced in acid environments. Using fluctuation analysis, Sigworth (1980*b*) has shown that the single-channel conductance in frog node sodium channels is reduced at pH 5 to about 40% of the normal value. His explanation for this is that hydrogen ions compete with sodium ions for the

Fig. 6.6. A sketch to show the main features of the sodium channel. (Based partly on diagrams by Armstrong, 1981, and Hille, 1984.)

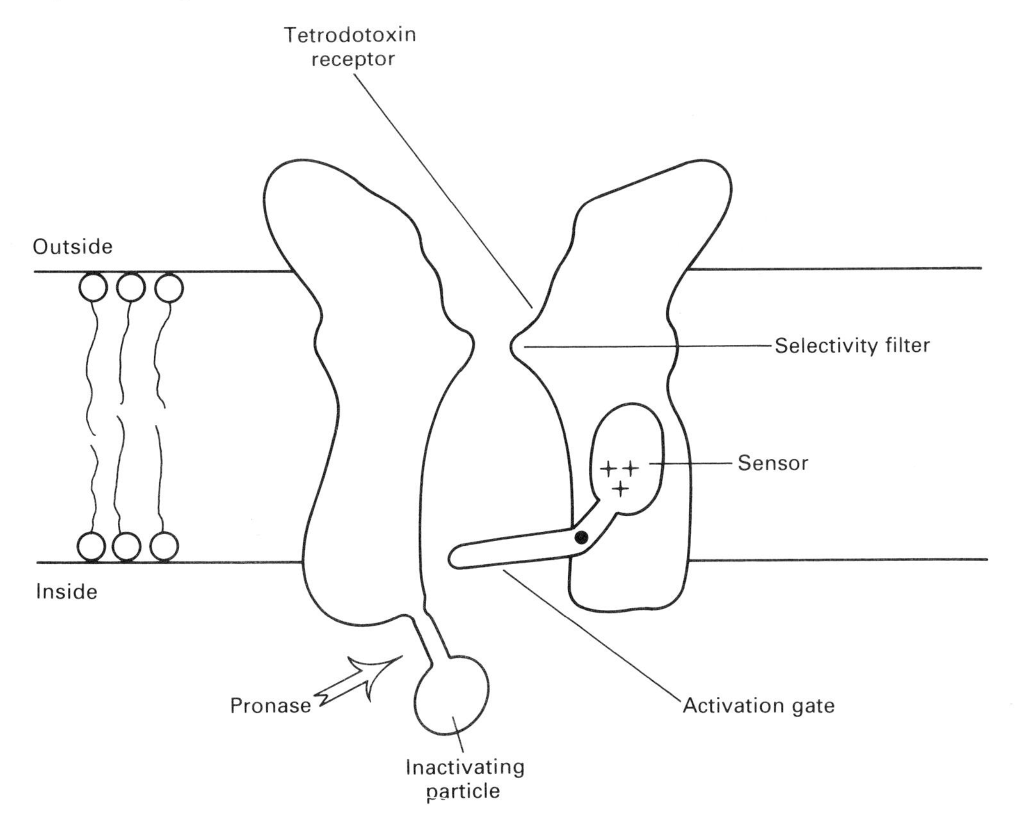

ion binding site in the selectivity filter (as suggested by Woodhull, 1973) but that the duration of the consequent block is so short that individual events cannot be resolved by the recording apparatus.

We saw in the previous chapter that external calcium ions may affect the current–voltage relations of squid axons. Yamamoto, Yeh and Narahashi (1984) found that single channel sodium currents in neuroblastoma cells are reduced by increased external calcium ion concentrations. This suggests that calcium ions may temporarily block the channels.

Modifiers of gating

The peptide toxins produced by scorpions and sea anemones cause hyperexcitation (and hence considerable pain) by preventing inactivation. The binding sites are on the outer surface of the channel, hence the toxin's action must involve a conformational change in the channel molecule in order to affect the inactivating particle on its inner side. Some American scorpions also possess toxins which alter the voltage dependence of the activation gate: the channels may remain open for some time after repolarization, resulting in repetitive production of action potentials (Cahalan, 1975).

Many neurotoxins are lipid-soluble substances which alter the activation-gating of the sodium channels so that they become open at the resting potential (Catterall, 1980). Examples include the plant alkaloids aconitine and veratridine, pyrethrins (natural insecticides from *Pyrethrum*, and their synthetic analogues), batrachotoxin from Columbian arrow-poison frogs, and the organochlorine insecticides DDT, aldrin and dieldrin.

Kinetics

The Hodgkin–Huxley equations which we saw in chapter 5 were constructed to describe the relations between conductance, voltage and time, as observed in voltage-clamped squid axons. They utilized a model in which each sodium channel was opened by three simultaneous events and blocked by a single event, so that the probability of the channel being open was given by m^3h (equation 5.10). This could be thought of as the movement of three activating m particles and one inactivation h particle to a particular place in the membrane. Note that the movements of the particles are independent of one another: this means that inactivation can occur independently of activation.

Fig. 6.7 gives a schematic diagram of the different possible states of the Hodgkin–Huxley model. Notice that, neglecting inactivation, there are three closed states and the sequence of opening the channel is

$$\text{closed} \overset{3\alpha}{\rightarrow} \text{closed} \overset{2\alpha}{\rightarrow} \text{closed} \overset{\alpha}{\rightarrow} \text{open}$$

The rate constants follow the descending 3:2:1

Fig. 6.7. Possible states of the Hodgkin–Huxley model for the sodium channel. (From Hille, 1984.)

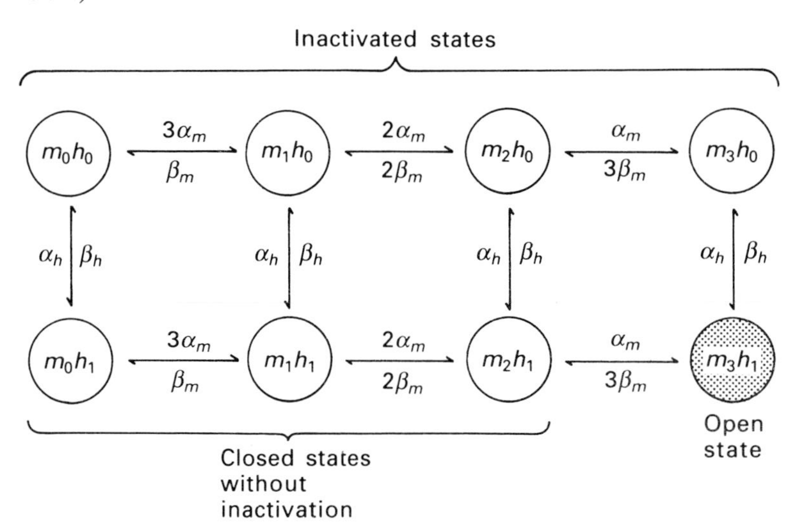

sequence because initially any of the three m particles can move, then two and finally one of them; the reverse sequence applies to the rate constants for closing.

It is possible to construct other models. If we assume that the three m gates are not independent of one another, then the rate constants for movement between states need no longer be in the 3 : 2 : 1 sequence. If we further assume that inactivation can only occur after opening, we get a rather simpler scheme that is known as the 'coupled inactivation model' (fig. 6.9*c*). Models of these types may produce predictions of the time course of conductance changes which are similar to those of the Hodgkin–Huxley model.

More information about inactivation comes from patch clamp experiments. Aldrich, Corey and Stevens (1983) found that the sodium channels of tissue-cultured neuroblastoma cells normally opened only once per depolarization and closed to the inactivated state. However, contrary results have been obtained by Vandenberg and Horn (1984), using cultured GH_3 cells (a rat pituitary cell line), where reopenings were frequent. Aldrich and his colleages also showed that some inactivation could occur without channel opening (fig. 6.8). This suggests that inactivation is partly but not completely coupled to activation.

Horn and Vandenberg (1984) obtained an extensive set of records of channel openings and closings in outside-out patches from cultured GH_3 cells. They fitted a variety of kinetic models to the data by a statistical procedure called the maximum likelihood method. This is a powerful method but requires a lot of computer time, hence the models they used were limited to five states with one open, one inactivated and up to three closed; even so they tested twentyfive different models. Three of these are shown in fig. 6.9, together with their predictions for the form of the histogram of times to first opening. Horn and Vandenberg concluded that acceptable models had to allow inactivation from closed states as well as from the open state (so ruling out the strictly coupled inactivation model), and that the irreversible-inactivation form of the Hodgkin–Huxley model was not satisfactory.

An acceptable scheme to describe the kinetics of change in the sodium channel must account for the properties of the gating current. The Hodgkin–Huxley model predicts that the sodium current following clamped repolarization after a brief depolarization (sometimes called the 'tail current') should decay three times faster than the 'off' gating current, but in fact the rates of decay of the two

Fig. 6.8. Averaged patch-clamp currents from neuroblastoma cells, to show inactivation. Trace *a* shows the average of 64 records produced in response to depolarizations of 70 mV. *b* is a similar trace in which an identical pulse was preceded by a depolarization of 50 mV for 7.5 ms and a reset pulse of 1 ms; the reduced size of the response shows that the prepulse has produced some inactivation. The individual traces from the set of which *b* was the average were then allotted to two groups: (1) where there were no channel openings during the prepulse: these are averaged in trace *c*; and (2) where there were such openings: these are averaged in trace *d*. Since trace *c* shows some inactivation (i.e. the response to the 70 mV depolarization is less than in *a*) we can therefore conclude that a channel can reach the inactivated state without necessarily passing through the open state. (From Aldrich and Stevens, 1983.)

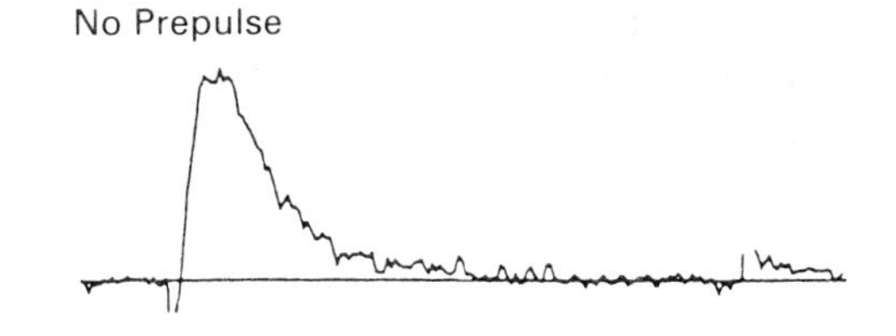

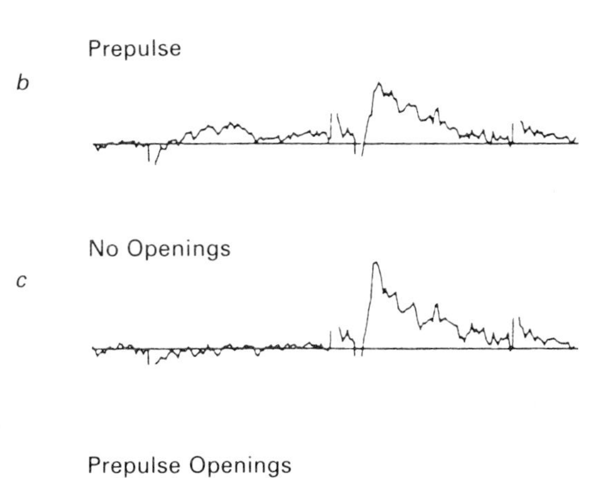

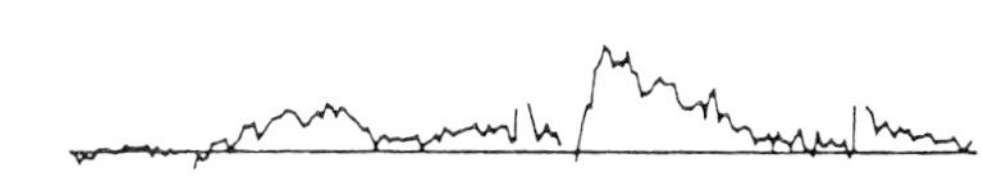

currents are very similar (fig. 5.29). Various other models have been suggested by Armstrong (1981), Aldrich *et al.* (1983) and Stimers, Bezanilla and Taylor (1985).

It is important to realize that Hodgkin and Huxley's splendid achievement in describing the action potential in terms of voltage-gated conductance changes is in no way diminished by the fact that their equations do not predict the form of the gating currents correctly.

Stimers *et al.* (1985) have produced a clear account of the relations between voltage, gating current and sodium channel activation that provides a template against which any kinetic scheme must be tested if it is to prove acceptable. Using squid giant axons treated internally with pronase

Fig. 6.9. Kinetic model-fitting and statistical analysis of sodium channel activity. The histogram shows the number of times that a sodium channel opened at a particular time after the onset of a clamped depolarization to −40 mV in a patch from a cultured GH_3 cell. The three curves show the predicted form of this relationship for three kinetic models of the channel; for each model the time constants were fitted by the maximum-likelihood procedure. The models are: *a*, the 'basic' model; *b*, an irreversible inactivation form of the Hodgkin–Huxley model; and *c*, a strictly coupled-inactivation model. (From Horn and Vanderberg, 1984.)

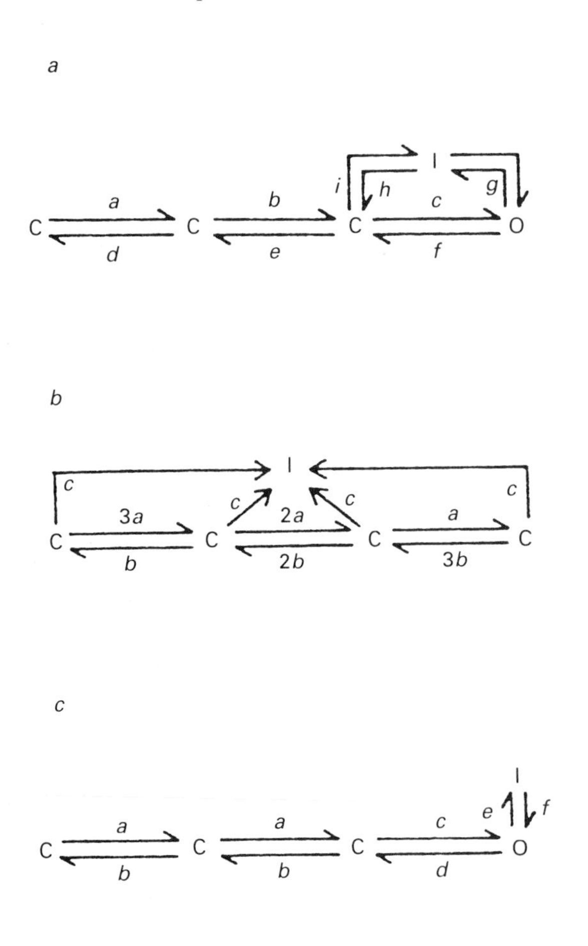

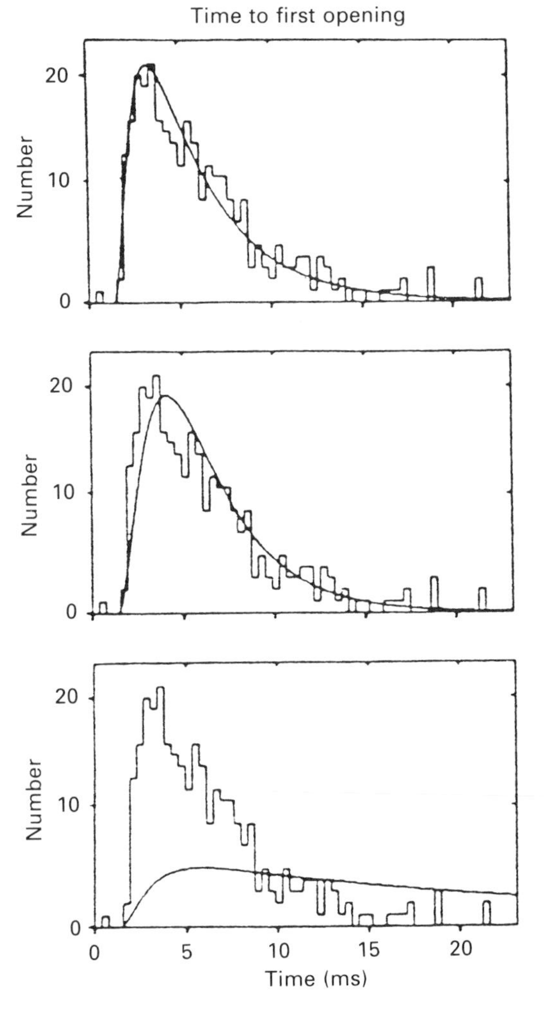

to remove inactivation, they measured gating currents and sodium currents over a wide range of membrane potentials. They wished to estimate the fraction of the channels in the open state from values of the peak sodium current. To do this they assumed that all the channels were opened by a clamped depolarization at +50 mV for 3.5 ms; then repolarization to different voltages allows measurement of the instantaneous sodium currents from the 'tail' currents (p. 72). Then the fraction F of the channels opened on depolarization to a particular voltage is given by the ratio of the peak inward current on depolarization to the instantaneous current at that voltage.

Some results of this work are shown in fig. 6.10, where the Q–V curve shows the proportion of gating charges moved and the F–V curve shows the fraction of channels opened. They confirm two predictions about gating currents: (1) the Q–V curve should always be above the F–V curve, since the gating charge must move before the channel can conduct, and (2) the two curves should saturate at the same potential, since when all the charge has moved all the channels should be open.

Fig. 6.10. Gating charge movement and sodium channel opening in squid axon. The Q–V curve shows the amount of gating charge movement (as a proportion of the maximum) following depolarization to different membrane potentials. The F–V curve shows the fraction of sodium channels opened by such depolarizations. The axons had been initially perfused with pronase to remove the complications of inactivation. The F–V curve was calculated from measurements of ionic currents as described in the text. (From Stimers *et al.*, 1985.)

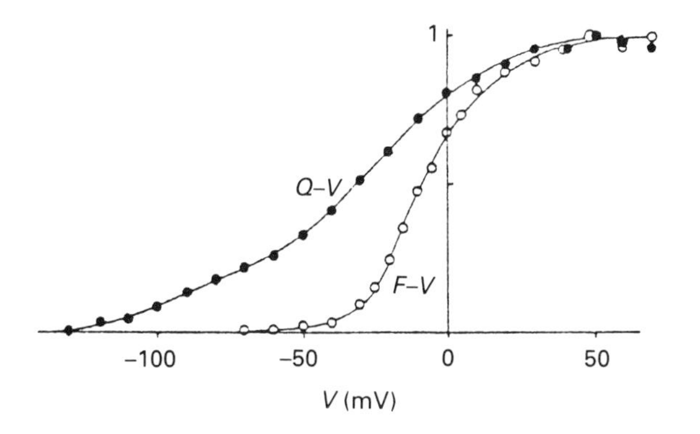

Counting the channels

Clearly it is of considerable interest to estimate how many sodium channels there are in an area of axon membrane. There are various ways of doing this. One method is to measure the binding of radioactive tetrodotoxin or saxitoxin to the nerve. The estimates vary from 35 sodium channels per square micrometre for the thin unmyelinated fibres of the garfish olfactory nerve to about 300 μm^{-2} for squid giant axons, and rather larger values for the nodes of myelinated axons (see table 6.1).

An alternative method is provided by the analysis of gating currents. Thus Keynes and Rojas (1974) found that the average maximum gating charge displacement was 1900 electronic charges per square micrometre. To utilize this figure we need to know how many gating charges there are per sodium channel. A minimum value for this can be calculated from the Boltzmann relation, expressed here as

$$F = \frac{1}{1 + \exp(-zeV/kT)}$$

where F is the fraction of channels open, e is the elementary unit charge, z is the number of charges per channel, V is the membrane potential (on a scale such that half the channels are open at $V = 0$), k is the Boltzmann constant and T is the absolute temperature. At 5 °C when V is measured in millivolts this becomes

$$F = \frac{1}{1 + \exp(-zV/24)}$$

which means that when F is small it increases e-fold for every $24/z$ millivolts change in membrane potential. Hodgkin and Huxley (1952*a*) found that the initial slope of the curve of peak sodium conductance against voltage was 3.9 mV per log unit (whence $z = 6$). Stimers *et al.* (1985), using pronase-treated squid axons and so allowing steady-state conductances to be measured, found a slope of 7 mV, suggesting $z = 4$. If we use a value of $z = 6$ and take 1900 as the number of charges moved per square micrometre, then the channel density is 316 μm^{-2}.

The third method of estimating channel density is to estimate the maximum sodium conductance per unit area and divide it by the single channel conductance as determined by patch clamping or fluctuation analysis. In frog myelinated fibres,

Table 6.1. *Estimates of sodium channel densities in various membranes*

Preparation	Channel densities (per square micrometre) estimated from		
	Toxin binding	Gating currents (charge/6)	Single channel conductance
Garfish olfactory nerve	35[1]		
Lobster walking leg nerve	90[1]		
Rabbit vagus	110[1]		
Squid giant axon	290[2]	317[3,4]	330[5] 92[6]
Squid fin nerve axon (16 μm)	94[2]		
Frog node		3000[7]	400–900[8]
Mammal node		2000–3000[9]	700[10]
Rabbit central fibre node	400–700[11]		
Frog muscle	371–557[12]	650[13]	140[14]

References: (1) Ritchie, Rogart and Strichartz, 1976; (2) Keynes and Ritchie, 1984; (3) Armstrong and Bezanilla, 1974; (4) Keynes and Rojas, 1974; (5) Conti *et al.*, 1975; (6) Bekkers, Greef and Neumcke, 1983; (7) Nonner, Rojas and Stämpfli, 1975*a*, *b*; (8) Sigworth, 1980*a*; (9) Chiu, 1980; (10) Neumke and Stämpfli, 1982; (11) Pellegrino and Ritchie, 1984; (12) Hansen Bay and Strichartz, 1980; (13) Collins, Rojas and Suarez-Isla, 1982*b*; (14) Calculated from Collins, Rojas and Suarez-Isla, 1982*a* and Sigworth and Neher, 1980.

for example, Conti *et al.* (1976*a*, *b*) found a typical maximum sodium conductance of 683 nS per node and a mean single channel conductance of 7.9 pS, giving about 86 500 channels per node. If the nodal membrane area was 50 μm^{-2} (and there is some uncertainty here) then the channel density would be 1730 μm^{-2}. Sigworth's (1980*a*) results give rather lower values, with 20 000–46 000 channels per node, corresponding to 400–920 channels per square micrometre.

Table 6.1 shows some estimates of sodium channel densities in various cells. There is evidently some variation in the various estimates, and the figures should not be regarded as being very precise. Nevertheless it is clear that we can make two useful generalizations: (1) densities are appreciably greater at nodes than in non-myelinated axons, and (2) in non-myelinated axons densities increase with increasing fibre diameter.

Is there an optimum density of the sodium channels? Clearly, conduction velocity will increase with increasing sodium conductance. But it will decrease with increasing membrane capacitance, and each channel, with its little store of moveable gating charges, must act as a small capacitance. At very high sodium channel densities, therefore, the effective membrane capacitance will be large enough to reduce the conduction velocity. Hence there is an optimum sodium channel density at which conduction velocity is maximal. This argument is developed by Hodgkin (1975), who calculates that the optimum density is in the range 500–1000 channels μm^{-2}, in reasonable agreement with the figures derived from experiments on squid axons.

Why should the sodium channel density be lower in axons of smaller diameter? Perhaps it is simply to reduce the quantity of sodium entering per impulse, all of which must later be pumped out using metabolic energy. This is much more of a problem for small diameter fibres because of their higher surface-to-volume ratio.

An important corollary of the differences in channel density in different axons is that calculations which assume that membrane characteristics are the same for fibres of different diameters (as on p. 42) are somewhat oversimplified.

It is interesting to notice that the density of sodium channels in unmyelinated axons is about one-tenth of that of sodium pumping sites (see p. 19). This ratio is presumably related to the much greater rate of flow of sodium ions through the channels than through the pumping sites.

The chemical nature of the sodium channel

The great specificity of the binding of tetrodotoxin and saxitoxin to sodium channels provides the basis for their biochemical isolation. Using radioactive toxins, all the biochemist has to do is to isolate the fraction of the cell protein which is radioactive. The proteins are extracted by using non-ionic detergents with some phospholipid present (Agnew *et al.*, 1978, 1983).

The electric organ of the electric eel *Electrophorus electricus* contains a large number of electrically excitable cells and so it is a good source of sodium channels. Agnew and his colleagues found that the sodium channel protein is a single large peptide with a number of carbohydrate residues attached, and a molecular weight of about 260 kDa. The carbohydrate accounts for 29% of the mass, half of it being sialic acid residues, which are negatively charged (Miller *et al.*, 1983). In rat brain the channel consists of three subunits: a large α chain similar to the *Electrophorus* chain at 260 kDa, and two smaller ones called the β1 and β2 chains, with molecular weights of about 39 and 37 kDa respectively (Hartshorne and Catterall, 1984).

The amino acid sequence of the sodium channel has been determined by Noda and his colleagues at Kyoto University, using the recombinant DNA techniques which we shall look at in chapter 8 (Noda *et al.*, 1984, 1986*a*). The electric eel protein chain is 1820 amino acids long and contains four homologous regions ('domains') which have very similar amino acid sequences (fig. 6.11). Within each of these domains are six segments (S1 to S6), each of which consists of 18–28 amino acid residues that are largely (but not entirely) hydrophobic and are probably arranged in an α-helix. These properties suggest that the six

Fig. 6.11. Parts of the amino acid sequence of the *Electrophorus* sodium channel, to show the four homologous domains. The four domains are aligned, with gaps introduced to maximise homology. The one-letter amino acid code is used. Boxes enclose identical or similar amino acids. The segments S1 to S6 are thought to be regions where the chain forms α-helices which traverse the membrane. (From Noda *et al.*, 1984.)

```
I    IRRGAI-RVFVNSAFNFFIM--FTIFSNCIFMTISNPPAWSK-----IVEYT---FTGIYTFEVIVKVLSRGFCIGHFTFLRDPWNWLDF   111- 189
II   LKKWVH-FVMMDPFTDLFIT--LCIILNTLFMSIEHHPMNES--FQSLLSAGNLVFTTIFAAEMVLKIIA-LDPYYYFQ---QTWNIFDS   555- 635
III  LRRTCYTIVEHDYF-ETFII--FMILLSSGVLAFEDIYIWRRRVIKVILEYADKVFTYVFIVEMLKWVAYGFKR-YFT---DAWCWLDF   989-1071
IV   VQGVVYDIV-TQPFTDIFIMALICINMVA--MMVESEDQSQV--KKDILSQINVIFVIIFTVECLLKLLA-LRQY-FFT---VGWNVFDF  1311-1390
                        S1                                      S2                      S3

I    SVVTMTYIT------EFIDL-----------RNVSALRTFRVLRALKTITIFPGLKTIVRALIESMKQMGDVVILTVFSLAVFTLAGMQL   190- 262
II   IIVSLSLLE-----LGLSNM-----------QGMSVLRSLRLLRIFKLAKSWPTLNILIKIICNSVGALGNLTIVLAIIVFIFALVGFQL   636- 709
III  VIVGASIMGITSSLLGYEEL-----------GAIKNLRTIRALRPLRALSRFEGMKVVVRALLGAIPSIMNVLLVCLMFWLIFSIMGVNL  1072-1150
IV   AVVVISIIG----LLLSDIIEKYFVSPTLFRVIRLARIARVLRLIRAAKGIRTLLF---ALMMSLPALFNIGLLLFLIMFIFSIFGMSN  1391-1472
                                             S4                                         S5

I    FMGNLRH-KCIRWPISNV-TL-DYE——(58)——NYDNFAWTFLCLFRLML-QDYWENLY---QMTLRAAGKSY-MVFFI--   263- 383
II   FGKNYKEYVC---KISDDCELPRWH——( 0)——MNDFFH-SFLIVFRALC-GEWIETMW---DCMEVGGVPMC-LAVYM--   710- 771
III  FAG--KFYRCI--NTTTDEILPVEE——(23)——NYDNAGMGYLSLLQVSTFKGWMDIMYAAVDSREVEDQPIYEINVYMYL  1151-1242
IV   FAYVKKQ------GGVDD-IF-NFE——(16)——GWDGLLLPTLNTGPPDC-DPDVEN--PGTDVRGNCGNPGKGITFFC--  1473-1548

I    ---MVIFLGSFYLINLILAVVAMAYEEQNQATLA---EAQEK   384- 419
II   ---MVIIIGNLVMLNLFLALLLSSFSSDNLSSIE---EDDEV   772- 807
III  YFVIFIVFGAFFTLNLFIGVIIDNFNRQKQKLGG---EDLFM  1243-1281
IV   ---SYIILSFLVVVNMYIAIILENFGVAQEESSDLLCEDDFV  1549-1587
           S6
```

segments traverse the membrane from one side to the other.

The N- and C-terminals of the chain are both on the cytoplasmic side of the membrane. The N-terminal end has 110 residues before domain I and the C-terminal end has 315 after domain IV. The major sites for glycosylation (the attachment of carbohydrate residues) are on the extracellular loops of the chain in domains I and III.

Noda *et al.* (1986*a*) were also able to determine the amino acid sequence of sodium channels from rat brain. They found two different molecular forms, with 2009 and 2005 amino acid residues per chain. The degree of sequence homology (i.e. the proportion of identical amino acids at corresponding positions in two protein chains) is 87% between the two forms and 62% between each of them and the *Electrophorus* sodium channel. The four homologous domains are similar in all three sodium channels, and all have hydropathicity profiles (see p. 19) with peaks corresponding to the S1 to S6 segments (fig. 6.12*a*).

Segments S1 to S3 have some negatively charged residues in their helices, whereas segments S5 and S6 are entirely nonpolar. Segment S4 is particular highly conserved in the different domains and the different sodium channel proteins. Its structure is ususual in that it contains four to eight arginine or lysine residues located at every third position; arginine and lysine are the most positively charged of all amino acid residues, whereas the other residues in the S4 sequences are all non-polar. Perhaps the S4 sequence achieves stability in the membrane by forming ion-pairs with the negative charges on the S1 and S3 segments.

Noda *et al.* (1986*a*, *b*) suggest that the four homologous domains are arranged around a central pore as is shown in fig. 6.12*c*. They suggest that the pore is bounded by the four S2 segments and is lined by their negative charges; this feature would presumably make the channel permeable to cations only.

The *Electrophorus* sequence has also been analysed by Greenblatt, Blatt and Montal (1985) and by Guy and Seetharamulu (1986), using computer graphics and model building. Their conclusions are generally in agreement with each other and with Noda and his colleagues, but with some distinct differences in detail. They both suggest, for example, that there are two extra short intramembrane segments between S5 and S6. Guy and Seetharamulu suggest that one of these combines with S4 in each domain to contribute to the pore lining, whereas Greenblatt and his colleagues think that it is lined by the S3 segments. But all three groups agree on the general conclusion that the four domains are grouped around a central pore.

A crucial question about the sodium channel must be the nature of its voltage sensor, whose movement provides the gating current. Segment S4, with its unusual array of positive charges, looks an attractive candidate; Catterall (1986*a*, *b*) has developed a prescient idea by Armstrong (1981) in proposing a 'sliding helix' model of gating based on this. Each positive charge in S4 is separated from its neighbours by about 5 Å, measured along the axis of the helix, and we may suppose that each one is loosely bound to a corresponding negative charge, probably on the S1 and S3 segments. Catterall suggests that the S4 segment responds to a depolarization of the membrane by rotating by about 60° and moving outward by about 5 Å, so as to bring each positive charge into alignment with the next negative charge (fig. 6.13). The net effect of this would be to expose a negative charge at the inner side of the membrane and a positive charge at the outer side, which is equivalent to moving one charge across the whole membrane.

Since there are four S4 segments in the molecule, Catterall's model implies a movement of four charges per channel. However, the S4 segments of domains III and IV contain six positive charges, whereas those of I and II contain four or five; Catterall suggests that the S4 segments of III and IV might move two steps up the ladder, as it were, so producing a total gating current of six charges per channel, the figure suggested by Hodgkin and Huxley (1952*d*).

Reconstitution in model membranes

Artificial phospholipid bilayers, in the form of either planar bilayers or liposomes, have been much used as models for cell membranes. Planar bilayer membranes are produced by coating a small hole in an insulating partition separating two chambers. Liposomes are multilayer bodies formed when phospholipids are suspended in

Fig. 6.12. Probable arrangement of transmembrane segments in the sodium channel. The hydropathicity profile of the amino acid sequence is shown in *a*, and structural predictions from this are shown in *b* and *c*. In *a* the averaged hydropathic index of amino acid residues $i-9$ to $i+9$ is plotted against i, where i is the amino acid number in the sequence. The bars show the positions of the homologous domains I to IV, and the boxes below them show the positions of the sections S1 to S6 that are thought to cross the membrane. The line of open boxes shows all sections of predicted α-helix or β-pleat structure. The bottom line shows the positions of the positively charged lysine and arginine residues as upward lines and the negatively charged aspartate and glutamate residues as downward lines. In *b* the four domains with their six segments are displayed linearly. N shows the position of the first amino acid residue and C shows that of the last. Probable sites for carbohydrate binding are shown by CHO. The view in *c* is perpendicular to the membrane, showing the four domains grouped around a central pore. The segments S1 to S6 in domains I to IV are shown by cross-hatching (S1), stipple (S2), diagonal hatching (S3), a plus sign (S4) and black (S5 and S6). (From Noda *et al.*, 1984, 1986.)

a

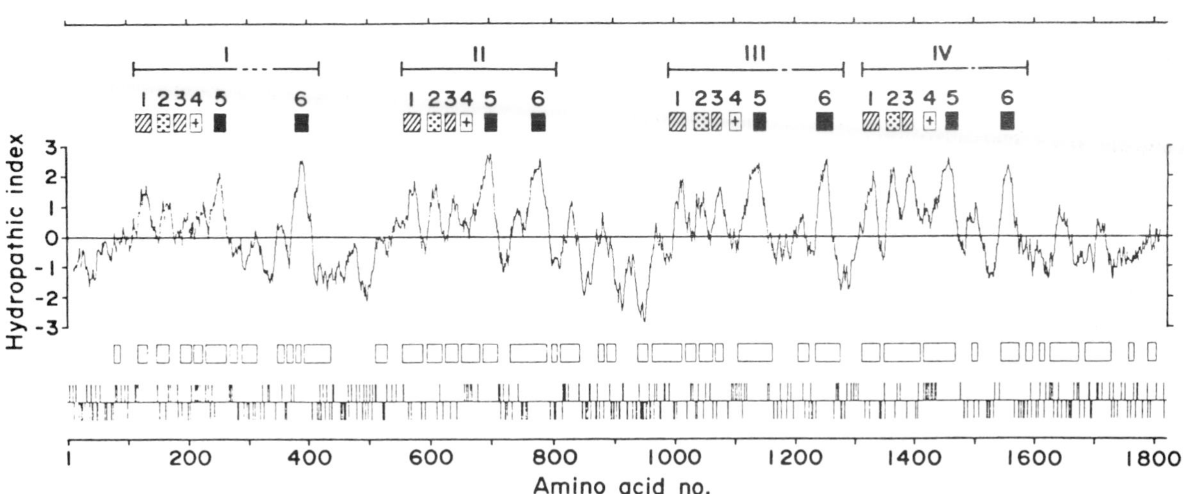

b

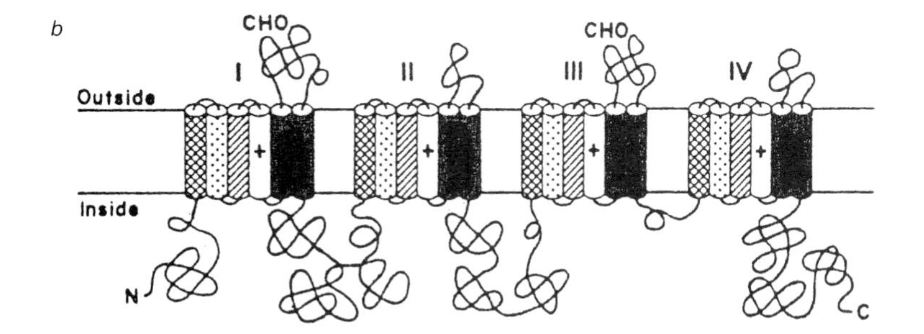

c

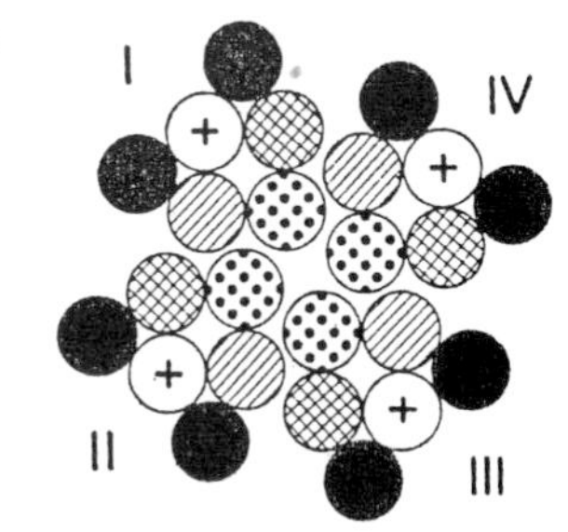

aqueous solutions; sonication produces small vesicles bounded by a single membrane.

Ionic channel proteins can sometimes be inserted into these artificial membranes and their properties studied (Miller, 1984, 1986). Thus Krueger, Worley and French (1983) fused membrane vesicles from rat brain with planar bilayers and observed current flow through single channels, and Hartshorne *et al.* (1985) did similar experiments with the purified protein. Tanaka, Eccleston and Barchi (1983) inserted purified rat muscle sodium channels into unilamellar vesicles and observed ionic fluxes by fast reaction techniques. All these studies used batrachotoxin to prevent inactivation of the sodium channels.

Some very pleasing experiments by Rosenberg, Tomiko and Agnew (1984) involved patch clamping of liposome membranes. Purified sodium channel protein from *Electrophorus* electroplax was incorporated into liposomes. Then a patch-clamp electrode was applied to the outside of a large vesicle. After the seal had formed, the electrode was briefly withdrawn into the air, with the result that the body of the liposome fell away to leave a small patch of the bilayer membrane on the end of the electrode.

Rosenberg and his colleagues were thus able to measure single-channel currents in response to brief depolarizations, just as in the experiments of Sigworth and Neher (p. 78), and with very similar results. The single-channel conductance was 11 pS, and the mean open time was about 1.9 ms.

Fig. 6.13. Catterall's sliding helix model of gating charge movement in the sodium channel. Segment S4 is presumed to form an α-helix crossing the membrance as shown on the left. Here the black circles represent the α-carbon atoms of the different amino acid residues and the open circles represent their side chains. Residues are indicated by the single letter code, with R showing the positively charged arginine; the rest are non-polar. The arginine residues thus form a helix of positive charges, shown on the right. The model proposes that these form ion pairs with an array of negative charges on other segments, and that depolarization allows an outward movement of the S4 segment by one step along this array. (From Catterall, 1986*a*.)

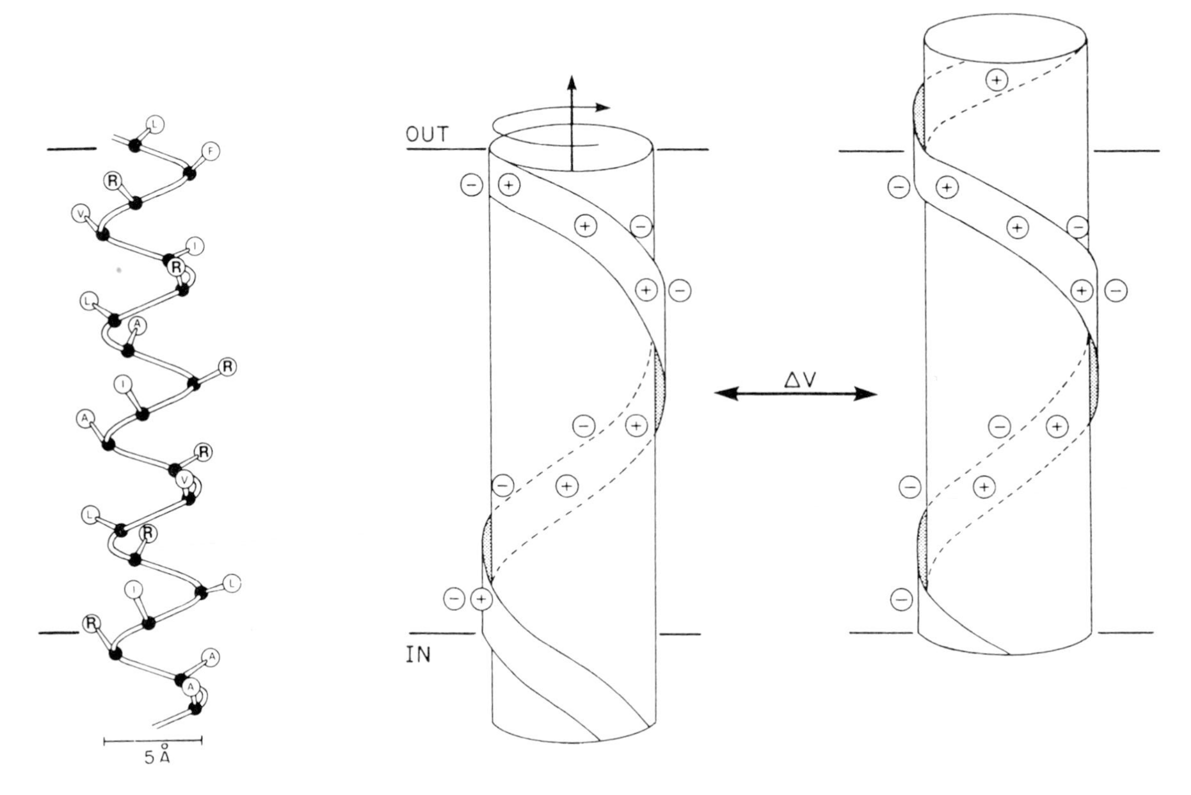

Inactivation was evident as a reduction in the number of channel openings at late times during a depolarizing pulse. In the presence of batrachotoxin there was no inactivation, the mean open time was 28 ms and the single-channel conductance increased to 25 pS. These experiments show conclusively that the single 260 kDa molecule, isolated from *Electrophorus* electroplax on the basis of its tetrodotoxin binding, is all that is required to make a functional sodium channel.

Expression in oocytes

During protein synthesis in animal cells, the nucleotide base sequence in the DNA of the appropriate gene is transcribed to form a complementary sequence in messenger RNA molecules, and this then acts as a template for translation to form the amino acid sequence of the protein. The oocytes of the African clawed toad *Xenopus laevis* provide an efficient translation system, and they can manufacture the corresponding proteins when mRNA is injected into them (Gurdon *et al.*, 1971). Since ionic channels can be detected by their electrophysiological effects, *Xenopus* oocytes can be used to investigate their properties in a readily accessible membrane system.

Voltage-gated sodium and potassium channels are not normally present in the oocyte membrane. But injection of mRNA from rat or human brains results, after a few days, in the production of such channels and their incorporation in the oocyte cell membrane (Gunderson *et al.*, 1983, 1984). Messenger RNA from brain tissue contains mRNAs for many different proteins; these can be partially separated by fractionation in a sucrose density gradient, and then the different fraction can be injected into oocytes. Fig. 6.14 shows some results of this procedure when applied to mRNA fractions from chick brain. A relatively heavy fraction produced a voltage-activated inward sodium current with a reversal potential near to +40 mV. A rather light fraction produced an outward current attributable to potassium ions.

We saw earlier how the Kyoto University group have sequenced cloned cDNA to determine the structure of the α-chain of rat brain sodium channels. They extended this work to make specific mRNA from the cDNA and injected it into *Xenopus* oocytes (Noda *et al.*, 1986*b*). Voltage-clamped depolarizations then gave transient inward currents sensitive to tetrodotoxin. These results imply that the α-chain alone is sufficient to form the voltage-gated channel. Perhaps the smaller β-chains serve some relatively minor accessory function.

Delayed rectifier potassium channels

So far we have designated the potassium channels opened by depolarization in squid axons as 'the' potassium channel. However, it turns out that there are potassium channels with rather different properties in some other cells, and hence

Fig. 6.14. Voltage-clamp records from *Xenopus* oocytes injected with messenger RNA fractions from chick brain. *a* shows inward currents produced in response to fraction 7 mRNA, at potentials from 0 to +40 mV in 10 mV steps. *b* shows outward currents produced in response to fraction 11, at potentials from −60 to +10 mV. There were 30 mRNA fractions from the sucrose density gradient, fraction 30 being the lightest. Clearly fraction 7 contains mRNA coding for sodium channels and fraction 11 contains that coding for potassium channels. (From Sumikawa *et al.*, 1984.)

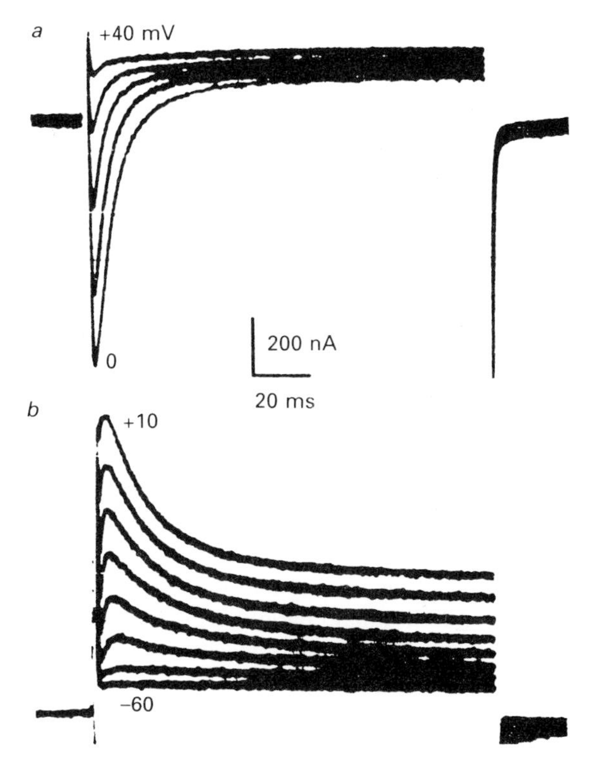

we need to distinguish between these different types. The potassium conductance increase which follows depolarization in nerve axons has been called 'delayed rectification' and so the channels involved have become known as delayed rectifier potassium channels.

Patch clamping

Single channel records of delayed rectifier potassium currents have been obtained from squid giant axons by Conti and Neher (1980). It is difficult to apply a patch-clamp electrode to the outside of the axon because it is closely surrounded by Schwann cells, so they made a special L-shaped pipette electrode to apply to the inner surface of the axon membrane. Other tricks included using a short internal perfusion with pronase to remove inactivation, external tetrodotoxin to eliminate sodium currents, and a perfusion solution of low ionic strength so as to give a good seal for the patch electrode. Thus with high external and low internal potassium ion concentrations, the reversal potential for the potassium current was +70 to +100 mV and the 'resting potential' was +50 to +70 mV. To record potassium currents, the membrane was clamped at −50 mV for a second and then depolarized by up to 120 mV; the ensuing currents were inward because of the reversed potassium ion gradient.

Fig. 6.15 shows Conti and Neher's recordings of patch currents at different membrane potentials. The trace at −31 mV shows a brief square pulse of inward current which, it seems reasonable to suggest, corresponds to the opening of one potassium channel for about 12 ms. More of these events occur at −25 and −19 mV, in accordance with our expectation that the probability of channel opening should increase as the membrane becomes more depolarized. At −8 and +17 mV the channel openings are clearly overlapping with each other to give a very 'noisy' trace; this is less evident at +34 mV. (There are probably two reasons for the reduction in the variance of the current flow at +34 mV: we would expect nearly all the channels in the patch to be open, and the current flow through each individual channel will be less since the potential is nearer to the potassium equilibrium potential.)

Close inspection of the individual events in fig. 6.15 (as in the −25 and −19 mV traces) shows that many of them contain brief interruptions, so that an event is a burst of a few short pulses. On average there were 2.8 short pulses per burst, each with a mean open time of 3.5 ms, and the mean duration of the interruptions was 1.3 ms. The mean overall duration of the burst was thus $(2.8 \times 3.5) + (1.8 \times 1.3) = 12$ ms.

What could these interruptions be caused by? Suppose there is a sequence of closed states leading to an open state:

$$C_n \rightleftharpoons \ldots \rightleftharpoons C_2 \rightleftharpoons C_1 \rightleftharpoons O. \qquad (6.4)$$

If the probability of C_1 returning to C_2 is lower than the probability of other changes in the system, then there is an appreciable chance that when the channel reaches C_1 from O it will revert to O rather than changing to C_2; such a situation will clearly produce bursts of the type seen here. Conti and Neher calculated that the rate constants necessary for such a situation are not compatible with the Hodgkin–Huxley equations, which would predict about 0.25 interruptions per burst on average, whereas actually there are 1.8. Similar bursting activity occurs in some other channels, and Colquoun and Hawkes (1982) have described the situation mathematically.

Fig. 6.15. Single-channel potassium currents recorded from squid axons at different membrane potentials. Further details in the text. (From Conti and Neher, 1980.)

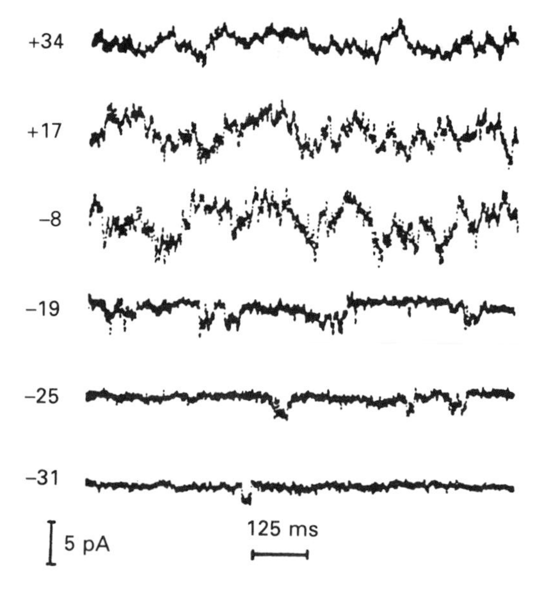

Conti and Neher calculate that the single channel conductance of the potassium channels in squid axons is about 10 pS. This is close to the value of 12 pS deduced from earlier investigations using noise analysis (Conti *et al.*, 1975). Using Hodgkin and Huxley's value of 36 mS cm^{-2} (i.e. 360 pS μm^{-2}) for the maximum potassium conductance, we get a channel density of 36 μm^{-2}. This is much lower than the corresponding figure for sodium channel density.

Gating currents

Potassium channel gating currents proved much more difficult to detect than did the gating currents for sodium channels. This arose from the much lower density of potassium channels in the membrane and also from their slower activation. Gilly and Armstrong (1980) noticed that the gating current produced by depolarizations to +20 mV or more showed a prominent slow phase, continuing well after the sodium ionic current had peaked but preceding the potassium ionic current. They discovered that the local anaesthetic dibucaine eliminates the sodium current (i.e. the sodium channels do not open) and much reduces the initial phase of the gating current, but the potassium current and the slow phase of the gating current are unaffected. So it seems likely that the slow phase is the potassium gating current.

Further investigations of the potassium gating current in squid axon have been made by White and Bezanilla (1985); see fig. 6.16. They found that the maximum amount of charge moved was around 490 e μm^{-2}, giving a charge of 7 e per channel if there are 70 channels μm^{-2} (the figure used by White and Bezanilla) or 13 e per channel if there are 36 channels μm^{-2} (as calculated above).

It is an interesting feature of the potassium gating current that the time constants for on and off currents are the same as those for the potassium ionic currents at the same voltage. This is not what one would expect from the Hodgkin–Huxley scheme, which predicts (from equation 5.6) that the time constant for potassium conductance should be related to the fourth power of that for gating. White and Bezanilla were able to model their results by using a kinetic scheme similar to that of Conti and Neher (equation 6.4), and making the rates of change progressively faster as one moves from left to right in that sequence.

Selectivity and block

If only potassium ions are moving through the potassium channels, then the reversal potential for the ionic current is equal to the equilibrium potential for potassium ions as in equation 3.4. If external potassium is now replaced by another ion X^+ then the amount of change in the reversal potential gives an estimate of the relative permeabilities (P) of X^+ and K^+; there will be no change if $P_X = P_K$ and a large change if P_X is much different from P_K. Using a modification of equation 3.5, it can be shown that

$$E_{r,X} - E_{r,K} = \frac{RT}{F} \log_e \frac{P_X\,[X]_o}{P_K\,[K]_o}$$

where $E_{r,K}$ is the reversal potential with external potassium ions and $E_{r,X}$ is the reversal potential with external X^+.

Fig. 6.16. Potassium gating current in a perfused squid axon. The internal solution contained impermeant organic cations to eliminate the potassium ionic current, the external solution contained Tris nitrate, tetrodotoxin and dibucaine to eliminate the sodium ionic current and reduce the sodium gating current. The gating currents were produced by a 6 ms depolarization from −110 to 0 mV, followed by a return to −60 mV. The ON response consists of a very brief sodium gating current followed by a more slowly changing potassium gating current. In the OFF response these two components are even more distinct and the sodium gating current (arrow) is smaller. (From White and Bezanilla, 1985.)

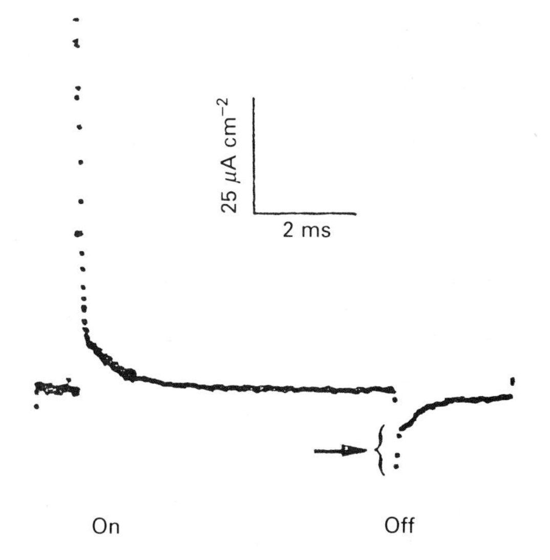

Hille (1973) used this technique to measure the relative permeabilities of different ions in the delayed rectifier potassium channels in frog nodes. His results were rather striking: only four ions, potassium, thallium, rubidium and ammonium, with diameters in the range 2.66–3.0 Å, could readily pass through the channel. Organic cations and caesium, all with diameters of 3.3 Å or more, were relatively impermeant. This suggests that the selectivity filter in the channel is a pore 3 to 3.3 Å in diameter. We can imagine a potassium ion sitting snugly in this pore, with the oxygen atoms lining the pore taking the place of the oxygen atoms of the water molecules that surrounded the ion in free solution. Lithium and sodium are smaller in diameter and so the walls of the pore cannot provide an adequate substitute for the water molecules of their hydration shells.

Delayed rectifier channels in squid axons are readily blocked by caesium and barium ions, and by TEA and other quaternary ammonium ions when applied from the inside of the membrane. The lipid-soluble compounds 4-aminopyridine, strychnine and quinidine cause block whether applied from the inside or the outside.

TEA has a diameter of about 8 Å, which is about the same as that of a hydrated potassium ion. Experiments by Swenson (1982), using an extensive range of quaternary ammonium compounds, suggest that the critical size about which block cannot occur is at a diameter of about 12 Å. This suggests that perhaps the channel has an inner 'mouth' of this size, which is large enough to receive hydrated potassium ions, and that TEA and the other quaternary ammonium ions fit into this.

Long-chain derivatives of TEA, such as triethylnonyl ammonium (C_9), can also cause block, perhaps with the TEA-head in the mouth and the long tail in a hydrophobic region of the membrane (Armstrong, 1971, 1975). A notable feature of C_9 block is that it is not present before the channels open but develops progressively during a clamped depolarization (fig. 6.17). This suggests that the site of blockage is inside the activation gate so that the blocking compound can only get access to it after the gate is opened.

The long-pore effect

Ussing (1949) showed that, if an ion X crosses the cell membrane solely under the influence of the chemical and electrical gradients across the membrane, and if the movement of any individual ion is independent of the movement of its neighbours, the ratio between the influx (M_i) and the efflux (M_e) is given by

$$\frac{M_i}{M_e} = \frac{[X]_o}{[X]_i} \exp(-EzF/RT), \qquad (6.5)$$

where the symbols in the exponent are as in (3.1).

Hodgkin and Keynes (1955*b*) measured the potassium fluxes of *Sepia* giants axons which had been poisoned with DNP so as to eliminate active transport effects. The axon was arranged so that its membrane potential could be altered by passing current through it at the same time as the potassium fluxes were measured. Thus it was possible to find the potential at which the fluxes were equal in each direction; if the flux ratio is given by (6.5), i.e.

Fig. 6.17. C_9 blockage of potassium channels in squid axon. The records show potassium ionic currents produced by clamped depolarizations to different membrane potentials; sodium currents were eliminated with tetrodotoxin. In *a* the axon was injected with triethylnonylammonium (C_9); this was not so in *b*. (From Armstrong, 1971.)

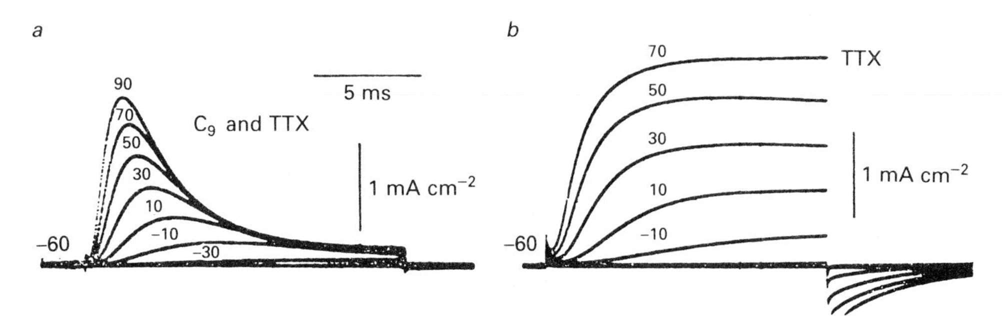

$$\frac{M_i}{M_e} = \frac{[\text{K}]_o}{[\text{K}]_i} \exp(-EF/RT)$$

or $$\frac{M_i}{M_e} = \exp\left(\frac{(E - E_K)F}{RT}\right), \quad (6.6)$$

the influx should be equal to the efflux when E is equal to E_K, E_K being calculated from (3.4). This expectation was fulfilled. This means that, in poisoned axons, potassium ions move solely under the influence of the concentration and electrical gradients across the membrane.

A most interesting discovery was made when the flux ratios at membrane potentials not equal to the potassium equilibrium potential were measured. Equation (3.11) can be written as

$$\log_{10} \frac{M_i}{M_e} = \frac{E - E_K}{58} \text{ mV}.$$

This implies that the flux ratio should change tenfold for a 58 mV change in membrane potential. But it was found (see fig. 6.18) that the relation was in fact much steeper than this, with a tenfold change in flux ratio for about a 23 mV change in membrane potential; it could be described by a modification of (6.6) such that

$$\frac{M_i}{M_e} = \exp \frac{(E - E_K)nF}{RT},$$

where n is 2.5. What does this mean? Hodgkin and Keynes suggested that potassium ions move through the membrane via pores whose length is appreciably greater than their diameter, so that a number of ions (n) can occupy a pore at the same time. Any one ion will then only pass through if it is hit n times in succession from the same side. This is much more likely to happen if the ion is moving in the same direction as the majority of the other ions. The phenomenon has become known as 'the long-pore effect'.

These results and those in the previous section lead us to a model for the delayed rectifier potassium channel (fig. 6.19). Like the sodium channel it has an activation gate at its inner end, opened by a sensor which contains a movable charge. Inside the activation gate there is a mouth at least 8 Å and perhaps 12 Å in diameter, where block by quaternary ammonium compounds can occur. This connects to a tunnel (responsible for

Fig. 6.18. The effect of membrane potential (plotted here as the difference between the membrane potential and the potassium equilibrium potential) on the potassium flux ratio in *Sepia* giant axons. (From Hodgkin and Keynes, 1955*b*.)

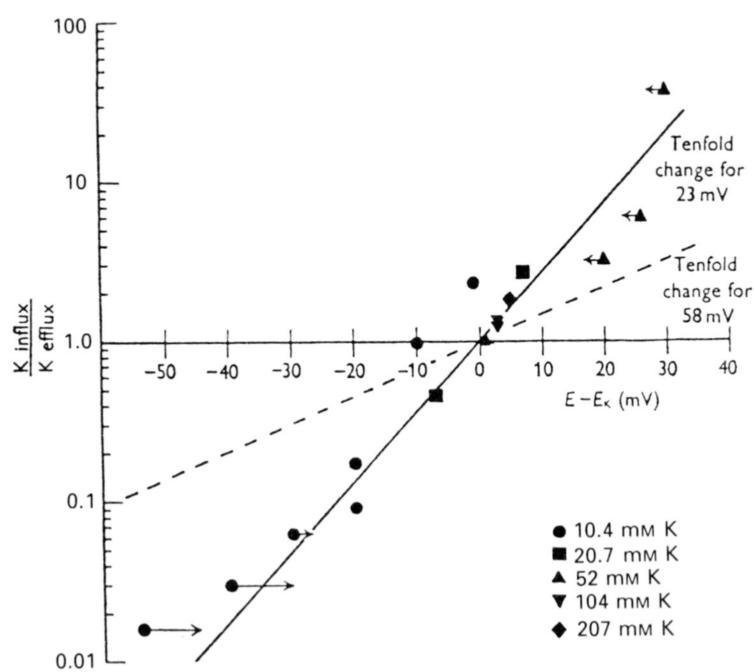

the long-pore effect) which contains a selectivity filter about 3.3 Å in diameter. In nodal channels (where external TEA can block) there may be an outer mouth as well.

Potassium inactivation

In the experiments by Hodgkin and Huxley on squid axons there was no evidence for any reduction of the potassium currents with time. The use of longer depolarizations showed that this did happen, so that the potassium conductance was reduced to about a third over a period of some seconds (Ehrenstein and Gilbert, 1966). Similar slow inactivation processes occur in the potassium channels of frog muscle (Adrian *et al.*, 1970) and frog nerve (Schwarz and Vogel, 1971). There is no evidence to show that this slow inactivation is operated by a separate inactivation gate such as occurs in the sodium channel.

Other voltage-gated potassium channels

There are a number of potassium channels which differ in their properties from the classical delayed rectifier type. Let us have a look at some of them.

A channels

Many molluscan central neurons produce long chains of action potentials when a steady depolarizing current is passed through them, the frequency being proportional to the current. Experiments by Connor and Stevens (1971*a*) on the sea slug *Anisodoris* showed that this behaviour is dependent on the presence of a second type of potassium channel. Depolarization results in a transient opening of these channels (i.e. they are rapidly activated and rapidly inactivated) so as to produce a brief outward flow of current. They have become known a A channels.

Fig. 6.20 shows how the total potassium current in an Archidoris neuron can be separated into two components, the delayed rectifier current (I_K) and the fast transient current (I_A). If the membrane is depolarized from -80 to -5 mV, the total current peaks rapidly and then falls back to a steady level. However, if the membrane is first held at -40 mV for 1.5 s, I_A inactivates so that I_K alone is seen on further depolarization to -5 mV. I_A is then given by the difference between the total trace and the I_K trace.

Crab walking-leg axons are unusual in showing repetitive firing at relatively low rates when current is passed through them (Hodgkin, 1948). They too depend upon the presence of fast transient potassium currents in addition to the usual delayed rectifier currents (Quinta-Ferreira *et al.*, 1982). Fast transient (A) currents have also been described in mammalian sympathetic and sensory neurons (Galvan and Sedlmeir, 1984).

Kasai and his colleagues (1986) have carried

Fig. 6.19. Sketch to show the probable features of the delayed rectifier potassium channel.

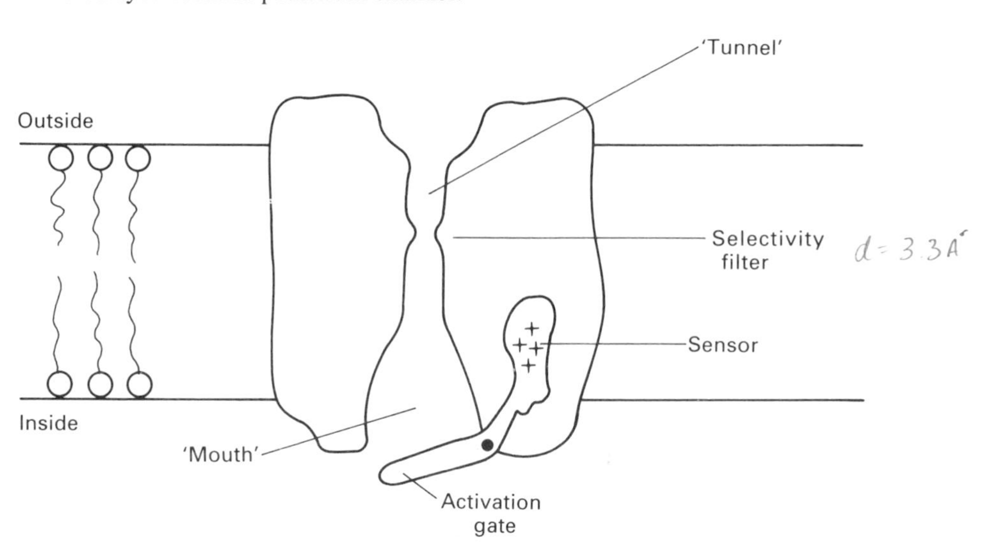

out patch-clamp studies on single A channels in mammalian sensory neurons in tissue culture. Patch clamping has the great advantage that one can investigate small patches of membrane which do not contain other channels: the test for A channels was simply that they should be completely inactivated by sustained depolarization. They found that the channels could be open at the resting potential (−60 mV) and rapidly became inactivated at potentials above −40 mV. The single-channel conductance was 20 pS.

Connor and Stevens (1971*b*) were able to compute the time-course of the repetitive action potentials in gastropod neurons by using a form of the Hodgkin–Huxley equations modified by the inclusion of the A current. As the cell starts to depolarize after an action potential, A channels are opened so that the outward flow of potassium ions prevents further depolarization. However, the A channels are inactivated after a short time so that the applied current is able to depolarize the membrane further, soon reaching the threshold for action potential production. The role of the A channels is thus to insert a pause between successive action potentials.

Molecular structure of A channels

It has been possible to isolate the proteins forming the sodium channel and the acetylcholine channel (chapter 8) from tissues where they occur in large quantities, such as the electric organs of some electric fish. Partial amino acid sequences have then provided a key to the production of complementary DNA for the whole protein. In the absence of tissues similarly rich in potassium channels, a different approach has had to be adopted.

The genetic structure of the fruit fly *Drosophila* has been under detailed scrutiny for much of the twentieth century. *Shaker* is a behavioural mutant which can be readily detected: the flies shake their legs when anaesthetised with ether. Voltage clamp studies on flight muscle cells show that the A currents are affected by mutations at the *Shaker* locus (Salkoff and Wyman, 1981). The Jans and their colleagues at the University of California, San Francisco, have therefore cloned genomic DNA from the *Shaker* locus so as to get at the structure of the A potassium channel proteins.

The *Shaker* locus is in the 16F region of the X chromosome. Papazian *et al.* (1987) cloned

Fig. 6.20. Potassium currents in an *Anisodoris* (sea slug) neuron under voltage clamp. Trace *a* shows the response to a depolarization from −80 to −5 mV: it consists of a fast transient current (I_A) plus the delayed rectifier current (I_K). Trace *b* shows the response when I_A has been inactivated by holding the membrane potential at −40 mV before depolarizing to −5 mV; it consists of I_K alone. Hence trace *c*, the difference between *a* and *b*, gives I_A. The sodium current, which inactivates within 5 ms, has been omitted from the diagram. (From Connor and Stevens, 1971*a*.)

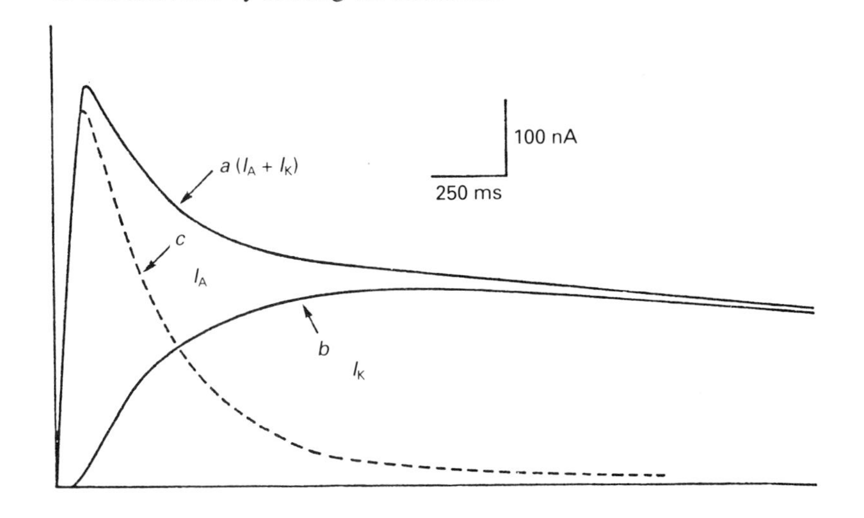

genomic DNA from this region and used it to probe complementary cDNA libraries made from fly head messenger RNA. They were thus able to prepare clones of *Shaker* genes and of cDNA coding for *Shaker* locus gene products. From this cDNA, Tempel *et al.* (1987) were able to deduce the amino acid sequence of a protein which was probably a component of the A potassium channel.

The putative A channel protein was 616 amino acids long with a molecular weight of 70 200 Da. It contained seven largely hydrophobic sections that were interpreted as membrane-crossing helices. Six of these (called H1 to H6) were of conventional structure, but there was one which was remarkably similar to the S4 section of the sodium channel: it had a positively charged arginine or lysine residue at every third position (fig. 6.21).

We shall see later that a similar S4-like section occurs in what is probably a voltage-gated calcium channel, but does not occur in ligand-gated channels such as the acetylcholine or glycine receptors. This fits very well with the idea that the S4-like section acts as the voltage sensor in voltage-gated channels, and suggests further that these channels are derived from a common evolutionary ancestor.

Further work showed that there are at least four different proteins produced from the *Shaker* locus, and that these are produced by alternative splicing of different sections of the primary transcript RNA to make different messenger RNAs (Schwarz *et al.*, 1988). These four proteins have a common central region, including most of the membrane-crossing sections, but different N- and C-terminal regions, as is shown in fig. 6.22.

The existence of these alternative A channel proteins raises questions about the nature of the A channels. A 70 kDa protein is too small to be a complete channel, as we can see by comparing it

Fig. 6.21. Amino acid sequences of the S4 segments of three voltage-gated channels: clone ShA of the potassium A channel from the *Shaker* locus in *Drosophila* (Tempel *et al.*, 1987), the *Drosophila* sodium channel (Salkoff *et al.*, 1987), and the rabbit muscle putative calcium channel (Tanabe *et al.*, 1987). Positively charged amino acids are indicated by shading. (From Miller, 1988.)

ShA -**Arg**-Val-Ile-**Arg**-Leu-Val-**Arg**-Val-Phe-**Arg**-Ile-Phe-**Lys**-Leu-Ser-**Arg**-His-Ser-**Lys**-Gly-Leu-Gln-

Na^+ -**Arg**-Val-Val-**Arg**-Val-Phe-**Arg**-Ile-Gly-**Arg**-Ile-Leu-**Arg**-Leu-Ile-**Lys**-Ala-Ala-**Lys**-Gly-Ile-**Arg**-

Ca^{2+} -**Lys**-Ile-Leu-**Arg**-Val-Leu-**Arg**-Val-Leu-**Arg**-Pro-Leu-**Arg**-Ala-Ile-Asn-Arg-Ala-**Lys**-Gly-Leu-**Lys**-

Fig. 6.22. Alternative splicing in the production of *Drosophila* A channel proteins. The diagram represents the amino acid sequences of four different proteins produced from the *Shaker* locus. There is a common central region which is combined with different end regions. End regions shaded in the same way have identical sequences. (From Miller, 1988, based on Schwarz *et al.*, 1988.)

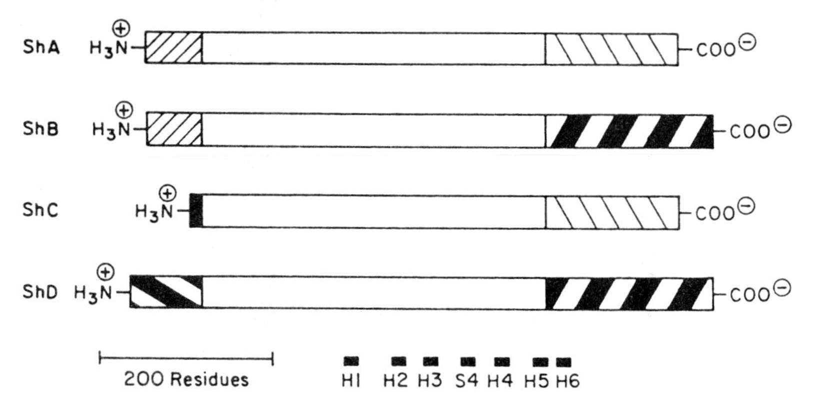

with the sodium channel (260 kDa, four domains and probably 24 membrane-crossing helices) and the acetylcholine channel (about 290 kDa, five subunits and probably 20 or 25 membrane-crossing helices, see chapter 8). It must be a component of a channel which consists in all of four or perhaps five similar units.

Is an A channel, then, composed of identical subunits, or can they (or must they) be different? A partial answer to this question has been provided by Timpe *et al.* (1988), using the *Xenopus* oocyte expression technique. They found that a single *Shaker* messenger RNA species was sufficient to make functional A channels in the oocytes. Furthermore the A currents produced by two of the different gene products had different rates of inactivation (fig. 6.23). These experiments show that functional A channels can be made from identical subunits and suggest that the single gene can be expressed to give a variety of different channel properties. They do not rule out the possibility that individual channels in the fly may include different subunits.

Inward rectifier channels

Katz (1949) discovered that in potassium-depolarized muscle the membrane conductance increased with hyperpolarization and decreased with depolarization. This means that potassium ions could flow into the cell much more easily than they could flow out. The effect is evident in fig. 3.17. Katz named the phenomenon 'anomalous rectification', but the alternative term 'inward rectification' is probably in more general use now.

Fig. 6.23. Expression of potassium A channels in *Xenopus* oocytes injected with *Shaker* mRNA a few days previously. The records are families of membrane currents produced by voltage-clamp depolarizations from a holding potential of -80 mV to levels between -30 and $+40$ mV. *a* shows the response from the ShA clone and *b* that from the ShB clone. (From Timpe *et al.*, 1988.)

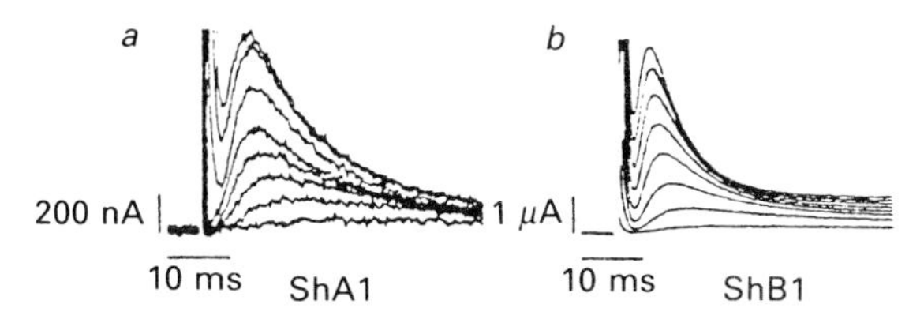

Patch-clamp experiments on the inward rectifier channels of rat myotubes (cultured embryonic muscle fibres) indicate a single channel conductance of about 10 pS with isotonic external potassium ion concentrations (Ohmori *et al.*, 1981). The conductance is reduced when the external potassium concentration is lower. The channel density is low, perhaps about one per square micrometre of surface membrane. Similar conclusions have been reached from fluctuation analysis experiments on adult frog muscle fibres (DeCoursey *et al.*, 1984).

Inward rectifier channels are present in vertebrate heart muscle fibres (Hutter and Noble, 1960). Their role here seems to be fairly clearcut: they provide a potassium conductance which contributes to the resting potential and which needs to be switched off during the plateau of the cardiac action potential (see p. 313). Their role in skeletal muscle, where they are largely confined to the T tubules (p. 295; Williams, 1976), is not so evident.

Calcium-activated potassium channels

Meech (1972, 1974) discovered that injection of calcium ions into molluscan neurons causes an increase in potassium conductance. Experiments by Gorman and Thomas (1978, 1980) showed how this calcium-activated potassium current ($I_{K(Ca)}$) is involved in the bursting behaviour of these cells (fig. 6.24). They used the calcium-sensitive dye arsenazo III, injected into the cell, to monitor the intracellular calcium concentration. During a burst of action potentials calcium ions enter the cell (via the calcium channels to be described in the next section) and so activate the K(Ca) channels. The resulting outward current terminates the burst and produces an interburst hyperpolarization. This dies away as the calcium ion concentration falls (presumably as a result of uptake by mitochondria and other intracellular organelles).

Calcium-activated potassium channels are found in a wide variety of cells (Schwarz and Passow, 1983). Barrett, Magleby and Pallotta (1982) have used the patch-clamp technique to measure their properties in cultured rat myotubes. They were able to demonstrate very clearly that the channels are activated both by a rise in internal calcium ion concentration and by depolarization.

The single-channel conductance of the K(Ca)

channels of rat myotubes was 220 pS at 20 °C, much higher than that of other ionic channels which we have so far discussed. Similar high conductances have been found in the K(Ca) channels of chromaffin cells and pituitary cells and also in the potassium channels of the sarcoplasmic reticulum of muscle. Latorre and Miller (1983) refer to these as 'maxi' potassium channels, and suggest that the high conductance arises from the 'tunnel' being much shorter than in the low conductance potassium channels. Low conductance K(Ca) channels also occur: Lux, Neher and Marty (1981) found values of 19 pS per channel from patch-clamp experiments on snail neurons.

Calcium channels

Appreciable calcium ion currents occur in many different types of cell, notably the muscle fibres of crustaceans and many other invertebrates, vertebrate heart muscle cells, the cell bodies of many invertebrate neurons (see fig. 6.24), nerve terminals, secretory cells, some sensory receptor cells and many egg cells (Hagiwara and Byerly, 1981). There may be some inflow of calcium ions through sodium channels (Baker *et al.*, 1971) and by the operation of the sodium–calcium exchange system (Requena *et al.*, 1986), but any appreciable inflow usually takes place through specific calcium channels (see Tsien, 1983; Stanfield, 1986; Miller, 1987).

Crustacean muscle fibres provided some of the first evidence for action potentials based on calcium ion flow. Hagiwara and Naka (1964), for example, used the large fibres of the barnacle *Balanus nubilis*, perfused internally with an isotonic potassium sulphate solution. They showed that the size of the all-or-nothing action potential was unaffected by removal of external sodium ions but was increased by a raised external calcium ion concentration.

Fig. 6.25 shows the calcium current in a squid giant axon presynaptic terminal during a clamped depolarization, after elimination of the sodium current with tetrodotoxin and the potassium current with TEA (Llinas *et al.*, 1981*a*). Llinas and his colleages fitted Hodgkin–Huxley type equations to their measurements of the calcium current and produced an estimate for the calcium conductance change during the action potential, as is shown in fig. 6.26. We shall see in chapter 7 that this calcium current is crucially important in the release of the transmitter substance from the presynaptic terminal.

The calcium conductance system in squid axon presynaptic terminals showed no evidence of any inactivation process. This is in contrast to many other calcium channel systems, where the calcium current inactivates after a short time. In some cases this inactivation appears to be voltage-dependant (Lux and Brown, 1984), in others it is

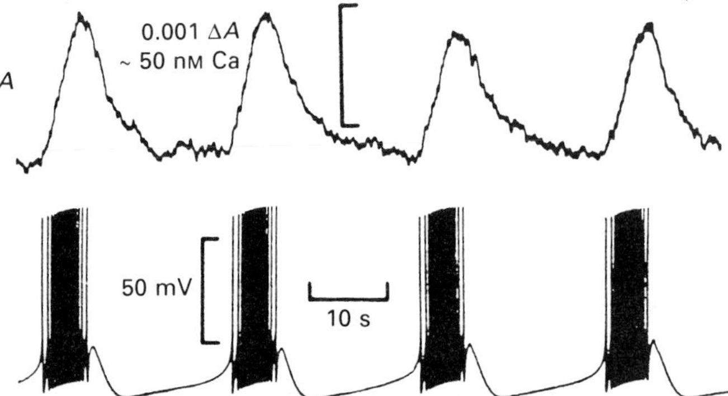

Fig. 6.24. Calcium ion concentration changes in an *Aplysia* pacemaker neuron. Electrical activity of the cell (bursts of action potentials at intervals of about 24 s) is shown on the lower trace. Changes in light absorbance, which reflect changes in intracellular calcium ion concentration since the cell was injected with the calcium-sensitive dye arsenazo III, are shown in the upper trace. The calcium ion concentration rises during the burst of action potentials and falls during the quiescent period. (From Gorman and Thomas, 1978.)

brought about by the action of internal calcium ions (Eckert and Tillotson, 1981; Eckert and Chad, 1984).

Various divalent cations, especially manganese, nickel, cobalt and cadmium ions, and also the trivalent lanthanum ions, block calcium channels. The organic compounds verapamil, D600 and dihydropyridines such as nifedipine and nitrendipine also cause block, and Bay K 8644 may prolong channel opening times; Reuter (1983) suggests that such compounds might be useful in treating certain types of heart disease. Batrachotoxin and veratridine, first known as blocking agents for sodium channels, will in fact block calcium channels, and batrachotoxin has a higher affinity for them than it has for sodium channels (Romey and Lazdunski, 1982). One of the toxins of the cone shell *Conus geographus* is a specific blocking agent for calcium channels (Kerr and Yoshikami, 1984).

Single calcium channel currents have been measured in a variety of cells, including molluscan neurons (Brown *et al.*, 1982), bovine chromaffin cells (Fenwick *et al.*, 1982) and chick sensory neurons (Nowycky *et al.*, 1985). They have a tendency towards the bursting behaviour seen in some potassium channels (p. 96). It has usually been necessary to use relatively high external calcium or barium ion concentrations in order to see them. Under these conditions the currents are usually in the range 0.5–2 pA at membrane potentials near −10 mV; thus Hagiwara and Ohmori (1983) estimate that the single channel conductance in GH_3 (pituitary) cells in tissue culture is about 7 pS. At lower calcium ion concentrations, which are nearer to the normal physiological situation, the single channel currents are correspondingly lower; in chromaffin cells at −12 mV, for example, the values are near 0.03 pA in 1 mM Ca^{2+} and 0.09 pA in 5 mM Ca^{2+}.

Patch-clamp records from chick sensory neurons in tissue culture suggest that there are three types of channel in these cells (Nowycky *et al.*, 1985). L channels produce long-lasting currents at large depolarizations, T channels provide transient currents at small depolarizations, and N channels require very negative potentials for removal of inactivation and large depolarizations for activation.

Fig. 6.26. Calculation of the calcium conductance (g_{Ca}) change during an action potential in squid giant axon terminal. Sodium (g_{Na}) and potassium (g_K) conductances are also shown. (From Llinas *et al.*, 1981*a*.)

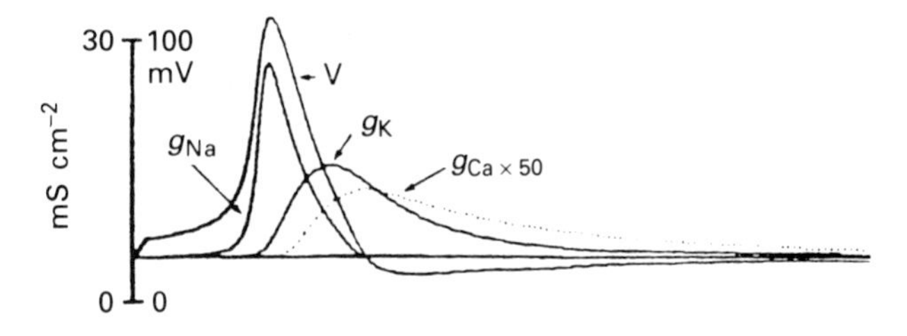

Fig. 6.25. Ionic currents in the presynaptic giant axon terminal at the squid giant synapse. Lower traces show the clamped depolarization, upper traces show ionic currents with inward currents as downward deflections. Trace *a* shows sodium and potassium currents after partial block with low concentrations of tetrodotoxin (TTX) and 3-aminopyridine (3-AmP). Trace *b* reveals the calcium current after total block of the sodium current by a higher concentration of TTX and of the potassium current by a higher concentration of 3-AmP and intracellular injection of TEA. Trace *c* is from another axon after treatment similar to that for trace *b*, again showing the inward calcium current, and trace *d* shows how this is blocked by 1 mM cadmium ions. (From Llinas *et al.*, 1981*a*.)

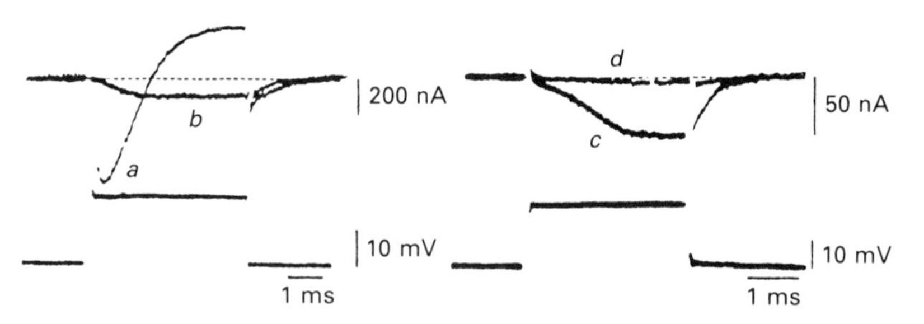

Table 6.2. *Properties of different types of calcium channel in chick dorsal root ganglion cells*

Channel type	T	N	L	
Activation threshold	−70	−10	−10	mV
Inactivation range	−100 to −60	−100 to −40	−60 to −10	mV
Single channel conductance (approx.)	9	13	25	pS
Cadmium block	weak	strong	strong	
Conus toxin block	weak	strong	strong	
Dihydropyridine sensitivity	no	no	yes	

Based on Miller, 1987.

The differences in the membrane potentials required for activation and inactivation allow the different channels to be preferentially opened (fig. 6.27). There are also differences in conductance and pharmacology, as is shown in table 6.2.

Some information on the nature of the calcium channels has been provided by the discovery of channels in frog muscle fibres which are permeable to monovalent cations only when the external calcium ion concentration is at a very low level. The time course of the monovalent ion currents during clamped depolarizations under these conditions was similar to that of the calcium ion currents at higher calcium concentrations. This led to the suggestion that the non-specific channels are simply calcium channels which contain no calcium ions (Almers *et al.*, 1984).

Fig. 6.27. Three types of calcium channels in cultured chick dorsal root ganglion cells. The records were obtained with the cell-attached patch-clamp method, with tetrodotoxin (to eliminate sodium channel activity) and 110 mM $BaCl_2$ in the recording pipette. (From Nowycky *et al.*, 1985.)

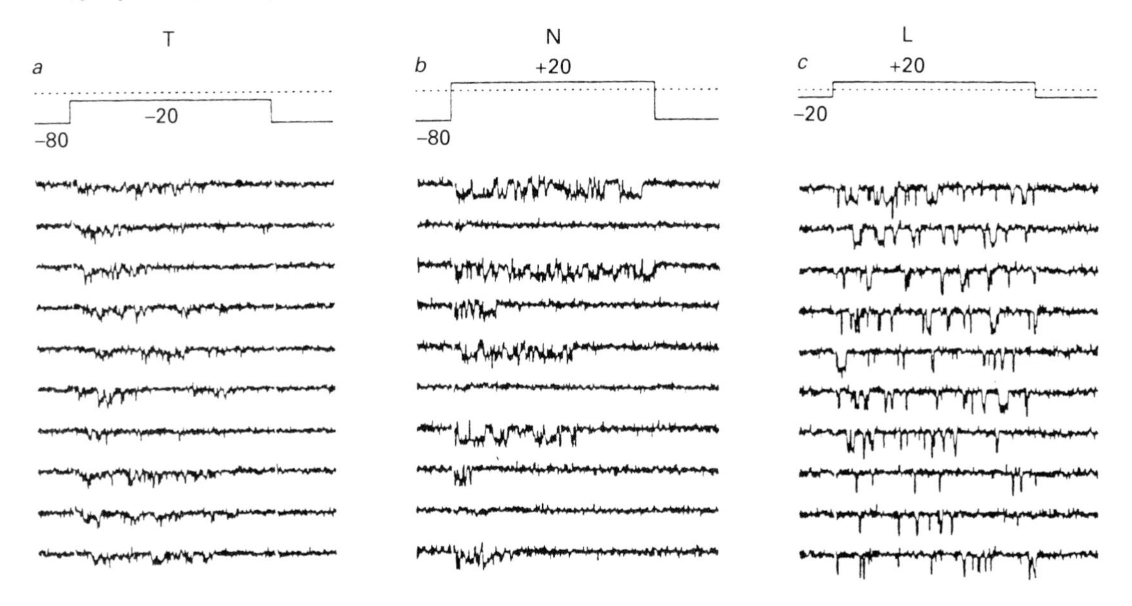

Fig. 6.28 shows how the inward currents through these channels vary with calcium ion concentration. At the higher calcium ion concentrations the currents decrease with decreasing calcium ion concentration until at concentrations between 10 μM and 1 mM there is very little current. This implies that the calcium channels are highly selective and that no other ion can carry the current. At lower concentrations, however, this selectivity is lost and sodium ions can readily flow through the channels, producing the currents shown on the left of fig. 6.28. So it seems clear that the selectivity of the channels depends on the presence of calcium ions.

Almers and McCleskey (1984) have produced a persuasive model of the calcium channel which accounts for these properties (fig. 6.29). They suggest that each channel has two binding sites with a high affinity for calcium ions. If one of these is occupied, then sodium and other ions are repelled and so cannot pass through the channel; the flow of calcium ions will also be low. But if both sites are occupied by calcium ions, the electrostatic repulsion between them will greatly increase the probability of one of the ions leaving the pore. Hence the rate of flow of calcium ions through the membrane is roughly proportional to the number of doubly occupied channels.

Molecular structure of the calcium channel

Mammalian skeletal muscles have calcium channels located in the T tubules. These are invaginations of the cell surface membrane which are concerned with carrying excitation from the cell surface into the interior of the muscle fibre (see chapter 13). The calcium channels are of the L type and bind dihydropyridines (DHP), hence it is possible to use their DHP-binding capability as an assay during their isolation and purification. For this reason it is perhaps more precise to refer to the purified protein (which is believed to be the calcium channel) as the DHP receptor.

Numa and his colleagues at Kyoto University were able to make complementary DNA coding for the DHP receptor in rabbit skeletal muscle by their usual methods of molecular biology (Tanabe *et al.*, 1987). The deduced protein sequence has 1813 amino acid residues and a calculated molecular

Fig. 6.28. Currents through calcium channels in frog muscle at different external calcium concentrations. Points show peak inward currents on depolarization to −20 mV (open symbols) or −5 to +7 mV (black symbols); sodium channels were blocked by tetrodotoxin. The curves were calculated by assuming that sodium ion flow through a channel could be blocked by one (continuous line) or two (dashed line) calcium ions. (From Almers *et al.*, 1984.)

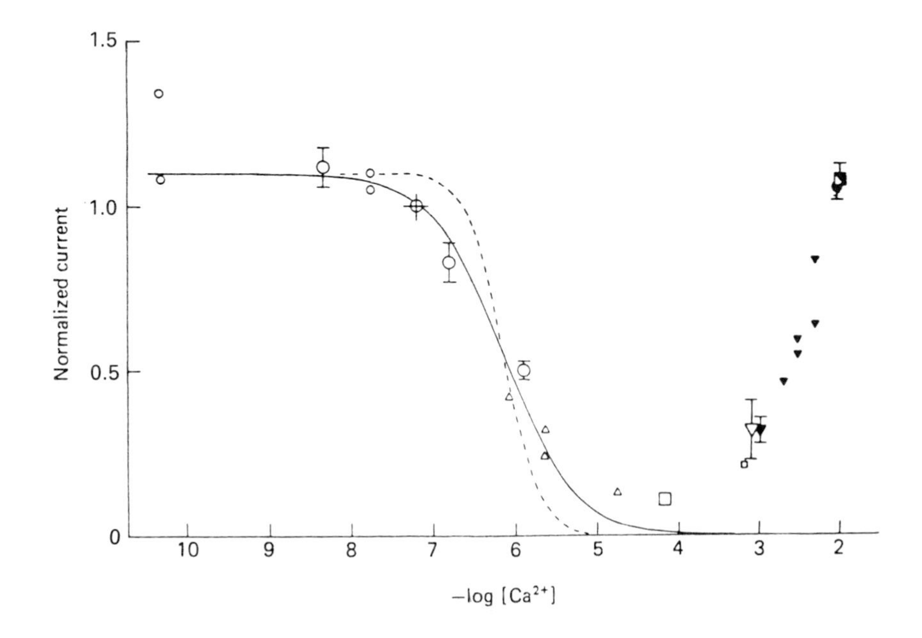

Fig. 6.29. Diagram to show a calcium channel with two binding sites. At low calcium ion concentrations there are no calcium ions bound in the channel so monovalent ions can pass through. At intermediate concentrations one calcium ion is bound in the channel, blocking passage to monovalent ions. At higher concentrations two calcium ions may be bound in the channel; electrostatic repulsion between them increases the probability of one of them leaving the pore, leading to an increase net flow of calcium ions. (From Almers *et al.*, 1986.)

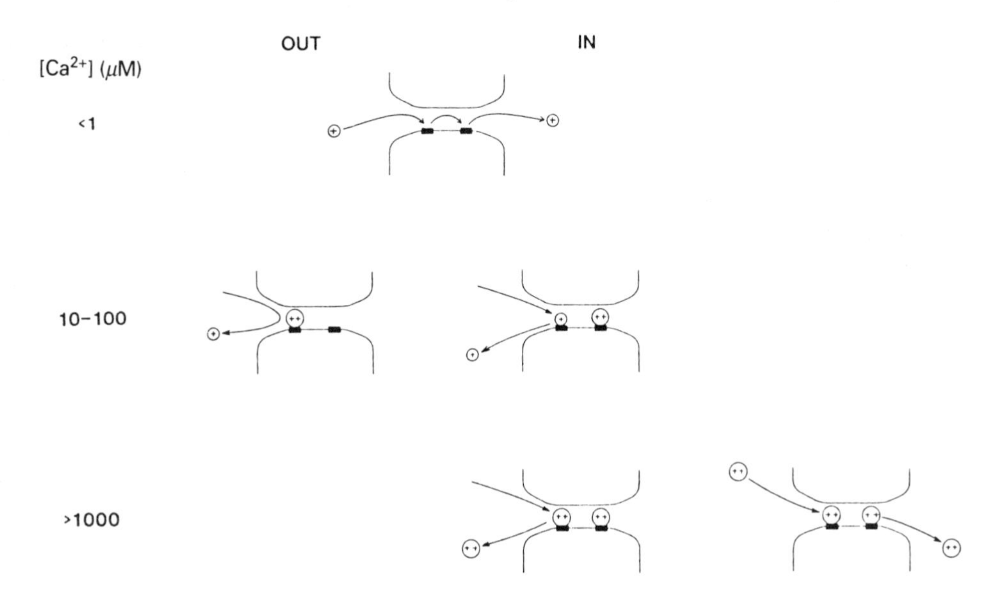

Fig. 6.30. Hydropathicity plot of the amino acid sequence of the DHP receptor from rabbit muscle, thought to be a calcium channel protein. Compare with the sodium channel plot shown in fig. 6.12*a*. (From Tanabe *et al.*, 1987.)

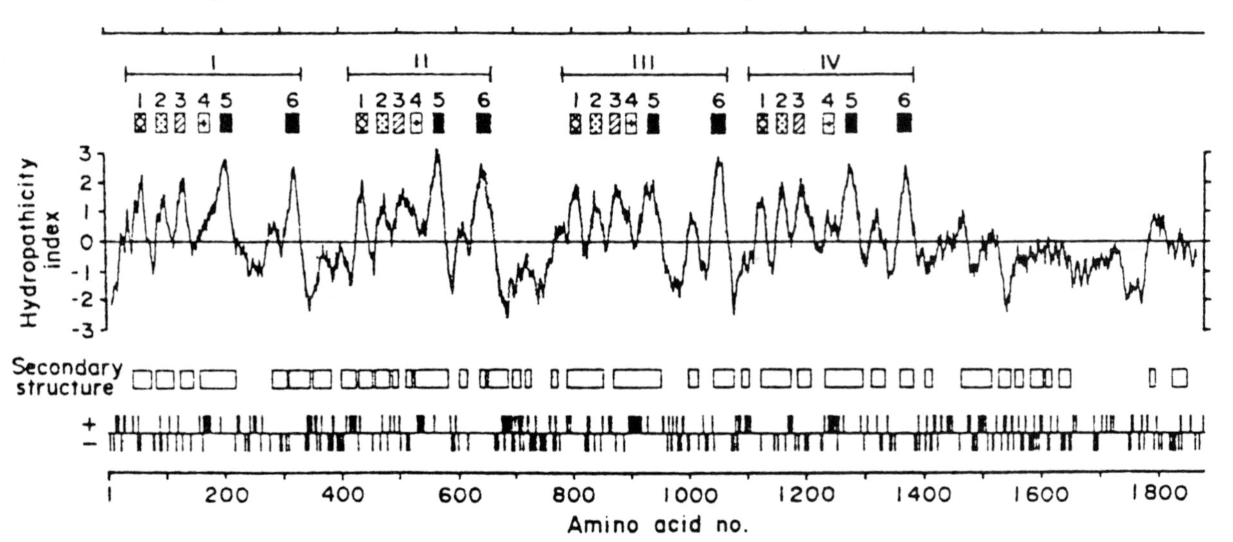

weight of 212 018 Da. The most striking feature of the sequence is its similarity to that of the sodium channel; 29% of the positions are occupied by identical residues and a further 26% by ones showing only conservative changes.

The hydropathicity plot of the deduced calcium channel sequence is shown in fig. 6.30. Just as in the sodium channel, there are four domains of high internal homology, each with six sections which appear to be membrane-crossing helices. The S4 section of each domain showed an unusual sequence with arginine or lysine every third residue, just like those of the S4 section in the sodium channel and the S4-like section in the potassium (A) channel (fig. 6.21). It looks very much as though the voltage-gated sodium and calcium channels are closely related in evolution.

What is the purpose of calcium channels? In some tissues, such as crustacean muscles, vertebrate heart muscle and some secretory cells, it is clear that calcium ion flow through these channels is the major vehicle of cell depolarization. In others this is not so. In the squid giant axon terminal, for example, the peak calcium current is less than a twentieth of the peak sodium current and occurs largely while the membrane potential is returning to its resting level; its purpose is rather to increase the calcium ion concentration inside the terminal, which is essential for the release of the transmitter substance, as we shall see in chapter 7. The intracellular calcium ion concentration is generally of great importance in the regulation of cellular activity, controlling such activities as secretion, neurotransmitter release, muscular contraction, enzyme activity, calcium activated potassium channel opening, and many others. We shall meet some of these functions in later chapters.

7
Neuromuscular transmission

Having dealt with the propagation of action potentials along the axons of nerve cells, we now have to consider how the information contained in nerve impulse sequences is passed on to other nerve cells and muscle cells, which brings us to the problem of synaptic transmission. This process is perhaps best understood in the case of the vertebrate neuromuscular junction, and so this chapter is largely concerned with experiments on this preparation. Later chapters deal with synapses between nerve cells and with some of the finer details of the process at the molecular level.

The nature of synaptic transmission

When the neuron theory had become generally accepted, at about the beginning of this century, the problem arose as to how excitation was transmitted across the gap between two neurons. Du Bois Reymond suggested in 1877 that the presynaptic cell could influence the post-synaptic cell either by electrical currents or by chemical mediators. We now know that transmission at the majority of synapses is brought about by release of a chemical *transmitter substance* from the presynaptic cell, although there are some cases of electrical transmission. The history of the establishment of this generalization provides a fascinating example of the development of a scientific theory, and we shall therefore briefly examine a few of the main steps in the story.

The first tentative evidence for the chemical transmission theory came from the work of Elliott in 1904. He showed that those mammalian muscles which are innervated by the sympathetic nervous system are responsive to adrenaline, whereas muscles which are not so innervated are not. In order to explain these observations he suggested that sympathetic nerves act by release of adrenaline from the nerve endings when a nerve impulse arrives at the periphery.

In 1914, Dale investigated the effects of the substance acetylcholine, originally isolated from ergot (a fungal infection of cereals) on various body functions. He found that it lowered the blood pressure in the cat, inhibited the heart beat in the frog, and caused contractions of frog intestinal muscles. He suggested that acetylcholine occurred naturally in the body, possibly acting as an antagonist to the effects of adrenaline, and, as an inspired guess, suggested that it was normally rapidly broken down by a hydrolytic enzyme; this would account for the fact that it had so far been impossible to isolate acetylcholine from the body.

The next stage in this story came from a classic experiment by Loewi (1921) on the frog heart. The heart normally beats spontaneously, but it can be inhibited by stimulation of the vagus nerve. Loewi found that the perfusion fluid from a heart which was inhibited by stimulation of the vagus would itself reduce the amplitude of the normal beat in the absence of vagal stimulation. Perfusion fluid from a heart beating normally did not have this effect. This means that stimulation of the vagus results in the release of a chemical substance (initially called the *Vagusstoff*), presumably from the nerve endings. The inhibitory action of vagal stimulation and of the vagusstoff was prevented in the presence of the alkaloid atropine. In this and other respects the pharmacological properties of the vagusstoff seemed to be similar to those of acetylcholine. It was later shown that the actions of the vagusstoff and acetylcholine could be potentiated by the alkaloid eserine, and that this substance acted by preventing the hydrolysis of acetylcholine by the enzyme acetylcholinesterase.

These experiments were leading to the conclusion that acetylcholine and the vagusstoff were one and the same substance; all that as needed for the general acceptance of this conclusion was a demonstration that acetylcholine does in fact occur in the body. This was provided by Dale and Dudley (1929), who isolated it from the spleens of cows and horses. Further advances came with the demonstrations that acetylcholine is released on stimulation from the sympathetic ganglia (Feldburg and Gaddum, 1934) and from the motor nerve endings in skeletal muscles (Dale *et al.*, 1936).

However, as the evidence for the chemical transmission theory accumulated, some difficulties arose in its application to all synapses. A number of neurophysiologists remained convinced that many cases of synaptic transmission must involve electrical current flow across the synapse. The principal objections to the chemical transmission theory arose from the rapidity of synaptic action (it was difficult to see how the transmitter substance could be released, produce its action and be destroyed all within the space of a few milliseconds) and from the conflicting evidence as to the action of acetylcholine in the central nervous system. These doubts were not finally resolved until the early nineteen-fifties, when it became possible to insert intracellular electrodes into the postsynaptic region (Fatt and Katz, 1951; Brock *et al.*, 1952), as we shall see in the following pages. Eccles (1976) gives the reasons why he originally believed in the electrical transmission theory, and provides a stimulating account of how he came to change his mind. Finally, a somewhat ironic footnote on the controversy over the electrical and chemical theories of synaptic transmission is provided by the discovery of some synapses in which the transmission process *is* by means of electric currents (see Bennett, 1985).

We begin our account of synaptic transmission by considering the structure and physiology of the neuromuscular junction in the 'twitch' fibres of vertebrate skeletal muscle. The situation here differs in some details from that found at other neuromuscular junctions and at synapses between neurons. Nevertheless, the detailed study of this particular synapse provides an excellent introduction to the physiology of chemically transmitting synapses in general.

The structure of the neuromuscular junction

A vertebrate 'twitch' skeletal muscle fibre, such as is found in the sartorius muscle of the frog, is a long cylindrical multinucleate cell, which is innervated at one or sometimes two points along its length by branches of a motor nerve axon. The structure of the neuromuscular junction as determined by light microscopy, used in conjunction with special staining techniques (see Couteaux, 1960), is shown in figs. 7.1 and 7.2. In the region of contact, the muscle fibre is modified to form the motor end-plate, which contains numbers of mitochondria and nuclei. The motor nerve axon loses its myelin sheath near the ending, and the terminal branches of the axon are partly sunken into 'synaptic gutters' as they spread over the

Fig. 7.1. Diagrammatic picture of a vertebrate 'twitch' muscle motor nerve terminal. In most cases a single motor axon innervates many more muscle fibres than the three shown here.

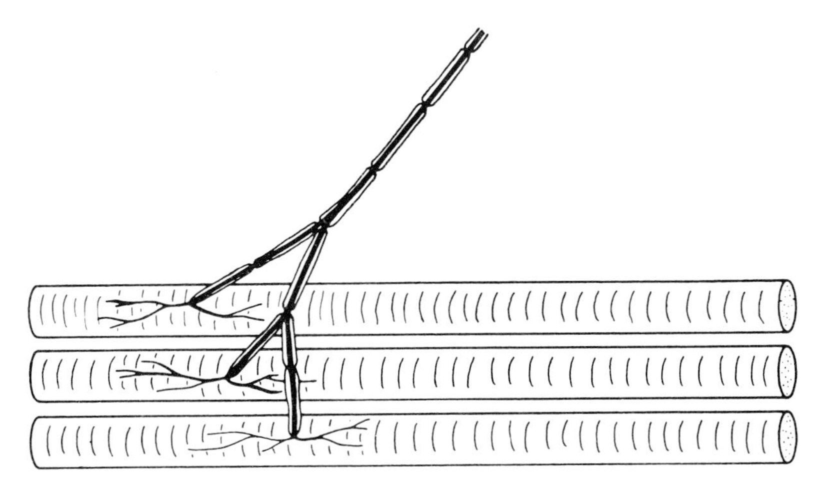

surface of the end-plate. Schwann cell (teloglia) nuclei can be seen in association with the unmyelinated terminals, but the precise distribution of their cytoplasm cannot be determined with the light microscope. A series of 'subneural lamellae' project into the end-plate cytoplasm under the terminal branches of the axon.

These observations have been extended by the use of electron microscopy (Birks *et al.*, 1960; Heuser and Reese, 1977), which shows that there is a space about 500 Å wide, the *synaptic cleft*, between the nerve terminal and the muscle cells (fig. 7.3). The 'subneural lamellae' are seen to be infoldings of the postsynaptic membrane. There is some extracellular material in the cleft, especially in these infoldings, which forms the basal lamina. The Schwann cell covers the nerve terminal in places where it is not in contact with the muscle cell. The terminal branches of the axon contain large numbers of small vesicles about 500 Å in diameter, which may be concentrated in regions opposite the folds in the postsynaptic membrane.

The release of acetylcholine by the motor nerve endings

In a very elegant series of experiments, Dale, Feldburg and Vogt (1936) showed that acetylcholine is released when the motor nerves of various vertebrate skeletal muscles are stimulated. Their experiments on the cat's tongue serve to illustrate some of their methods and conclusions. In most cases the superior cervical ganglia had been removed some weeks previously, so that the hypoglossal nerve would now contain no fibres from the sympathetic system, the only efferent fibres being the motor axons of the tongue muscle. The blood vessels to the tongue were now prepared so that a perfusion fluid (Locke's solution saturated with oxygen at 37 °C) could be passed into the lingual artery via the carotid arteries and collected from the jugular vein after its passage though the capillaries of the tongue. A small quantity of eserine was added to the perfusion fluid so as to prevent the breakdown of acetylcholine by acetylcholinesterase.

The quantities of acetylcholine released in an experiment of this kind were too small to be identified and measured by ordinary chemical methods, and so some type of bioassay had to be used. Acetylcholine in very dilute concentrations causes contraction of the dorsal longitudinal muscles of leeches (after sensitization by eserine) and produces a fall in the blood pressure of anaesthetized cats. Thus if the perfusate also produced these effects, and particularly if the same concentration of acetylcholine was needed to mimic the effects of the perfusate in different tests, then there is a strong indication that the perfusate contained acetylcholine.

Now consider fig. 7.4, which shows the results of one of the experiments on the cat's tongue. Each record in the diagram is a smoked drum trace of the contraction of a strip of leech muscle in response to perfusate from the tongue (diluted by 50%) or a solution containing a known concentration of acetylcholine. Trace *a* shows the absence of response to perfusate obtained before stimulation of the hypoglossal nerve. Trace *b* shows the contraction produced by acetylcholine at a concentration of 10^{-8} w/v. Trace *c* shows the response to perfusate obtained during a period of stimulation at 5 shocks per second, traces *d* and *e* the responses to perfusate obtained 2 and 20

Fig. 7.2. The fine structure of a vertebrate motor end-plate and nerve ending, seen in longitudinal section, as determined by light microscopy. mf, myofibril; m.n., muscle nucleus; ax, axon; tel, Schwann cell (teloglia) nucleus; sarc, sarcoplasm; my, myelin sheath. (From Couteaux, 1960.)

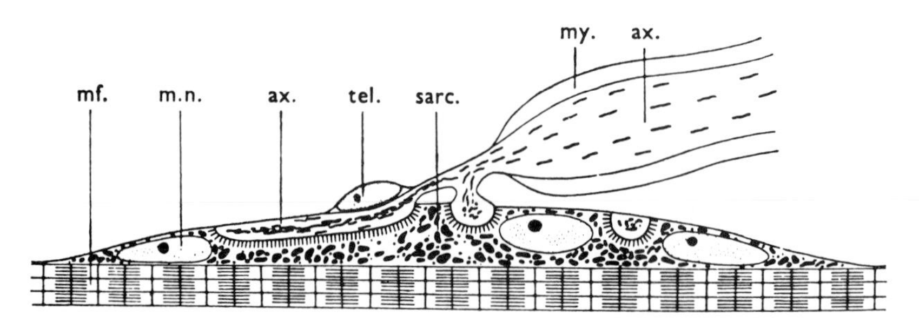

Fig. 7.3. Electron micrograph of a frog neuromuscular junction. The axon terminal (A) is seen in longitudinal section. It contains mitochondria (Mi) and numerous synaptic vesicles (V), and is covered by a Schwann cell (S). Collagen fibres (Co) can be seen over the Schwann cell. The muscle fibre (Mu) is separated from the axon terminal by the synaptic cleft (C), which contains some darkly staining material. The muscle fibre postsynaptic membrane is indented to form junctional folds (F), and is underlaid by dense material at the top of them, where the acetylcholine receptors are concentrated. Presynaptic active zones (Z) occur opposite some of the junctional folds; notice the slight protusion of the presynaptic membrane, the dense cytoplasmic material and the concentration of synaptic vesicles there. Magnification 40 000 times, i.e. 1 mm is equivalent to 250 Å. (Photograph kindly supplied by Professor J. E. Heuser.)

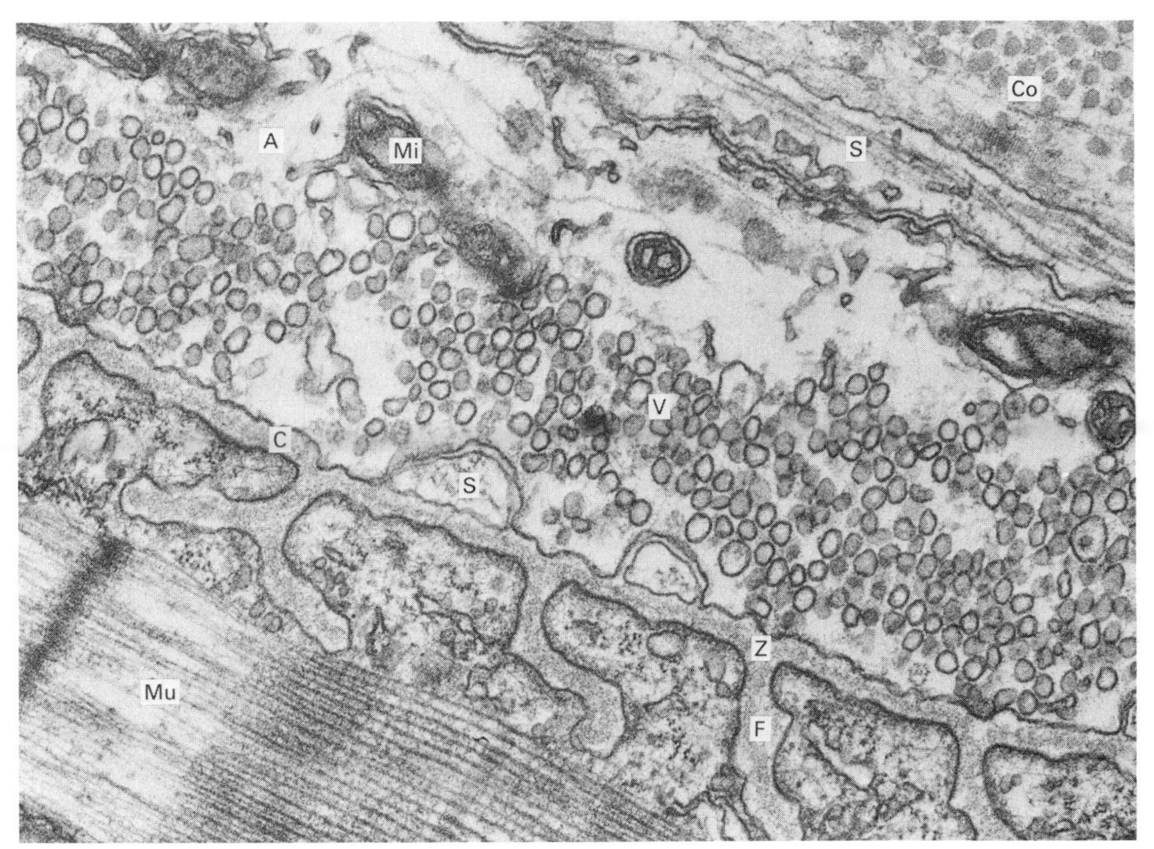

Fig. 7.4. Contractions of eserinized leech muscle in response to acetylcholine and perfusates from cat tongue muscle under various conditions. Explanation in text. (From Dale *et al.*, 1936.)

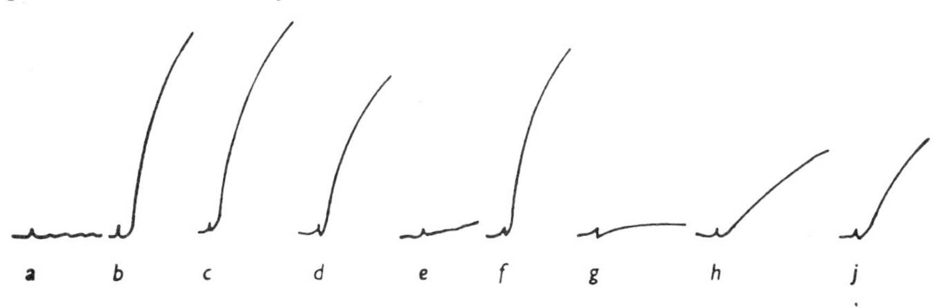

minutes later. Trace *f* shows the response to perfusate obtained during a second period of stimulation; trace *g*, 20 minutes later. Finally trace *h* shows a smaller respone to perfusate obtained from a third period of stimulation, which can be compared with trace *j*, the contraction produced by acetylcholine at a concentration of 2.5×10^{-9} w/v.

This experiment shows that a substance which causes leech muscle to contract is released on stimulation of the nerves supplying the cat's tongue. It was also found that the perfusate from a period of stimulation reduced the blood pressure in the cat, and that this effect was increased in the presence of eserine and abolished by atropine; in all these respects the released substance was similar to acetylcholine. Furthermore, the concentrations of acetylcholine required to mimic the effects of the perfusate on cat blood pressure and on the leech muscle were the same. It follows that acetylcholine must have been present in the perfusion fluid obtained during a period of stimulation.

Where does this acetylcholine come from? At this stage of the investigation there were three possibilities: the acetylcholine could be released from the motor nerves, from the sensory nerves, or from the muscle fibres. In order to resolve this question, Dale and his colleagues used a similar perfusion technique on leg muscles of the dog and the frog. Stimulation of the sensory nerves via the dorsal roots (which were cut proximally to the electrodes so as to avoid reflex excitation of the motor nerves) did not induce release of acetylcholine, whereas stimulation of the motor nerves via the ventral roots did. Direct stimulation of the muscles also produced acetylcholine release, but of course it is impossible to stimulate a normal muscle without also stimulating the motor nerve endings in the muscle. However, stimulation of a muscle which had been denervated ten days previously, or stimulation of a muscle after transmission had failed following prolonged stimulation of the motor nerves, did not result in release of acetylcholine. Hence acetylcholine is released from the motor nerve endings on stimulation.

This conclusion was very interesting in itself, but, in terms of the chemical transmission theory, it had to be shown that the release of acetylcholine was relevant to the excitation of the muscle, and was not merely some incidental by-product of nervous activity. Earlier work had shown that some contraction could be obtained from muscles after injection of acetylcholine, but only when the amount injected was very high. Brown, Dale and Feldburg (1936) reinvestigated the problem, paying particular attention to the details of the perfusion technique, so as to ensure that the administered doses of acetylcholine came into contact with all the fibres in the muscle as rapidly as possible. In this way they were able to show that injections of as little as 2 μg of acetylcholine could produce contractions reaching as high as that in a twitch produced by stimulation of the motor nerves. This response was abolished in the presence of curare, a South American arrow poison which blocks neuromuscular transmission. Injections of eserine caused the contractions produced by single stimuli to the motor nerves to be enhanced and prolonged, as in repetitive stimulation. Hence it was concluded that the action of acetylcholine on the muscle fibre is an essential step in the transmission process.

The end-plate potential

We have now to seek an answer to the question: what is this action of acetylcholine on the muscle fibre? Electrical activity in the region of the end-plate, produced by nervous stimulation, was first observed by Göpfert and Schaefer in 1938 and was subsequently investigated by a number of other workers (e.g. Kuffler, 1942). The general conclusion to be reached from these initial experiments was that the first postsynaptic electrical event is a depolarization (recorded as a negative potential with external electrodes) of the muscle fibre membrane, known as the *end-plate potential* (EPP) or postsynaptic potential (PSP). This initial depolarization is restricted to the end-plate region of the muscle fibre, but is normally of sufficient size to elicit an all-or-none action potential which propagates along the length of the fibre.

The end-plate potential was first investigated with intracellular electrodes by Fatt and Katz (1951), using the frog sartorius muscle. Fig. 7.5 gives an idea of the experimental arrangement. The microelectrode is filled with a concentrated solution of potassium chloride (see p. 7) and is inserted into the muscle fibre in the end-plate region. A suitable amplifier then measures the voltage between the tip of the electrode and another electrode in the external solution, so giving the membrane potential. It is much easier to study the nature of the EPP if it is reduced in size by partial block with curare, so that the complicating effects of the propagated action potential are absent. Under these conditions, the EPP is a brief depolarization with a rapid rising phase and a much slower falling phase, as is shown in the uppermost record of fig. 7.6.

When the electrode is inserted at increasing distances from the end-plate, the electrical change becomes reduced in size and its time course is lengthened (fig. 7.6). This is what one would expect if the EPP is a result of a brief 'active' phase of ionic current flow across the membrane at the junctional region, followed by a passive electronic spread and decay of the charge (or rather, deficiency of charge, since the membrane, is depolarized) across the membrane. That is to say, the form of the EPP should be in accordance with the equations used by Hodgkin and Rushton (p. 37) in the application of the core-conductor theory to nerve axons. A relatively simple way of testing this idea is to see if the total charge on the membrane (which is proportional to the integral of voltage with respect to distance) decays exponentially, as is predicted by (4.10). Fig. 7.7 shows that it does. A more elaborate test was provided by assuming

Fig. 7.6. The end-plate potential in curarized frog sartorius muscle, recorded by means of an intracellular microelectrode inserted at different distances from the neuromuscular junction. The figures to the left of each record indicate the distance (mm) between the electrode and the motor nerve ending; the time scale at the bottom is in milliseconds. (From Fatt and Katz, 1951.)

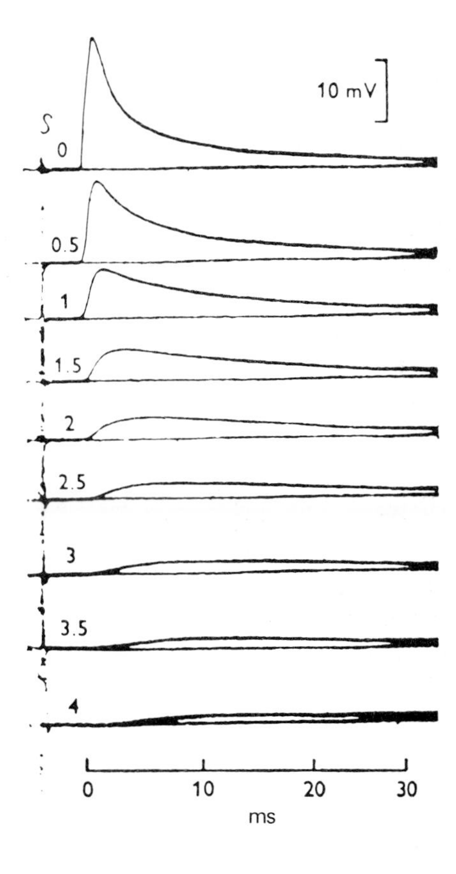

Fig. 7.5. Schematic diagram to show the arrangement for intracellular recording of postsynaptic responses at the vertebrate neuromuscular junction.

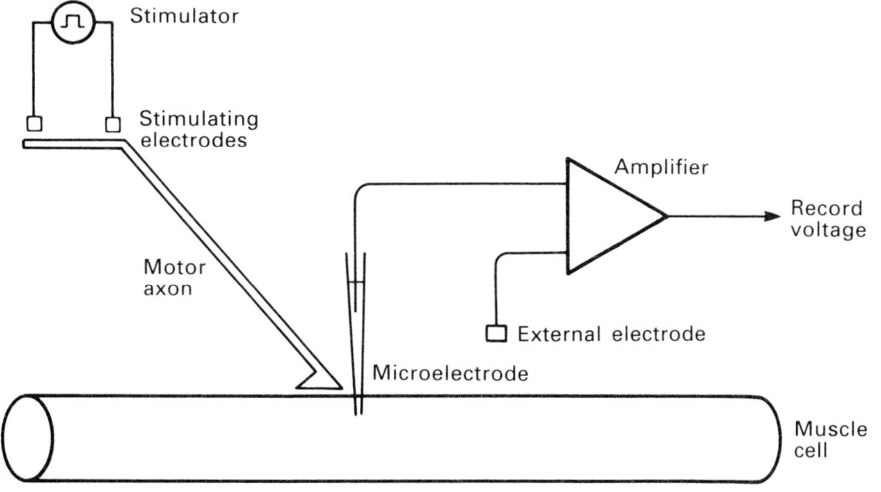

that the idea was correct, calculating the resistance and capacitance of the membrane from the form of the EPP, and then comparing the values obtained with those derived from measurements using applied current pulses; similar estimates were obtained in each case.

This analysis also allows the amount of charge displaced during an EPP to be calculated. The peak value in fig. 7.7 is 3.2×10^{-3} V cm; this will give the peak charge displacement in coulombs if multiplied by the membrane capacitance of a unit length of muscle fibre. With a membrane capacitance of 6 μF cm^{-2} and a fibre diameter of 135 μm, this works out as 2.54×10^{-7} μF cm^{-1}, giving a charge transfer of about 8×10^{-10} C, which is equivalent to a net movement of about 8×10^{-15} moles of univalent ions. These calculations refer to the EPP of curarized muscle, so that in the normal condition the action of the neuromuscular transmitter must result in an ionic movement of at least three times this amount at the end-plate.

The ionic current at the end-plate was later measured directly by Takeuchi and Takeuchi (1959), using a voltage-clamp technique (fig. 7.8). The results (shown in fig. 7.9) were fully in agreement with the scheme suggested by Fatt and Katz.

Fig. 7.7. Diagram to show the exponential decay of charge associated with the end-plate potential. The ordinate shows the integral of voltage with respect to distance (derived from the records of fig. 7.6), plotted on a logarithmic scale. (From Fatt and Katz, 1951.)

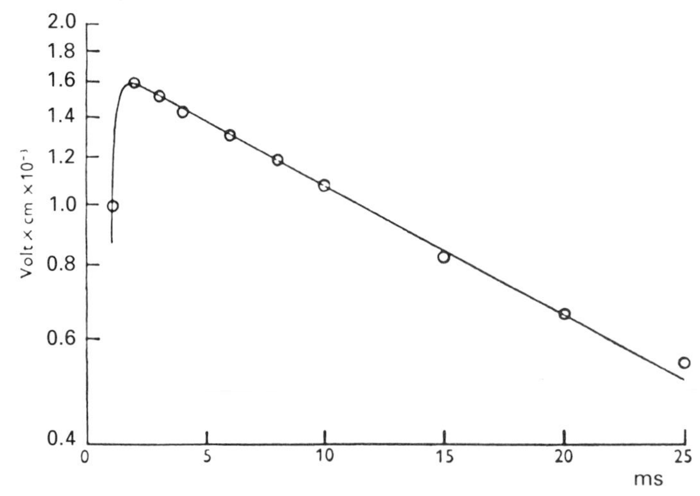

Initiation of the action potential

If a frog sartorius fibre is stimulated electrically so as to depolarize the membrane sufficiently, an action potential arises in the depolarized region and is propagated along the length of the fibre. This action potential is similar to that of non-myelinated axons (although the time course is longer) in that it shows the familiar phenomena of threshold, all-or-nothing response and refractoriness, and is produced by a specific regenerative increase in the permeability of the cell membrane to sodium ions (Nastuk and Hodgkin, 1950).

In the normal muscle, the depolarization produced by the EPP is sufficient to cross the threshold for production of an action potential. However, Fatt and Katz showed that the form of an action potential produced in response to nervous stimulation and recorded at the end-plate (the *N* response) is different from that recorded at distances away from the end-plate (fig. 7.10). It is also different from responses to direct stimulation of the muscle fibre (*M* responses) recorded at the end-plate (fig. 7.11). Fig. 7.10 shows the transitional stages between the two types that occurs at points near the end-plate. What is the reason for these differences? It is obvious that the inflection in the rising phase of the *N* response at about −50 mV is caused by the action potential 'taking off' from the rising phase of an EPP as the membrane potential crosses the threshold. The later differences between *N* and *M* responses were at first more difficult to interpret; the peak of the *N* response is lower than that of the *M* response and there is a hump on the falling phase of the *N* response. Fatt and Katz suggested that these phenomena were caused by the effect of the conductance change at the end-plate on the form of the normal action potential.

The action of acetylcholine on the end-plate membrane

If the transmitter substance at the neuromuscular junction really is acetylcholine, then it should be possible to show that acetylcholine produces

Fig. 7.8. Arrangement for voltage–clamp measurements on neuromuscular transmission. The output of the feedback amplifier passes through a microelectrode so as to keep the membrane potential (measured via the other microelectrode) constant, at a level set by the signal voltage. The potential across the resistor *R* is proportional to the current flowing through the membrane. The muscle fibre is in ringer solution containing curare.

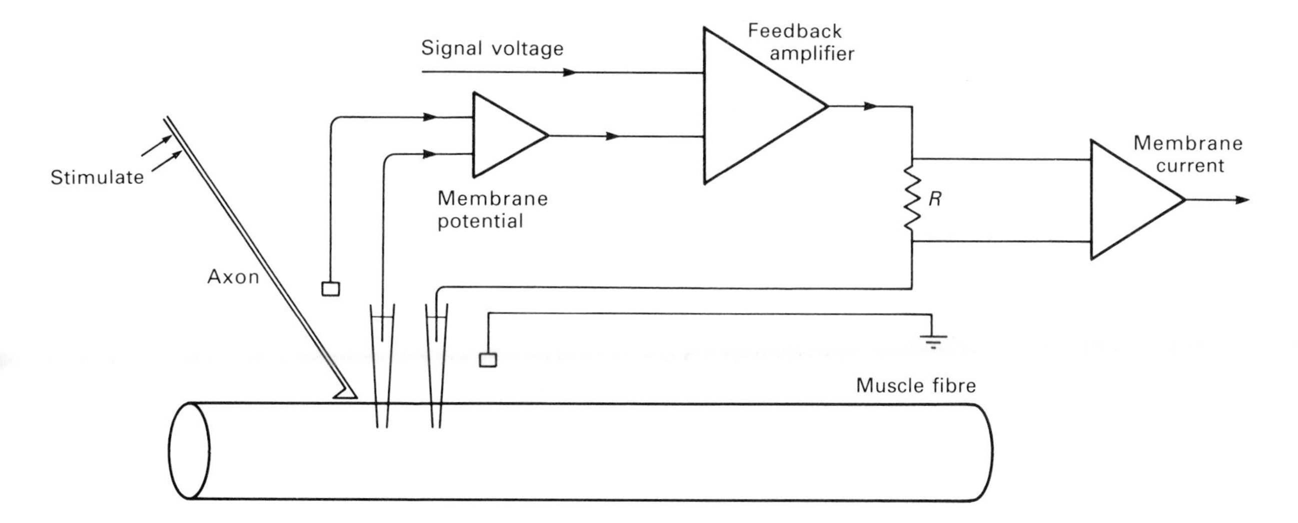

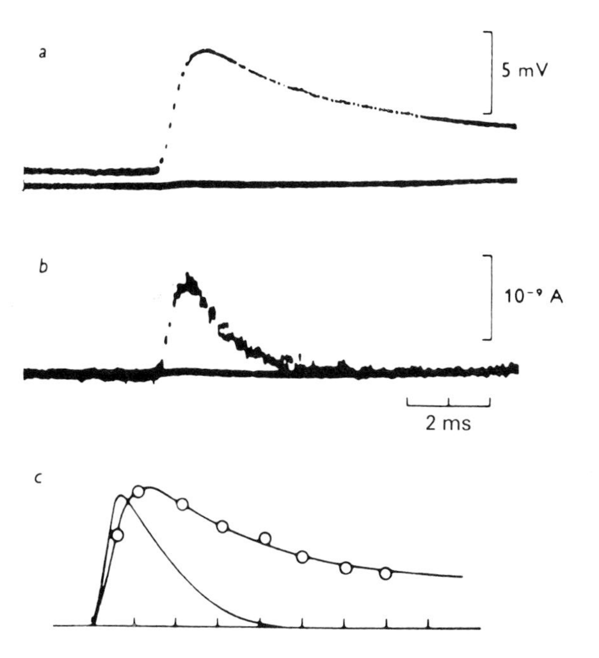

Fig. 7.9. Voltage-clamp analysis of the end-plate potential in a curarized frog muscle. Trace *a* shows the EPP, recorded in the usual manner. Trace *b* shows the end-plate current (EPC) recorded from the same end-plate with the membrane potential clamped at its resting level. The continuous lines in *c* show the actual EPP and EPC superimposed; the circles show the expected EPP as calculated from the EPC, assuming that the muscle fibre has a time constant of 25 ms and an effective resistance of 320 kΩ. (From Takeuchi and Takeuchi, 1959.)

Fig. 7.10. The initiation of the muscle action potential at the motor end-plate. Traces obtained by inserting a microelectrode at different distances (shown in the inset) along the length of a frog muscle fibre, and recording the response to stimulation of the motor nerve. Membrane potentials recorded as changes from the resting level. (From Fatt and Katz, 1951.)

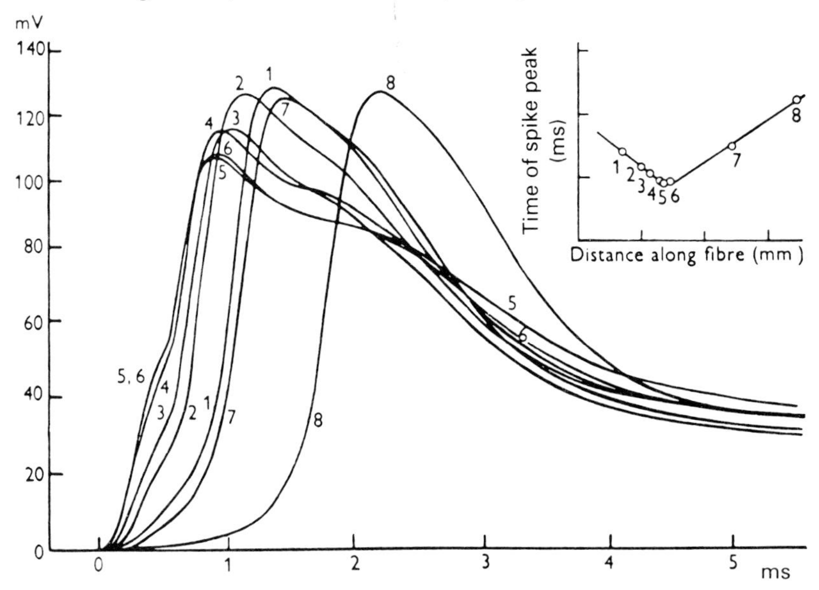

Fig. 7.11. Muscle action potentials recorded intracellularly from the end-plate region in response to direct electrical stimulation of the muscle fibre (trace *M*) and in response to stimulation of the motor nerve (record *N*). The voltage shows the change with respect to the resting potential; the broken line shows where the potential across the membrane is zero. (From Fatt and Katz, 1951.)

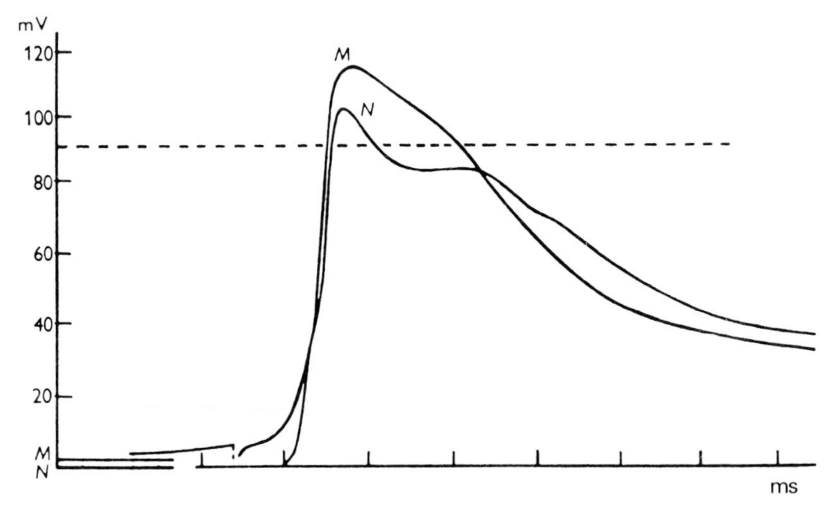

electrical changes at the end-plate similar to those following nervous stimulation. But how is the acetylcholine to be applied to the end-plate? As we have seen, the normal action of the transmitter substance is a brief, impulsive event lasting no more than a few milliseconds, yet it is not possible to limit the action of externally applied acetylcholine to less than a few seconds when using normal perfusion techniques. The answer to this problem was provided by an ingenious technique developed by Nastuk (1953) and further used by del Castillo and Katz (1955) and others. Acetylcholine ionizes in solution so that the acetylcholine ion is positively charged. Consequently it will move electrophoretically if placed in an electric field. Thus after filling a glass micropipette (of the same order of size as a glass microelectrode) with acetylcholine, it is possible to eject it from the tip of the pipette for very brief times by passing a pulse of electrical current outward through the pipette. This technique is known as *ionophoresis*. It is usually necessary to pass an inward 'braking' current through the pipette in order to prevent outward diffusion of acetylcholine in the absence of an ejecting pulse.

The experimental arrangement used by del Castillo and Katz is shown in fig. 7.12. When the acetylcholine pipette was brought near the end-plate, an ejecting pulse caused a depolarization of the muscle fibre membrane, which could initiate propagated action potentials if it was sufficiently large (fig. 7.13). On the other hand, if the pipette was inserted into the muscle fibre, the acetylcholine ejecting pulse merely caused an electrotonic potential change, which is what would occur even if there were no acetylcholine in the pipette.

These experiments show that acetylcholine, when applied to the outside of the end-plate membrane, causes a depolarization similar to that

Fig. 7.13. Responses of a muscle fibre to ionophoretic application of acetylcholine to the end-plate. The upper traces show the membrane potential of the fibre, the lower traces monitor the current in the ionophoresis circuit. (From del Castillo and Katz, 1955.)

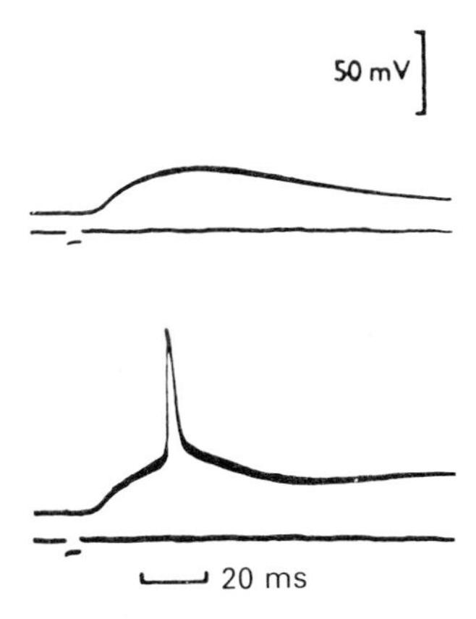

Fig. 7.12. Arrangement for ionophoretic application of acetylcholine at the neuromuscular junction. The intracellular microelectrode is inserted in the end-plate region and connected to the circuit on the right to record the membrane potential. The ionophoresis circuit is shown on the left: a pulse generator is connected to an extracellular micropipette which contains a solution of acetylcholine (ACh).

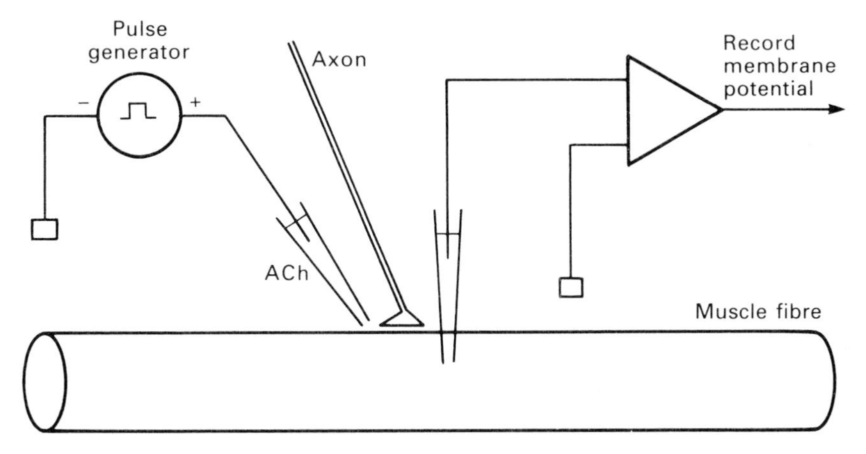

produced during neuromuscular transmission. The longer time course of the rising phase of the acetylcholine potential was ascribed to the much longer distance from the sensitive sites: a few microns instead of 500 Å. Later workers have obtained responses that are much more rapid.

Acetylcholine acts via particular molecular sites known as *receptors*, which occur on the outside of the postsynaptic membrane in the end-plate region. The receptors have been isolated in recent years and we now know quite a lot about their structure and properties, as we shall see in the next chapter. Probably two acetylcholine molecules are bound to each receptor in order to activate it. When this happens a channel is momentarily opened, so permitting ionic current flow across the membrane. End-plate potentials, and the responses to acetylcholine applied by ionophoresis, are produced by currents flowing simultaneously through large numbers of channels.

Localization of the acetylcholine receptors

Responses to acetylcholine in ionophoresis experiments can only be obtained when it is applied to the end-plate region of the muscle fibre. This suggests that the acetylcholine receptors are localized in or near the subsynaptic region. Just how localized they are has been shown in some elegant experiments by Kuffler and Yoshikami (1975*a*). They used the proteolytic enzyme collagenase to free the nerve terminals from the muscle fibres, so enabling acetylcholine to be applied directly to the subsynaptic membrane. They found, as expected, that the subsynaptic membrane had a high sensitivity to acetylcholine; but the sensitivity fell to as little as two per cent of this within a distance of 2 μm once the ionophoresis pipette was moved over the extrasynaptic membrane.

The localization of acetylcholine receptors has also been determined histochemically. A most useful substance for this is α-bungarotoxin, a polypeptide constituent of the venom of the Formosan snake *Bungarus multicinctus*; it causes neuromuscular block by binding tightly to the acetylcholine receptors (see Lee, 1970). Using toxin made radioactive with either tritium or iodine-125 it is possible to show by autoradiography that the toxin rapidly binds to the postsynaptic membrane in the end-plate region (Barnard *et al.*, 1971; Fertuck and Salpeter, 1974). Fertuck and Salpeter found that binding occurs at the crests of the postsynaptic membrane between the junctional folds and for about 2000 Å into the folds. This distribution of binding sites corresponds with the occurrence of electron-dense thickenings of the postsynaptic membrane. By counting the grains of silver in the autoradiographs, Fertuck and Salpeter calculated that there are about 30 000 sites per square micrometre (±27%) in these regions in mouse muscle. If there are two binding sites per ionic channel (see chapter 8) then the channel density will be half this at about 15 000 per square micrometre. This density is considerably higher than that of sodium channels in nerve axon membranes.

Acetylcholine receptors have also been seen by electron microscopy of neuromuscular junctions and electric organs. A variety of techniques has been used, including negative staining of membrane fragments, freeze-fracture and freeze-etching of quick-frozen material (Rosenbluth, 1974; Cartaud *et al.*, 1973, 1978; Heuser and Salpeter, 1979; Hirokawa and Heuser, 1982). The results (fig. 7.14) show arrays of particles on the subneural regions (but not in the depths of the junctional folds of the neuromuscular junction) at densities in the region of 10 000 per square micrometre. Hirokawa and Heuser showed that the subsynaptic density at the neuromuscular junction is in fact a submembranous meshwork which appears to anchor the acetylcholine receptors in the membrane to underlying filaments of the cytoskeleton.

Desensitization

If a muscle is soaked in a solution containing acetylcholine, the initial period of excitation is followed by neuromuscular block which can only be removed by washing away the acetylcholine. Thesleff (1955) showed that such application causes an initial depolarization which then dies away, and introduced the term *desensitization* to describe this effect. The situation was further investigated by Katz and Thesleff (1957*a*) using the ionophoretic application technique. Both barrels of a double-barrelled pipette were filled with an acetylcholine solution; steady ('conditioning') currents were passed through one and brief pulses of current through the other. When acetylcholine was continuously applied by switching on the 'conditioning' current, the resulting depolarization gradually faded, and the response to 'pulses' of acetylcholine also declined, as is shown in fig. 7.15. Cessation of the conditioning current was followed by a recovery in the response to acetylcholine pulses.

The ionic basis of the end-plate potential

Del Castillo and Katz (1954*a*) found that the EPP increased in size if the muscle cell membrane was hyperpolarized at the end-plate region, and decreased if it was depolarized. This observation provided further evidence for the view that the EPP is produced by an increase in membrane conductance. But it was not clear whether or not this conductance increase occurred for all ions. The problem was solved by Takeuchi and Takeuchi (1960), using a voltage-clamp method on frog sartorius muscle fibres. It is not possible to clamp the membrane potential of a whole muscle fibre since one cannot insert an intracellular wire electrode along its length. However, using glass intracellular microelectrodes, the membrane potential can be clamped at the end-plate region, as long as no large currents flow through the adjacent unclamped membrane. As we have seen (fig. 7.9), this technique allows the end-plate current (EPC) to be measured. The principle of the Takeuchis' experiments was to determine the relations between the

Fig. 7.15. A series of records showing desensitization to acetylcholine in a frog muscle fibre. In each case the lower trace monitors the ionophoretic current and the upper trace shows the membrane potential of the muscle fibre. Acetylcholine was applied ionophoretically from a double-barrelled micropipette. Brief 'test' pulses were applied through one barrel (producing the dots on the lower traces and the vertical deflections on the upper traces), and constant 'conditioning' currents were applied through the other barrel at the time indicated. The strength of the 'conditioning' current is lowest in the top record and highest in the bottom one. (From Katz and Thesleff, 1957*a*.)

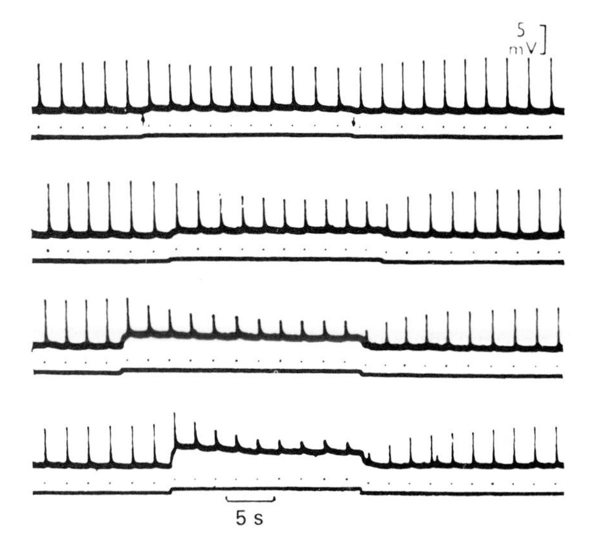

Fig. 7.14. Array of acetylcholine receptors on the electrocyte postsynaptic membrane in the electric ray *Torpedo*. Notice the tendency for the receptors to form rows of four abreast, and that each receptor consists of a number of subunits around a central hollow. The picture is of a platinum replica of the surface of a fragment of postsynaptic membrane, quick-frozen and freeze-etched. Magnification 270 000 times, i.e. 1 mm is equivalent 35 Å. (Photograph kindly supplied by Professor J. E. Heuser, from Heuser and Salpeter, 1979.)

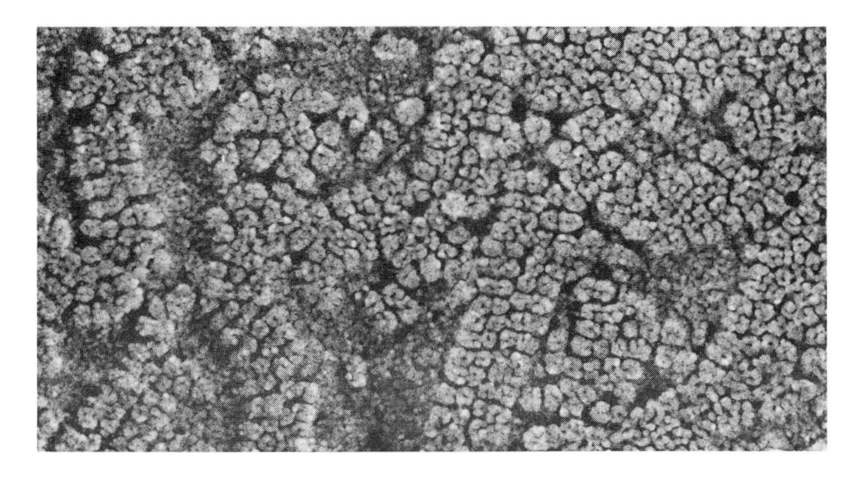

size of the EPC and the membrane potential in solutions containing different ionic concentrations.

First, consider fig. 7.16*a*. In a solution containing 3×10^{-6} g ml^{-1} curare, the EPC varies linearly with membrane potential over the measured range of -50 to -120 mV (it was not possible in these experiments to clamp the membrane at depolarizations greater than about -50 mV). When the line drawn through these points is extrapolated, it cuts the membrane potential axis at -15 mV. This means that the EPC would be zero at this value, and negative (i.e. in the opposite direction) beyond it; hence this point (-15 mV) is called the *reversal potential*. If the curare concentration is increased, the values of EPC are reduced, but the reversal potential is not altered.

Fig. 7.16*b* shows the results of a similar experiment, in which the relation between the EPC and the membrane potential in a curarized muscle fibre was compared in two solutions containing different sodium ion concentrations. It is evident that reduction of the sodium ion concentration makes the reversal potential more negative, and thus it follows that the sodium conductance of the end-plate membrane must increase during the EPC. Similar experiments were carried out on the effect of variations in potassium ion concentration (fig. 7.16*c*). Again, it was found that the reversal potential is dependent upon the potassium ion concentration, and therefore the potassium conductance of the end-plate membrane must also increase during the EPC. However, with drastic changes in chloride concentration, using glutamate as a substitute, there was no change in reversal potential, and therefore there is no change in the chloride conductance of the end-plate membrane during the EPC (fig. 7.16*d*).*

We can conclude that the transmitter substance, acetylcholine, acts by increasing the permeability of the subsynaptic membrane to cations, so that sodium ions flow inward and potassium ions flow outward. Notice that, in contrast with the situation in a propagated action potential, the *ratio* of the sodium and potassium conductances does not vary with time, and moreover the conductances are not initiated by changes in membrane potential, so that the potential change constituting the EPP is not regenerative.

These conclusions tell us some important things about the ionic channels which are opened during neuromuscular transmission. They differ fundamentally from the channels involved in the nerve action potential in that they are opened not by changes in membrane potential but by binding of the transmitter substance: they are ligand-gated, not voltage-gated. Thus the linear relations shown in fig. 7.16 show that, for any one of the straight lines in that diagram, the number of channels opened is constant at different membrane potentials. Curare blocks the channels, so the difference between the two lines in fig. 7.16*a* is that there are fewer channels open at the higher curare concentration. Each individual channel allows both sodium and potassium ions to pass through, but not chloride ions. Many channels are open at the peak of the end-plate current, but the number falls steadily to zero over the next few milliseconds. We shall look further at the nature and properties of these channels in chapter 8.

It was originally supposed that the conductance changes occurring during the EPP were totally independent of the membrane potential. Later work showed that the end-plate current decays more rapidly at less negative membrane potentials. For example, Kordas (1969) found that the half-time of decline of the end-plate current was 1.6 ms when the membrane potential was clamped at -120 mV, but only 0.6 ms at $+40$ mV (see fig. 7.17). Similar results were obtained by Magleby and Stevens (1972).

Acetylcholinesterase

The enzyme acetylcholinesterase hydrolyses acetylcholine to form choline and acetic acid. Nonspecific ('pseudo') cholinesterases occur in the bloodstream; these are not as active in hydrolysing acetylcholine as is 'true' cholinesterase, and can be distinguished by the fact that they are able to hydrolyse butyrylcholine. Acetylcholinesterase occurs in very high concentration at the neuromuscular junction. The first evidence for this came from the work of Marnay and Nachmansohn (1938), using a manometric estimation technique on frog sartorius

* Although the reversal potential is unchanged in fig. 7.16*d*, it is obvious that the EPCs recorded were much higher in the absence of chloride. The reason for this is that the glutamate solution contained a higher calcium concentration, which would increase the amount of transmitter released (p. 136), and also a lower curare concentration.

Fig. 7.16. The effect of membrane potential on the end-plate current of a frog muscle fibre, determined by means of a voltage-clamp technique, under various conditions. *a*: Effect of different curare concentrations. *b*: Effect of different external sodium ion concentrations. *c*: Effect of different external potassium ion concentrations. *d*: Effect of different external chloride concentrations. Notice particularly the changes in the reversal potential brought about by changes in sodium and potassium ion concentrations, and the absence of such changes when the curare or chloride concentrations are altered. (From Takeuchi and Takeuchi, 1960).

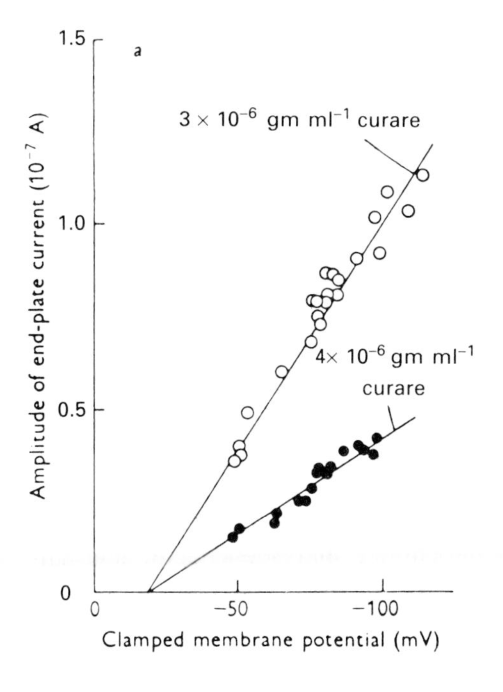

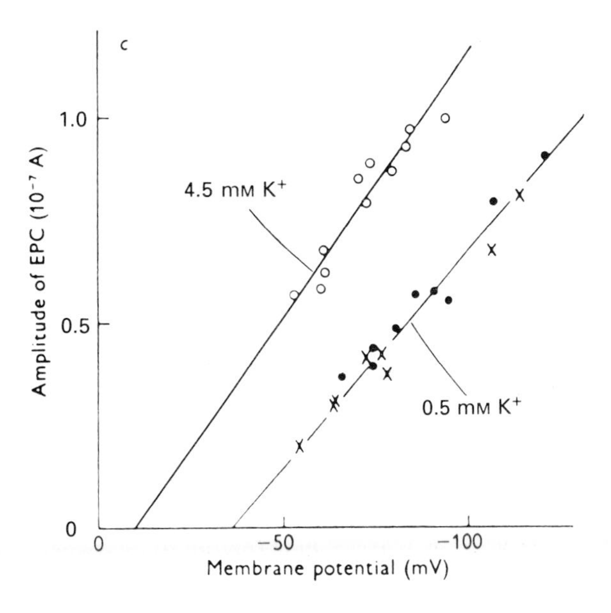

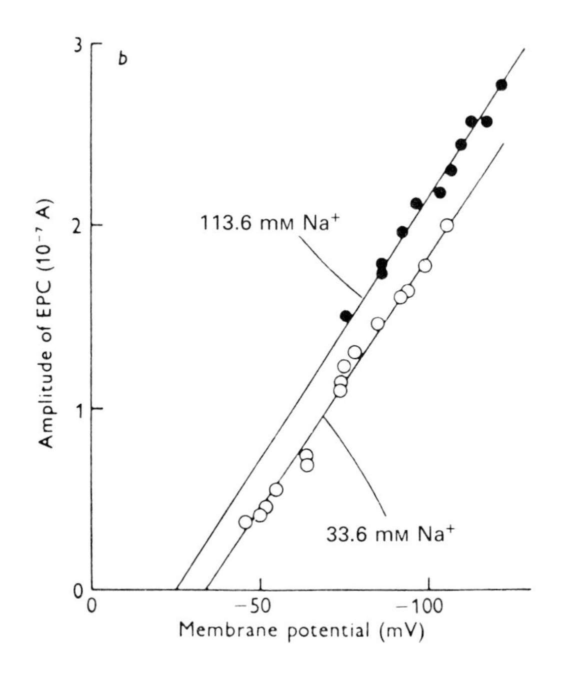

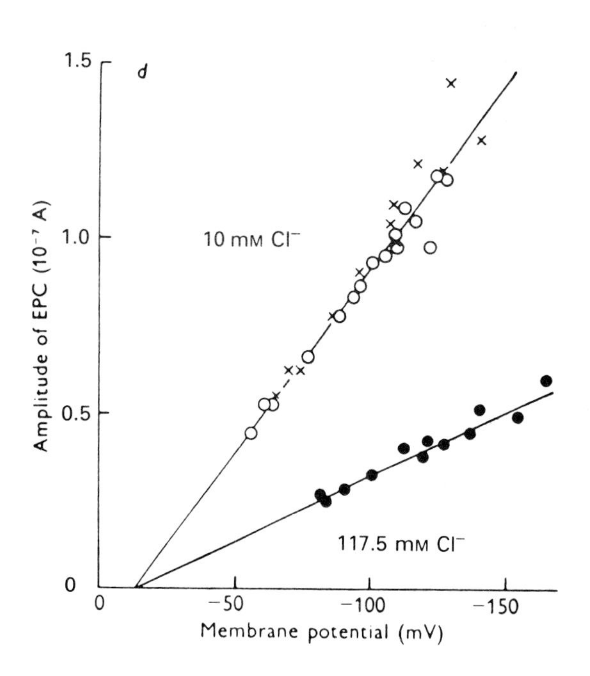

muscle. A portion of the tissue under investigation was ground up in Ringer solution, and acetylcholine was added. The acetic acid formed by the enzymic hydrolysis of the acetylcholine released carbon dioxide from the bicarbonate in the Ringer, and the volume of this carbon dioxide was then measured manometrically. They found that portions of the muscle from regions which contained no nerve endings would hydrolyse 0.14 mg of acetylcholine per 100 mg of tissue per hour, whereas in the region with the greatest number of nerve endings the rate was about 0.8 mg. Since the neuromuscular junctions would only constitute a very small proportion of the latter sample, it is evident that the acetylcholinesterase concentration there must be very high. It was further shown that the concentration in the sciatic nerve was not sufficient to account for this high concentration at the junctional region, which must therefore be a specific property of the junction. Marnay and Nachmansohn calculated that there was sufficient acetylcholinesterase at each neuromuscular junction to hydrolyze about 10^{-14} moles of acetylcholine in 5 ms; this is more than enough to deal very rapidly with the amount of acetylcholine released per impulse.

Fig. 7.17. The effect of membrane potential on the time course of the end-plate current. Frog muscle fibres were treated with glycerol to prevent contraction; this allows the membrane potential to be clamped at values less than the normal threshold for contraction without damage to the fibre. Notice the reversal potential at about −5 mV, and notice particularly that the time course of the fall of the end-plate current is much longer at more negative membrane potentials. (From Kordas, 1969.)

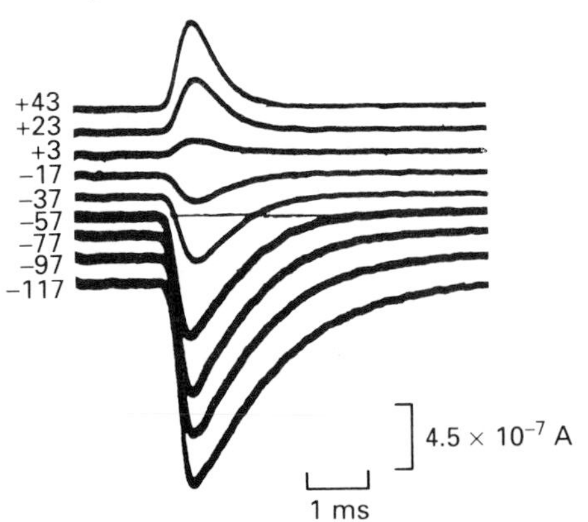

The breakdown of acetylcholine by acetylcholinesterase can be inhibited by a number of compounds. The most well-known of these is eserine (physostigmine), an alkaloid extracted from the calabar bean. The EPP is greatly prolonged in the presence of eserine and may initiate a series of action potentials in the muscle fibre. Many organophosphorus compounds also inactivate cholinesterases, a capability which has formed the basis for their use as war gases and insecticides; examples are diisopropylphosphorofluoridate (DFP) and tetraethylpyrophosphate (TEPP).

The localization of acetylcholinesterase at the neuromuscular junction can be determined by histochemical means. The essentials of this technique, as developed by Koelle and Friedenwald (1949), are as follows. Fresh-frozen material (or material lightly fixed with formaldehyde) is incubated in a medium containing acetylthiocholine in the presence of magnesium ions and copper glycinate. The enzyme hydrolyses the acetylthiocholine to thiocholine and acetic acid, and the thiocholine then complexes with the copper glycinate. On treatment of this complex with ammonium sulphide, copper sulphide is deposited at the enzymic site, and can be easily seen under the microscope. The results showed that acetylcholinesterase is located in the terminal gutters of the postsynaptic membrane.

Salpeter and her colleagues (1978) used tritiated DFP as a marker to localize acetylcholinesterase using electron microscope autoradiography. They found that the silver grain density was high over the synaptic cleft and especially high over the junctional folds. In mouse fast extraocular muscle they estimate that there are about 2500 acetylcholinesterase sites per square micrometre of postsynaptic membrane.

There is now good evidence that the acetylcholinesterase molecules are in the extracellular material (the basal lamina) of the synaptic cleft and junctional folds. Hall and Kelly (1971) found that treatment of endplates with proteolytic enzymes would release acetylcholinesterase into the perfusion fluid. After collagenase treatment neuromuscular transmission still occurred but the postsynaptic responses had a longer time course, comparable with that after eserine treatment.

McMahan, Sanes and Marshall (1978) found that the basal lamina left after degeneration of the muscle and nerve following injury would still stain histochemically for acetylcholinesterase. The enzyme was found where the synaptic gutters had been, with dense bars at 1 μm intervals, corresponding to the positions of the junctional folds.

The quantal nature of transmitter release

Miniature end-plate potentials

With a microelectrode inserted into a frog sartorius muscle fibre in the end-plate region, and using a high-gain low-noise amplifier, Fatt and Katz (1952) observed the occurrence, in the resting muscle, of small fluctuations in membrane potential of much the same shape as normal EPPs, but only about 0.5 mV in height (fig. 7.18). These *miniature end-plate potentials* could only be found at the end-plate region, were reduced in size by curare, and increased in the presence of prostigmine (an anticholinesterase similar to eserine in its action). They were not found in muscle that had been denervated two weeks previously. Statistical analysis of the time intervals between successive miniature EPPs showed that their occurrence was randomly distributed in time. These observations suggested that the miniature EPPs were caused by the action of acetylcholine released spontaneously from the motor nerve ending.

Fig. 7.18. A series of membrane potential records from a frog neuromuscular junction showing miniature end-plate potentials. (From Fatt and Katz, 1952.)

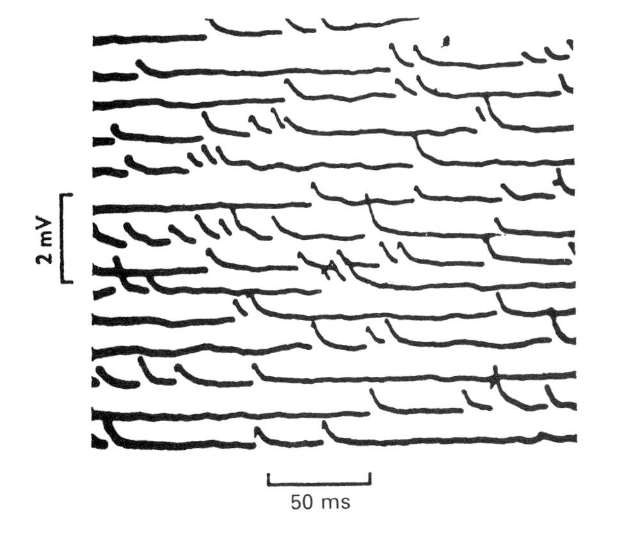

How much acetylcholine is involved in the production of a single miniature EPP? The most obvious suggestion is perhaps just one molecule. Fatt and Katz argued that this could not be so, however, for the following reasons. First, if one molecule could produce a depolarization of about 0.5 mV, then not more than 1000 at the most would be needed to produce an EPP of 50 mV, and therefore the amount of acetylcholine released per impulse would be no more than 10^{-21} moles. In fact it is much larger than this (Krnjević and Mitchell, 1961, estimated that each nerve terminal in rat diaphragm muscle releases about 10^{-17} moles of acetylcholine per impulse), so that each miniature EPP must be produced by several thousand molecules of acetylcholine. Secondly, the response to externally applied acetylcholine is a smooth depolarization, not a series of unitary events about 0.5 mV in height.

Kuffler and Yoshikami (1975*b*) have provided a more direct estimate of the number of molecules of acetylcholine per miniature EPP. They used preparations in which the subsynaptic membrane was exposed, by removing the nerve terminals from frog or snake muscles which had been treated with the enzyme collagenase so as to loosen their connective tissue sheaths. They then measured the size of the responses to ionophoretic applications of acetylcholine. The amount of acetylcholine ejected from the pipette during each ionophoretic pulse was determined by means of an ingenious bioassay: several thousand such pulses were passed into a droplet of Ringer solution which was then applied to an end-plate under a layer of fluorocarbon oil, and its depolarizing effect was compared with that of a droplet of the same volume of acetylcholine solution of known concentration. They found that about 6000 molecules of acetylcholine were required to produce a depolarization of 1 mV. The average size of a miniature EPP in their experiments was about 1.5 mV, and hence somewhat less than 10 000 molecules were involved in its production.

A most interesting property of these miniature EPPs is that their frequency can be altered by procedures which we would expect to change the membrane potential of the presynaptic nerve terminals. Liley (1956) investigated these effects at the neuromuscular junctions in the rat diaphragm. If depolarizing currents were passed through an external electrode placed near the nerve terminals, the frequency of discharge increased; conversely, it decreased if the current was hyperpolarizing (fig. 7.19*a*). The frequency also increased if the external potassium ion concentration was raised above 10 mM; in this case the increase can be related to an estimate of the presynaptic membrane potential, which cannot be measured directly since the terminals are too fine (fig. 7.19*b*).

We have seen that the size of the miniature EPPs can be altered by curare and anticholinesterases; it also changes with postsynaptic membrane potential in the same way as does the EPP. In the frog sartorius, the size of miniature EPPs varies inversely with the diameter of the muscle fibre; thus, since large fibres have a smaller total membrane resistance than small fibres, it would seem that the currents associated with the miniature EPPs (and therefore the amounts of acetylcholine released) are more or less constant (Katz and Thesleff, 1957*b*). The discharge frequency can be altered by changing the osmotic pressure of the bathing solution or (in mammals) the magnesium or calcium ion concentrations, as well as by polarization of the nerve terminals. These observations lead to an important generalization: as Katz (1962) puts it, 'the *frequency* of the miniature potentials is controlled entirely by the conditions of the *pre*synaptic membrane, while their *amplitude* is controlled by the properties of the *post*synaptic membrane'.

The quantal nature of the end-plate potential

It has long been known that an excess of magnesium ions blocks neuromuscular transmission. Del Castillo and Engbaek (1954) showed that this effect was due to a reduction of the quantity of acetylcholine released per impulse, so that the EPP was very much smaller. Calcium ions antagonized this action, i.e. the degree of reduction of acetylcholine release was dependent upon the ratio of the

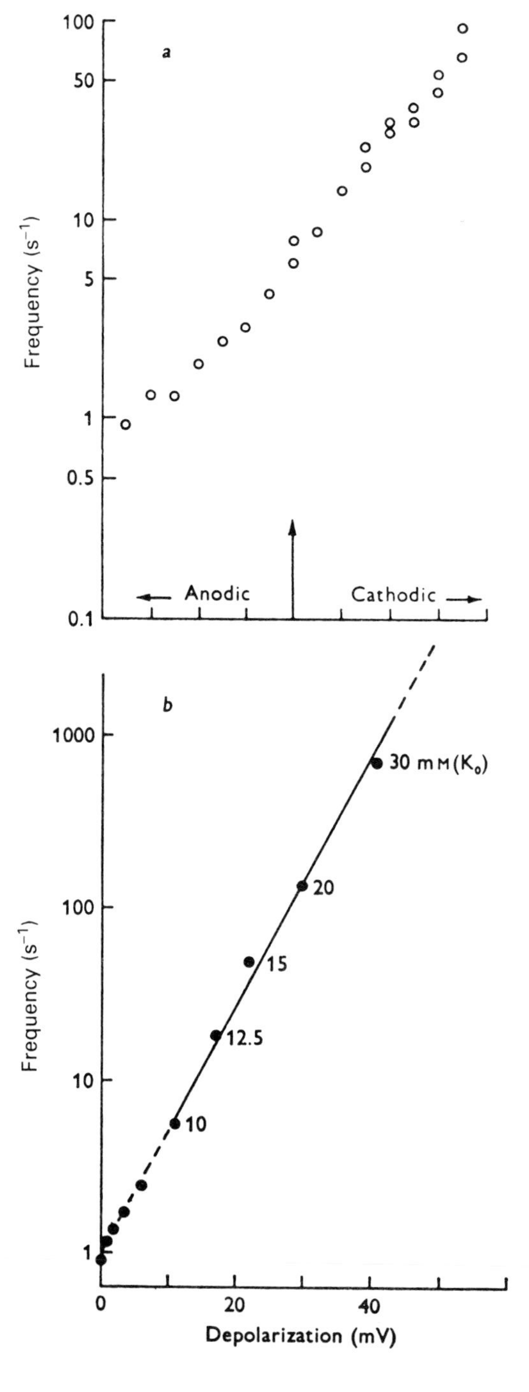

Fig. 7.19. The effect of the presynaptic membrane potential on the frequency of miniature end-plate potentials in a rat diaphragm muscle fibre. *a*: The effect of applying various electrical currents (measured in arbitrary units) to the nerve ending. *b*: The effect of increasing the external potassium ion concentration; the abscissa shows the estimated depolarizations of the presynaptic nerve membrane produced by this treatment. (*a* from Liley, 1956; *b* from Katz, 1962, derived from results given by Liley, 1956.)

magnesium and calcium ion concentrations. When the block produced by this treatment was intense, it was found that the size of successive EPPs may fluctuate in a step-wise manner (fig. 7.20). This led to the suggestion that acetylcholine is released from the motor nerve terminals in discrete 'packets' or *quanta*, that the normal EPP is the response to some hundreds of these quanta, and that miniature EPPs are the result of spontaneous release of single quanta.

An extension of the quantal release hypothesis is that there is a large population of quanta in each nerve terminal, each one of which has a small probability of being released by a nerve impulse. This idea was tested by del Castillo and Katz (1954*b*) by means of a statistical analysis applied to a large number of nervous stimuli. The experiments were performed on a frog toe muscle in a solution containing a high magnesium ion concentration so as to reduce the quantal content of each response. Let x be the number of quanta comprising any one response, and let m be the mean of all values of x. Let N be the total number of events measured, and let n_0, n_1, n_2, etc., be the number of events in the classes $X = 0, 1, 2$, etc.* Then the relative frequencies (n/N) of the different values of x should be given by the terms of a Poisson distribution, i.e.

$$\frac{n_x}{N} = e^{-m} \cdot \frac{m^x}{x!}. \qquad (7.1)$$

The easiest way of finding m in this equation is to take the case when $x = 0$, when we get

$$\frac{n_0}{N} = e^{-m}$$

or

$$m = \log_e (N/n_0), \qquad (7.2)$$

i.e. m is given by the natural logarithm of the reciprocal of the proportion of events in which transmission fails. There is also another way of finding m. Since we suspect that each miniature EPP corresponds to one quantum, the mean number of quanta per response should be given by

$$m = \frac{\text{mean amplitude of EPP}}{\text{mean amplitude of miniature EPPs}} \qquad (7.3)$$

Thus we have two independent ways of determining m, from (7.2) and (7.3), and if the hypothesis is correct, they should agree. Fig. 7.21. shows that they do.

Now consider the situation when x is 1 or more. The relative frequencies of EPPs of different sizes should be described by a Poisson distribution, as in (7.1). A slight complication arises here, however, since the size of the spontaneous miniature EPPs varies somewhat (fig. 7.22*a*), and this has to be taken into account when calculating the frequency distribution of EPP amplitudes. Fig. 7.22*b* shows that the curve calculated in this way is in good agreement with the actual distribution of EPP amplitudes.

Does this analysis apply to normal EPPs, where m is large? Martin (1955) showed that it

Fig. 7.20. Quantal fluctuations in end-plate potential size after magnesium poisoning. Spontaneous miniature potentials can be seen in some of the records. (From del Castillo and Katz, 1954*b*.)

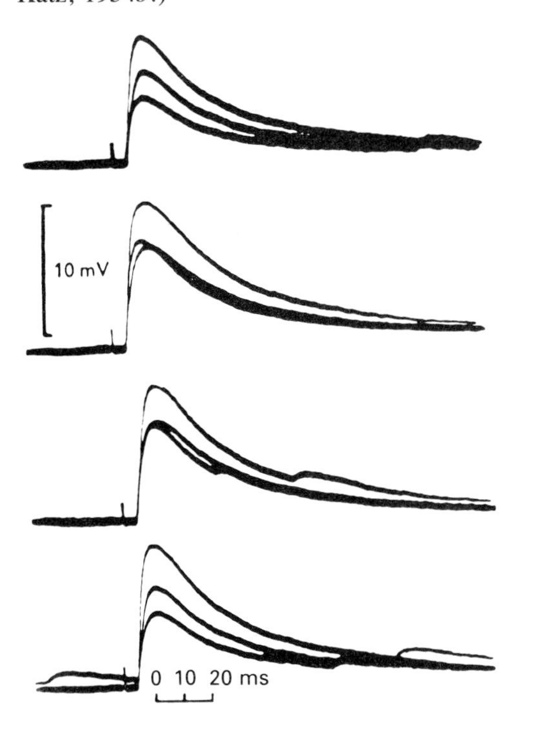

* An analogy may help the non-mathematical reader with the terminology here. Suppose a large class of students attends a long series of lectures, and each student has a constant low probability of sneezing during a lecture. Then x is the number of sneezes in any particular lecture, m is the mean of x, N is the total number of lectures, n_0 the number of lectures with no sneezes, n_1 the number with one sneeze, and so on.

does, as long as (7.3) is modified to take account of the non-linear addition of quantal changes. For example, it follows from Ohm's law that a change in conductance which produces a depolarization of 0.5 mV when the membrane potential is −95 mV will only produce a depolarization of three-quarters of that amount (i.e. 0.375 mV) at −75 mV (assuming that the reversal potential is −15 mV). From these considerations, Martin calculated that a normal EPP of about 50 mV is produced by the action of about 250 quanta.*

We can now see the importance of the experiments showing that the frequency of spontaneous discharge of miniature EPPs rises when the presynaptic terminal is depolarized. If the line in fig. 7.19*b* is extrapolated to, say, 90 mV (so that, to put it another way, we are assuming that the probability of discharge of a quantum of actylcholine is logarithmically proportional to the depolarization of the presynaptic terminal membrane), the discharge frequency would be about 10^6 per second. Since the presynaptic action potential only lasts for a fraction of a millisecond, this is about the right order of magnitude to account for the release of the number of quanta needed to produce an EPP.

But why should acetylcholine be discharged from the nerve ending in quantal units of 10^3 to 10^4 molecules? Fig. 7.3 shows that the nerve terminal contains large numbers of vesicles about 500 Å in diameter. These vesicles were first observed by de Robertis and Bennett (1954) and Palade and Palay (1954), and have since been found in the presynaptic terminals of all chemically transmitting synapses where they have been looked for. Del Castillo and Katz (1956) suggested that they contain the transmitter substance, and that the discharge of the contents of one vesicle into the synaptic cleft corresponds to the release of one quantum of the transmitter. This suggestion is known as the 'vesicle hypothesis'.

* The Poisson distribution is a limiting case of the binomial distribution when the probability of any individual event is low, as in quantal release in high-magnesium solutions. Consequently one might expect that quantal release would be distributed binomially under more normal conditions. Some investigations have found results in agreement with this expectation (e.g. Wernig, 1975; Bennett *et al.*, 1977), but there are difficulties in relating the binomial parameters to our understanding of what is happening in the nerve terminal (Silinsky, 1985).

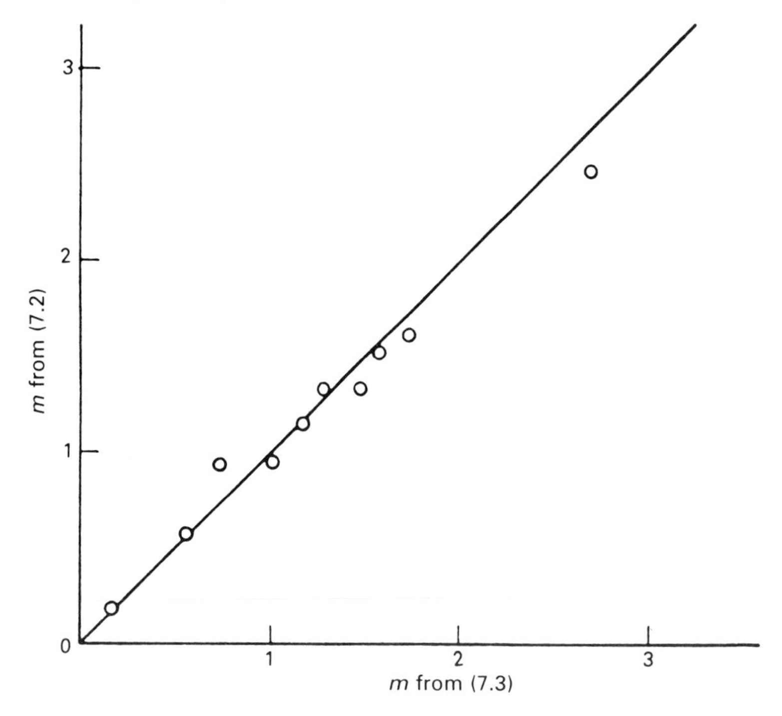

Fig. 7.21. Graph showing the agreement in the values of m derived from (7.2) and (7.3). Explanation in text. (From del Castillo and Katz, 1954*b*.)

Contents of the vesicles

If the vesicle hypothesis is correct it should be possible to isolate vesicles and show by chemical means that they actually do contain acetylcholine. Much work on this problem has been done by Whittaker and his colleagues. Homogenization of a suitable tissue (such as guinea-pig brains) followed by centrifugation enables a subcellular fraction containing fragments of nerve endings to be isolated. The fragments are known as synaptosomes and contain synaptic vesicles. The vesicles can be released from the synaptosomes by rupturing them in a hyposmotic solution. Such extracts from homogenized brains contain quantities of a variety of transmitter substances, including acetylcholine (Whittaker *et al.*, 1964).

Fig. 7.22. Statistical analysis of the quantal components of transmission at a frog neuromuscular junction. *a*: Histogram showing the variation in size of spontaneous miniature end-plate potentials; the continuous curve is a normal distribution curve. *b*: Frequency distribution histogram of the sizes of responses to nervous stimulation in the same muscle fibre as *a*, after magnesium block. The ordinate shows the number of occurrences of potentials in a particular size range; the abscissa shows the size of the potentials, the unit of measurement being the mean size of the spontaneous potentials, from *a*. The arrows (to show the expected number of failures of transmission) and the continuous curve were calculated from a Poisson distribution, modified to take account of the variability in size of the quantal units as seen in *a*. (From del Castillo and Katz, 1954b.)

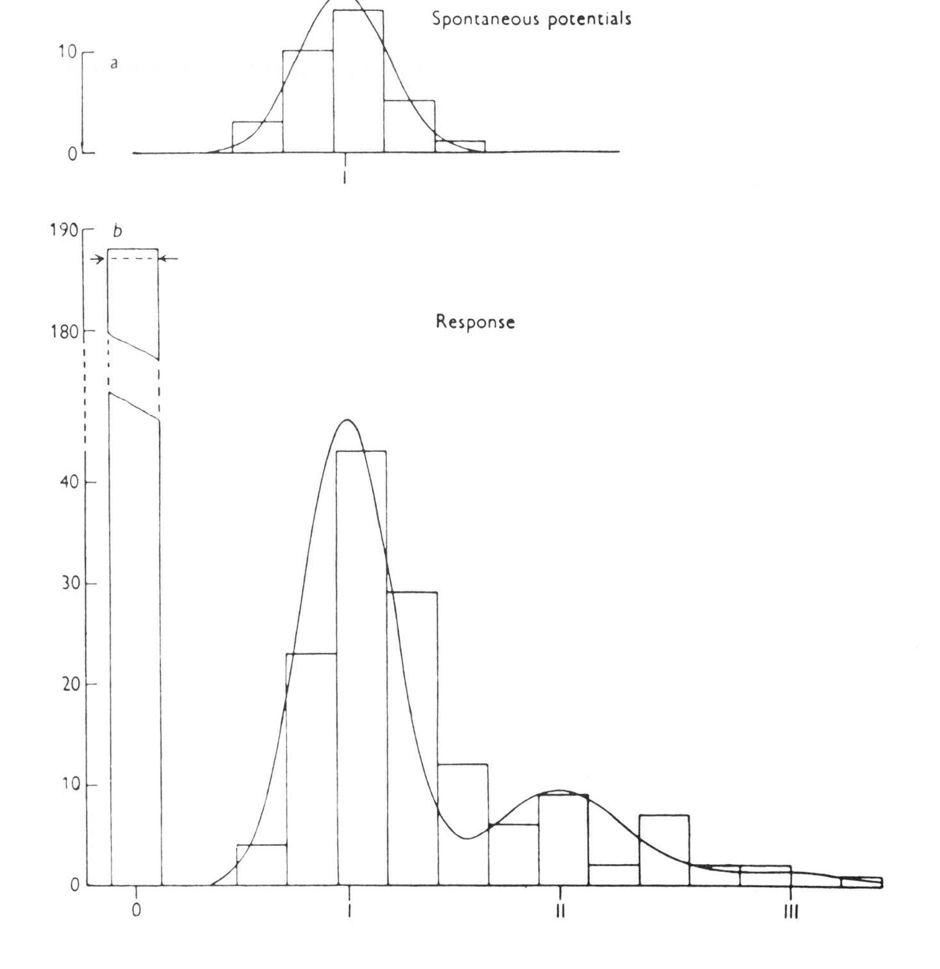

The electric organ of the electric ray *Torpedo* is very richly innervated with cholinergic nerve terminals and provides an excellent source of nearly pure fractions of acetylcholine-containing vesicles (Whittaker and Zimmermann, 1974; Tashiro and Stadler, 1978; Ohsawa *et al.*, 1979; Whittaker, 1984). The acetylcholine content of each vesicle is in the region of 200 000 molecules. Electric organ vesicles are larger than those at the frog neuromuscular junction: the outside diameter is about 84 nm instead of 50 nm and so the volume they contain (allowing 4 nm for membrane thickness) is nearly six times greater. Since the osmotic concentration of frog tissues is about one fifth of that in rays (Schmidt-Nielsen, 1983), we might expect frog vesicles to contain only one thirtieth the amount of acetylcholine in *Torpedo* vesicles, i.e. about 7000 molecules. This figure is in reasonable agreement with the estimates of quantal content from physiological measurements or chemical determinations (Kuffler and Yoshikami, 1975*b*; Miledi *et al.*, 1983).

The vesicles also contain adenosine triphosphate (ATP), about one molecule for every seven molecules of acetylcholine (Dowdall *et al.*, 1974). It is not yet clear what the physiological function of this ATP is, although we do know that it is released into the synaptic cleft along with the acetylcholine, as is shown by collecting perfusates from rat muscle during stimulation (Silinsky, 1975).

Preparations of vesicles also contain phospholipids and membrane-bound proteins, including proteoglycans, which are proteins to which large quantities of acid polysaccharides are attached. The polysaccharide chains may serve to balance the ionic charges of the vesicle contents.

Vesicle discharge and recycling

Del Castillo and Katz (1956) suggested that each synaptic vesicle discharges its contents by fusion of its membrane with the presynaptic plasma membrane as is shown in fig. 7.23. Membrane fusion of this type has been seen in secretory cells and the process is called 'exocytosis'. Electron microscopy showed that the discharge sites are not simply distributed at random but are localised to 'active zones' (Couteaux and Pecot-Dechavassine, 1974; Heuser, 1976). These are seen as bands of dense material and double rows of intramembranous particles, with a row of vesicles on each side, opposite the junctional folds of the postsynaptic membrane (fig. 7.24). This means that the vesicles are positioned where release of their contents will be most effective, opposite the concentrations of acetylcholine receptors near the openings of the junctional folds.

Direct evidence for the fusion of the vesicular and presynaptic membranes comes from quick-freezing experiments (Heuser *et al.*, 1979; see also Torri-Tarelli *et al.*, 1985). The muscle is mounted on a device which will slam it onto a block of copper which has been cooled to −269 °C by liquid helium, with its nerves being stimulated a few milliseconds beforehand. Transmitter release was enhanced with 4-aminopyridine, which blocks potassium channels and so produces a much

Fig. 7.23. Dell Castillo and Katz's hypothesis about the quantal release of transmitter. Reactive sites on the vesicles and the presynaptic membrane are shown as dots. Release of transmitter occurs when two sites meet, and the probability of this happening increases greatly on depolarization. (From Katz, 1969.)

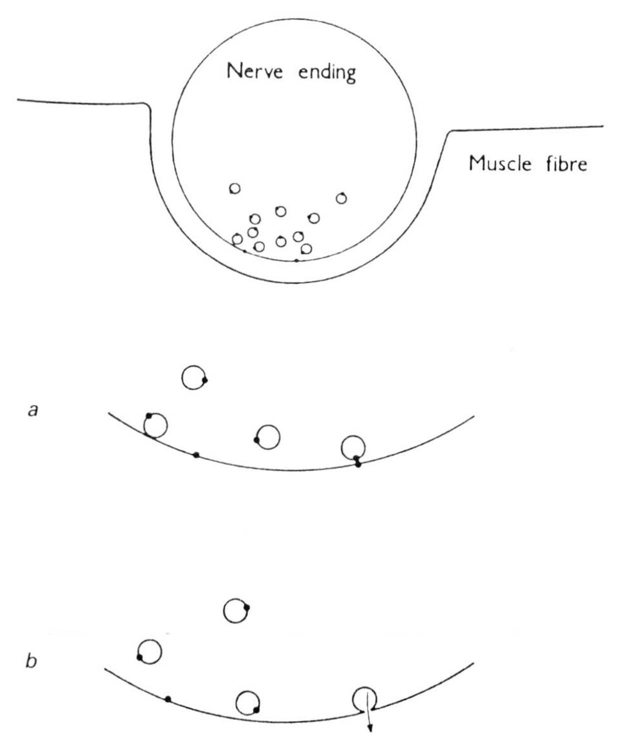

prolonged action potential in the nerve terminal. Freeze-fracture replicas showed vesicles caught in the act of exocytosis (fig. 7.25), in numbers comparable to the enlarged quantal content of the EPP.

What happens to the membrane of the vesicle after the release of its contents? Heuser and Reese (1973) found that the population of synaptic vesicles in the nerve terminal became temporarily depleted after a period of repetitive stimulation. After one minute of stimulation of the nerve at 10 per second there was a 30% loss of vesicles together with a corresponding increase in the area of the presynaptic membrane. However, after 15 min of such stimulation numerous membrane-bound vacuoles (cisternae) appeared in the terminal, whose surface area largely accounted for the 60% loss in synaptic vesicles. After a period of rest these cisternae disappeared and the vesicles reappeared. Heuser and Reese concluded that there is a recycling of the vesicular membrane after exocytosis.

More evidence for this view came from the use of horse-radish peroxidase as a marker in the extracellular space. This enzyme oxidizes a variety of aromatic compounds to produce reaction products which combine with osmium tetroxide to produce electron-dense material, so its position in the cell can be detected by electron microscopy. After stimulation it appeared inside the nerve terminal in vesicles and cisternae away from the active zones, indicating uptake of plasma membrane and some extracellular material ('endocytosis'). Some of these vesicles are coated with fibrous material (probably the protein clathrin), and endocytosis probably begins with the formation of invaginations of the presynaptic membrane called coated pits. There may also be some invagination of non-coated membrane when stimulus rates are high, as is shown in fig. 7.26 (Miller and Heuser, 1984).

By treatment of the neuromuscular junction with albumin solution before fixation with osmium tetroxide, Gray (1978) was able to show that the synaptic vesicles in the nerve terminals are closely associated with microtubules. These run either parallel to the active zones or across them, hence it

Fig. 7.24. Diagram to show part of the structure of the frog neuromuscular junction. The postsynaptic membrane is thrown into junctional folds at right angles to the long axis of the axon terminal branch. Acetylcholine receptors are arranged in this membrane at the crests and upper parts of these folds, but are absent from their deepest regions. Acetylcholinesterase is found in the intercellular material (the basal lamina, not shown) at the base of these folds. The presynaptic membrane shows active zones: there are dense bars opposite the folds and synaptic vesicles line up on each side of them, probably along microtubules.

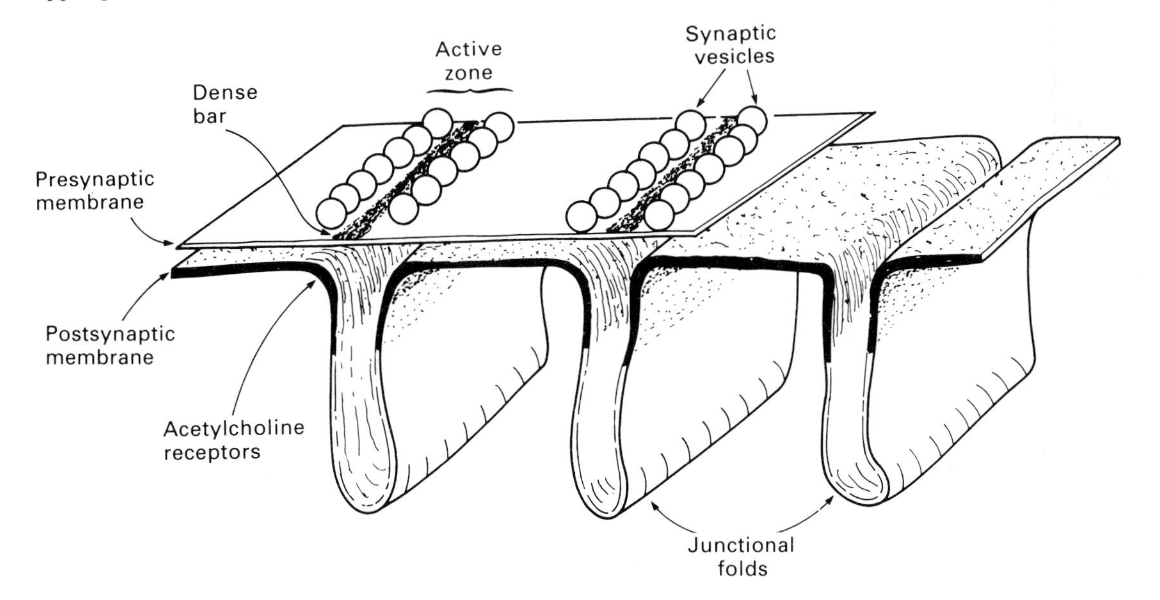

looks as though they may be involved in the transport of vesicles to their release sites.

Presynaptic events producing transmitter release

We have seen that the arrival of a nerve impulse at the motor nerve terminal results in the release of a number of quantal packets of acetylcholine from the terminal. The next question to be raised is the nature of the connecting links between these two phenomena.

Fig. 7.25. Synaptic vesicle exocytosis caught in the act by quick freezing of a frog neuromuscular junction. The muscle was slammed into a copper block cooled to −269 °C by liquid helium, then freeze-fractured at −105 °C, etched (i.e. ice on the surface was allowed to evaporate) and then a replica of the exposed surface was made by evaporation of platinum and carbon onto it. The photographs show transmission electron micrographs of such replicas, showing the cytoplasmic face of the presynaptic membrane. The double row of membrane particles in each case represents an active zone. The nerve was stimulated just before cooling: 3 ms before in *a* and 5 ms in *b*. The 'holes' in *b* are thought to be openings into synaptic vesicles that were frozen while they were discharging their contents into the synaptic cleft. Magnification 143 000 times, i.e. 1 mm is equivalent to 70 Å. (Photographs kindly supplied by Professor J. E. Heuser, from Heuser *et al.*, 1979.)

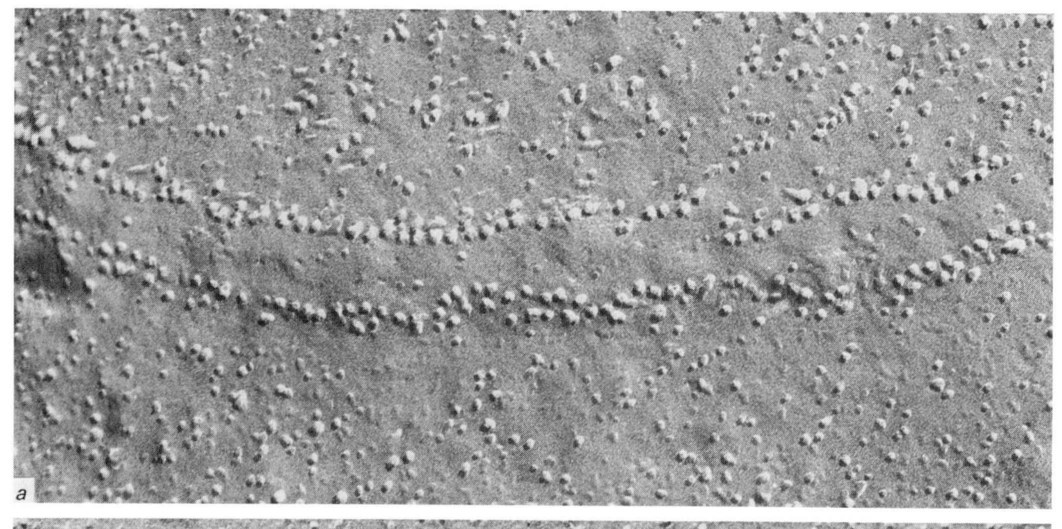

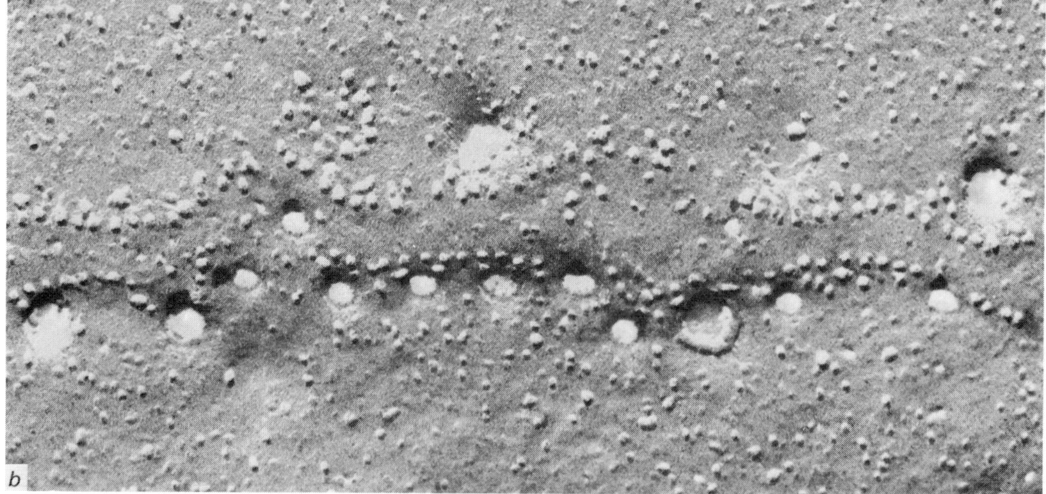

Synaptic delay

When an impulse arrives at the motor nerve ending, there is a short delay before the EPP arises in the muscle fibre. This delay is a general property of chemically transmitting synapses, and has been investigated at the frog sartorius neuromuscular junction by Katz and Miledi (1965*a, b*). The essence of their technique was to use an extracellular microelectrode applied closely to a point on the terminal so as to be able to make 'focal recordings' of the presynaptic action potential and the postsynaptic EPP from localized points at the junction.

One possible explanation of synaptic delay was that the action potential in the motor nerve axon does not invade the terminal endings, so that the depolarization there would be a slower process, caused by electrotonic spread from the myelinated region of the axon. However, Katz and Miledi found that the presynaptic spike recorded from the myelinated terminals occurred at later times after the stimulus as the recording electrode was moved distally, and also that antidromic* impulses could be set up in the axon by focal stimulation of the terminals. Hence they concluded that the action potential in the axon does invade the terminals.

The focal recording technique was then applied to measuring accurately the extent of the delay at any particular point; the method Katz and Miledi used for this is extremely ingenious. The preparation was kept in a calcium-free Ringer solution containing 1 mM magnesium ions; this completely blocked neuromuscular transmission but did not prevent presynaptic impulse conduction. The recording microelectrode was filled with a 0.5 M solution of calcium chloride, so that diffusion of calcium ions from the electrode (which could be controlled electrophoretically) allowed transmission to occur at, and only at, the point of recording.

* An *antidromic* impulse is one travelling in the opposite direction to that which normally occurs in the animal. An *orthodromic* impulse is one which travels in the same direction as naturally occurring impulses.

Fig. 7.26. Recycling of vesicle membrane at the frog neuromuscular junction. Vesicles discharge their contents at the active zones, where their membranes fuse with the presynaptic membrane. Vesicular membrane (including intramembrane particles) is retrieved via coated pits to form coated vesicles, which may fuse with cisternae from which synaptic vesicles are reformed (black arrows). After intensive stimulation uncoated pits may form, retrieving excess membrane but not selecting intramembrane particles (white arrows). (After Miller and Heuser, 1984.)

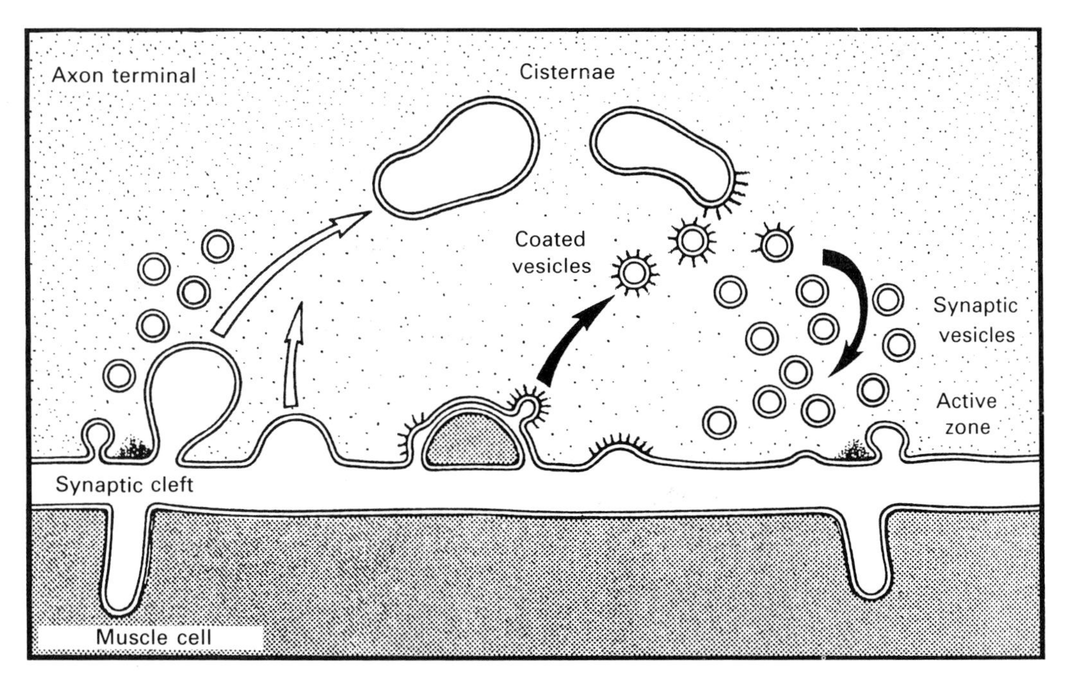

Thus the focal electrode recorded the presynaptic action potential at a point on the nerve terminal and the EPP produced immediately below it; there could be no interference from transmitter action at distances away from this point, where presynaptic depolarization would occur at different times. Fig. 7.27*a* shows the form of the record obtained by this means, and fig. 7.27*b* shows a number of successive actual records. The size of the EPPs produced fluctuates in a step-wise manner and may fail completely, as is to be expected from the quantal release hypothesis. More interesting is the fact that the synaptic delay is shown to be variable. By measuring the delays of a large number of responses it was possible to measure the probability of transmitter quanta being released at various times after the presynaptic impulse. The results of this experiment are given in fig. 7.28, showing that the minimum synaptic delay is about 0.5 ms at 17 °C.

We now have the problem of what the nature of the synaptic delay is. Katz and Miledi considered three possibilities: delay between depolarization of the axon terminals and release of acetylcholine, the time taken for the acetylcholine to diffuse across the synaptic cleft, and the time between the impact of acetylcholine on the postsynaptic membrane and the subsequent depolarization. They showed that the time needed for acetylcholine molecules to diffuse the distance of 500 Å across the synaptic cleft must be very small, and certainly less than 0.05 ms. When acetylcholine was applied ionophoretically to the endplate, the consequent postsynaptic depolarization began within 0.15 ms of the start of the ejecting current pulse. Hence it was concluded that at least 0.3 ms, the major part of the synaptic delay, is needed for the release of acetylcholine from the presynaptic nerve terminals.

The depolarization of the presynaptic terminal

As mentioned in chapter 5, the puffer fish poison tetrodotoxin blocks nervous conduction. A frog sartorius nerve-muscle preparation treated with tetrodotoxin will not therefore show the normal transmission process when the nerve is stimulated in the normal manner, because the axon and terminal membranes have been made inexcitable. However, Katz and Miledi (1967*a*) showed that if the terminals of such a preparation are briefly depolarized by a current pulse applied via a microelectrode, an end-plate potential is elicited in the muscle fibre. Thus the depolarization of the

Fig. 7.27. Records obtained from a frog neuromuscular junction with the focal recording technique, details of which are given in the text. A downward deflection indicates negativity at the focal electrode. *a*: Form of a typical record, showing the components of the response and the method of measuring the synaptic delay. *b*: Part of a series of such responses; two traces are superimposed on each record. (From Katz and Miledi, 1965*b*, by permission of the Royal Society.)

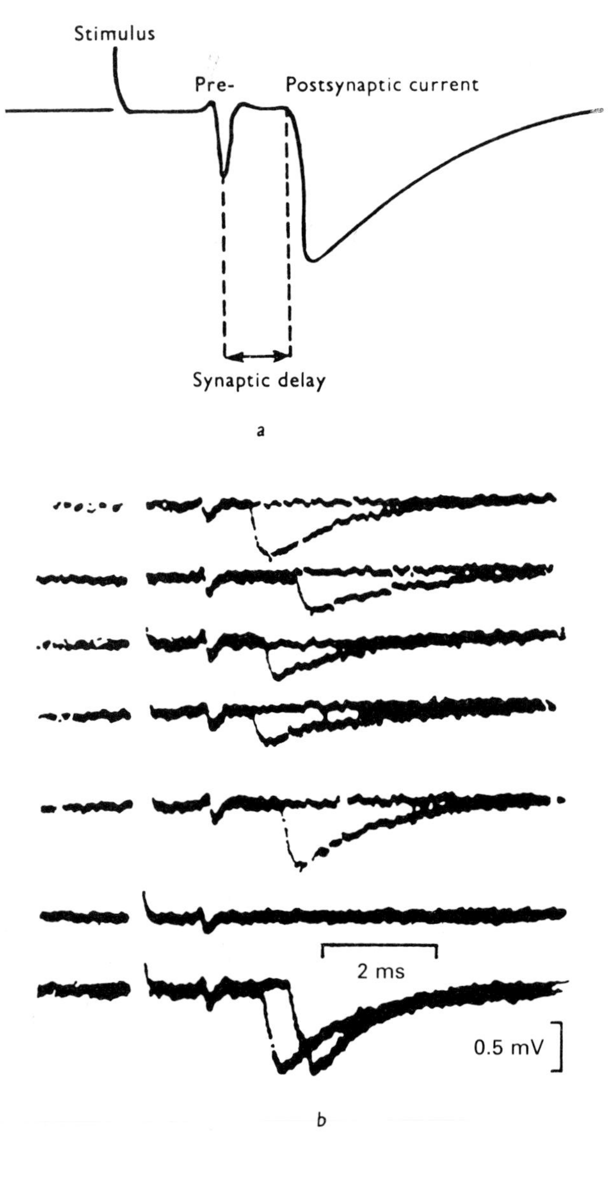

terminal membrane, however it is produced, appears to be the effective agent in producing transmitter release.

The giant synapse in the stellate ganglion of the squid has very distinct advantages for the study of presynaptic events in that the presynaptic axon is large enough for an intracellular microelectrode to be inserted into it (fig. 7.29). Under normal conditions, an action potential in the presynaptic fibre produces an action potential in the postsynaptic fibre, with a delay of about 0.4 ms at 21 °C. However, if the preparation is fatigued by repetitive stimulation, or synaptic block is induced by magnesium ions or by hyperpolarization of the postsynaptic fibre, a postsynaptic potential can be recorded from the postsynaptic fibre (Bullock and Hagiwara, 1957; Hagiwara and Tasaki, 1958).

Fig. 7.28. Frequency distribution of synaptic delays in a large number of responses of the type shown in fig. 7.27. (From Katz and Miledi, 1965*b*, by permission of the Royal Society.)

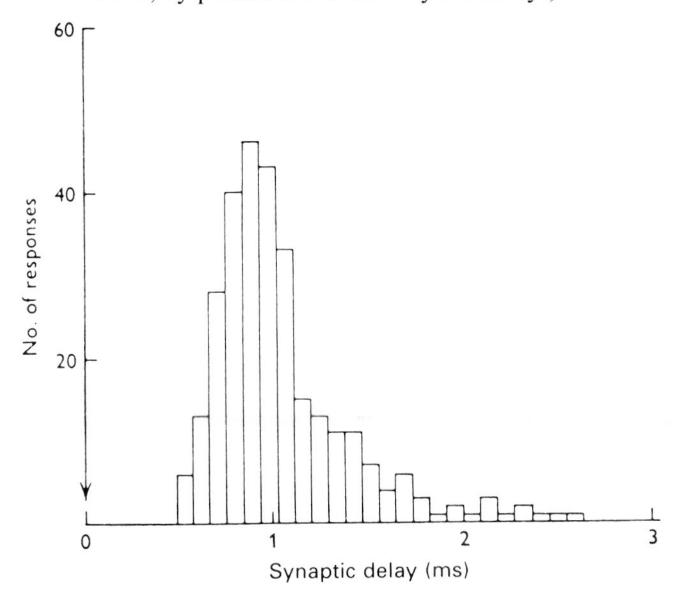

Fig. 7.29. The anatomy of the squid stellate ganglion, showing the positioning of electrodes for recording pre- and postsynaptic activity in the giant axons. (From Bullock and Hagiwara, 1957.)

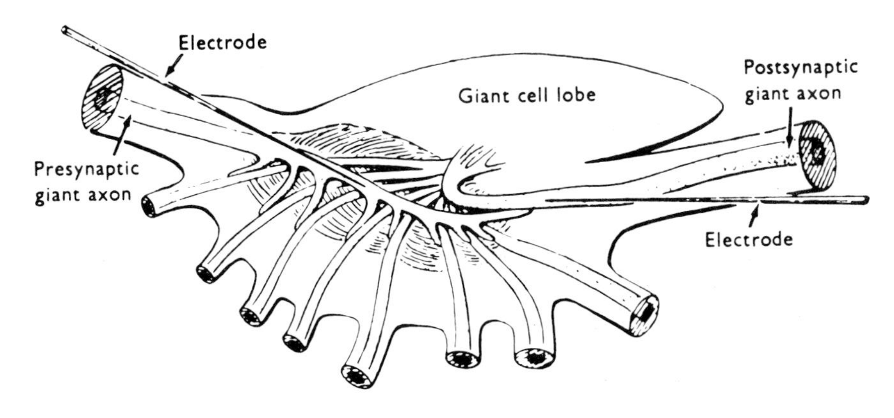

Katz and Miledi (1966) determined the input–output relations of the squid giant synapse using tetrodotoxin to eliminate the nerve action potentials. Presynaptic depolarizations of 40 mV or more produced postsynaptic responses, as is shown in fig. 7.30.

The role of calcium ions

Synaptic transmission is dependent upon calcium ions in the external medium and is, as we have seen, much reduced when the magnesium:calcium ratio is high. This suggests that calcium ions may be involved in the release of transmitter. Direct evidence for this idea has been provided by Miledi (1973), using the squid giant synapse. Because of the large size of the presynaptic axon he was able to inject calcium ions into it via an intracellular micropipette. This procedure caused a depolarization of the postsynaptic axon, indicating that raising the presynaptic intracellular calcium ion concentration causes release of the transmitter substance.

We have seen in the previous chapter that axon terminals are rich in calcium channels which are opened by depolarization (Llinas *et al.*, 1981*a*). One might therefore suppose that transmitter release is triggered by calcium flowing in through these channels, which are themselves opened by the presynaptic action potential. This idea can be tested by applying calcium ions electrophoretically to the terminals of a tetrodotoxin-treated preparation in calcium-free Ringer solution

Fig. 7.30. The relation between presynaptic membrane potential and postsynaptic response in the squid giant synapse. Currents of various strengths were passed through the presynaptic terminal, while microelectrodes recorded the pre- and postsynaptic membrane potentials. The preparation was treated with tetrodotoxin to eliminate action potentials. (From Katz and Miledi, 1966.)

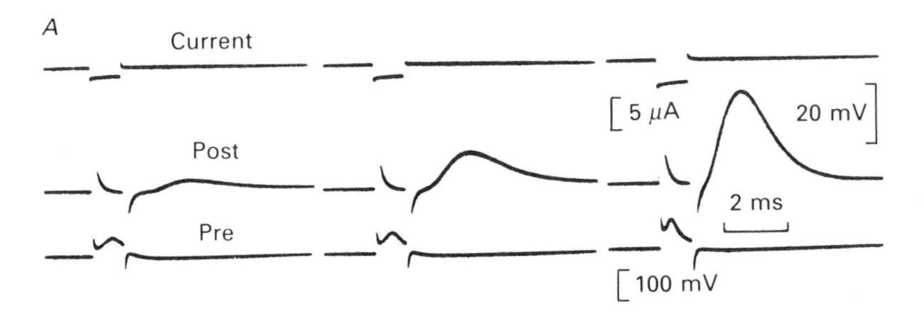

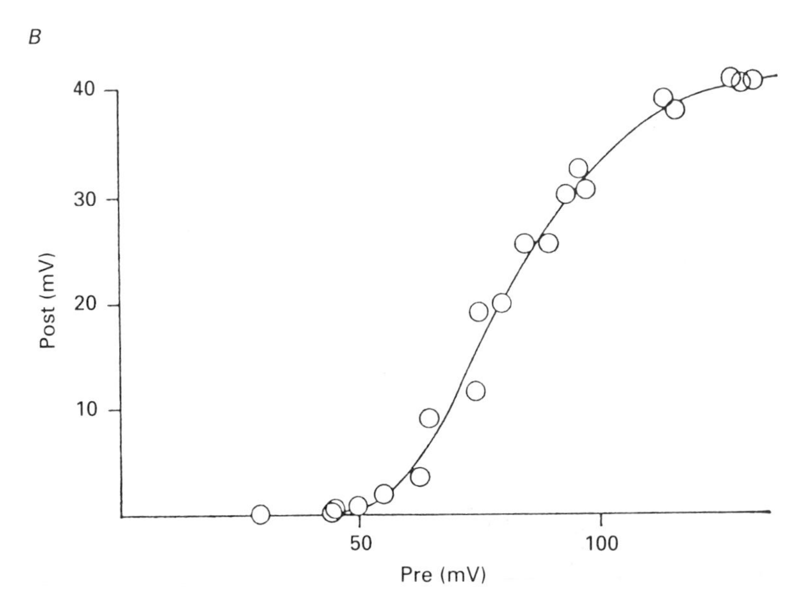

at various times before and after the application of depolarizing pulses (Katz and Miledi, 1967*b*). It was found that application of calcium ions before the depolarizing pulse (the interval between the two pulses could be as little as 50 μs) resulted in transmission, whereas application of calcium ions after the depolarizing pulse did not.

Del Castillo and Katz supposed that quantal release occurs when calcium ions react with a substance X and that the release rate should be proportional to the concentration of the intermediate CaX. However, the size of the EPP is a nonlinear function of the external calcium concentration, so Dodge and Rahamimoff (1967) proposed that several calcium ions needed to act cooperatively to produce release, perhaps by the formation of an intermediate Ca_nX, in which case

$$\text{EPP amplitude} = K([Ca^{2+}]_{out})^n$$

or

$$\text{log amplitude} = \log K + n\log[Ca^{2+}]_{out} \quad (7.4)$$

where K is a constant.

Dodge and Rahamimoff found that the maximum value of n at the neuromuscular junction was 3.8, suggesting that four calcium ions cooperate to cause release of a quantum. Cohen and van der Kloot (1985) suggest that (7.4) is a simplification and that the actual value of n is much greater than four, although this cannot normally be detected. Llinas, Steinberg and Walton (1981*b*) found a linear relation between calcium entry and transmitter release in squid axon terminals.

Just how the calcium promotes release of vesicles once it is inside the terminal is not yet clear. Perhaps electrostatic effects are involved (Silinsky, 1985), perhaps enzymic cascades and actomyosin-like proteins (Moskowitz and Puszkin, 1985).

Facilitation and depression

If a curarized rat diaphragm muscle is repetitively stimulated through its motor nerve, it is found that successive EPPs decline in size, as is shown in fig. 7.31*a*. This phenomenon is known as *neuromuscular depression.* However, if the calcium ion concentration is lowered, the reverse effect, known as *neuromuscular facilitation*, occurs (fig. 7.31*b*). The mechanisms of these effects were investigated by del Castillo and Katz (1954*c*), using a frog toe muscle. In solutions with high magnesium concentrations, they found that the proportion of failures of transmission decreased during a train of stimuli. If the stimuli were paired, the number of failures following the second stimulus was less than that following the first stimulus. These experiments indicate that facilitation is a presynaptic phenomenon; it is produced by an increase in the probability of discharge of acetylcholine quanta. When transmission was blocked by curare (which, of course, does not itself reduce the number of quanta released per impulse), depression was observed, and it was found that the later EPPs in a train showed fluctuations in amplitude. Here, again, it is evident that depression is mainly a presynaptic phenomenon, although it is possible that desensitization may be important in some cases after prolonged stimulation.

What is facilitation caused by? Katz and Miledi (1968) suggested that some of the calcium entry during the first nerve impulse remains active and adds to that entering with the second impulse so as to produce a larger response. They tested this idea using the ionophoretic application of calcium ions, and found that facilitation is much enhanced if calcium is present during the first nerve impulse.

Fig. 7.31. Facilitation and depression in curarized rat diaphragm muscle, stimulated at 120 shocks s^{-1}. *a*: External calcium ion concentration 2.5 mM, showing depression. *b*: External calcium ion concentration 0.28 mM, showing facilitation. (From Lundberg and Quilisch, 1953.)

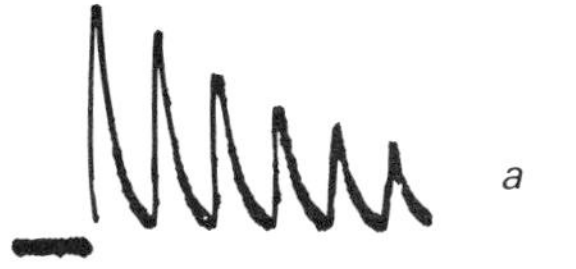

The fourth-power relation between calcium entry and release has an interesting implication for the 'residual calcium' hypothesis. Suppose the first impulse raises the calcium level of the active zones to some value q, from which it declines rapidly to $0.2\,q$ and then much more slowly. The release rate at $0.2\,q$ will be proportional to $(0.2)^4\,q$, i.e. about 1/600 of its value at q. A second impulse will raise the calcium level to $1.2\,q$ and so the release rate will be proportional to $(1.2)^4\,q$, approximately twice its value for the first impulse (Katz and Miledi, 1968).

More recent analyses suggest that there are at least four different facilitatory processes, separable by their time courses and their differential responses to strontium and barium ions (Magleby and Zengel, 1982). Facilitation proper is seen as a two-stage process, with recovery time constants of tens and hundred of milliseconds respectively. Augmentation has a recovery time constant of a few seconds, and potentiation (previously known as post-tetanic potentiation: Lloyd, 1949; Hubbard, 1963) lasts for several seconds to some minutes. Augmentation and potentiation are normally only seen after a series of presynaptic impulses.

Non-quantal release of acetylcholine

In addition to the relatively massive release of acetylcholine from motor nerve endings on stimulation, there is also a continuous low-level release from resting muscles (Mitchell and Silver, 1963; Fletcher and Forrester, 1975). In rat diaphragm muscles, for example, Fletcher and Forrester found values of 0.65 pmole min^{-1} over a 30-minute period.

Can the spontaneous release of quanta, producing miniature EPPs, account for this effect? Assume that each quantum contains 10 000 molecules, that the frequency of miniature EPPs is 3 per second, and that there are 1000 end-plates per muscle. Then the amount of acetylcholine in spontaneous quantal release should be $(3 \times 60 \times 10\,000 \times 1000)$, i.e. 1.8×10^9 molecules / min^{-1} or about 0.03 pmole min.^{-1} This is less than a twentieth of the actual value. Using a slightly different approach, Vizi and Vyskocil (1979) conclude that quantal release represents not more than 1% of the total release of acetylcholine from the resting muscle.

Does this steady 'leak' of acetylcholine come from the resting presynaptic terminal and would it produce any postsynaptic depolarization? Katz (1969) considered that any such depolarization would be very small and likely to be undetected, but a closer look by Katz and Miledi (1977) did produce results. They used muscles treated with the anticholinesterase DFP to enhance the effect, and applied a massive dose of curare to the endplate by ionophoresis. This produced a small hyperpolarization of about 40 μV for a few seconds, which we can attribute to a temporary block of the acetylcholine receptors.

In further experiments, Katz and Miledi (1981) concluded that stimulation does not produce any increase in the rate of nonquantal leakage of acetylcholine. The effect may be an accidental consequence of the presence of acetylcholine in the terminal, but it is also possible that it may serve some functional purpose in the development or maintenance of the nerve-muscle interconnection.

8

The nicotinic acetylcholine receptor

In the previous chapter we saw that the postsynaptic response at the neuromuscular junction is mediated by discrete molecular entities, the acetylcholine receptors, to which acetylcholine binds, and that the response itself consists of ionic flow through the ionic channel. Structural studies, which we shall look at later in this chapter, show that the acetylcholine binding sites and the ionic channel are part of the same macromolecular unit, the acetylcholine receptor.

More precisely, we should call this receptor the *nicotinic* acetylcholine receptor, since there is another type (the muscarinic acetylcholine receptor, described in chapter 10) which has rather different properties. Nicotinic receptors are activated by nicotine and blocked by curare, whereas muscarinic receptors are activated by muscarine and blocked by atropine.

In this chapter we consider first physiological studies, which have concentrated largely on the gating and properties of the ionic channel, and then the structure of the receptor as a whole, as revealed by biochemical and molecular biological approaches.

Single-channel responses

We have seen that acetylcholine receptors are located on the postsynaptic membrane at the neuromuscular junction, and we have assumed (p. 120) that the end-plate current is the sum of the currents through a large number of individual channels each associated with an acetylcholine receptor. Can we obtain information about these single-channel currents? Some answers to this question began to appear in the 1970s, as a result of the two powerful techniques of noise analysis and patch clamping. We have already met these methods in our study of voltage-gated channels (chapter 6), but they were used early in their development to study the acetylcholine-gated channels involved in neuromuscular transmission.

Let us start with some simple assumptions. Del Castillo and Katz (1957) suggested that receptor activation is a two-stage process in which the acetylcholine (A) first combines with the receptor (R) to form a complex AR and that this then undergoes a conformational change which opens the ionic channel. The channel-open state is represented by AR*. We assume that these reactions are reversible and are governed by different rate constants (k_1, k_{-1}, α and β), to give an overall reaction scheme

$$\underset{\text{closed}}{\mathrm{A+R}} \underset{k_{-1}}{\overset{k_1}{\rightleftharpoons}} \underset{\text{closed}}{\mathrm{AR}} \underset{\alpha}{\overset{\beta}{\rightleftharpoons}} \underset{\text{open}}{\mathrm{AR^*}} \qquad (8.1)$$

More recently, as we shall see later, it has become clear that two molecules of acetylcholine combine with each receptor molecule. The reaction scheme then becomes

$$\underset{\text{closed}}{\mathrm{2A+R}} \underset{k_{-1}}{\overset{2k_1}{\rightleftharpoons}} \underset{\text{closed}}{\mathrm{A+AR}} \underset{2k_{-2}}{\overset{k_2}{\rightleftharpoons}} \underset{\text{closed}}{\mathrm{A_2R}} \underset{\alpha}{\overset{\beta}{\rightleftharpoons}} \underset{\text{open}}{\mathrm{A_2R^*}} \qquad (8.2)$$

The factor of two for the rate constants k_1 and k_{-2}, which both lead to the intermediate AR, arises because there are two possible forms of AR according to which of the two binding sites is occupied.

Schemes (8.1) and (8.2) imply that the ionic channel is a two-state system: it can be either open or closed, and the conductance when it is open is constant. We cannot predict what the duration of any particular opening by a single channel will be since the rate constants are probabilistic rather than determinate. But the schemes do imply that the probability of an open channel closing in any one time interval is constant. This means that if we measure a large number of channel openings, their durations (the time between opening and closing) will be exponentially distributed with a mean value of $1/\alpha$.

Acetylcholine noise

The depolarization produced by acetylcholine is the sum of the effects of a very large number of random collisions between acetylcholine molecules and receptor molecules. We would expect there to be random fluctuations in the rate of these molecular collisions, which would produce corresponding fluctuations – or 'noise' – in the resulting depolarization. Katz and Miledi (1970, 1972) found that such fluctuations are indeed detectable, as can be seen in fig. 8.1. The phenomenon has also been investigated by Anderson and Stevens (1973) with the voltage-clamp technique; this allowed the noise (measured as fluctuations in current) to be measured at different membrane potentials, independently of the acetylcholine concentration. Let us examine some of their results.

We can estimate the single-channel conductance by measuring the variance of the current. This conclusion depends upon a simple relation between the mean and the variance in a binomial distribution, as described in chapter 6. Suppose there are a total of N channels of which r are open at any one time. Suppose the probability of any particular channel being open is p. Then

$$\text{mean value of } r = Np$$

and

$$\text{variance of } r = Np(1-p).$$

The membrane conductance G will be the number of open channels (r) times the single-channel conductance γ, i.e.

$$G = \gamma r$$

Fig. 8.1. End-plate 'noise' produced by acetylcholine. The upper trace shows the membrane potential recorded at high gain from the end-plate region of a frog muscle fibre, with no acetylcholine (ACh) present. The lower trace is a similar record obtained when acetylcholine was applied by diffusion from a nearby micropipette. (From Katz and Miledi, 1972.)

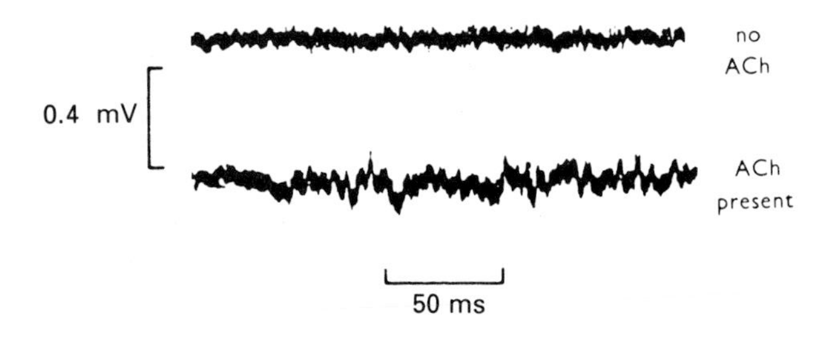

so

$$\text{mean value of } G = \gamma Np$$

and

$$\text{variance of } G = \gamma^2(\text{variance of } r) = \gamma^2 Np(1-p)$$

hence

$$\frac{\text{variance of } G}{\text{mean of } G} = \gamma(1-p).$$

If p is very low (as it would be in low acetylcholine concentrations) this simplifies to give

$$\frac{\text{variance of } G}{\text{mean of } G} = \gamma. \tag{8.3}$$

Fig. 8.2 shows some results obtained by Anderson and Stevens. There is clearly a linear relation between the mean and the variance of the membrane conductance. The slope of this gives a value for the single channel conductance of 19 pS. In another fibre where a particularly good voltage clamp was achieved, a value of 32 pS was obtained.

Fig. 8.2. Determination of the single-channel conductance from acetylcholine noise measurements. Each point is derived from a voltage-clamp record of the current through the membrane, from which the mean and the variance of the membrane conductance (G) can be determined. The slope of the relationship gives the single-channel conductance, from (8.3). (From Anderson and Stevens, 1973.)

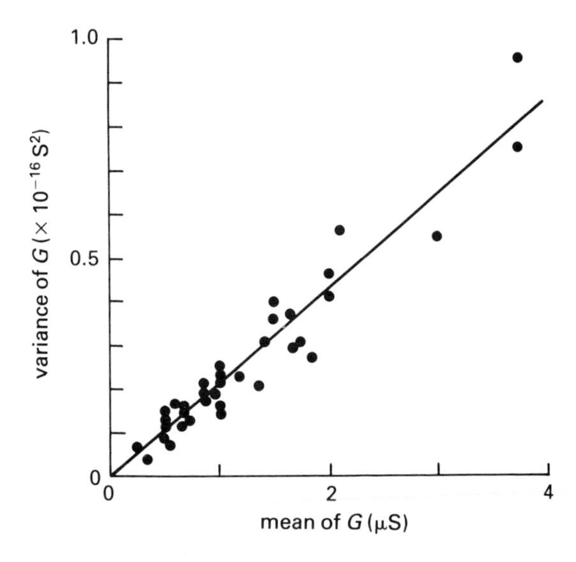

Information about the duration of channel openings can be derived from a frequency analysis of the noise. The current fluctuations are digitised and subjected to Fourier analysis by computer. This leads to a spectral density curve (or 'power density spectrum') in which the density at any particular frequency is the sum of the squared amplitudes of the sine and cosine components of the current fluctuations at that frequency. The spectral density curve then shows the contribution of different frequencies to the noise, as is shown in fig. 8.3. It can also – and this is its really useful property – give us some information about the time characteristics of the individual events which cause the fluctuations.

(Should there be any readers who regard the last sentence as nonsensical magic, I invite them to consider the Analogy of the Fleas. Suppose the platform of a small rapid-reading balance is placed in a box in which there are a large number of fleas jumping about at random. Each flea remains still for a short period of time (for an average of 100 ms, let us say) between jumps. Clearly the number (and hence weight) of fleas on the platform will fluctuate, and so the continuous record of weight will be 'noisy'. If we measure the weight of fleas on the platform every second we will find that successive measurements are essentially independent of one another because almost all the fleas on the platform at one particular time will have jumped off when we look again one second later. Hence the variance of the difference between successive weights recorded at one-second intervals will be maximal. The same will apply for measurements taken at 10-second intervals. Another way of describing this situation is to say that the low frequency components of the noise are large. However, if we measure at intervals of 1 ms, each successive measurement will be very close to the previous

Fig. 8.3. Spectral density curve (or 'power density spectrum') of current fluctuations produced at the end-plate by acetylcholine. The membrane potential was clamped at −60 mV. The curve is drawn according to (8.4) with $\alpha' = 132\ \mathrm{s}^{-1}$. (From Anderson and Stevens, 1973.)

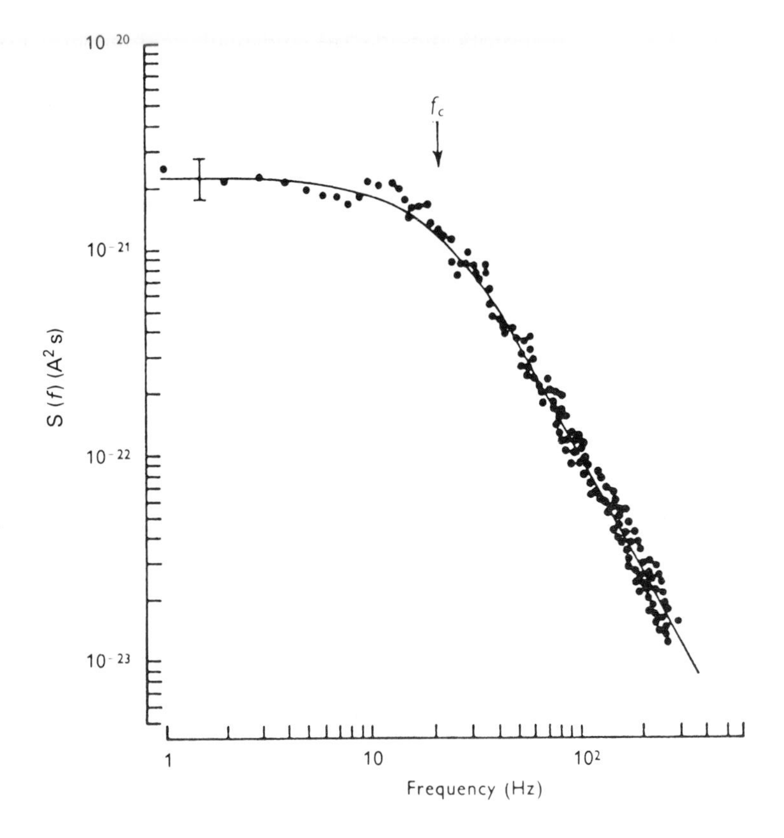

one, because only about one hundredth of the fleas will have jumped off in the meantime. Hence the variance of the differences between successive weights recorded at 1 ms intervals will be low, that is to say the high-frequency components of the noise are small. At intermediate frequencies we will see corresponding intermediate values for the components of the noise, and so a graph of these components against frequency might well look like that in fig. 8.3. Notice that the contribution of any particular high frequency to the noise depends upon the time for which the fleas are still between jumps. If we warm them up, for example, so that this time is reduced to 10 ms, then measurements at 1 ms intervals will show larger differences between successive measurements because now about one tenth of the fleas will have jumped off in the meantime instead of only about one hundredth. Clearly there must be information in the frequency characteristics of the noise which can tell us about the length of time between jumps. For fleas read acetylcholine molecules, for platform read postsynaptic membrane and for weight read current. Now read on.)

If we assume that the acetylcholine noise is produced by a system such as that described in (8.1), in which openings and closings occur at random with an exponential distribution of opening times, then it can be shown that the form of the spectral density curve should be given by

$$S(f) = \frac{S_0}{1 + (2\pi f/\alpha')^2} \tag{8.4}$$

where f is the frequency, S_0 is a constant (the value of $S(f)$ when $f = 0$) and α' is a rate constant. (Anderson and Stevens used the rate constant α from (8.1) here, implying that it can be measured from the spectral density curve, but we shall see later that this view is probably not correct, and so we shall here use α' instead.) This type of relation is known as a Lorentzian curve: $S(f)$ is maximal and the curve is flat at low frequencies, whereas at high frequencies $S(f)$ falls in proportion to $1/f^2$ (see fig. 8.3). Further details of the theory of noise analysis are give by Stevens (1972), Neher and Stevens (1977) and DeFelice (1981).

We can measure the apparent mean duration of channel opening from the spectral density curve. Let f_c be the frequency at which the spectral density falls to half its maximum value. Then

$$S(f_c) = \frac{S_0}{2} = \frac{S_0}{1 + (2\pi f_c/\alpha')^2}.$$

Hence

$$(2\pi f_c/\alpha')^2 = 1$$

and so

$$f_c = \alpha'/2\pi.$$

In fig. 8.3, for example, the value of f_c is 21 Hz, giving a rate constant α' of $0.13\ \mathrm{ms}^{-1}$. The apparent mean channel opening time is $1/\alpha'$, i.e. 7.6 ms in this case. Anderson and Stevens found that f_c became larger (and thus the apparent mean opening time became shorter) at more positive membrane potentials and at higher temperatures.

Patch-clamp records

The patch-clamp technique, which we have met in chapter 6, was invented by Neher and Sakmann (1976*a*) and first used by them to examine single channel currents through acetylcholine receptors. It had been known for some time that denervation results in a remarkable change in vertebrate muscle fibres: the whole surface becomes sensitive to acetylcholine, as if the occurrence of receptors outside the end-plate region were normally inhibited in some way by the presence of the motor axon (Axelsson and Thesleff, 1959). Neher and Sakmann pushed the polished tip of a micropipette electrode containing a low concentration of acetylcholine against the membrane of a denervated muscle fibre and so, by using a voltage-clamp circuit, were able to measure the current flow across a small patch of membrane. Since the density of the channels is much lower than at the end-plate, only a relatively small number of channels would be included in the patch and so the activity of individual channels could be recorded.

Neher and Sakmann found that each channel produces a square pulse of current lasting for up to a few milliseconds, as is shown in fig. 8.4. The single-channel currents were 2.2 pA at a membrane potential of −80 mV and 3.4 pA at −120 mV. This change of 1.2 pA per 40 mV implies that the current would be zero at −7 mV, which is thus the reversal potential. We can calculate the single-channel conductance γ from the relation

$$\gamma = i/(E - E_r)$$

where i is the single-channel current, E is the membrane potential and E_r is the reversal potential. Here, at $-80\,\text{mV}$,

$$\gamma = 2.2/0.073$$
$$= 30\,\text{pS}.$$

The first patch-clamp records, then, were in pleasing agreement with the results of fluctuation analysis. But, with improvements in techniques (Hamill *et al.*, 1981) and more extensive observations, it was not long before complications arose. Openings occurred in bursts (and sometimes the bursts were grouped in clusters), there were brief closings and brief openings, and sometimes the single channel conductance was not always the same size.

One of the first of the new observations was that single 'openings' of the channel are frequently interrupted by brief closed periods or 'gaps' (Colquhoun and Sakmann, 1981), such as is shown in fig. 8.5. This means that the events are actually bursts of openings. The phenomenon had actually been predicted by Colquhoun and Hawkes (1977, 1981) on the basis of scheme (8.1): if β is not much less than k_{-1} then AR is quite likely to revert to AR* instead of dissociating to A + R (fig. 8.6). A similar argument can be applied to scheme (8.2). This means that the results of noise analysis must be reinterpreted. If the gaps within bursts are short, then the apparent opening times measured by noise analysis are actually burst length times, and the rate constant α in (8.1) or (8.2) cannot be equated to the parameter α' in (8.3).

Colquhoun and Sakmann (1985) made some careful high-resolution measurements of the time

Fig. 8.4. Single-channel currents from acetylcholine receptors at the frog muscle end-plate. Traces show the responses to acetylcholine (ACh) and its analogues suberyldicholine (SubCh), decan-1,10-dicarboxylic acid dicholine ester (DecCh) and carbachol (CCh). (From Colquhoun and Sakmann, 1985.)

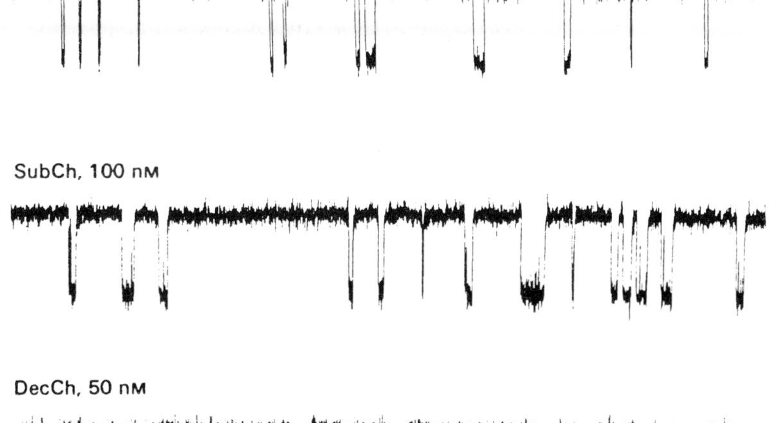

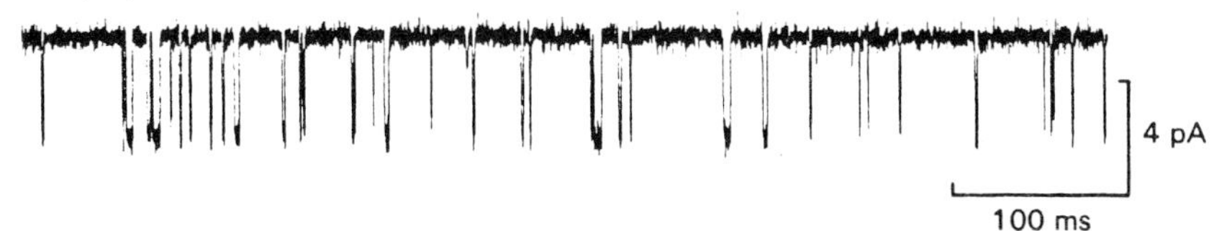

characteristics of the end-plate channel openings in frog muscle fibres, and used these to estimate some of the rate constants in scheme (8.2). An open channel A_2R^* can close only by reverting to A_2R, and the rate constant for this is α. Hence

$$\text{mean duration of channel opening} = 1/\alpha. \qquad (8.5)$$

The rate constant for the departure from the A_2R state in scheme (8.2) will be the sum of the rate constants for conversion to A_2R^* and to AR,

Fig. 8.5. Single-channel current at the frog neuromuscular junction, produced by acetylcholine, showing a burst of two openings separated by a brief gap. The two traces show the same event at low and high time resolution. (From Colquhoun and Sakmann, 1985.)

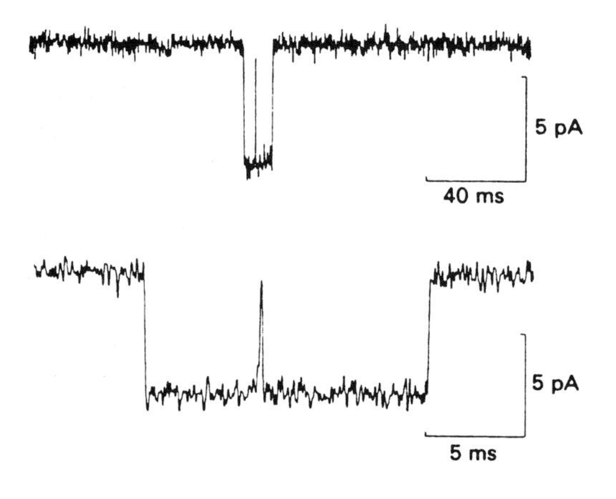

Fig. 8.6. Diagram to show how the transitions between three states, as in scheme (8.1), give rise to bursts of channel openings. (From Colquhoun and Hawkes, 1982.)

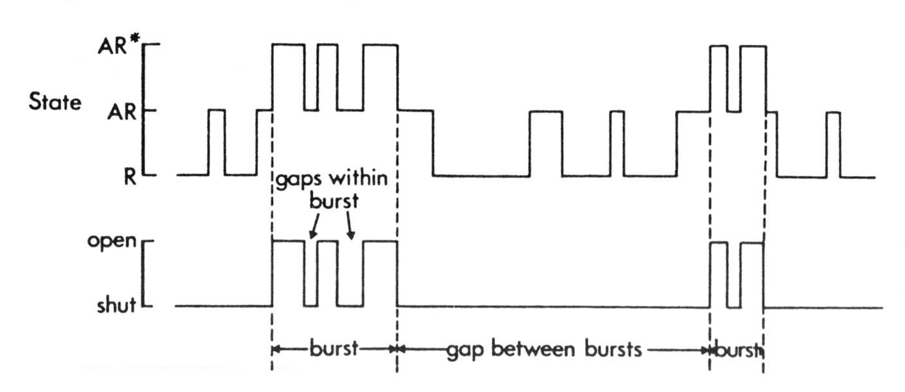

hence the time constant will be the reciprocal of this, i.e.

$$\text{mean gap duration} = 1/(\beta + 2k_{-2}). \qquad (8.6)$$

The number of gaps in a burst will be dependent upon the relative values of β and k_{-2}. If β is greater than k_{-2} the chances of A_2R reverting to A_2R^* will be relatively high and so there will be on average more openings (and so more gaps) in the burst; in fact

$$\text{mean number of gaps per burst} = \beta/k_{-2}. \qquad (8.7)$$

Equations (8.5) to (8.7) show that the rate constants in scheme (8.2) can be calculated from measurements of the time characteristics of the single channel currents. There is a technical difficulty here since we cannot measure all of the channel closings: some of them are too brief for the recording apparatus. This leads to two errors. Firstly, the apparent mean duration of gaps in bursts will be too long. Secondly, two openings separated by a very brief gap will be recorded as a single long opening, and so the mean value for the open time will be too long. Colquhoun and Sakmann met this problem by fitting an exponential distribution to all the brief gaps that they could record, and assuming that the number of even shorter gaps was as predicted from this distribution. They could thus work out the total number of brief gaps and so, knowing the total time during bursts, calculate the mean duration of channel closing.

After applying these corrections, Colquhoun and Sakmann produced values of 1.4 ms for the mean open time duration, 20 μs for the mean gap duration and 1.9 for the mean number of gaps per burst. From these they calculated that $\alpha = 714\,s^{-1}$, $\beta = 30\,600\,s^{-1}$, and $k_{-2} = 8150\,s^{-1}$.

These values for the rate constants imply that the conformational change from A_2R to A_2R^* is energetically favoured, so that a channel which has two molecules of acetylcholine bound to it will spend most of its time in the open condition. This, of course, is just what is required for effective functioning (Colquhoun and Sakmann, 1983).

There are other aspects of patch clamp records which show that scheme (8.2) is not the whole story. Colquhoun and Sakmann (1985) found that a small proportion of the gaps in a burst are appreciably longer than the rest; they were named intermediate gaps. A proportion of the channel openings are much shorter than the rest. Colquhoun and Sakmann suggest that these could be caused by channel opening sometimes occurring after only one acetylcholine molecule has been bound. In line with this idea, brief openings are relatively more frequent at low acetylcholine concentrations; but this cannot be the whole story since some still occur at high concentrations. Sine and Steinbach (1987), using a tissue-cultured cell line, did not find any such concentration effect, and propose that the brief openings arise from an alternative state ($A_2Q \rightleftharpoons A_2Q^*$) which has shorter opening times.

A number of workers have reported various 'subconductance states', in which the single-channel conductance is some fraction of the normal value (Hamill and Sakmann, 1981; Auerbach and Sachs, 1983; Colquhoun and Sakmann, 1985). These states seem to be common in cultured cells, but do also occur in mature frog muscle; Colquhoun and Sakmann found that 1–2% of all bursts contained a partial closure to either 18% or 71% of the normal open channel conductance level.

In conclusion, we can explain many of the features of the patch-clamp records of single-channel currents by means of the relatively simple model of receptor behaviour described by scheme (8.2), but there are other features which imply that the actual situation is rather more complicated.

Selective permeability of the acetylcholine channel

The experiments of the Takeuchis, referred to in the previous chapter, showed that the acetylcholine channel was selectively permeable to cations and that both sodium and potassium ions could pass through. Later work confirmed these conclusions and extended them to calcium and magnesium ions (Takeuchi, 1963*a*, *b*; Linder and Quastel, 1978; Lewis, 1979).

Some organic cations can also pass through the acetylcholine channel, and this opens the possibility of investigating its dimensions. Hille and his colleagues (Dwyer *et al.*, 1980) measured the permeability of frog muscle end-plate channels to a wide variety of organic cations with this end in view. They used frog muscle fibres with their ends cut and open to sodium fluoride solutions so that sodium was the principal internal cation. The end-plate region was in a separate pool isolated with vaseline (petroleum jelly) from the cut ends; it was perfused with a solution containing either 114 mM NaCl or 114 mM XCl, where X^+ is the organic cation. Acetylcholine was applied to the end-plate via a micropipette by ionophoresis. A voltage-clamp circuit held the membrane potential at various desired levels.

By applying acetylcholine at different membrane potentials it is easy to measure the reversal potential. In fig. 8.7, for example, the reversal potential for sodium is +4 mV and that for Tris is −39 mV, a difference of −43 mV. Let us see how such measurements can tell us something about the permeabilities of the channel to different ions.

Under the conditions of the experiments, with sodium as the only internal permeant cation, the reversal potential for sodium ions is given by

$$E_{r,Na} = \frac{RT}{F} \ln \frac{P_{Na}[Na]_o}{P_{Na}[Na]_i}$$

and that for the cation X^+ is

$$E_{r,X} = \frac{RT}{F} \ln \frac{P_X[X]_o}{P_{Na}[Na]_i}$$

where P_{Na} and P_X are permeability coefficients as in (3.5). Then the change in reversal potential on substituting X^+ for sodium ions is

$$\Delta E_r = E_{r,X} - E_{r,Na}$$

$$= \frac{RT}{F} \ln \frac{P_X[X]_o}{P_{Na}[Na]_o}.$$

Thus, if $[X]_o$ is equal to $[Na]_o$,

$$\Delta E_r = 24.66 \ln (P_X/P_{Na}) \quad \text{mV}$$

or

$$P_X/P_{Na} = \exp(\Delta E_r/24.66). \quad (8.7)$$

If we apply (8.7) to our value of ΔE_r for Tris, we get

$$P_{Tris}/P_{Na} = \exp(-43/24.66) = 0.175.$$

Dwyer and his colleagues measured the values of P_X/P_{Na} for a large number of organic cations, with the results shown by the open circles in fig. 8.8. Cations with permeability coefficients greater than P_{Na} were all smaller than 5 Å in diameter, and there was a fairly steady drop in permeability with increasing diameter, reaching zero at about 7 Å. They concluded that the acetylcholine channel must have a cross-sectional area of at least 40 Å^2 and suggested that a square 6.5 Å × 6.5 Å, as in fig. 8.9, would provide a suitable model. Since we now know that the acetylcholine receptor is composed of five subunits (p. 148), a pentagonal cross-section might be more appropriate; a regular pentagon with a height of 7 Å would have an area of 41 Å^2.

Hille and his colleagues also measured the relative permeabilities of the end-plate membrane to various metal ions, using essentially the same techniques (Adams *et al.*, 1980). The results are shown as the black symbols in fig. 8.8. Permeability ratios for the alkali metal ions fell in the sequence Cs > Rb > K > Na > Li from $P_{Cs}/P_{Na} = 1.42$ to $P_{Li}/P_{Na} = 0.87$. This order is the same as that of the mobilities of the ions in free solution. The thallous ion has an unexpectedly high P_{Tl}/P_{Na} ratio of 2.51.

The permeabilities for divalent cations are lower, and decrease in the order Mg > Ca > Ba > Sr. They are affected by the concentration of the ion; P_{Ca}/P_{Na} is 0.22 at 20 mM $CaCl_2$ but only 0.16 at 80 mM.

The relatively low selectivity between different metal ions and the relatively high values for single channel conductance suggest that the ionic channel is a water-filled pore. The reduction of divalent cation permeability at higher concentrations might be explained if there are negative charges on the external surface near the mouth of the channel (Lewis, 1979; Dwyer *et al.*, 1980). Such charges could also account for the impermeability of the channel to anions.

Fig. 8.7. Reversal potential measurements with organic cations substituted for external sodium at the neuromuscular junction. The records show currents under voltage clamp at different membrane potentials (separated by 7 mV steps) in response to brief pulses of acetylcholine. (From Dwyer *et al.*, 1980.)

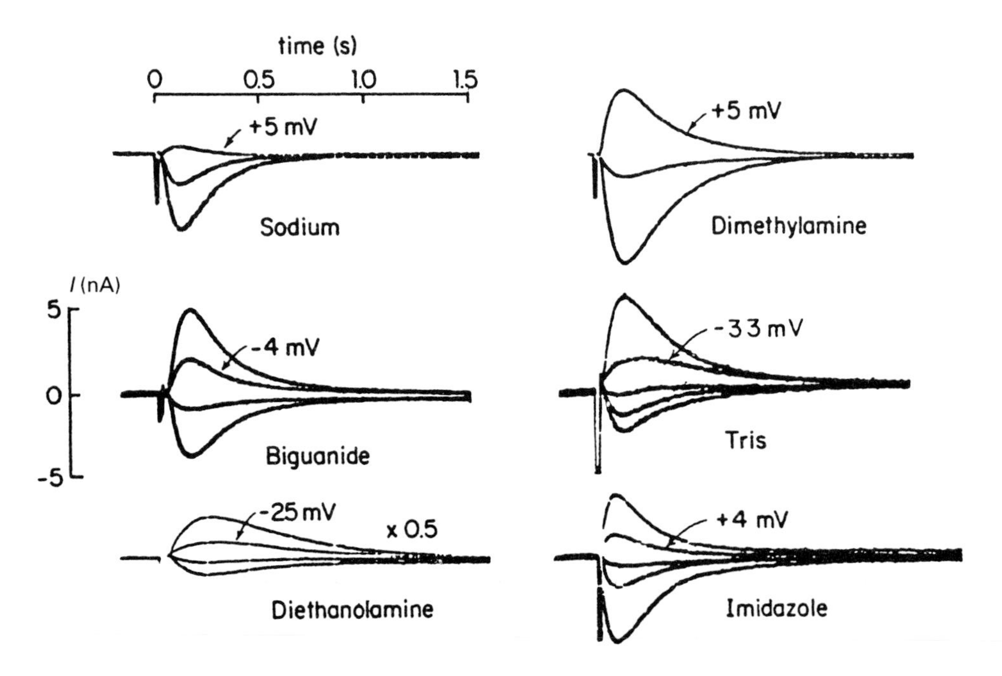

Structure of the acetylcholine receptor

The richest known source of acetylcholine receptors is the electric organ of the electric ray *Torpedo*, one kilogram of which contains over 100 mg. Lower concentrations are present in the electric organ of *Electrophorus*, the electric eel, and in vertebrate muscle fibres. Acetylcholine receptors have been isolated from all these sources, but the *Torpedo* electric organ remains the biochemists' favourite.

In order to isolate receptors from *Torpedo* electric organ it is first necessary to homogenise the tissue. Then fractionation by centrifugation in a sucrose density gradient separates the receptor-rich subsynaptic membranes, which are relatively dense. Further purification leads to small vesicles in which the receptors are embedded outside-out in the vesicular membrane. Treatment with detergents releases the receptors from the membrane components, and they can then be further purified by affinity chromatography on α-bungarotoxin columns.

Fig. 8.8. Relations between ionic diameter and the relative permeability of acetylcholine channels at the frog muscle end-plate. The three curves represent different theoretical models which assume a cylindrical pore with a diameter of 7.4 Å. (From Adams *et al.*, 1980.)

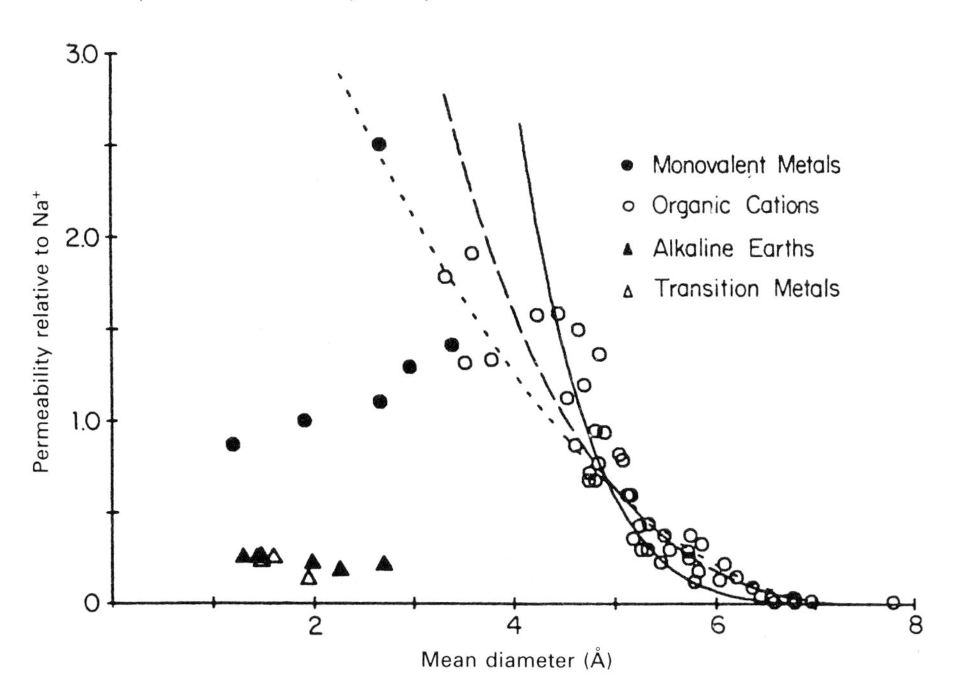

The molecular weight of the acetylcholine receptor is probably in the region of 285 to 290 kDa (Popot and Changeux, 1984); rather lower values were found in the earlier determinations. Electrophoresis on sodium dodecyl sulphate (SDS) gels shows that the receptor contains four different polypeptide chains, with apparent molecular weights of 40, 50, 60 and 65 kDA, known

Fig. 8.9. Hypothetical cross-sections of three types of ionic channels in frog nerve and muscle, based on their permeabilities to ions of different sizes. The sodium and potassium (delayed rectifier) channels occur in nerve axons at the nodes of Ranvier. (From Dwyer *et al.*, 1980.)

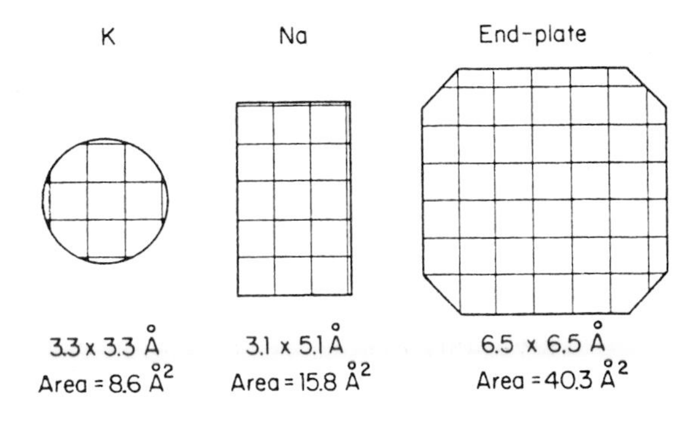

respectively as the α, β, γ and δ subunits (Weill *et al.*, 1974; Raftery *et al.*, 1976). Later it became clear that there were two α chains and one of each of the others in each receptor (Reynolds and Karlin, 1978; Lindstrom *et al.*, 1979; Raftery *et al.*, 1980). The binding sites for acetylcholine are located on the α chains (Reynolds and Karlin, 1978), hence it seems that there are at least two binding sites per receptor.

Primary and secondary structure

Knowledge of the amino acid sequence of the protein chains is clearly necessary before we can understand how the receptor works. Raftery and his colleagues (1980) used an automatic sequencing system to determine the first 54 amino acids in each of the four subunits. They found that each subunit was distinct from the others, but that there was considerable homology between them: if any two subunits were compared then 35% or more of the amino acids in corresponding positions were the same.

The next step was to determine the full amino acid sequences of the receptor subunits. With the advent of DNA cloning techniques it became easier to obtain the sequence of bases in the nucleic acids coding for the protein and then to deduce the amino acid sequence from this, than to determine the amino acid sequence directly by analysis of the protein. Laboratories in Japan, Britain, France and the United States attacked the problem more or less simultaneously (Noda *et al.*, 1982*a*; Sumikawa *et al.*, 1982; Devillers-Thiery *et al.*, 1983; Claudio *et al.*, 1983). Work by the Kyoto University group was particularly productive, so that they had soon published the amino acid sequences of all four subunits (Noda *et al.*, 1982*a*, 1983*a, b*; Numa *et al.*, 1983; see also Numa, 1986). Let us have a brief look at how this was done.

The normal sequence of information transfer in protein synthesis by cells can be stated as 'DNA makes RNA makes protein'. The genetic information in the appropriate section of the nucleotide base sequence of the DNA in the nucleus is first transcribed to form a complementary sequence of RNA bases on a primary transcript. This is then modified to give messenger RNA (mRNA), which in turn is translated into the amino acid sequence forming the protein. The first step, therefore, is to isolate mRNA from the tissue. This contains a mixture of different mRNAs, less than 1% being of the desired type.

Next, a complementary DNA (cDNA) strand is formed on each mRNA molecule by the action of reverse transcriptase, an enzyme found in retroviruses. The mRNA is then removed with alkali and the cDNA made double-stranded with the enzyme DNA polymerase I from the bacterium *Escherichia coli.* The cDNA molecules are inserted into plasmids (loops of double-stranded DNA which self-replicate in bacteria) which contain genes for antibiotic resistance, and the plasmids are taken up by *E. coli* cells (fig. 8.10). The bacteria are then plated on a medium containing antibiotic, which ensures that only those cells containing plasmids will grow. Thus a number of small colonies of bacteria are produced, each one being a clone which contains many copies of its own particular piece of the cDNA. In their determination of the primary structure of the α subunit, Noda and his colleagues (1982) used 2.3 μg of mRNA to produce a library of about 200 000 clones of cDNA.

Only a few of these clones will contain cDNA coding for the desired protein, so there is a need to select these particular ones out of the mass. For this screening programme, we need a probe which has something in common with the sequence we are seeking. Noda and his colleagues synthesized small pieces of DNA corresponding to the amino acid residues 25 to 29 and 13 to 18 of the α subunit in the sequence determined by Raftery *et al.* (1980), and labelled them with radioactive phosphorus. These probes would hybridize only with complementary sequences in the DNA library. A nitrocellulose filter is pressed onto each plate of *E. coli* clones, bringing samples of the clones away with it on removal (fig. 8.11). The samples are then lysed and heated to expose single-stranded cDNA and the radioactive probes are applied to them: binding occurs only with complementary sequences in the cDNA, and excess probes are washed away. The position of the probes on the filter is then detected by autoradiography; this indicates which clones contain the required cDNA and so they can be removed from the original plate and cultured so as to produce sufficient cDNA for sequencing (a few nanograms is enough). Noda and his colleagues found 20 clones that would hybridize with both their

Fig. 8.10. Production of clones of cDNA in *E. coli* from *Torpedo* electric organ mRNA.

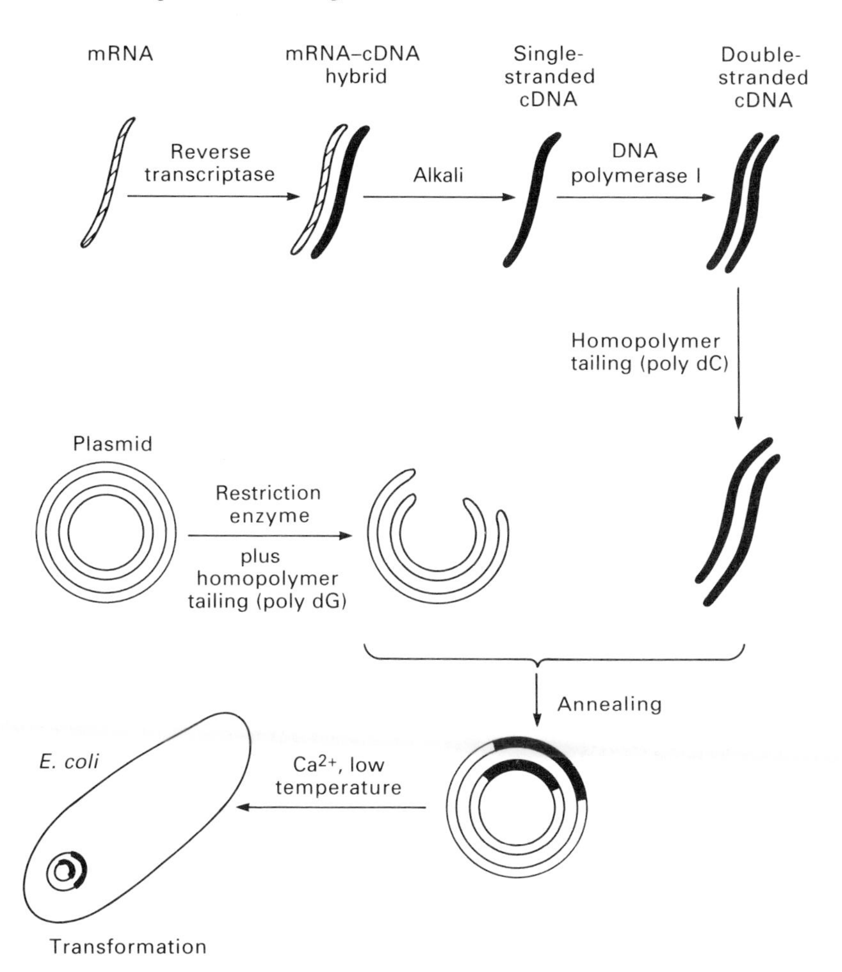

Fig. 8.11. Screening of cDNA clones from *Torpedo* electric organ for the acetylcholine receptor α subunit. Further details in the text.

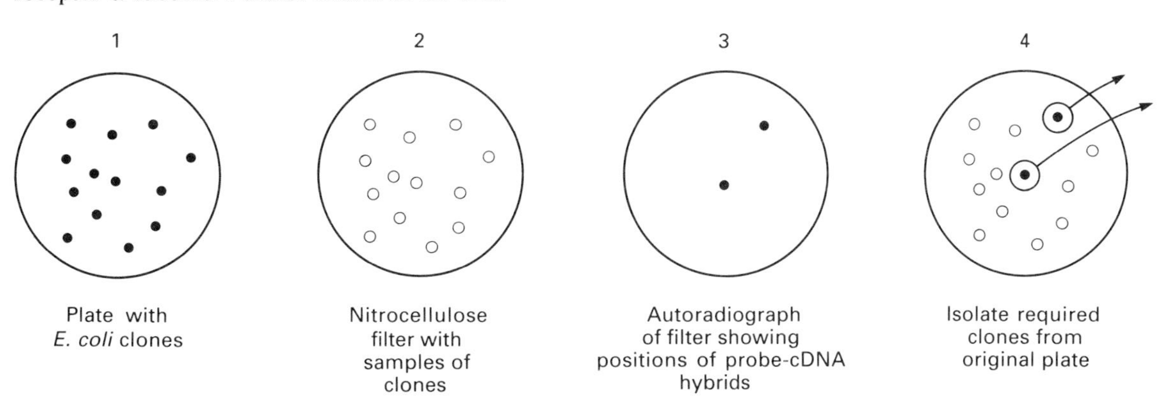

5'---CAGCUAUCAGCUGUCGCUGAGACAGGUGGCAUAAGAGUGGAACAGAGAGUUGAAAAGGCAGGAAACUGGCUUAUCUCUUCACUAGAAAAGAGCUGAACACAGAAGUCCAGAAGAU
-240 -220 -200 -180 -160

-20
Met Ile Leu Cys Ser Tyr Trp His Val Gly Leu Val
CUAACAAGUUCAUCGUUUAGUUAUUAGAAGUGGCAGAUUUGCUUGAAAAGCCAAUUAUUGAAAGCUGAAGA AUG AUU CUG UGC AGU UAU UGG CAU GUA GGG UUG GUG
-140 -120 -100 -80 -60 -40

-10 -1 1 10
Leu Leu Leu Phe Ser Cys Cys Gly Leu Val Leu Gly Ser Glu His Glu Thr Arg Leu Val Ala Asn Leu Leu Glu Asn Tyr Asn Lys Val
CUA CUG UUA UUU UCG UGU UGU GGU CUG GUA CUA GGU UCU GAA CAU GAA ACA CGU UUG GUU GCU AAU UUA UUA GAA AAU UAU AAC AAG GUG
-20 -1 1 20 40

20 30 40
Ile Arg Pro Val Glu His His Thr His Phe Val Asp Ile Thr Val Gly Leu Gln Leu Ile Gln Leu Ile Ser Val Asp Glu Val Asn Gln
AUU CGU CCA GUG GAG CAU CAC ACC CAC UUU GUA GAU AUU ACA GUG GGG CUA CAG CUG AUA CAA CUC AUC AGU GUG GAU GAA GUA AAU CAA
60 80 100 120 140

50 60 70
Ile Val Glu Thr Asn Val Arg Leu Arg Gln Gln Trp Ile Asp Val Arg Leu Arg Trp Asn Pro Ala Asp Tyr Gly Gly Ile Lys Lys Ile
AUU GUG GAA ACA AAU GUG CGC CUA AGG CAG CAA UGG AUU GAU GUG AGG CUU CGC UGG AAU CCA GCC GAU UAU GGU GGA AUU AAA AAG AUC
160 180 200 220

80 90 100
Arg Leu Pro Ser Asp Asp Val Trp Leu Pro Asp Leu Val Leu Tyr Asn Asn Ala Asp Gly Asp Phe Ala Ile Val His Met Thr Lys Leu
AGA CUG CCU UCU GAU GAU GUU UGG CUG CCA GAU UUA GUU CUG UAC AAC AAU GCU GAU GGU GAU UUU GCC AUU GUU CAC AUG ACC AAA CUG
240 260 280 300 320

110 120 130
Leu Leu Asp Tyr Thr Gly Lys Ile Met Trp Thr Pro Pro Ala Ile Phe Lys Ser Tyr Cys Glu Ile Ile Val Thr His Phe Pro Phe Asp
CUU UUG GAU UAU ACG GGA AAA AUA AUG UGG ACA CCU CCA GCA AUC UUC AAA AGC UAU UGU GAA AUU AUU GUA ACA CAU UUC CCA UUU GAU
340 360 380 400

140 150 160
Gln Gln Asn Cys Thr Met Lys Leu Gly Ile Trp Thr Tyr Asp Gly Thr Lys Val Ser Ile Ser Pro Glu Ser Asp Arg Pro Asp Leu Ser
CAA CAA AAU UGC ACU AUG AAG UUG GGA AUC UGG ACG UAC GAU GGG ACA AAA GUU UCC AUA UCC CCG GAA AGU GAC CGU CCG GAU CUG AGU
420 440 460 480 500

170 180 190
Thr Phe Met Glu Ser Gly Glu Trp Val Met Lys Asp Tyr Arg Gly Trp Lys His Trp Val Tyr Tyr Thr Cys Cys Pro Asp Thr Pro Tyr
ACA UUU AUG GAA AGU GGA GAG UGG GUA AUG AAA GAU UAU CGU GGA UGG AAG CAC UGG GUG UAU UAU ACC UGC UGU CCU GAC ACU CCU UAC
520 540 560 580

200 210 220
Leu Asp Ile Thr Tyr His Phe Ile Met Gln Arg Ile Pro Leu Tyr Phe Val Val Asn Val Ile Ile Pro Cys Leu Leu Phe Ser Phe Leu
CUG GAU AUC ACC UAC CAU UUU AUC AUG CAG CGU AUU CCU CUU UAU UUU GUU GUG AAU GUC AUC AUU CCU UGU CUG CUU UUU UCA UUU UUA
600 620 640 660 680

230 240 250
Thr Gly Leu Val Phe Tyr Leu Pro Thr Asp Ser Gly Glu Lys Met Thr Leu Ser Ile Ser Val Leu Leu Ser Leu Thr Val Phe Leu Leu
ACU GGA UUA GUA UUU UAC UUA CCA ACU GAU UCA GGU GAG AAG AUG ACU UUG AGU AUU UCC GUU UUG CUG UCU CUG ACU GUG UUC CUU CUG
700 720 740 760

260 270 280
Val Ile Val Glu Leu Ile Pro Ser Thr Ser Ser Ala Val Pro Leu Ile Gly Lys Tyr Met Leu Phe Thr Met Ile Phe Val Ile Ser Ser
GUU AUU GUU GAG CUG AUC CCC UCA ACU UCC AGC GCU GUG CCU UUG AUU GGC AAA UAC AUG CUU UUU ACA AUG AUU UUU GUC AUC AGU UCA
780 800 820 840 860

290 300 310
Ile Ile Ile Thr Val Val Val Ile Asn Thr His His Arg Ser Pro Ser Thr His Thr Met Pro Gln Trp Val Arg Lys Ile Phe Ile Asp
AUC AUC AUU ACU GUU GUU GUA AUU AAU ACU CAC CAU CGC UCU CCA AGU ACA CAU ACA AUG CCA CAA UGG GUA CGA AAG AUC UUU AUU GAU
880 900 920 940

320 330 340
Thr Ile Pro Asn Val Met Phe Phe Ser Thr Met Lys Arg Ala Ser Lys Glu Lys Gln Glu Asn Lys Ile Phe Ala Asp Asp Ile Asp Ile
ACU AUA CCC AAU GUU AUG UUU UUC UCA ACA AUG AAA CGA GCU UCU AAG GAA AAG CAA GAA AAU AAG AUA UUU GCU GAU GAC AUU GAU AUC
960 980 1,000 1,020 1,040

350 360 370
Ser Asp Ile Ser Gly Lys Gln Val Thr Gly Glu Val Ile Phe Gln Thr Pro Leu Ile Lys Asn Pro Asp Val Lys Ser Ala Ile Glu Gly
UCU GAC AUU UCU GGA AAG CAA GUG ACA GGA GAA GUA AUU UUU CAA ACA CCU CUC AUU AAA AAU CCA GAU GUC AAA AGU GCU AUU GAG GGA
1,060 1,080 1,100 1,120

380 390 400
Val Lys Tyr Ile Ala Glu His Met Lys Ser Asp Glu Glu Ser Ser Asn Ala Ala Glu Glu Trp Lys Tyr Val Ala Met Val Ile Asp His
GUC AAA UAU AUU GCA GAG CAC AUG AAG UCU GAU GAG GAA UCA AGC AAU GCU GCA GAG GAA UGG AAA UAU GUU GCA AUG GUG AUU GAU CAC
1,140 1,160 1,180 1,200 1,220

410 420 430
Ile Leu Leu Cys Val Phe Met Leu Ile Cys Ile Ile Gly Thr Val Ser Val Phe Ala Gly Arg Leu Ile Glu Leu Ser Gln Glu Gly
AUU CUG CUG UGU GUC UUC AUG CUG AUU UGU AUA AUU GGU ACA GUU AGC GUG UUU GCU GGC CGU CUC AUU GAA CUC AGU CAA GAG GGC UAA
1,240 1,260 1,280 1,300

AUCUUCAUUGUGAGCAAAAAAGGCAAUACUGGAAUAAGGGAUGGAUAUCACUCCACAGAAAAGAUGUGUGGGUUUAGUGUUGCAAUUGUAGUCUGUUUUAUGAGAUAUAUAGUUUGCUUU
1,320 1,340 1,360 1,380 1,400 1,420

GUUUUACAAUGAAAUGUACUUAAGGUAUUUGAAUAUGUAAAAAAAAGUAAUGAAAUAACAGUAAGUGAAAAAUGUUAUUAUGCAAGUACCUGAAACGUGUAAUAAGUGGAACAACUUUUU
1,440 1,460 1,480 1,500 1,520 1,540

AAUACAUUACAUAAAAGUAAGCAAAAAAUAAGUUUAACAAAUUAUGAGGGUAGUCAUUUGAAAUGUAACAGAGAAAUGAAAAUUAUUAGAAAUAUAAACAGUAUAUAUUAAGUUAAACAA
1,560 1,580 1,600 1,620 1,640 1,660

AGUUAAUCCAUUCUUUUAUAUCCAAAGUUUGUAUUAUACAUUUAGAAGUGUAGUUCUAUUGUAUAAUUUAAAGUAUGUUUUACAGAUCAUUAAUAAAUAUUCAAUGCAUUACU---3'
1,680 1,700 1,720 1,740 1,760 1,780

probes, and they picked the two largest of these for sequencing.

DNA can be cut by restriction endonucleases, bacterial enzymes that act only at particular sites. Using a number of different endonucleases it is possible to cut the cDNA into a series of smaller units whose position on the DNA strand is approximately known. These pieces can themselves be cloned using a bacteriophage vector in *E. coli* cells. The nucleotide sequences of these various pieces are then determined by one of the standard methods (Maxam and Gilbert, 1980). Then by matching up the sequences of the different pieces the nucleotide sequence of the whole cDNA strand can be worked out.

Fig. 8.12 shows the base sequence of *Torpedo* α subunit mRNA as determined by this method, and the amino acid sequence of the protein deduced from this. In protein synthesis the mRNA is read from the 5′-terminal end until the sequence AUG is reached: this indicates the beginning of the protein chain and also codes for the amino acid methionine. From then on successive base triplets act as codons for the various amino acids in the sequence until a stop codon (UAA in this case) is reached, when the protein chain is complete.

The protein chain of the α subunit, as determined by amino acid analysis by Raftery *et al.* (1980), begins with the sequence Ser–Glu–His–Glu–Thr, which does not occur until the 25th residue in fig. 8.12. The first 24 residues probably form a 'signal sequence' which serves to start the insertion of the protein chain into the membrane and is removed after it has reached its final position. Hence in numbering the residues the first 24 are given negative numbers. The mature protein thus contains 437 residues with serine at the N-terminal (position 1) and glycine at the C-terminal. It has a calculated molecular weight (neglecting glycosylation and other modifications of the amino acid chain) of 50116.

Opposite
Fig. 8.12. Base sequence of the mRNA coding for the α subunit of the acetylcholine receptor in the electric organ of *Torpedo californica*, and the amino acid sequence predicted from this. Numbering of the amino acid sequence starts at the first residue (serine) of the mature protein. (From Noda *et al.*, 1982.)

The mature β, γ and δ subunits contain 469, 489 and 501 amino acids, with calculated molecular weights of 53 681, 56 279 and 57 565 respectively. Their sequences show marked homologies with that of the α subunit (fig. 8.13). This suggests that the four subunits are similar in structure and that the genes encoding for them are descended from a single common ancestor (Noda *et al.*, 1983*a*). The hydrophobicity profiles for each subunit are very similar (fig. 8.14), and show four regions of 18 or more residues which look as though they would form hydrophobic α-helices. These regions (M1 to M4) are long enough to cross the membrane and it seems highly likely that they do just that.

Fig. 8.15 shows a model for all the subunits, produced by Finer-Moore and Stroud (1984). They suggest that there is a fifth membrane-spanning α-helix between M3 and M4, labelled MA in the diagram. It contains both charged and uncharged residues and so is amphipathic rather than hydrophobic, hence it may well form part of the lining of the ionic channel. Guy (1984) came to the same conclusion.

The long section from the N-terminal to the beginning of M1 is apparently all outside of the cell. It contains sites for glycosylation (attachment of sugars) and disulphide crosslinking. In the α subunit it contains the α-bungarotoxin and acetylcholine binding sites (Kao *et al.*, 1984; Lennon *et al.*, 1985).

Young *et al.* (1985) made a synthetic peptide corresponding to the last 16 residues of the δ subunit and raised antibodies to it. They showed that these antibodies were attached to the cytoplasmic side of receptor-rich membranes from *Torpedo* electric organ. Similar results were obtained for the other subunits by Lindstrom *et al.* (1984). Hence the C-terminal end of the chain is cytoplasmic. This conclusion provides evidence in favour of the model with five membrane-crossing segments as shown in fig. 8.15, and against the initial view that there were just four such sections.

Quaternary structure

Postsynaptic membranes of *Torpedo* electric organ show arrays of receptors which seem to consist of a number of subunits arranged round a central pit (Cartaud *et al.*, 1978; fig. 7.14). The

$\alpha_2\beta\gamma\delta$ structure suggests that there are five subunits and that the receptor should therefore show an approach to pentamerous symmetry.

How are the different subunits arranged? Some information has been provided by using monoclonal antibodies to the α subunits (Fairclough *et al.*, 1983). Whole antibodies are complex molecules about the same size as the whole acetylcholine receptor, but their Fab fragments (which contain the binding sites) are smaller and do not combine with each other. Electron microscopy of negatively stained preparations of receptors to which Fab fragments had been bound could thus show the relative positions of the two subunits: they were always separated by another subunit.

The structure of biological macromolecules can be determined by X-ray diffraction if three-dimensional crystals of sufficient size can be prepared. This has been done for haemoglobin,

Fig. 8.13. The amino acid sequences of the four subunits of *Torpedo californica*, as determined by Noda *et al.* (1982, 1983*a, b*) and arranged by Finer-Moore and Stroud to show homologies. The single-letter code for amino acids (table 1.1) is used. Notice that the numbering here is not always the same as in the individual units (as in fig. 8.3, for example), because of the need to incorporate gaps so as to maintain the maximum homology between the different subunits. Notice the high degree of homology in the sections M1 to M4, which are thought to be hydrophobic membrane-crossing α-helices. The section MA shows the amphipathic section that may form the lining of the ionic channel. (From Fairclough *et al.*, 1983.)

```
           -24                     -1
ALPHA      MILCSYWHVGLVLLLFSCCGLVLG
BETA       MENVRRMALGLVVMMALALSGVGA
GAMMA           MVLTLLLIICLALEVRS
DELTA      MGNIHFVYLLISCLYYSGCS   G

1
SEHETRLVANLL EN YNKVIRPVEHHTHFVDITVGLQLIQLISVDEVNQIVETNVRLRQQWIDVRLRWNPADYGGIKKIRLPSDDV
SVMEDTLLSVLF ET YNPKVRPAQTVGDKVTVRVGLTLTNLLILNEKIEEMTTNVFLNLAWTDYRLQWDPAAYEGIKDLRIPSSDV
ENEEGRLIEKLL GD YDKRIIPAKTLDHIIDVTLKLTLTNLISLNEKEEALTTNVWIEIQWNDYRLSWNTSEYEGIDLVRIPSELL
VNEEERLINDLLIVNKYNKHVRPVKHNNEVVNIALSLTLSNLISLKETDETLTSNVWMDHAWYDHRLTWNASEYSDISILRLPPELV

        100
WLPDLVLYNNADGDFAIVHMTKLLLDYTGKIMWTPPAIFKSYCEIIVTHFPFDQQNCTMKLGIWTYDGTKVSIS      PESDRP
WQPDIVLMNNNDGSFEITLHVNVLVQHTGAVSWQPSAIYRSSCTIKVMYFPFDWQNCTMVFKSYTYDTSEVTLQ   HALDAKGERE
WLPDVVLENNVDGQFEVAYYANVLVYNDGSMYWLPPAIYRSTCPIAVTYFPFDWQNCSLVFRSQTYNAHEVNLQLS     AEEGEA
WIPDIVLQNNNDGQYHVAYFCNVLVRPNGYVTWLPPAIFRSSCPINVLYFPFDWQNCSLKFTALNYDANEITMDLMTDTIDGK DYP

                   200
     DLSTFMESGEWVMKDYRGWKH WVYYTCCPD TPYLDITYHFIMQRIPLYFVVNVIIPCLLFSFLTGLVFYLPTDSG EKM
VKEIVINKDAFTENGQWSIEH KPSRKNW RSD DP  S YEDVTFYLIIQRKPLFYIVYTIIPCILISILAILVFYLPPDAG EKM
VEWIHIDPEDFTENGEWTIRH RPAKKNYNWQLTKDD TDFQEIIFFLIIQRKPLFYIINIIAPCVLISSLVVLVYFLPAQAGGQKC
IEWIIIDPEAFTENGEWEIIH KPAKKN IYPDKFPNGTNYQDVTFYLIIRRKPLFYVINFITPCVLISFLASLAFYLPAESG EKM
                                                     M1

                             300
TLSISVLLSLTVFLLVIVELIPSTSSAVPLIGKYMLFTMIFVISSIIITVVVINTHHRSPSTHTMPQWVRKIFIDTIPNV
SLSISALLAVTVFLLLLADKVPETSLSVPIIIRYLMFIMILVAFSVILSVVVLNLHHRSPNTHTMPNWIRQIFIETLPPFLWIQRPV
TLSISVLLAQTIFLFLIAQKVPETSLNVSLIGKYLIFVMFVSMLIVMNCVIVLNVSLRTPNTHSLSEKIKHLFLGFLPKYLGMQLEP
STAISVLLAQAVFLLLTSQRLPETALAVPLIGKYLMFIMSLVTGVIVNCGIVLNFHFRTPSTHVLSTRVKQIFLEKLPRILHMSRAD
      M2                             M3
                                      400
                             MFFSTMKRASKEKQENKIFADDIDISDISGKQVTGEVIFQTPLIKNP DV
TTPSPD           SKPTIISRANDEYFIRKPA    GDFVCPVDNARVAVQPERLFSEMKWHLNG          LTQPVTLPQDL
SEETPEKPQ PR RRSSFGIMI KA EEYILKKPRSELMFEEQKDRHGLKRVNK MTSDIDI       GTTVDLYKDLANFAP EI
ESEQPDWQNDLKLRRSSSVGYISKAQ EYFNIKSR    SELMFEKQSERHGLVP RVTPRIGFGNNNEN        IAASDQLHDEI

                                               500
KSAIEGVKYIAEHMKSDEESSNAAEEWKYVAMVIDHILLCVFMLICIIGTVSVFAGRLIELSQEG
KEAVEAIKYIAEQLESASEFDDLKKDWQYVAMVADRLFLYVFFVICSIGTFSIFLDASHNVPPDNPFA
KSCVEACNFIAKSTKEQNDSGSENENWVLIGKVIDKACFWIALLLFSIGTLAIFLTGHFNQVPEFPFPGDPRKYVP
KSGIDSTNYIVKQIKEKNAYDEEVGNWNLVGQTIDRLSMFIITPVMVLGTIFIFVMGNFNHPPAKPFEGDPFDYSSDHPRCA
          MA                                 M4
```

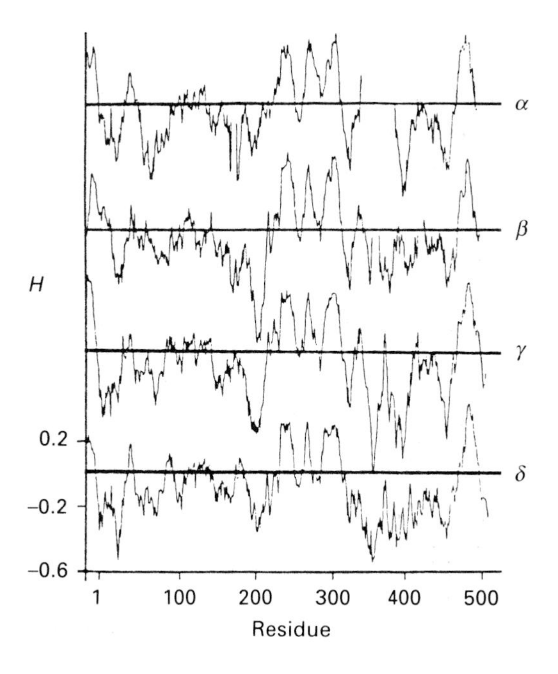

Fig. 8.14. Hydrophobicity plots for the four subunits of the acetylcholine receptor. The plots show the average of seven adjacent residues, and the vertical scale shows the average hydrophobicity as determined by Kyte and Doolittle (table 1.1). Residue numbering as in fig. 8.13. (From Finer-Moore and Stroud 1984.)

Fig. 8.15. Diagram showing the predicted secondary structure of all four of the subunits of the acetylcholine receptor. The numbering of residues is as in fig. 8.13. The dashed lines show regions which are not present in all subunits. Cysteine residues (which may take part in disulphide bonds) are indicated by S and the Greek letter for the subunit in which they occur. Stars show potential glycosylation sites. (From Finer-Moore and Stroud, 1984.)

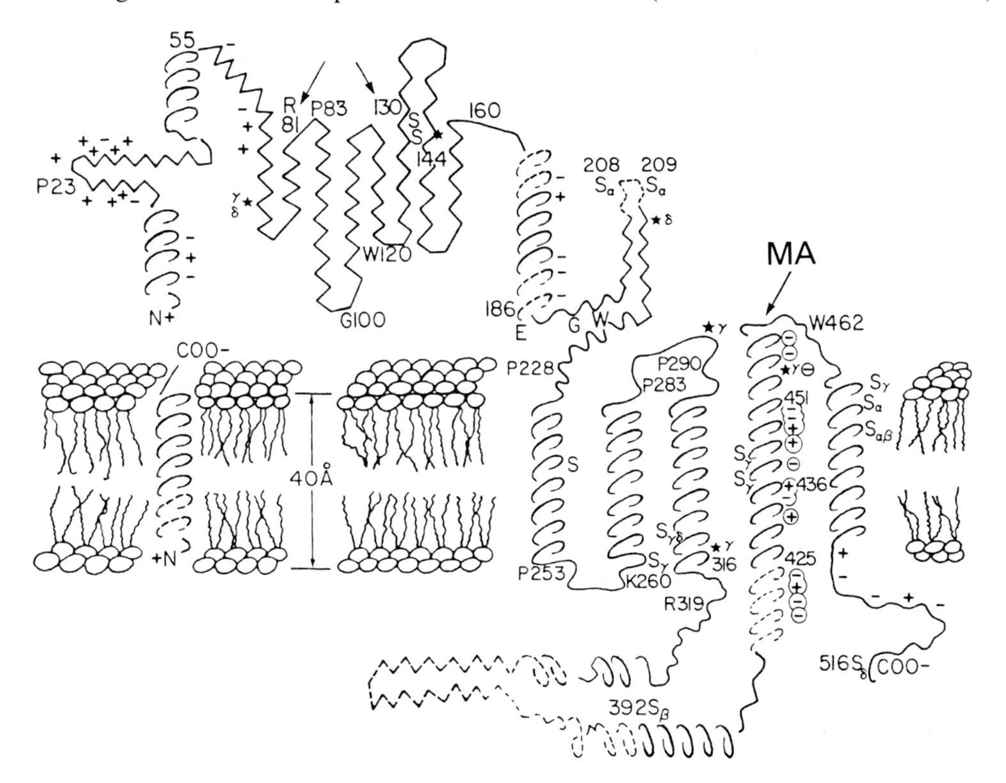

lysozyme and a number of other proteins. Two-dimensional arrays may have a sufficient degree of order in them for analogous methods to be used. Electron micrographs of such '2-D crystals' can be subjected either to optical diffraction or to computerized image analysis. In order to obtain some idea of the 3-D structure of the molecules, the crystalline array must be viewed at different angles in the electron microscope, and the image digitized and subjected to Fourier analysis. Details of the process are given by Amos, Henderson and Unwin (1982).

Brisson and Unwin (1985) have applied this technique to *Torpedo* acetylcholine receptors. They isolated membrane vesicles containing receptors and found that these would aggregate to form tubular crystals if left in buffer solutions for some weeks. The tubes flattened when placed on the carbon support grid for electron microscopy, and were then either frozen or negatively stained. Electron microscopy showed regular arrays of receptors, and image analysis produced the results shown in fig. 8.16.

It is clear from fig. 8.16 that the acetylcholine receptor really is a pentamerous structure with a hole in the middle, at least on the outer (synaptic) side of the membrane. The resolution of the method was insufficient to show the ionic channel itself in the membrane-crossing and cytoplasmic parts of the molecule.

Expression in the oocyte membrane

The african clawed toad *Xenopus laevis* commonly lives in ponds in tropical Africa. It adapts readily to life in the laboratory, where it has been used for a variety of purposes, from pregnancy testing to the study of how genes are expressed in development. The oocytes are found in the ovaries of adult females. They are large cells which are waiting for the hormonal stimulus that will make them develop into mature eggs ready for fertilization.

Xenopus oocytes possess the normal translation machinery and so will respond to the injection of mRNA from other sources by making the protein for which it codes (Gurdon *et al.*, 1971; Gurdon, 1974). They have been much used in recent years to investigate the properties of various neurotransmitter and channel proteins.

Fig. 8.16. Structure of the acetylcholine receptor from *Torpedo* electric organ as determined by quantitative electron image analysis. *a* shows a section through the receptor in a plane perpendicular to the membrane; the contour lines show electron densities. *b* shows the projected densities in the plane of the membrane. (From Brisson and Unwin, 1985.)

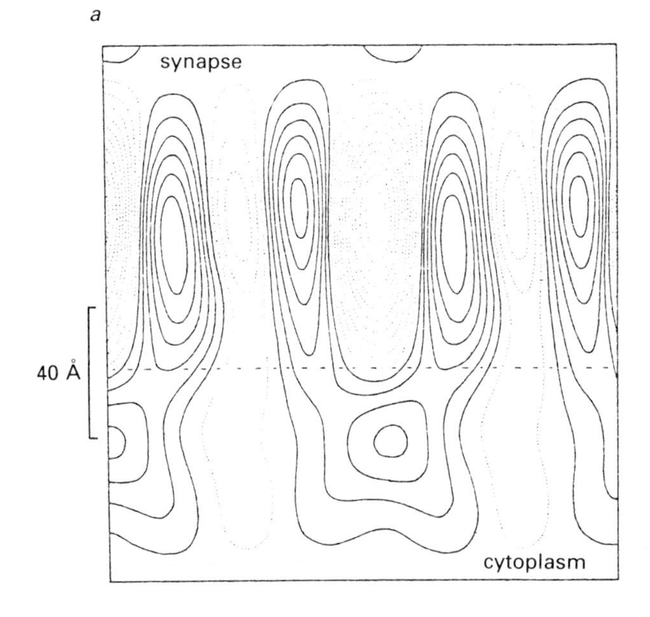

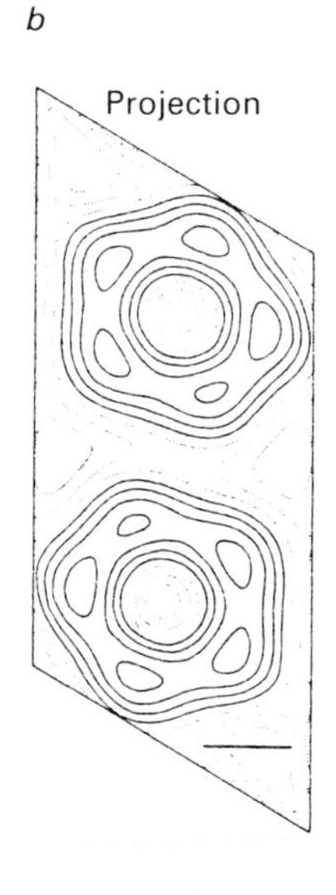

Sumikawa *et al.* (1981) showed that injection of mRNA from *Torpedo* electric organ into *Xenopus* oocytes resulted in the production of acetylcholine receptor molecules, detected by their binding of α-bungarotoxin. Are such molecules functionally effective as receptors? In order to provide an answer to this question, Barnard, Miledi and Sumikawa (1982) injected oocytes with *Torpedo* electric organ mRNA and then tested them two days later by applying acetylcholine by ionophoresis and observing the electrical response.

Xenopus oocytes normally contain some acetylcholine receptors, but these are muscarinic and so can be blocked by atropine, which has no effect on the nicotinic receptors produced by the injected mRNA. Application of acetylcholine to noninjected oocytes produces the muscarinic response (fig. 8.17*a*), which is rather irregular in form and delayed for some seconds after the start of the ionophoretic pulse. Fig. 8.17*b* shows the response of the injected oocytes: it is a smooth inward current which begins almost simultaneously with the onset of the ionophoresis current. It occurs in the presence of atropine and is blocked by curare, hence it is nicotinic. It has a reversal potential near to 0 mV and appears to be largely caused by the flow of sodium and potassium ions. Injection of acetylcholine into the inside of the oocyte did not produce any response. These experiments show that the mRNA contains all the information for the cell to make mature functional acetylcholine receptors that are correctly oriented in the membrane.

The *Xenopus* oocyte system has also been used by the Kyoto University group to see which of the different subunits are necessary for the making of a functional receptor (Mishina *et al.*, 1984). Starting from their clones of *E. coli* containing plasmids with the cDNA coding for one of the subunits, they constructed an artificial recombinant plasmid which was capable of entering a cell line of tissue-cultured monkey cells and which also contained the receptor cDNA. Following this process of transfection, the cultured cells produced relatively large quantities of the mRNA coding for the receptor subunit. In this way, mRNAs specific for the four different subunits were obtained, and they could then be injected into oocytes in various combinations.

The results were clearcut. After injection of mRNAs specific for all four subunits, the oocytes produced appreciable inward currents when acetylcholine was applied by ionophoresis in the presence of atropine. With only three subunits, there was normally no response.* This shows clearly that all four subunits are required for normal acetylcholine receptor function. In contrast, only the α subunit was required for α-bungarotoxin binding.

* In a few cases small responses were seen after injection of α, β and γ, or α, β and δ chains only. White *et al.* (1985) also obtained small responses when the δ chain was omitted. This may mean that the γ and δ chains can to a small extent substitute for each other.

Fig. 8.17. Responses of *Xenopus* oocytes to acetylcholine applied by ionophoresis. Trace *a* shows the muscarinic response, a delayed and rather irregular depolarization, in an oocyte which had been injected with a little water as a control. Trace *b* is a voltage-clamp record of the nicotinic response, a relatively rapid and smooth inward current, in an oocyte which had been injected with *Torpedo* electric organ mRNA two days previously. (From Barnard *et al.*, 1982.)

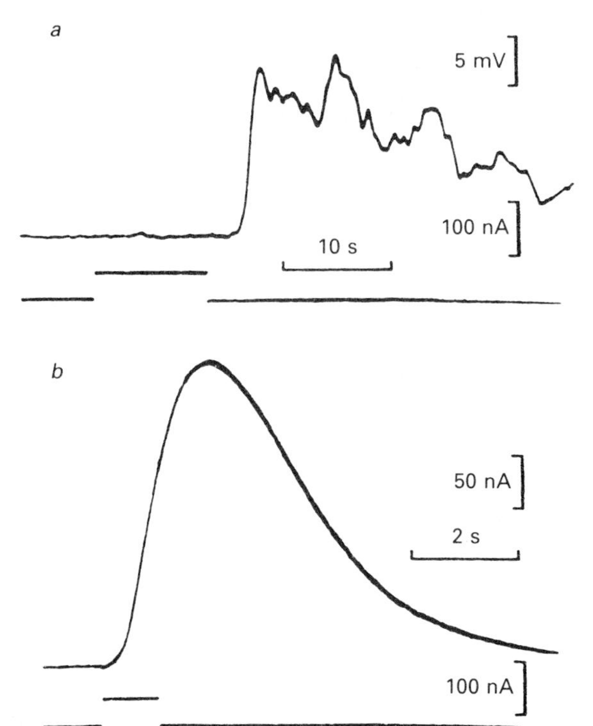

Modifying the receptors

The next step in this saga of the interaction of molecular biology with physiology was to modify the primary structure of the protein chains at particular sites and to see what effect this had on the functioning of the receptor. This procedure is called site-directed mutagenesis. It should enable us to get some idea of where particular functions are located in the protein molecule.

Using the standard tricks of the gene-manipulator's trade, Mishina *et al.* (1985) prepared mutant forms of the cDNA coding for the acetylcholine receptor α subunit. These were of two types: deletion mutants, which would produce a chain with a particular short sequence of amino acids missing (being replaced by 1–3 extraneous amino acids), and point mutants, leading to the replacement of a particular amino acid by another one. Then mRNAs were prepared from these mutant cDNAs by transcription, using an *in vitro* system this time. The mutant α subunit mRNAs were then injected into *Xenopus* oocytes together with normal β, γ and δ mRNAs in the ratio 2:1:1:1, and the response of the oocytes to acetylcholine or their binding of α-bungarotoxin was measured.

The results of these experiments (table 8.1) serve to confirm and extend our ideas about the structure and functioning of the receptor molecule. Table 8.1 shows that cytoplasmic mutations (those between M3 and MA and near the C-terminal) have relatively little effect on toxin binding or the response to acetylcholine. Deletions in the transmembrane segments abolish the response to acetylcholine, presumably by destroying the integrity of the ionic channel, and reduce the binding. Point mutations in the external region of the molecule, between the N-terminal and M1, show the importance of glycosylation (which requires the asparagine residue at 141) and the sulphydryl links that are presumably made by the cysteines at 128, 142, 192 and 193 (the numbering here follows fig. 8.12, not fig. 8.13). These point mutations probably have direct or indirect effects on the structure of the acetylcholine binding site.

Single channel currents can be measured by applying the patch-clamp technique to injected oocytes. Sakmann and his colleagues (1985) have used this method to look at the roles of the different acetylcholine receptor subunits in ion transport and gating. Messenger RNAs from the α, β, γ and δ subunits were injected in the ratio 2:1:1:1. Single channel currents from bovine (cow) receptors were of longer average duration than those from *Torpedo* receptors.

Table 8.1. *Effects of site-directed mutagenesis in the α subunit on the properties of the acetylcholine receptor, as expressed in Xenopus oocytes*

	Response to acetylcholine	α-bungarotoxin binding
Point mutations		
Asn 141	zero	zero
Cys 222	normal	slightly reduced
Cys 128 or 142	zero	zero
Cys 192 or 193	zero	reduced
Deletions		
M1	zero	zero
M2, M3 or M4	zero	reduced
MA	zero	reduced
between M3 and MA	normal or reduced	normal or reduced
near C-terminal	reduced	reduced

Based on Mishina *et al.* (1985).

Then hybrid receptors were produced by injecting oocytes with bovine mRNA for one of the subunits and *Torpedo* mRNA for the others. The results are shown in table 8.2. The figures for the binding of radioactive α-bungarotoxin give an indication of the density of the receptors in the oocyte membrane. Clearly the densities of bovine and *Torpedo* receptors are much the same, as are those of the hybrid receptors in which the α or δ chains are bovine and the others are from *Torpedo*. The very low figures for hybrids with bovine β or γ chains suggest that hybrids of this type cannot combine to form proper receptors in the membrane.

The single channel conductances of the functional receptors in table 8.2 are similar but the average durations of their openings are very different. It is particularly noticeable that the hybrid with the bovine δ chain has a long average opening time like that of the whole bovine receptor; the corresponding α hybrid has an intermediate average

Fig. 8.18. Responses of foetal and adult acetylcholine receptors in bovine muscle. The traces on the left show single-channel currents from (*a*) foetal and (*b*) adult muscle. Those on the right show currents from *Xenopus* oocytes which had been previously injected with mRNA coding for (*c*) α, β, γ and δ subunits or (*d*) the α, β, δ and ε subunits. Clearly the difference between foetal and adult muscle receptors lies in the substitution of the ε for the γ subunit. (From Mishina *et al.*, 1986.)

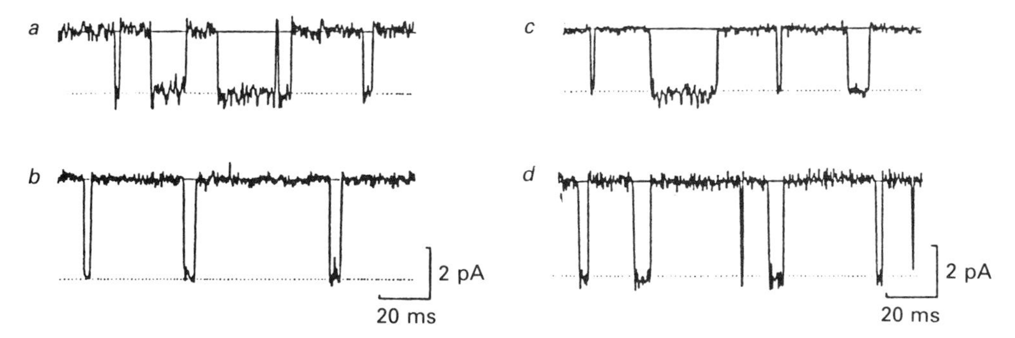

Table 8.2. *Properties of Torpedo, cow and hybrid acetylcholine receptors, as expressed in Xenopus oocytes*

Subunit constitution	Surface α-bungarotoxin binding (fmol per oocyte)	Whole cell current (nA)	Single channel conductance (pS)	Mean single channel open time (ms)
$\alpha_T\beta_T\gamma_T\delta_T$	7.6	23	42	0.6
$\alpha_c\beta_c\gamma_c\delta_c$	7.9	1600	40	7.6
$\alpha_c\beta_T\gamma_T\delta_T$	10.4	80	42	2.5
$\alpha_T\beta_c\gamma_T\delta_T$	0.1	<1	—	—
$\alpha_T\beta_T\gamma_c\delta_T$	0.3	<1	—	—
$\alpha_T\beta_T\gamma_T\delta_c$	5.4	1400	35	8.6

Simplified from Sakmann *et al.* (1985).

opening time. Thus it seems that the δ chain is particularly important in determining the duration of channel opening, perhaps by affecting the rate constant α in scheme (8.2).

The next step was to use genetic manipulation techniques to construct chimaeric δ subunit cDNAs which contained partly the *Torpedo* sequence and partly the bovine sequence (Imoto *et al.*, 1986). These were used to make chimaeric δ subunit mRNAs which were then injected into oocyts together with *Torpedo* α, β and γ subunit mRNAs. At low divalent cation concentrations (less than 1 mM) the single channel conductance for *Torpedo* receptors is higher than that for bovine ones (mean values were 87 and 62 pS respectively). With a bovine δ chain and the other chains from *Torpedo*, the lower value was observed. With chimaeric chains, single channel conductances were either in the higher (*Torpedo*) or lower (bovine) ranges. The crucial part of the molecule seemed to be the M2 helix and the external sequence connecting the M2 and M3 helices: low conductances were found when this section was bovine and high ones when it was from *Torpedo*.

Receptors at different sites

There are not only species differences in the molecular nature of acetylcholine receptors, but also differences at different sites within the body and during development. In adult vertebrate muscles, the receptors are largely confined to the neuromuscular junction, but in foetal muscles and in adult muscles after denervation they are also found in the nonjunctional membrane. These non-junctional receptors have lower conductances and longer opening times than the junctional receptors (Neher and Sakmann, 1976*b*).

Takai *et al.* (1985) isolated a cDNA from bovine muscle which produced an mRNA coding for a new acetylcholine receptor subunit, which they called the ε (epsilon) chain. This was more simular to the γ chain than to the others. Injection of α, β, δ and ε subunit mRNAs into oocytes produced functional receptors (Mishina *et al.*, 1986). Measurements of their single channel currents showed that the receptors with the γ subunit produced responses like those of foetal muscle (conductance 40 pS, mean duration 7.2 msec), whereas those with the ε subunit produced responses like adult muscle (conductance 60 pS, mean duration 2.3 ms), as is shown in fig. 8.18.

Nicotinic acetylcholine receptors occur in sympathetic ganglion cells and at various sites in the brain. Boulter *et al.* (1986) have isolated a cDNA clone from rat brain which encodes for a protein similar to the muscle receptor α subunit. Its main difference is that it is 37 amino acids longer, with most of the extra length being between M3 and MA.

9
Synapses between neurons

Synapses between neurons are of two physiologically distinct types: those in which transmission is chemical in nature as at the neuromuscular junction, and those in which the presynaptic cell directly excites the postsynaptic cell by means of electric current. We shall first consider the properties of some representative chemically transmitting synapses.

The structure of chemically transmitting synapses (fig. 9.1) shows some variation in detail, but two particular features are common to all of them: synaptic vesicles are found in the presynaptic nerve

Fig. 9.1. Diagrams to show the variety of structure found in chemically transmitting synapses in the mammalian central nervous system. *a*: Two types of synapse between axon terminals and a dendrite. *b*: Synapse between an axon terminal and a dendritic spine. *c*: Two types of synapse between axon terminals and a neuron soma. *d*: A synapse whose postsynaptic spine is invaginated into the presynaptic terminal. Type 1 synapses (p. 174) are shown in *b* and on the left in *a*; type 2 synapses are shown in *c* and on the right in *a*. (From Whittaker and Gray, 1962.)

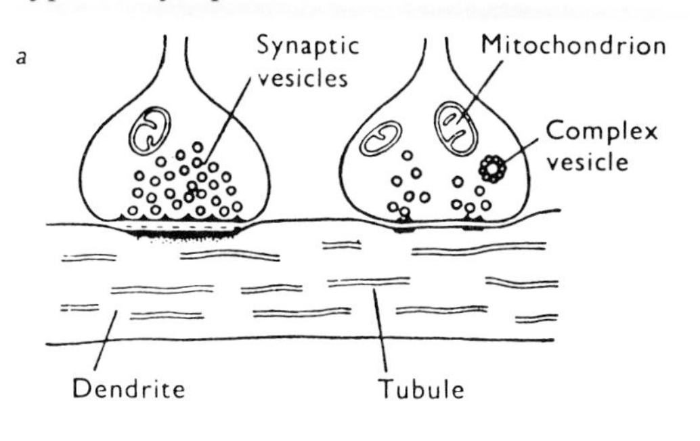

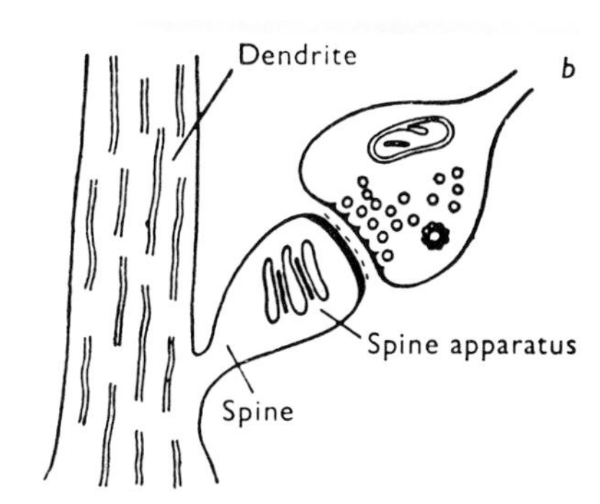

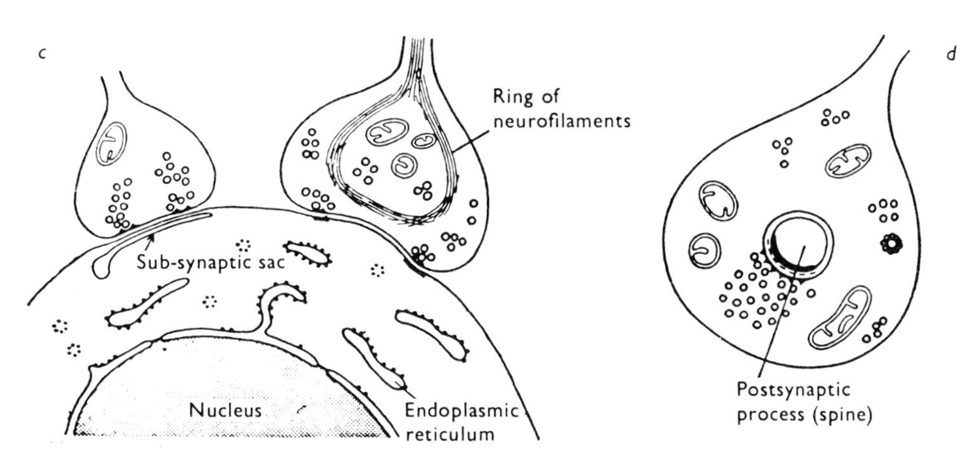

ending, and the pre- and postsynaptic cells are separated by an extracellular space (the synaptic cleft) about 200 Å across. In addition, the pre- and post-synaptic membranes are usually thickened in places, and it seems probable that these thickened regions are particularly associated with the transmission process. On the inner side of the presynaptic membrane is a geometrical array of dense material called the presynaptic grid, which seems to be associated with a set of microtubules whose function is to direct the vesicles to the active zones (Gray, 1983). A variety of subcellular components may occur in different types of synapse (see Gray and Guillery, 1966; Gray, 1974; Peters *et al.*, 1976), but their functional significance is not always clear.

Synaptic excitation in mammalian spinal motoneurons

Motoneurons are neurons which directly innervate skeletal muscle fibres. In mammals, the cell bodies of the motoneurons innervating the limb muscles lie in the ventral horn of the spinal cord, and their axons pass out to the peripheral nerves via the ventral roots. The cell body, or soma, is about 70 μm across, and extends into a number of fine branching processes, the dendrites (fig. 9.2). The surface of the soma and dendrites is covered with small presynaptic nerve endings (terminal boutons), showing the typical features of a chemically transmitting synapse. Some of these terminal

Fig. 9.2. The dendritic field of a cat spinal motoneuron. The motoneuron was injected with horseradish peroxidase and its structure then reconstructed from stained serial sections. The circles show sites of synaptic contact with one particular presynaptic fibre, shown in more detail in fig. 9.9. (From Redman and Walmsley, 1983*a*.)

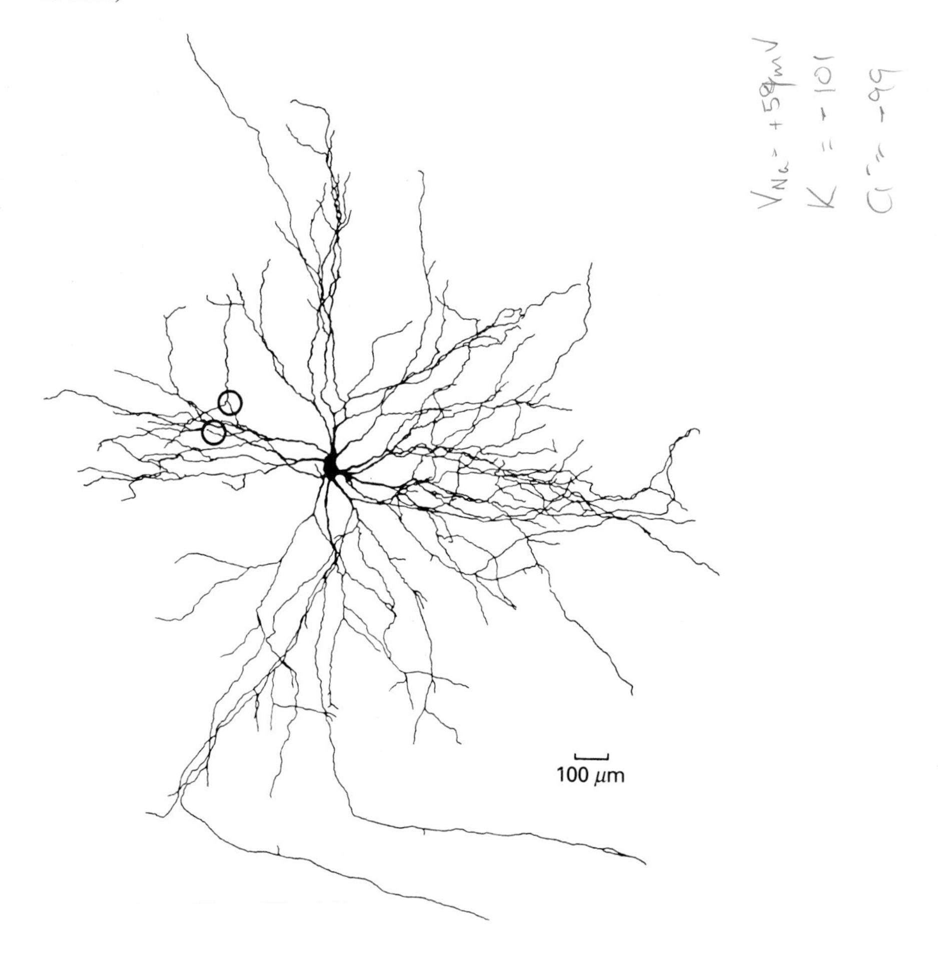

$V_{Na} = +58$ mV
K = −101
Cl = −99

boutons are the endings of group Ia fibres from stretch receptors (muscle spindles) in the muscle which the motoneuron innervates. Stretching the muscle excites these group Ia fibres, which may then excite the motoneurons supplying the muscle so that it contracts. This system is known as a *monosynaptic reflex* (fig. 9.3).

Most of our knowledge of the synaptic responses of motoneurons comes from investigations in the lumbar region of the spinal cord of the cat, using intracellular electrodes. This technique was first developed by Eccles and his colleagues (Brock *et al.*, 1952). The cat is anaesthetized and its spinal cord transected in the thorax. The spinal cord is exposed, a patch of the tough sheath round the cord removed, and the microelectrode inserted through this region until it reaches a motoneuron. Maps of the positions of various motoneuron groups are available, prepared by observing the chromatolytic changes following section of the nerve supplying a particular muscle (e.g. Romanes, 1951), so that the micro-electrode can be placed in approximately the desired position. The identity of a motoneuron is accurately established by stimulating the motor nerve fibres of a particular muscle and observing the monosynaptic response in the motoneuron. In many experiments the ventral roots are cut so that stimulation of a peripheral nerve does not result in antidromic stimulation of the motoneurons.

Motoneurons have a resting potential of about −70 mV. Depolarization of the membrane by about 10 mV results in the production of an action potential which propagates along the axon to the nerve terminals. The results of a number of experiments on the effects of injection of various ions into motoneurons via microelectrodes indicate that the ionic basis of the resting and action potentials is much the same as in squid axons. That is to say, the resting potential is slightly less than the potassium equilibrium potential, the action potential is caused primarily by a regenerative increase in sodium permeability, and the ionic concentration gradients necessary for these potentials are maintained by an active extrusion of sodium ions.

Excitatory postsynaptic potentials

The responses shown in fig. 9.4 were recorded by means of a microelectrode inserted into a motoneuron supplying fibres in the biceps or semitendinosus muscles. They were produced by stimulating the nerve supplying these muscles with single shocks. The size of the mass response of the sensory nerves was recorded from the dorsal roots, as shown in the inset records of fig. 9.4 (the form of these dorsal root responses shows that all the sensory fibres stimulated were of about the same conduction velocity – all group Ia fibres, in fact). The

Fig. 9.3. Anatomical organization of the monosynaptic stretch reflex system. *a* is grossly simplified; there are very many stretch receptors and afferent and efferent neurons associated with each muscle. *b* is slightly less simplified, showing how the afferent fibres branch to synapse with different members of the motoneuronal pool.

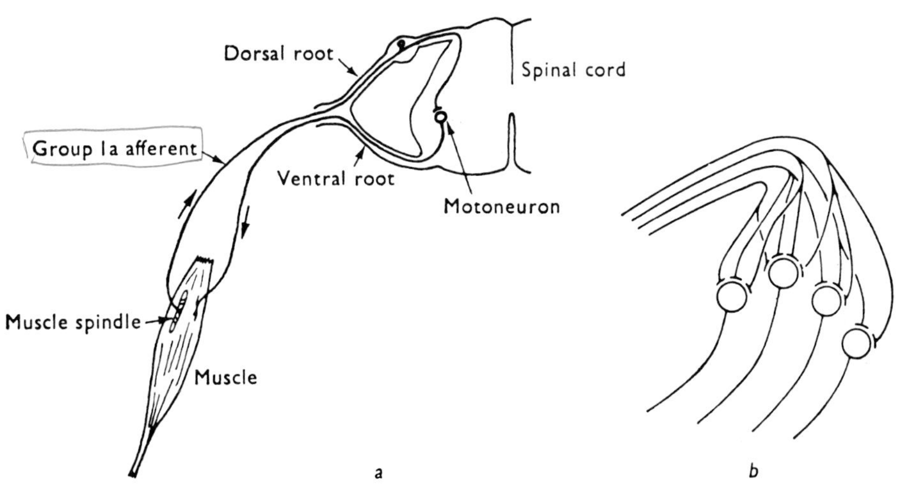

ventral roots were cut so as to avoid antidromic stimulation of the motoneurons. Each record was obtained by superimposition of faint traces from about forty separate responses to stimuli of the same intensity.

The form of these responses, which are known as *excitatory post-synaptic potentials* (EPSPs), is much the same as that of the end-plate potential in a curarized frog sartorius muscle fibre: a fairly rapid rising phase followed by a slower decay which follows an approximately exponential time course. Notice, however, that the size of the response is proportional to the stimulus intensity, and therefore to the number of presynaptic fibres which are active. This property is known as *spatial summation.* If the EPSP is large enough, a propagated action potential is set up (fig. 9.5). If a second EPSP is produced a short time later, the total response is greater; thus two successive EPSPs may be able to produce an action potential whereas either alone could not do so. This phenomenon is called *temporal summation.*

Fig. 9.4. Excitatory postsynaptic potentials recorded intracellularly from a cat biceps–semitendinosus motoneuron in response to stimuli of increasing intensity (from *a* to *c*) applied to the group Ia afferent fibres from the muscle. The inset records, taken at constant amplification, show the size of the dorsal root responses. (From Coombs *et al.*, 1955*c*.)

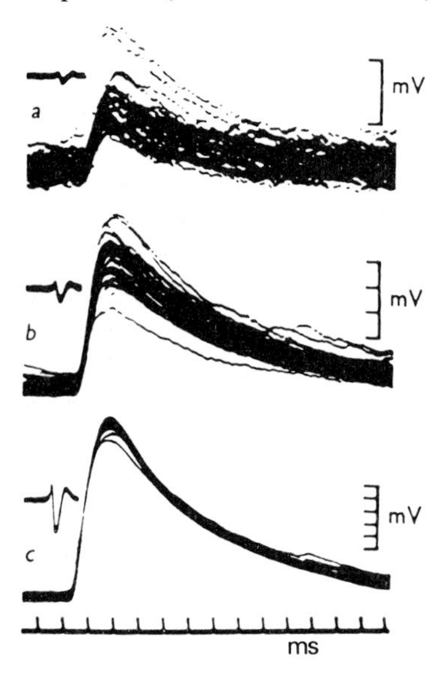

If the cell membrane is progressively depolarized, the EPSP decreases in size and eventually becomes reversed in sign; the reversal potential in cat motoneurons is about 0 mV (fig. 9.6). With small depolarizations an EPSP which is submaximal at the normal membrane potential may cross the threshold for production of an action potential, but larger maintained depolarizations make the cell electrically inexcitable. Hyperpolarization of motoneurons reduces the time constant of the falling phase of the EPSP and does not change the size of the EPSP.

The ionic mechanism of the EPSP in motoneurons has not been investigated with the precision that has been applied to the ionic basis of the EPP. The reason for this is mainly that it is technically very difficult to change the external ionic environment of motoneurons. However, the results described in the previous paragraph indicate that the excitatory transmitter substance causes a simultaneous and non-regenerative increase in the permeability of the postsynaptic membrane to more than one ion.

Perhaps the best evidence that motoneuronal EPSPs are produced by chemical transmission is the fact that they are reversed in sign when the membrane potential is made positive. Voltage-clamp experiments show a similar reversal of the excitatory postsynaptic current (Finkel and Redman, 1983). Further evidence is provided by

Fig. 9.5. Initiation of an action potential by the EPSP in a cat gastrocnemius motoneuron. The stimulus intensity to the afferent nerve was increased in the order *a* to *d*, with the result that the EPSP is of sufficient size to produce an action potential in *b* to *d*, and does so progressively earlier in these cases. (From Coombs *et al.*, 1957*b*.)

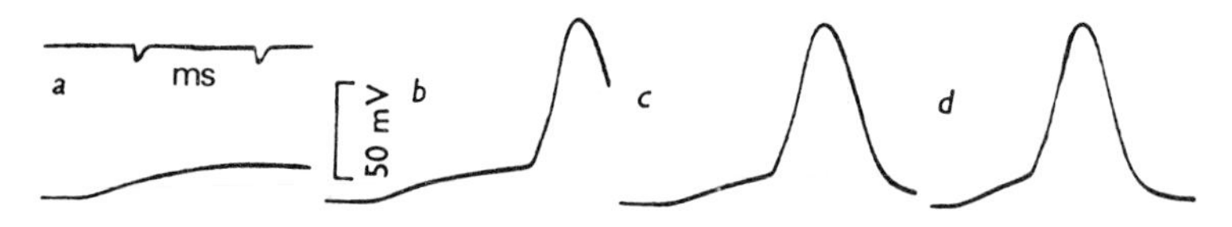

the presence of synaptic vesicles in terminal boutons. The transmitter substance is probably glutamic acid (p. 212).

Components of the EPSP

Fig. 9.4 shows that the motoneuronal EPSP arises from the activity of several afferent fibres. What would the response to individual afferent fibres be? Could we distinguish responses at single terminal boutons, or see quantal responses arising from the release of single vesicles? How many ionic channels are opened by such responses?

Some answers to these questions have been produced by Redman and his colleagues. Jack, Redman and Wong (1981) looked at the responses to activity in single presynaptic axons: they cut away the dorsal root to leave just a thin filament in which only one group Ia fibre from a particular muscle was active. We can call the response to stimulation of a single presynaptic fibre a unitary EPSP. These unitary EPSPs were only a few hundred microvolts in size and were superimposed upon a noisy background (fig. 9.7*a*). A large number (800) of these EPSPs were recorded and stored in digital form for computer analysis following stimulation of the same presynaptic fibre. Fig. 9.7*b* shows the average form of the EPSPs and fig. 9.7*e* is a histogram showing their frequency distribution. This can be compared with the frequency distribution of the background noise voltage (fig. 9.7*f*).

The histogram in fig. 9.7*e* is wider than that of the noise and somewhat skewed. This means that there must be some variation in the EPSPs in addition to that attributable to the background noise. If we assume that this variation is discrete rather than continuous then it is possible, by using some rather sophisticated statistical computing, to determine its frequency distribution. The result, shown in fig. 9.7*g*, implies that the EPSPs in this experiment fluctuate between four discrete peak voltages. This conclusion is strengthened by the analysis shown in fig. 9.8, in which the histogram and the average EPSP are reconstructed from the four different components.

The amplitudes of the four components in this experiment differ by about 100 μV and the smallest one is 302 μV, so it looks as though they consist of 3, 4, 5 and 6 quantal units respectively. Other unitary EPSPs consisted of fewer or (rarely) more quantal units. We might expect that these quantal units correspond to the release of single vesicles, or the activity of single synaptic boutons, or perhaps (if one vesicle is released per bouton) both. We need some more detailed anatomical information to resolve this question, and this has been provided by Redman and Walmsley (1983*a, b*).

Redman and Walmsley recorded unitary

Fig. 9.6. Effect of membrane potential on the size of the EPSP in a cat motoneuron. The membrane potential was set at the values shown to the left of each set of records by passing current through one barrel of a double-barrelled intracellular electrode; the other barrel was used to record the membrane potential. The traces at −42 and −60 mV show the initiation of action potentials. (From Coombs *et al.*, 1955*c*.)

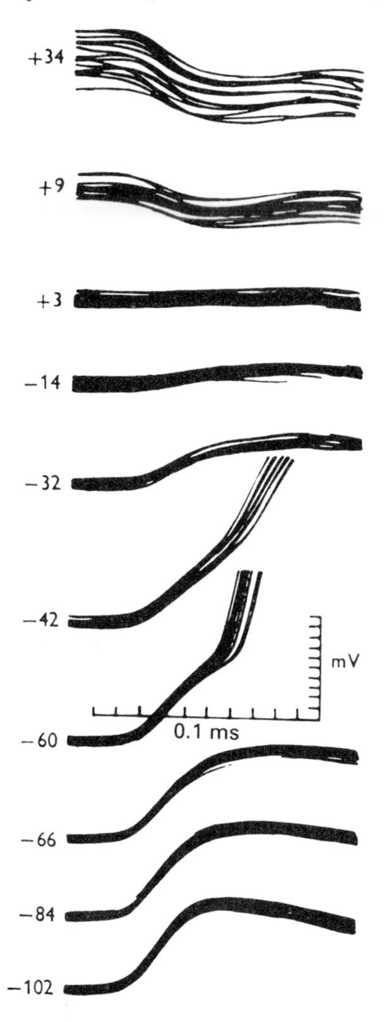

EPSPs from single motoneurons and analysed them in the same way as in the earlier experiments. However, here the recording microelectrode was filled with a solution containing horseradish peroxidase (HRP). This could be injected into the motoneuron by ionophoresis and later stained for light or electron microscopy, so as to reveal the dendritic field of the motoneuron. HRP was also used to stain the terminals of the group Ia fibre that produced the unitary EPSP,

Fig. 9.7. Components of unitary EPSPs in a spinal motoneuron. 800 responses to stimulation of a single group Ia fibre were recorded in digitized form on disc; the records in *a* show four of them. *b* shows the averaged time course for all 800 responses and *c* is the standard deviation of these responses. 800 similar traces were obtained in the absence of stimulation, so as to measure the background noise of the system; *d* shows three of these records. *e* is a histogram of the sizes of the 800 EPSPs. *f* is a similar histogram showing the frequency distribution of the noise, with the dotted line as a Gaussian curve fitted to it. *g* shows the deconvolved probability graph by which *e* can be modelled as a combination of four discrete events combined with the Gaussian curve. (From Jack *et al.*, 1981.)

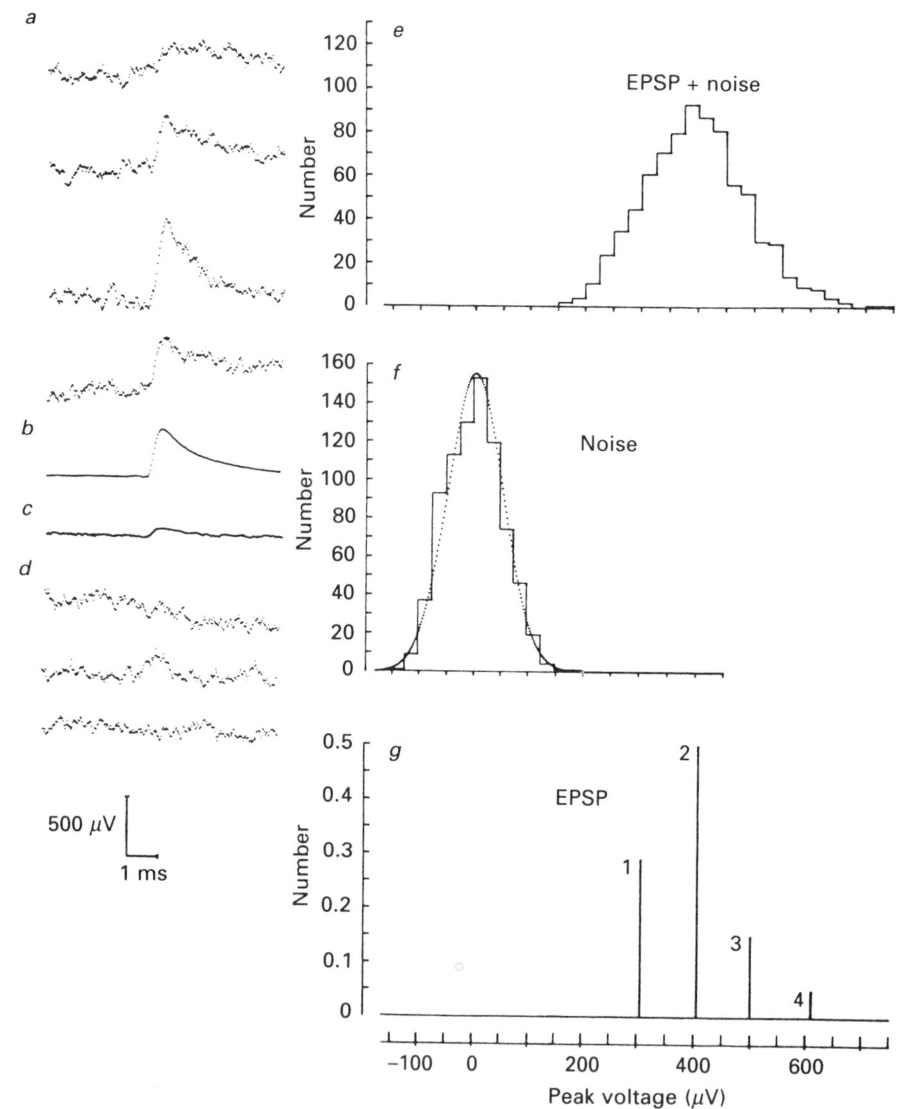

by applying it to the cut end of the dorsal root filament.

One of Redman and Walmsley's results is shown in fig. 9.9. Here the presynaptic terminal makes contact with the motoneuron at four boutons, three on one branch of a dendrite and one on the other. The average of 1600 unitary EPSPs is shown in fig. 9.9*a*. Statistical analysis gives the

Fig. 9.8. Reconstructing the average unitary EPSP from its four components. The histogram *a* is the one shown in fig. 9.7*e*. The four dotted curves show the contributions of the individual components as determined in fig. 9.7*g*: thus curve 1 is a normal (Gaussian) distribution with mean 302 μV (from line 1 in fig. 9.7*g*), S.D. 51.4 μV (as in the noise curve in fig. 9.7*f*) and amplitude such that its cumulative probability is 0.29 (again from line 1 in fig. 9.7*g*). The EPSP labelled 1 in *b* is the average of responses attributable to this component. Curves 2, 3 and 4 are derived similarly from fig. 9.7*g* and *f*, corresponding to the average EPSPs 2, 3 and 4. Addition of curves 1 to 4 gives the dashed curve in *a*, which fits the histogram very well. Weighted averaging of the EPSP components 1 to 4 gives the reconstructed average EPSP *c*, which is effectively identical with the actual average EPSP *d*. (From Jack *et al.*, 1981.)

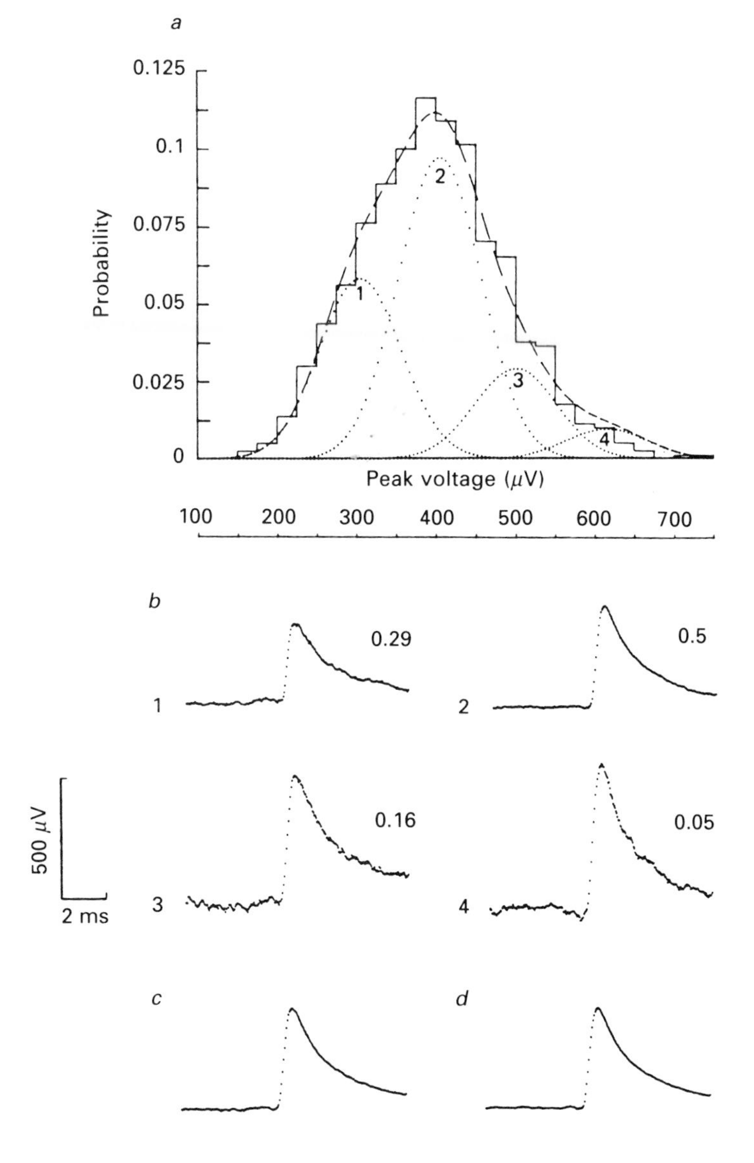

two components of this as unitary EPSPs with amplitudes of 322 and 443 μV (fig. 9.9*b–d*), implying quantal contents of 3 and 4 respectively. Redman and Walmsley's interpretation of this is that the 322 μV EPSP results from transmission at three boutons at which failures never occur, and that the 443 μV EPSP results from transmission at all four boutons. The all-or-nothing nature of the response at individual boutons suggests that the contents of just one vesicle are released per bouton. It is very pleasing that these delicate and time-consuming experiments have provided such agreement

Fig. 9.9. Components of the unitary EPSP produced at an identified motoneuronal synapse. After the experiment the motoneuron and the presynaptic group Ia fibre were filled with horseradish peroxidase so that their structure could be seen, as described in the text. The motoneuron is the same one as in fig. 9.2; its outline is shown as a dotted line here. The preterminal branches of the group Ia fibre are shown as a continuous line, with the four synaptic contacts (boutons) arrowed. Unitary responses were analysed as in fig. 9.7. *a* shows the averaged EPSP (upper trace) and its standard deviation (lower trace). Deconvolution produced the fluctuation pattern for the peak amplitude of the EPSP as shown in *b*. *c* and *d* show the two component EPSPs and their standard deviations. (From Redman and Walmsley, 1983*a* and *b*.)

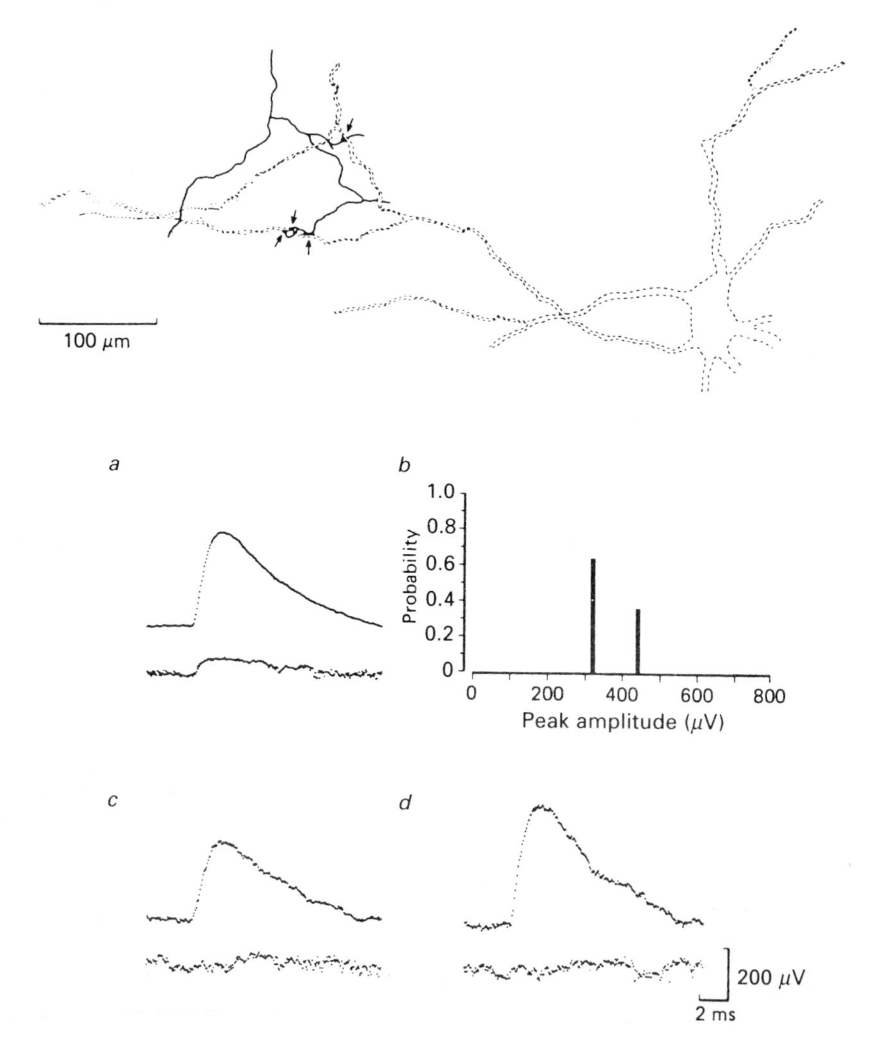

between the microcircuitry of the synaptic contact and the physiological response in the motoneuron.

Fig. 9.9 also shows the standard deviation of the component EPSPs. It is very small, giving a coefficient of variation (the standard deviation divided by the mean) of less than 5%. This compares with a value of around 30% for the corresponding measure in miniature end-plate potentials, as is evident from fig. 7.22*a*. Why should the quanta seen at motoneurons be so much less variable than those seen at the neuromuscular junction? Redman and his colleagues suggest that the effect may be caused by saturation of the receptors on the subsynaptic membrane, so that the contents of one vesicle are more than sufficient to open all the postsynaptic transmitter-activated channels at any particular bouton. Variation in vesicle size would not then be reflected in variation in the postsynaptic response. Even if two vesicles were released from one bouton there would be no increase in the postsynaptic response.

Another curious feature also emerges from these experiments. The electrical behaviour of the dendritic tree can be modelled using cable theory (Rall, 1962, 1977; Jack *et al.*, 1975). When an EPSP, recorded in the soma, results from a brief injection of current as a result of synaptic action some distance away along a dendrite, it should be smaller and have a longer time course than one resulting from the same current injected near to or at the soma (compare figs. 4.13 amd 7.4). Redman and Walmsley (1983a) found that the time to peak of the EPSP does indeed increase with increasing distance of the synapse from the soma in the way predicted. But the quantal components of EPSPs with longer time courses are not less in amplitude than those with shorter time courses (Jack *et al.*, 1981). This surprising result might mean that the presynaptic vesicles are larger in boutons synapsing on the more distal regions of the dendrites; this seems highly improbable. A much more likely explanation is that the density of receptors and their channels on the postsynaptic membrane is greater the farther the distance from the soma.

The size of the quantal component of the EPSP is about 100 μV. Voltage-clamp measurements on soma synapses show that this is brought about by a peak current of about 330 pA, corresponding to a conductance of about 5 nS (Finkel and Redman, 1983). If the single channel conductance is 21 pS (as in the glutaminergic synapses on crayfish muscle), then 240 channels would be opened. However, if it is 125 pS (as in locust muscle), only 40 channels would be opened. For distant dendritic synapses we have to multiply these figures by ten, giving 400–2400 channels opened per unitary EPSP. These figures are compatible with the suggestion that the contents of a single vesicle, which we might expect to include several thousand transmitter molecules, are sufficient to saturate all the receptors at one bouton.

Initiation of the action potential

As is shown in fig. 9.5, an EPSP of sufficient size elicits an action potential in the motoneuron. The form of this action potential is rather more complicated than that in the axon at some distance from the site at which it is initiated; as recorded in the soma, the action potential consists of three components. These are (1) the M spike, representing activity in the *m*yelinated region of the axon and recorded by a microelectrode placed in the soma as a small depolarization of 1–3 mV; (2) the IS spike, representing activity in the *i*nitial *s*egment of the axon, which is not surrounded by a myelin sheath, and recorded in the soma as a depolarization of 30–40 mV; and (3) the SD spike, representing activity in the *s*oma and *d*endrites, and therefore recorded as a full-sized action potential of 80–100 mV.

This analysis (due to Coombs *et al.*, 1955*a*, 1957*a*, *b*) was derived mainly by examination of the effects of the soma membrane potential on the form of an antidromic action potential produced by stimulation of the ventral roots. The results of such an experiment are shown in fig. 9.10*a*. At the resting potential (−80 mV in this case) only the IS spike is seen. It is reasonable to assume that the size of the IS spike *in the initial segment* is about 100 mV, but the electrotonic currents set up by this activity only produce a depolarization of about 40 mV in the soma, where it is recorded; since this is still below the threshold for the regenerative response of the soma and dendrites, no SD spike arises. However, if the soma is depolarized below about −78 mV, the electrotonic currents produced by activity in the initial segment are now sufficient to excite the soma, so an SD spike arises from the IS spike, with decreasing latency at lower membrane potentials (compare the traces obtained at

−77 and −63 mV in fig. 9.10*a*). Many motoneurons are sufficiently excitable to produce SD spikes in the absence of any depolarization prior to the IS spike. If the soma is hyperpolarized beyond about −82 mV (a procedure which must also hyperpolarize the initial segment to some extent), the IS spike disappears, and we are left with a very small deflection, the M spike (shown in the traces obtained at −87 mV in fig. 9.10*a*), which seems to be caused by the electrotonic currents produced in the soma by the action potential in the myelinated region of the axon.

How does this analysis relate to the action potentials initiated orthodromically by excitatory synaptic action? An initial IS component can be seen at the beginning of the SD spike produced by an EPSP of sufficient size (fig. 9.11*b*). This implies that the initial segment is the site of impulse initiation in orthodromic excitation (fig. 9.12). The reason for this is that the threshold is very much lower in the initial segment, so that it can be preferentially excited by the depolarization produced by the EPSP. The IS spike then itself excites both the soma and the myelinated axon. Confirmation of this view is provided by the experiments of Coombs *et al.* (1957*b*) in which the M spike of a single neuron (produced in response to orthodromic stimulation) was recorded from a thin bundle of fibres (a 'filament') in the ventral root while the soma was hyperpolarized. It was found that blockage of the SD spike alone did not affect production of an action potential in the axon, whereas blockage of the IS spike always eliminated the axonal response. Furthermore, the action potential in the ventral root filament followed the IS spike by about the time that one would expect from the conduction velocity of the motor fibres,

Fig. 9.10. Analysis of action potentials recorded in the motoneuronal soma following antidromic stimulation. *a*: Effect of soma membrane potential on the response. *b*: Diagram of a motoneuron showing the probable sites of the SD, IS and M spikes. (From Eccles, 1957.)

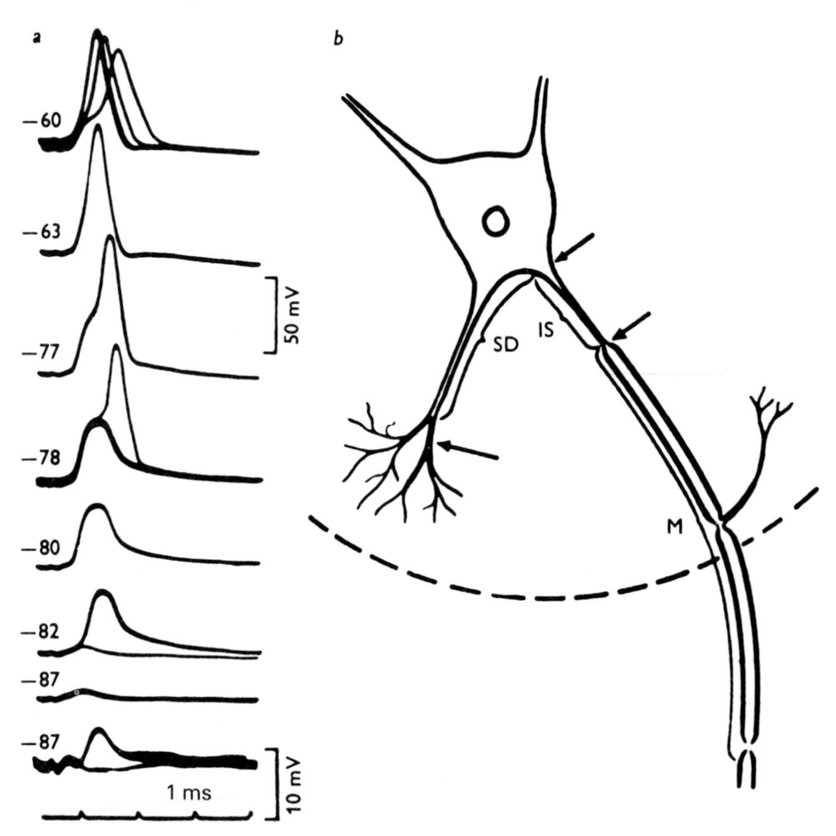

Fig. 9.11. Action potentials recorded intracellularly from cat motoneurons after (*a*) antidromic and (*b*) orthodromic (monosynaptic) stimulation. The lower traces are electrically differentiated records of the upper traces, i.e. they show the rate of change of membrane potential with time. (From Coombs *et al.*, 1957*b*.)

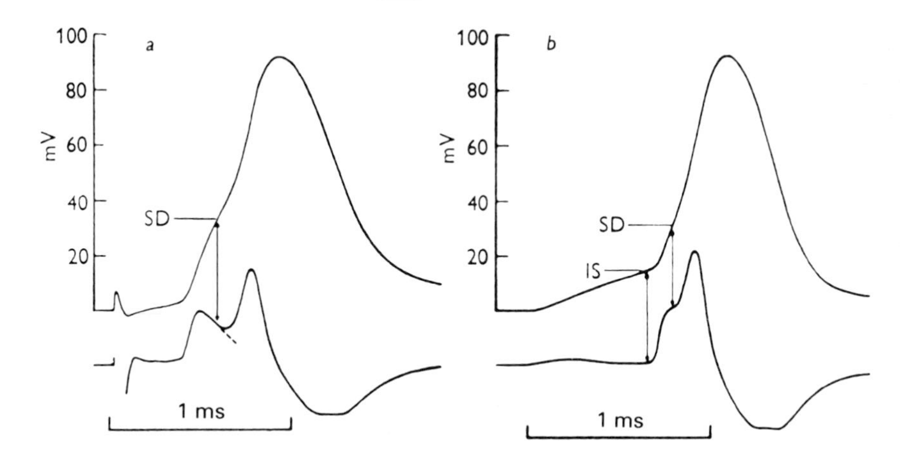

Fig. 9.12. Spike initiation in vertebrate spinal motoneuron. The arrows in the diagram show how the initial segment is depolarized by the local circuit currents set up by excitatory synaptic action in the soma and dendrites. Also shown are the sequences of events in orthodromic and antidromic activation.

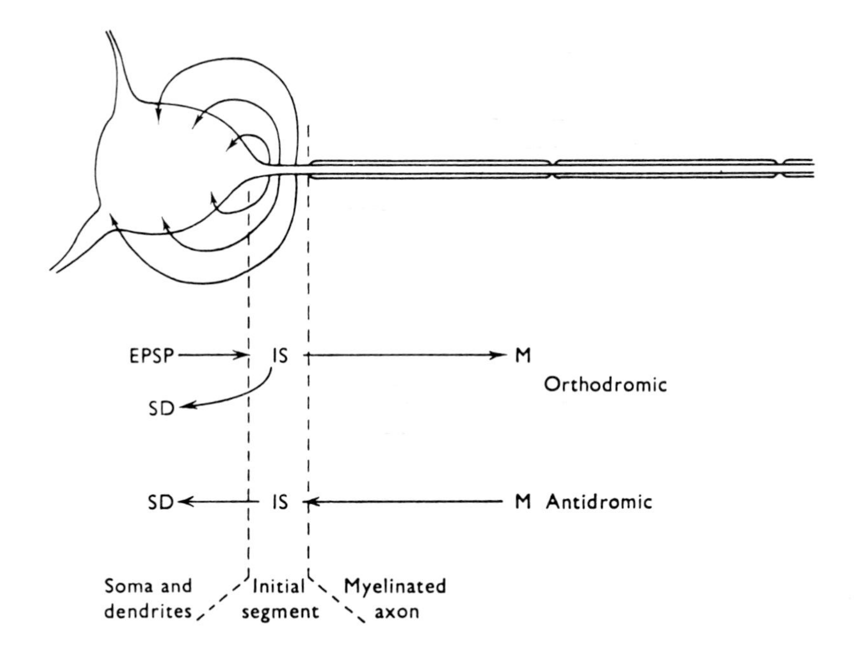

whereas if the SD spike initiated the peripheral action potential, the conduction velocity would have to be impossibly fast (fig. 9.13).

Fig. 9.13. Evidence that the action potential in the axon of a motoneuron is initiated by the IS spike and not by the SD spike. The upper trace shows the action potential recorded intracellularly from a posterior biceps-semitendinosus motoneuron sona following monosynaptic activation; the lower trace shows the response recorded from a small bundle of fibres in the ventral root. The arrow marks the time at which these action potentials in the ventral root must have been initiated centrally. (From Coombs *et al.*, 1957*b*.)

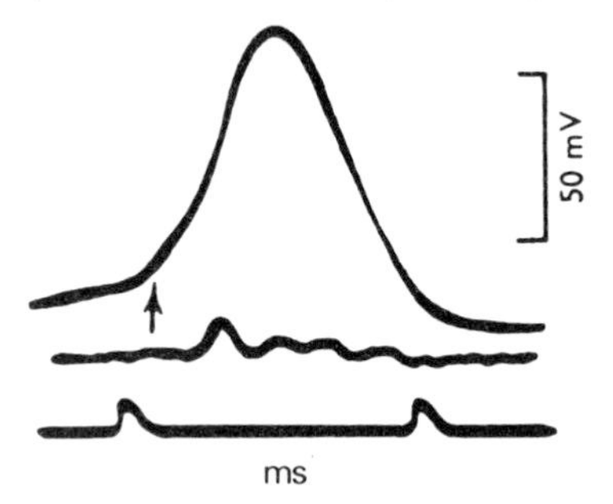

Fig. 9.14. The direct inhibitory pathway. This diagram is much simplified. Many afferent, inhibitory and efferent neurons are involved at each stage; each inhibitory interneuron is innervated by several afferents, and itself innervates several motoneurons.

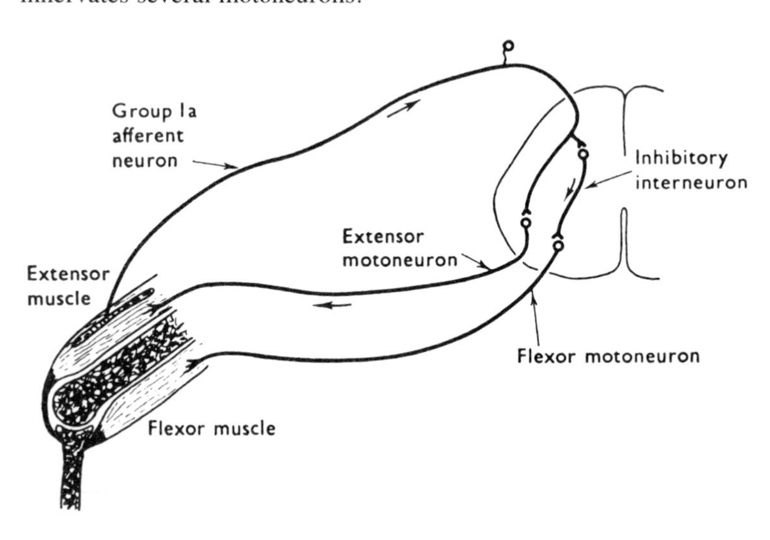

Inhibition in mammalian spinal motoneurons

If the contraction of a particular limb muscle is to be effective in producing movement, it is essential that those muscles which oppose this action, the *antagonists*, should be relaxed. In the monosynaptic stretch reflex systems of mammals, this is brought about by *inhibition* of the motoneurons of the antagonistic muscles, through a system known as the direct inhibitory pathway. Fig. 9.14 shows the arrangement of neurons in this pathway. Group Ia afferent fibres from the stretch receptors in a particular muscle (an extensor in fig. 9.14) synapse in the ventral horn with motoneurons innervating that muscle and, to a lesser extent, with its synergists (i.e. other muscles acting in a similar way). These afferents also synapse, in a region called the intermediate nucleus, with small interneurons which themselves innervate the motoneurons of antagonistic muscles. It is these interneurons which exert the inhibitory action on the motoneurons of antagonistic muscles. This inhibitory action can thus be examined by inserting a microelectrode into a motoneuron and stimulating the group Ia afferents from an antagonistic muscle. Fig. 9.15 shows the results of such an experiment: the records were taken from a biceps-semitendinosus (flexor) motoneuron in response to group Ia afferent volleys from the quadriceps muscle (an extensor). The responses consist of small hyperpolarizing potentials known as *inhibitory postsynaptic potentials*, or IPSPs.

Apart from the fact that it is normally hyperpolarizing rather than depolarizing, the shape of the IPSP is similar to the shapes of the EPSP and the EPP. Hence we might expect it to be produced by a similar mechanism, i.e. by a brief change in the ionic conductance of the membrane, which causes the initial hyperpolarization, followed by a passive decay of the charge on the membrane capacitance. This idea can be verified by using a voltage-clamp technique (Araki and Terzuolo, 1962), as is shown in fig. 9.16. Observation of the effects of the membrane potential on the size of the IPSP provides a further test of the hypothesis. It is found (Coombs *et al.*, 1955*b*) that the IPSP increases on depolarization, but decreases and then reverses in sign on hyperpolarization (fig. 9.17).

The question now arises, to which ions does the postsynaptic membrane alter its permeability during the action of the inhibitory transmitter substance? As we have seen, this type of problem is most readily solved by observing the effects of altering the ionic concentration gradients across the membrane. This was done by Coombs and his colleagues (1955*b*); since it is very difficult to change the external environment of spinal motoneurons, they altered the ionic concentrations inside the cell by injecting various ions through one barrel of a double microelectrode, the other barrel being used to measure membrane potentials. After chloride injection, the reversal potential always moved to a more depolarized level, so that the IPSP at the resting potential became a depolarizing response of increased size (fig. 9.18). This result implies that the potential at the peak of the IPSP is at least partially determined by the Nernst equation for chloride ions, i.e. the inhibitory transmitter substance (which is probably glycine) causes an increase in the permeability of the postsynaptic membrane to chloride ions.

Interactions between inhibitory and excitatory postsynaptic potentials

We have seen that an EPSP which is sufficiently large will cause the production of a propagated action potential in the axon of a motoneuron. If an IPSP is induced so as to coincide approximately with the EPSP, the depolarization produced by the EPSP is reduced, so that the membrane potential does not cross the threshold and no action potential ensues (fig. 9.19). When the EPSP is too small to elicit an action potential we can see the effects of interaction with the IPSP more clearly (fig. 9.20). If the EPSP follows the IPSP by more than about 2 ms, it is not reduced in size (as measured from foot to peak) but the membrane potential at its peak is more negative than usual (*b–f*, fig. 9.20). At much shorter intervals, when the

Fig. 9.16. Time course of the inhibitory transmitter action in motoneurons. The upper record shows an IPSP, the lower record shows the current flow when the membrane potential was clamped at the resting level. (From Araki and Terzuolo, 1962.)

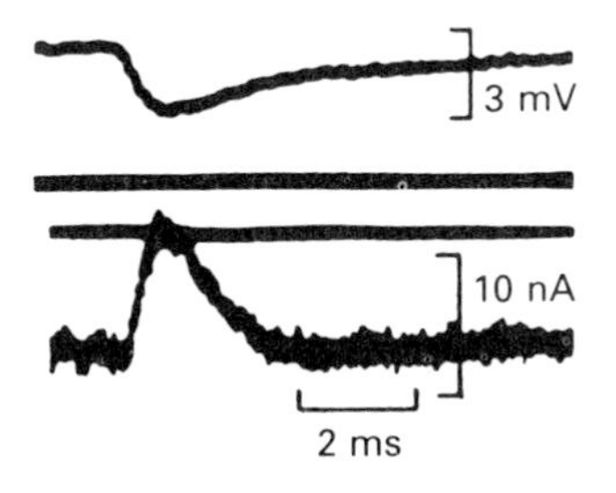

Fig. 9.15. Inhibitory responses in a cat motoneuron (biceps–semitendinosus) to afferent volleys from the antagonistic muscle (quadriceps). Stimulus intensity increases from *a* to *f*. The upper trace shows the afferent volley recorded from the dorsal root. (From Coombs *et al.*, 1955*d*.)

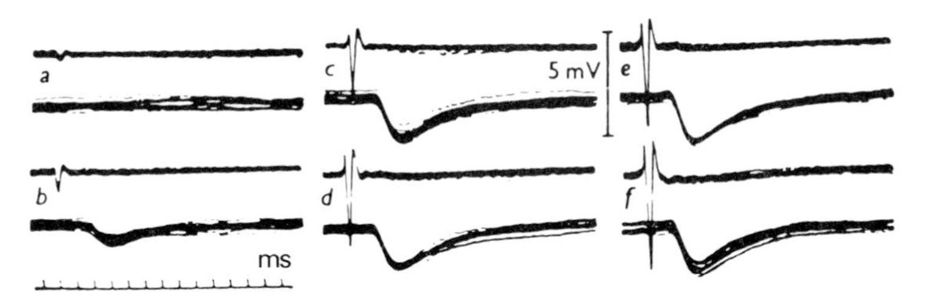

EPSP and IPSP are practically coincident (*g*–*i*), the EPSP is reduced in size; this is because the increase in chloride conductance reduces the effects of the excitatory transmitter. If the IPSP occurs after the peak of the EPSP (*j* and *k*) it does not, of course, affect the peak depolarization.

Thus there are two effects of inhibition: one, caused by the hyperpolarization of the membrane, which lasts for the whole of the IPSP, and the other, caused by an antagonism between the inhibitory and excitatory synaptic currents, which lasts only for the duration of the conductance change which occurs during the rising phase of the IPSP. This analysis accords very well with the effect of an inhibitory volley on the

Fig. 9.17. The effect of membrane potential on the magnitude of the IPSP in a cat motoneuron. *a*: A series of records of the IPSP at different membrane potentials. *b*: These results (and others from the same cell) are expressed in graphical form; the arrow indicates the resting potential. (From Coombs *et al.*, 1955*b*.)

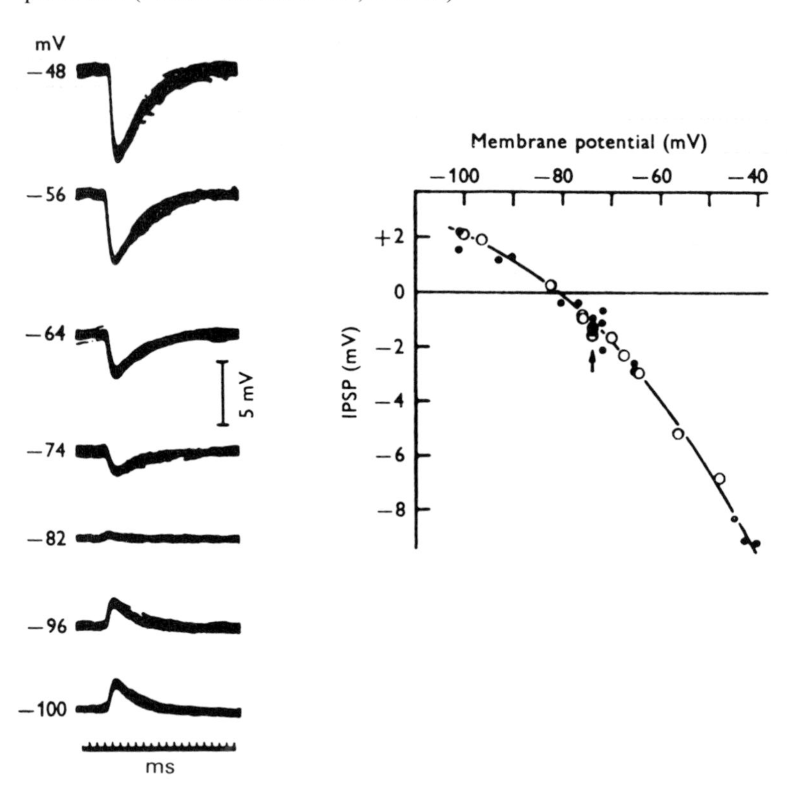

Fig. 9.18. The effect of increasing the internal chloride concentration on the IPSP of a cat motoneuron. The records were obtained by inserting a micropipette electrode filled with 3 M KCl into the cell. Record *a* was obtained immediately after insertion, records *b* and *c* at successively later times. Notice the change in the IPSP following diffusion of chloride out of the electrode. Record *d* was obtained immediately after *c*, but with the membrane potential set at a lower level (−27 mV, instead of −59 mV). (From Coombs *et al.*, 1955*b*.)

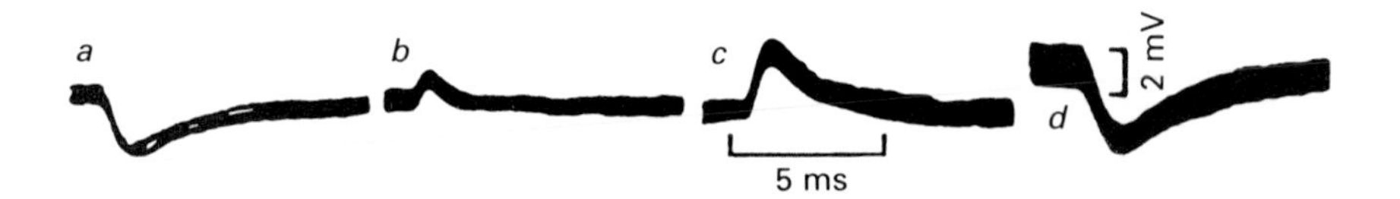

monosynaptic response recorded from a ventral root. It is found (fig. 9.21) that the compound action potential in the ventral root is very much reduced in size when the excitatory volley occurs up to 2 ms after the inhibitory volley, but thereafter the degree of inhibition is such as one might expect from the hyperpolarization constituting the IPSP.

The motoneuron is in a sense a decision-making device. The decision to be made is whether or not to 'fire', that is to say whether or not to send an action potential out along the axon towards the muscle. If the incoming excitatory synaptic action is sufficiently in excess of the incoming inhibitory action, the resulting depolarization will cross the

Fig. 9.19. Interactions between inhibitory and excitatory responses in a cat biceps–semitendinosus motoneuron. The lower records show motoneuron membrane potentials, the upper records monitor group Ia afferent responses in the sixth lumbar dorsal root. The excitatory response alone, produced by stimulating the biceps–semitendinosus nerve, is shown in *j* (the peak of the resulting action potential is not shown in these records). The inhibitory response alone, produced by stimulating the quadriceps nerve, is shown in *i*, *a* to *h* show responses to stimulation of both nerves, the excitatory stimulus following the inhibitory stimulus after the times (ms) shown by each record. The conduction time in the inhibitory pathway is about 1 ms longer than that in the excitatory pathway, so that the EPSP occurs before the IPSP in records *f* to *h*. (From Coombs *et al.*, 1955*d*.)

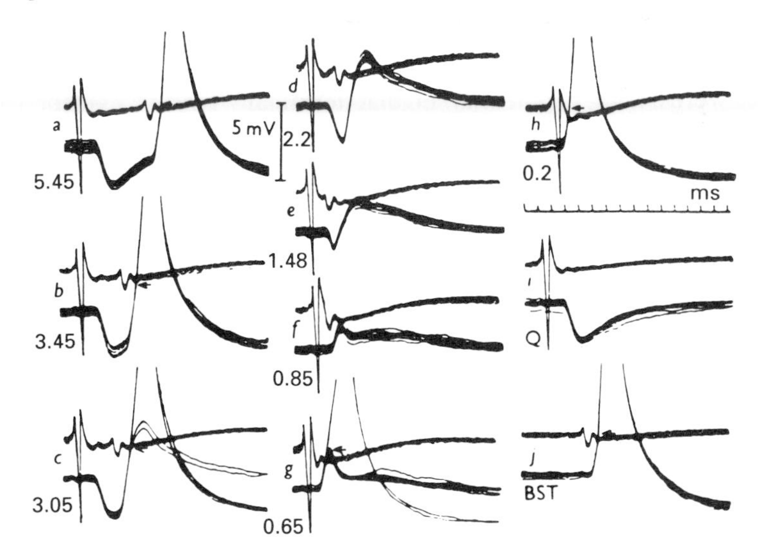

Fig. 9.20. Interaction between inhibitory and excitatory responses in a cat motoneuron. The experimental procedure was similar to that in fig. 9.19, except that the excitatory stimulus was insufficient to produce an action potential in the motoneuron. Trace *a* shows the IPSP alone and trace *l* shows the EPSP alone. Traces *b* to *k* show superimposed records of (i) the EPSP alone and (ii) EPSP plus IPSP at various intervals between the two. Further details in the text. (From Curtis and Eccles, 1959.)

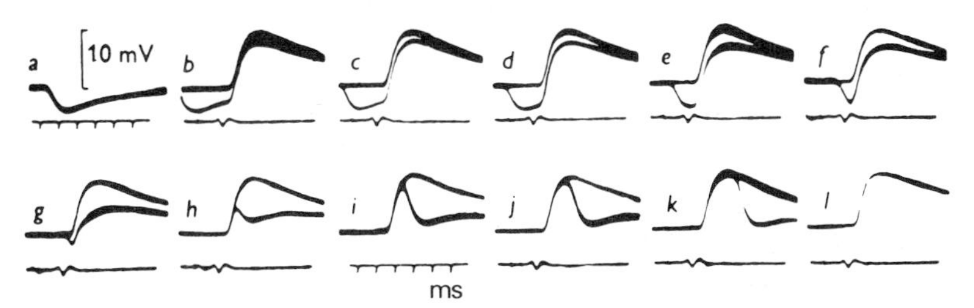

threshold for production of an action potential and the motoneuron will 'fire'. But a reduction in synaptic excitation or an increase in synaptic inhibition will make the membrane potential more negative so that it drops below the threshold and the motoneuron ceases firing. We should remember that the motoneuron receives excitatory and inhibitory inputs from many sources, so that, for example, a 'decision' based on inhibition from group Ia fibres from an antagonistic muscle may be 'overruled' by excitatory inputs from neurons descending from the brain.

Round and flat vesicles

In studying the fine structure of the cerebral cortex by electron microscopy, Gray (1959) was able to discern two structural types of synapse. One of them, called Type 1, showed more marked postsynaptic thickenings and wider synaptic clefts than the other, called Type 2 (see fig. 9.1). It soon became evident that, in those cases where their function was known, Type 1 synapses were excitatory and Type 2 were inhibitory. Then Uchizono (1965) made a remarkable discovery. In aldehyde-fixed sections of the cerebellar cortex, he found that the synaptic vesicles in known excitatory synapses were spherical or 'round' whereas those of inhibitory synapses were ellipsoidal or 'flat'. A similar distinction between round and flat vesicles can be seen in nerve terminals in the spinal cord; presumably they also correspond to excitatory and inhibitory endings.

It seems very likely that 'flat' vesicles are in fact 'round' in life, and simply become 'flat' as a result of osmotic changes occurring during the fixation process (Peters *et al.*, 1976). In freeze-fractured material all vesicles appear to be spherical (Akert *et al.*, 1972). But clearly there must be some difference between the two types before fixation. Whatever this difference is, it does provide another useful tool for the neuroanatomist.

Presynaptic inhibition

The amount of transmitter released from a nerve terminal is related to the size of the action potential in that terminal. This is evident from studies on neuromuscular transmission and on squid giant synapses (fig. 7.30). If the nerve

Fig. 9.21. The effect of an inhibitory volley on the size of the monosynaptically excited compound action potential recorded in the ventral root. *a* shows the inhibitory (quadriceps) volley recorded in the dorsal root (upper trace), and also the IPSP (lower trace) that this produces in the biceps–semitendinosus motoneuron. The time scale of these records is the same as the abscissa of the graph *b*. The ordinate of *b* shows the size of the compound action potential in the ventral root, with the response in the absence of inhibition taken as 100%. The two curves in *b* were obtained from different sizes of afferent volley. The broken lines indicate the time courses of that part of the inhibitory process which can be attributed directly to the hyperpolarization of the IPSPs. (From Araki *et al.*, 1960.)

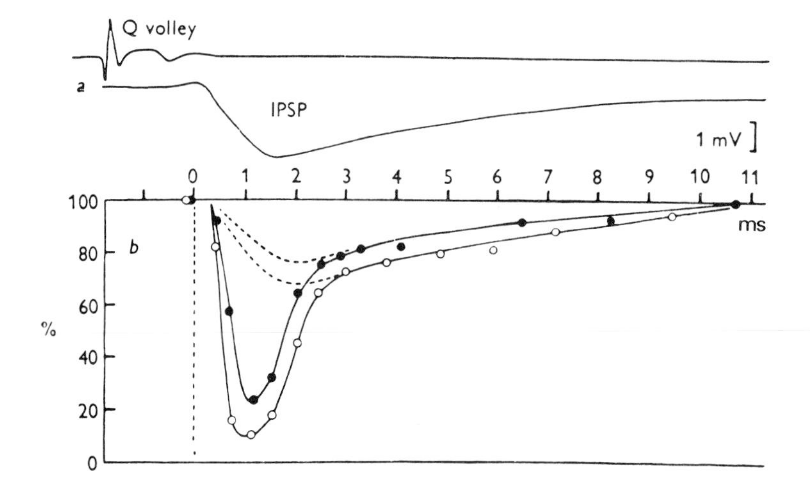

terminal is slightly depolarized, the size of the action potential will be correspondingly reduced, partly because it will arise from a less negative level of membrane potential, and partly because the depolarization will cause slight increases in the potassium conductance and sodium inactivation of the nerve terminal membrane. Consequently, a depolarization of the nerve terminal by a few millivolts will considerably reduce the amount of transmitter released, and will therefore also reduce the size of the postsynaptic response produced by activity in that terminal.

Now if there were some mechanism in the nervous system whereby such a depolarization of a presynaptic terminal could be brought about by the activity of a second neuron, then this second neuron would be capable of reducing the responses in the postsynaptic cell elicited by the action of the first neuron. Such a mechanism would thus be inhibitory, and since it would act presynaptically, we could call the phenomenon *presynaptic inhibition*, to contrast it with the postsynaptic inhibition mechanism which we have already examined.

There is good evidence that presynaptic inhibition does in fact occur in the mammalian spinal cord (Frank and Fuortes, 1957; Eccles *et al.*, 1961; Schmidt, 1971). Consider the experiment whose results are shown in fig. 9.22. The EPSPs shown in records *a–d* were obtained from a gastrocnemius motoneuron in response to a single stimulus exciting the group Ia fibres in the gastrocnemius–soleus nerve. When the stimulus was preceded by stimulation of the group I fibres in the posterior biceps–semitendinosus nerve, as in records *b–d*, the EPSP was depressed in size, although there was no evidence of any inhibitory synaptic action on the motoneuron itself. This depression, or inhibitory action, began about 2.5 ms after the inhibitory volley entered the cord, reached a maximum after about 15 ms, and lasted for over 200 ms (fig. 9.22*e*).

This experiment shows that it is possible for the monosynaptic EPSP to be depressed by neuronal action without there being any IPSP in the motoneuron, and therefore suggests that the inhibitory action takes place presynaptically. But it does

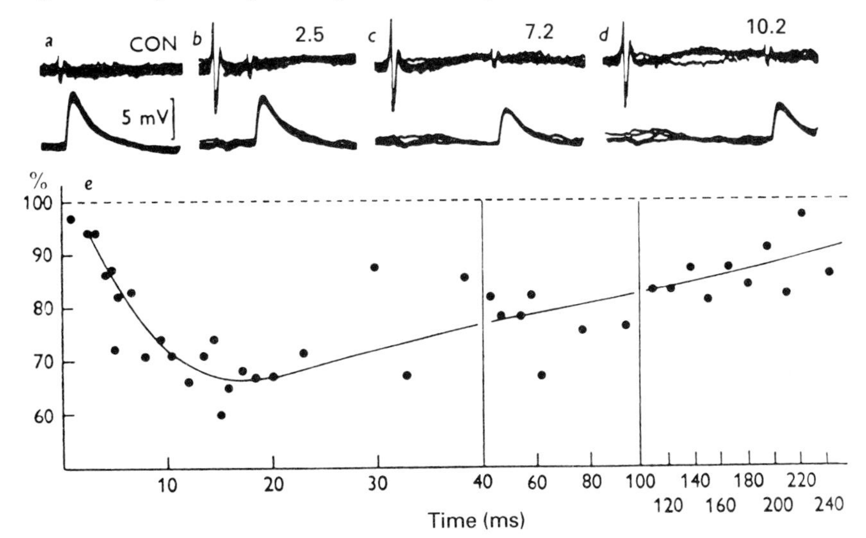

Fig. 9.22. Presynaptic inhibition in the spinal cord of the cat. The lower traces in *a* to *d* show EPSPs recorded from a gastrocnemius motoneuron, produced by stimulation of the gastrocnemius–soleus nerve. The upper traces show extracellular records from one of the dorsal roots. In *b* to *d*, the excitatory volley was preceded by a volley in the posterior biceps–semitendinosus nerve; the figures above each trace give the time intervals between the two volleys. The graph *e* shows the depression of the EPSP (expressed as a percentage of its size in the absence of inhibition) at various times after the inhibitory volley. (From Eccles *et al.*, 1961.)

not tell us anything about how the inhibitory action is brought about. It has been known for some time (e.g. Barron and Matthews, 1938) that dorsal root volleys are followed by a depolarization (the *dorsal root potential*) which spreads electrotonically along the same or adjacent dorsal roots. Eccles, Magni and Willis (1962) showed that dorsal root afferent fibres subjected to presynaptic inhibition are in fact depolarized, and that the depolarization follows a time course which is apparently identical with that of the inhibitory action. Thus it seems that presynaptic inhibition is brought about by depolarization of the afferent nerve terminals in the way suggested at the beginning of this section.

A central delay of 2.5 ms is longer than would be expected in a pathway with a single synapse only, so it is probable that the presynaptic inhibitory pathway involves one or two interneuronal stages (fig. 9.23*a*). This could account for the long duration of presynaptic inhibition, if each interneuron fires many times when it is excited. Fig. 9.23*b* shows, in diagrammatic form, the suggested anatomical basis of presynaptic inhibition; 'serial synapses' of this type have in fact been seen in electron micrographs of the spinal cord (Gray, 1962).

Slow synaptic potentials

In the sympathetic nervous system of vertebrates there is a chain of ganglia lying near to the spinal cord. These contain the cell bodies of the postganglionic fibres which terminate on smooth muscle or gland cells (fig. 10.1). The preganglionic fibres arise in the spinal cord and form synapses with the cell bodies of the postganglionic fibres in the ganglia.

Bullfrog sympathetic ganglia contain B cells and C cells, the B cells being the larger. Each is innervated by preganglionic fibres which form numerous synaptic boutons on the neuronal soma. They show a variety of different types of synaptic activity (Kuffler, 1981; Adams *et al.*, 1986), as is shown in fig. 9.24.

A single stimulus applied to the preganglionic fibres produces a fast EPSP in both B and C cells, and this may be large enough to produce an action potential in the postganglionic fibres (fig. 9.24*a*). The response is blocked by curare and can be mimicked by acetylcholine. The size of the fast EPSP is linearly related to membrane potential, with a reversal potential at about −5 mV (Nishi and Koketsu, 1960). All this suggests that the mechanism of production of the fast EPSP is very similar to that at the neuromuscular junction: it is mediated by nicotinic acetylcholine receptors in which a cation-selective ionic channel opens when acetylcholine is bound to it.

In B cells a slow EPSP with a much longer time course occurs after the fast EPSP (fig. 9.24*c*). Similar responses are seen after application of acetylcholine to the ganglion. The slow EPSP occurs in the presence of curare or other nicotinic blocking agents but is itself blocked by atropine,

Fig. 9.23. Diagram to show the suggested anatomical basis of presynaptic inhibition: *a*, neuronal connections; *b*, synaptic structure.

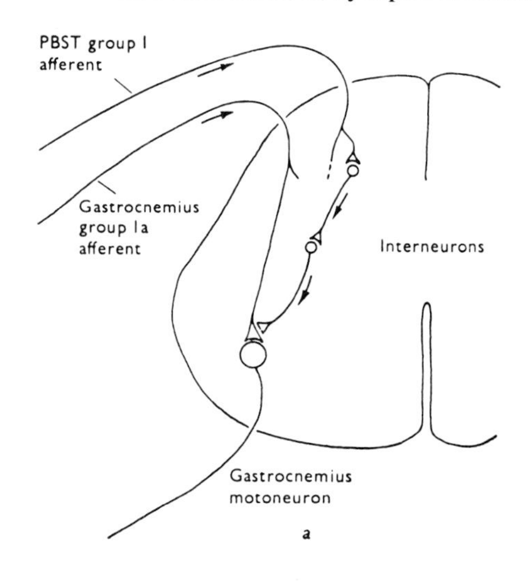

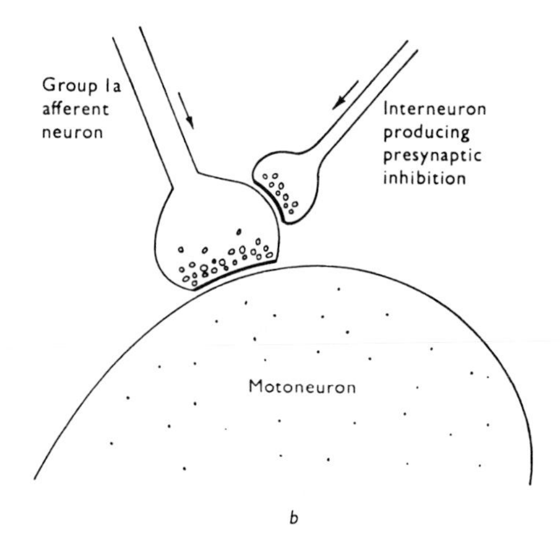

hence the receptors which mediate it are muscarinic.

Further investigation showed that the mechanism of the slow EPSP is quite different from that of the fast EPSP in that it is produced by a closing of ionic channels, not an opening of them (Kobayashi and Libet, 1970; Weight and Votova, 1970). Evidence for this conclusion is shown in fig. 9.25, in which fast and slow EPSPs are elicited at different membrane potentials by passing currents through a microelectrode in the postsynaptic cell. The fast EPSP decreases in size with depolarizing (positive) currents, reversing at a potential near to −5 mV. The slow EPSP, however, increases in size with depolarizing currents, and is reduced by hyperpolarizing currents, with a reversal potential at about −88 mV.

It is very striking in fig. 9.25 that the fast EPSP is always a voltage change towards the reversal potential (just as in figs. 9.6 and 9.17) whereas the slow EPSP is always a movement away from it. This implies that, while there is an increase in membrane conductance during the fast EPSP, there is a decrease in conductance during the slow EPSP. The reversal potential for the slow EPSP is very near to the equilibrium potential for potassium ions and is unaffected by changes in chloride equilibrium potential. Hence we can conclude that the slow EPSP is brought about by a temporary closure of some of the potassium channels in the cell.

Later experiments showed that the particular potassium channels that are closed are responsible for a potassium ion current (the M-current) which is activated rather slowly by depolarizations beyond −60 mV (Brown and Adams, 1980; Adams *et al.*, 1982; Adams and Brown, 1982). In some cells there is also an inward current which may be produced by movement of sodium ions; it is associated with a conductance increase and so must involve the opening of another set of channels (Jones, 1985). The functional purpose of the slow EPSP may be to increase the excitability of the neuron over relatively long time periods.

In C cells the fast EPSP is followed by a slow, hyperpolarizing IPSP. This IPSP is muscarinic with a latency of 50 ms or more and lasts for a few seconds (fig. 9.24*b*). It has a reversal potential

Fig. 9.24. Fast and slow synaptic responses in frog 10th sympathetic ganglion neurons. *a*, fast EPSP produced by a single preganglionic stimulus, and (right) a stronger stimulus excites more preganglionic fibres giving a larger EPSP which is sufficient to produce an action potential. *b*, slow IPSP in a C neuron, produced in response to a burst of stimuli applied to the spinal nerves; fast EPSPs were blocked with the curare-like compound dihyro-β-erythroidine. *c*, slow EPSP produced by brief repetitive stimulation of the sympathetic chain. *d*, the late slow EPSP produced by repetitive stimulation of the spinal nerves. Note the different time scales. (From Kuffler, 1980.)

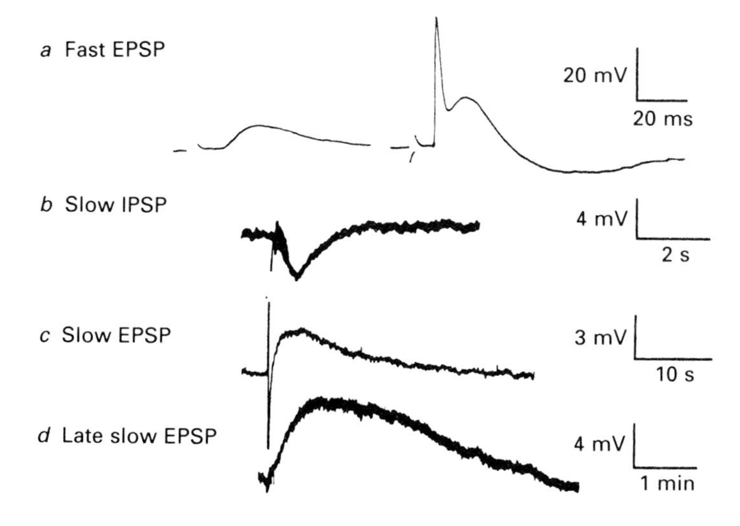

around −102 mV; this is unaffected by chloride ion concentration but strongly dependent upon the external potassium ion concentration with a slope of 55 mV per tenfold change (Dodd and Horn, 1983). Hence the slow IPSP is produced by the opening of channels selective for potassium ions.

The final event in this complex sequence of postsynaptic responses is a depolarization which lasts for a few minutes after a prolonged period of repetitive stimulation of the preganglionic fibres. This was discovered by Nishi and Koketsu (1968) and called by them the late slow EPSP (fig. 9.24*d*). It persists in the presence of both nicotinic and muscarinic blocking agents and is thus not cholinergic. There is good evidence that the transmitter substance is a peptide similar to luteinising hormone-releasing hormone, LHRH (Jan *et al.*, 1979; Kuffler, 1980; Jan and Jan, 1982). The ionic mechanism appears to be similar to that of the slow EPSP: closure of the M-current potassium channels and, in some cells, an additional inward current which may involve sodium ion flow (Jones, 1985).

The late slow EPSP is found in both C and B cells but only C cells possess presynaptic boutons that show immunoreactivity to LHRH. Hence it seems likely that the LHRH-like substance which is released from terminals on C cells can diffuse to the B cells many micrometres away. In accordance with this idea the time course of the late EPSP is faster in C cells than in B cells (Jan and Jan, 1982).

Slow potentials are widely distributed. Their time course, and especially their long latency, could be explained if channel opening or closing

Fig. 9.25. Effects of applied currents on fast and slow EPSPs in frog sympathetic ganglion cells. *a* shows a series of fast EPSPs each produced by a single preganglionic stimulus, with depolarizing (+) or hyperpolarizing (−) currents applied to the postsynaptic cell. *b* shows a similar series of slow EPSPs, produced by a burst of repetitive stimuli; the fast EPSPs were blocked with nicotine. The resting potential was −65 mV; reversal potentials were approximately 0 mV for the fast EPSP and −88 mV for the slow EPSP. (From Weight and Votova, 1970.)

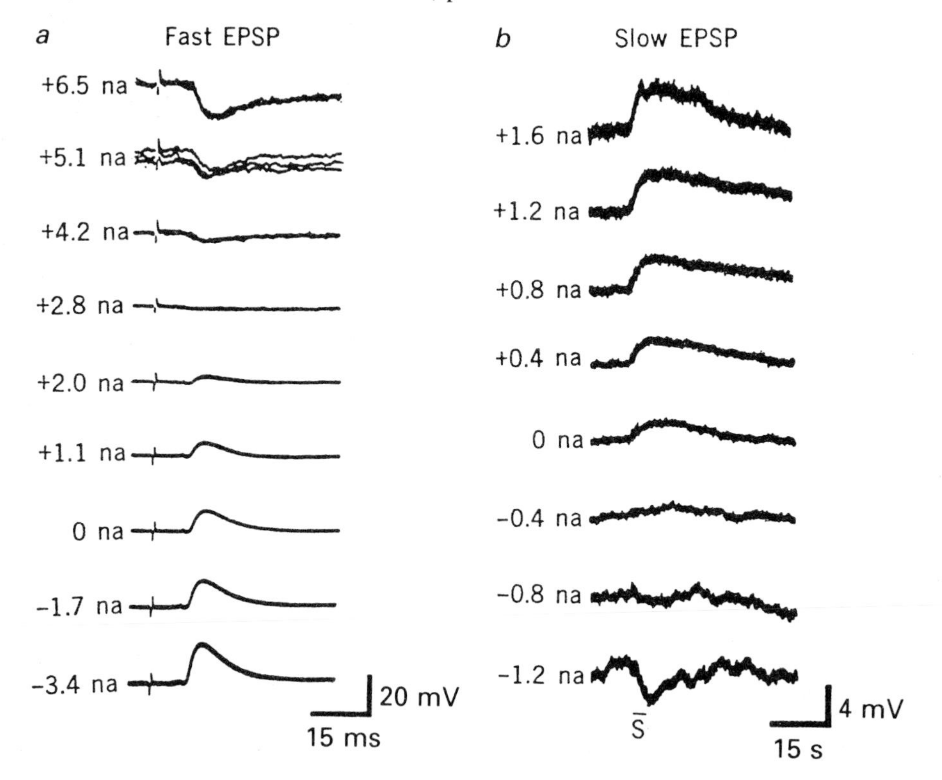

is mediated not by a direct combination of the transmitter with the channel but by an indirect process involving intermediate steps between binding at the receptor and the response of the channel. This suggests that cytoplasmic 'second messengers' are involved (Greengard, 1976; Hartzell, 1981); we shall look at some of the details of these systems in the next chapter. Fig. 10.2 outlines the different model systems for fast and slow responses to receptor activation.

Electrically transmitting synapses

An excitatory electrically transmitting (or electrotonic) synapse is one in which the postsynaptic cell is directly excited by the electrotonic currents accompanying an action potential in the presynaptic axon. Electron microscopy of electrotonic synapses shows regions where the intercellular space between the two cells is much narrower than usual, being about 2 nm instead of 20 nm. These regions are known as 'gap junctions'. As we shall see later, they contain channels which provide direct connections between the pre- and postsynaptic cells, so that current can flow readily from one cell to the other.

In order for electrotonic transmission to be effective, the size of the postsynaptic cell must not be much greater than that of the presynaptic cell, otherwise the postsynaptic current density will be insufficient to cause excitation. (Kuffler *et al.*, 1984, put this point well in drawing an analogy with heat flow: a hot knitting needle cannot heat up a cannonball.) Hence if transmission is to occur between a small axon and a large postsynaptic cell (as at the neuromuscular junction, for example), then it must be chemical in nature.

The advantages of electrical transmission are that it does not involve the complex apparatus of the chemical transmitter mechanism and that the synaptic delay can be almost negligible.

Septal synapses in invertebrate giant nerve fibres

Many annelids and crustaceans possess giant fibres in the ventral nerve cord which are multicellular units, each being divided by a transverse septum in each segment. The giant fibres of the earthworm, mentioned in chapter 4, are of this type. The physiology of the transmission process across the septa was investigated by Watanabe and Grundfest (1961) in the lateral giant fibres of the crayfish. They found that currents passed through a microelectrode into one of the component cells of a fibre would readily cross the septum to cause a potential change in its neighbour. Nevertheless, the resistance of the septum is not negligible, so that its presence must slightly reduce the conduction velocity in the whole fibre.

Electrical interconnections between neurons

A number of cases are now known in which there is some electrical interaction between neurons that frequently fire in synchrony. One example of this is in the two giant cells which occur in the segmental ganglia of leeches (Hagiwara and Morita, 1962). In fig. 9.26*a*, the upper trace shows the response of one of the cells to depolarizing current applied via an intracellular microelectrode, and the lower trace shows the electrical changes which occur simultaneously in the other cell. Fig. 9.26*b* shows the very similar results obtained from a similar experiment on two adjacent pyramidal cells in the mammalian brain.

Similar electrotonic interconnections occur between pacemaker cells in the lobster heart ganglion (Hagiwara *et al.*, 1959), between supermedullary neurons in the puffer fish (Bennett, 1960), between spinal electromotoneurons in mormyrid electric fish (Bennett *et al.*, 1963), the motoneurons of toadfish sound-producing muscles (Bennett, 1966) and elsewhere (Bennett, 1977). Bennett points out that such electrotonic coupling is much used in synchronizing the activity of neurons innervating effector organs which show marked pulsatile activity. It is essential, for example, that all the cells of the lobster heart, the mormyrid electric organ or the toadfish sound-producing muscle should fire at the same time. In many cases, the impedance of the junction increases at higher frequencies, so that brief action potentials have less effect on an adjacent cell than do slow subthreshold potentials.

The giant motor synapses of the crayfish

The flexor muscles in the abdomen of the crayfish are innervated by motor fibres of large diameter which can be excited by impulses in the lateral giant fibres of the nerve cord. This system provides the physiological basis of the 'tail-flick' escape response of the animal. Furshpan and Potter

(1959) investigated the transmission process at this synapse using microelectrodes inserted into the pre- and postsynaptic axons, as is shown in fig. 9.27. Fig. 9.28 shows the presynaptic and postsynaptic responses to orthodromic stimulation. Notice particularly that the postsynaptic response begins almost simultaneously with the presynaptic action potential; this suggests that the transmission process is electrical in nature. However, antidromic stimulation of the motor fibre produces only a very small response (less than 1 mV) in the lateral giant fibre, and so there can be no transmission of an impulse across the synapse in the reverse direction.

In further experiments, depolarizing and hyperpolarizing currents were applied through microelectrodes inserted into the axons on either side of the synapse. The results, shown in fig. 9.29, were very interesting. Depolarizing currents passed into the presynaptic axon produced corresponding changes in the postsynaptic axon, whereas hyperpolarizing curents did not; but depolarizing currents passed into the postsynaptic axon did not produce changes in the presynaptic axon, whereas hyperpolarizations did. This indicates that the synaptic membrane is a rectifier: current can only flow from the pre-fibre to the

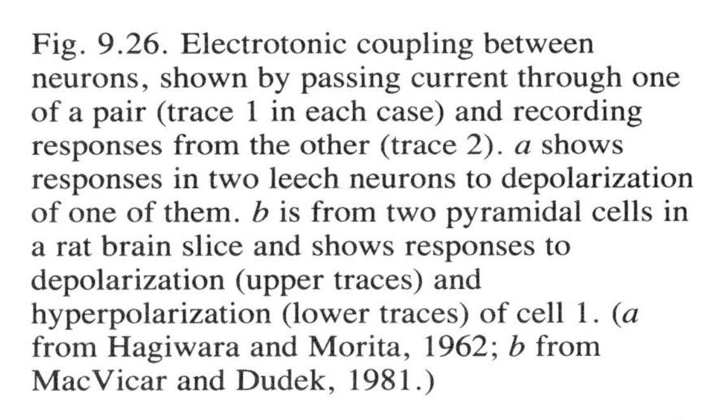
Fig. 9.26. Electrotonic coupling between neurons, shown by passing current through one of a pair (trace 1 in each case) and recording responses from the other (trace 2). *a* shows responses in two leech neurons to depolarization of one of them. *b* is from two pyramidal cells in a rat brain slice and shows responses to depolarization (upper traces) and hyperpolarization (lower traces) of cell 1. (*a* from Hagiwara and Morita, 1962; *b* from MacVicar and Dudek, 1981.)

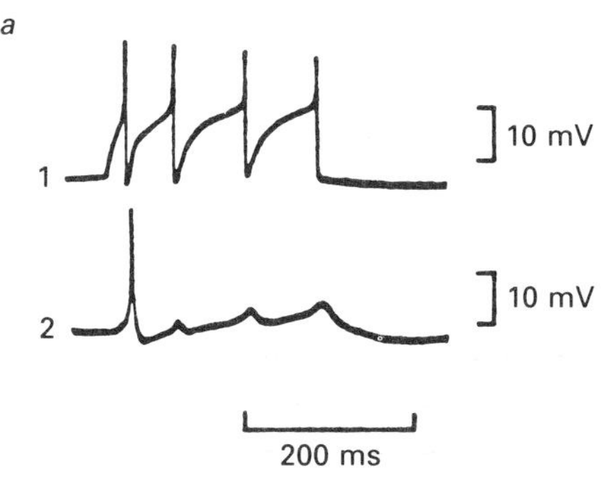

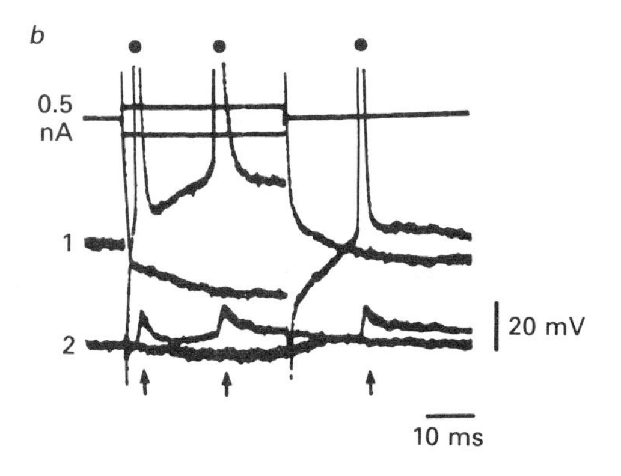

Fig. 9.27. Part of the abdominal nerve cord of a crayfish, to show the position of the electrically-transmitting synapse between the lateral giant fibre and the large motor axon in the third root. (From Furshpan and Potter, 1959.)

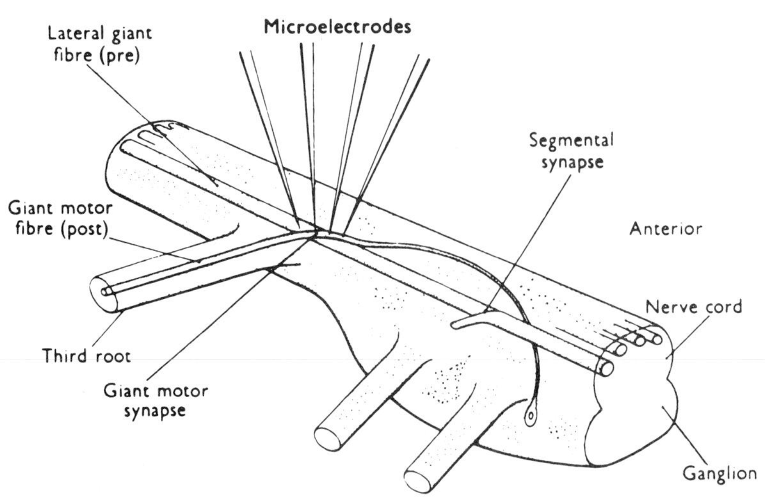

Fig. 9.28. Pre- and postsynaptic action potentials at the crayfish giant motor synapse. Two pairs of records are shown, at different amplifications. The upper trace of each pair is the presynaptic response. The kink in the postsynaptic response (at the arrow) shows the level at which the postsynaptic response crosses the threshold. Notice the negligible latency between the times of onset of the two responses. (From Furshpan and Potter, 1959.)

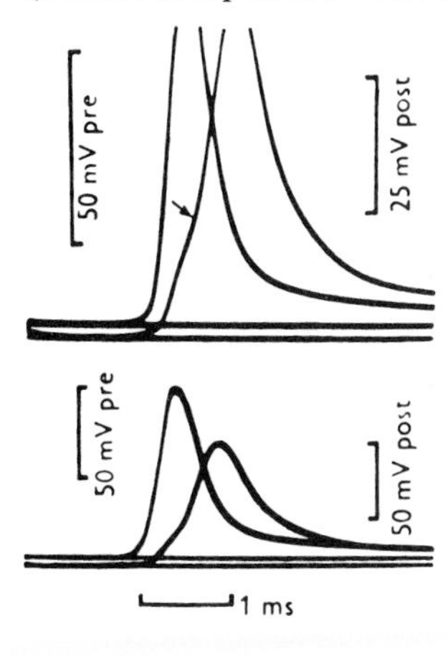

post-fibre, and not in the opposite direction. The rectifying properties of the junction are essential to the animal, since the motor fibre can also be excited (by chemical transmission) by other neurons; if there were no rectification at the junction, such excitation would always produce excitation of the lateral giant fibres and the crayfish would be continually subject to irrelevant escape responses.

Gap junction channels

Gap junctions are regions of fairly close apposition between the cell membranes of adjacent cells; the intercellular space is narrowed from about 20 nm to 2–3 nm. They are called 'gap junctions' to distinguish them from 'tight junctions', where the cell membranes make contact so that there is no extracellular space between the cells. The distinction was made by Ravel and Karnovsky (1967), who used electron microscopy to show that lanthanum salts could permeate between the two membranes at gap junctions but not at tight junctions. They also found that there was some material

Fig. 9.29. The rectifying properties of the crayfish giant motor synapse. Curent is passed through the membrane of one of the fibres, and the membrane potential in both fibres is recorded. *a* and *b*: Responses in the pre- and post-fibres to current through the pre-fibre membrane; notice that only depolarizing currents cross the synapse. *c* and *d*: Responses in the pre- and post-fibres to current through the post-fibre membrane; notice that only hyperpolarizing currents cross the synapse. (From Furshpan and Potter, 1959.)

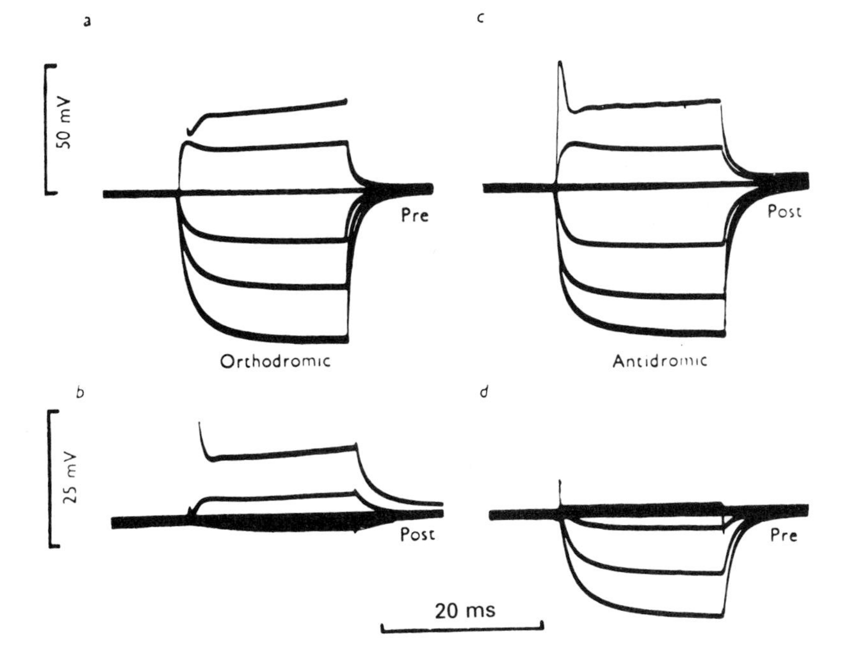

crossing the gap and that sections in the plane of the membranes showed particles with subunits in hexagonal arrays.

Gap junctions occur at regions where cells are known to be in electrical connection with each other, such as in liver cells and heart muscle and at electrotonic synapses (see Bennett and Spray, 1985). Tight junctions are found where sheets of cells separate distinct fluid compartments, as in kidney tubules and intestinal epithelium.

Structural studies, using X-ray diffraction or electron microscopy of liver cell membranes, show that each gap junction channel is composed of a pair of hexamers ('connexons'), one in each of the two apposed membranes, as is shown in fig. 9.30 (Makowski *et al.*, 1977, 1984; Unwin and Zampighi, 1980; Unwin and Ennis, 1983, 1984).

Some gap junction channels are closed by alterations in transmembrane voltage (fig. 9.31), some by pH changes and some by increases in cytoplasmic calcium ion concentration. The sensitivity of channel gating to these factors differs considerably in different tissues and in different animals (Spray *et al.*, 1985). Unwin and Zampighi (1980) suggested that gating is produced by an alteration in the inclination of the six subunits in the membrane. An alternative model suggests that there is a gating particle in the cytoplasmic side of the hexamer, which may move so as to open the channel (Makowski, 1985; Sosinsky *et al.*, 1986).

Gap junctions allow the passage of much larger molecules than will pass through most ionic channels. Thus Kanno and Loewenstein (1964) found that fluorescein (molecular weight 376) diffused fairly freely from one cell to another in *Drosophila* salivary gland, but would not diffuse out of the cell into the external medium. Spray, Harris and Bennett (1979) investigated the

Fig. 9.30. The structure of gap junctions isolated from mouse liver cells, as deduced from X-ray diffraction studies. (From Makowski *et al.*, 1984.)

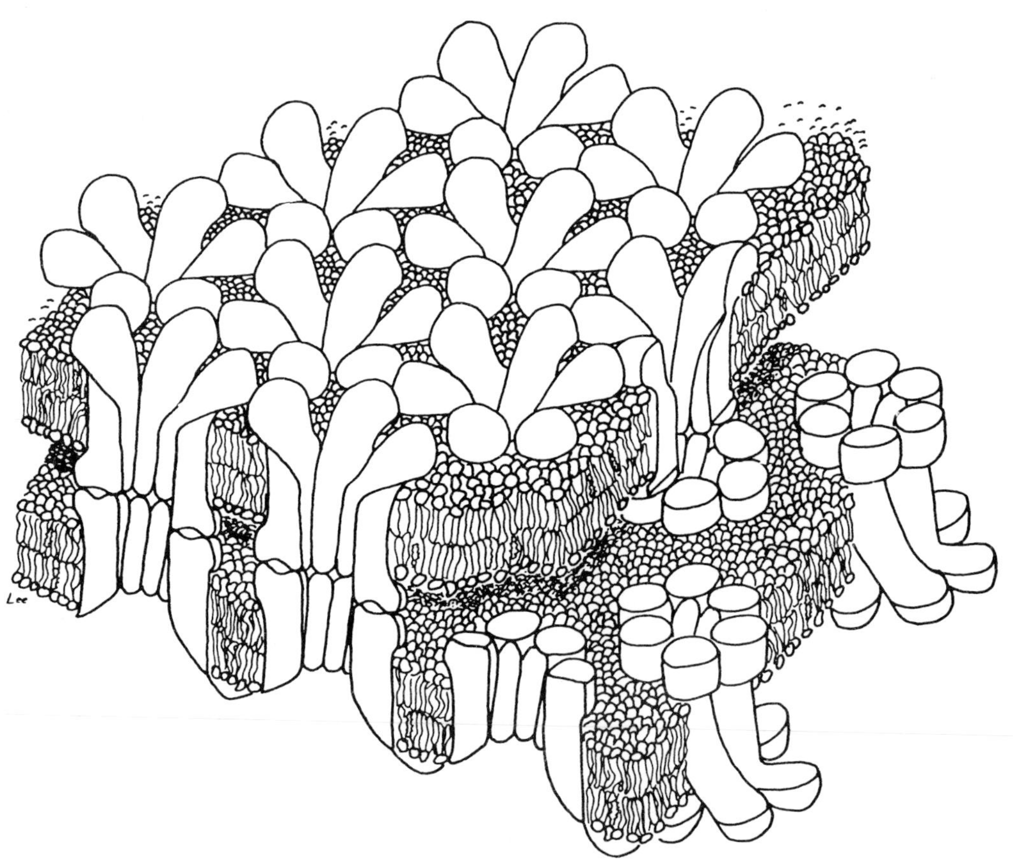

electrotonic coupling between pairs of blastomeres from early amphibian embryos. They found that the junctional conductance was much reduced if the voltage across the junction was more than a few millivolts in either direction. Permeability to the dye lucifer yellow was altered in the same way by such voltage changes, implying that current flow and dye movement take place via the same channels.

Experiments by Verselis *et al.* (1986) showed that gap junction permeability to tetraethylammonium (TEA) ions was proportional to the junctional conductance, which again confirms the view that current flow and passage of small molecules take place through the same channels. Since the TEA ion (diameter 8.5 Å) is larger than the potassium ion (hydrated diameter 4.6 Å), which is the main current carrier, this result also suggests that single gap junction channels are either open or closed: if the channel could become partly closed one would expect that the permeability to the larger TEA ion would be disproportionately reduced.

Fig. 9.31. Proposed reaction scheme for the voltage-sensitive gating of amphibian blastomere gap junction channels. It is suggested that the gate in connexon 1 is closed when the voltage across the junction reaches some threshold value, whereas the gate in connexon 2 is closed when the voltage reaches the same value in the opposite direction. When the voltage gradient is below threshold in either direction, both gates are open. (From Spray *et al.*, 1985.)

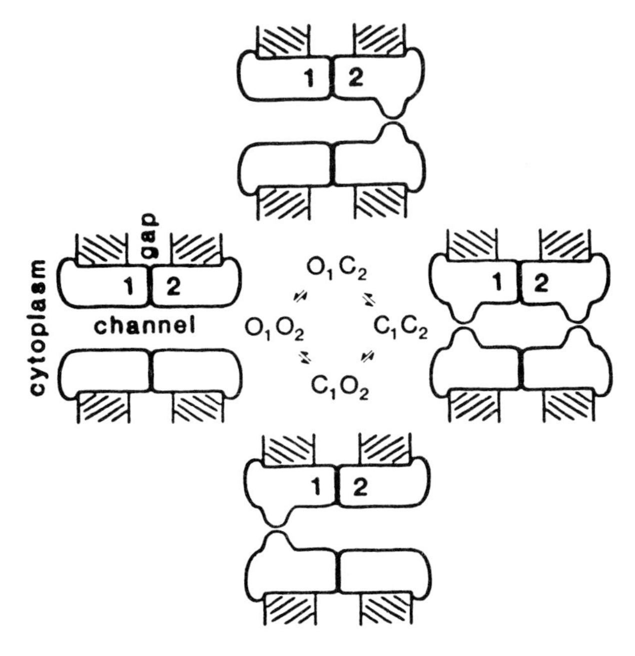

Loewenstein and his colleagues have used fluorescent probes of different sizes to estimate the diameter of the gap junction channel (Flagg-Newton *et al.*, 1979; Schwarzmann *et al.*, 1981; Loewenstein, 1981). They concluded that the limiting diameter of the channels in mammalian cells is between 16 and 20 Å, so that substances with molecular weights up to about 1000 can pass through them. Channels in insect cells are rather larger, with a limiting diameter between 20 and 30 Å, allowing substances with molecular weights up to about 1800 to pass through.

Single channel currents at gap junctions have been measured in pairs of rat lachrymal gland cells by Neyton and Trautmann (1985). They calculated that the conductance of a single open channel was about 120 pS.

The function of gap junctions in non-excitable cells is still not altogether clear (see Sheridan and Atkinson, 1985). One possibility is that they are involved in the transmission of intracellular signal molecules, such as cyclic AMP, between cells. Results compatible with this idea have been produced by Lawrence, Beers and Gilula (1978), using a mixture of two different cell types, rat ovarian granulosa cells and mouse heart muscle cells, in tissue culture. Gap junctions form between the two cell types. Follicle stimulating hormone (FSH) stimulates granulosa cells and noradrenalin stimulates heart cells, by a cyclic AMP mechanism in both cases. Application of FSH to a mixed culture causes the beating rate of the heart cells to rise within minutes, whereas it has no effect on heart cells cultured alone. It looks as though the FSH produces cyclic AMP in the granulosa cells and that this then diffuses through the gap junctions to increase the activity of the heart cells.

In electrotonic synapses, the gap junctions clearly serve as routes for current flow between one cell and another. Undue voltage sensitivity, of the type seen in blastomeres, would prevent such current flow, hence it is not surprising to find that it is absent in symmetrical junctions such as the septal synapses of crayfish abdominal lateral giant fibres (Johnston and Ramon, 1982). The rectifying synapses, however, may well have the connexons on one side of the junction voltage-sensitive while those on the other are not, as seems to occur in the giant-motor axon synapse of the crayfish (Jaslove and Brink, 1986). The rate constants for closing of the channel are much more rapid here than in the symmetrically voltage-sensitive channels of blastomeres.

An inhibitory synapse operating by electrical transmission

Many fishes possess a large pair of interneurons, the Mauthner fibres, whose cell bodies lie in the brain and whose axons pass down the spinal cord. The space around the axon hillock is bounded by a dense layer of glial cells and extracellular material called the axon cap; this contains the terminals of a number of interneurons (fig. 9.32). Furukawa and Furshpan (1963) recorded an external hyperpolarizing potential (EHP) with a microelectrode placed inside the axon cap but outside the Mauthner cell, especially in response to stimulation of the contralateral Mauthner axon (fig. 9.33*a*). Firing of the Mauthner cell is inhibited during the course of the EHP, and it is therefore inhibitory in function.

The EHP is unaffected by injection of tetrodotoxin into the axon cap (Faber and Korn, 1978). The terminals of the interneurons ending within the axon cap are unmyelinated and so it seems likely that the action potential does not propagate into them. Local circuit currents in the terminals within the cap will therefore produce a brief positive-going potential change, seen as the

Fig. 9.33. The external hyperpolarizing potential (EHP) produced in a Mauthner cell in response to stimulation of the axon of the other cell. *a* is an extracellular record with the tip of the microelectrode within the axon cap, *b* is an intracellular record from near the axon hillock. Since the reference electrode was distant in each case, the potential across the membrane is the difference between traces *a* and *b*. The EHP is followed by an IPSP, which is depolarizing in this case, probably as a result of chloride leakage from the microelectrode. (From Furukawa and Furshpan, 1963.)

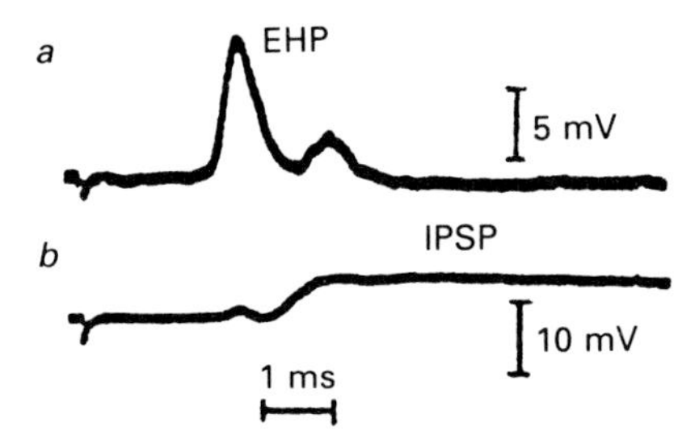

Fig. 9.32. Mauthner cells in the brain of fishes, showing how the interneurons causing electrical inhibition are innervated by collateral branches from their axons.

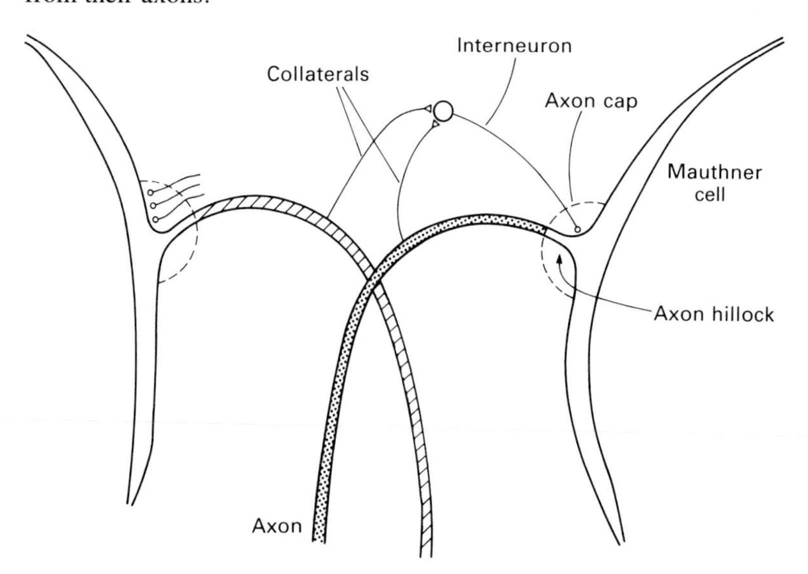

EHP. This will increase the potential across the Mauthner cell membrane, given by the difference between the traces *a* and *b* in fig. 9.33, and so result in inhibition. Fig. 9.34 shows the likely currents involved.

The function of Mauthner fibres is to elicit the startle response of the fish. This consists of a very rapid flexion of the body, brought about by a massive synchronous contraction of the muscles on that side, followed by a tail-flip which drives the fish forward (Eaton *et al.*, 1977). Clearly it is most important that only one of the two Mauthner fibres should fire at one time, otherwise the fish's muscles would contract on both sides at once. Hence the need for effective and immediate inhibition of one cell by the other when it fires. The EHP is followed by a conventional chemically transmitted IPSP (fig. 9.33*b*), hence the functional value of the electrical inhibition may well lie in the rapidity with which inhibition becomes effective.

Fig. 9.34. The origin of the EHP in Mauthner cells. The action potential in the interneuron travels only as far as the outside of the axon cap; inward flow of current at this point produces the local circuit currents shown in the diagram.

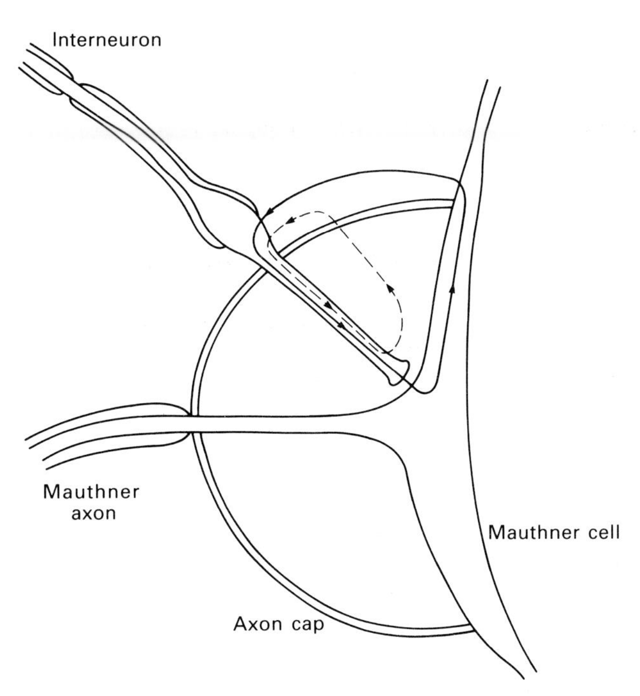

10
Neurotransmitters and their receptors

In the previous three chapters we have seen that chemical transmission at synapses involves the release of a transmitter substance from the nerve terminals and its combination with receptors on the postsynaptic cell. In this chapter we shall look rather briefly into the nature of some of these neurotransmitters and their receptors.

Evidence on the role of acetylcholine in transmission at parasympathetic nerve endings accumulated during the 1920s, a role that was extended to neuromuscular transmission and transmission from sympathetic preganglionic fibres in the 1930s. We have seen some of the evidence for this in chapter 7.

At this time there were indications that the neurotransmitter at postganglionic sympathetic nerve endings might be the hormone adrenaline (alternatively called epinephrine) or a similar substance; it was later shown to be noradrenaline (norepinephrine). In 1933, Dale introduced the terms 'cholinergic' to describe nerve fibres using acetylcholine as the transmitter and 'adrenergic' for fibres releasing adrenaline or a similar substance (fig. 10.1).

Dopamine (a precursor of noradrenaline) and serotonin became strong candidates for transmitters in the 1950s. Soon after this the effects of glutamate and some other amino acids on neurons were discovered, and they gradually gained acceptance as genuine neurotransmitters. In the 1970s a new class of neurotransmitters, the neuropeptides, was discovered; Hökfelt and his colleagues counted over twenty of these in 1980, and the list of possible candidates has more than doubled since then. Finally there is evidence that adenosine and related purines are also neurotransmitters. Table 10.1 provides a summary of the range of different transmitters. It is very likely that the list will be added to in the future.

Neurotransmitters act by combining with receptors. When the receptor molecule is part of an ionic channel the response is direct and rapid, as in the nicotinic acetylcholine receptor. Slower

Fig. 10.1. Cholinergic and adrenergic neurons in the peripheral nervous system of mammals. The diagram is oversimplified in a number of ways. CNS, central nervous system; ACh, acetylcholine; NorAd, noradrenaline.

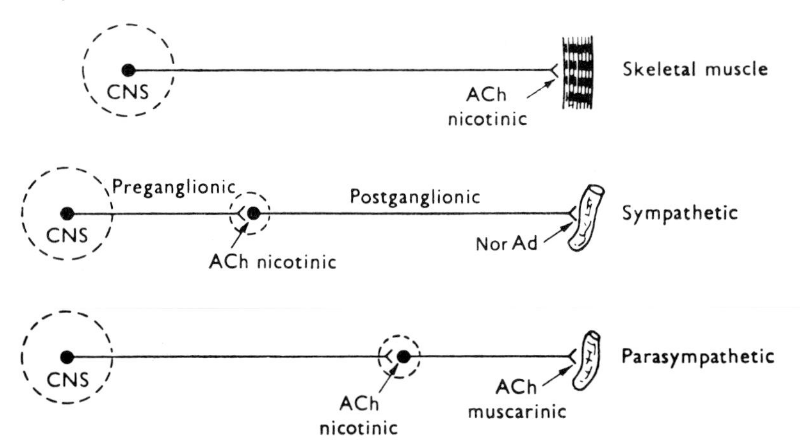

Table 10.1 *Some neurotransmitters*

Acetylcholine
Catecholamines
Noradrenaline
Adrenaline
Dopamine
Serotonin
Histamine
Amino acid transmitters
γ-amino butyric acid (GABA)
Glycine
Glutamate
Aspartate
Purines
ATP
Adenosine
Neuropeptides
Met-enkephalin
Substance P
CGRP
etc. (see table 10.5)

responses are produced when the receptor acts via a G protein either to produce a cytoplasmic second messenger or to activate an ionic channel (fig. 10.2).

There are cases when neurons release substances which affect the process of synaptic transmission brought about by a different transmitter substance. They may influence transmitter synthesis or release, or they may act postsynaptically by altering the sensitivity of the postsynaptic receptors. Such substances are called neuromodulators. Presynaptic terminals may possess receptors for their own neurotransmitter ('autoreception') or for others (Starke, 1981).

The elucidation of the nature of the transmitter involved at any particular synapse is not an easy task. Paton (1958) suggested that the following criteria should be satisfied before a transmitter role is accepted for a substance:

1. The presynaptic neuron should contain the substance and be able to synthesize it.
2. The substance should be released on stimulation of the presynaptic axons.
3. Application of the substance to the postsynaptic cell should reproduce the effects of normal transmission.
4. The action of the substance on the postsynaptic cell should be affected by competitive blocking agents in the same way that synaptic transmission is.

Paton originally included a fifth criterion in this list, that an enzyme capable of destroying the substance should exist in the vicinity of the synapse (such as, for example, acetylcholinesterase at the neuromuscular junction). Since then it has become evident that there may be other methods of limiting the duration of transmitter action, such as diffusion or desensitization (Gerschenfeld, 1966) or reabsorption (Brown, 1965).

The application of these criteria can be illustrated by reference to the hypothesis that acetylcholine is the transmitter substance at vertebrate skeletal neuromuscular junctions, as follows:

Criterion 1. The direct demonstration that motor nerves can synthesize acetylcholine is difficult, because of the relative sparsity of nerve endings in muscular tissue. However, synthesis of acetylcholine has been clearly demonstrated in sympathetic ganglia (Birks and MacIntosh, 1961), and the synaptic vesicles isolated from the nerve terminals in *Torpedo* electric organ contain acetylcholine (Whittaker, 1984).

Criterion 2. Release of acetylcholine from motor nerves on stimulation was shown by Dale, Feldburg and Vogt, as described on p. 112.

Criterion 3. The ionophoretic experiments of del Castillo and Katz and others (p. 119) show that acetylcholine has the same effect on the postsynaptic membrane as does the transmitter substance. Takeuchi (1963*a*) showed that the reversal potential of the responses to ionophoretically applied acetylcholine is the same as that of the EPP and is similarly affected by changes in the external sodium and potassium ion concentrations.

Criterion 4. We have seen that curare blocks neuromuscular transmission by reducing the size of the EPP. Similarly, curare antagonizes the action of externally applied acetylcholine.

It is much more difficult to establish whether or not a particular neurotransmitter is involved at particular synapses in the central nervous system, since a very large number of synapses and a wide variety of different neurotransmitters are involved. The criteria needed have been discussed by Werman (1966) and Orrego (1979). Bradford (1986) emphasizes the importance of demonstrating the presence of specific receptors for the proposed neurotransmitter in the tissue containing the synapse.

The situation is further complicated by the phenomenon of co-transmission, in which two neurotransmitters (often a 'classical' transmitter

Fig. 10.2. Direct and indirect actions of neurotransmitters on ionic channels. *a* shows the direct action which occurs when the ionic channel is an integral part of the receptor, as in the nicotinic acetylcholine receptor. In *b* and *c* the receptor molecule is not directly linked to an ionic channel but acts via a G protein. In *b* the G protein acts directly on the channel to open or close it. In *c* the G protein activates an enzyme which generates a second messenger such as cyclic AMP which itself then alters the state of the channel.

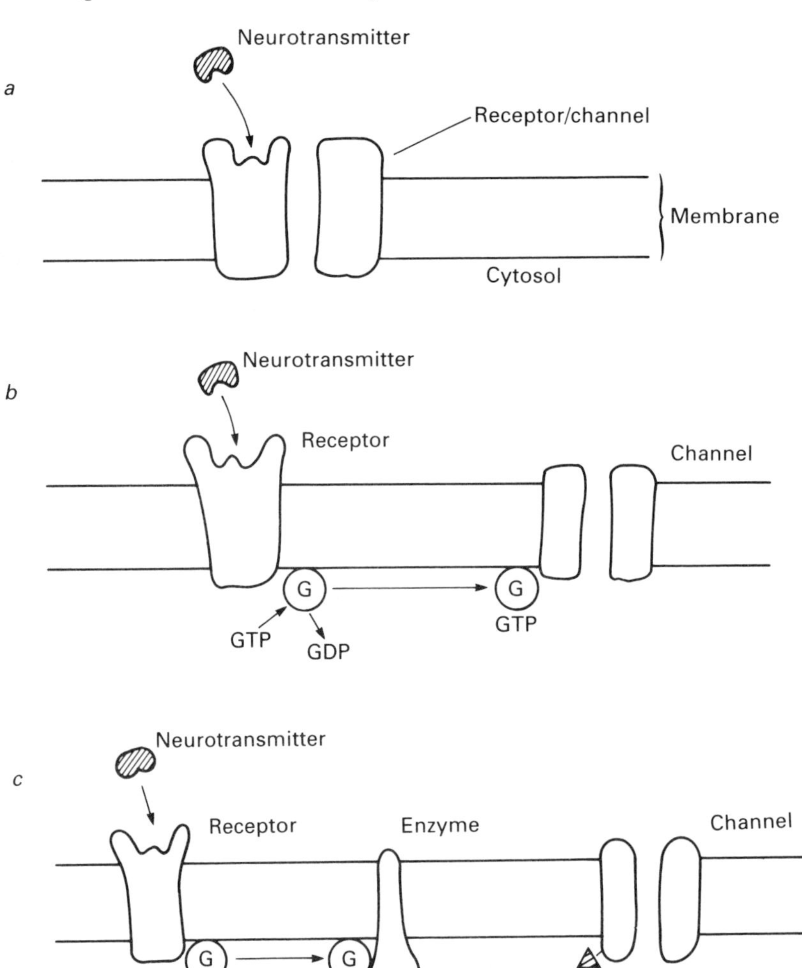

and a peptide) are released simultaneously from the same nerve terminal (Cuello, 1982; Campbell, 1987).

Second messengers and G proteins

We have seen in the previous chapter that the slowness of certain synaptic potentials could perhaps be explained by the production of cytoplasmic 'second messengers', rather than by direct activation of ionic channels. The second messenger concept was first introduced to describe the role of cyclic adenosine monophosphate (cyclic AMP or cAMP) in hormone action (Sutherland and Rall, 1960; Sutherland *et al.*, 1965). In skeletal muscle, for example, adrenaline causes an increase in intracellular cAMP levels which leads ultimately to the breakdown of glycogen. Binding of the hormone to its receptor leads to activation of the membrane-bound enzyme adenylate cyclase, which converts ATP to cyclic AMP. The cyclic AMP combines with a cAMP-dependent protein kinase which then serves to phosphorylate target proteins, so altering their activity and producing some physiological or metabolic response. Protein phosphorylation seems to provide a major general mechanism for the control of intracellular events (Cohen, 1982; Nestler and Greengard, 1984).

A second messenger system of this type can clearly involve some considerable amplification. One or two transmitter molecules may elicit a cascade of responses, so that many second messenger molecules are produced and perhaps more target protein molecules phosphorylated.

One example of this type of system occurs in the sensory neurons of the marine mollusc *Aplysia* (Siegelbaum *et al.*, 1982). Application of the transmitter substance serotonin produces a slow EPSP which is caused by the closure of a set of potassium channels. Injection of cyclic AMP into the cell also closes these channels and it therefore seems likely that the serotonin is acting via a cyclic AMP second messenger system of the type shown in fig. 10.2*c*.

The link between the receptor and adenylate cyclase is mediated by a protein which binds guanosine triphosphate (GTP) to become active. This belongs to a family called G proteins or N proteins (Rodbell, 1980; Gilman, 1984, 1987; Stryer and Bourne, 1986). G proteins have been described as 'cytoplasmic shuttles' moving between receptor and effector molecules on the internal surface of the plasma membrane (Chabre, 1987).

Adenylate cyclase may be stimulated or inhibited by different transmitters, which act via separate receptors and separate G proteins, G_s and G_i (fig. 10.3). G_s acts directly to stimulate the adenylate cyclase, but there seem to be two ways in which G_i might inhibit it: either by a direct action on the enzyme (Gilman, 1984), or indirectly by

Fig. 10.3. The cyclic AMP second messenger system. Adenylate cyclase (AC) converts ATP to cAMP when activated by the G protein G_s. G_s is itself activated when the appropriate neurotransmitter combines with its receptor R_s. Inhibition of AC occurs when combination of a different neurotransmitter with its receptor R_i occurs, so activating the inhibitory G protein G_i. The diagram does not show the dissociation of the G proteins on activation and the inhibitory action may be less direct than indicated here.

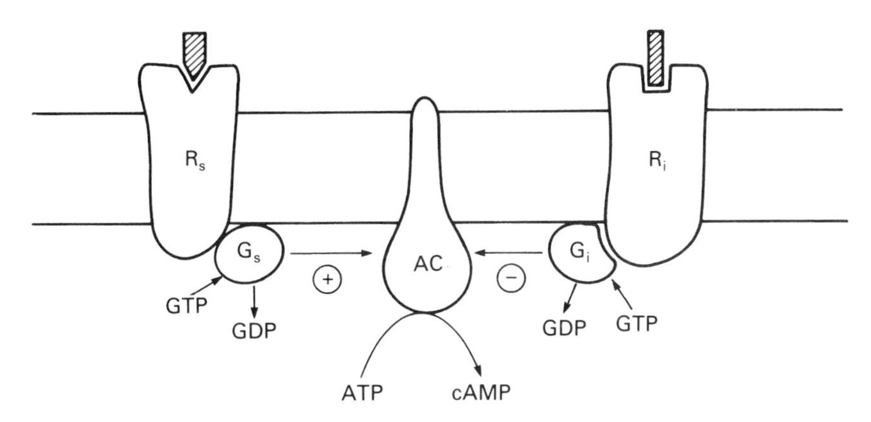

preventing the activation of G_s by the excitatory receptor (Levitski, 1986).

G proteins consist of three subunits, the α, β and γ chains. Activation of the receptor molecule (following binding of the neurotransmitter or hormone) promotes the binding of GTP to the α chain of the G protein, displacing GDP. This causes the α chain to dissociate from the other two chains, and it then activates the adenylate cyclase or other target protein. This activation ceases when the GTP is hydrolysed to GDP by the α chain, which then reassociates with the βγ complex. Experiments on G protein systems frequently use analogues of GTP which cannot be hydrolysed, such as guanosine (γ-thio) triphosphate, so as to maintain the activating effect on the target protein. Another approach utilises the effects of two bacterial toxins: cholera toxin maintains the stimulatory proteins G_s in the active condition, whereas pertussis toxin prevents G_i and some other G proteins from becoming active.

Another G protein, called G_o, has been found in large amounts in the brain. In a tissue culture line of neuroblastoma × glia hybrid cells, G_o is activated by a peptide and inhibits voltage-gated calcium channels (Hescheler *et al.*, 1987). It is possible that its main function in nervous tissue is also via regulation of calcium channel activity; the inhibition of voltage-dependent calcium channels in cultured chick embryonic dorsal root neurons involves a G protein which could well be G_o (Holz *et al.*, 1986). G proteins are also involved in sensory transduction in the vertebrate retina (transducin, or G_T; see chapter 18) and in vertebrate olfaction (Pace *et al.*, 1985).

Second messenger systems other than the cyclic AMP system exist. They include the cyclic GMP system in the retina (chapter 18) and the phosphatidylinositol system (Berridge and Irvine, 1984; Berridge, 1985, 1986).

Phosphatidylinositol (PI) is a phospholipid found largely in the inner leaflet of the cell membrane lipid bilayer. Unlike other phospholipids it is readily phosphorylated, to give phosphatidylinositol-4,5-bisphosphate (PIP_2). This is hydrolysed by the membrane-bound enzyme phosphodiesterase to give two second messengers, inositol 1,4,5-trisphosphate (IP_3) which is water-soluble and so cytoplasmic, and diacylglycerol (DG), which remains in the membrane. IP_3 promotes release of calcium ions from the endoplasmic reticulum in a variety of cell types (Streb *et al.*, 1983; Berridge and Irvine, 1984), and opens calcium channels in the plasma membrane of *Xenopus* oocytes (Parker and Miledi, 1987). DG activates the membrane-bound enzyme protein kinase C, which will then phosphorylate a target protein and so alter its activity. Target proteins for protein kinase C include calcium, potassium and chloride channels (Nishizuka, 1984; Kaczmarek, 1986, 1987).

The PI system is activated by acetylcholine in *Xenopus* oocytes. We have seen that acetylcholine acts on muscarinic receptors in the oocyte membrane to produce a slow response involving the opening of chloride channels (fig. 8.17*a*). Injection of IP_3 produces a very similar response (Oron *et al.*, 1985) which, it turns out, is brought about by the opening of calcium channels in the plasma membrane; the resulting increase in cytoplasmic calcium ion concentration then opens the membrane chloride channels (Parker and Miledi, 1987). This provides a nice example of the PI system being activated by a neurotransmitter. The link between neurotransmitter or hormone and the activation of the phosphodiesterase is provided by another G protein, G_p (Cockroft and Gomperts, 1985; Cockroft, 1987). Fig. 10.4 provides a summary of the system.

In heart muscle it is clear that the muscarinic acetylcholine receptors are coupled to potassium channels via a G protein (G_k) without the intervention of a second messenger system (Pfaffinger *et al.*, 1985; Yatani *et al.*, 1987). Systems of this type (fig. 10.1*b*) may well occur elsewhere. It would be interesting to know how many channels are activated by one receptor.

The action of drugs on neuromuscular transmission

From the evidence presented in chapter 7, we can safely assume that the transmitter substance at the vertebrate skeletal neuromuscular junction is acetylcholine. In order for acetylcholine to act effectively as the neuromuscular transmitter it has to be synthesized in the motor nerve, be released from it on stimulation, react with the receptor sites on the postsynaptic membrane so as to increase the ionic permeability of the membrane, and finally it must be broken down by acetylcholinesterase.

Some of the choline resulting from the hydrolysis is recycled by being taken up into the presynaptic terminal to be made into acetylcholine again. The different drugs which affect the transmission process may act at any of the stages in this sequence of events.

Acetylcholine is synthesized in the nerve terminals from choline and acetyl coenzyme A, by means of the enzyme choline acetyltransferase. The enzyme is inhibited by 4-naphthylvinyl pyridine, which thus blocks the synthesis of acetylcholine.

Choline is present in or near the synaptic cleft as a result of the breakdown, by acetylcholinesterase, of the acetylcholine previously released by nerve impulses. From here it is rapidly taken up into the nerve terminals for use in acetylcholine synthesis. The uptake process is specifically inhibited by the drug hemicholinium-3.

Botulinum toxin, obtained from the bacterium *Clostridium botulinum*, prevents the release of acetylcholine from the presynaptic terminals. Black widow spider venom has an almost opposite effect: it produces a massive discharge of acetylcholine quanta and a corresponding reduction in the numbers of vesicles present in the terminals (Longenecker *et al.*, 1970). Neuromuscular block ensues as a result of exhaustion of the acetylcholine content of the nerve terminals.

Drugs which act on the postsynaptic membrane in much the same way as acetylcholine does are called *agonists*, whereas those which reduce or prevent the action of acetylcholine are called *antagonists*. Molecules of each of these types are thought to act by combination with the receptor sites at which acetylcholine itself acts. The formulae of acetylcholine and some of its agonists are shown in table 10.2. Many agonists can cause neuromuscular block by desensitization; in some cases (e.g. decamethonium) this action is more marked than the depolarization. Succinylcholine has been used as an immobilizing agent for large wild mammals, since the blocking effect is only temporary as the compound is gradually destroyed by the nonspecific cholinesterase in the blood.

It has been known since the work of Claude Bernard in 1857 that *curare*, the substance used by South American Indians to poison the tips of their arrows, acts at the neuromuscular junction. It is an antagonist to acetylcholine. Crude curare is a mixture of substances obtained from various plants; there are three varieties, known as pot-curare, tube-curare and calabash-curare. One of the active constituents of tube curare is the alkaloid D-tubo-

Fig. 10.4. The phosphatidylinositol second messenger system. The enzyme phosphodiesterase is activated by the G protein G_p which itself is activated when the appropriate neurotransmitter combines with its receptor. Phosphodiesterase hydrolyses phosphatidylinositol-4,5-bisphosphate (PIP_2) to give two second messengers, diacylglycerol (DG) and inositol trisphosphate (IP_3). Then DG activates protein kinase C (which will then phosphorylate some target protein) and IP_3 causes release of calcium ions from the endoplasmic reticulum.

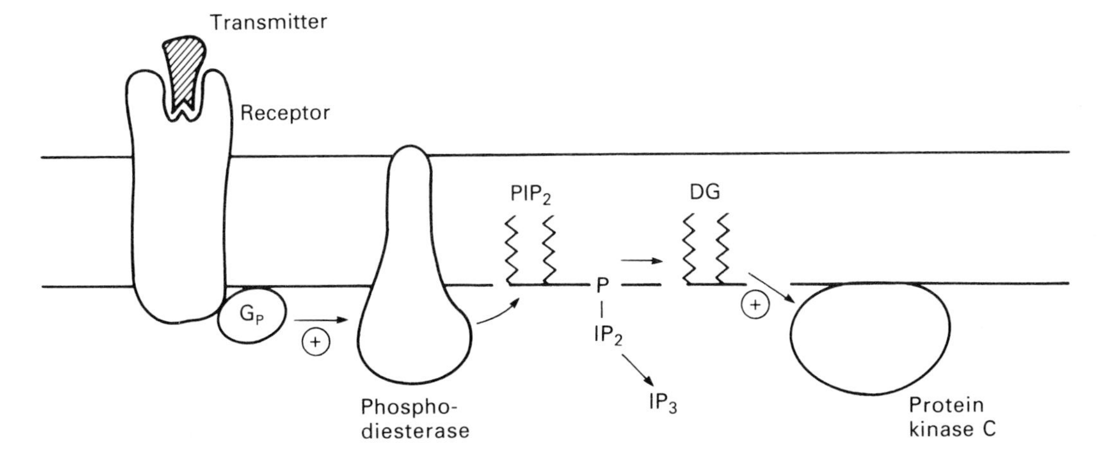

Table 10.2. *Nicotinic agonists of acetylcholine*

Acetylcholine	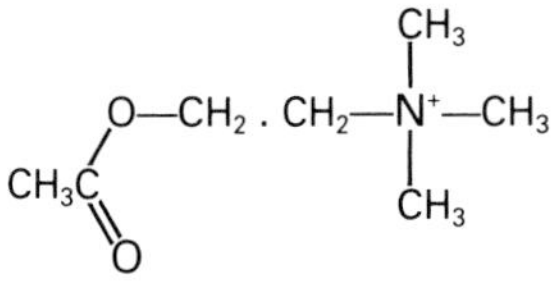 $CH_3C(=O)O-CH_2.CH_2-N^+(CH_3)_3$
Choline	$HO-CH_2CH_2N^+(CH_3)_3$
Carbachol	$H_2N.CO.O.CH_2CH_2N^+(CH_3)_3$
Succinylcholine	$CH_2CO.O.CH_2CH_2N^+(CH_3)_3$ $\vert$ $CH_2CO.O.CH_2CH_2N^+(CH_3)_3$
Decamethonium	$(CH_3)_3N^+-(CH_2)_{10}-N^+(CH_3)_3$
Murexine	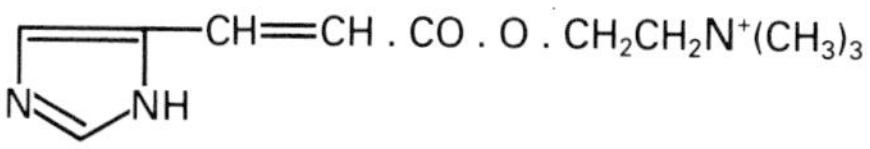 (imidazole, N, NH)$-CH=CH.CO.O.CH_2CH_2N^+(CH_3)_3$
Nicotine	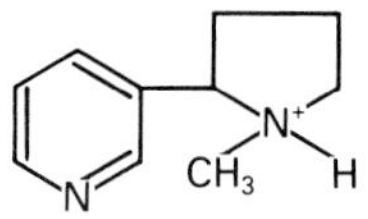 (pyridine ring, N)–(pyrrolidine ring, N^+, CH_3, H)

curarine (fig. 10.5), which is the most widely used of the various curare extracts. Some of the calabash-curare alkaloids are extremely effective neuromuscular blocking agents; C-alkaloid E, for instance, effectively blocks transmission in frog sartorius muscle at 1/200th the concentration of D-tubocurarine required to produce the same effect. β-*Erythroidine* is an alkaloid derived from the American leguminous plant *Erythrina*. Unlike most neuromuscular blocking agents it is active when taken orally. Various snake venoms are also powerful antagonists of acetylcholine; we have seen in chapter 7 how α-bungarotoxin has proved to be very useful in this respect.

Fig. 10.5. D-tubocurarine, an acetylcholine antagonist at nicotinic synapses.

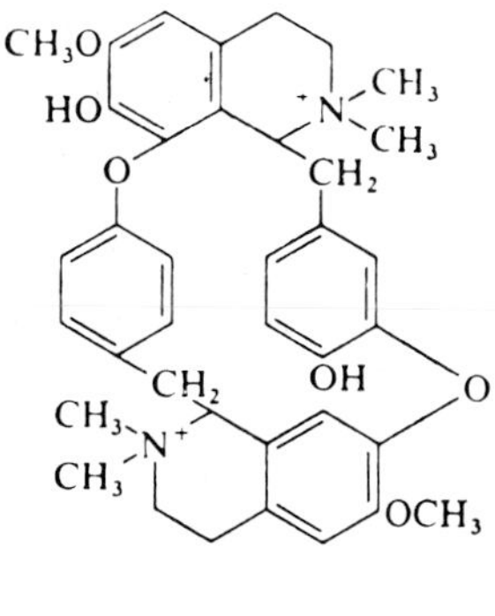

Finally we must consider the drugs which inhibit the breakdown of acetylcholine by acetylcholinesterase. The most well-known of these is *eserine* (physostigmine), an alkaloid which can be extracted from the Calabar bean. Compounds with similar structure and activity are miotene, neostigmine (prostigmine) and edrophonium (table 10.3). In the presence of these compounds, the EPP is greatly prolonged and may initiate a series of action potentials in the muscle fibre. The effect of these compounds can be removed by washing the preparation in Ringer solution, when the inhibitor–enzyme complex breaks up; edrophonium seems to be particularly loosely bound. A number of organophosphorus compounds inactivate acetylcholinesterase. Examples are diisopropylphosphorofluoridate (DFP) and tetraethylpyrophosphate (TEPP). The complexes between DFP or TEPP and acetylcholinesterase are very stable,

Table 10.3. *Some acetylcholinesterase inhibitors*

Eserine	$CH_3 . NH . CO . O$— [ring structure]; CH_3; N; N; CH_3 CH_3
Miotene	$CH_3 . NH . CO . O$—[benzene ring]—$CH(CH_3)N(CH_3)_2$
Neostigmine	$(CH_3)_2N . CO . O$—[benzene ring]—$N^+(CH_3)_3$
Edrophonium	HO—[benzene ring]—$N^+(CH_3)_2$; C_2H_5
DFP (diisopropyl-phosphorofluoridate)	$(H_3C)_2CH—P(F)(\rightarrow O)—CH(CH_3)_2$
TEPP (tetraethyl pyrophosphate)	$(C_2H_5O)_2P(\rightarrow O)—O—P(\rightarrow O)(OC_2H_5)_2$

so that the effects of the inhibitor cannot be removed by washing. Organophosphorus compounds have been used as war gases and are much used as insecticides (examples are parathion, malathion and dichlorvos); their biological activity in each case is via their anticholinesterase action.

Other cholinergic synapses

Besides being the transmitter substance at voluntary motor nerve endings in vertebrates, acetylcholine is also released at the endings of preganglionic sympathetic fibres, and pre- and postganglionic endings of parasympathetic fibres (fig. 10.1). The pharmacology of the junctions involved in the fast EPSPs of autonomic ganglia is generally similar to that of the skeletal neuromuscular junction; in particular, nicotine mimics the actions of acetylcholine and curare is a competitive antagonist. The pharmacology of parasympathetic postganglionic endings, and of the slow responses in autonomic ganglion cells (chapter 9), is rather different, in that the roles of nicotine and curare are here taken by muscarine and atropine respectively. The two types of actions, and the receptors which promote them, are accordingly known as nicotinic and muscarinic.

Muscarine (fig. 10.6) is obtained from *Amanita muscaria*, a mushroom sometimes eaten in the past for its dionysiac properties (Graves, 1958). Pilocarpine and arecoline (the active principle in betel nuts) have similar actions, although pilocarpine also has weak nicotinic properties.

The deadly nightshade (*Atropa belladonna*) contains the alkaloid L-hyoscyamine; this readily racemizes to form DL-hyoscyamine, which is known as *atropine* (fig. 10.7). Atropine is an antagonist to acetylcholine (and its agonists) at synapses where its action is muscarinic. It has been used since classical times as a cosmetic, because it causes dilation of the pupils by blocking the synapses of parasympathetic fibres on the iris sphincter muscle. Synthetic antagonists include pirenzepine and quinuclidinyl benzilate (QNB).

Renshaw cells

Mammalian motoneurons are subject to a form of negative feedback via small interneurons in the spinal cord. These are called Renshaw cells. They are excited by synaptic action from collateral branches of the motor axons and cause inhibition of the neighbouring motoneurons. Their function seems to be to prevent motoneurons from firing at an unnecessarily high frequency. We have unusually good evidence as to the nature of the transmitter at the synapses between motor axon collaterals and Renshaw cells.

Renshaw cells can be easily activated by antidromic stimulation of motor nerves. The response of a Renshaw cell to a single antidromic volley produced by stimulation of a lumbar ventral root, as recorded by an extracellular microelectrode in the vicinity of the cell, is shown in fig. 10.8*a*. Eccles, Eccles and Fatt (1956) investigated the action of various drugs on this system by injecting them into the arterial blood supply of this part of the spinal cord. *b* and *c* (fig. 10.8) show the

Fig. 10.6. Muscarine, an agonist of acetylcholine at postganglionic parasympathetic synapses.

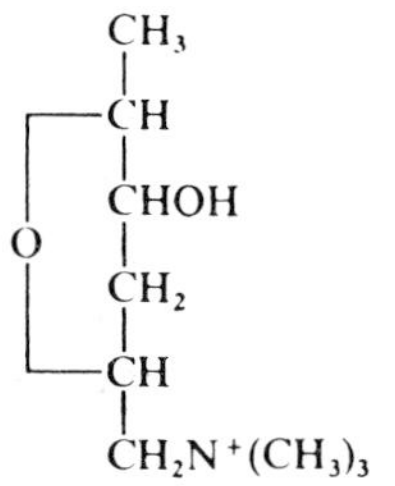

CH_3
|
CH ─┐
| │
CHOH O
| │
CH_2 │
| │
CH ─┘
|
$CH_2N^+(CH_3)_3$

Fig. 10.7. Atropine, an antagonist of acetylcholine at muscarinic synapses.

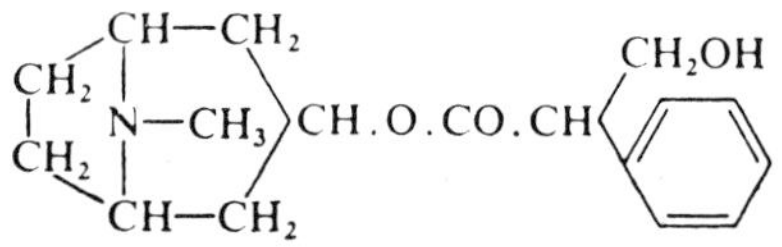

CH—CH_2 / CH_2, CH_2 / N—CH_3 / CH—CH_2 ⟩CH.O.CO.CH(CH_2OH)(C₆H₅ ring)

Fig. 10.8. Responses of a Renshaw cell, as recorded by an extracellular microelectrode, to antidromic motor nerve volleys and injections of acetylcholine (ACh), nicotine, eserine and dihydro-β-erythroidine (DHE). Explanation in text. (From Eccles *et al.*, 1956.)

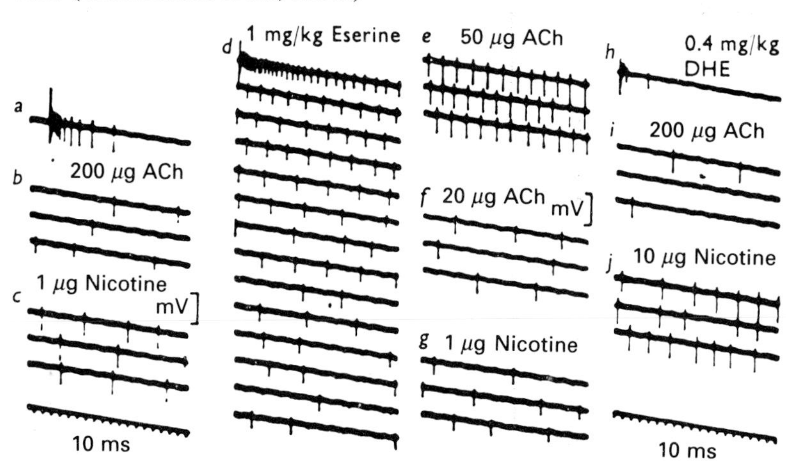

spontaneous activity following injection of acetylcholine and nicotine. After injection of eserine (*d*–*g*), the response to stimulation of the ventral root is enormously prolonged (*d*) and the response to acetylcholine is greatly increased (*e* and *f*). The response to nicotine (*g*) is unchanged, as one would expect, since it is not destroyed by acetylcholinesterase. After treatment with the cholinergic blocking agent dihydro-β-erythroidine, the response to a ventral root stimulus is much reduced (*h*) and the responses to injected acetylcholine and nicotine (*i* and *j*) are also reduced. No effects were seen after injection of drugs, other than acetylcholine, that contain a quaternary nitrogen atom, and are therefore ionized in solution (such as succinylcholine, D-tubocarine and neostimine); however, Curtis and R. M. Eccles (1958*a, b*) were later able to show that responses to these drugs could be obtained if they were ionophoretically ejected from micropipettes in the immediate vicinity of the cells. This suggests that there must be some diffusion barrier closely surrounding the cells. The micropipettes used by Curtis and his colleagues have five barrels (fig. 10.9); the central barrel is filled with 4 M sodium chloride and used as an extracellular recording electrode, and the barrels are each filled with a different drug, so that the actions and interactions of four drugs on any one cell can be examined.

It is clear that the evidence for cholinergic transmission at these Renshaw cell synapses satisfies the third and fourth of Paton's criteria. Since we are sure that the motoneuron releases acetylcholine at the neuromuscular junction, we may take the first criterion as satisfied also. The collection of the transmitter substance from the vicinity of the motoneuron collateral endings after stimulation (to satisfy the second of Paton's criteria) is obviously a much more difficult task, but Kuno and Rudomin (1966) showed that antidromic stimulation of motor nerves in the cat causes release of acetylcholine in the spinal cord. The synapses of motoneurons on skeletal muscles and Renshaw cells are interesting in that they provide evidence in favour of the idea, described by Eccles as 'Dale's principle', that any single neuron always releases the same transmitter substance or substances at its different endings.

The nicotinic action at the motoneuron collateral synapses is not the only action of acetylcholine on Renshaw cells, as was shown by Curtis and Ryall (1964). They found that the excitatory action of acetyl-β-methylcholine was unaffected by the nicotinic blocking agent dihydro-β-erythroidine,

Fig. 10.9. Multi-barrelled electrodes used for microionophoresis experiments on single neurons in the central nervous system. (from Curtis, 1965.)

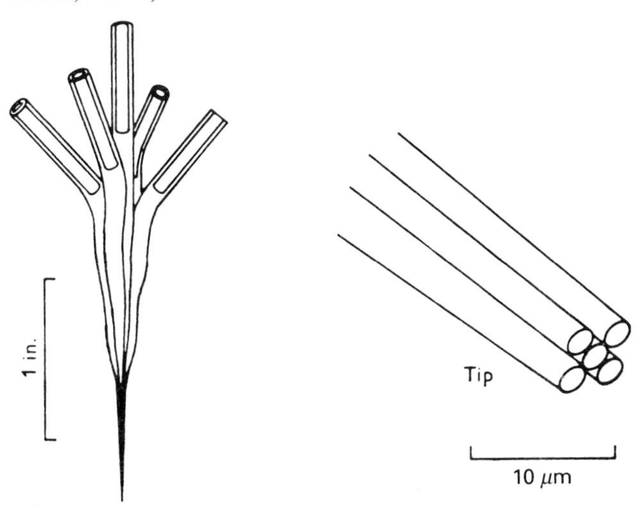

whereas it was blocked by atropine (fig. 10.10). Thus there are some muscarinic receptors on the Renshaw cell, as well as the nicotinic receptors activated by motoneuron collaterals.

Cholinergic synapses in the mammalian brain

Using the ionophoretic application technique, Krnjević and Phillis (1963*a, b*) were able to demonstrate the presence of acetylcholine-sensitive cells with muscarinic receptors in the cerebral cortex. In an investigation of the sensitivity of brain stem neurons to acetylcholine, Bradley and Wolstencroft (1965) showed that some of the sensitive neurons were excited by acetylcholine, whereas others were inhibited. The receptors on the inhibited neurons were entirely muscarinic, whereas those on excited neurons had both muscarinic and nicotinic properties.

New techniques have increased our knowledge of the distribution of cholinergic neurons in the brain (Cuello and Sofroniew, 1984; Kelly and Rogawski, 1985). One technique is to purify the acetylcholine-synthesizing enzyme choline acetyltransferase (ChAT) so as to prepare antibodies to it, and then use these antibodies for localization of the ChAT by immunocytochemistry, using a 'sandwich' of rabbit and goat antibodies. In the method used by Kimura *et al.* (1980) sections of fixed brain were incubated with Fab fragments of rabbit antibodies to ChAT. Then goat antibodies to rabbit antibodies were attached, and finally rabbit antibodies to the enzyme peroxidase, bound to the peroxidase itself; the goat antibody thus holds the two types of rabbit antibody together. Peroxidase will oxidize the compound 3,3′-diaminobenzidine in the presence of hydrogen peroxide to produce a brown electron-dense product, which can then be identified under the microscope. The results showed that cholinergic neurons are located at certain particular sites in the brain. For example, there are some cholinergic cell bodies at the base of the forebrain that send axons into the cerebral cortex, but there are no cholinergic neuron cell bodies in the cortex itself.

An alternative method is to try and localize acetylcholine receptors by using radioactive binding agents. Binding of radioactive QNB is much more extensive than that of nicotinic ligands, indicating that the great majority of brain acetylcholine receptors are muscarinic. Two separate types of nicotinic receptors occur, as indicated by cDNA cloning of their α chains. Their distributions have been determined by hybridization of radioactive mRNAs prepared from the cDNAs (Goldman *et al.*, 1987).

Invertebrates

Tauc and Gerschenfeld (1961, 1962) provided convincing evidence that acetylcholine acts both as an excitatory and as an inhibitory transmitter substance in the cells of *Aplysia* ganglia.

Fig. 10.10. Evidence for muscarinic acetylcholine receptors on Renshaw cells. The records show the firing frequency of the cells in response to ionophoretic application of acetylcholine (ACh) and acetyl-β-methylcholine (ACβMe). *a*: Control responses. *b*: Responses during simultaneous application of dihydro-β-erythroidine, which blocks nicotinic receptors. *c*: Responses after injection of atropine, which blocks muscarinic receptors. (From Curtis and Ryall, 1964.)

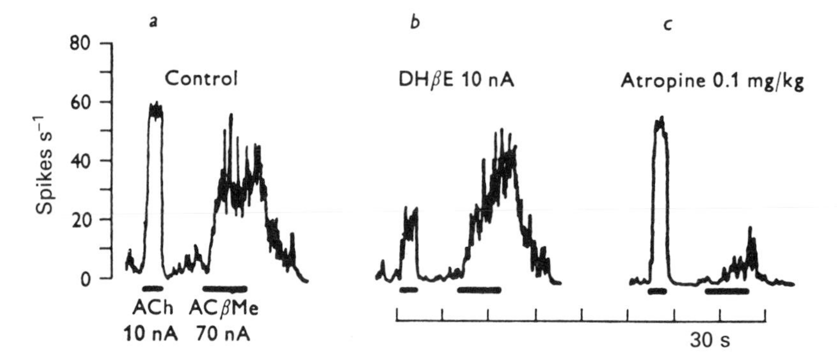

Application of acetylcholine to cells of one type (D-cells) causes depolarization (fig. 10.11), whereas application to the other type (H-cells) causes a hyperpolarization with a reversal potential equal to that of the IPSP. These actions, and also the EPSPs in D-cells and the IPSPs in H-cells, are antagonized by curare and atropine and potentiated by eserine. The excitatory input to H-cells is not cholinergic. Tauc and Gerschenfeld suggested that interneurons which would simultaneously induce excitatory and inhibitory actions in different post-synaptic neurons might exist, and Strumwasser (1962) showed this to be so. It was later shown that some of the D-cells ('D_{inhi}-cells') are inhibited by a non-cholinergic input, and it is particularly interesting that in these cells the IPSP does not involve an increase in chloride conductance, whereas the IPSP of H-cells does (Gerschenfeld and Chiarandini, 1965).

In view of the effectiveness of nicotine and of anticholinesterases as insecticides, it is not surprising to find that acetylcholine is probably the transmitter at some synapses in the insect central nervous system (Callec, 1974; Satelle, 1985). Lummis and Satelle (1985) measured the binding of radioactive α-bungarotoxin and QNB to extracts of cockroach nerve cord. They concluded that nicotinic receptors outnumber muscarinic receptors by about seven to one, the reverse of the situation in the vertebrate brain.

Some invertebrate peripheral synapses appear to be cholinergic. The neuromuscular transmitter in annelids and holothurians appears to be acetylcholine; indeed, as we have seen (p. 112), contraction of the dorsal longitudinal muscle of the medicinal leech has been used as a bioassay for acetylcholine. Molluscan hearts are inhibited by acetylcholine.

Muscarinic receptors

The postsynaptic potentials produced by the activation of muscarinic receptors are much slower than those involving nicotinic receptors, as is illustrated in figs. 8.17 and 9.24. It seems clear that the muscarinic receptor does not include an ionic channel in its structure, but acts indirectly via G protein activation, resulting in either modulation of an ionic channel or the production or inhibition of a cytoplasmic second messenger.

Muscarinic receptors have been purified from pig brain and other sources: they are single proteins with a molecular weight of about 70 kDa (Haga *et al.*, 1985). This implies that the molecule is of insufficient size to enclose an ionic channel: compare the sodium channel (*ca.* 260 kDa) and the nicotinic acetylcholine channel (5 subunits, total *ca.* 290 kDa).

Good evidence for the association of the muscarinic receptor with G proteins is provided by reconstitution experiments by Haga and his colleagues (Haga *et al.*, 1985, 1986). Purified receptors from pig brain and purified G proteins were mixed with phospholipid vesicles. This would allow the proteins to be reconstituted in a membrane environment free of other membrane proteins. Reconstituted G_i alone has some GTPase activity, which is not affected by the muscarinic agonist carbachol. Reconstituted receptor alone

Fig. 10.11. The effect of brief electrophoretic applications of acetylcholine on the membrane potential and firing frequency of *Aplysia* neurons. The D-cell is depolarized by acetylcholine, resulting in an increase in the firing frequency. The H-cell is hyperpolarized, resulting in a temporary cessation of activity. (From Tauc and Gerschenfeld, 1961.)

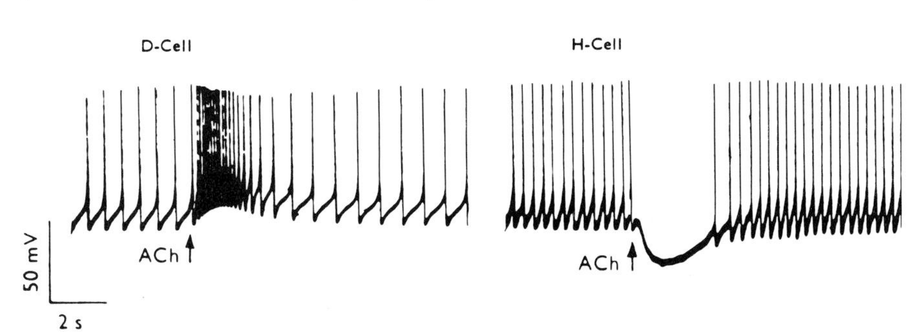

has no GTPase activity. When both proteins are reconstituted together, the GTPase activity is much increased by carbachol. This shows that G_i is indeed activated when the muscarinic receptor is activated. We would therefore expect that activation of muscarinic receptors linked to G_i would inhibit the production of cyclic AMP.

Similar results were obtained with another G protein, G_o, which shows that in this case a single receptor type is capable of interacting with two separate G proteins.

Not all muscarinic receptors have precisely the same pharmacological properties. Those in sympathetic ganglia have a high affinity for the antagonist pirenzepine, whereas those in heart muscle have a much lower affinity (Hammer and Giachetti, 1982). On the basis of this distinction the two types have been called M_1 and M_2 receptors. Another distinction appears in the action of agonists on chick embryonic heart cells: both carbachol and oxotremorine inhibit the production of cyclic AMP, but only carbachol stimulates phosphoinositol hydrolysis (Brown and Brown, 1984).

The molecular structure of M_1 muscarinic receptors from pig brain has been deduced from the complementary DNA sequence by the Kyoto University group (Kubo *et al.*, 1986*a*). Their methods were similar to those they used for the nicotinic acetycholine receptor, as described in chapter 8. The protein consists of 460 amino acid residues, giving a molecular weight of 51 416. The difference between this figure and the 70 000 derived from SDS gel measurements is probably largely accounted for by the carbohydrate content of the receptor protein.

Analysis of the amino acid sequence by the Kyte and Doolittle method (p. 19) shows a hydropathicity profile which has seven hydrophobic segments (fig. 10.12). These presumably represent seven α-helical sections each of which traverses the membrane. There are two probable glycosylation sites at asparagine residues near the N-terminal, which is therefore probably on the external side of the membrane. This would imply that the long section between the hydrophobic segments V and VI is on the cytoplasmic side, and the presence of possible sites for phosphorylation in this section is in accordance with this.

Similar hydropathicity profiles are found in two other membrane proteins important in the functioning of excitable cells, rhodopsin (the photosensitive pigment in the eye) and the β-adrenergic receptor. Kubo and his colleagues extended this exciting finding by making a detailed comparison of the amino acid sequences of the three proteins (fig. 10.13). They found that 30% of the muscarinic acetylcholine receptor sequence is identical with that of the β-adrenergic receptor and 23% with that of rhodopsin. This suggests that the genes encoding for all three proteins are derived from a common ancestor.

Kubo *et al.*, (1986*b*) have also determined the

Fig. 10.12. Hydropathicity profile of the muscarinic acetylcholine receptor. The graph shows the average hydropathicity index (values from Kyte and Doolittle) of the 19 amino acids from $i - 9$ to $i + 9$, plotted against i, where i represents the amino acid number. Black bars show the positions of the putative transmembrane segments I to VII. Arrows show possible N-glycosylation sites in the N terminal region. (From Kubo *et al.*, 1986*a*.)

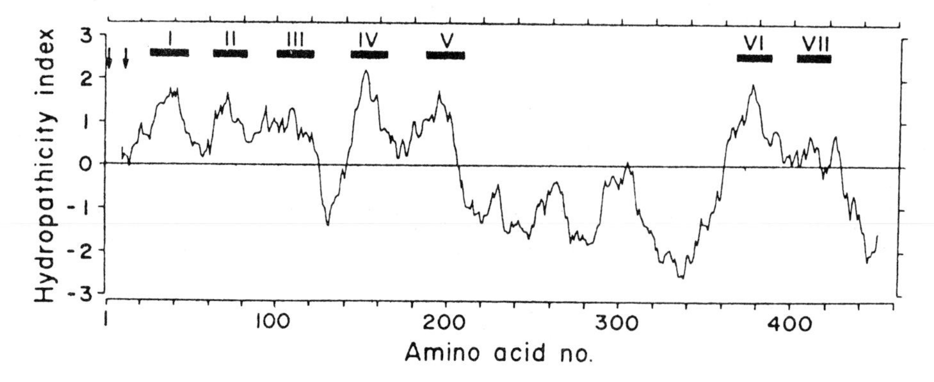

amino acid sequence of the M_2 muscarinic receptors from pig heart. The M_1 and M_2 sequences are not identical, but they are highly homologous, with 60% of the amino acids being identical. The two receptors must be coded for by different genes which are closely related to each other.

Injection of the mRNAs specific for the two receptors into *Xenopus* oocytes shows that they respond differently to acetylcholine, as is shown in fig. 10.14 (Fukuda *et al.*, 1987). The M_1 receptors produce relatively large oscillatory inward currents which are completely eliminated by injection of the calcium-chelating agent EGTA into the oocyte. The M_2 receptors produce smaller currents which are of two components, the first being not oscillatory and not abolished by EGTA injection, whereas the second is opposite in both these respects. The reversal potential of the oscillatory component was about -25 mV, suggesting that chloride channels were involved. The reversal potential for the smooth current was about $+10$ mV, not affected by chloride ion concentration, but sensitive to changes in the concentration of sodium or potassium ions. It looks as though M_1 receptors, in the oocyte membrane environment, act via the phosphatidylinositol system to release calcium ions which then open chloride channels. M_2 receptors, it seems, act more directly on cation-selective channels.

Catecholamines

The catecholamines are an important group of physiologically active compounds which contain the catechol group, a benzene ring with two adjacent hydroxyl groups. The hormone *adrenaline* (epinephrine) is released by the adrenal glands of mammals and serves to prepare the animal for action; the rate and amplitude of the heart-beat increase, constriction of the blood-vessels (except in the muscles) occurs, the spleen contracts, the pupils dilate, intestinal movements are inhibited, etc. Adrenaline is synthesized from the amino acid

Fig. 10.13. Alignment of the amino acid sequences of pig muscarinic acetylcholine receptor (mAChR), hamster β-adrenergic receptor (βAR) and cow opsin. The one-letter amino acid notation is given in table 1.1. Sets of identical residues are enclosed with solid lines and sets of conservative residues with dashed lines. Bars show the positions of the supposed transmembrane segments I to VII in the mAChR. Stars show probable N-glycosylation sites in the N terminal region. Opsin is the protein component of the visual pigment rhodopsin. (From Kubo *et al.*, 1986*a*.)

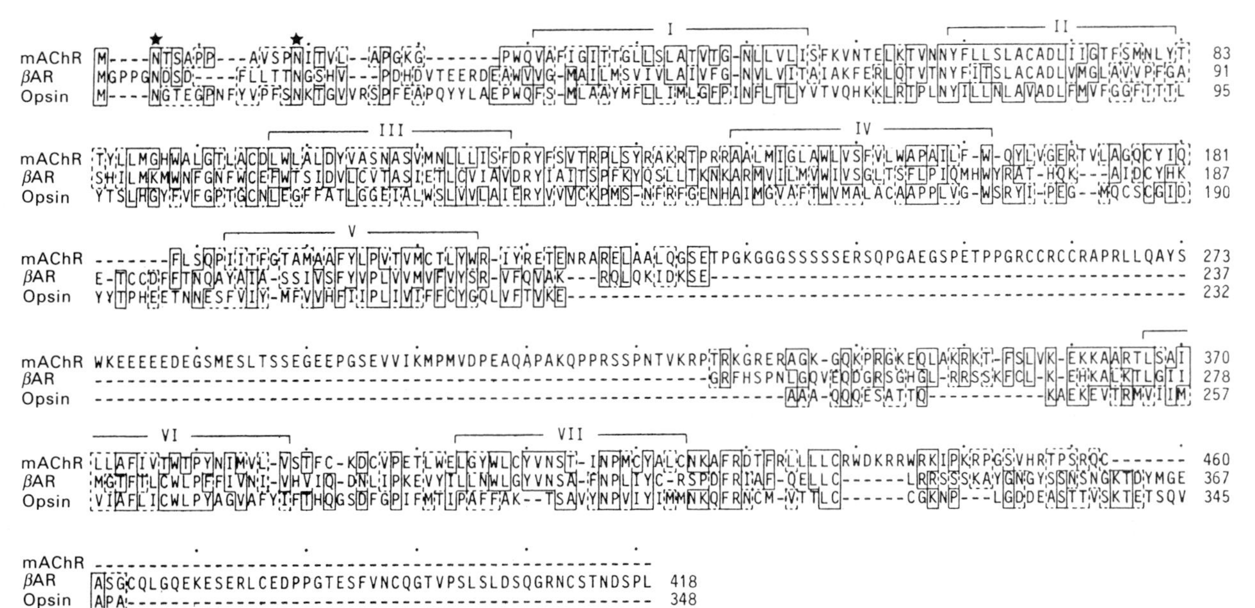

tyrosine via a series of other catecholamines, dopa, dopamine and noradrenaline, as is shown in fig. 10.15.

As mentioned earlier, Elliott in 1904 suggested that adrenaline was released from postganglionic sympathetic nerves on stimulation. In later years, a number of discrepancies between the action of adrenaline and sympathetic stimulation were discovered, so the sympathetic transmitter substance became known by the noncommittal name 'sympathin'. Later, von Euler (see von Euler, 1955) established that in most cases the sympathetic transmitter is not adrenaline but noradrenaline. The results of one of von Euler's experiments, on an extract of bovine spleen nerves, is shown in fig. 10.16. He found that 0.08 g of extract was equivalent to about 0.6 μg of noradrenaline in its action on cat blood pressure, and that 0.4 g of extract was equivalent to about 5 μg of noradrenaline in its action on the rectal caecum of the fowl; thus in the two reactions 1 g of extract is equivalent to 7.5 and 12.5 μg of noradrenaline respectively. However, the adrenaline equivalents for 1 g of extract work out as 17.5 μg for the cat blood pressure assay but only 0.25 μg for the fowl rectal caecum assay. Since the figures for adrenaline are so different, we must conclude that most of the active agent of the extract is not adrenaline.

Fig. 10.14. Responses of M_1 and M_2 muscarinic receptors in *Xenopus* oocytes. Oocytes had been previously injected with mRNA specific to M_1 (upper traces) or M_2 receptors (lower traces), made by transcription from the appropriate cloned cDNA. Bars show duration of application of acetylcholine to the oocyte. Traces show currents under voltage clamp at −70 mV, before (*a*) and after (*b*) injection of the calcium-chelating agent EGTA into the oocyte. The experiment shows that the response of the M_1 receptors is oscillatory and requires internal calcium ion, whereas the major component of the M_2 receptor response is smooth and does not. (From Fukuda *et al.*, 1987.)

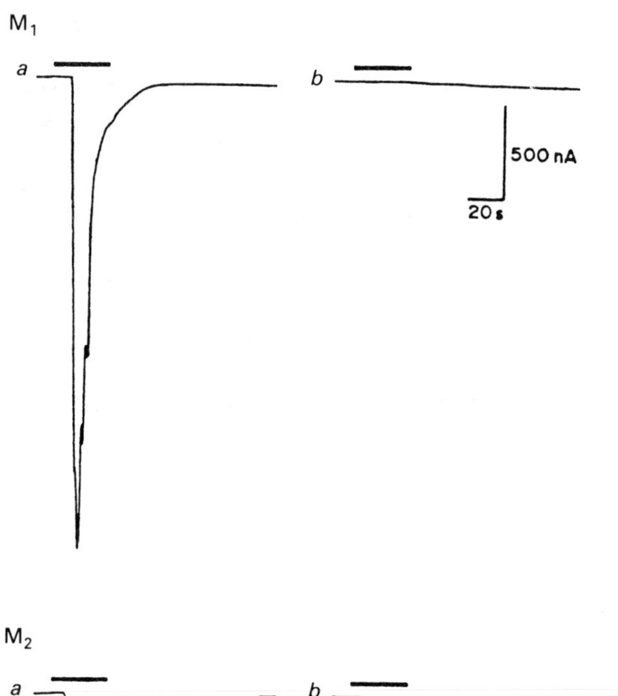

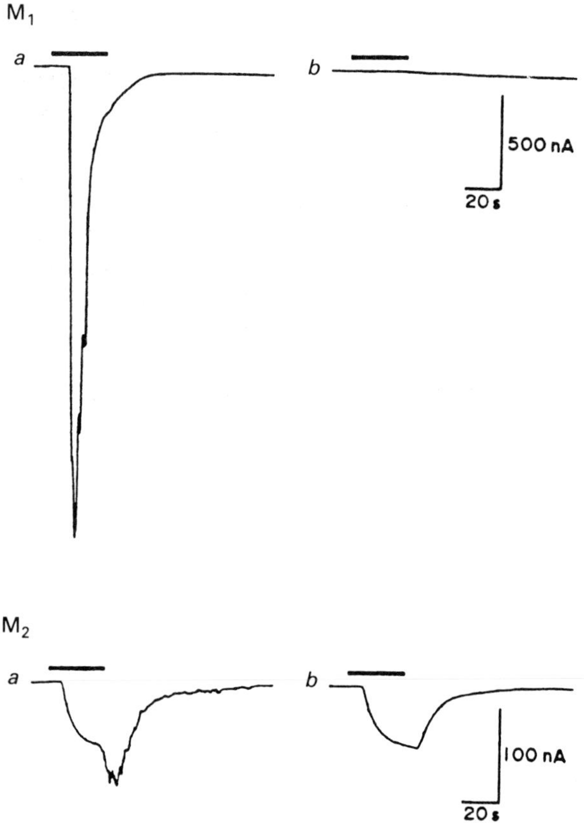

Fig. 10.15. The synthesis of adrenaline from tyrosine.

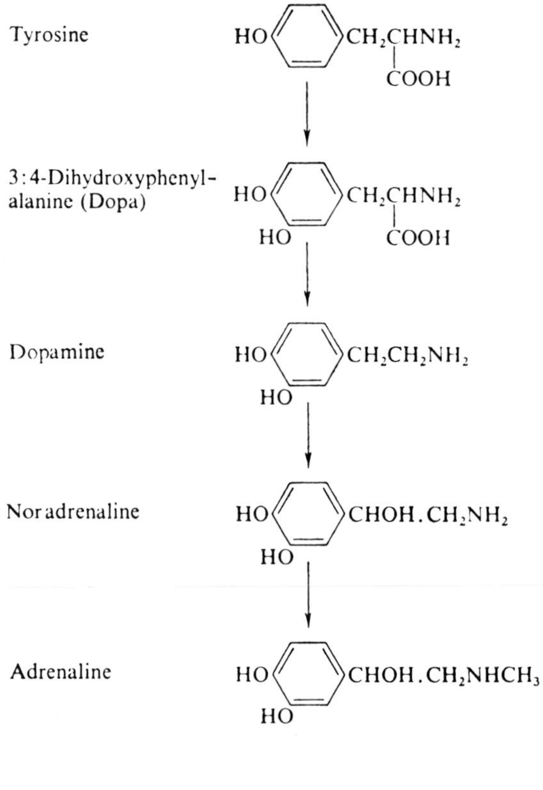

Tyrosine — HO–C_6H_4–$CH_2CH(NH_2)COOH$

↓

3:4-Dihydroxyphenylalanine (Dopa) — (HO)$_2$–C_6H_3–$CH_2CH(NH_2)COOH$

↓

Dopamine — (HO)$_2$–C_6H_3–$CH_2CH_2NH_2$

↓

Noradrenaline — (HO)$_2$–C_6H_3–$CHOH.CH_2NH_2$

↓

Adrenaline — (HO)$_2$–C_6H_3–$CHOH.CH_2NHCH_3$

Neurons in which noradrenaline is the transmitter substance are fairly widespread in the mammalian central nervous system. Our evidence for this is based largely on a useful technique in which freeze-dried sections of brain tissue are exposed to hot formaldehyde vapour; the catecholamines are then converted to substances which fluoresce strongly under ultraviolet light (Carlsson *et al.*, 1962; see Carlsson, 1987). Neurons in which dopamine appears to be the transmitter are also present.

Adrenergic receptors

Ahlquist (1948) compared the activity of the adrenergic agonist isoproterenol (isoprenaline) with those of adrenaline and noradrenaline on a variety of tissues. He found that in some cases (some blood vessels, the pregnant uterus, nictitating membrane, ureter) isoproterenol was much less active than the other compounds; he suggested that the adrenergic receptors in these tissues should be called α-receptors. For other tissues (other blood vessels, nonpregnant uterus, heart) isoproterenol was more active; these were described as having β-receptors. There are different competitive blocking agents (antagonists of noradrenaline) at the two types: α-receptors are blocked by phentolamine whereas β-receptors are blocked by propanolol, for example.

Table 10.4. *Agonists and antagonists at adrenergic receptors*

Receptor type	Agonists	Antagonists
α_1 and α_2		Phentolamine Phenoxybenzamine
α_1	Phenylephrine	Prazosin Corynanthine
α_2	Clonidine	Yohimbine Rauwolscine
β_1 and β_2	Isoproterenol	Propanolol
β_1	Dobutamine	Practolol
β_2	Salbutamol	Butoxamine

Later work resulted in further subdivisions of these receptor types, largely on the basis of their responsiveness to different agonists or antagonists. Thus β_1-receptors are responsive to dobutamine and not responsive to salbutamol, whereas the reverse holds for β_2-receptors. α_1-receptors are blocked by prazosin and corynanthine, whereas α_2-receptors are blocked by yohimbine and rauwolscine (see table 10.4).

There is good evidence that α_1-receptors are linked to the phosphatidylinositol second messenger system. In rat liver cells, for example, treatment with adrenaline causes rises in the cytoplasmic concentrations of IP_3 and calcium ions (Charest *et al.*, 1985; Exton, 1985). α_2-Receptors, on the other hand, act in a variety of tissues to inhibit adenylate cyclase via activation of the inhibitory protein G_i (Bylund and U'Prichard, 1983).

β-Receptors are also linked to adenylate

Fig. 10.16. Experiment to show that an extract of beef splenic nerves contains predominantly noradrenaline rather than adrenaline. The records show the effects of particular quantities of adrenaline (Adr.), noradrenaline (Nor-adr.) and the splenic nerve extract (Spl. n.) on the arterial pressure of a cat (left) and on the length of the isolated rectal caecum of a hen (right). (From von Euler, 1955.)

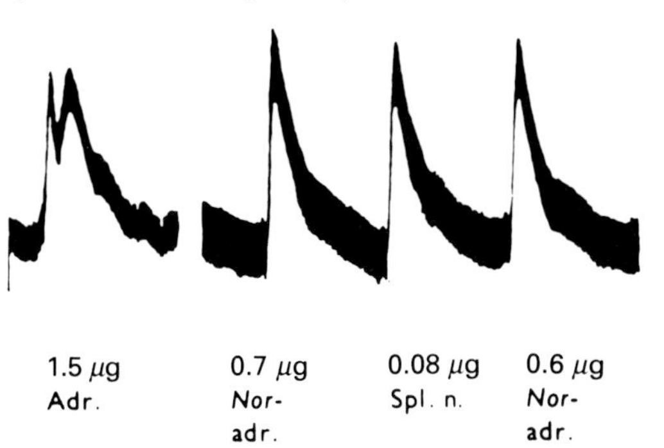

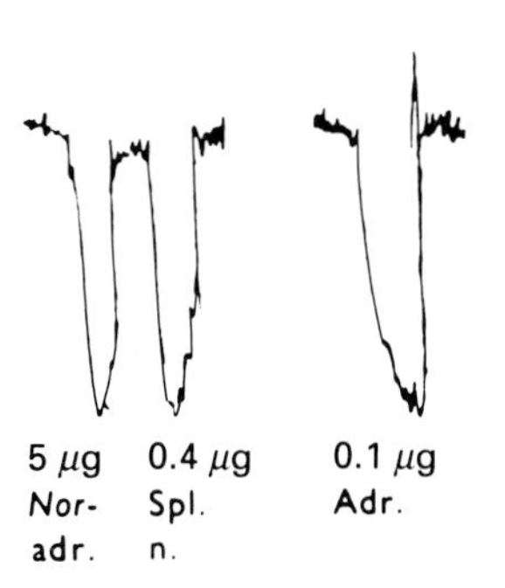

cyclase, but via the stimulatory protein G_s (see Levitzki, 1986). Thus activation of β-receptors leads to an increase in the cytoplasmic concentration of the second messenger cyclic AMP. A nice test of this conclusion is provided by the reconstitution experiments by May *et al.* (1985). They purified the three necessary proteins for this system: the β_1-adrenergic receptor (from turkey erythrocytes), the G_s protein (from rabbit liver) and the enzyme adenylate cyclase (from cow brain). These were then restored to their membrane lipid environment by reconstitution in phospholipid vesicles. Adenylate cyclase activity (the ability to make cAMP from ATP) was stimulated 2.6-fold by β-adrenergic agonists. This stimulation was dependent upon GTP and was blocked by β-adrenergic antagonists.

Complementary DNA cloning has been used to determine the amino acid sequence of the β_2-receptors in hamsters (Dixon *et al.*, 1986) and man (Kobilka *et al.*, 1987) and of the β_1-receptors in turkey erythrocytes (Yardon *et al.*, 1986). The three sequences are very similar to each other. It is a remarkable fact, however, that they show considerable homology with the visual pigment rhodopsin and also (as was discovered soon afterwards) with the muscarinic acetylcholine receptor (see fig. 10.13). It seems clear that, as in rhodopsin and bacteriorhodopsin, there are seven hydrophobic membrane-spanning sections in the molecule, as is shown in fig. 10.17.

Adrenergic vesicles

The adrenergic neurons of the sympathetic nervous system have non-myelinated axons which terminate in various effector organs. The terminals branch, and have a large number of swollen portions called *varicosities* which are packed with mitochondria and vesicles. Most of the vesicles have electron-dense cores and are hence known as 'granular vesicles'. They are of two sizes, large (diameter up to 100 nm) and small (diameter 40–50 nm). The small ones are predominant in the varicosities, whereas the large ones are also found in the cell body and axon.

Both types of vesicle contain noradrenaline and ATP. The large vesicles contain the proteins chromogranin A and dopamine β-hydroxylase (the enzyme which converts dopamine to noradrenaline) and a number of others, together with

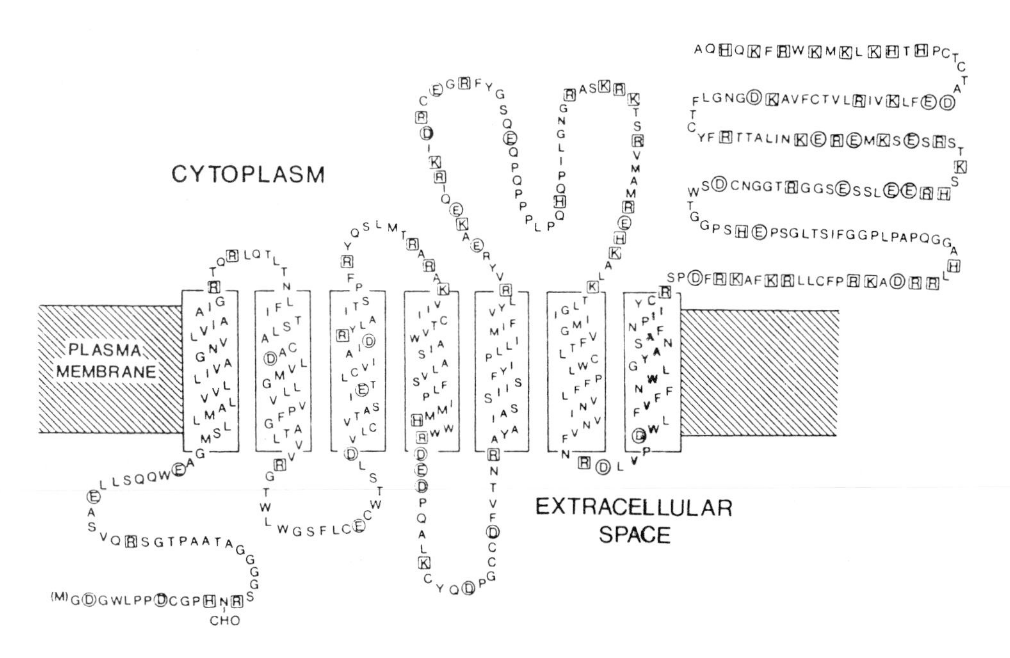

Fig. 10.17. Amino acid sequence and proposed orientation of the β-adrenergic receptor. The N terminal is shown at the left in the extracellular space. Positively charged amino acid residues are shown in squares, negatively charged ones in circles. (From Yarden *et al.*, 1986.)

enkephalins and other opioid peptides which presumably are acting as co-transmitters. Klein and Lagercrantz (1981) calculate that small vesicles contain 700–1000 molecules of noradrenaline whereas large ones contain 8000–16 000.

In some electron micrographs there are also some 'agranular vesicles' present; it seems likely that these are derived from small granular vesicles by loss of some of their contents.

Presynaptic receptors

Noradrenergic neurons have their activity modulated by noradrenaline and a number of other substances via presynaptic receptors (Langer, 1974, 1981). Evidence for this conclusion comes from experiments in which the amount of noradrenaline released at a synapse is measured in the presence of drugs which act at adrenergic receptors. In the dog heart, for example, Yamaguchi, De Champlain and Nadeau (1977) found that the amount of noradrenaline released after stimulating sympathetic nerve fibres is increased in the presence of phenoxybenzamine (an α-receptor antagonist) and reduced in the presence of clonidine (an α_2-receptor agonist). Thus there appear to be α_2-adrenergic receptors on the presynaptic varicosities, whose action is to inhibit noradrenaline release. It looks as though there is some negative feedback control of the amount of transmitter released.

Other presynaptic receptors occur. Inhibition of noradrenaline release may be brought about by acetylcholine (via muscarinic receptors), dopamine, 5-hydroxytryptamine or opioid peptides. Noradrenaline release can be increased via β-adrenergic receptors, angiotensin II receptors (angiotensin II is a peptide hormone involved in the control of blood pressure) and nicotinic acetylcholine receptors. The mechanisms of these effects could perhaps involve the control of the presynaptic intracellular calcium ion concentration.

The fate of released noradrenaline

We have seen that in vertebrate skeletal muscles acetylcholine combines with receptors for a short time but is then rapidly hydrolysed by acetylcholinesterase. The primary activity of released catecholamines is similarly a brief combination with postsynaptic receptors, but thereafter there are several possibilities.

Adrenaline and noradrenaline are enzymically inactivated in two main ways: by oxidative deamination (by the mitochondrial enzyme monoamine oxidase) or by methylation of a hydroxyl group (by the cytoplasmic enzyme catechol-O-methyl transferase). Both these enzymes are intracellular and hence can only act on the catecholamines after they have been taken up by cells. This situation is in marked contrast with that at cholinergic synapses, where the extracellular localization of acetylcholinesterase in the synaptic cleft enables it to inactivate the transmitter substance immediately and effectively.

Clearly then, enzymic inactivation cannot be the prime method of terminating the effects of noradrenaline on the postsynaptic membrane. It became evident that many tissues, and especially adrenergic nerve endings, are able to take up catecholamines from the extracellular fluid. Thus Whitby, Axelrod and Weil-Malherbe (1961) found that tritium-labelled noradrenaline was rapidly concentrated in adrenergic nerve endings. Iversen (1963, 1967) found that the adrenergic nerve terminals in a perfused rat heart were capable of clearing the entire extracellular space of the heart in about ten seconds. He calculated that this uptake system would therefore remove noradrenaline from the immediate vicinity of the terminals in a matter of milliseconds, thus constituting a most effective method of inactivating the transmitter (Iversen, 1971). It is also an economical method, since the noradrenaline is accumulated intracellularly in the synaptic vesicles, from which it can be released again in further synaptic transmission.

It is possible to fool the uptake process by feeding it with an analogue of noradrenaline or one of its metabolites. The analogue is then taken up into the nerve terminal, stored in the synaptic vesicles and then released into the synaptic cleft when a presynaptic nerve impulse arrives. Such compounds are known as *false transmitters*.

Noradrenaline is also taken up by a variety of non-neural cells in the body. Thus noradrenaline may be removed from the synaptic region by diffusion to the blood stream or by uptake into the postsynaptic cell. In each case it is usually enzymically inactivated by catechol-O-methyl transferase or monoamine oxidase.

Drugs active at adrenergic synapses

In view of the variety of processes occurring at adrenergic synapses (summarized in fig. 10.18), it is not surprising that many different drugs affect transmission in some way. Such drugs may affect the synthesis, release or re-uptake of catecholamines, may inhibit their intracellular enzymatic breakdown, or may act as agonists or antagonists at pre- or postsynaptic receptors. It is common for a drug to act in more than one way.

Table 10.4 gives some examples of adrenergic agonists and antagonists. Propranolol and other 'β-blockers' are clinically useful in the treatment of high blood pressure. Clonidine produces a brief rise in blood pressure, probably by stimulating α_2-receptors in the arterioles, followed by a prolonged fall which is thought to be caused by activation of the presynaptic α_2-receptors on adrenergic neurons in the brain (Katzung, 1987). Isoproterenol and salbutamol are β-receptor agonists which are useful as bronchodilator agents in the treatment of asthma. Isoproterenol may produce side-effects by stimulating the β_1-receptors in the heart, whereas salbutamol does not since it is selective for β_2-receptors.

Amphetamine promotes the release of transmitter at noradrenergic and dopaminergic synapses in the brain. Cocaine blocks the re-uptake mechanism at the same group of synapses. Both these substances produce sensations of euphoria and so have been used as drugs of abuse. Drugs used as antidepressants include imipramine (a re-uptake

Fig. 10.18. Sites of action of drugs at a peripheral adrenergic synapse. 1, Propagation of the action potential into the presynaptic varicosity; 2, synthesis of noradrenaline (NA); 3, uptake of noradrenaline into presynaptic vesicles; 4, deamination by the enzyme monoamine oxidase (MAO); 5, release of noradrenaline; 6, binding to postsynaptic receptors; 7, binding to α_2 presynaptic receptors; 8, reuptake into the presynaptic terminal; 9, inactivation by the enzyme catechol O-methyl transferase; 10, loss of noradrenaline to the circulation.

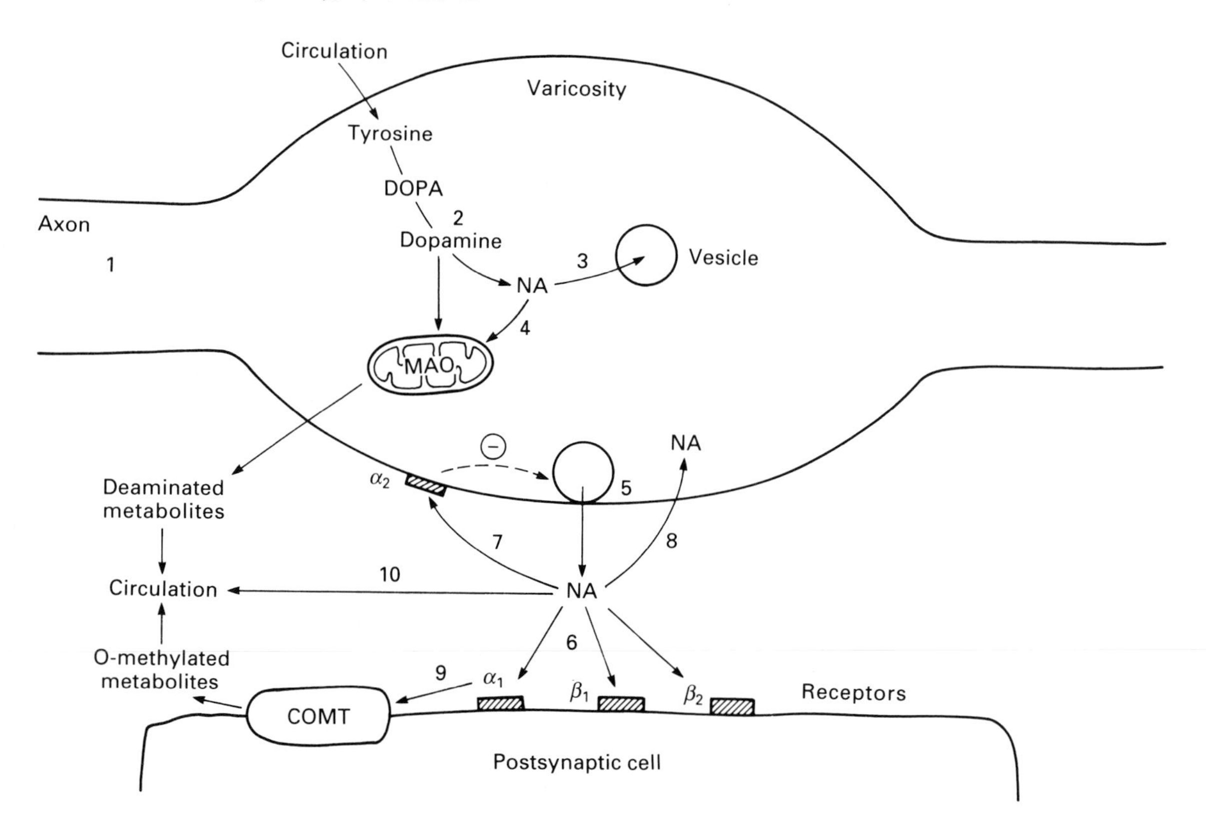

inhibitor) and iproniazid (an inhibitor of monoamine oxidase).

Guanethidine and reserpine cause depletion of the noradrenaline content of adrenergic vesicles, and have been used as agents for reducing the blood pressure.

Dopamine

Dopamine is a precursor of noradrenaline (fig. 10.15), but it also acts as a neurotransmitter. The discovery in 1961 that particular neurons in the brain contain appreciable quantities of dopamine, and that these same neurons were the ones which degenerate in Parkinson's disease, set the scene for a dramatic advance in the treatment of the disease (see Hornykiewicz, 1973). Administration of L-dopa, the immediate precursor of dopamine, produced a rapid relief of the symptoms. Malfunction of dopaminergic neurons has also been implicated in schizophrenia, but the evidence here is much less conclusive (Owen *et al.*, 1985).

Postsynaptic dopamine receptors have been divided into two types, D_1 and D_2 (Kebabian and Calne, 1979). Stimulation of D_1 receptors activates adenylate cyclase and so increases cyclic AMP formation, whereas stimulation of D_2 receptors does not. D_2 receptors also occur presynaptically on the terminals of dopaminergic and other neurons. Antipsychotic drugs such as chlorpromazine and others act as antagonists at D_2 receptors.

Serotonin

Serotonin (5-hydroxytryptamine, 5-HT) is an indoleamine (fig. 10.19). Indoleamines and catecholamines are sometimes grouped together as monoamines.

Relatively large quantities of serotonin are found in blood serum and platelets, and in the chromaffin cells of the gut lining. It causes potentiation of the heartbeat in molluscs, and this effect was used by Twarog and Page (1953) to show that there is an appreciable concentration of serotonin in the brain.

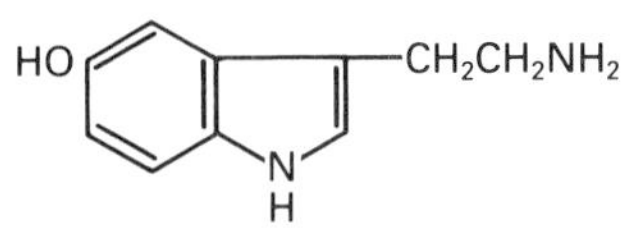

Serotonin has a wide variety of effects on nerve cells (see Vandermaelen, 1985). Inhibitory effects depend largely on increases in membrane potassium conductance, and many excitatory effects seem to be caused by a decrease in potassium conductance. Some effects are produced via changes in acetylcholine or noradrenaline release, suggesting that presynaptic serotonin receptors are involved.

Different types of serotonin receptors occur. Peroutka, Lebowitz and Snyder (1981) measured the ability of various different drugs to displace radioactive serotonin and radioactive spiperone from binding sites on rat brain membrane fractions. They found that inhibition of serotonin binding was proportional to the degree of inhibition of serotonin-sensitive adenylate cyclase activity. However, inhibition of spiperone binding was not related to adenylate cyclase activity but was proportional to the inhibition of serotonin-induced head twitches in rats. They suggested that the two types of response involved two different receptor types. 5-HT_1 receptors, with a high affinity for serotonin, normally mediated neuronal inhibition and were linked to adenylate cyclase. 5-HT_2 receptors, with a higher affinity for spiroperidol, showed no linkage to adenylate cyclase and were thought to be responsible for excitatory responses.

5-HT_1 receptors have been further subdivided into 5-HT_{1A} and 5-HT_{1B} subtypes according to their affinity for various compounds. Alternative classifications exist for peripheral neuron receptors, for which powerful new blocking agents have been discovered (Richardson *et al.*, 1985).

Substances active at serotoninergic synapses include the ergot alkaloids methysergide and lysergic acid diethylamide (LSD). Methysergide is used in the treatment of migraine. LSD is a hallucinogen; there is some evidence that it is an agonist at 5-HT_2 receptors (see Jacobs, 1987).

Serotonin is synthesized from the amino acid tryptophan and can be deaminated by the enzyme monoamine oxidase. In the pineal organ (which contains large quantities of serotonin) it is converted into the hormone melatonin.

Fig. 10.19. Serotonin (5-hydroxytryptamine).

HO —CH$_2$CH$_2$NH$_2$ N H

Histamine

The main concentrations of histamine in the body are in the mast cells and basophil leucocytes, from which it is released to act as a 'local hormone'. Histamine has a marked effect on some smooth muscles, increases capillary permeability, and stimulates sensory nerve endings, leading to sensations of pain and itching. It also occurs as a neurotransmitter in the brain (see Schwartz *et al.*, 1986). It is synthesized from the amino acid histidine, using the enzyme histidine decarboxylase.

Postsynaptic and effector cell receptors are of two types, H_1 and H_2 (Black *et al.*, 1972). Mepyramine and the 'antihistamines' (such as promethazine) used in the treatment or allergic reactions and in preventing motion sickness, are antagonists at H_1 receptors, whereas burimamide is an antagonist at H_2 receptors. Presynaptic histamine receptors form a third class (H_3), both in the brain (Arrang *et al.*, 1983, 1987) and at peripheral autonomic nerve endings (Ishikawa and Sperelakis, 1987).

Amino acid transmitters

Gamma-aminobutyric acid (GABA) and glycine act as fast inhibitory transmitters in the mammalian central nervous system, whereas glutamate and probably aspartate and others are responsible for fast excitatory transmission. Their depressant and excitatory actions were first discovered by using the ionophoresis technique on central neurons (Curtis and Watkins, 1960; Curtis *et al.*, 1960). Nevertheless it was first thought that they were not neurotransmitter substances released from nerve terminals and acting on postsynaptic receptors, but that they acted via some more generalized effect on membrane excitability. This view was revised with the discovery of particular blocking agents for amino acid action, such as strychnine for glycine and bicuculline for GABA, suggesting that there are specific receptors for these substances. Krnjević (1986) gives a pithy summary of the development of research in this field.

Gamma-aminobutyric acid (GABA)

There is excellent evidence that GABA is the transmitter mediating peripheral inhibition in crustaceans. Kuffler and Edwards (1958) showed that the action of the inhibitory transmitter substance on crustacean stretch receptor cells could be imitated by GABA. Both synaptic inhibition and GABA produce a hyperpolarization or depolarization according as to where the membrane potential is set, and the reversal potential is the same for each. Changes in external chloride concentrations produce similar changes in the two responses. Both responses are depressed by picrotoxin.

A similar imitation of synaptic inhibition by GABA is seen in the peripheral inhibition of crustacean and insect muscle fibres; the responses involve an increase in chloride conductance (Takeuchi and Takeuchi, 1965; Usherwood and Grundfest, 1965). Kravitz, Kuffler and Potter (1963) showed that the inhibitory fibres from the legs of lobsters contain appreciable quantities of GABA while the excitatory fibres do not. And finally Otsuka *et al.* (1966) found that GABA appeared in a perfusate of the lobster claw opener muscle when the inhibitory axon was stimulated. Thus, at crustacean peripheral inhibitory junctions, GABA fulfils all of Paton's criteria for the identification of a transmitter substance.

Some of the compounds active at GABA-ergic synapses are shown in fig. 10.20. GABA receptors are widely distributed in the mammalian nervous system. There are two distinct types, called $GABA_A$ and $GABA_B$ (Hill and Bowery, 1981). $GABA_A$ receptors are largely postsynaptic in the brain, but are responsible for presynaptic inhibition on the spinal cord. Muscimol (obtained from the mushroom *Amanita muscaria*) and isoguvacine are agonists and bicuculline (from the herb *Corydalis*) and picrotoxin are antagonists (Curtis *et al.*, 1970). Activation of postsynaptic $GABA_A$ receptors produces fast IPSPs which involve the opening of chloride channels.

$GABA_B$ receptors have baclofen as an agonist and are insensitive to bicuculline (Bowery *et al.*, 1980). In some cases they appear to be located presynaptically, but in others they are clearly postsynaptic (Newbery and Nicoll, 1984). $GABA_B$ responses appear to be slower than $GABA_A$ responses and may be linked to adenylate cyclase (Hill *et al.*, 1984; Karbon *et al.*, 1984). The rest of this section relates to $GABA_A$ receptors only.

Krnjević and Schwarz (1967) applied GABA by ionophoresis to cells in the cat cerebral cortex. This produced hyperpolarizations similar to IPSPs. Both responses were converted to depolarizations by injection of chloride ions into the cell, and the reversal potential of the GABA response was always the same as that of the IPSP. They concluded that GABA was probably the inhibitory transmitter substance and that it acted by opening ionic channels permeable to chloride ions.

Much information about the nature of GABA-activated chloride channels has been obtained by Bormann, Hamill and Sakmann (1987), by application of the whole cell and patch clamp techniques to mammalian spinal neurons in tissue culture. The whole-cell clamp (fig. 6.1*a*) allows the internal ionic concentrations to be varied because they equilibrate with the solution in the recording electrode. Bormann and his colleagues found that the reversal potential of the response to GABA is given by the chloride equilibrium potential, hence the GABA channel is permeable to chloride ions. They also determined the reversal potential with different anions internally, and from this they calculated the permeability ratios for a variety of anions with respect to chloride. The results (fig. 10.21) show that permeability falls with increasing ionic diameter, reaching zero at about 5.6 Å. This suggests that the internal diameter of the GABA channel is slightly smaller than that of the nicotinic acetylcholine channel (7.4 Å), as determined in the essentially similar experiments by Dwyer *et al.* (chapter 8).

Bormann and his colleagues also obtained single GABA channel records using the patch-clamp technique. Different conductance states were evident, with conductances of 44, 30, 19 and 12 pS, the 30 pS state being by far the most common. With thiocyanate as the permeant anion, the main conductance level was reduced from 30 to 22.5 pS. With mixtures of chlorides and thiocyanate, however, channel conductance was still lower, reaching a minimum of 12 pS at 16% thiocyanate (fig. 10.22). This rather surprising phenomenon is known as the anomalous mole fraction effect; it has been seen in other channels (Almers and McCleskey, 1984) and is interpreted as evidence that there are at least two interacting ion binding sites in each channel.

(An inexact analogy may help in thinking about this anomalous mole fraction effect. Consider a junior school sports day in which parents (chloride ions) and children (thiocyanate ions) take part: the parents can run faster than the children because they have longer legs. In a three-legged race people run in pairs with their adjacent legs tied together, and success goes to those who can synchronize their leg movements. Child–child pairs will still lose to parent–parent pairs. But parent–child pairs will be the slowest of all because it will be very difficult for them to move their legs in synchrony.)

Analysis of how we might expect two-ion channels to behave involves Eyring rate theory and some complex calculations with an uncomfortably large number of constants to which values have to

Fig. 10.20. Some substances which combine with $GABA_A$ receptors.

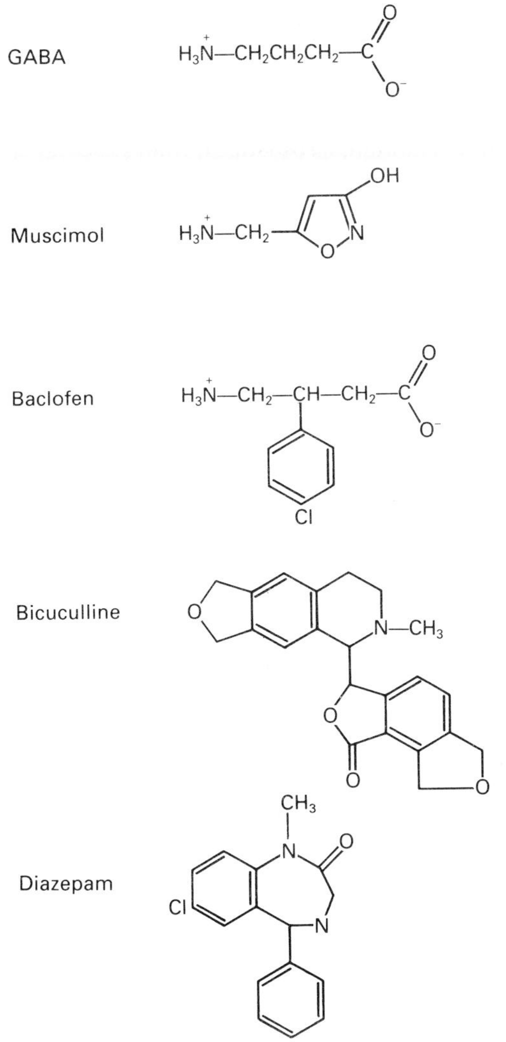

GABA

Muscimol

Baclofen

Bicuculline

Diazepam

be assigned. But the results are in good agreement with actuality (fig. 10.22) and provide some justification for the model shown in fig. 10.23.

Interest in $GABA_A$ receptors was much enhanced by the discovery that they are modulated by two types of drugs widely used in medical practice as central nervous system depressants. These are the benzodiazepines such as diazepam (the tranquillizer Valium) and the barbiturates (Choi *et al.*, 1977; Macdonald and Barker, 1978; Nicoll *et al.*, 1975; Barker and Ransom, 1978). These compounds enhance the inhibitory action of GABA. Fluctuation analysis suggests that diazepam increases the frequency of channel openings whereas phentobarbitol increases their average open lifetime (Study and Barker, 1981). It has been estimated that the consumption of benzodiazepines in the USA is over 8000 tons per year (Tallman *et al.*, 1980).

There has been much discussion as to the nature of the links between the benzodiazepine receptor, the barbiturate receptor and the $GABA_A$ receptor–channel complex. Binding and other studies suggested that these were closely coupled,

Fig. 10.21. Permeability of the GABA-activated chloride channel to various anions. The oscilloscope traces *a* to *c* show whole-cell currents of cultured mouse spinal neurons following application of GABA while held under voltage clamp at different membrane potentials. *a* shows the responses when the extracellular and intracellular solutions had an identical chloride concentration of 145 mM; the reversal potential is very close to 0 mV. *b* and *c* show similar responses when the cell was internally dialysed with a solution containing 140 mM formate (*b*) or propionate (*c*) and 4 mM chloride; the reversal potential is lower for formate and much lower for propionate, showing that the channels are less permeable to these ions than to chloride. In *d* the permeabilities of the different ions with respect to chloride, as calculated from the reversal potential experiments, are plotted against their diameters, so giving an estimate of the size of the channel. (From Bormann *et al.*, 1987.)

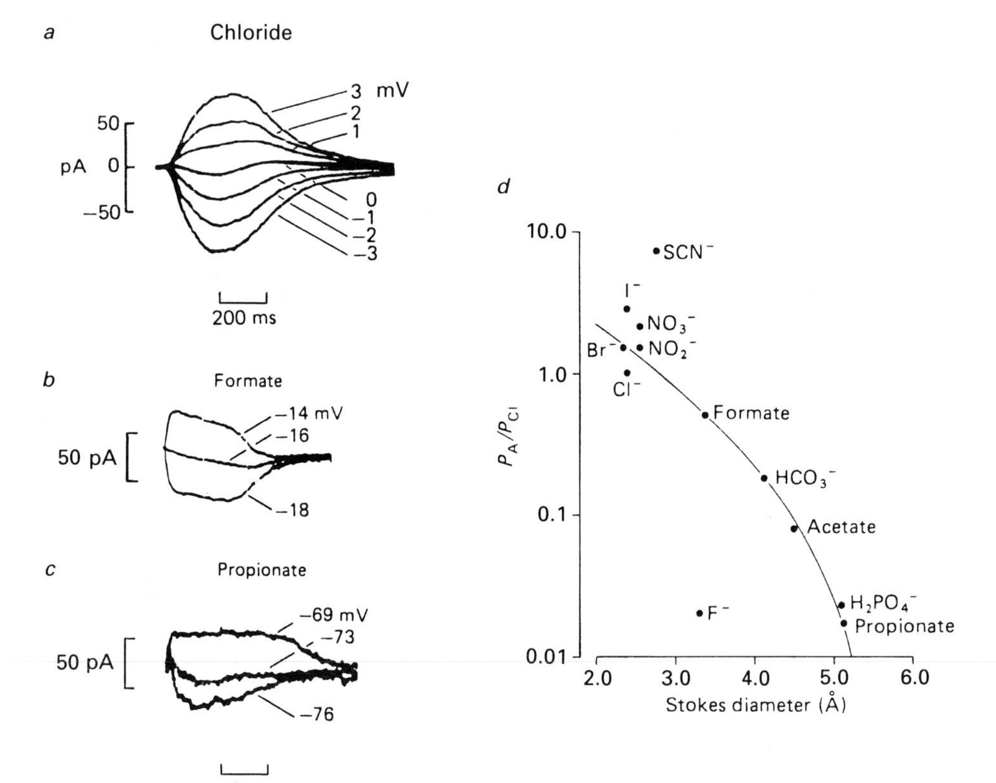

but most models implied that the different components of the system were different protein molecules (Olsen, 1982; Turner and Whittle, 1983). The situation has been greatly clarified by the application of the techniques of molecular biology to the problem by Barnard and his colleagues (Schofield *et al.*, 1987), whose results we shall now consider.

The $GABA_A$ receptor can be purified from the cow cerebral cortex by using benzodiazepine affinity chromatography (Sigel and Barnard, 1984). It consists of two pairs of protein chains forming an $\alpha_2\beta_2$ complex; the α chains bind benzodiazepine and the β chains bind GABA (Mamalaki *et al.*, 1987).

The methods of sequencing used by Schofield and his colleagues were similar to those used for the nicotinic acetylcholine receptor and described in chapter 8. Some polypeptide sequences were determined from the purified receptor, and these were used to construct synthetic oligonucleotide probes. Brain cDNA libraries were then screened with these for hybridizing sequences, with the

Fig. 10.22. The anomalous mole-fraction effect in GABA-activated chloride channels of cultured mouse spinal neurons. The points show the single channel conductance at −70 mV with different proportions of thiocyanate in a chloride–thiocyanate mixture; solutions were the same on each side of the membrane and the total anion concentration was always 145 mM. The continuous line shows the expected relation for a channel with a single binding site for which the two anions compete. The dashed line shows that for a channel with two such sites. (From Bormann *et al.*, 1987.)

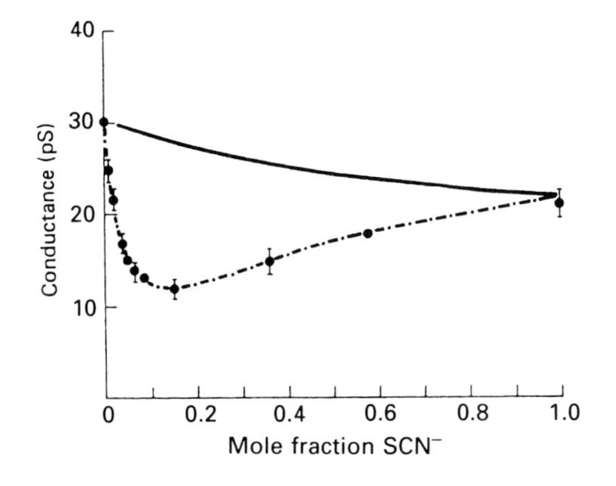

Fig. 10.23. Schematic model of the GABA-activated chloride channel in cultured mouse spinal neurons. The diagram in *a* is drawn to scale and shows two fixed positive charges to which the chloride ion can bind. The profile in *b* shows the free-energy changes for a chloride ion passing through the open channel in the absence of a potential across the membrane. Energy barriers have to be crossed on entering and leaving the channel. Wells correspond to the positions of the two binding sites, and there is another barrier between them. The profile is calculated according to Eyring rate theory using values suitable to fit the experimental results. (From Bormann *et al.*, 1987.)

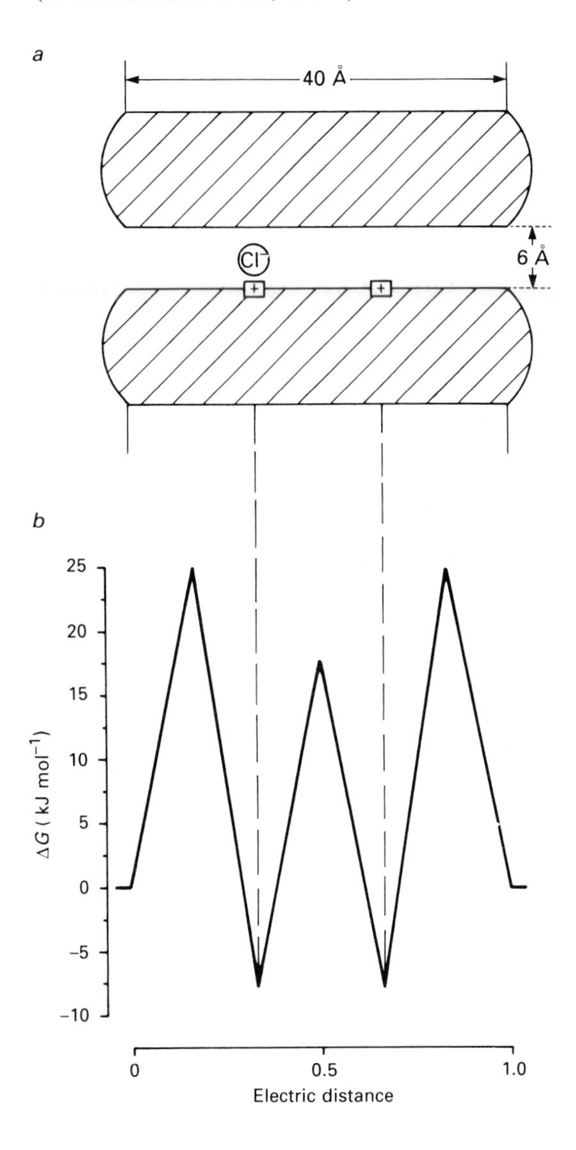

result that two were found, coding for the α and β chains of the receptor. From the nucleotide sequences of these two cDNAs the amino acid sequences of the two receptor subunits were deduced. The mature α chain has 429 amino acid residues, molecular weight (excluding added carbohydrate residues) 48 800, and the β chain has 449 residues, molecular weight 51 400.

The two subunit chains have 35% identity in their amino acid residues and their hydropathy profiles are very similar (fig. 10.24), showing a high degree of homology between them. In each chain there are four hydropathic regions which seem to be transmembrane α-helices. A schematic model of the receptor is shown in fig. 10.25. It assumes that the four subunits form a tight group with the chloride channel in the middle, its walls formed by one or two of the membrane-spanning sections from each subunit.

Fig. 10.24. Hydropathicity profiles of the primary sequences of cow $GABA_A$ receptor α and β subunits. The ordinate shows the summed hydropathicity indices for windows of 17 residues from $i - 8$ to $i + 8$, the abscissa shows the position of the corresponding amino acid i in the sequence. Black bars show the presumed signal sequence at the left and the four probable membrane-crossing segments. The lower graph shows, for comparison, a similar plot for the cow nicotinic acetylcholine receptor α subunit. (From Schofield *et al.*, 1987.)

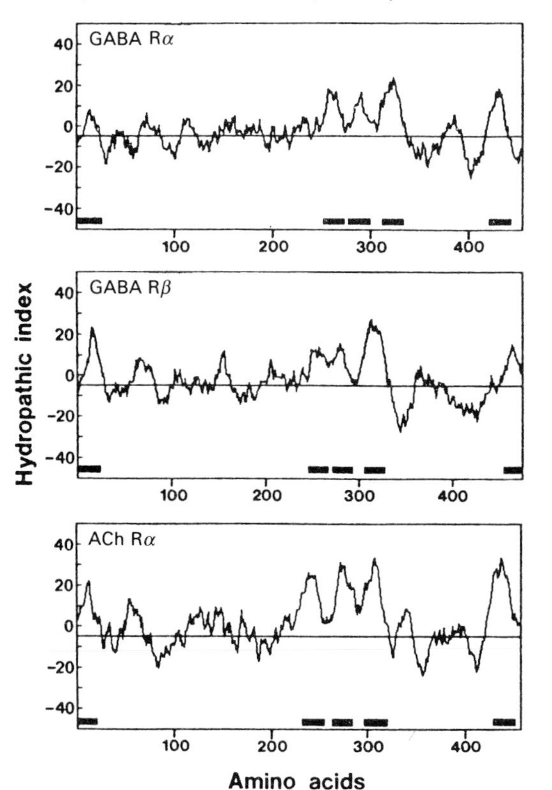

An exciting and remarkable feature of these results is that there is some consider similarity between the amino acid sequences of the GABA receptor subunits and those of the nicotinic acetylcholine receptor. The M1 to M4 transmembrane sequences are similarly placed in the two receptor types, but there is no MA sequence in the GABA receptor subunits. It looks as though both receptors are derived from some common evolutionary ancestor. Together with the glycine receptor (p. 212) they form a family of neurotransmitter receptors with fast intrinsic ionic channels. They are quite separate from another major receptor family which contains the β-adrenergic receptor, the muscarinic acetylcholine receptor and rhodopsin; here there is no intrinsic ionic channel, the whole receptor is formed from just one protein chain with seven membrane-crossing regions, and the receptors produce their effects via activation of G proteins. It will be interesting to see if and where the $GABA_B$ receptor fits into this pattern.

Does the $\alpha_2\beta_2$ complex constitute the entire $GABA_A$ receptor system? Schofield and his colleagues prepared α and β subunit messenger RNAs by transcription from the corresponding cDNAs inserted into plasmids, and these mRNAs were then injected into *Xenopus* oocytes. A few days later, application of GABA produced inward currents brought about by appreciable increases in membrane conductance. The reversal potential for these currents was −27 mV, the equilibrium potential for chloride ions. Such responses were never produced after the oocyte had been injected with the mRNA for either the α chain or the β chain alone.

The inward chloride currents produced by application of GABA to injected oocytes were blocked by bicuculline or picrotoxin and potentiated by chlorazepate (a benzodiazepine) or barbiturates. This implies that the complex contains all of the receptor sites involved in the activation and modulation of the GABA receptor.

Glycine

It is likely that the transmitter mediating postsynaptic inhibition on motoneurons in the spinal cord is glycine. This was first suggested by Aprison and Werman (1965), as a result of their demonstration that the local concentration of glycine in the spinal cord is much higher in those regions where inhibitory interneurons occur. Electrophoretic application of glycine produces hyperpolarization of spinal motoneurons and strychnine blocks this action (Werman and Aprison, 1968; Curtis, 1969). Thus glycine appears to meet the first, third and fourth of Paton's criteria.

The amino acids β-alanine and taurine act as agonists at glycinergic synapses and the alkaloid strychnine (obtained from the Indian tree *Strychnos nux-vomica*) is an antagonist. The blockage of the inhibitory synapse on motoneurons by strychnine appears to account for its convulsive effects (Bradley *et al.*, 1953). Tetanus toxin, however, produces its convulsive effect by presynaptic action on inhibitory interneurons, preventing the release of the transmitter substance.

In the previous section we looked at some work by Bormann *et al.* (1987) on the GABA-activated channels of cultured mammalian spinal neurons. They investigated the properties of the glycine-activated channels at the same time, and

Fig. 10.25. The topology of the $GABA_A$ receptor in the postsynaptic membrane. There are four subunits in all, only two of which (one α and one β) are shown in this diagram. The four membrane-crossing helices in each subunit are shown as cylinders. Triangles show possible sites for N-glycosylation on the extracellular side. The cysteines which form disulphide bridges are shown by C. The site for cAMP-dependent phosphorylation in the β subunit is shown by P. Charges are indicated on residues close to or within the membrane. We presume that the ionic channel is lined by a number (perhaps five, according to Bormann *et al.*, 1987) of the sixteen membrane-crossing helices, whose arrangement must therefore be rather more close-packed than shown here. (From Schofield *et al.*, 1987.)

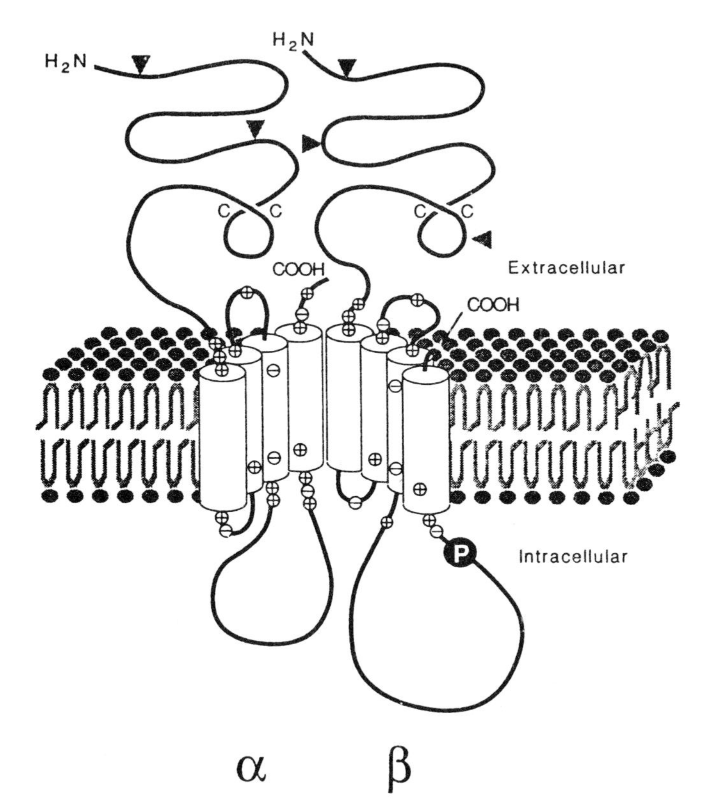

found that the two channels are very similar. The reversal potential for whole-cell currents in glycine-activated channels changed by 56 mV per tenfold change in internal chloride ion concentration and was unaffected by changes in potassium ion concentration. This seems to rule out the possibility (Eccles, 1964) that the IPSP in the direct inhibitory pathway involves an increase in permeability to potassium ions, unless another different type of channel is also involved.

Permeability studies on the glycine-activated chloride channels suggested that their internal diameter is 5.2 Å, just a little smaller than for the GABA channel. Multiple conductance states like those of the GABA channel were observed in patch-clamp experiments, with conductances of 46, 30, 20 and 12pS, but in the glycine channel the 46 pS state was the most common. The great similarity between the properties of the glycine and GABA channels suggests that the molecular structure of the channel lining is likely to be similar in the two cases.

Purification of the glycine receptor and sequencing of one of its subunits has been achieved by Betz and his colleagues (Betz, 1987; Grenningloh *et al.*, 1987). The receptor contains two proteins that are glycosylated and have integral membrane components, with molecular weights of 48 and 58 kDa. There is also a 93 kDa protein associated with the cytoplasmic domains of the complex. The 48 kDa protein binds strychnine and has had its amino acid sequence determined by cDNA cloning. The mature subunit has 421 amino acid residues and is homologous in structure with GABA receptor and nicotinic acetylcholine receptor subunits. Like them it has four hydrophobic transmembrane segments, in similar positions in the molecule, and a cysteine–cysteine disulphide linkage with the intervening 13 residues highly conserved, that occurs in the extracellular region between the N terminal and the M1 segment.

Excitatory amino acids

Curtis, Phillis and Watkins (1960) found that glutamic, aspartic and cysteic acids cause excitation of spinal motoneurons when applied ionophoretically. Later studies extended these results to a wide variety of neurons and a considerable number of other acidic amino acids. It now seems likely that glutamate is the major mediator of fast synaptic transmission in the vertebrate central nervous system.

Most glutamate receptors directly activate ionic channels selective for cations. They appear to be of two or three different types, distinguished by their responses to different agonists (Watkins and Evans, 1981). The most clearly defined are those which are activated by N-methyl-D-aspartate (NMDA) and blocked by D-2-amino-5-phosphonovalerate (APV); they are called NMDA receptors. Other types are unaffected by these substances but are activated either by quisqualic acid or by kainic acid; these Q and K receptors are often described as 'non-NMDA receptors'. Antagonists of both types include kynurenic acid and *cis*-2,3-piperidine dicarboxylic acid (PDA). Some of these various compounds are shown in fig. 10.26.

Neurons grown in tissue culture are much more accessible than when they are embedded in the central nervous system. The value of this accessibility can be seen in the development of our understanding of a rather unusual feature of NMDA receptor channels. Ionophoresis experiments on spinal motoneurons *in vivo* led to the surprising conclusion that glutamate apparently produced a decrease in membrane conductance (Engberg *et al.*, 1979). Voltage-clamp experiments on neurons in tissue culture allowed the current–voltage curve of the glutamate-induced current to be measured (as in fig. 10.27), and it was found that it had a region of negative slope conductance (MacDonald *et al.*, 1982). The explanation for this came from experiments in which magnesium was removed from the bathing solution: the current–voltage relation was now much more conventional, suggesting that magnesium ions can block the channels at negative membrane potentials but that this block is removed by depolarization (Nowak *et al.*, 1984). Patch-clamp experiments are in line with this explanation: magnesium causes 'flickering' (suggesting intermittent blockage of the channel) at negative membrane potentials (fig. 10.28). Non-NMDA channels do not show this sensitivity to magnesium ions.

MacDermott and her colleagues (1986) have shown that calcium ions can pass through NMDA channels. They used voltage-clamp methods with cultured mouse embryo spinal neurons, and measured the intracellular calcium ion activity with the

Fig. 10.26. Some compounds active at glutaminergic synapses.

Glutamate

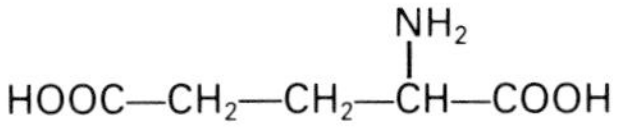

$$HOOC-CH_2-CH_2-\overset{NH_2}{\overset{|}{C}}H-COOH$$

Aspartate

$$HOOC-CH_2-\overset{NH_2}{\overset{|}{C}}H-COOH$$

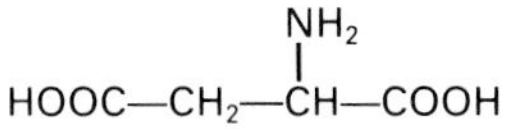

N-methyl-D-aspartate (NMDA)

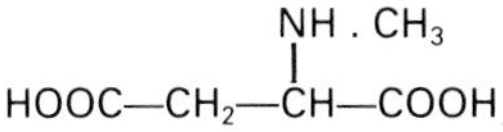

$$HOOC-CH_2-\overset{NH\,.\,CH_3}{\overset{|}{C}}H-COOH$$

D-2-amino-5-phosphonovalerate (APV)

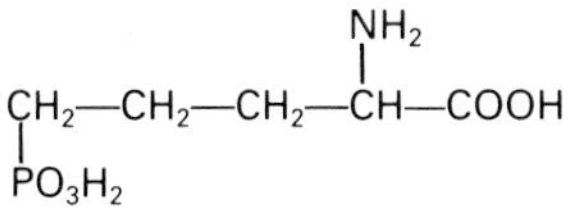

$$\underset{PO_3H_2}{\underset{|}{C}}H_2-CH_2-CH_2-\overset{NH_2}{\overset{|}{C}}H-COOH$$

Quisqualic acid

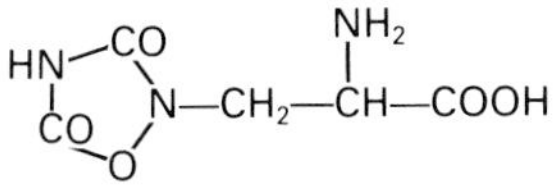

HN—CO, CO—O ring; N—CH_2—CH(NH_2)—COOH

Kainic acid

$H_3C-C{=}CH_2$; CH_2-CH ; $CH-CH_2-COOH$; $HN-CH\,.\,COOH$

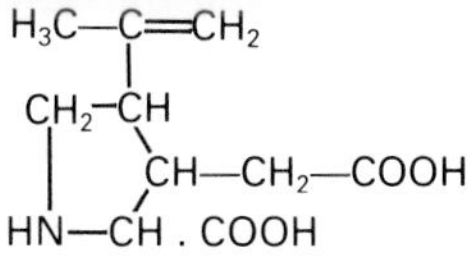

Cis-2,3-piperidine dicarboxylic acid (PDA)

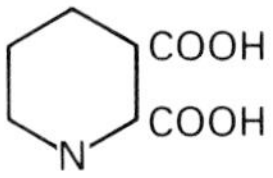

N ring; COOH; COOH

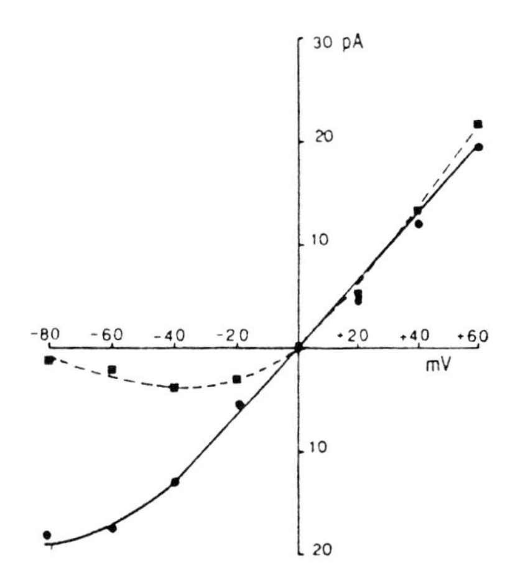

Fig. 10.27. Voltage dependence of the glutamate-induced current in cultured embryonic mouse brain cells. In a magnesium-free solution (continuous line) the relation is approximately linear between +60 and −60 mV. With 0.5 mM Mg (broken line) the currents are much reduced at negative membrane potentials and the curve has a negative slope between −80 and −40 mV. (From Nowak *et al.*, 1984.)

calcium-sensitive dye arsenazo III. At −60 mV application of NMDA caused an increase in internal calcium ion activity, and this was blocked by magnesium ions. Does this mean that the calcium entered via the NMDA channels? It could not have entered via voltage-gated calcium channels because they are not activated at this membrane potential. Another possibility is that the calcium might be released from internal stores as a result of inositol trisphosphate production. But application of NMDA at +60 mV (where the ionic current through the channel is outward) does not produce a rise in internal calcium, showing that activation of the NMDA receptor on its own is not sufficient to produce the response. Further evidence for calcium flow through the channel is given by experiments on the reversal potential for the ionic current produced by NMDA: it is altered by changes in external calcium ion concentration, whereas the response to kainate is not (fig. 10.29).

Fig. 10.28. Effect of magnesium ions on single channel currents induced by glutamate in cultured embryonic mouse brain cells. In the absence of magnesium (*a*) the timing of the currents is largely unaffected by membrane potential. With a low magnesium ion concentration (*b*) the currents are similar to the Mg-free situation at +40 mV, but at −60 mV they are much interrupted. This suggests that a magnesium ion can block a glutamate-activated channel when it is held in position by a negative membrane potential. (From Nowak *et al.*, 1984.)

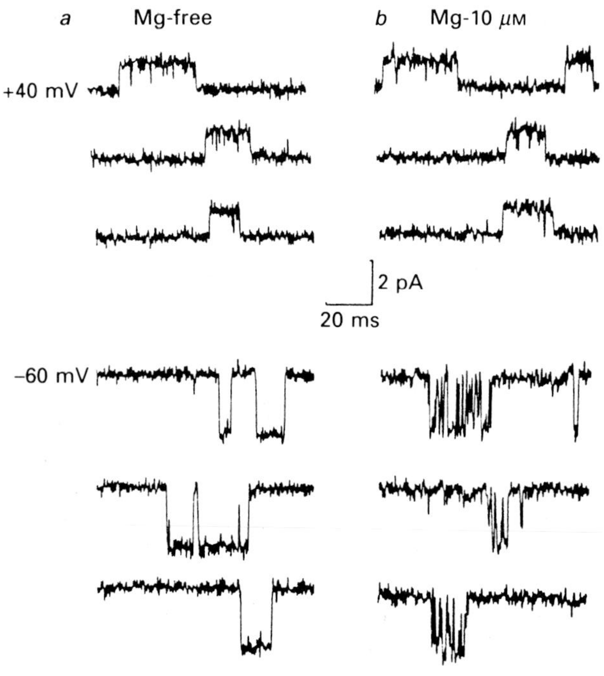

Patch-clamp analyses show that glutamate-activated channels from mammalian brain neurons display at least four different conductance levels and that individual channels may readily change from one conductance level to another (Jahr and Stevens, 1987; Cull-Candy and Usowicz, 1987), as is shown in fig. 10.30. NMDA tends to activate the higher conductance levels and to keep the channels open longer, but there is clearly overlap between NMDA and non-NMDA responses. Only the larger conductance levels are blocked by magnesium ions. The authors have suggested that there is just a single type of channel which can switch between NMDA and non-NMDA characteristics, the probability of adopting any particular level being affected by the nature of the agonist. A less radical conclusion is that the two types of channel have similar structures and so can adopt a similar variety of conductance states (Ascher and Nowak, 1987). This idea seems to be more in accordance with the different distributions of NMDA and non-NMDA receptors in the brain (Cotman *et al.*, 1987).

There has been much interest in the possibility that NMDA receptors are sometimes involved in long-term changes in synaptic responsiveness. The phenomenon of long-term potentiation occurs in certain cells of the hippocampus: a brief burst of repetitive stimulation produces enhanced postsynaptic responses for up to several hours afterwards (Bliss and Lomo, 1973). The NMDA antagonist APV prevents this effect (Collingridge *et al.*, 1983). APV also produces changes in rat learning when injected into the ventricles of the brain: spatial learning is greatly reduced while visual learning is unaffected (Morris *et al.*, 1986).

Because of their blockage by magnesium ions, it seems likely that NMDA receptors are only active when the membrane potential is depolarized, such as might occur during a burst of synaptic activation of non-NMDA channels. It has been suggested that opening of NMDA channels allows calcium to enter the cell and that this calcium then brings about long-term changes in cellular responsiveness, perhaps by activating a protein kinase

Fig. 10.29. NMDA channels are permeable to calcium ions whereas non-NMDA channels are not. The upper set of traces shows currents produced by cultured mouse spinal cord neurons under voltage clamp at different membrane potentials in response to NMDA or kainic acid (KA) at different calcium ion concentrations. In 1 mM Ca^{2+} (*a*) the reversal potential for both NMDA and KA responses is about 0 mV. In 20 mM Ca^{2+} (*b*) the reversal potential for the NMDA response is now +17 mV, whereas that for the KA response is unchanged. The lower set of traces shows the increase in internal calcium ion concentration (upper traces, showing change in absorbance of the calcium-sensitive dye arsenazo III) following activation of NMDA channels (*c*); application of KA (*d*) at a level sufficient to produce a similar current flow (lower traces) does not have this effect. (From MacDermott *et al.*, 1986.)

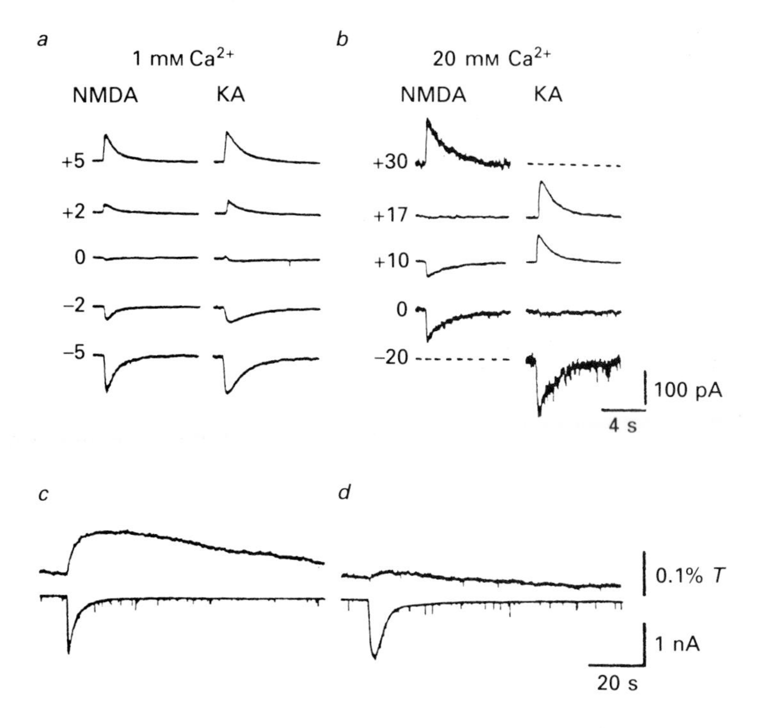

Fig. 10.30. Multiple-conductance channels activated by excitatory amino acids. The traces show patch-clamp records of single-channel currents in cultured mouse cerebellar neurons in the presence of (*a*) aspartate and (*b*) quisqualate. (From Cull-Candy and Usowicz, 1987.)

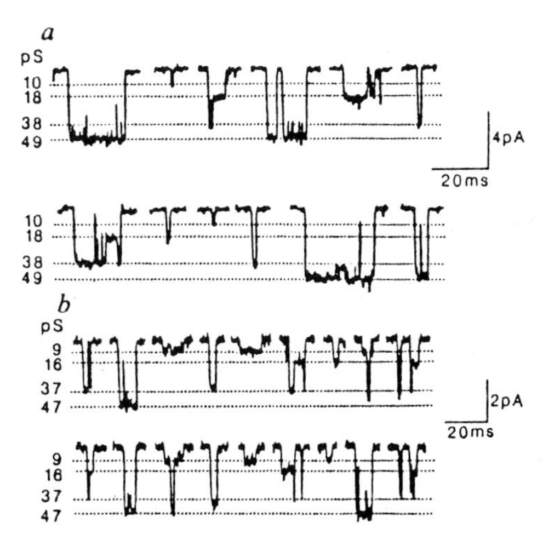

(Miller and Kennedy, 1986; Stevens, 1986; Collingridge and Bliss, 1987). A process of this nature might well be involved in the cellular basis of learning and memory. It is also possible that NMDA receptors are involved in epilepsy; the voltage-sensitivity of the channels makes them likely to be involved in positive feedback processes, and NMDA antagonists prevent induced seizures in mice (Croucher *et al.*, 1982; Herron *et al.*, 1986).

Injection of glutamate into newborn mice causes obesity, brain lesions and other developmental disturbances (Olney, 1969). Similar excitotoxicity is known by other substances which activate glutamate receptors. It may be that the brain damage associated with stroke, hypoglycaemic coma, epilepsy or neurodegenerative disease (such as Huntingdon's chorea) is brought about in part by excessive activation of NMDA receptors. If this is so it might be desirable to use NMDA blocking agents in the treatment of these conditions (Rothman and Olney, 1987).

A different type of glutamate receptor has been described by Sugiyama, Ito and Hirono (1987). They found that inward currents could be induced by glutamate and quisqualate in *Xenopus* oocytes injected with rat brain mRNA, but that these currents were slow and irregular in form with the delayed onset typical of the slow responses mediated by second messengers. The responses were much reduced by pertussis toxin, suggesting the involvement of G proteins. Injection of inositol trisphosphate into the oocyte produced similar responses. Injection of EGTA (a calcium-chelating agent) inhibited both the quisqualate–glutamate responses and the IP_3 responses, suggesting that the currents are produced by calcium-activated chloride channels.

Some sensitive individuals have suffered from sensations of burning, facial pressure, chest pain and sometimes headache as a result of eating Chinese food. This 'Chinese restaurant syndrome' is caused by the monosodium glutamate that is used as a flavour enhancer (Schaumburg *et al.*, 1969). Recognition of this effect, and of the possibility of excitotoxic effects, has led to the withdrawal of monosodium glutamate as an additive from many canned baby foods.

Glutamate is the excitatory neuromuscular transmitter in arthropod skeletal muscles (Takeuchi and Takeuchi, 1964; Usherwood and Cull-Candy, 1975). Extrajunctional glutamate channels from locust muscles have a rather high single-channel conductance, at about 150 pS (Cull-Candy *et al.*, 1980). Specific NMDA receptors appear to be absent.

Like other amino acid neurotransmitters, glutamate is removed from the synaptic cleft by an efficient uptake system in the nerve terminals (Storm-Mathisen and Iversen, 1979; Iversen and Bloom, 1972). It can thus be re-packaged in the synaptic vesicles and used again.

Purines

The structure of some biologically active purines is shown in fig. 10.31. We have seen that adenosine triphosphate (ATP) forms part of the contents of cholinergic synaptic vesicles and that it is released into the synaptic cleft along with the acetylcholine, so it is not too surprising to find that in some cases ATP itself may be a neurotransmitter. Burnstock and his colleagues (1970) found that

Fig. 10.31. Some biologically active purines.

ADENOSINE

CAFFEINE

CYCLIC AMP

ATP

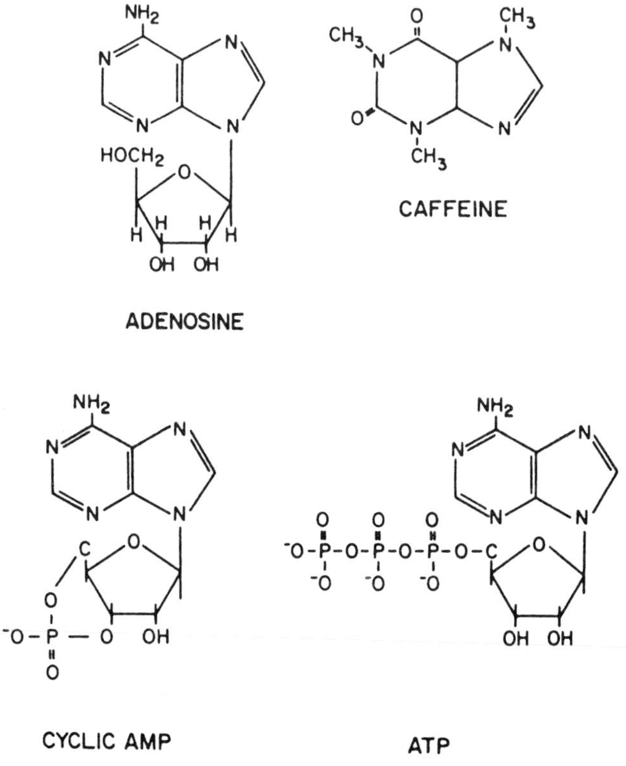

some smooth muscle responses to nerve stimulation persist in the presence of both cholinergic and adrenergic blocking agents, and that they could be imitated by the application of ATP; this led to the idea of 'purinergic' transmission (Burnstock, 1972). It is now clear that a number of other neurotransmitters are also involved in the control of the gut musculature, but the concept of purinergic transmission remains viable (Burnstock, 1981; Su, 1983). ATP-activated channels that are permeable to calcium and sodium ions have been demonstrated in smooth muscle (Benham and Tsien, 1987.)

There is also evidence that adenosine and ATP may act as transmitters or neuromodulators in the central nervous system (Stone, 1981; McAfee and Henon, 1985). It may be that the stimulants caffeine (from coffee) and theophylline (from tea) act by blocking adenosine receptors.

Neuropeptide transmitters

Neuropeptides are single chains of amino acids produced in nerve cells by the standard methods of protein synthesis. Some years ago it was generally believed that all of the few secreted neuropeptides were released into the blood stream and so acted as hormones. More recently it has become evident that there are many more neuropeptides than was originally suspected and that the majority of them act as neurotransmitters, often in addition to a role as hormones (table 10.5). Let us see how this change of viewpoint has happened.

Certain cells of the hypothalamus produce secretory materials which pass along their axons to be released into the circulation at their terminals in the pituitary gland. This process is called neurosecretion, and the released substances are neurohormones (Scharrer and Scharrer, 1945; Bargmann and Scharrer, 1951). The first of the hypothalamic neurohormones to have their structures determined were vasopressin and oxytocin (see du Vigneaud, 1956); they are each peptides with nine amino acid residues (fig. 10.32). Vasopressin increases the blood pressure and oxytocin produces contraction of the uterus.

Von Euler and Gaddum (1931) obtained an alcoholic extract from horse brain and gut which produced marked lowering of the blood pressure when injected into rabbits and made rabbit intestinal muscle contract. Von Euler later showed that

Table 10.5. *Some neuropeptides in the mammalian nervous system*

Pituitary peptides
Corticotropin (ACTH)
Growth hormone (GH)
Lipotropin
α-Melanocyte-stimulating hormone (α-MSH)
Oxytocin
Vasopressin
Circulating hormones
Angiotensin
Calcitonin
Insulin
Gut hormones
Cholecystokinin (CCK)
Gastrin
Motilin
Pancreatic polypeptide (PP)
Secretin
Substance P
Vasoactive intestinal polypeptide (VIP)
Opioid peptides
Dynorphin
β-Endorphin
Met-enkephalin
Leu-enkephalin
Kyotorphin
Hypothalamic releasing hormones
Corticotropin-releasing factor (CRF)
Luteinizing-hormone-releasing hormone (LHRH)
Somatostatin
Thyrotropin-releasing hormone (TRH)
Miscellaneous peptides
Bombesin
Bradykinin
Calcitonin gene related peptide (CGRP)
Carnosine
Neuropeptide Y
Neurotensin
Proctolin
Substance K

the activity of the extract was destroyed by trypsin, suggesting that the active component (by now called 'substance P') was a peptide. It was not until many years later that the precise structure of substance P was determined (Chang *et al.*, 1971).

Relatively high levels of substance P are found in the dorsal horn of the spinal cord. The possibility that it might be the transmitter at some afferent endings was discussed in the fifties and sixties (e.g. McLennan, 1963); further evidence for this idea was later provided from a variety of sources and it is now well established (Go and Yaksh, 1987). In the early seventies, then, substance P appeared to be the only neuropeptide likely to be a neurotransmitter. This view did not survive investigations arising from the awesome human effects of the products of the opium poppy.

Opiates are drugs that have powerful actions on the nervous system; they include morphine, which is of great importance in the relief of pain, and heroin, well known as an addictive narcotic. Pert and Snyder (1973) used a tritium-labelled antagonist which could be displaced by competition with opiates to demonstrate the presence of specific opiate receptors in the brain. But what normally activates these receptors? This question initiated a search for endogenous opioid substances, which was fulfilled by the discovery and sequencing of two of the opioid peptides, met-enkephalin and leu-enkephalin, by Hughes *et al.* (1975) and Simantov and Snyder (1976). Then followed a great burst of research on brain neuropeptides, so that the number of them to be considered as possible candidates for neurotransmitter status is now well over 30 (Iversen, 1984) and may be as many as 100 (Scharrer, 1987).

The detection of neuropeptides in the nervous system has been greatly aided by the development of new techniques. Immunohistochemistry uses antibodies to detect particular peptides. Radio-labelled peptides can be used to localise their receptors (Kuhar *et al.*, 1986). Molecular cloning methods have produced amino acid sequences for peptide precursors, and specific antibodies can be made against peptides synthesized from them (Lynch and Snyder, 1986). Nucleic acid probes can be used for hybridization with mRNAs coding for particular peptides (Schwartz and Costa, 1986).

Many peptidergic neurons contain a non-peptide neurotransmitter as well. One of the first demonstrations of this was by Hökfelt *et al.* (1977) for sympathetic ganglion cells, using the fluorescent antibody technique. A thin section of tissue was washed first with a solution containing sheep antibodies to somatostatin, and then with fluorescent rabbit antibodies to sheep antibodies. An adjacent section, mostly passing through the same cells, was similarly stained for dopamine-β-hydroxylase, the enzyme which converts dopamine to noradrenaline. Hökfelt and his colleagues found that many of the cells (a majority in the inferior mesenteric ganglion) were fluorescent in

Fig. 10.32. The amino acid sequence of some neuropeptides. The NH_2 groups indicate amidation of the C-terminal residue. pGlu indicates pyroglutamate: an N-terminal glutamate residue has been modified to form an internal peptide bond.

met-enkephalin	Ty-Gly-Gly-Phe-Met
leu-enkephalin	Tyr-Gly-Gly-Phe-Leu
angiotensin II	Asp-Arg-Val-Tyr-Ile-His-Pro-Gly-NH
arg-vasopressin	Cys-Tyr-Phe-Gln-Asn-Cys-Pro-Arg-Gly-NH
lys-vasopressin	Cys-Tyr-Phe-Gln-Asn-Cys-Pro-Lys-Gly-NH
oxytocin	Cys-Tyr-Ile-Gln-Asn-Cys-Pro-Leu-Gly-NH
LHRH	pGlu-His-Trp-Ser-Tyr-Gly-Leu-Arg-Pro-Gly-NH
dynorphin	Tyr-Gly-Gly-Phe-Leu-Arg-Arg-Ile-Arg-Pro-Lys-Leu-Lys-Trp-Asp-Asn-Glu
substance P	Arg-Pro-Lys-Pro-Gln-Gln-Phe-Phe-Gly-Leu-Met-NH
β-endorphin	Tyr-Gly-Gly-Phe-Met-Thr-Ser-Glu-Lys-Gln-Thr-Pro-Leu-Val-Thr-Leu-Phe-Lys-Asn-Ala-Ile-Val-Lys-Asn-Ala-His-Lys-Lys-Gly-Gln
CGRP	Ser-Cys-Asn-Thr-Ala-Thr-Cys-Val-Thr-His-Arg-Leu-Ala-Gly-Leu-Leu-Ser-Arg-Ser-Gly-Gly-Val-Val-Lys-Asp-Asp-Phe-Val-Pro-Thr-Asp-Val-Gly-Ser-Glu-Ala-Phe-NH

Table 10.6. *Some examples of the coexistence of peptide and non-peptide ('classical') neurotransmitters in the same neuron*

Classical transmitter	Peptide	Location
Dopamine	Neurotensin	Ventral midbrain
	CCK	Ventral midbrain
Noradrenaline	Neuropeptide Y	Medulla oblongata
5-HT	Substance P	Medulla oblongata
	Enkephalin	Medulla oblongata
Acetylcholine	VIP	Cortex
	Substance P	Pons
GABA	Somatostatin	Thalamus, cortex
	CCK	Cortex
	Enkephalin	Retina

Simplified after Hökfelt *et al.*, 1986.

both sections and must therefore have contained both somatostatin and noradrenaline. Since then a considerable number of such cases of coexistence have been described (Hökfelt *et al.*, 1982; Campbell, 1987); some of them are listed in table 10.6.

Synthesis

Non-peptide neurotransmitters like acetylcholine and noradrenaline are synthesized in the nerve terminal by specific enzymes, and there are usually mechanisms for their reuptake from the extracellular space. Neither of these systems is available for neuropeptides, which are synthesized, usually as precursors, by the normal ribosomal protein synthesis system in the cell body. This means that the peptide or its precursor must be transported along the axon to the terminal, usually in large granular storage vesicles, before it can be released (fig. 10.33). The concentrations of neuropeptides are typically in the range of some picomoles per gram of brain, perhaps 1000 times less than the monoamines and about 100 000 times less than the amino acid neurotransmitters (Hökfelt *et al.*, 1980).

It seems to be the common pattern that neuropeptides and peptide hormones are formed from longer-chain protein precursor molecules. Molecular cloning methods have allowed the sequences of many of these precursors to be determined (see Lynch and Snyder, 1986). Preproenkephalin A, for example, is the translation product of a gene which is expressed in the adrenal gland and in the brain. It contains 268 amino acid residues of which the first 24 probably form a signal sequence which is removed once the proenkephalin is formed in the endoplasmic reticulum (Noda *et al.*, 1982). There are four sequences of met-enkephalin and one of leu-enkephalin in the complete chain (fig. 10.34).

Each of the enkephalin sequences in proenkephalin is bounded at each end by a pair of basic residues, lysine or arginine, which form the sites at which endopeptidases of the trypsin type will be able to break the peptide chain. Some rather larger sequences correspond to peptides found in the adrenal gland; it seems likely that processing proceeds further in the brain to give the smaller opioids as end products. Two brain peptides of intermediate size, metorphamide and amidorphin, are amidated at their C-terminal ends (Weber *et al.*, 1983; Seizinger *et al.*, 1985); this is done by the breakage of a glycine residue.

Similar results have been obtained for other neuropeptide precursors. Thus proopiomelanocortin contains the sequences for γ-melanocyte stimulating hormone (γ-MSH), adrenocorticotropic hormone (ACTH) and β-lipotropin (β-LPH), and further splitting can occur so that ACTH produces α-MSH and β-LPH produces γ-LPH and β-endorphin. The large peptides are produced in the anterior lobe of the pituitary, but processing proceeds farther in the brain, to give the smaller ones (see Douglass *et al.*, 1984).

Another way of generating diversity occurs when the same gene produces different mRNAs as a result of alternative RNA splicing. In the thyroid the calcitonin gene produces an mRNA which codes for the hormone calcitonin, but in the hypothalamus most of the mRNA has a partly different sequence and so codes for another peptide, referred to as the calcitonin gene-related peptide or CGRP (Amara *et al.*, 1982). The probable sequence of events is shown in fig. 10.35.

The existence of CGRP was predicted from the mRNA sequence, and it is highly satisfactory that it was later shown to have many of the properties expected of a neurotransmitter. It was localized by immunocytochemistry to particular parts of the brain such as the trigeminal ganglion, using antisera to a synthetic peptide based on the pre-

Fig. 10.33. Diagram to show the differences between a peptidergic neuron (right) and one releasing a non-peptide ('classical') neurotransmitter. (From Hökfelt *et al.*, 1980.)

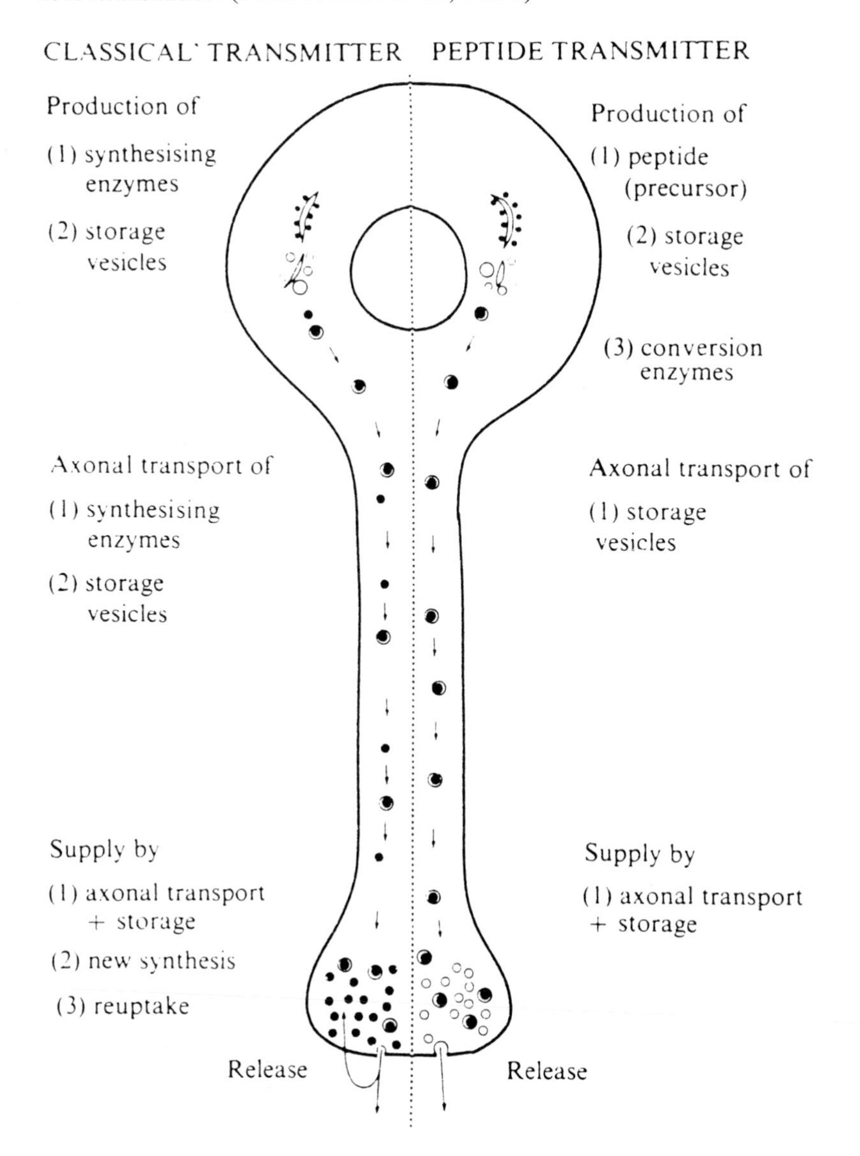

dicted sequence (Rosenfled *et al.*, 1983). It was released from cultured trigeminal ganglion cells by depolarization with potassium ions in the presence of calcium (Mason *et al.*, 1984). Synthetic CGRP produced an increase in the heart rate when injected into rats and an increase in the noradrenaline level in the blood when perfused into the cerebral ventricles (Fisher *et al.*, 1983). Immunocytochemistry shows that CGRP is present in nerves supplying the guinea pig heart, and stimulation of these nerves produces acceleration of the heart beat (Saito *et al.*, 1986).

Substance P is another peptide whose gene can produce two different mRNAs by alternative splicing. The precursor gives rise either to substance P alone or to substance P and another predicted peptide called substance K because of its similarity in sequence to the amphibian peptide kassinin (Nawa *et al.*, 1984).

Receptors and postsynaptic responses

There are at least three different types of opiate receptors (Zukin and Zukin, 1981; North,

Fig. 10.34. Processing of preproenkephalin A, precursor of the enkephalins and a number of other opioid peptides. Proteases usually split the molecule at sites where there are two acidic residues, arginine (R) and/or lysine (K). Sometimes cleavage can occur at a single arginine residue next to a glycine (G) residue, the latter being then fractured to leave an amide (NH_2) group. (From Lynch and Snyder, 1986.)

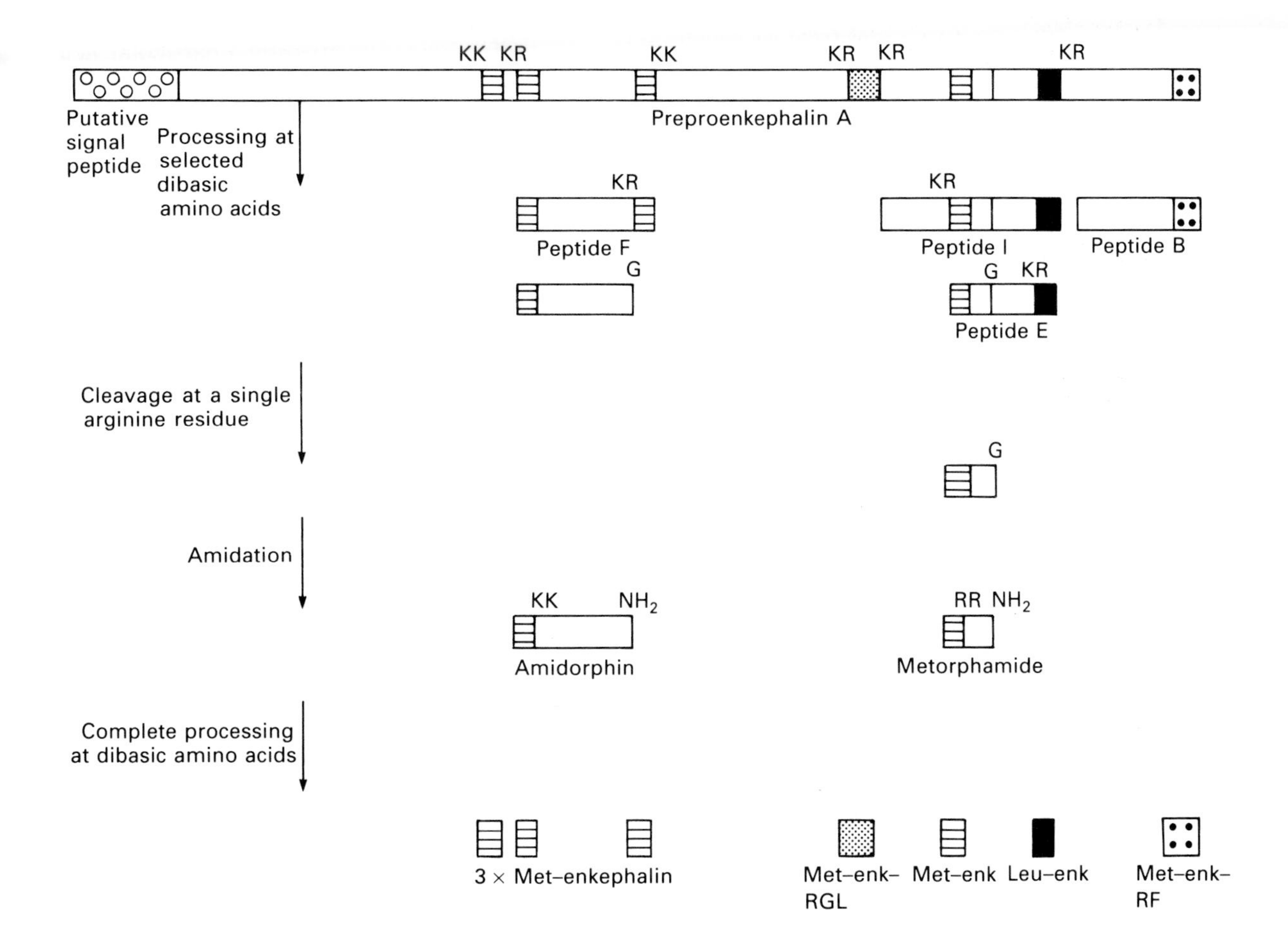

1986): μ (mu) receptors are most specific for morphine, δ (delta) receptors bind the enkephalins most readily, and κ (kappa) receptors are specific for dynorphin. Metorphamide and amidorphin combine with both μ and δ receptors but not with κ receptors. The antagonist naloxone is effective at all three types, but especially so at δ receptors.

Opioids produce hyperpolarization by activation of receptors in certain cells in rat brain slices, by means of an increase in potassium conductance which is also activated by α_2-adrenergic receptors (North and Williams, 1985). In guinea pig mesenteric plexus neurons, μ and κ receptors occur both pre- and postsynaptically; activation of μ receptors opens potassium channels whereas activation of κ receptors seems to reduce excitability by closing calcium channels (Cherubini and North, 1985).

In chapter 9 we examined the slow synaptic responses of frog sympathetic ganglion cells. The late slow EPSP (fig. 9.24*d*) is produced by the neuropeptide LHRH and involves a reduction in membrane potassium conductance by closure of the M-current potassium channels (Jones, 1985). The relative slowness of this and other responses to neuropeptides suggests that they usually act via second messenger systems. Opioid action is in some cases associated with cAMP production (West and Miller, 1983). Concentrations of neuropeptides are relatively low in comparison with non-peptide transmitters, hence the amplification inherent in second messenger systems may be a necessary feature of neuropeptide action.

Inactivation

Various different peptidases (enzymes which hydrolyse peptides) are widespread in their distribution in the body and are capable of hydrolysing peptides which they might come into contact with. But it may be quite difficult to show which of them, if any, is effective in inactivating a particular neuropeptide transmitter at a particular site. The best

Fig. 10.35. Tissue-specific expression of the calcitonin gene. The gene consists of a DNA sequence with at least six exons (regions which code for mRNA). Transcripts of these exons are spliced together in two different ways to make two different mRNAs, so that calcitonin is produced in the thyroid and predominantly CGRP (calcitonin gene-related peptide) in the brain. (From Amara *et al.*, 1982.)

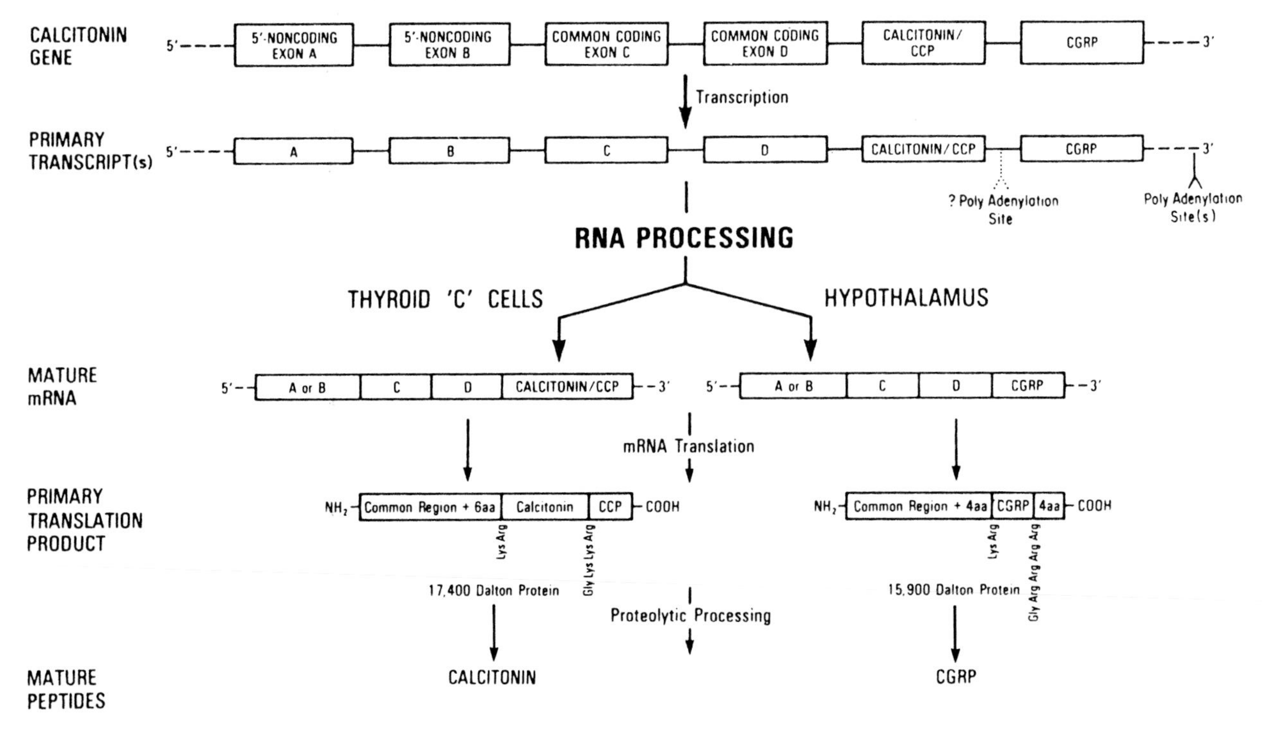

candidate seems to be one called enkephalinase, which hydrolyses the enkephalins (Schwartz *et al.*, 1981; Lynch and Snyder, 1986). Autoradiography using a radioactive specific inhibitor of the enzyme shows that its distribution in the brain is similar to that of the μ and δ opioid receptors (Waksman *et al.*, 1986).

11

The mechanics and energetics of muscle

In this and the next three chapters we shall examine some of the properties of muscle cells. The function of muscle cells is to shorten and develop tension. Thus the end product of cellular activity can be measured with considerable precision, by mechanical measurement of the change in length or tension or both. Such a contraction* must obviously involve the consumption of energy, some of which may appear as heat.

In this chapter and the next two we shall mainly be concerned with the properties of rapidly contracting vertebrate skeletal muscles, such as frog sartorius and the rabbit psoas. Skeletal muscles are activated by motoneurons, as we have seen in previous chapters. Their cells are elongate and multinuclear and the contractile material within them shows cross-striations, hence skeletal muscle is a form of *striated muscle*. Some of the special properties of other muscles will be examined in chapter 14.

The normal stimulus for the contraction of a muscle fibre in a living animal is an impulse in the motor nerve by which it is innervated. The sequence of events following the nerve impulse is shown schematically in fig. 11.1. We have examined stages 1 to 4 of this sequence (the excitation processes) in previous chapters. This chapter is concerned with some of the overall consequences of contraction (stage 6); details of the cellular mechanisms involved in stages 5 and 6 are considered in the following chapters.

* In muscle physiology, the word 'contraction' is usually used to denote mechanical activity of the muscle, whether or not this includes shortening. If we hold the length of a muscle constant, it may still 'contract' in this sense, the contraction being seen as tension development.

Fig. 11.1. The control sequence leading to contraction in a vertebrate 'twitch' muscle fibre.

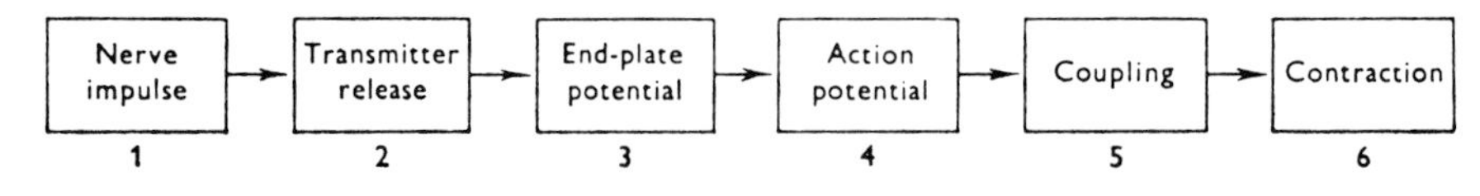

Anatomy

Skeletal muscle fibres are multinucleate cells (fig. 11.2) formed by the fusion of numbers of elongated uninucleate cells called myoblasts. Mature fibres may be as long as the muscle of which they form part, and 10–100 μm in diameter. The nuclei are arranged around the edge of the fibre. Most of the interior of the fibre consists of the

Fig. 11.2. Diagram to show the arrangement of fibres in a vertebrate striated muscle. The cross-striations on the myofibrils can be seen with light microscopy. After Schmidt-Neilsen (1983).

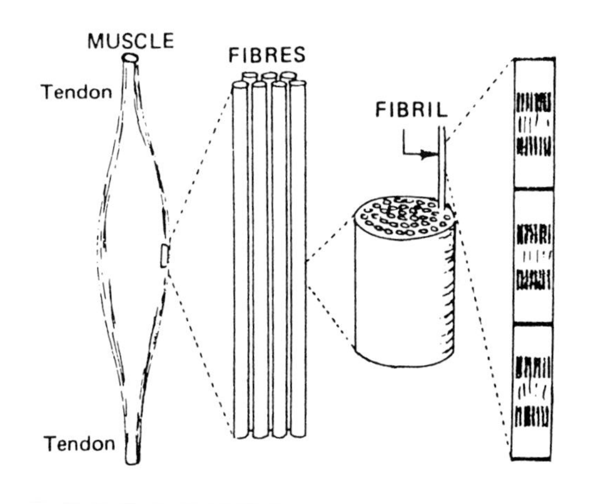

protein filaments which constitute the contractile apparatus, grouped together in bundles called *myofibrils*. The myofibrils are surrounded by cytoplasm (or *sarcoplasm*), which also contains mitochondria, the internal membrane systems of the sarcoplasmic reticulum and the T system, and a fuel store in the form of glycogen granules and sometimes a few fat droplets. We shall examine the structure of the contractile apparatus and the internal membrane systems in more detail in the following chapters.

The muscle fibre is bounded by its cell membrane, sometimes called the sarcolemma, to which a thin layer of connective tissue (the endomysium, formed of collagen fibres) is attached. Bundles of muscle fibres are surrounded by a further sheet of connective tissue (the perimysium) and the whole muscle is contained within an outer sheet of tough connective tissue, the epimysium. These connective tissue sheets are continuous with the insertions and tendons which serve to attach the muscles to the skeleton.

Mammalian muscles have an excellent blood supply, with blood capillaries forming a network between the individual fibres. Sensory and motor nerve fibres enter the muscle in one or two nerve branches. The sensory nerve endings include those on the muscle spindles (sensitive to length), in the Golgi tendon organs (sensitive to tension), and a variety of free nerve endings in the muscle tissue, some of which are involved in sensations of pain.

In mammals the γ motoneurons provide a separate motor nerve supply for the muscle fibres of the muscle spindles, while the bulk of the muscle fibres are supplied by the α motoneurons. Each α motoneuron innervates a number of muscle fibres, from less than ten in the extraocular muscles (those which move the eyeball in its socket) to over a thousand in a large limb muscle. The complex of one motoneuron plus the muscle fibres which it innervates is called a *motor unit.* Since they are all activated by the same nerve cell, all the muscle fibres in a single motor unit contract at the same time. Muscle fibres belonging to different motor units may well contract at different times, however.

Most mammalian muscle fibres are contacted by a single nerve terminal, although sometimes there may be two terminals originating from the same nerve axon. Muscle fibres of this type are known as *twitch fibres*, since they respond to nervous stimulation with a rapid twitch. In the frog and other lower vertebrates, another type of muscle fibre is also commonly found, in which there are a large number of nerve terminals on each muscle fibre. These are known as *tonic fibres*, since their contractions are slow and maintained. There are some tonic fibres in the extraocular muscles of mammals, and also in the muscles of the larynx and the middle ear.

Mechanical properties

The mechanical properties of muscles are best investigated with isolated muscle or nerve–muscle preparations such as the gastrocnemius or sartorius in the frog. The sartorius muscle is attached at one end to the pelvic girdle, and at the other end to the tibia at the knee joint. Contraction of the muscle moves the leg forward and flexes it at the knee. For physiological experiments, the muscle can be removed intact from the animal by cutting the tendons, or the bones to which they are attached, which can then be connected to the recording apparatus.

We have seen in chapter 7 that the fibres of frog twitch muscles are electrically excitable. Thus a muscle can be simulated by direct application of electric shocks, as well as via its motor nerve. For many experiments it is more convenient to use such direct stimulation, often via a number of electrodes spaced along the length of the muscle so that all parts are activated simultaneously.

When muscles contract they exert a force on whatever they are attached to (this force is equal to the *tension* in the muscle) and they shorten if they are permitted to do so. Hence we can measure two different variables during the contraction of a muscle: its length and its tension. Most often one of these two is maintained constant during the contraction. In *isometric* contractions the muscle is not allowed to shorten (its length is held constant) and the tension it produces is measured. In an *isotonic* contraction the load on the muscle (which is equal to the tension in the muscle) is maintained constant and its shortening is measured.

Sometimes the tension is measured while length changes, linear or sinusoidal, are imposed on the muscle. Contractions during constant velocity length changes may be called isovelocity responses.

Various other types of load can be used. An *elastic load* causes the shortening of a muscle to be proportional to its tension. An *inertial load* consists of a mass which is accelerated by contraction of the muscle. Loads of these types, although very commonly those which the muscle contracts against in nature, are not much used in experiments on the mechanics of muscle, since they do not allow either the length or the tension to be extrinsically determined during the course of the contraction.

Isometric contractions

In isometric contractions both ends of the muscle are fixed so that it cannot shorten, and we measure the contraction as a change in tension. This means that one end of the muscle has to be attached to a suitable tension-recording device. A simple method is to use a lever pivoted on a torsion wire, so that its deflection is proportional to the force applied to it. The deflection of the lever can be recorded by a revolving smoked drum on which the tip of the lever writes, or by a beam of light and a photoelectric cell. Unfortunately such a system is not very stiff (the muscle must always shorten to some extent in order to move the lever) and has appreciable inertia.

Better isometric recording devices are much stiffer and have negligible inertia. One method is to use a special triode valve whose anode projects through a stiff spring diaphragm. Movement of the anode then alters the current flowing through the valve, so the force applied to the anode peg can be displayed as a voltage change on an oscilloscope. Another useful method is to attach the muscle to a steel bar of suitable stiffness which has semiconductor strain gauges bonded onto it. The resistance of the strain gauges then varies with muscle tension (fig. 11.3). For recording the tension from single muscle fibres, more sensitive devices are required, such as a silicon beam strain gauge system or a variable capacitance (see Woledge *et al.*, 1985).

The tension produced by a muscle is a force, and is therefore measured in newtons or in grams weight (1 kg weight is equivalent to 9.80 N). Fig. 11.4 shows the time course of the tension development in isometric contractions. A single stimulus produces a rapid increase in tension which then decays; this is known as a *twitch*. The duration of the twitch varies from muscle to muscle, and decreases with increasing temperature. For a frog

Fig. 11.4. Isometric contractions. *a*, response to a single stimulus, producing a twitch; *b*, response to two stimuli, showing mechanical summation; *c*, response to a train of stimuli, showing an 'unfused tetanus'; *d*, response to a train of stimuli at a higher repetition rate, showing a maximal fused tetanus.

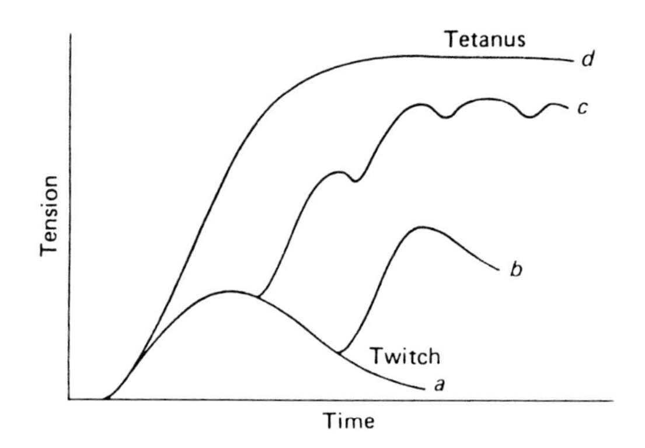

Fig. 11.3. An isometric lever system for measuring the force exerted by a muscle without allowing it to shorten. Semiconductor strain gauges are bonded to a steel bar (*a*), and form two arms of a resistance bridge connected to a battery (*b*).

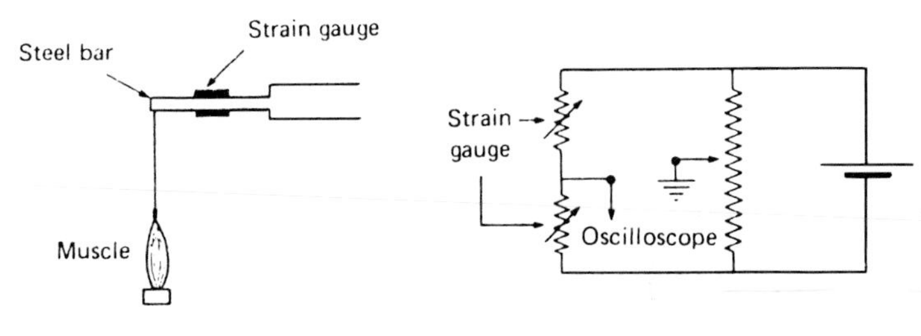

sartorius at 0 °C, a typical value for the time between the beginning of the contraction and its peak value is about 200 ms, tension falling to zero again within 800 ms. If a second stimulus is applied before the tension in the first twitch has fallen to zero, the peak tension in the second twitch is higher than that in the first; this effect is known as mechanical summation. Repetitive stimulation at a low frequency thus results in a 'bumpy' tension trace. As the frequency of stimulation is increased, a point is reached at which the bumpiness is lost and the tension rises smoothly to reach a steady level. The muscle is then in *tetanus*, and the minimum frequency at which this occurs is known as the *fusion frequency*.

A resting muscle is resistant to stretching beyond a certain length, so that it is possible to determine a passive length–tension curve, as is shown in fig. 11.5. The full isometric tetanus tension of the stimulated muscle is also dependent on length, as is shown in the 'total active tension' curve in fig. 11.5. The difference between the two curves is known as the 'active increment' curve; notice that this passes through a maximum at a length near to the maximum length in the body, falling away at longer or shorter lengths.

When a muscle in isometric tetanus is suddenly shortened, the tension falls abruptly, and then rises to a new maximum level which is determined by the isometric length–tension relation (fig. 11.6). This type of experiment is known as a *quick release* (Gasser and Hill, 1924). The amount by which the tension falls is roughly proportional to the length change during the quick release, up to the point at which the tension falls to zero. With quick releases of greater extents, there is an increasing delay between the time of release and the redevelopment of tension. The time course of the redevelopment of tension after the quick release is similar to, but not usually identical with, that of the development of tension at the beginning of the

Fig. 11.6. The results of an isometric quick-release experiment. See text for details.

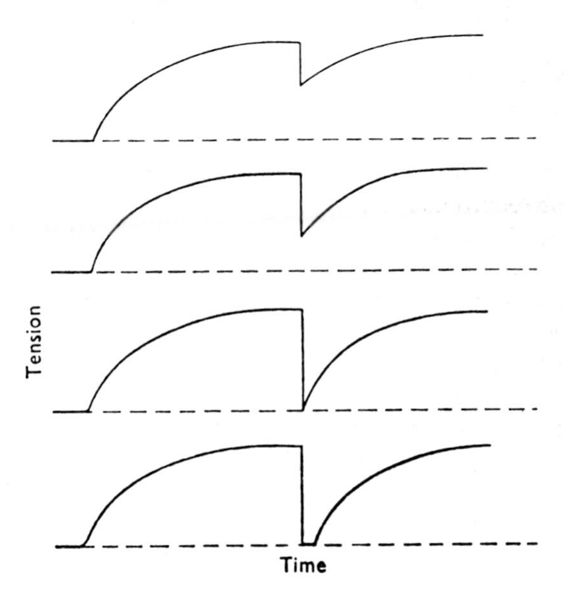

Fig. 11.5. The length–tension relation of a muscle.

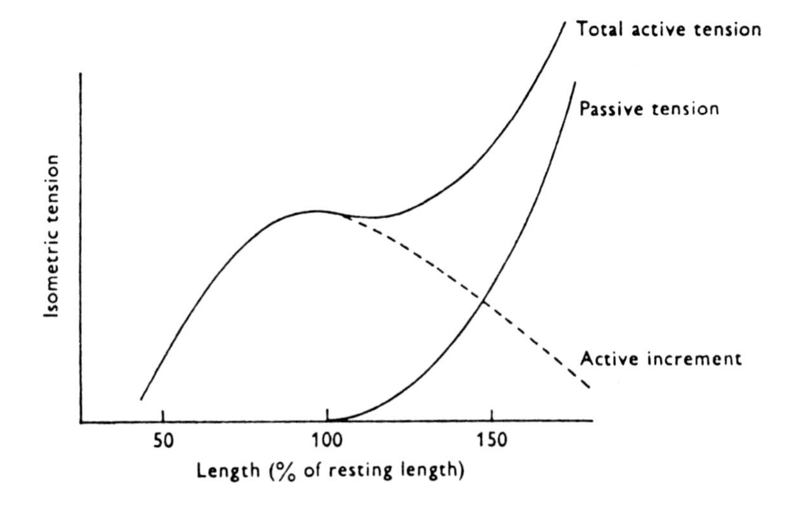

period of stimulation. The converse of this type of experiment occurs when the muscle is stretched during contraction.

Isotonic contractions

In the measurement of isotonic contractions, the tension exerted by the muscle is maintained constant (usually by allowing it to lift a constant load) and its length change is measured. A *free-loaded* contraction is one in which the muscle is loaded in the resting state and then stimulated. This type of measurement is not much used, since the initial length of the muscle will vary with different loads. In an *after-loaded* contraction, the muscle is not loaded at rest, but must lift a load in order to shorten.

The apparatus used for measuring an after-loaded contraction is shown schematically in fig. 11.7. The muscle is attached to a light, freely pivoted isotonic lever so that it must lift a weight when it shortens. The movement of the lever can be recorded by means of a smoked drum or by the interruption of a beam of light focused on a photocell. When the muscle is relaxed, the lever rests against a stop, so that the resting muscle does not have to support the load. If the stop were not there the muscle would take up longer initial lengths with heavier loads, which would make it more difficult to interpret the results of experiments with different loads.

Fig. 11.8*a* shows what happens when the muscle has to lift a moderate load while being stimulated tetanically. The tension in the muscle starts to rise soon after the first stimulus, but it takes some time to reach a value sufficient to lift the load, so that there is no shortening at first and the muscle is contracting isometrically. Eventually the tension becomes equal to the load and so the muscle begins to shorten; the tension remains constant during this time and the muscle contracts isotonically. It is noticeable that initially the velocity of shortening during the isotonic phase is constant, provided that the muscle was initially at a length near to its maximum length in the body. As the muscle shortens further, however, its velocity of shortening falls until eventually it can shorten no further and shortening ceases. When the period of simulation ends, the muscle is extended by the load as it relaxes until the lever meets the after-load stop, after which relaxation becomes isometric and the tension in the muscle falls to its resting level.

If we repeat this procedure with different loads (as in fig. 11.8*b*), we find that the contractions are affected in three ways.

Fig. 11.7. Isotonic lever, arranged for use with after-loaded contractions. The insert shows the photocell system used to record the position of the lever.

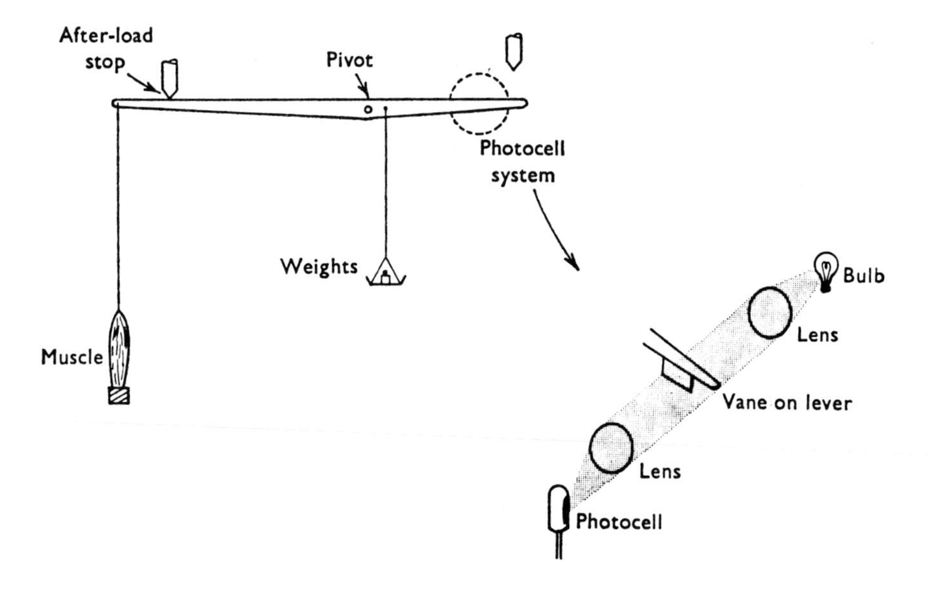

(1) The delay between the stimulus and the onset of shortening is longer with heavier loads. This is because the muscle takes longer to reach the tension required to lift the load.

(2) The total amount of shortening decreases with increasing load. This is because the isometric tension falls at shorter lengths (fig. 11.5) and so the more heavily loaded muscle can only shorten by a smaller amount before its isometric tension becomes equal to the load. Fig. 11.9 illustrates this point.

(3) During the constant-velocity section of the isotonic contraction, the velocity of shortening decreases with increasing load. It becomes zero when the load equals the maximum tension which can be reached during an isometric contraction of the muscle at that length.

Notice that the first two of these observations are essentially predictable from what we already know about isometric contractions. We can quantify the third observation by plotting the initial velocity of shortening against the load to give a *force–velocity curve*, as in fig. 11.15.

An alternative way of measuring isotonic contractions is by means of the *isotonic release* method. The apparatus used for this is shown in fig. 11.10. The muscle is attached to an isotonic lever as in fig. 11.7, but the lever can here be held up against the contraction of the muscle by means of a stop which can be immediately withdrawn at any desired time during the contraction. The results of a typical experiment are shown in fig. 11.11. At the beginning of the stimulation period the muscle contracts isometrically and the tension rises to its maximum value. When the release relay stop is withdrawn, the tension falls abruptly to a lower level which is equal to the value of the load. The change in length seems to consist of two components: first an abrupt shortening which is coincident with the change in tension, and secondly a steady isotonic shortening at more or less constant

Fig. 11.8. After-loaded isotonic tetanic contractions. *a* shows the length and tension changes during a single contraction, with shortening as an upward deflection of the length trace. *b* shows the initial length changes in contractions against different loads.

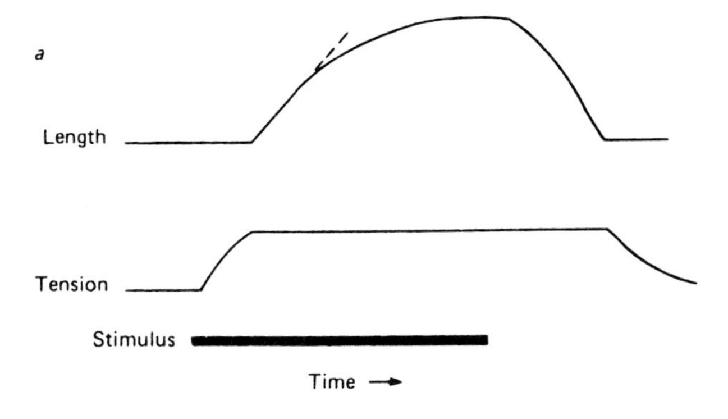

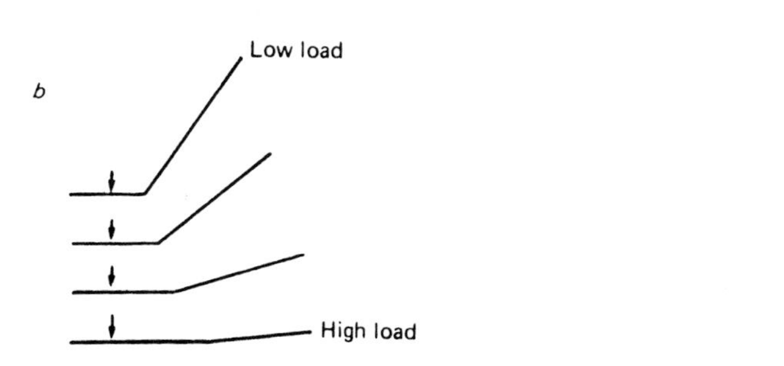

Fig. 11.9. Diagram showing why it is that a lightly loaded muscle can shorten further than a heavily loaded one. Starting from point *a* on the length axis, the muscle contracts isometrically until its tension is equal to the load it has to lift, and then it shortens until it meets the isometric length–tension curve. With a heavy load (P_1) this occurs at x, with a lighter load (P_2) it occurs at y. Notice that point x can also be reached by an isometric contraction from point b. (When starting from a much extended length, a muscle may in practice stop short of point x when lifting load P_1; this is probably caused by inequalities in the muscle.)

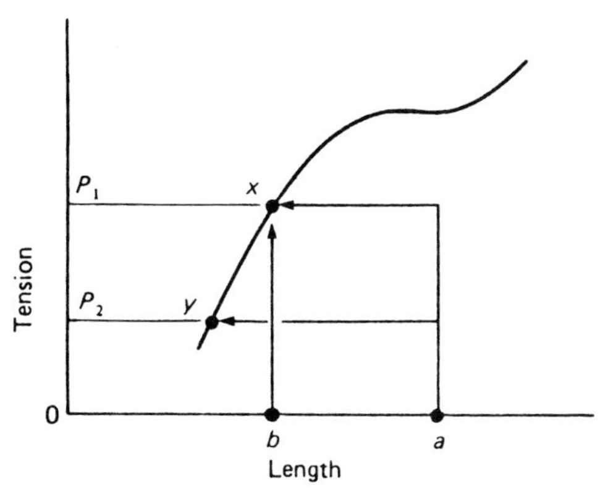

velocity. Since the mass of the lever and the load can never be negligible, the lever system possesses inertia, and so there is usually some oscillation at the change-over between these two phases. With increasing loads (fig. 11.11*b*), the initial length change is less, and the velocity of shortening during the isotonic phase is less. The initial phase of length change thus behaves as an elasticity, since its extent is proportional to the change of tension. The isotonic phase shows the same relation between force and velocity as is seen in afterloaded contractions.

The onset of activity

A number of changes in the properties of a muscle occur just after the stimulus and before contraction begins. These changes are not fully understood.

The time between the stimulus and the beginning of the rise in tension during an isometric twitch is known as the *latent period.* It lasts about 15 ms in a frog sartorius at 0 °C and decreases with increasing temperature. If the muscle is stretched so that there is some resting tension, the use of very sensitive recording methods reveals that there is a brief decrease in tension before the large increase

Fig. 11.11. Isotonic releases. *a*: Diagram of the length and tension changes during an isotonic release; shortening is shown upwards on the length trace. *b*: Length changes during a series of isotonic releases against different loads. Shortening is again shown upwards, the dots on the grid lines are at 1 ms intervals, and the figures opposite each trace indicate the load (*g*) on the muscle after release. From a frog sartorius muscle, with maximum isometric tension 32 g. (*b* from Jewell and Wilkie, 1958.)

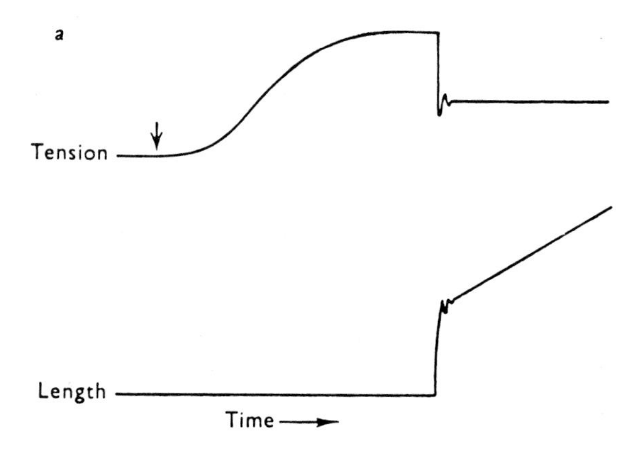

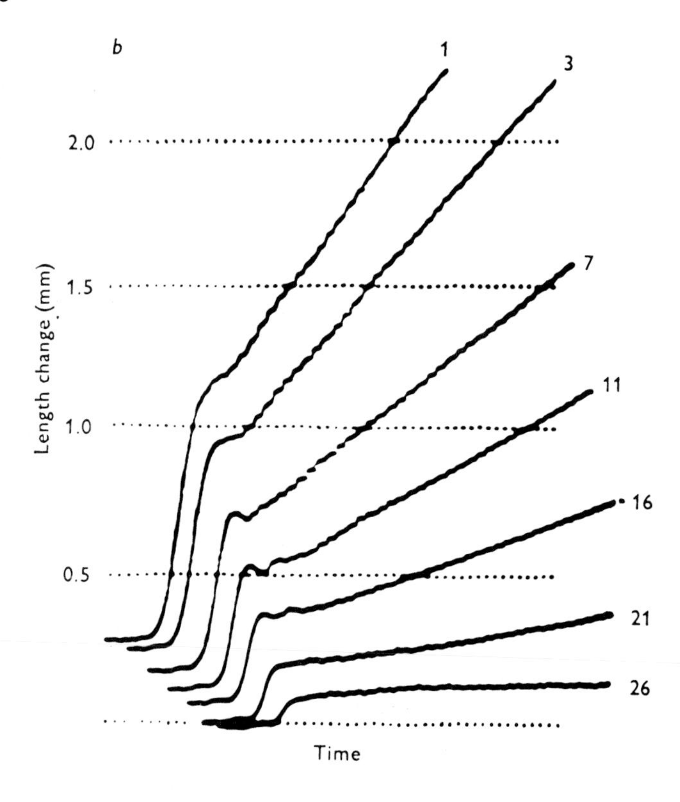

Fig. 11.10. Isotonic lever arranged for isotonic releases. The system is as in fig. 11.7 except that the lever is held up by a stop mounted on a relay. When the relay is activated, the stop is immediately withdrawn so that the muscle can now shorten.

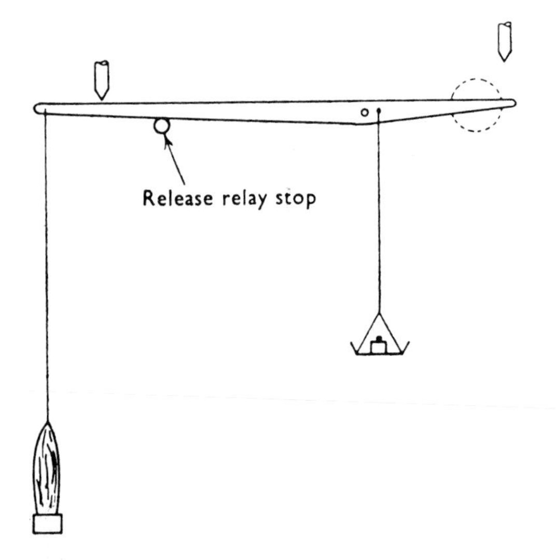

that constitutes the contraction begins. This change is known as *latency relaxation* (Sandow, 1944). The amount of light diffracted by a muscle begins to change during the latent period, at about the time that latency relaxation begins (D. K. Hill, 1953). The resistance of the muscle to stretching begins to rise about half-way through the latent period (A. V. Hill, 1950*b*).

The heat production of muscle

Almost all chemical reactions, and a number of physical changes, are accompanied by the evolution or absorption of heat. Contracting muscles produce heat, and the measurement of this heat production is of considerable interest in that we might expect it to be related to the chemical reactions involved in the contraction process. Furthermore, measurement of the heat production is an essential prerequisite in any attempt to measure the energy expended by a contracting muscle. This energy ($-\Delta E$) is the sum of the work done by the muscle on the load (w) and the heat produced during the contraction (h), or

$$-\Delta E = h + w. \qquad (11.1)$$

The minus sign on the left-hand side of (11.1) is a matter of thermodynamic convention.

The technical problems involved in the measuring the heat production of muscles are considerable, since the temperature changes involved may be extremely small. The methods used by A. V. Hill and his colleagues involve the use of a thermopile. If an electrical circuit is made up of two different metals, an electromotive force is produced if the temperature of one of the junctions is different from that of the other; such a device is called a thermocouple. A thermopile consists of a large number of thermocouples in series, and is therefore more sensitive than a single thermocouple. In Hill's experiments, the temperature changes produced by a moist muscle resting against the 'hot' junctions of the thermopile were measured by means of a galvanometer. The deflection of the galvanometer was determined by fixing a mirror to the galvanometer axis and projecting a thin beam of light via the mirror either onto photosensitive paper fixed to a revolving drum or onto a photocell whose output was displayed on an oscilloscope (fig. 11.12). Further details of the techniques for measuring heat production are given by Woledge *et al.* (1985).

The main features of the heat production during isometric contractions were determined by Hill and Hartree in 1920. Later work has been

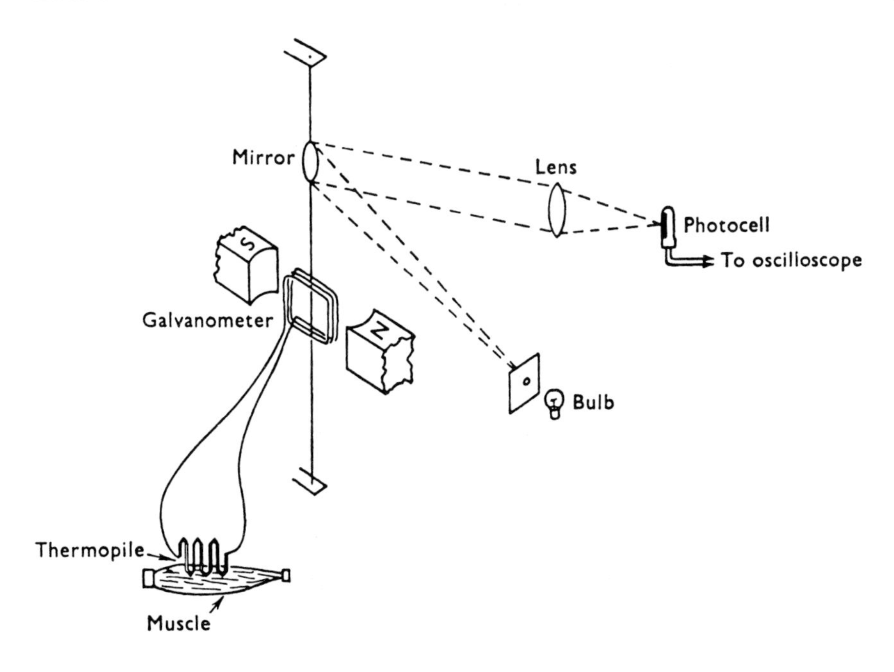

Fig. 11.12. Schematic diagram to show one method of measuring the heat production of a muscle.

much concerned with increasing the accuracy and time resolution of the measurements involved, and with the heat changes associated with shortening or lengthening of the muscle.

In resting muscle, various metabolic processes occur throughout the life of the muscle, and these processes are associated with the liberation of heat. This resting heat is about 10 mJ g^{-1} min^{-1} in frog sartorius muscles at 20 °C (Hill and Howarth, 1957). The nomenclature of the heat changes produced by contracting muscles has varied somewhat since the original formulation by Hill and Hartree; we shall adopt the scheme given by Kushmerick (1983), in which the heat produced during contraction and relaxation is called the *initial heat*, and the heat produced after the muscle has relaxed is called the *recovery heat*.

In an isometric tetanus, the initial heat consists of three phases (fig. 11.13). The *activation heat* is a burst of heat production produced soon after the onset of stimulation. It is thought to accompany the processes whereby the force-producing reactions are switched on; we shall see in the next chapter that these processes involve calcium ion movements within the muscle cells. The *maintenance heat* is produced steadily throughout a tetanus and is thought to reflect largely the force-producing reactions themselves. The *relaxation heat* is produced while the muscle is relaxing, and is probably largely due to the degradation of mechanical work within the muscle.

Fig. 11.13. The rate of heat production of frog sartorius muscle during an isometric tetanus. The muscle was stimulated for a period of 1.2 s. From Hill and Hartree (1920).

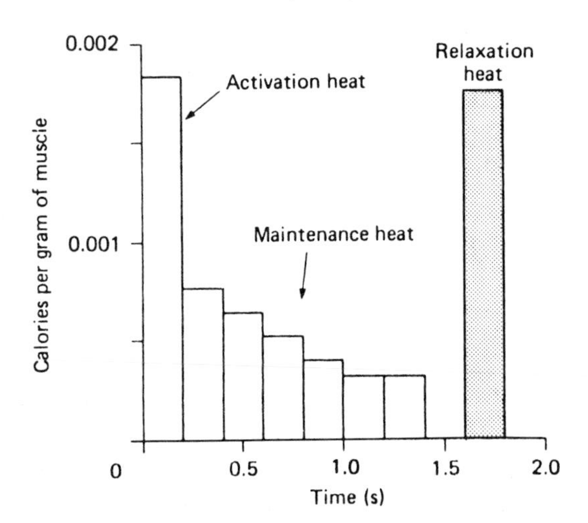

Elastic bodies usually undergo heat changes when they are stretched or allowed to shorten. Most substances, such as steel or wood, absorb heat when stretched and release heat when released; active muscle shows this type of behaviour (Hill, 1953; Woledge, 1961, 1963). Rubber, on the other hand, releases heat when stretched and cools on release; this type of behaviour is shown by resting muscle.

When an active muscle shortens, an extra amount of heat is released, in addition to the activation heat. This is called the *heat of shortening*, and we shall examine its properties in the following section. When a contracting muscle is stretched, its heat production is greatly reduced, so that it is actually absorbing energy during the period of stretch.

The high-energy compounds which have been broken down during contraction have to be resynthesized. For a long period after the end of a contraction, the heat production of a muscle is higher than the normal resting level. This heat production is known as the *recovery heat*, and is thought to correspond to the respiratory processes involved in resynthesis of the high-energy phosphate compounds; it is greatly reduced in the absence of oxygen.

A. V. Hill's analysis

So far we have described some of the mechanical and thermal properties of contracting muscles without providing any theoretical scheme which will account for these properties and relate them to each other. The search for such a scheme has occupied muscle physiologists for much of this century (see Hill, 1965) and has still not yet been achieved. One of the best analyses was that produced by A. V. Hill in 1938, which has greatly influenced much of the work performed in this field since that time. Hence we must examine Hill's work in some detail, although, as we shall see, a number of later observations indicate that his analysis is incomplete.

Most early theories of the mechanism of muscular contraction suggested that, on stimulation, the muscle became effectively a stretched elastic body. But the force exerted by a purely elastic body when stretched is determined by its length, and is independent of the velocity of shortening. Thus the observation that, during isotonic shortening, force is determined by velocity indicates that the 'elastic' theory cannot be correct. It was then suggested (Gasser and Hill, 1924; Levin and Wyman, 1927) that some of the elasticity of the muscle is damped by an internal viscosity. The idea was that some of the tension in the elasticity was used up internally in working against the viscosity. More tension would be so absorbed at high velocities, leaving less to appear at the ends of the muscle.

The visco-elastic theory was generally accepted until 1938, in spite of the fact that Fenn (1923, 1924) had provided evidence to show that it must be incorrect. Fenn measured the shortening of a muscle under various loads and also measured its heat production. He was then able to calculate the total energy release (from (11.1)), and found that more energy is released when the muscle is allowed to shorten during the contraction. This is not what one would expect from the elastic and visco-elastic theories, which postulate that the muscle contains a fixed amount of potential energy at the beginning of a contraction, which must be converted into heat and work during the contraction. The extra release of energy on shortening is frequently known as the 'Fenn effect'.

Let us now consider some of the details of A. V. Hill's analysis (Hill, 1938). The first stage in this work was the experimental measurement of heat production during isotonic contractions, when it immediately became obvious that extra heat production (known as the heat of shortening) occurs if the muscle is allowed to shorten. Fig. 11.14 shows the results of some of these experiments. In fig. 11.14*c*, isotonic releases were performed at different times after the beginning of a period of stimulation; the muscle was allowed to shorten by the same amount against the same load in each case. It is obvious from the diagram that the heat of shortening does not vary appreciably with time. Fig. 11.14*a* shows the heats of shortening when the muscle was allowed to shorten for different distances against a constant load after an isotonic release; the heat production increased with increasing shortening. Fig. 11.14*b* shows that the amount of the heat of shortening was apparently independent of the load when the muscle shortened over the same distance against various loads, although the rate of heat production (and, of course, the velocity of shortening) decreased with increasing loads. From these results, it was concluded that the extra heat released during shortening is proportional to the distance shortened, i.e.

$$\text{heat of shortening} = a \cdot x, \tag{11.2}$$

where x is the distance shortened and a is a constant having the dimensions of a force. Now the work done by the muscle in an isotonic contraction is given by the product of the force exerted and the distance shortened, i.e.

$$\text{work} = P \cdot x,$$

where P is the load. The extra energy released by the muscle during an isotonic contraction (E_e) is given by

$$\begin{aligned} E_e &= \text{heat of shortening} + \text{work} \\ &= P \cdot x + a \cdot x \\ &= (P + a)x. \end{aligned}$$

Therefore the *rate* of extra energy liberation is given by

$$\frac{dE_e}{dt} = (P + a)V \tag{11.3}$$

where V is the velocity of shortening.

The next stage in the analysis was to determine how the rate of extra energy liberation varied with the load. Knowing a, this could be determined by measuring the velocity of shortening (V) at different loads (P), and then putting the appropriate values in (11.3). Hill found that the rate of extra energy liberation decreased linearly with increasing tension, i.e.

$$(P + a)V = b(P_0 - P), \tag{11.4}$$

where P_0 is the isometric tension and b is a constant with the dimensions of a velocity.

Equation (11.4) can be expressed differently by the following procedure. Adding $b(P + a)$ to both sides, we get

$$(P + a)V + b(P + a) = b(P_0 - P) + b(P + a),$$

i.e.

$$(P + a)(V + b) = b(P_0 + a) = \text{constant}. \tag{11.5}$$

Equation (11.5) is a hyperbola in which the tension produced by the muscle is related to its velocity of shortening. Hence, if the analysis is correct, the equation should fit the force–velocity relation of the muscle as determined by purely mechanical measurements, and, in particular, the values of a and b determined from such measurements should agree with those derived from measurements of the heat of shortening and the rate of extra energy release during shortening. Hill concluded that this was so: fig. 11.15 shows the force–velocity relation fitted by (11.5).

Hill's 1938 paper consists of three parts. The first part deals with the experimental methods used in determining the heat production and mechanics of contraction, and the second part contains the elegant experiments and analysis outlined above. In the final section, Hill proposed a new mechanical model of contracting muscle to replace the visco-elastic model. He suggested that a contracting muscle consists of two components: (1) a *contractile component*, whose properties are determined by the force–velocity relation, in series with (2) a *series elastic component*, whose properties can be described by a force–extension curve. This idea is illustrated diagrammatically in fig. 11.16.

Fig. 11.14. Heat production in a frog sartorius muscle when it was allowed to shorten during a tetanus at 0 °C, to show the heat of shortening. The curves are galvanometer deflections proportional to the amount of heat produced by the muscle. For the upper traces (*a*), the muscle was released at 1.2 s after the onset of stimulation and allowed to shorten various distances (*B*, 1.9; *C*, 3.6; *D*, 5.2 mm) against a constant load (3 g) before being stopped. For the middle traces (*b*), the muscle was released at 1.2 s after the onset of stimulation and allowed to shorten a constant distance (5.2 mm) against various loads (*F*, 24.9; *G*, 13.9; *H*, 5.7; *J*, 3.0 g). For the lower traces (*c*), the muscle was released at various times (*L*, 1.8; *M*, 0.95 s; *N*, immediately) after the onset of stimulation and allowed to shorten a certain distance (9.1 mm) against a constant load (2.9 g). Traces *A*, *E* and *K* show galvanometer deflections during isometric contractions. (From Hill, 1938, by permission of the Royal Society.)

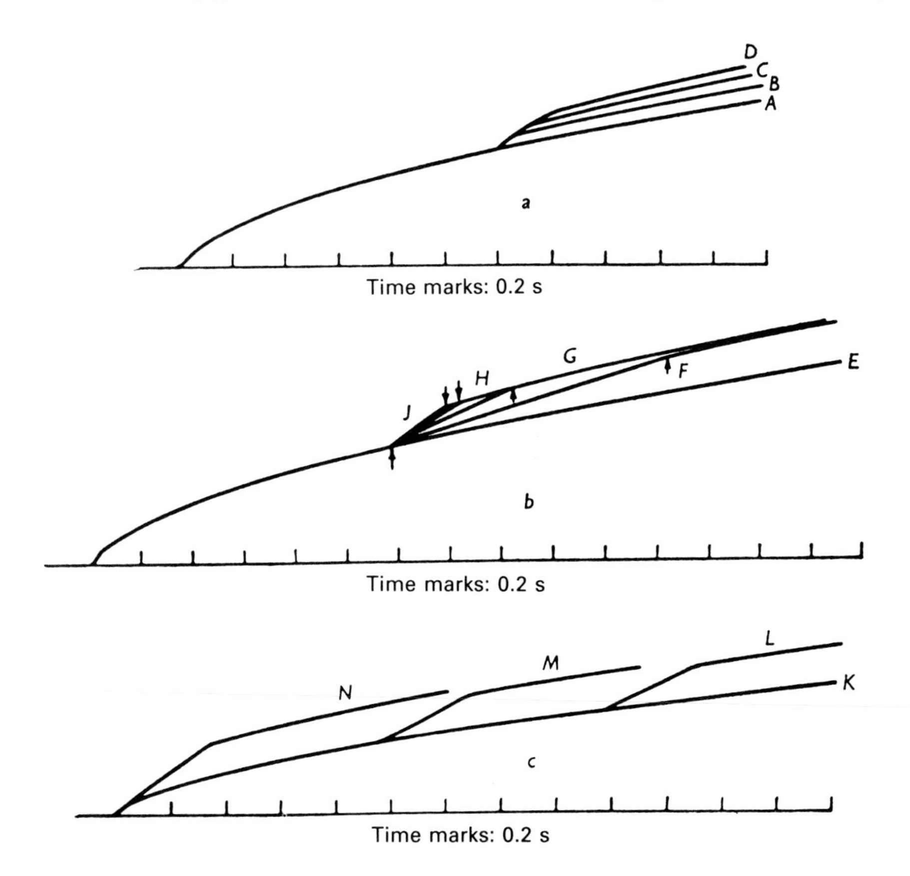

Fig. 11.15. The force–velocity curve of a frog sartorius muscle at 0 °C. The experimental points were determined from after-loaded contractions as in fig. 11.3. The curve is drawn according to (11.5) with $a = 14.35$ g, $b = 1.03 \text{ cm s}^{-1}$, and $P_0 = 65.2$ g. (From Hill, 1938, by permission of the Royal Society.)

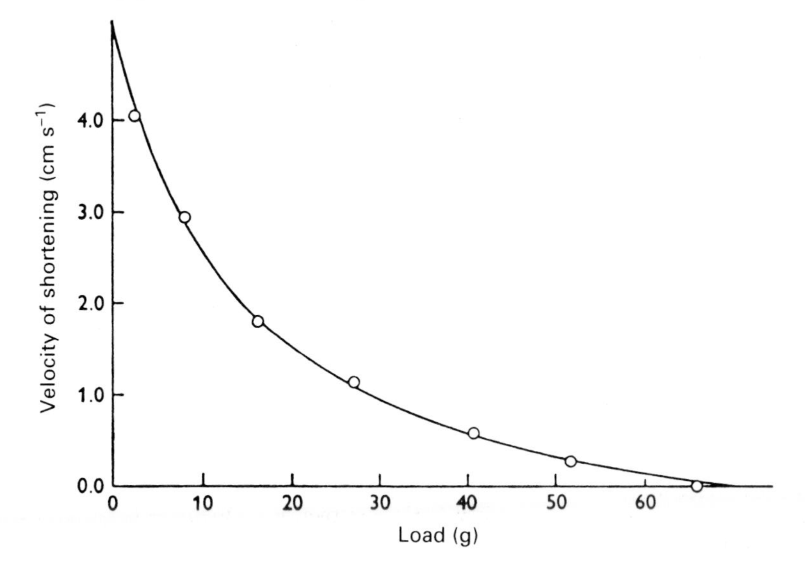

Fig. 11.16. A. V. Hill's two-component model of contracting muscle. The properties of the contractile component are determined by its force–velocity curve, and the properties of the series elastic component are determined by its force–extension curve.

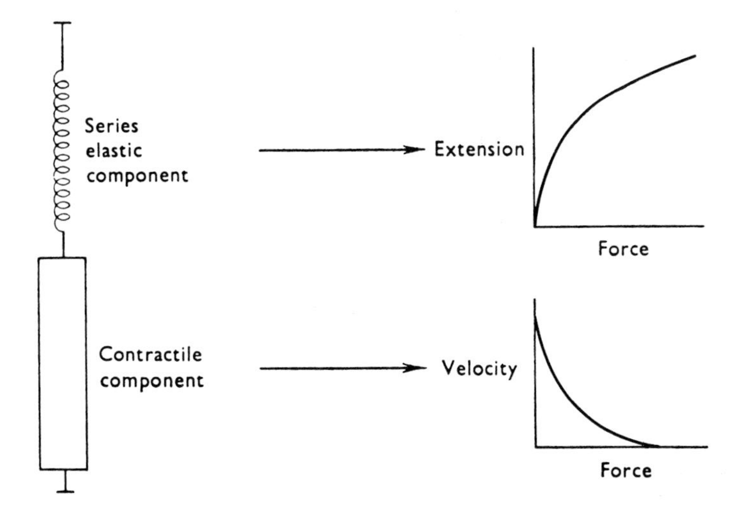

Now let us see how this model is able to explain some of the mechanical properties of tetanically contracting muscles. Firstly, consider the time-course of the rise of tension during an isometric tetanus (fig. 11.4). In the resting condition, there is no tension on the muscle, therefore the series elastic component must be slack. With the onset of stimulation, the contractile component begins to shorten; since the tension is zero, it will initially shorten at its maximum velocity. But since the ends of the muscle are fixed, the shortening of the contractile component must produce a corresponding extension of the series elastic component, and so the tension will begin to rise according to the force–extension curve. This increase in tension results in a reduction in the velocity of shortening of the contractile component (according to the force–velocity curve), so the series elastic component will be extended less rapidly, and hence the tension will rise more slowly. These interactions will continue until the tension has reached its full isometric level (P_0), when the shortening velocity is zero and the series elastic component is fully extended.

The model can also provide an explanation for the quick release phenomena observed by Gasser and Hill (fig. 11.6). The quick release allows an abrupt shortening of the series elastic component, and therefore an abrupt fall in tension. Because of this fall in tension (which will be roughly proportional to the extent of the release), the contractile component now begins to shorten at a velocity determined by the new tension. In so doing, it will extend the series elastic component again, and so the tension will again rise to its isometric level in the same way that it did at the beginning of the period of stimulation. The new isometric tension may be slightly different from the old one, since the muscle is now at a shorter length.

During after-loaded isotonic contractions (fig. 11.8), the series elastic component is at constant length and tension during the period of shortening, so that the velocity of shortening of the whole muscle is equal to that of the contractile component. In an isotonic release (fig. 11.11), the sudden reduction in tension from isometric to the value of the isotonic load causes an abrupt shortening of the series elastic component. Thereafter the series elastic component is at constant length and the muscle shortens at a velocity determined by the force–velocity relation.

The elastic elements

We must now consider the nature of the series elastic component. Part of this series elasticity is external to the muscle, being the elasticity of the connections between the muscle and the recording system. It is desirable to reduce this external elasticity as far as possible (by using steel wire or chain connections rather than cotton thread, for example) but it can never be entirely eliminated. Of the remaining series elasticity, about half is located in the muscle tendons and the rest in the muscle cells themselves (Jewell and Wilkie, 1958); we shall see later, in considering Huxley and Simmons' model of the contractile mechanism, just where in the muscle cells this is. The force–extension curve of the series elastic component can be measured in various ways; one of the most reliable methods involves measuring the extent of the elastic length change during isotonic releases.

At long lengths, the two-component model shown in fig. 11.16 is inadequate, since it does not account for the resting tension exerted by a stretched muscle. The model is then modified by including a third element, the *parallel elastic component*, as is shown in fig. 11.17. This is probably mainly composed of the connective tissue sheaths of the muscle, but it is possible that components of the muscle cells are also involved.

Fig. 11.17. Modification of the two-component model at long lengths to include a third component, the parallel elastic component, which is responsible for the resting tension of a stretched muscle.

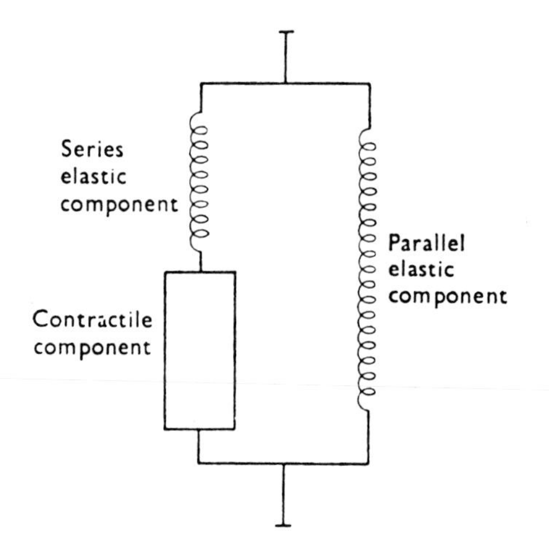

Alternative force–velocity equations

Hill's equation for the force–velocity relation was originally applied to muscles at rest length only. Abbott and Wilkie (1953) measured the force–velocity relation at various lengths below the rest length, and found that the results could be described by a simple modification of Hill's equation, so that, at any particular length,

$$(P + a)(V + b) = b(P'_0 + a), \qquad (11.6)$$

where P'_0 is the full isometric tension at that length. The constants a and b did not appear to change with length. Equation (11.6) is merely an extension of (11.5).

The first force–velocity equation was that given by Fenn and Marsh (1935):

$$P = P_0 e^{-\alpha V} - kV,$$

where a and k are constants. An equation of somewhat similar form is that provided by Aubert (1956):

$$P = A\, e^{-V/B} \pm F,$$

where B and F are constants, and $A = P_0 + F$. Notice that Aubert's equation introduces a 'frictional element', F, which decreases the tension on shortening and increases it on lengthening. Both these equations are largely empirical; they are not connected with any particular theory of muscular contraction.

Limitations of Hill's analysis

It is clear that Hill's analysis provides an elegant theoretical scheme which fits many of the features of muscular contraction. However, a number of discrepancies between the predictions of the theory and experimental observations have been discovered; we must now examine some of these discrepancies.

The effect of load on the heat of shortening

In Hill's original experiments, it appeared that the heat of shortening was independent of the load, being determined only by the distance shortened. Hill (1964*a*) later made a very careful reinvestigation of this point, and concluded that the heat of shortening increases with increasing loads. The relation between the heat of shortening (now represented by α) and the load P is given by

$$\alpha/P_0 = 0.16 + 0.18\, P/P_0.$$

This conclusion invalidates the derivation of (11.5) from thermal measurements, but does not, of course, affect its applicability as an empirical description of the force–velocity relation.

The time course of the rise in tension during an isometric tetanus

According to Hill's model, the time course of the rise in tension during an isometric tetanus is precisely determined by the force–velocity curve of the contractile component and the force–extension curve of the series elastic component, as we have already seen (p. 236). In mathematical terms,

$$\frac{dP}{dt} = \frac{dP}{dx} \cdot \frac{dx}{dt}$$

or

$$\frac{dP}{dt} = V\frac{dP}{dx} \qquad (11.7)$$

where V is given by the force–velocity curve and dP/dx is the slope of the force–extension curve. Integration of (11.7) with respect to P gives

$$t = \int_0^{P_t} \frac{1}{V \cdot dP/dx}\, dP, \qquad (11.8)$$

where P_t is the tension at time t after the beginning of stimulation.

This relation has been very carefully examined by Jewell and Wilkie (1958). They determined experimentally the force–velocity and force–extension curves, and then calculated the expected form of the time course of tension development in an isometric contraction from (11.8) using numerical integration. This calculated curve could then be compared with the actual responses (both at the start of stimulation and after a quick release) of the same muscle. The results are shown in fig. 11.18. It is clear that the tension does not rise as rapidly as the theory predicts; the time taken for the tension to reach a given level is about 80% longer than expected for the initial development of tension, and about 50% longer than expected after a quick release. Obviously, something must be wrong with the theory; Jewell and Wilkie suggested that the velocity of shortening is determined by the 'past history' of the muscle, as well as by the force acting on it at a particular time.

Transient changes in velocity of shortening after isotonic releases

During an isotonic release, there is a period of oscillation between the abrupt shortening ascribed to the series elastic component and the steady shortening that follows, as can be seen in fig. 11.11*b*. Some of this oscillation is caused by the inertia of the lever system used for recording the length change. Any such inertial oscillation in length must also be accompanied by an oscillation in tension; thus any investigation of this phenomenon must include measurements of the tension of the muscle as well as its length. Podolsky (1960) carried out experiments of this type on sartorius muscles, and showed that the shortening velocity does not reach a steady value until some time after the oscillation in tension has stopped (fig. 11.19). Related results were obtained by Aidley (1965*b*) from frog rectus abdominis muscles, and by Armstrong, Huxley and Julian (1966) using single frog

Fig. 11.18. The time-course of the rise in tension during an isometric tetanus. The broken line shows the final maximum tension. Curve *a* is calculated from the force–velocity relation of the contractile component and the force–extension curve of the series elastic component by means of (11.8). Curve *b* is the actual time-course of tension development after a quick release. Curve *c* is the actual time-course of tension development at the beginning of a period of stimulation. (From Jewell and Wilkie, 1958.)

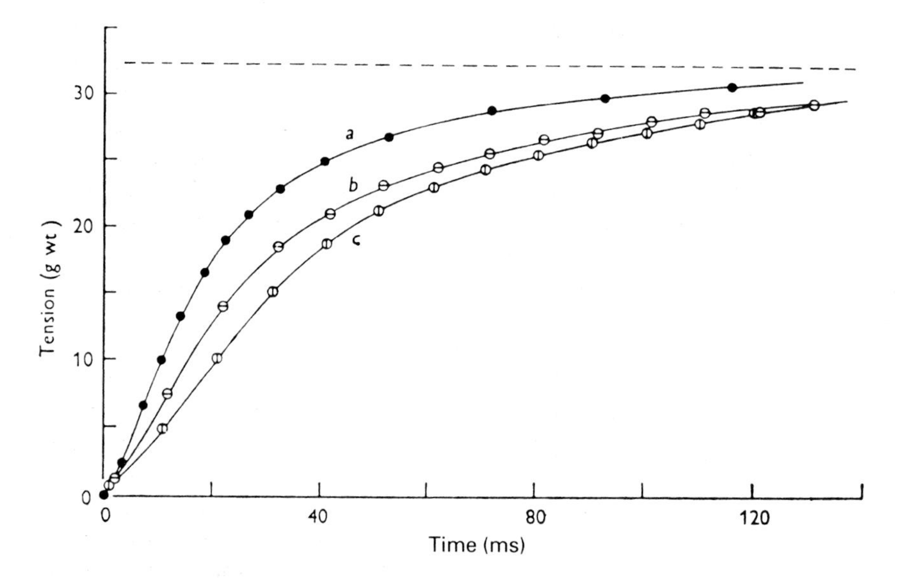

Fig. 11.19. Isotonic releases of frog sartorius muscle, showing transient velocity changes immediately after the reduction in tension. Shortening is shown as a downward deflection of the upper traces. The loads on the muscle after release, expressed as fractions of P_0, were: *a*, 0.84; *b*, 0.69; *c*, 0.55. (From Podolsky, 1960.)

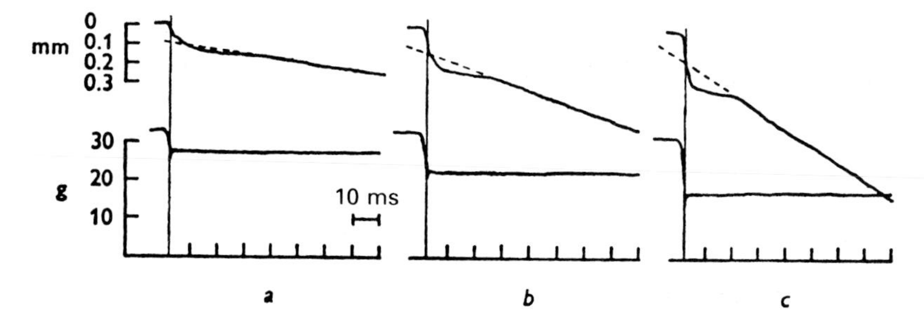

semitendinosus fibres. We shall examine the details of these transients, as measured by A. F. Huxley and his colleagues, in the following chapter.

All these results imply that the force–velocity relation is not instantaneously obeyed when the tension changes; shortening velocities seem to be higher than usual immediately after a fall in tension. In other words, the properties of the contractile component are not completely described by the steady-state force–velocity relation.

The analysis of twitches

We have so far been mainly concerned with the dynamics of muscle during repetitive stimulation, when contraction is tetanic. We must now consider what happens during the response of the muscle to a single electrical stimulus.

Isometric and isotonic twitches

A possible explanation of the form of the isometric twitch in terms of A. V. Hill's two-component model is as follows (Hill, 1949). In the resting condition, there is no tension in the muscle, and the series elastic component is slack. Soon after the stimulus, the contractile component begins to shorten. This extends the series elastic component (since the ends of the muscle are fixed) and so the tension begins to rise in accordance with the force–extension curve of the series elastic component. This rise in tension causes the shortening velocity of the contractile component to fall in accordance with the force–velocity curve and therefore the rate of rise in tension decreases. For a short time, the time course of tension development is indistinguishable from that at the start of a tetanus, but soon the twitch tension does not rise so rapidly as the tetanus tension. The reason for this is that the activity of the contractile component is less, in other words, its force–velocity relation is no longer maximal – the tension at any particular shortening velocity is less than that in a tetanus. Another way of stating this idea is to say that the tension which could be exerted by the contractile component if it were neither shortening nor lengthening (denoted by P_i and termed the 'active state') begins to fall. At the peak of the twitch, the series elastic component is at its fullest extent and the contractile component has just stopped shortening, so that the tension of the muscle is equal to P_i. But now P_i continues to fall, so that the contractile component is no longer capable of sustaining the tension in the series elastic component, so the series elastic component shortens, restretching the contractile component in the process; hence the tension begins to fall, and eventually reaches zero.

This explanation contains a number of assumptions. A major one is that the contractile component assumes the properties that it possesses during maintained stimulation very soon after the onset of contraction. Hill's evidence for this assumption depends upon experiments in which a muscle was rapidly stretched shortly after a stimulus. It was found (fig. 11.20) that tension rose

Fig. 11.20. Tension changes in a toad sartorius muscle following a rapid stretch applied just after a single stimulus. Trances *b* to *e* show the effects of stretches applied 34 ms after the stimulus. In *a*, the stretch was applied 70 ms before the stimulus. The final length of the muscle was the same in each case. (From Hill, 1949, by permission of the Royal Society.)

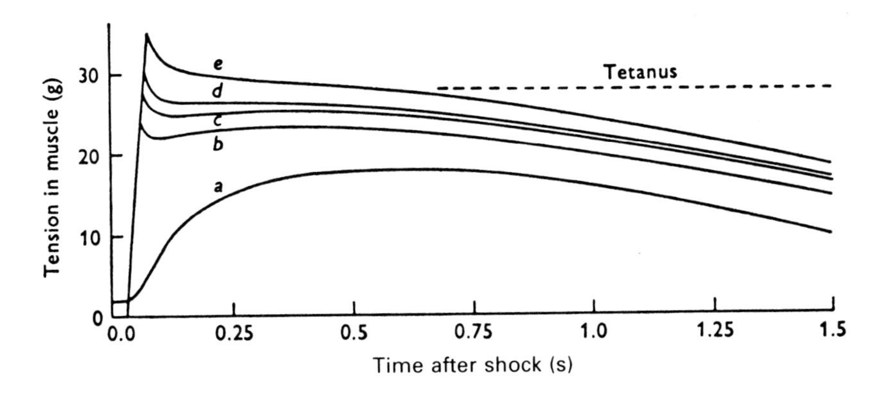

rapidly during the stretch and fell a little immediately afterwards; after this the change in tension depended upon the extent of the stretch. For long stretches (trace *e* in fig. 11.20) the tension continued to fall. For shorter stretches (*b* and *c*), the tension rose for a short time, indicting that the contractile component was still able to shorten at these high tensions at this time. When the stretch was just sufficient to bring the tension up to the maximum tetanic tension, the tension remained steady for a short time before falling as the muscle relaxed (*d*). This indicates that the contractile component is capable of maintaining the full tetanic isometric tension at very short times after the onset of contraction.

In isotonic afterloaded twitches, shortening begins when the tension is sufficient to lift the load. The rate of shortening falls off until the peak of the twitch, when the load begins to extend the muscle again.

The 'active state'

A. V. Hill represented his ideas about what happens during an isometric twitch in the form shown in fig. 11.21. The curve P_n is the tension during a normal twitch, P_s is that in a stretch which is just sufficient to cause the tension to remain constant for a short time after the end of a stretch (trace *d* in fig. 11.20); the tension at the beginning of this curve is equal to P_0, the maximum isometric tension in a tetanus. P_i, as we have seen, represents the tension which could be exerted by the contractile component if it were neither shortening nor lengthening. Originally, Hill used to term 'active state' merely to refer to the condition which a contracting muscle is in (in contrast to the 'resting state'), but the term later acquired quantitative connotations, being applied to the curve P_i in fig. 11.21.

The precise form of the P_i curve could not be determined from Hill's experiment; that given in fig. 11.21 is merely an intelligent guess. An attempt to perform a more accurate determination was made by Ritchie (1954). His method involved the measurement of the maximum tension following a quick release at various times during an isometric twitch. At these points, the tension is neither rising nor falling, therefore the contractile component is neither shortening nor lengthening, and therefore the tension is equal to P_i. Thus P_i can be plotted against time to give an 'active state' curve.

Is the time course of the 'active state' independent of length and tension changes? Consider fig. 11.22, which shows length changes following isotonic releases against the same load at different times during a twitch (Jewell and Wilkie, 1960). At late times after the stimulus (at 0.6 s, for example), the muscle is still shortening after a late release, but is lengthening after an early release. This indicates that, at this time, the 'active state' is greater than the load after the late releases but less than the load after the early releases. In other words, if the

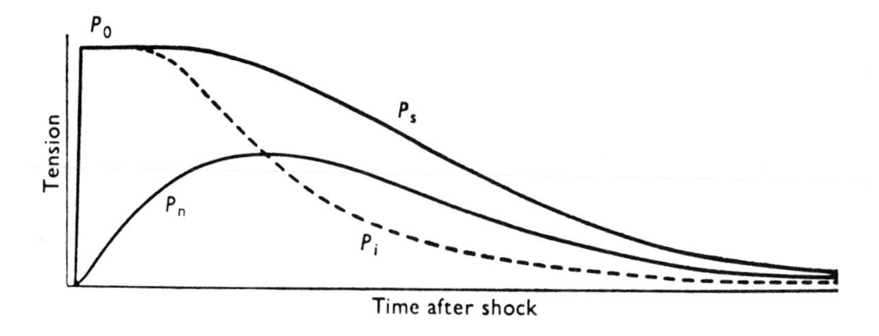

Fig. 11.21. The mechanical changes during a muscle twitch, according to the 'active state' concept. P_n, time course of the tension change during an isometric twitch. P_s, time course of the tension changes during a twitch in which the muscle is rapidly stretched, by an amount just sufficient to extend the series elastic component to the length which it is during the plateau of an isometric tetanus, soon after the stimulus. This results in the tension attained immediately after the stretch being equal to the maximum tetanic isometric tension, P_0. P_i is the 'active state' curve deduced from these results. (From Hill, 1949, by permission of the Royal Society.)

muscle is allowed to shorten, the duration of the 'active state' lessens. Further evidence that this is so was provided by Pennyquick (1964), from mechanical measurements, and by Hill (1964*b*), from heat measurements.

The results of these experiments lead to a rather curious situation. The time course of the 'active state' can only be measured by means of experiments in which the length and tension of the muscle are changed; indeed, as Pringle (1960) pointed out, the 'active state' is defined in such a way that it must be assumed that it is unaffected by changes in length and tension. Yet we have seen that such changes do alter the 'active state'! We must conclude that the concept of the 'active state' leads more to confusion than enlightenment.

A satisfactory substitute for the concept of the active state, that is to say a quantitative measure of the 'activity' of the muscle, is not yet available. A promising start was made by Julian (1969, 1971; Julian and Sollins, 1973), using some of the concepts developed in chapter 12. He was able to produce computer predictions of the behaviour of muscles during a twitch that agree closely with actuality.

The nature of the energy source

We have seen that Hill's two-component model provides an approximate description of the mechanical properties of contracting muscle, and his estimates of total energy release provide a quantitative description of the Fenn effect. But a description of muscle solely in terms of its gross output is necessarily incomplete, since it does not answer one of the major questions about muscle: how is chemical energy converted into work?

Two major lines of investigation have approached this question. Firstly there has been a search for the fuel for muscular contraction and attempts to relate energy output to chemical breakdown. This became increasingly fruitful from 1927 onwards. Secondly, beginning in the 1950s, structural studies have led to considerable advances in our understanding of the mechanisms involved in the contractile process. Both these paths have drawn upon the mechanical and thermal studies we

Fig. 11.22. An experiment which appears to indicate that the time course of the 'active state' is affected by length and tension changes in the muscle. Responses of a frog sartorius muscle at 0 °C to a single stimulus. Trace *a* shows the isometric twitch. Trace *b* shows an isotonic after-loaded twitch with a very light load. The other traces show the length changes occurring after isotonic releases against the same light load at different times after stimulation. (From Jewell and Wilkie, 1960.)

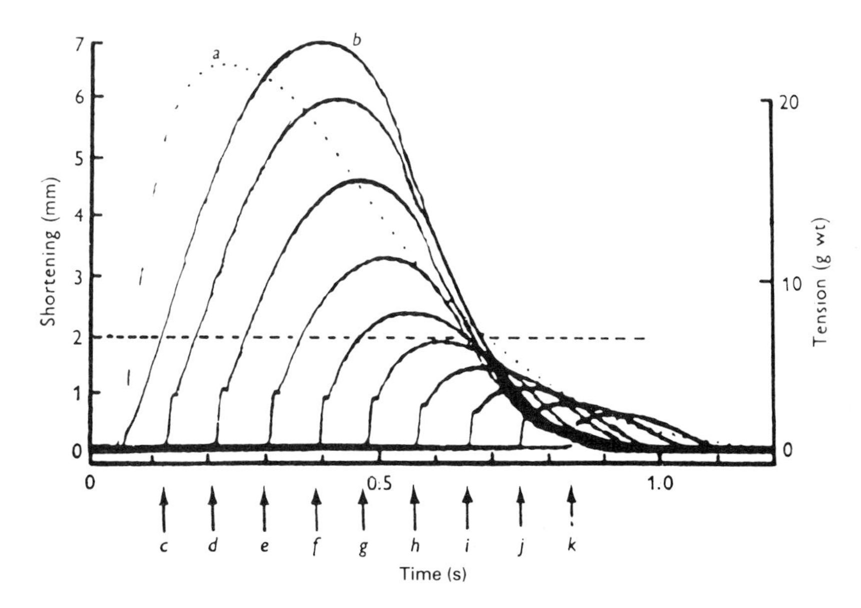

have described so far. For the remainder of this chapter we consider chemical change; the nature of the contractile process forms the subject of chapter 12.

The energy for muscular contraction is ultimately derived from the chemical energy released by the oxidation of food substances. For our purposes it is more pertinent to enquire what the *immediate* source of the energy for contraction is. Quantitative studies on this question began with the work of Fletcher and Hopkins in 1907. They showed that muscles can continue to contract in the absence of oxygen, and that they produce lactic acid under these conditions. It was later shown by Meyerhof and others that this lactic acid is formed by the breakdown of glycogen, which itself is formed by polymerization of glucose derived from carbohydrates in the food. Until about 1930, then, it was generally believed that the immediate source of energy for contraction was the reaction which results in the formation of lactic acid; this was known as 'the lactic acid theory'.

The next step was the discovery of 'phosphagen', soon shown to be creatine phosphate, and the demonstration that it was broken down during contraction (Eggleton and Eggleton, 1927; Fiske and Subbarow, 1927). In a crucial series of experiments, Lundsgaard (1930*a, b*) determined the concentrations of lactic acid and creatine phosphate in muscles poisoned with iodoacetate and held in anaerobic conditions. He found that stimulation and contraction did not cause the production of lactic acid under these conditions, whereas creatine phosphate was broken down in proportion to the tension produced by the muscle. These results clearly disproved the lactic acid theory.

The final stage in this 'revolution in muscle physiology', as it was called by A. V. Hill in 1932 (see Hill, 1965), was the discovery of adenosine triphosphate (ATP) and its role as a carrier of energy within the cell (Lohmann, 1929, 1934; Meyerhof and Lohmann, 1932; Lipmann, 1941). This led to an understanding of the basic scheme of energy production and utilization in cells (fig. 11.23) which is now familiar to all students of cell biology. A fascinating account of these discoveries, and of much else in the history of muscle research, is given by Needham (1971).

If the scheme in fig. 11.23 is correct, then, as A. V. Hill (1950*a*) pointed out in his 'challenge to biochemists', it should be possible to show that ATP is broken down during the contraction of living muscle cells. The general technique used is to stimulate a muscle, very rapidly freeze it so as to prevent any further biochemical changes, and then determine how much of the substance one is interested in is present. A similar determination is performed on an unstimulated muscle, usually the equivalent muscle on the opposite leg of the same animal. Using these techniques, it was shown by a number of workers that contraction leads to the breakdown of creatine phosphate in muscles poisoned with iodoacetic acid (which prevents glycolysis, and hence prevents resynthesis of creatine phosphate). This could mean either that creatine phosphate is the immediate source of energy for contraction, or that creatine phosphate reacts with adenosine diphosphate (ADP) to give ATP and creatine, ATP being the immediate energy source.

Fig. 11.23. Diagram to show, very schematically, the respiratory energy production in an animal cell. Oxygen is needed to convert pyruvic acid to carbon dioxide and water. Many invertebrates use arginine phosphate instead of creatine phosphate for storage of high energy phosphate.

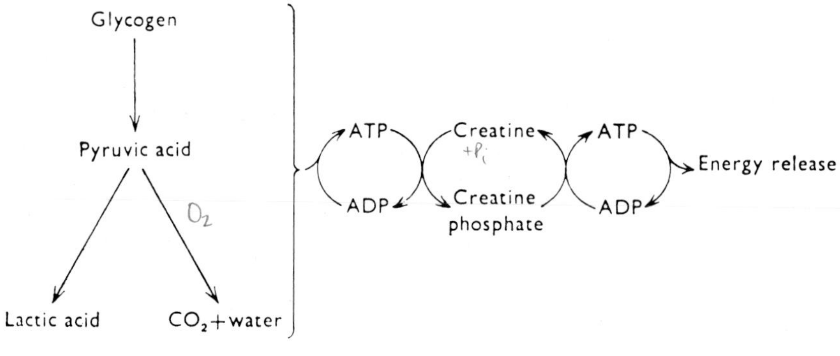

Table 11.1 *Changes in the ATP content of frog sartorius muscles after treatment with fluorodinitrobenzene. The stimulated muscles were frozen in liquid propane at the peak of a single isotonic twitch at 0 °C. The control muscles were unstimulated. ATP contents are given in micromoles per gram of muscle*

Muscle pair	Control ATP	Twitch ATP	Difference in ATP
1	2.55	2.39	−0.16
2	3.38	3.25	−0.13
3	2.55	2.53	−0.02
4	2.82	2.58	−0.24
5	3.25	3.12	−0.13
6	2.93	2.57	−0.36
7	3.38	3.20	−0.18
8	2.16	2.01	−0.15
9	3.16	2.98	−0.18
10	2.51	2.40	−0.11
11	1.90	2.09	+0.19
12	2.90	2.40	−0.50
13	3.56	2.64	−0.92
Mean ± SE	2.85 ± 0.14	2.63 ± 0.10	−0.22 ± 0.07

From Infante and Davies, 1962.

The substance 1-fluoro-2,4-dinitrobenzene (FDNB) blocks the action of the enzyme creatine phosphotransferase, so that ADP cannot be rephosphorylated to ATP by the breakdown of creatine phosphate. Using this substance, Davies and his colleagues were able to show that breakdown of ATP occurs during contractions of frog rectus abdominis (Cain *et al.*, 1962) and sartorius (Infante and Davies, 1962) muscles. The results of their experiments on sartorius muscles are given in table 11.1. The average amount of ATP lost during the rising phase of each twitch was 0.22 μmoles per gram of muscle. Is this sufficient to account for the energy used in the contraction? In these experiments, the muscle shortened against a light load, and the average value for the work done in each contraction was 17.4 $g\,cm\,g^{-1}$; this is equivalent to $1.71 \times 10^{-3}\,J\,g^{-1}$. The heat of hydrolysis of the terminal phosphate bond in the conversion of ATP to ADP is probably about 47 $kJ\,mole^{-1}$ (Homsher, 1987). Hence the dephosphorylation of 0.22 μmoles of ATP should make available about 10^{-2}J. The amount of ATP broken down is therefore more than enough to account for the work done in a twitch; the excess energy appears as heat, we assume.

Energy balance during contraction

The first law of thermodynamics states the principle of the conservation of energy: energy is neither created nor destroyed in a chemical reaction. In contracting muscle, chemical energy is converted into heat and work, and hence it should be possible to show that the energy available from chemical breakdown is equal to the heat released plus the work done. If we cannot account for the energy release (heat plus work) by estimating the energy available from the known chemical reactions in the muscle, then there must be some unknown reaction or reactions present. Energy balance experiments therefore provide a clear test of our understanding of the overall chemical changes that take place in contracting muscle. The subject has been well reviewed by Curtin and Woledge (1978), Kushmerick (1983), Woledge *et al.* (1985) and Homsher (1987).

In thermodynamic terms, it is useful to refer to the enthalpy change of reaction, ΔH. At constant pressure P, this is related to the energy change ΔE by

$$\Delta H = \Delta E + P\Delta V$$

where ΔV is the change in volume. If ΔV is zero (as is very nearly the case in contracting muscle) the enthalpy change is effectively the same as the energy change. We may thus rewrite (11.1) more precisely as

$$-\Delta H = h + w$$

If there are n reactions occurring in the contracting muscle, then we have to add the enthalpy changes for all n of them together to get the total enthalpy change. Thus if ΔH_i is the molar enthalpy change for the ith reaction (i.e. the heat of reaction per mole of one of the reactants) and ξ_i (xi$_i$) is the extent of the reaction (i.e. the number of moles of that reactant which are converted), then the total enthalpy change is given by

$$h + w = -\sum_{i=1}^{n} \xi_i \cdot \Delta H_i.$$

Let us first examine a piece of work by Wilkie (1968) that at first sight provided a clearcut correspondence between energy release and chemical breakdown. Wilkie used frog sartorius muscles which were poisoned with iodoacetate (to prevent glycolysis) and nitrogen (to prevent oxidative phosphorylation). Under these conditions we would expect any ADP formed from ATP breakdown to be immediately rephosphorylated by the breakdown of creatine phosphate, so that change in creatine phosphate content should represent the net chemical energy consumption of the muscle. Wilkie therefore measured the amount of creatine phosphate in individual muscles after they had performed a contraction in which he had also measured the work done by the muscle and the heat it produced. He also measured the creatine phosphate content of the unstimulated muscle from the other leg of the frog so as to estimate the change in creatine phosphate content due to activity. He repeated this experiment for a variety of different types of contraction: isometric and isotonic, twitches and tetani. The results are shown in fig. 11.24. Clearly the breakdown of creatine phosphate is directly proportional to the sum of the heat and work produced by the muscle, 46.4 kJ of energy being released per mole of creatine phosphate broken down.

Does all of this energy come from the breakdown of creatine phosphate? Although the figure of 46.4 kJ mol^{-1} (or 11.1 kcal mol^{-1}) did at first appear to be in reasonable agreement with the expected values for the heat of hydrolysis of creatine phosphate, doubts soon arose. Woledge (1973), as a result of careful calorimetric measurements, concluded that the expected value should be about 34 kJ mol^{-1}.

In another investigation, Kushmerick and Davies (1969) measured the ATP breakdown during the contraction phase of muscles poisoned

Fig. 11.24. The relation between energy production (heat plus work) and creatine phosphate (ΔPC) breakdown in frog sartorius muscles poisoned with iodoacetate and nitrogen. Each point represents a determination on one muscle after the end of a series of contractions. The symbols indicate different types of contraction (isometric twitches and tetani, isotonic tetani in which the muscle was stretched by the load). (From Wilkie, 1968.)

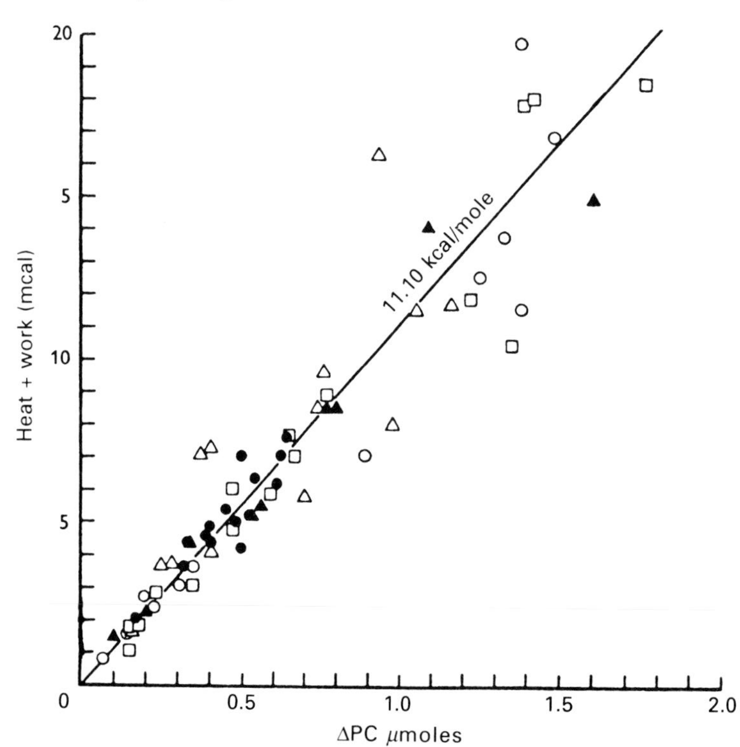

with DTNB. They found that here also the chemical changes could not account for all the heat produced by the muscle. The difference between the energy that could be explained in terms of creatine phosphate and ATP breakdown and the total observed energy release became known as 'unexplained energy'.

It was desirable to determine the time course of the unexplained energy release. In order to do this it was necessary to freeze muscles at different times during the contraction cycle, and hence to use a very rapid freezing technique. Kretschmar and Wilkie (1969) designed a rather dramatic apparatus to do this: the muscle was hit simultaneously from each side by two aluminium hammers which until that moment had been sitting in liquid nitrogen. Thus the creatine phosphate content of muscles could be determined at different times during a period of contraction, and compared with the heat production of other muscles. Using this method Gilbert *et al.* (1971) found that the initial rate of heat production was notably higher than could be accounted for by chemical breakdown. Curtin and Woledge (1979) reached similar conclusions.

In a 5 s isometric tetanus, the unexplained energy accounted for about 28% of the total energy release, but thereafter little extra unexplained energy was produced (fig. 11.25).

At this point the faint-hearted might begin to doubt whether the first law of thermodynamics applies to frog muscles, but Paul (1983) has produced clear evidence that it does. He stimulated muscles so as to produce an isometric tetanus for 3 s every 256 s over a period of some minutes. During this time they reached a steady-state condition in which the heat produced per cycle was constant; this implies that the recovery processes had proceeded to the same extent after each tetanus. Oxygen consumption measurements provided an estimate of

Fig. 11.25. Enthalpy production during isometric tetanuses of different durations in frog muscle. The solid symbols show the observed enthalpy (heat plus work), as measured at different times during a 15 s tetanus; the heat was measured with a thermopile and the work was calculated as that done against the series elasticity. The open symbols show the explained enthalpy, obtained from measurements of the breakdown of creatine phosphate and ATP. The difference between the two curves is the unexplained enthalpy. (From Curtin and Woledge, 1979.)

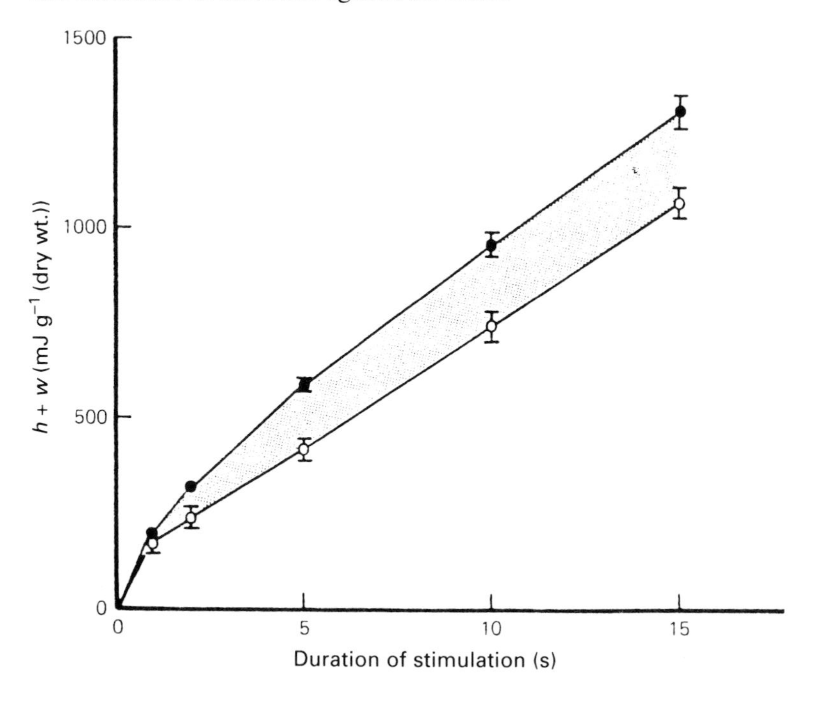

the total chemical change during both contraction and recovery phases. Measurements of creatine phosphate and ATP levels (by using the cooled hammer method) in experimental and control muscles allowed the chemical change during a tetanus to be determined.

Paul found that that total observed enthalpy change, measured as heat production (averaging 181.5 mJ g^{-1} per cycle of tetanus plus recovery) was precisely explained by the total chemical change as measured by oxygen consumption (180.6 mJ g^{-1}). The excess unexplained enthalpy during the contractions (the enthalpy difference between the heat production and the creatine phosphate breakdown) was thus balanced by a corresponding deficiency during the recovery periods. Apart from restoring our confidence in the laws of nature, this experiment provides further evidence for the reality of the reactions producing the unexplained enthalpy and shows that they are reversed during the recovery period.

What is the nature of these reactions? One way of approaching this question is to try to see whether or not they are associated with the contractile mechanism itself or with the coupling process (the intracellular control processes which switch contraction on and off). We shall see in the next chapter that when muscles are greatly stretched they produce little or no tension, but the intracellular calcium movements which form part of the coupling process still occur. Unexplained enthalpy is still produced at these long lengths, so it is unlikely to be associated with the contractile mechanism (Curtin and Woledge, 1981; Homsher and Kean, 1982).

We shall also see later (chapter 13) that the coupling process involves release of calcium ions from the sarcoplasmic reticulum into the sarcoplasm. Some of these calcium ions combine with the protein troponin, which forms part of the contractile mechanism. Frog muscles also contain appreciable quantities of the calcium-binding protein parvalbumin (see Wnuk *et al.*, 1982). Calcium is therefore likely to bind to both these proteins soon after the onset of stimulation.

Homsher *et al.* (1987) discuss the amount of heat that would be released by these reactions. The concentration of parvalbumin in frog muscle is about 0.4 μmol g^{-1}, and each molecule has two binding sites for calcium. There are 0.2 μmol g^{-1} of similar binding sites on troponin so in all there are about 1 μmol of calcium binding sites per gram of muscle. Displacement of magnesium by calcium from these proteins results in a heat release of about 31 kJ per mole of calcium bound, so calcium binding in the muscle could account for a heat release of about 31 mJ g^{-1}, in good agreement with the size of the unexplained enthalpy. Thus it seems likely that the isometric unexplained enthalpy is largely a product of calcium movements associated with the coupling process.

Another form of unexplained enthalpy occurs during rapid shortening. Homsher, Irving and Wallner (1981) found that the extra energy release on shortening (Hill's heat of shortening plus work) was not accounted for by the chemical breakdown of high energy phosphate (creatine phosphate plus ATP) during the period of shortening. The unexplained enthalpy production under the conditions used (a 30% shortening in 0.3 s) was about 6 mJ per gram of muscle. This was balanced by a corresponding deficiency of heat production during the isometric contraction, which continued immediately after the end of the shortening period. It looks as though there is some exothermic change in the contractile mechanism during rapid shortening and that this change to some extent precedes the breakdown of high-energy phosphate.

A. V. Hill (1960) tells how a newspaper report in 1928 alleged that he had solved the Mystery of Life. We must regard this view as somewhat overoptimistic, but it is clear that his successors have been making some further progress towards that distant goal.

12
The contractile mechanism of muscle

One of the most characteristic features of vertebrate skeletal muscles is that microscopic examination reveals that they are striated. Suitable optical techniques or staining methods show that bands of light and dark material alternate along the length of the myofibrils, and that these bands are aligned across the breadth of the fibre.

Detailed descriptions of these striation patterns, and sometimes observations on how they altered with changes in fibre length, were made by a number of nineteenth century microscopists. But, as A. F. Huxley (1980) points out, this knowledge was disregarded and further structural studies were largely neglected in the first half of the twentieth century. The advances in muscle understanding during this time arose largely from biochemical and physiological studies, and the nature of the striation pattern seemed to have no relevance to these approaches.

All this changed with the advent of the sliding filament theory in 1954. Quite suddenly muscle fine structure made sense in terms of function. The search for structural detail as the means of interpreting physiological and biochemical observations began afresh, with new and increasingly sophisticated methods. As a result we now have some exciting glimpses of the molecular activity that underlies muscular contraction. But let us first put the sliding filament theory in its context by taking a look at the biochemical and structural background from which it emerged.

The myofibril in 1953

The contractile machinery of striated muscle cells consists of a small number of different proteins which are aggregated together in filaments. The two major components of this system are the proteins *actin* and *myosin*, which interact with each other to produce the contraction.

Biochemistry

The name myosin was first used by Kühne in 1864 to describe the protein which he isolated by saline extraction from frozen frog muscle. But pure myosin was not obtained until the 1940s, when Straub (1943), working with Szent-Györgyi in Hungary separated it from the second major protein, which he called actin. A number of minor constituents have been discovered since then; they serve either to modify the interaction between actin and myosin or in determining the structural organization of the system.

Myosin is a rather complex protein with a molecular weight of about 500 000. One of its most important properties is that it is an ATPase, i.e. it will enzymatically hydrolyse ATP to form ADP and inorganic phosphate (Engelhardt and Ljubimowa, 1939). This reaction is activated by calcium ions, but inhibited by magnesium ions. Treatment with the proteolytic enzyme trypsin splits the myosin molecule into two sections known as light meromyosin and heavy meromyosin; of these, only heavy meromyosin acts as an ATPase (Mihaly and Szent-Györgi, 1953).

Isolated *actin* exists in two forms: G-actin, a more or less globular molecule of molecular weight about 42 000, and F-actin, a fibrous protein which is a polymer of G-actin. Neither form has any ATPase activity.

If solutions of actin and myosin are mixed, a great increase in viscosity occurs, due to the formation of a complex called *actomyosin*. Actomyosin is an ATPase but, unlike myosin ATPase, it is activated by magnesium ions. 'Pure' actomyosin (a mixture of purified actin and purified myosin) will split ATP in the absence of calcium ions, but 'natural' actomyosin (an actomyosin-like complex which can be extracted from minced muscle with strong salt solutions) can only split ATP if there is a low concentration of calcium ions present. In the absence of calcium ions, addition of ATP to a solution of natural actomyosin results in a decrease in viscosity, suggesting that the actin–myosin complex becomes dissociated.

These properties of actomyosin solutions can be paralleled by the properties of glycerol-extracted muscle fibres, which are prepared by soaking a muscle in cold 50% glycerol for a period of weeks (Szent-Györgyi, 1949). Such treatment removes most of the sarcoplasmic material from the fibres, leaving only the contractile structures. In the absence of ATP the fibres do not contract, but cannot be extended without exerting considerable force; they are said to be in *rigor*. This stiffness is caused by the formation of cross-linkages between the actin and myosin, and corresponds to the formation of actomyosin when solutions of actin and myosin are mixed. On addition of ATP in the presence of magnesium ions the fibres become readily extensible. This 'plasticizing' action of ATP is a result of the breakage of the cross-links, and corresponds to the dissociation of actomyosin in solution by ATP. Finally, if calcium ions are added to the glycerinated fibres in the presence of magnesium and ATP, ATP is split and the fibre contracts; this corresponds to the calcium-activated splitting of ATP by 'natural' actomyosin.

Structure

The striation pattern of myofibrils, as seen by conventional light microscopy after fixing and staining, or by phase-contrast, polarized light or interference microscopy, is shown in fig. 12.1*a*. The two main bands are the dark, strongly birefringent *A* band and the lighter, less birefringent *I* band. These bands alternate along the length of the myofibril. In the middle of each *I* band is a dark line, the *Z* line. In the middle of the *A* band is a lighter region, the *H* zone, which is sometimes bisected by a darker line, the *M* line.* The unit of length between two *Z* lines is called the *sarcomere*.

The structural basis of the myofibrillar striation pattern was obscure until the advent of the techniques of electron microscopy and thin sectioning, and their application to muscle by H. E. Huxley and Jean Hanson (Huxley, 1953*a*, 1957; Hanson and Huxley, 1953, 1955). They found that

* The origins of these letters used in the description of the striation pattern are seen in the original names: *I*sotropisch, *A*nisotropisch, *Z*wischenscheibe, *H*ensen's disc, and *M*ittlemembran. These names are now no longer used.

Fig. 12.1. The striation pattern of a myofibril (*a*), and its structural basis; interdigitating arrays of thick and thin filaments (*b*).

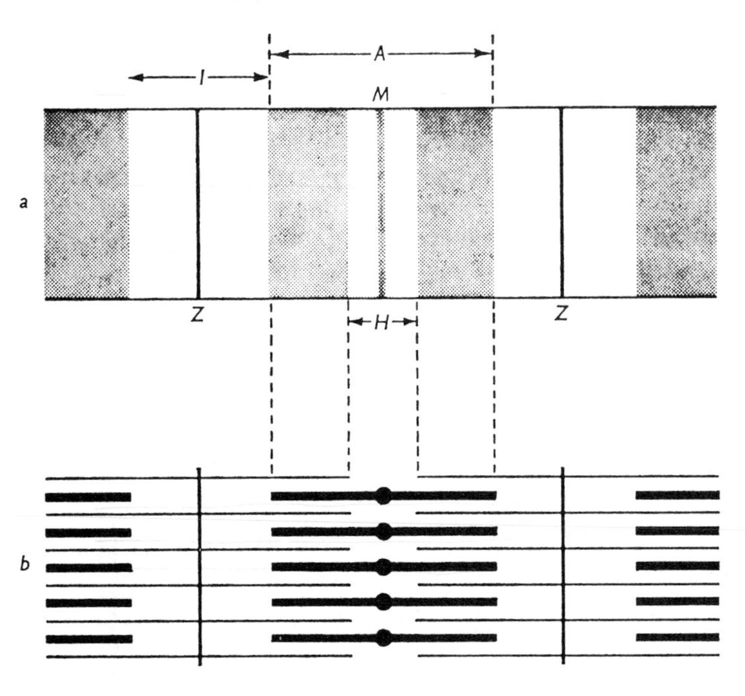

the myofibrils are composed of two interdigitating sets of filaments, about 50 and 110 Å in diameter. The thin filaments are attached to the *Z* lines and extend through the *I* bands into the *A* bands. The position of the thick filaments is coincident with that of the *A* band. The *H* zone is that region of the *A* band between the ends of the two sets of thin filaments, and the *M* line is caused by cross-links between the thick filaments in the middle of the sarcomere. Because of their positions, the thick filaments are sometimes called *A* filaments and the thin filaments are called *I* filaments. This arrangement can be seen in the electron micrographs shown in figs. 12.2 and 12.3, and, diagramatically, in fig. 12.1*b*.

As we have seen, the major part of the myofibrillar material consists of two proteins, *actin* and *myosin*, and the interaction between these two seems to be the chemical basis of muscular contraction. Consequently, it was desirable to determine the localization of actin and myosin in the myofibrils, and to see if their distribution was related to that of the thick and thin filaments discovered by electron microscopy. The question was investigated by Hanson and Huxley (1953, 1955) using isolated myofibrils from a muscle which had been previously extracted with glycerol; glycerol extraction removes most of the sarcoplasmic material from rabbit psoas muscles, leaving only the contractile structures. The appearance of such a myofibril, viewed by phase-contrast microscopy, is shown in fig. 12.4*a*. After treatment with a 0.6 M solution of potassium chloride containing some pyrophosphate and a little magnesium chloride, the dark material of the *A* bands disappeared, as is shown in fig. 12.4*b*. This solution had previously been used to extract myosin from minced muscle, and it was therefore concluded that the *A* filaments are composed of myosin. The myosin-extracted fibre was then treated with a 0.6 M potassium iodide solution, which was known to extract actin from muscle. This removed the substance of the *I* bands (fig. 12.4*c*), showing that the *I* filaments are composed of actin. It would seem that the *Z* lines, which were not affected by the extraction of myosin and actin, are composed of some other substance.

Transverse sections through the *A* band in the region of overlap show that the filaments are arranged in a hexagonal array so that each myosin filament is surrounded by six actin filaments, and each actin filament is surrounded by three myosin filaments (fig. 12.5). Hence there are twice as many actin filaments as there are myosin filaments. Cross-sections through the *H* zone show only myosin filaments, and cross-sections through the *I* band show only actin filaments. Previous to these

Fig. 12.2. Thick longitudinal section of a frog sartorius muscle fibre, showing the striation pattern as seen by electron microscopy. Magnification 28 000 times. (Photograph kindly supplied by Dr H. E. Huxley.)

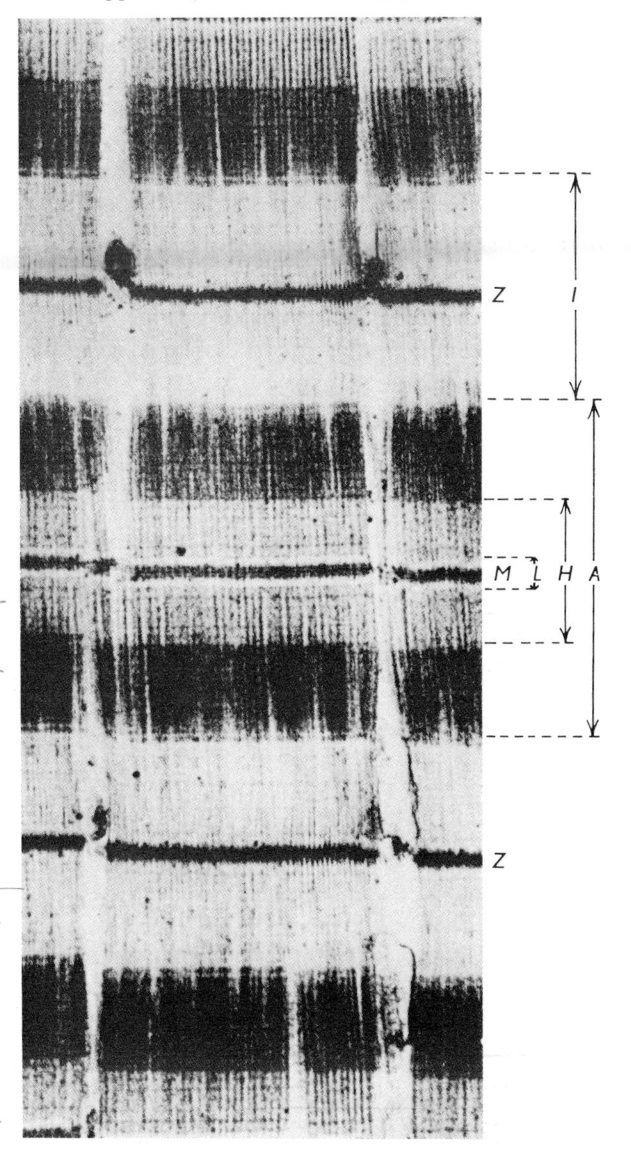

electron microscope studies, H. E. Huxley (1953*b*) had obtained evidence for a hexagonal array of two types of filaments from equatorial measurements of low-angle X-ray diffraction patterns. Taking his results in conjunction with the electron micrographs, it was deduced that the centre-to-centre distance between each myosin filament in the array was about 450 Å at rest length (this distance increases at shorter lengths and becomes less if the muscle is stretched). In electron micrographs the distance may appear to be less than this, as a result of the shrinkage which occurs during fixation and embedding.

High-magnification electron micrographs of glycerol-extracted muscles (fig. 12.3) show that the thick filaments are covered with projections (H. E. Huxley, 1957). In the ovelap region, these projections may be joined to the thin filaments, and so they are known as *cross-bridges*. There are no

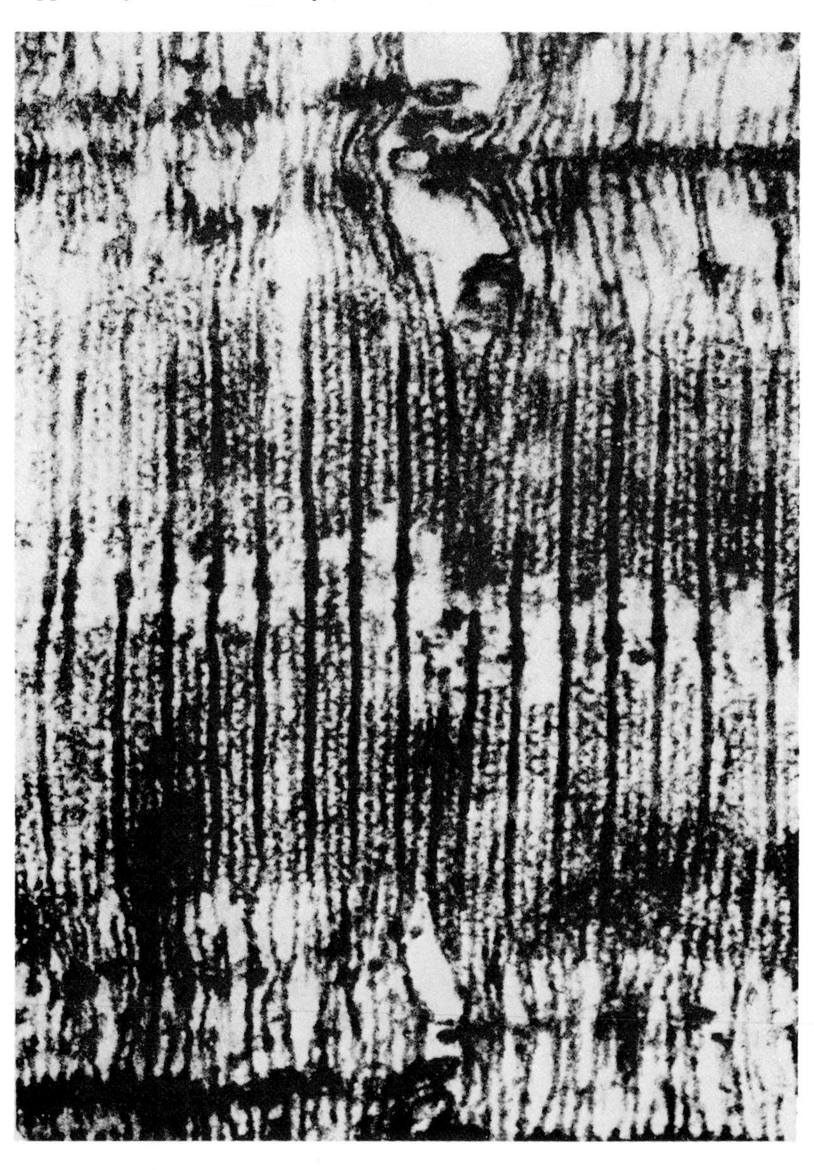

Fig. 12.3. Thin longitudinal section of a glycerol-extracted rabbit psoas muscle fibre. Notice, particularly, the cross-bridges between the thick and thin filaments. (Photograph kindly supplied by Dr H. E. Huxley.)

projections in the very middle of the filaments; this produces a light region, variously known as the '*L* zone', 'pseudo *H* zone', or '*M* region', about 0.15 μm long, in the middle of the sarcomere (fig. 12.2).

The sliding filament theory

Prior to 1954, most suggestions as to the mechanism of muscular contraction involved the coiling and contraction of long protein molecules, rather like the shortening of a helical spring. In that year, the *sliding filament theory* was independently formulated by H. E. Huxley and J. Hanson (from phase-contrast observations on glycerinated myofibrils) and by A. F. Huxley and R. Niedergerke (using interference microscopy of living muscle fibres). In each case the authors showed that the *A* band does not change in length either when the muscle is stretched or when it shortens actively or passively. This observation, interpreted in terms of the interdigitating filament structure described in

Fig. 12.4. Diagram showing the appearance of a stretched myofibril from glycerol-extracted rabbit muscle, as viewed by phase-contrast microscopy. *a*: Before treatment; *b*: after extraction of myosin; *c*: after extraction of actin. (Drawn from a photograph in Hanson and Huxley, 1955.)

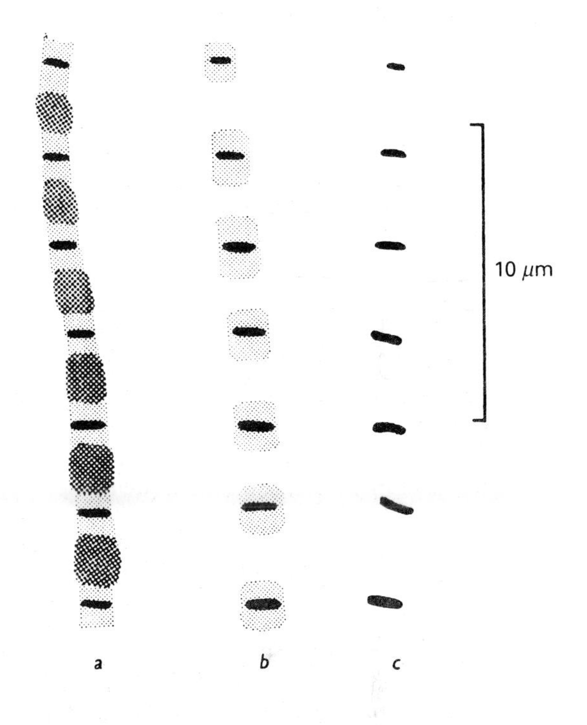

Fig. 12.5. The filament array in vertebrate striated muscle. *a* shows a longitudinal view, *b* to *e* show transverse sections at various levels: *b*, in the *I* band near to the *Z* line; *c*, in the overlap region of the *A* band; *d*, in the *H* zone; and *e*, at the *M* line. The transverse sections show the thick (myosin, M) and thin (actin, A) filaments and, in *e*, the *M* filaments (mf). (From Squire, 1981.)

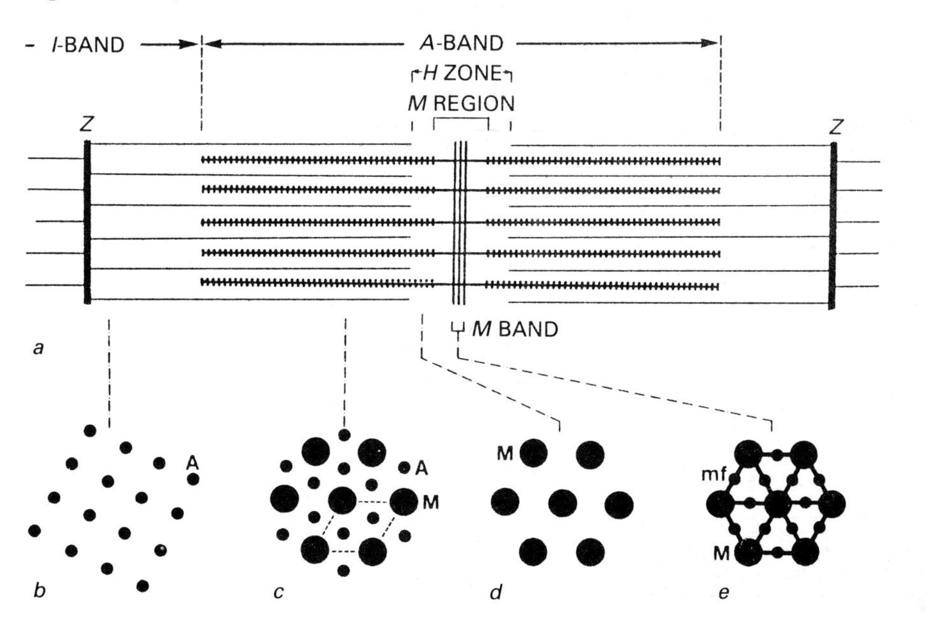

the previous section, suggests that contraction involves sliding of the *I* filaments between the *A* filaments, with the lengths of both sets of filaments remaining unchanged (fig. 12.6).

What makes the filaments slide past each other? Various proposals have been made in answer to this question (see A. F. Huxley, 1974, 1980). The most generally accepted view is that sliding is caused by a series of cyclic reactions between the projections on the myosin filaments and the active sites on the actin filaments. Each projection is first attached to the actin filament to form a cross-bridge, then it moves or pulls on the actin filament, and finally it lets go, moving back to attach to another site further along the actin filament. The cross-bridges thus act as 'independent force generators', to use A. F. Huxley's useful phrase.

Let us have a look at some of the evidence for this theory. There are two points to be established: firstly, that the filaments remain constant in length during shortening or contraction of the muscle, and secondly that independent force generators provide the motive force for the sliding movement.

The lengths of the A *and* I *filaments*

We have met a number of examples of the use of electron microscopy in previous chapters, but it may be as well to consider at this point some of the difficulties encountered in making accurate measurements with the technique. The study of the structure of cells and tissues usually involves the preparation of thin sections of the material. The tissue is first fixed in a suitable fixative such as a solution of osmium tetroxide or glutaraldehyde. It is then dehydrated in alcohol and embedded in a substance such as methacrylate ('Perspex') or various epoxy resins (such as 'Araldite') so as to form a relatively hard block. This block is then sectioned on a special microtome to provide sections of the order of 0.02–0.05 μm thick.

The electron microscope can only produce an image of an object if parts of the object are 'electron dense', i.e. if they are opaque to electrons. In practice, this means that the tissue has to be 'stained' by depositing heavy metal atoms upon it; some suitable stains are osmium textroxide, potassium permanganate, lead hydroxide and uranyl acetate. In the final electron micrograph, therefore, one observes the distribution of heavy metal atoms on a thin section of tissue which has been fixed, dehydrated and embedded. Hence it is as well to check that the structures seen after any particular treatment are compatible with those seen after other procedures (use of a different fixative, for example) or deduced from the use of other techniques, such as light microscopy or X-ray diffraction.

The preparatory procedures used in electron microscopy frequently result in shrinkage of the tissue involved. Hence it is no easy matter to determine the lengths of the *A* and *I* filaments in living or glycerinated muscles. However, Page and Huxley (1963) investigated this problem in a piece of work that is a most beautiful example of mensurative electron microscopy. They began by showing that the sarcomere length of unrestrained muscles (measured by an optical diffraction technique) decreased during the fixation and dehydration processes, but was unaffected by the embedding process. These effects could be eliminated if the muscle was fixed at each end, so this procedure was followed in the rest of their experiments.

By homogenizing glycerinated fibres in a suitable medium, it was possible to obtain '*I* segments' consisting of a *Z* line and a set of *I* filaments. These were examined by negative staining and shadow-casting; both of these techniques can be used without fixation or dehydration in alcohol,

Fig. 12.6. The structural changes in a sarcomere on shortening, according to the sliding filament theory.

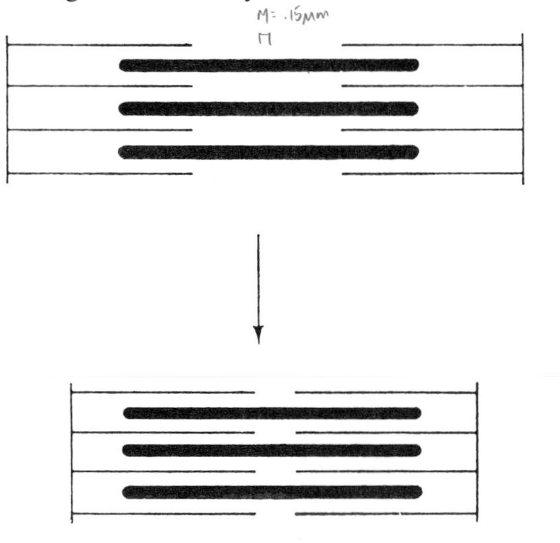

and so shrinkage from these causes could be eliminated. Shadowed segments and those stained with sodium phosphotungstate were 2.05 μm long; those stained with uranyl acetate were slightly shorter. These results on their own can only give a rather tentative estimate of the *I* filament length; but the value so obtained must be 2.05 μm.*

Particular care was used in the electron microscopy of longitudinal sections of muscles: the blocks were cut at right angles to the longitudinal axis so as not to shorten the filaments by the pressure of the microtome knife, the effectiveness of this procedure was checked by measuring the width of the sections and the width of the block, the magnification of the electron microscope was calibrated at frequent intervals, and allowances were made for the shrinkage of the photographic prints.

Muscles were fixed in osmium tetroxide when contracting isometrically under the influence of electrical stimulation or solutions containing a high potassium ion concentration. The filament lengths were 2.01–2.05 μm for the *I* filaments (except at a sarcomere length of 3.7 μm, when the average *I* filament length was 1.98 μm), and 1.56–1.61 μm for the *A* filaments (fig. 12.7). These results indicate that very little shrinkage of the filaments occurs during osmium fixation of active muscles, in spite of the fact that appreciable shrinkage occurs during osmium fixation of resting muscles. Page and Huxley suggested that the reason for this is that the *A* and *I* filaments are cross-linked during activity, whereas in the resting condition they are able to shrink by sliding past each other. At a sarcomere length of 3.7 μm, the filaments will not be overlapping, therefore no cross-bridges can be formed, which accounts for the shrinkage of the *I* filaments at this length.

The conclusion from this set of experiments, illustrated diagrammatically in fig. 12.7, is that the lengths of the filaments do not change in the resting or active muscles, whatever the length of the sarcomere. In frog striated muscles the *I* filaments are 2.05 μm long and the *A* filaments are 1.6 μm long.

* The term '*I* filament length' is here used to indicate the total length of *I* substance on each side of a *Z* line, including the thickness of the *Z* line, not the length of *I* substance on only one side of a *Z* line.

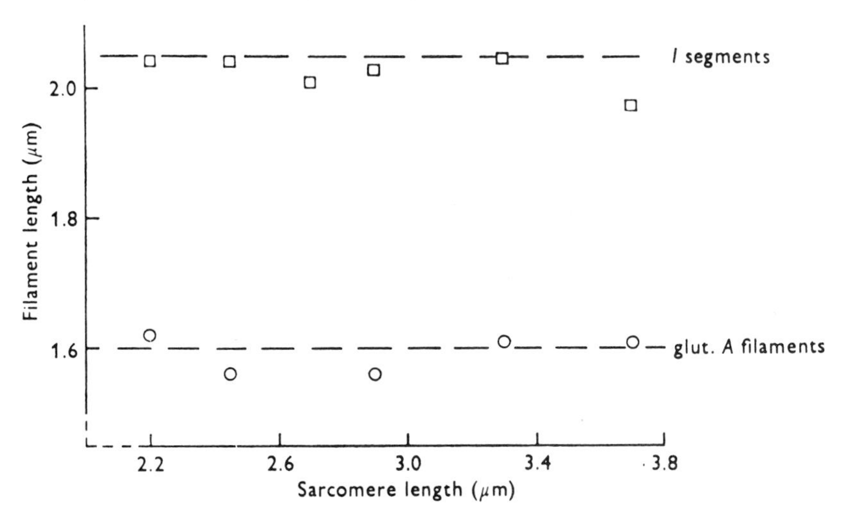

Fig. 12.7. The lengths of the *A* filaments (circles) and *I* filaments (squares) in frog muscle fixed in osmium tetroxide during isometric tetani at various sarcomere lengths. The broken lines show the lengths of isolated *I* segments and the lengths of the *A* filaments in glutaraldehyde-fixed muscles. (From Page, 1964, by permission of the Royal Society.)

X-ray diffraction measurements

X-rays are a type of electromagnetic radiation; they differ from visible light in that their wavelength is extremely short, being about 1 Å. The resolution of an optical microscope is ultimately limited by the wavelength of the light which it uses. Thus if we could make a microscope which used X-rays instead of visible light, we should be able to examine very small structures indeed. This cannot be done, however, since it is not possible to focus X-rays. But the light in an optical microscope passes through two stages between the object and its image; first it is scattered by the object, then it is focused by the lenses of the microscope. X-rays are also scattered by objects. The essence of the X-ray diffraction technique is that the pattern produced by this scattering is interpreted by mathematical means so as to indicate what a focused image would look like. In practice, it is only possible to do this if we possess certain information about the X-rays and if the scattering pattern shows a sufficient degree of regularity. A regular pattern is produced only if there are corresponding regularities in the object, such as are produced by the regular packing of atoms and molecules in crystals or fibres. It would be inappropriate here to go any further into the theory of X-ray diffraction methods; the interested reader should consult, for example, the book by Wilson (1966).

The arrangement for an examination of a muscle fibre by the X-ray diffraction technique is shown schematically in fig. 12.8. If we were to draw a line on the photographic plate parallel to the fibre axis and passing through the point at which the undiffracted X-ray beam hits the plate, such a line would lie along the *meridian* of the pattern. A similar line passing through the undiffracted beam but at right angles to the fibre axis would lie along the *equator* of the pattern. Spots or lines produced on the plate by diffracted X-rays are called 'reflections'. Reflections lying on the meridian are produced by regularly repeating structures spaced along the fibre axis. Reflections lying on the equator are produced by regularly repeating structures in the transverse plane of the fibre. Structures which repeat in directions other than axially or radially (as, for example, in helically arranged units) produce 'off-meridional reflections', which may be arranged in a series of lines, parallel to the equator, known as 'layer lines'.

Each set of repeating structures produces a series of reflections on the photographic plate at regular distances from the line of the original X-ray beam, and their intensity usually decreases at increasing distances from this line. The first member of a series of this type (i.e. that nearest to the centre of the pattern) is known as a first-order reflection, the second is known as a second-order reflection, and so on. The distance between reflections of successive orders (which is equal to the distance between the first-order reflection and the centre of the pattern) is inversely related to the distance between the repeating units in the fibre. Hence low-angle patterns (patterns in which the emerging rays do not diverge much from their original direction) give information about structures repeating at relatively long distances, such as might be seen with the electron microscope, and wide-angle patterns give information about repeating structures which are closer together.

Wide-angle X-ray diffraction measurements on muscle show the α-helix pattern, irrespective of

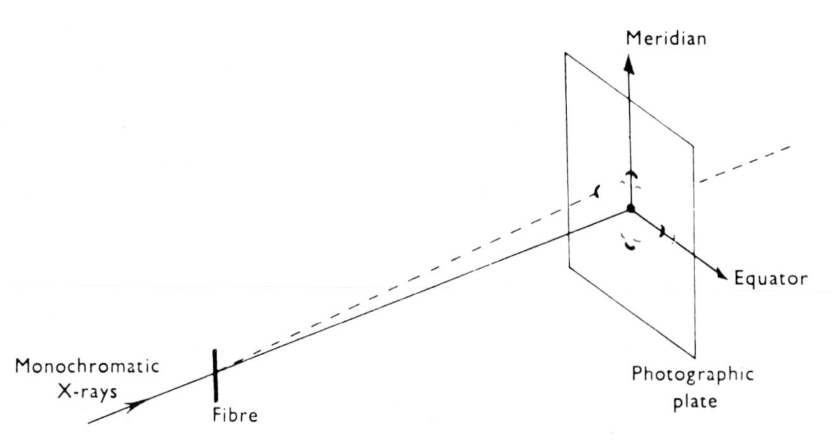

Fig. 12.8. Schematic diagram to show the arrangement for making low-angle X-ray diffraction observations on fibres. The dotted line shows one of the many diffracted rays, producing a reflection where it hits the photographic plate.

the length of the muscle (see H. E. Huxley, 1960). This α-helix pattern appears to be mainly derived from myosin, and its constancy at different lengths implies that there is little change in the lengths of the myosin molecules during stretching or contraction. This observation is in accordance with the sliding filament theory, although it does not provide any strong evidence in its favour.

Much more conclusive evidence is provided by low-angle X-ray diffraction measurements of meridional reflections, which indicate the distances between repeating units along the axis of the myofibrillar filaments. H. E. Huxley (1953*b*) showed that such measurements on resting muscles were independent of the length of the muscle. This is just what one would expect from the sliding filament theory; if, on the other hand, the filaments did shorten during shortening of the muscle, we would expect the distance between their repeating units to decrease.

It is obviously desirable that this conclusion should be confirmed for actively contracting muscles. This was done by two groups of workers, in London (Elliott *et al.*, 1965) and in Cambridge (H. E. Huxley *et al.*, 1965). The chief technical difficulty of this type of experiment is that the exposure times for X-ray diffraction experiments may have to be a matter of hours, whereas it is impossible to excite a vertebrate striated muscle for more than a few seconds at a time without fatiguing it. The problem was overcome by stimulating the muscle at intervals and, using refined X-ray diffraction methods, only passing the X-ray beam through the muscle when the isometric tension was above a certain level. In each case it was found that the axial spacings during isometric contractions were not significantly different from those in the resting muscle and were unaffected by the length of the muscle.

More precise measurements by Huxley and Brown (1967) showed that there is in fact a slight *increase* in the axial spacings derived from the myosin filaments during an isometric contraction. Their measurements gave the value of the principal myosin subunit spacing as 143 ± 0.1 Å in the resting muscle and 144.6 ± 0.1 Å during contraction, an increase of just over 1%. There was no significant change in the actin subunit spacing during contraction.

Do the cross-bridges move?

The sliding filament theory postulates that the contractile force is mediated via the cross-bridges originating on the myosin filaments. This suggests that the cross-bridges should move during the contraction, and there is some evidence that this is so. Reedy, Holmes and Tregear (1965) found that glycerol-extracted insect flight muscles produced different X-ray diffraction and electron micrograph patterns according to whether they were in rigor (p. 248) or were relaxed. They concluded that the projections from the myosin filaments in resting muscle stick out at right angles from the myosin filaments, but that in rigor they become attached to the actin filaments and move through an angle of about 45° so as to pull the actin filaments towards the centre of the sarcomere (fig. 12.9).

Changes in the orientation of the cross-bridges in living muscles during isometric contractions are apparent from the X-ray diffraction measurements of Huxley and Brown. This conclusion is based on the observation that, during contraction, the intensity of the off-meridional reflection at 429 Å is only about 30% of that in resting muscle (fig. 12.10), whereas the intensity of the meridional reflection at 143 Å is about 66% of that in resting muscle. This implies that the helically repeating structures on the myosin filament becomes much less regularly arranged than do the axially repeating structures, and therefore that the outer portions of the myosin reflections move (in an asynchronous fashion) with respect to their bases.

Further evidence for cross-bridge movement during contraction comes from X-ray diffraction measurements on insect asynchronous (or 'fibrillar') muscle (Tregear and Miller, 1969). The physiological properties of these muscles are dealt with in chapter 14, suffice it to say here that their glycerinated fibres will perform oscillatory work when subjected to sinusoidal length changes at the right frequencies in the presence of ATP and calcium ions. Tregear and Miller measured the fluctuations in the intensity of the 145 Å meridional reflection during such oscillations at different frequencies by means of a rather ingenious technique. A metal plate was placed in the X-ray beam, with a rectangular aperture positioned so as to let through only the diffracted rays comprising the 145 Å meridional reflection. Behind the aperture was a proportional counter to measure the X-rays passing through it, and the output of this counter was recorded on an eight-channel scaler which scanned synchronously with the length oscillation. In this way the results shown in fig. 12.11 were obtained.

At low frequencies, the diffraction intensity of the spot varied in time with the variations in tension, being low when tension was high and high when it was low, as is shown in fig. 12.11. A diminution in the intensity of a meridional reflection must be produced by a decrease in the degree of axial order of structures occurring every 145 Å in the muscle. One way in which this might occur is by axial displacements of the mass of each cross-bridge, such as would be produced by bending.

Tregear and Miller's experiments provided some degree of time resolution in measuring the intensity of particular X-ray reflections, by utilizing the oscillatory properties of glycerinated insect asynchronous flight muscle. Much more precise measurements have since been made on frog muscles by H. E. Huxley and his colleagues. These were made possible by using the intense X-ray sources available from high-energy electron accelerators such as the electron-positron storage ring DORIS near Hamburg, coupled with the development of sophisticated electronic position-sensitive X-ray detectors (see Huxley and Faruqi, 1983). Such techniques have made it possible to measure X-ray reflection intensities during the course of a contraction for time periods as short as 1 ms in some cases.

Figs. 12.12 and 12.13 show time-resolved observations on frog sartorius muscle twitches (Huxley *et al.*, 1982). There is a marked fall in the intensity of the 429 Å off-meridional layer-line during the twitch. The time course of the change closely parallels that of the rise and fall in tension, preceding it by a few milliseconds. This is inter-

Fig. 12.9. Diagram, derived from X-ray diffraction and electron microscopy measurements, to show the positions of the cross-bridges of glycerinated insect flight muscle in the relaxed state (*b*) and in rigor (*a*). (From Reedy *et al.*, 1965.)

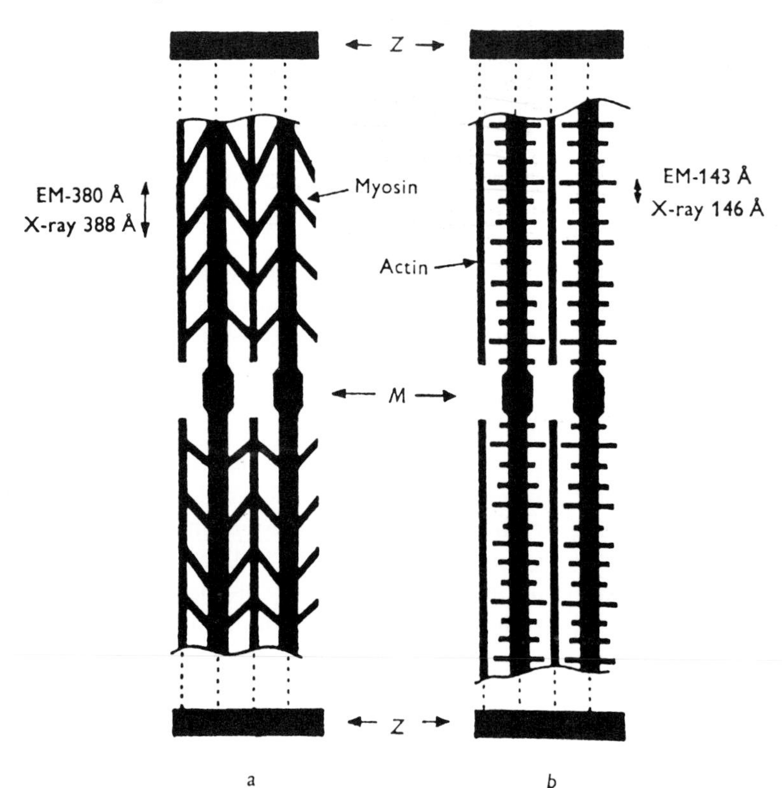

preted as a disordering of the helical arrangement of the cross-bridges as they become attached to the actin filaments and pull on them.

The relation between sarcomere length and isometric tension

If the sliding filament theory is correct, and the cross-bridges act as independent force generators during contraction, then the isometric tension should be proportional to the degree of overlap of the filaments. In terms of fig. 11.5, the decline in the active increment curve at longer lengths should be caused by the reduced degree of overlap (and therefore smaller number of cross-bridges) between the two types of filaments as the muscle is extended (A. F. Huxley and Niedergerke, 1954). The way to test this suggestion is to measure the active increment length–tension curve, with the abscissa as sarcomere length rather than the length of the whole muscle, and compare it with the lengths of the *A* and *I* filaments. However, this turns out to be not so easy as one might think. A. F. Huxley and Peachey (1961) found that the sarcomere length in single frog muscle fibres was not constant along the length of the fibre, being shorter at the ends. At long lengths there may be overlap of the filaments in the sarcomeres at the ends of the fibre, but not in those in the middle, so that the ends contract at the

Fig. 12.10. X-ray diffraction patterns of living frog sartorius muscle at rest (*a*) and during isometric contractions (*b*). *c* is a schematic diagram showing the reflections apparent in *a*. Notice the disappearance of the system of layer lines based on 429 Å during contraction. The 72 Å and 143 Å meridional reflections remain (the latter is not evident in this reproduction), as does the 59 Å actin reflection. (*a* and *b* from Huxley, 1972.)

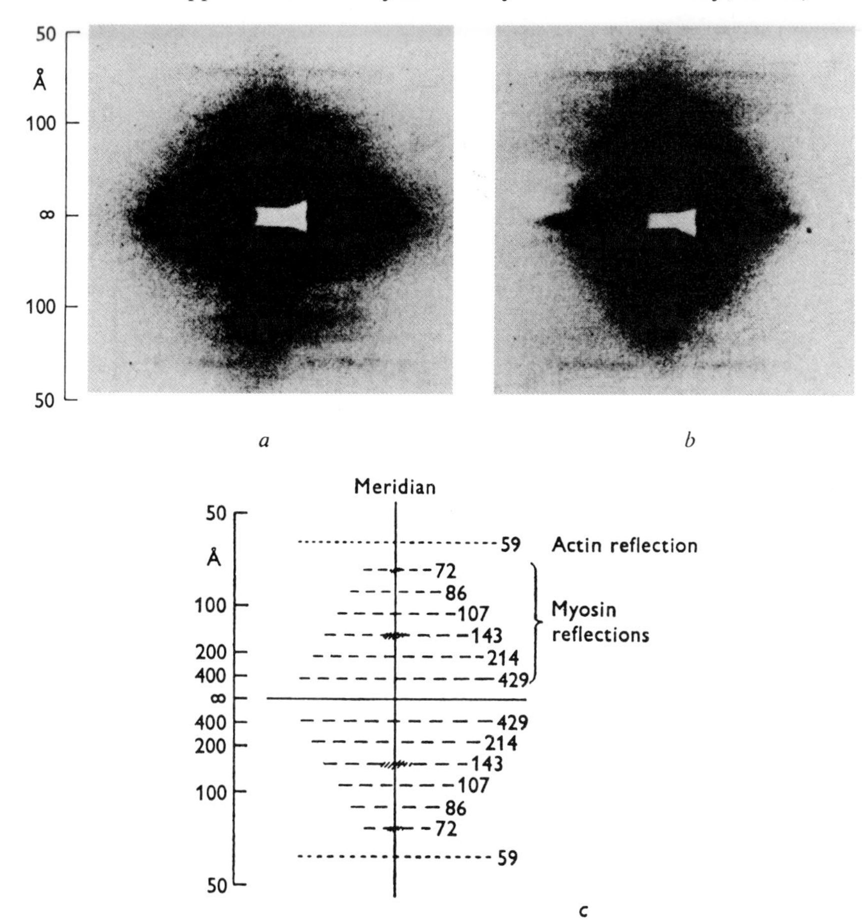

Fig. 12.11. X-ray diffraction measurements on a bundle of glycerol-extracted flight muscle fibres from a giant water bug, *Lethocerus cordofanus*, showing changes in intensity (black circles and full line) of the 145 Å meridional reflection during sinusoidal oscillation at 4.8 Hz in the presence of ATP and calcium ions. The dashed line shows the tension change. (From Tregear and Miller, 1969.)

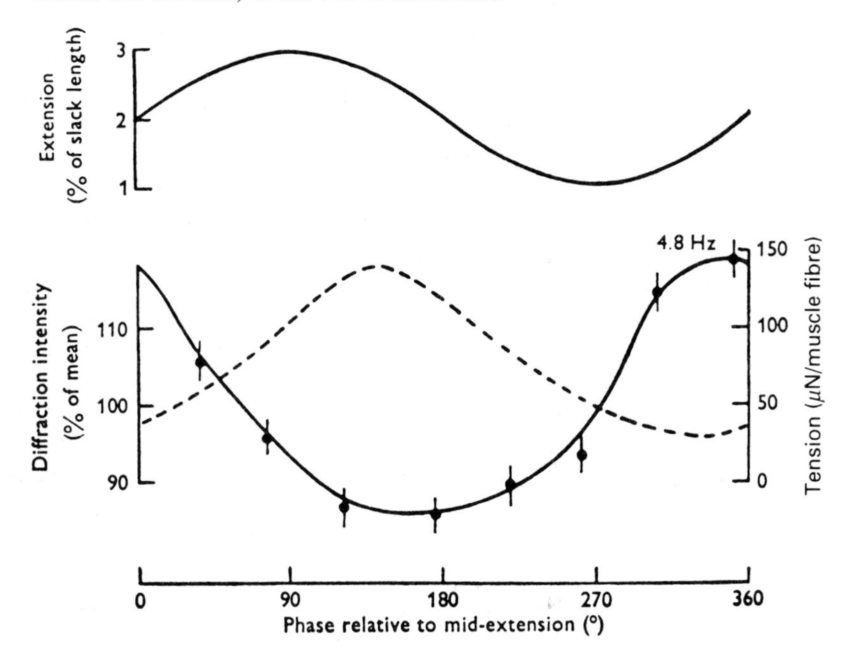

Fig. 12.12. Intensity of the 429 Å off-meridional layer line during an isometric contraction as measured by time-resolved X-ray diffraction using synchrotron radiation. The upper curve (squares) shows the intensity of the layer line, measured as counts per five milliseconds in single twitches at 10 °C, the lower curve (asterisks) shows the corresponding isometric tension. The stimulus was at 42.5 ms. (From Huxley *et al.*, 1982.)

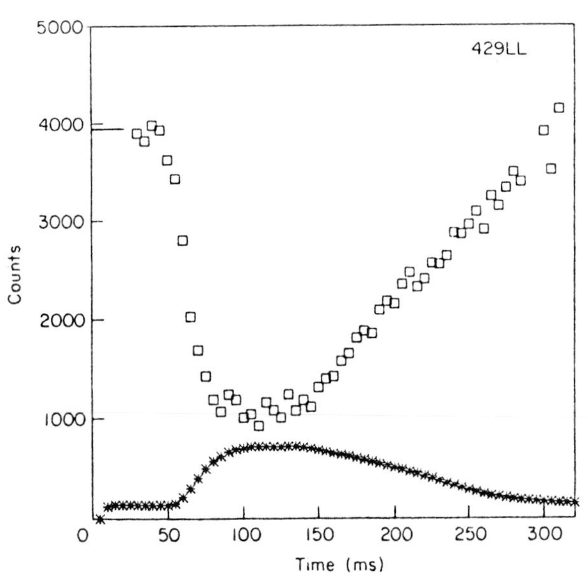

Fig. 12.13. The data of fig. 12.12 plotted as a percentage of the maximum change. Notice that the change in layer line intensity (squares) preceeds the tension change (asterisks) by about 10 ms at the start of the contraction. (From Huxley *et al.*, 1982.)

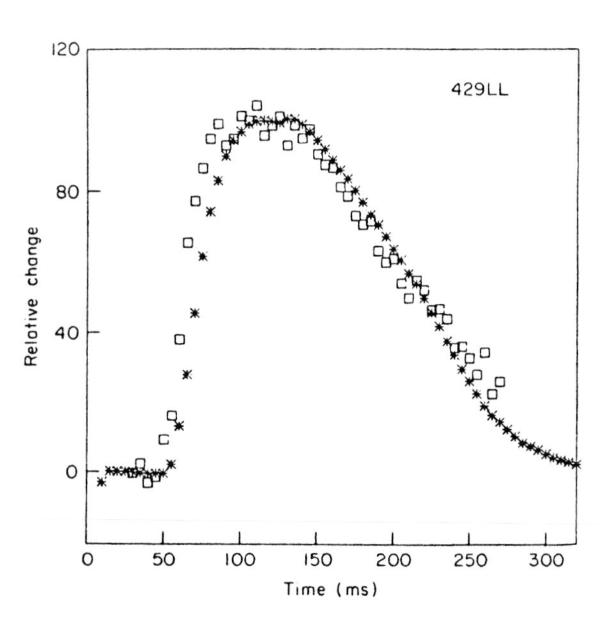

expense of the middle section if the fibre is fixed at each end. In order to overcome this difficulty, it is necessary to hold a small portion of the fibre (in which the sarcomere length *is* constant) at a constant length while its tension is measured. This has been done by A. F. Huxley and his colleagues (Gordon *et al.*, 1966*a, b*), whose work we shall now consider.

Before examining the apparatus used in these experiments, let us consider a useful electronic feedback device known as a 'spot follower' (Fig. 12.14). This consists of a cathode-ray tube, a photocell, two lenses, and a movable vane placed between the lenses. The object of the system is to ensure that the movement of the spot follows exactly the vertical movement of the edge of the vane, however irregular this may be. This is achieved by feeding the output of the photocell into an amplifier connected to the *Y* plates of the cathode-ray tube. The system is arranged so that if the photocell can 'see' more than half the spot, the spot is raised, and if it can see less than half the spot, it is lowered (remember that the spot on the screen moves in the opposite direction to its image on the vane). Thus the movement of the spot follows (inversely) the movement of the edge of the vane. It follows that the output of the photocell, since it is this that directly determines the position of the spot, must be proportional to the position of the edge of the vane.

Fig. 12.14. A simple 'spot follower' circuit. Explanation in text.

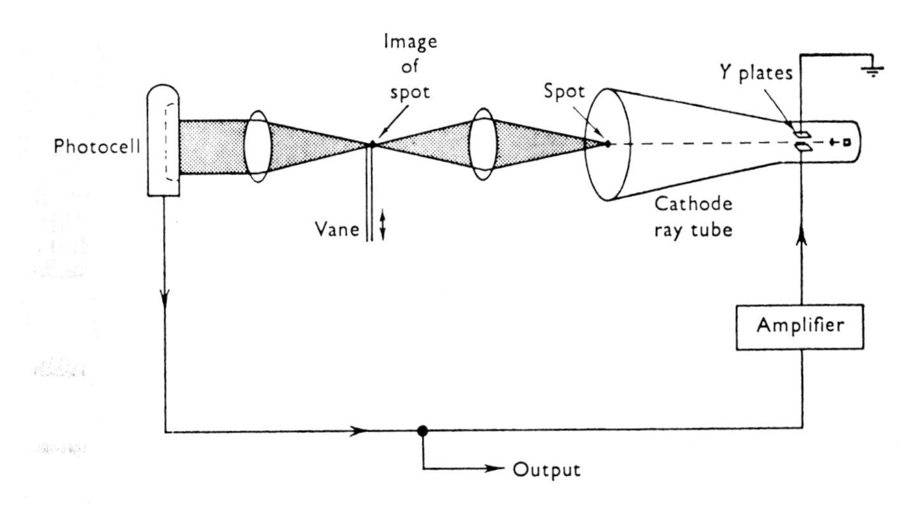

Now let us examine some of the details of the apparatus used by Gordon *et al.* (fig. 12.15). A single muscle fibre is used, mounted on a microscope stage, and stimulated electrically. It is connected via its tendons to a tension transducer valve at one end and to the arm of a galvanometer at the other. The galvanometer arm moves according to the force exerted on it by the muscle fibre and the current in its coil (the usual zeroing spring is not present). Two small pieces of gold leaf are attached to the fibre with grease, to act as markers. The position of these markers is chosen so that the sarcomere length (observed through the microscope) does not vary along the length of the fibre between them. Below the microscope is a double-beam cathode-ray tube, mounted vertically. The substage lens is positioned so that images of the two spots on the cathode ray tube screen can be focused on an edge of each marker. Light from these images is then collected separately onto two photocells.

The output of the right-hand photocell is fed to the amplifiers controlling the position of both spots. Thus if both markers move to the right or left by the same amount, the follower system works so that both spots move correspondingly, and there is therefore no change in the output of the left-hand photocell. But if the distance between the markers alters, the left-hand photocell (and its spot-follower loop) is activated, and the right-hand spot (which is focused on the left hand marker) moves independently of the left-hand spot. The output of the left-hand photocell then indicates the distance between the markers, shown as the box marked 'length' in fig. 12.15.

Besides the spot-follower feedback loops, there is another feedback loop, indicated as the 'length-regulator loop' in the diagram. The output of the left photocell ('length') is fed to the main amplifier, which drives the galvanometer. This circuit is arranged so that the galvanometer moves so as to bring the output of the photocell back to its mid-position, i.e. so as to keep the 'length' (the distance between the markers on the muscle fibre) constant. This, of course, is just what the apparatus is required to do. The length can be set at different values by altering the 'length input signal', since the input to the main amplifier is the difference between this and the 'length' signal.*

* This account is rather simplified as also is fig. 12.15. The apparatus also has facilities for (i) a tension-regulator loop so that the tension can be held constant while the length changes, (ii) switching between length regulation and tension regulation according to the magnitude of either of these parameters, and (iii) forcing either length or tension to follow an input command signal.

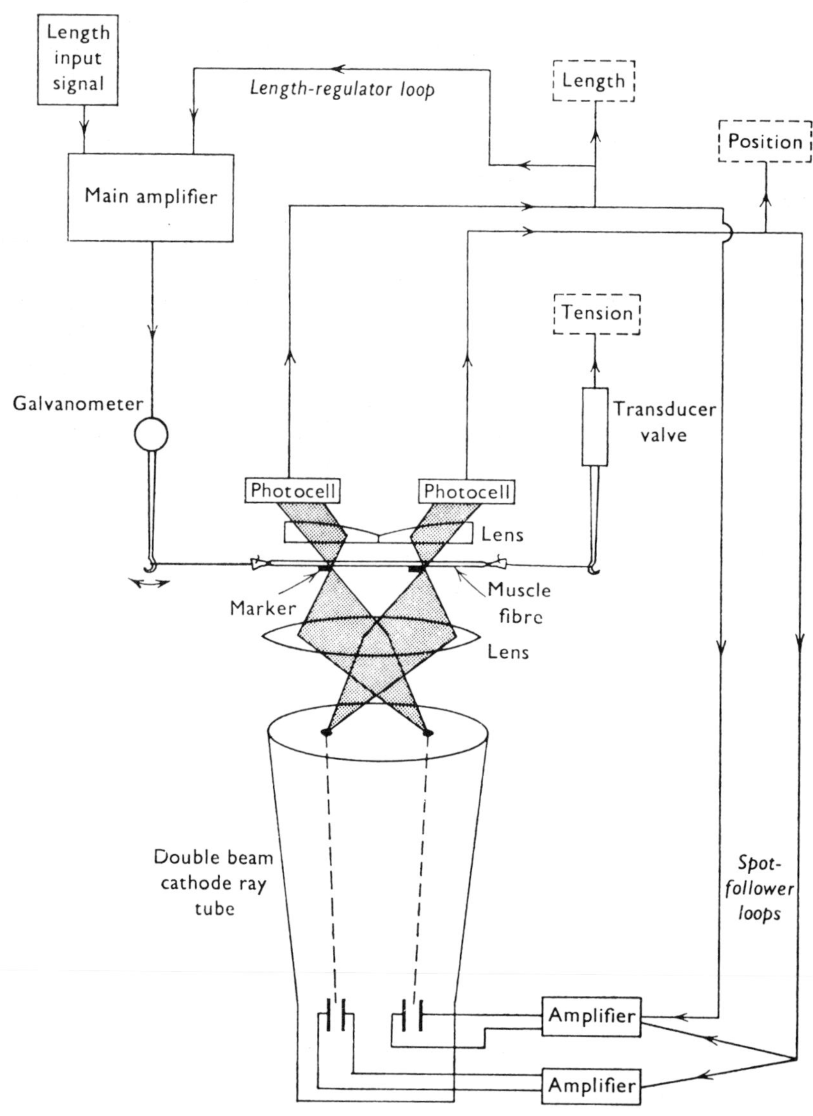

Fig. 12.15. Apparatus used for experiments on the mechanics of portions of a muscle fibre. Simplified; explanation in text. (Based on Gordon *et al.*, 1966*a*.)

A single experiment with this apparatus consists of a series of isometric tetani obtained at different sarcomere lengths; for technical reasons it is not possible to measure isometric tension over the whole range of sarcomere lengths in any one experiment. The results of these experiments are summarized in fig. 12.16. It is evident that the length–tension diagram consists of a series of straight lines connected by short curved regions. There is a 'plateau' of constant tension at sarcomere lengths between 2.05 and 2.2 µm. Above this range, tension falls linearly with increasing length; the projected straight line through most of the points in this region passes through zero at 3.65 µm, but there is in fact a very slight development of tension at this point. Below the plateau, tension falls gradually with decreasing length down to about 1.65 µm, then much more steeply, reaching zero at about 1.3 µm.

According to the sliding filament theory, the isometric tension is directly proportional to the number of cross-bridges that can be formed between the *A* and *I* filaments, less any internal resistance tending to extent the sarcomeres. Thus, at long lengths, the tension ought to be proportional to the degree of overlap of the *A* and *I* filaments. In order to see whether the length–tension diagram fits with this prediction we need to know the dimensions of the filaments and the position of the cross-bridge-forming projections on the myosin filaments. From the measurements of Page and Huxley (1963), we know that the *A* filaments are 1.6 µm long (symbol *a* in fig. 12.17) and the *I* filaments, including the *Z* line, are 2.05 µm long (*b*). The middle region of the *A* filaments, which is bare of projections and therefore cannot form cross-bridges, (*c*) is 0.15 to 0.2 µm long and the thickness of the *Z* line (*z*) is about 0.05 µm.

Now let us see if the length–tension diagram shown in fig. 12.16 can be related to these dimensions, starting at long sarcomere lengths and working through to short ones. Fig. 12.17*b* shows the points at which qualitative changes in the relations between the elements of the sarcomere occur. These stages are numbered 1 to 6, and the corresponding lengths are shown by the arrows in fig. 12.16. Above 3.65 µm (1) there should be no cross-bridges, and therefore no tension development. In fact there is some tension development up to about 3.8 µm; the reason for this is not clear. Between 3.65 µm and 2.25–2.2 µm (1 to 2) the number of cross-bridge increases linearly with decrease in length, and therefore the isometric tension should show a similar increase. It does. With further shortening (2 to 3) the number of cross-bridges remains constant and therefore there should be a plateau of constant tension in this region; there is. After stage 3, we might expect there to be some increase in the internal resistance to shortening, as the *I* filaments must now overlap; after stage 4 the *I* filaments from one half of the sarcomere might interfere with the cross-bridge formation between the *A* and *I* filaments in the other half of the sarcomere. These effects would be

Fig. 12.16. The isometric tension (active increment) of a frog muscle fibre, measured as a percentage of its maximum value, at different sarcomere lengths. The numbers 1 to 6 refer to the myofilament positions shown in fig. 12.17*b*. (From Gordon *et al.*, 1966*b*.)

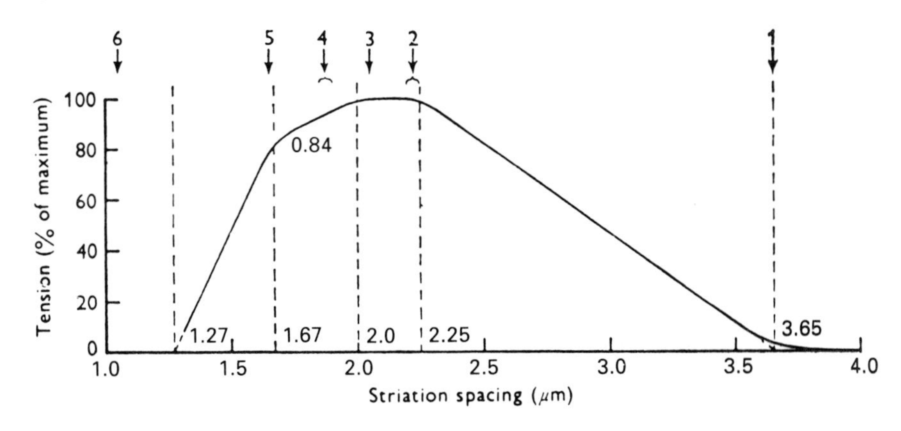

expected to reduce the isometric tension. This does in fact occur, although no extra reduction corresponding to stage 4 is detectable. At 1.65 μm (5), the *A* filaments hit the *Z* lines, and therefore there should be a considerable increase in the resistance to shortening; it is found that there is a distinct kink in the length–tension curve at almost exactly this point, after which the tension falls much more sharply. The curve reaches zero tension at about 1.3 μm, before stage 6 (when the *I* filaments would have hit the *Z* line) is reached.

It would be difficult to find a more precise test of the sliding filament theory than is given by this experiment, and the theory obviously passes the test with flying colours. In addition, the results tell us something about the mode of action of the cross-bridges. The filament array in the myofibrils is a constant-volume system, so the transverse distance between the actin and myosin filaments must decrease as the muscle is stretched. However, the fact that the decrease in isometric tension with increasing sarcomere length above 2.2 μm is linear means that the tension per cross-bridge is independent of length. Therefore the tension produced per cross-bridge is independent of the transverse distance between the filaments. We shall return to this point later (p. 279).

Fig. 12.17. *a*: Myofilament dimensions in frog muscle. *b*: Myofilament arrangements at different lengths; the letters *a, b, c* and *z* refer to the dimensions given in part *a*. The sarcomere lengths corresponding to the positions labelled 1 to 6 are indicated by the arrows in fig. 12.16. (From Gordon *et al.*, 1966*b*.)

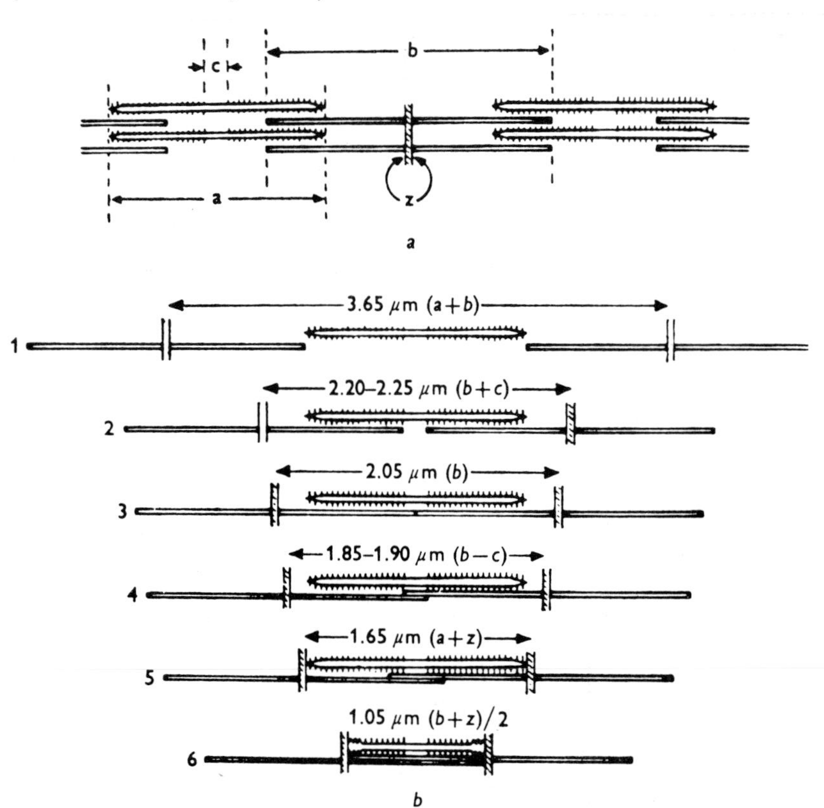

Structure of the contractile machinery

The success of the sliding filament theory has led to further questions. What makes the filaments slide and how is the force produced? The search for answers has involved a great deal of work on the fine detail of myofibrillar structure, including especially studies of the protein molecules and how they fit together to form the thick and thin filaments. Let us have a selective look at some of this work; more detailed accounts are given by Craig (1986), Haselgrove (1983), Huxley (1983),

Table 12.1. *Myofibrillar proteins of vertebrate skeletal muscle*

Protein	Molecular weight ($\times 10^3$)	Content wt%	Localization (band)
Contractile			
Myosin	520	43	*A*
Actin	42	22	*I*
Regulatory			
Troponin	70	5	*I*
Tropomyosin	33 × 2	5	*I*
M-Protein	165	2	*M* line
C-Protein	135	2	*A*
α-Actinin	95 × 2	2	*Z* line
Cytoskeletal			
Titin	2800	10	*A–I*
Nebulin	750	5	*I*

Minor structural proteins less than 1% of the total proteins are omitted.
(Data from Maruyama, 1986.)

Sheterline (1983) and Squire (1981, 1986). Table 12.1 gives a summary of the main proteins of the myofibril.

Myosin

Myosin molecules can be seen by electron microscopy by using the shadow-casting technique (Huxley, 1963; Slayter and Lowey, 1967; Elliott and Offer, 1978). They are tadpole-like structures with a 'tail' about 1500 Å long and two 'heads' at one end (fig. 12.18). Under suitable conditions they are able to aggregate to form filaments, as we shall see later.

The myosin molecule consists of a number of different protein chains. These can be dissociated from one another by treatment with suitable reagents such as disulphide-link reducing agents, alkaline conditions and sodium dodecyl sulphate. The mixture of the resulting complexes (of protein chains to which dodecyl sulphate ions are attached) can then be separated by electrophoresis in a column of polyacrylamide gel; the negatively charged complexes move towards the anode. The gel acts as a three-dimensional net which impedes the passage of large complexes more than that of small ones. This enables the complexes to be separated from one another, and the distance which a particular complex moves is fairly precisely related to its molecular weight: the lighter the protein chain, the further its dodecyl sulphate complex moves. Fig. 12.19 shows an application of the method to muscle proteins.

This technique has been used by a number of workers to investigate the complex nature of the myosin molecule (Weeds and Lowey, 1971; Lowey and Risby, 1971; see also Bagshaw, 1982 and Lowey, 1986). It is found that the molecule consists of two heavy chains (each of molecular weight about 200 000) and four light chains of three different types (fig. 12.20). Two of the light chains have a molecular weight of about 18 000 and are detachable from the whole molecule by DTNB (5,5′-dithiobis-(2-nitrobenzoate), a substance which breaks disulphide linkages in proteins) and

Fig. 12.18. Rotary shadowed myosin molecules from scallop striated muscle. Vertebrate striated myosin molecules are identical in appearance. Magnification 100 000. (Photograph kindly supplied by Dr L.-L. Y. Frado.)

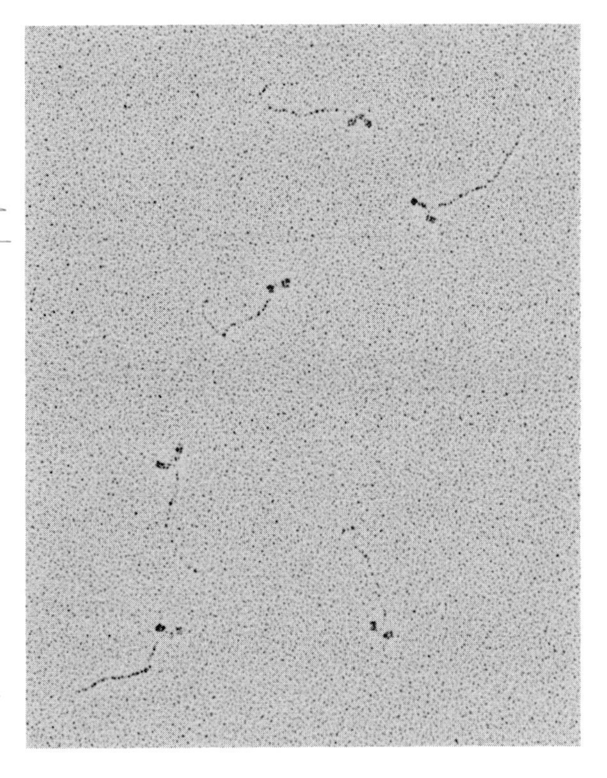

Fig. 12.19. Separation of muscle proteins by gel electrophoresis in the presence of SDS. Proteins are from solubilized rabbit skeletal myofibrils in (*a*), and rabbit fast skeletal myosin in (*b*). The proteins become separated according to their molecular weights, with heavy molecules at the top in the photograph and lighter ones towards the bottom. (Photograph kindly supplied by Dr C. R. Bagshaw, after Bagshaw, 1982.)

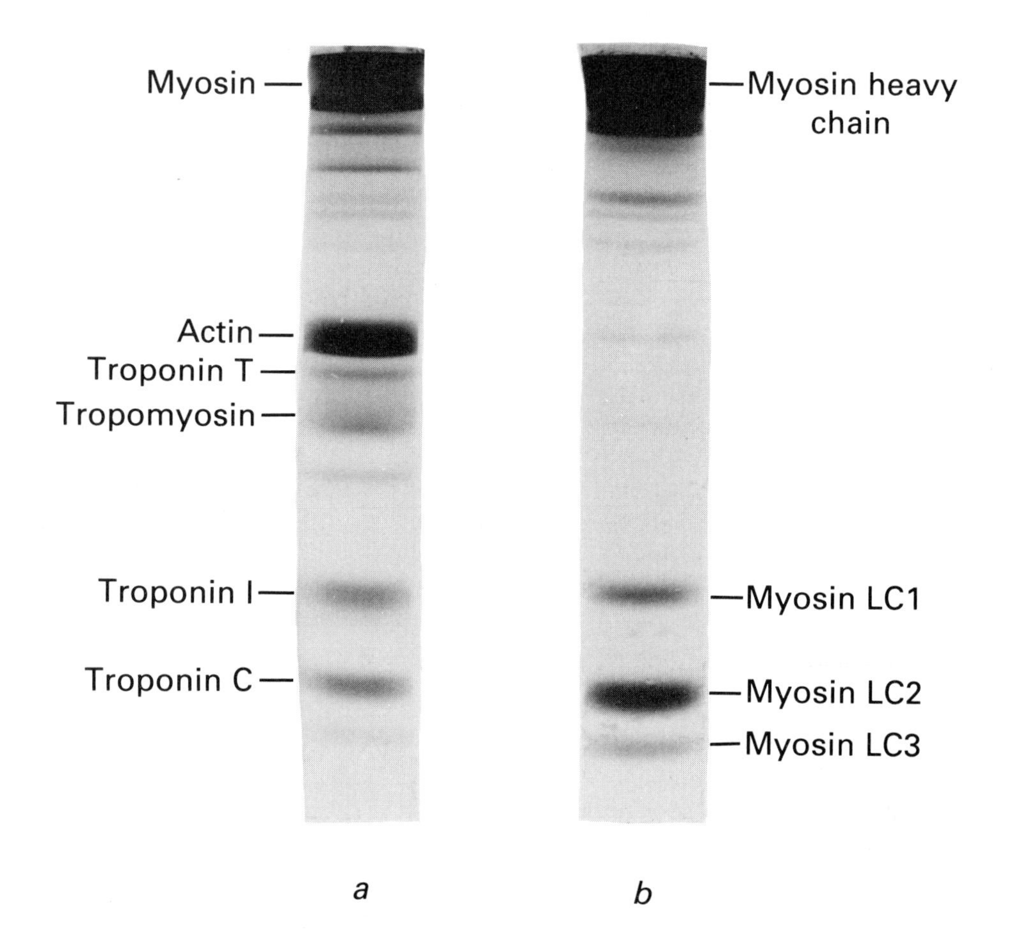

are known as DTNB light chains. The other two light chains are separable from the whole molecule by treatment with alkali; they are of two types, known as the A1 and A2 light chains, with molecular weights of 21 000 and 17 000 respectively. Alternatively, the light chains may be named from the order of their molecular weights as the LC1 (A1), LC2 (DTNB) and LC3 (A2) chains. Different types of muscle fibre may have different light chains present.

The myosin molecule can be split by proteolytic enzymes into various subfragments (Mihaly and Szent-Gyorgyi, 1953; Lowey *et al.*, 1969). Mild digestion by trypsin produces two sections known as light meromyosin (LMM) and heavy meromyosin (HMM); HMM acts as an ATPase and binds to actin, but LMM does not. Examination of the meromyosins by electron microscopy shows that HMM has two more or less globular 'heads' and a short 'tail', whereas LMM is a rod-like molecule. HMM can be further split into subfragments, two globular S1 portions and a rod-like S2 portion, by digestion with papain. The S2 rod ('long S2') can be split again to remove a 'hinge' region, leaving a section called 'short S2'. LMM molecules will aggregate to form filaments under suitable conditions, but neither HMM nor its constitutent subfragments will.

The bulk of the S2 fragment and LMM consists of α-helix, and the two heavy chains are wound round each other to give a coiled-coil structure (Lowey *et al.*, 1969). The S1 heads are separate from each other and 'globular' in structure; electron micrographs show that each is an elongated club-shaped molecule about 190 Å long, attached to the S2 rod via a narrow neck. Each head can apparently swing freely about its junction with the rod and there is also a 'hinge' in the rod about 430 Å from the head-rod junction (Elliott and Offer, 1978).

The amino acid sequence of the myosin heavy chain from the nematode worm *Caenorhabditis* has been determined by gene cloning techniques (Karn *et al.*, 1983). It shows considerable homologies with the corresponding sequences later determined for vertebrate skeletal and smooth muscle (Strehler *et al.*, 1986; Yanagisawa *et al.*, 1987). Hence the myosin molecule is, in evolutionary terms, a rather conservative one. The mammalian skeletal muscle heavy chain has 1939 amino acids, with a molecular weight of 223 900. The N-terminal sequences are appropriate to a globular head region, and the C-terminal rod shows the extended α-helix structure which would be expected from a coiled-coil dimer.

The charge distribution in the rod section, evident from the amino acid sequence, shows some very interesting regularities (McLachlan and Karn, 1982; McLachlan, 1984). Firstly, there are those connected with the coiled-coil structure. Many α-helical coiled-coil proteins have a characteristic pattern whereby hydrophobic residues occur alternately every three or four residues, so producing a seven-residue repeat (Crick, 1953). We can thus designate the residues in this repeating sequence as *a b c d e f* and *g*, with *a* and *d* as the hydrophobic residues. Successive *a* and *d* residues then lie on a zigzag line down the length of the

Fig. 12.20. Schematic diagram of the myosin molecule. The two heavy chains are shown as thick black lines; in the rod region they are coiled-coil α-helices. The A1 and DTNB light chains occur in the S1 heads. (From Lowey, 1986.)

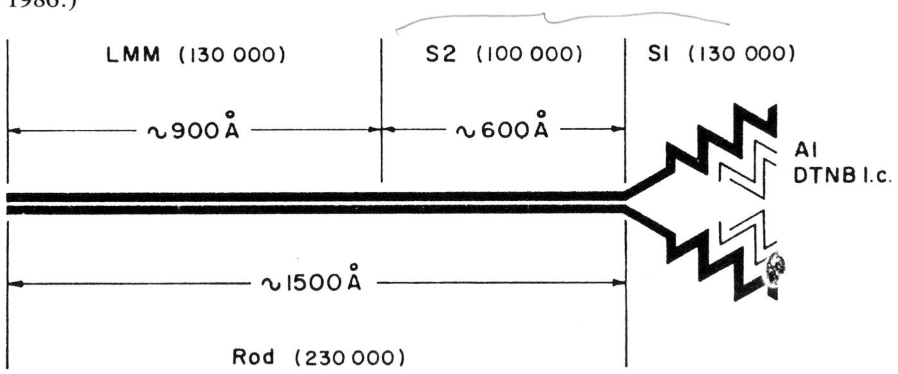

Fig. 12.21. Cross section of the rod section of the myosin molecule. Successive amino acid residues in the two α-helices are labelled *a* to *g*, and of these normally only *a* and *d* are hydrophobic. Consequently the *a* and *d* residues of the two chains associate with each other to form the coiled coil structure. The diagram is drawn as if there were 3.5 residues per turn in each α-helix; in fact there are about 3.6, so that the two helices coil around each other. (From McLachlan and Karn, 1982.)

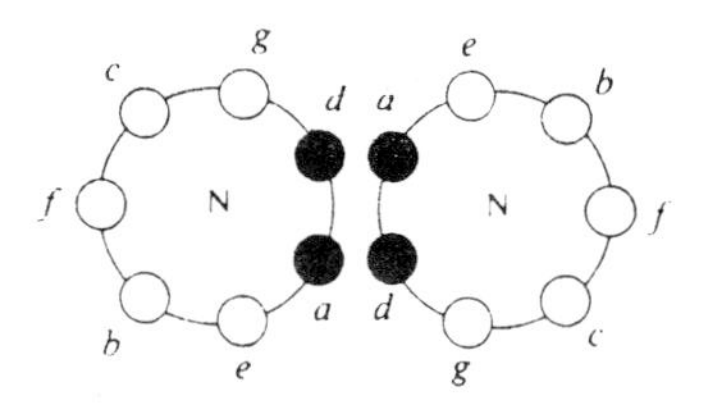

α-helix, so that the two helixes can lie with their lines of hydrophobic residues in mutual contact (fig. 12.21).

Acidic and basic residues are clustered on the outside of the coiled-coil, and so arranged that there is a regular repeating pattern along its entire length (fig. 12.22). The regular bands of charged residues, alternating positive and negative, suggest that strong interactions will arise if two molecules (i.e. two coiled-coils) are placed side by side. If they are precisely aligned with each other, or if they are displaced by steps of 28 residues, there will be a strong repulsion between them since similar charges will be opposite to each other. If one molecule is displaced by 14 residues, however, then the two molecules will attract each other since the negative charges in one will be opposite the positive charges in the other, and vice versa.

Fig. 12.22. Distribution of different types of amino acid residue on the surface of the supercoil of the two α-helices in the myosin rod. A section of coil 56 residues long, containing two repeats of the averaged 28-residue repeating sequence, has been projected onto a cylinder and unrolled so that each section of the sequence appears twice side-by-side, first from one helix and then from the other. The sequence reads across from right to left with a slight downward slant and a pitch of 3.5 residues per turn of each α-helix. The zig-zag lines connect the residues *a* and *d* on the side of the supercoil. On the left the radii of the circles represent the average net charge of the residues in that position (white circles for positive charge, black for negative), on the right they represent the average hydophobicity of the residues. Notice the repeating bands of positive and negative charge alternating along the length of the molecule. (From McLachlan and Karn, 1982.)

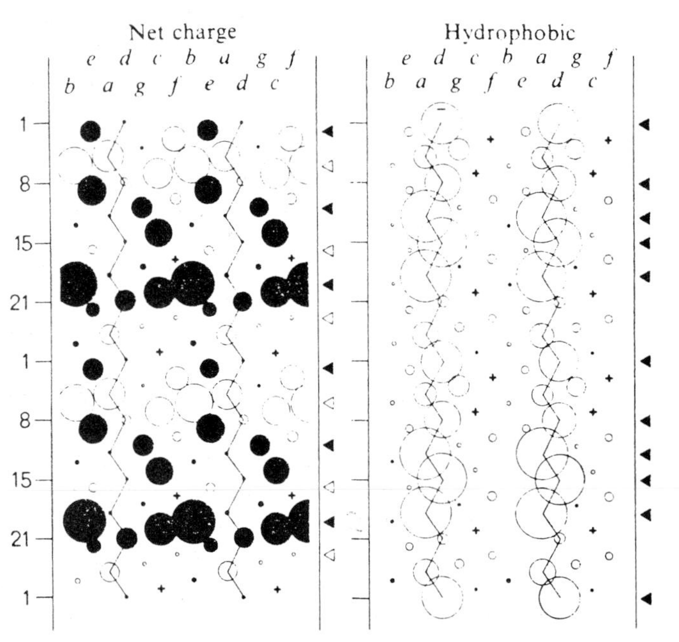

The same thing will happen at the similar displacements of 42, 70, 98 etc, i.e. at displacements of $14n$ residues, where n is an odd number. Successive 28-residue sections are not precisely identical in their amino acid sequence, so there may be some variation in the degree of attraction between adjacent molecules at different values of n; McLachlan and Karn calculate that the attraction is particularly strong at $n = 7$, and also at $n = 21$ and $n = 35$, i.e. at displacements of 98, 294 and 490 residues. We might therefore expect myosin molecules to pack alongside each other with displacements of 98 residues in their chain positions. The helical rise in an α-helix is 1.485 Å per residue, so 98 residues occupy a length of 145.5 Å. This is in most attractive agreement with the 143 Å X-ray meridional reflection attributed to myosin which, as we shall see, probably represents the stagger of successive triplets of myosin molecules along the myosin filament (fig. 12.25).

The regular 28-residue cycle is interrupted at four places in the chain by the insertion of an extra residue; McLachlan and Karn call these 'skip residues'. Their presence probably makes the coiled-coil slightly unwound, so that it has a longer pitch than usual, near these locations. We would expect this to have consequences for the packing of the molecules in the thick filaments.

The hydrophobic core of the coiled-core seems to be somewhat weakened by the inclusion of polar residues at the a and d positions in a short section from positions 316 to 327 of the rod, corresponding to the link between short S2 and the hinge region. There may be thus be some flexibility in the rod at this point. However, there is no indication that the whole of the hinge region may assume a non-helical form, as is suggested in some theories of cross-bridge action (see p. 285).

The amino acid sequence shows that the S1 head is a much less regular structure than the rod; it contains only short sections with the α-helix structure. The S1 heavy chain can be readily split in two places by proteolytic enzymes, to give 25 kDa, 50 kDA and 20 kDa fragments, reading in that order from the N-terminal. These fragments appear to correspond to separate domains in the intact molecule (Yanagisawa *et al.*, 1987).

Photoaffinity labelling is a method for identifying the active sites of enzymes. An analogue of the substrate is prepared that contains a photoreactive group that will bind irreversibly to the nearest part of the enzyme molecule when it is activated by a flash of light. So the analogue reaches the enzymic site and is then bound in place by photoactivation. Using this technique, various ATP analogues have been shown to bind to the 25 kDa domain and the N-terminal end of the 50 kDa domain, suggesting that the ATPase site involves these regions (Okamoto and Yount, 1985).

The approximate location of the ATPase site has been found by electron microscopy, using the egg-white protein avidin to bind to a suitable ADP analogue which was itself bound to myosin by photoaffinity labelling; it lies towards the outer end of the S1 head, about 140 Å from the head-rod junction and about 40 Å from the actin binding site (Sutoh *et al.*, 1986; Tokunaga *et al.*, 1987). This agrees with the location of two antibodies which bind to the 25 kDa unit and inhibit the ATPase activity (Winkelmann and Lowey, 1986). The 20 kDa peptide can be removed from S1 without loss of the actin-activated ATPase activity (Okamoto and Sekine, 1987).

The thick filaments

H. E. Huxley (1963) managed to isolate myosin filaments by homogenizing portions of glycerinated fibres. The filaments were then examined by electron microscopy, using the negative-staining technique. The most noticeable feature of these filaments was the presence of fairly regularly spaced projections on them, which almost certainly correspond to the projections and cross-bridges seen in thin sections of glycerinated muscle fibres. In the middle of each filament was a section, 0.15–0.2 µm long, from which these projections were absent, and which must correspond to the 'pseudo *H* zone' of intact muscle fibres.

We have seen that the myosin molecule is a tadpole-like structure with two 'heads' attached to a rod-like 'tail' section. A most interesting feature of these molecules, discovered by Huxley (1963), is that they are able to aggregate, under suitable conditions, to form filaments. The 'artificial filaments' so formed are of varying lengths, but all otherwise show the same general structure as the isolated natural filaments, including the projection-free region in the middle. Huxley suggested that the 'tails' of the myosin molecules become attached to each other to form a filament as is shown in fig. 12.23, with the 'heads' projecting from the body of the filament. Notice particularly that this type of arrangement accounts for the bare region in the middle. It also has two crucial implications for the design of the contractile apparatus: (1) the polarity of the myosin molecules is reversed in the two halves of the filament, and (2) the particularly reactive regions of the myosin molecule, the ATPase site and the actin binding site, are placed on the outside of the filament at the ends of the projections from it.

The analysis of the amino acid sequence of the myosin rod by McLachlan and Karn (1982), referred to in the previous section, fits very well with Huxley's model of filament structure. Fig. 12.24 shows how we would expect adjacent molecules to be staggered by 143 Å with respect to one another.

In his pioneer investigation on the the low-angle X-ray diffraction pattern of living muscle, H. E. Huxley (1953*b*) observed a series of reflections corresponding to an axial repeat distance of about 420 Å. Using more accurate methods,

Fig. 12.23. H. E. Huxley's suggestion as to how the myosin molecules aggregate to form an *A* filament with a projection-free shaft in the middle and reversed polarity of the molecules in each half of the sarcomere. (From Huxley, 1971.)

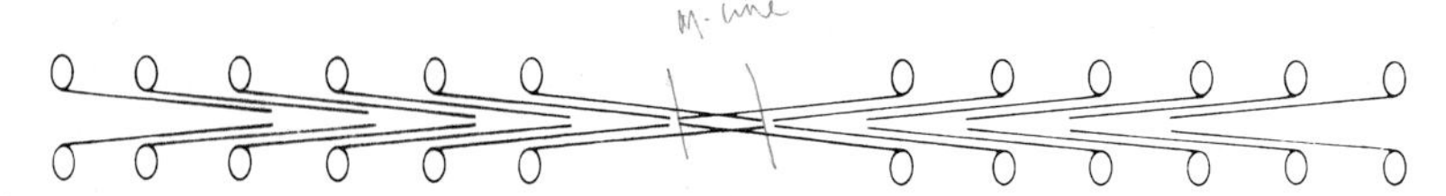

Fig. 12.24. Longitudinal packing of myosin molecules in the thick filament, as suggested by the amino acid sequence of the rod section of the molecule. *a* shows the bipolar arrangement in the middle of the filament. *b* shows the parallel array of staggered myosin rods. 28-residue repeats are shown as circles in rods 0 to 2, with the first 12 repeats (white circles) being in the short S2 section. Skip residues (see text) are shown by vertical lines, and the half-stagger between them is emphasized in rods 3 to 12. Successive rods are shown displaced by 98 residues. Vertebrate thick filaments probably have three such arrays in each cross section. (From McLachlan and Karn, 1982.)

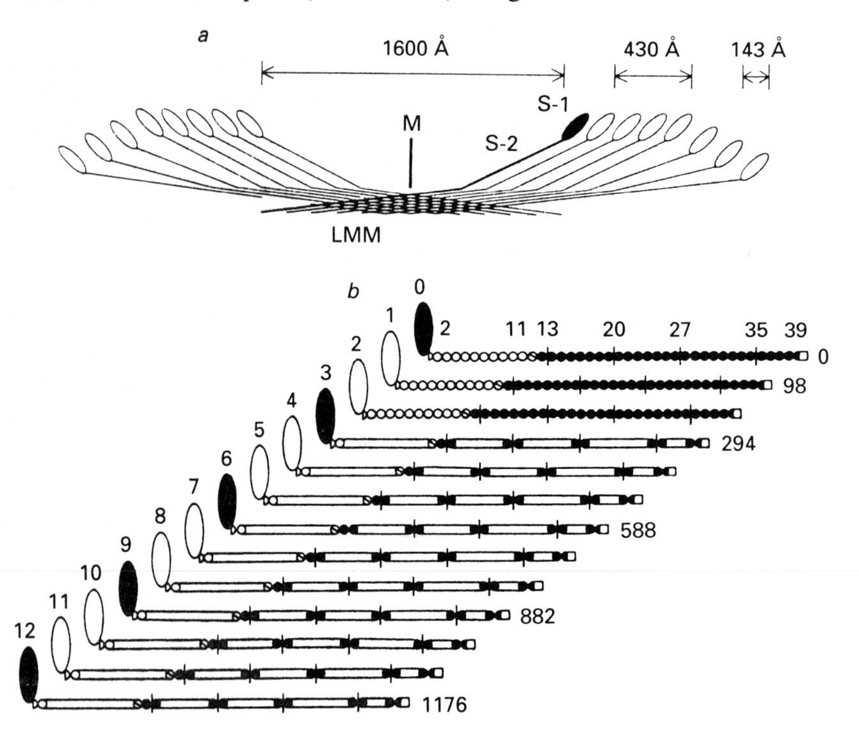

Worthington (1959) observed a strong reflection at 145 Å with a fainter one at 72 Å, and suggested that these corresponded to the third and sixth orders of a 435 Å repeat distance. These reflections did not correspond to those obtained from actin (which could also be seen as a separate series), and Worthington therefore suggested that they were produced by myosin.

Huxley and Brown (1967) made a detailed investigation of the X-ray diffraction pattern of living muscle. The strong meridional reflection at 143 Å (equivalent to Worthington's 145 Å), together with the off-meridional layer lines based on 429 Å (Worthington's 435 Å) led them to suggest that the cross-bridges emerge from the filament in pairs every 143 Å, there being a rotation of 60° between each successive pair, so that projections oriented in the same direction occur every 429 Å. An alternative model was proposed by Squire (1974), in which a 'crown' of three cross-bridges emerged every 143 Å and there was a rotation of 40° between each sucessive crown, as is shown in fig. 12.25.

In principle it is possible to distinguish between these different models by measuring the number of molecules in each filament. This demands very careful extraction techniques so that all the myosin, and nothing but the myosin, is measured. Application of such techniques was initially inconclusive, with different investigators producing answers of two, three or even four myosin molecules per crown.

Fig. 12.25. Diagram showing the arrangement of cross-bridges on the myosin filament in vertebrate striated muscle. (From Offer, 1974.)

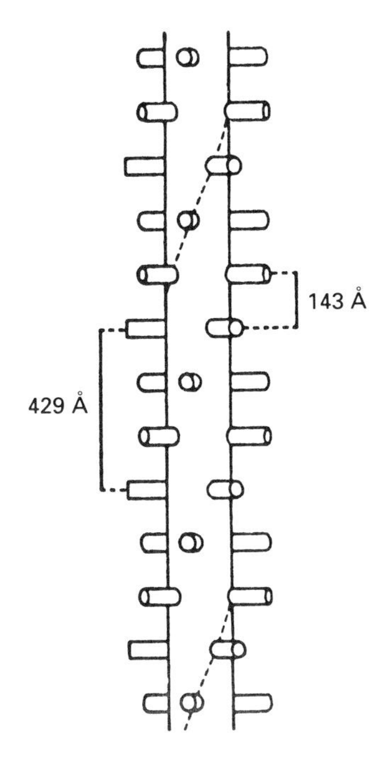

A fresh approach was provided by experiments involving scanning transmission electron microscopy (Lamvik, 1978; Reedy *et al.*, 1981; Knight *et al.*, 1986). Measurements of electron scattering by individual filaments provide an estimate of filament mass. Thus Knight and his colleagues found that the mass per half filament of rabbit thick filaments was 91.7 MDa, spread along a length of 8170 Å. If the myosin crowns occur every 143 Å along the filament from the edge of the bare zone to the tip, then there will be about 52 crowns per half filament. If we further assume that 6% of the mass is proteins other than myosin (see later), then the mass per crown is $91.7 \times 0.94/52$, i.e. 1.66 MDa. Since the mass of the myosin molecule is 0.52 MDa there are 1.66/0.52 i.e. 3.19 myosins per crown. This suggests that there are indeed three myosins per crown, as indicated in fig. 12.25.

Do cross-bridges occur regularly at all the 143 Å levels along the length of the filament, or are there some gaps in the sequence? Craig and Offer (1976) attempted to answer this question by labelling glycerinated rabbit fibres with antibodies to the S1 subfragment. They found that there was dense labelling throughout the A band except at the centre and along a narrow stripe near its ends. They concluded that the cross-bridges extend in a regular sequence from the end of the bare zone to the end of the filament except for a single gap between the second and third cross-bridges from the end (fig. 12.26). Notice that in this model there are only 49 crowns per half filament; adoption of this value would produce a slight revision in the calculations in the previous paragraph.

A regular helix of the type shown in fig. 12.25, with a repeat distance of 429 Å and an axial displacement of 143 Å between each crown of cross-bridges, should give an X-ray diffraction pattern with off-meridional layer lines based on 429 Å and on-meridional reflections at 143 Å, 71.5 Å and so on. There should be no on-meridional reflections at 216 and 107 Å. But there are: they are known as 'forbidden' reflections (see Haselgrove, 1983). Perhaps the distance between successive crowns is not always 143 Å, but could be in some repeating sequence such as 161, 161, 107 (Yagi *et al.*, 1981) or 160, 120, 150 (Squire *et al.*, 1982).

Vertebrate thick filaments contain some accessory proteins, called C-, X- and H-proteins, which bind to myosin filaments at regular intervals (Offer *et al.*, 1973; Starr and Offer, 1983). Examination of isolated *A* bands using negative staining shows a series of 11 stripes about 430 Å apart, with the first one at the edge of the bare zone (Craig, 1977). Antibody staining reveals that some of these stripes are sites of attachment of the accessory proteins (Bennett *et al.*, 1986). The existence of these accessory proteins raises many questions. Are they important in maintaining the structure of the myson filament? Do they modulate actin-myosin interactions? Why are they only present in the inner two-thirds of the *A* band?

Actin

G-actin is a globular protein of molecular weight about 42 000. Each molecule normally contains one molecule of bound ATP or ADP and one bound calcium ion. In salt solutions of physiological ionic strength, G-actin containing bound ATP will polymerize to form F-actin, during which the ATP is split to leave bound ADP in its place (Szent-Györgyi, 1951).

As well as forming the main protein of the thin filaments in striated and smooth muscles, actin occurs in a wide variety of other animal and plant cells, where it plays an important role in cell motility and the cytoskeleton. Actin from rabbit skeletal muscle contains a sequence of 375 amino acids (Elzinga *et al.*, 1973; Vandekerckhove and Weber, 1978*a, b*; Korn, 1982). Actins from heart muscle, vascular smooth muscle and non-muscle tissues show only very small differences in the sequence, indicating that the actin molecule is highly conserved in evolution. Perhaps this is because actin has to bind to a large number of different proteins (including, in skeletal muscle, other actin molecules, myosin, tropomyosin, troponin, *Z* line proteins and perhaps others), so that the possibilities for change may be strictly limited.

G-actin forms a crystalline complex with pancreatic DNase 1 and so it is possible to get some idea of the structure of the molecule by X-ray diffraction methods. The results show that the molecule is elongated (67 × 40 × 37 Å) and separated into two domains, one somewhat larger than the other, with a cleft between them (Kabsch *et al.*, 1985).

The thin filaments of all muscles consist principally of the polymerized form of actin, F-actin. X-ray diffraction patterns of F-actin were first obtained by Selby and Bear (1956), using dried muscles from the clam *Venus*. They concluded that the actin monomers were arranged either in a net-like structure or in helices (topologically a helix is similar to a cylindrical net); the pitch of the helix model was 350 or 406 Å, corresponding to 13 or 15 G-actin monomers respectively. Using negative staining and other techniques, Hanson and Lowy (1963) concluded that F-actin consists of two chains of monomers connected together in a double helical form, as is shown in figs. 12.30*a* and 12.27. X-ray diffraction measurements by Huxley and Brown (1967) suggested that the pitch of the long helix was probably 2 × 370 Å, with about 13.5 monomers per turn.

A more recent model of F-actin has been produced by Egelman and DeRosier (1983; see also Egelman and Padrón, 1984, Egelman, 1985, and Amos, 1985). This places less emphasis on the idea that there are two intertwined chains of monomers: binding occurs between adjacent

Fig. 12.26. Distribution of cross-bridges in half a myosin filament, as deduced from antibody staining experiments. (From Craig and Offer, 1976*a*.)

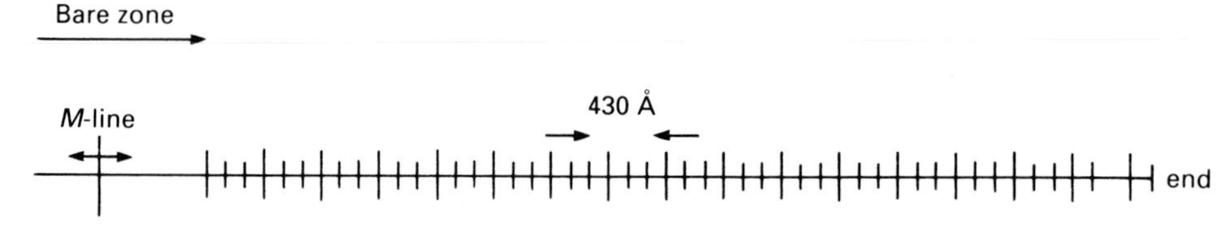

monomers only, and not between successive monomers on the same long-pitch strand. Each elongated subunit projects nearly (but not quite) at right angles to the long axis of the filament (fig. 12.28). Successive subunits are rotated by about 167° around the axis of the filament, and displaced by 27.3 Å along it. Image analysis of isolated filaments suggests that there is considerable variation in the value of the long-pitch, which implies that adjacent subunits may be able to rotate to some extent about the filament axis, perhaps by $\pm 10°$ (Egelman *et al.*, 1982). This idea receives further support from studies in the thin filaments of insect flight muscle in rigor (Taylor *et al.*, 1984).

Tropomyosin and troponin

Tropomyosin was first isolated by Bailey (1948) and its amino acid sequence was determined by Stone *et al.* (1974). The molecule consists of two α-helical chains which intertwine to form a coiled-coil structure of the type which we have seen in the myosin rod. It has a molecular weight of $2 \times 33\,000$ and length of about 400 Å.

The role of tropomyosin began to emerge with the discovery of a second accessory protein, troponin, by Ebashi and his colleagues (Ebashi and Kodama, 1966*a, b*; Ebashi *et al.*, 1968, 1969). The responsiveness of 'natural' actomyosin, and of glycerinated muscle fibres, to calcium ions depends on the presence together of both tropomyosin and troponin. Addition of tropomyosin and troponin to purified actomyosin systems inhibits

Fig. 12.27. Helical geometry of the actin filament based on the 'double string of beads' model originally proposed by Hanson and Lowy. Successive monomers can be arranged on two primitive helices, a right-handed one whose pitch is about 51 Å and a left-handed one at about 59 Å. This gives the appearance of two intertwining helices whose pitch is 2×375 Å. The axial distance between the centres of adjacent monomers (such as 5 and 6) is 27.3 Å, hence that between successive monomers on one of the long-pitch helices (such as 5 and 7) is 54.6 Å. X-ray diffraction diagrams (e.g. fig. 12.10) show relatively strong off-meridional reflections at 51 and 59 Å. (From Egelman *et al.*, 1982.)

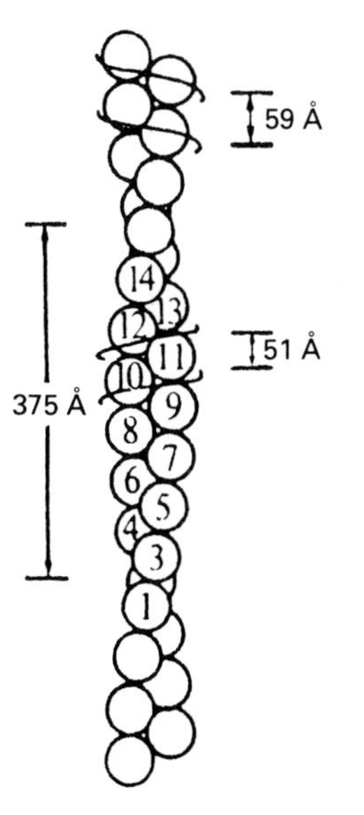

Fig. 12.28. Egelman and DeRosier's model of F-actin structure. Each actin monomer is represented as a dumbbell with its long axis at 75° to the long axis of the filament. (Redrawn by S. Mitchell from Amos, 1985, after Egelman and DeRosier, 1983.)

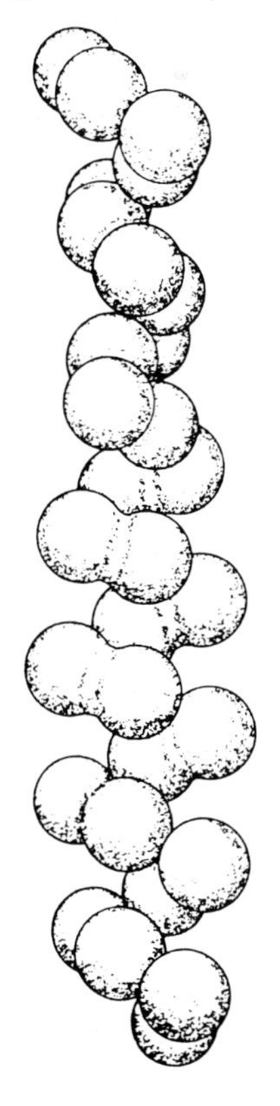

their ATPase activity in the absence of calcium, but does not do this if calcium ions are present. Thus troponin and tropomyosin serve to sensitize the actomyosin ATPase to the presence of calcium ions. We shall see in the following chapter (where we shall examine the properties of troponin in more detail) that this effect is crucial to the control of contractile activity in the living muscle cell.

Electron micrographs show a periodicity of about 400 Å in the *I* band, which is greatly enhanced after staining with ferritin-labelled antibody to troponin (Ohtsuki *et al.*, 1967). In their study of the X-ray diffraction patterns of living muscle, Huxley and Brown (1967) found a marked meridional reflection at 385 Å, and they concluded that this arose from the thin filament. Thus it seems likely that this 385 Å reflection represents the distance between the points at which troponin is attached via tropomyosin molecules to the actin backbone of the filament. Ebashi and his colleagues therefore suggested that the tropomyosin molecules lie end-to-end along the grooves between the two strings of actin monomers, each tropomyosin with a troponin molecule attached to it, as is shown in fig. 12.29.

The ratio of actin to tropomyosin to troponin molecules is 7:1:1 (Greaser *et al.*, 1973). If each tropomyosin molecule binds to seven actin monomers on one of the long-pitch actin helices, then the distance between successive tropomyosin molecules will be $(7 \times 55) = 385$ Å. The amino acid sequence of tropomyosin fits very well with this idea: there are 14 pseudo-repeats containing similar sequences of about 200 amino acids (Parry, 1975; McLachlan and Stewart, 1976). McLachlan and Stewart suggest that the molecule could lie alongside the F-actin filament and bind at 55 Å intervals to seven successive actin monomers on one of the actin long-pitch helices; the 14 pseudo-repeats perhaps correspond to two sets of alternative binding sites, which can be locked on or off by rotating the tropomyosin molecule by about a quarter of a turn about its long axis. We shall see in the next chapter that this idea that the tropomyosin molecule can adopt two alternative positions in the thin filament is crucial to modern ideas of how the interaction between actin and myosin is controlled in the living muscle.

Decorated thin filaments

In his study of the structure of muscle molecules and myofibril filaments with the negative staining technique, H. E. Huxley (1963) discovered a fruitful example of actin–myosin interaction. Under suitable conditions individual myosin molecules or their HMM or S1 subunits will attach to thin filaments or F-actin to produce 'decorated' thin filaments (fig. 12.30). It was immediately obvious that such filaments are polarized; the overall appearance shows a series of 'arrowheads' which always point in the same direction on any one filament, away from the *Z* line.

The detailed structure of decorated filaments has been investigated by computational analysis of electron micrographs using a three-dimensional reconstruction technique developed by DeRosier and Klug (1968) and used by them to determine the structure of the tail of the T4 bacteriophage. The principle of the method is essentially similar to that whereby it is possible to determine the structure of molecules by X-ray analysis of their crystals. Electron micrographs of negatively stained material are transmission images in which the total density through the three dimensional object is projected onto a two-dimensional plane. From a number of such images obtained with different orientations of the object (or, in the case of a helical object, from

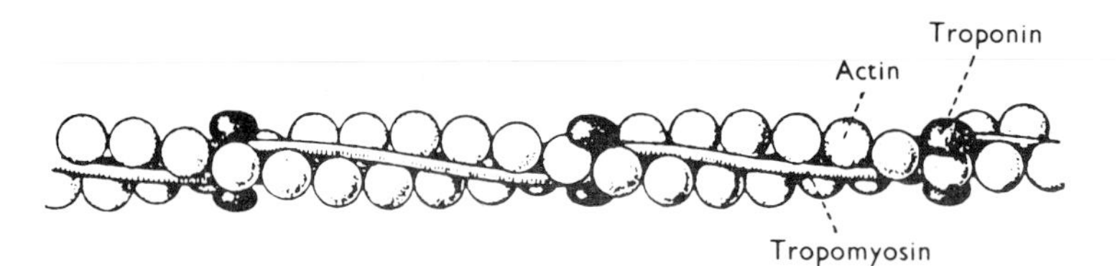

Fig. 12.29. A model for the structure of the thin filament, showing the probable position of the tropomyosin and troponin components. (From Ebashi *et al.*, 1969.)

images at different levels in the helix) it is possible to reconstruct the appearance of the original three-dimensional object. In order to do this the electron micrograph is first scanned by a microdensitometer which feeds its measurements into a computer. The computer then carries out a Fourier transformation on these measurements (so producing the information that would have been obtained from an optical diffraction pattern of the electron micrograph) and then combines transformations from different orientations to produce maps of the three-dimensional distribution of density. Further details of the method are given by Amos *et al.* (1982).

This technique has been applied to the structure of decorated thin filaments by Moore, Huxley and DeRosier (1970), Taylor and Amos (1981), and Vibert and Craig (1982). The 'arrowheads' are caused by the binding of one S1 head to each actin monomer. Their periodicity (each arrowhead is about 370 Å long) thus reflects the periodicity of the actin long-pitch helices. The S1 heads are tilted relative to the filament axis as is shown in fig. 12.31.

M line, Z line and other proteins

At the *M* line, each thick filament is connected to its neighbours by a series of '*M* bridges'. There are also '*M* filaments', which lie halfway between pairs of thick filaments and parallel to them and are contacted by the *M* bridges as is shown in fig. 12.32 (Knappeis and Carlsen, 1968; Luther and Squire, 1978). These structures must serve to maintain the thick filaments in their hexagonal array.

The *M* line contains two well-characterized proteins: creatine kinase (creatine phosphotransferase), the enzyme which catalyses the transfer of phosphate from creatine phosphate to ATP, and *M*-protein or myomesin (Turner *et al.*, 1973; Strehler *et al.*, 1983). However, Woodhead and Lowey (1983) were unable to detect any associations between these two proteins either with each other or

Fig. 12.30. Negatively stained actin filaments from vertebrate striated muscle. In (*a*) the filaments are as isolated from the muscle, in (*b*) they have been 'decorated' with molecules of vertebrate myosin subfragment 1. Magnification 270 000 for (*a*) and 260 000 for (*b*). (Photographs kindly supplied by Dr R. Craig.)

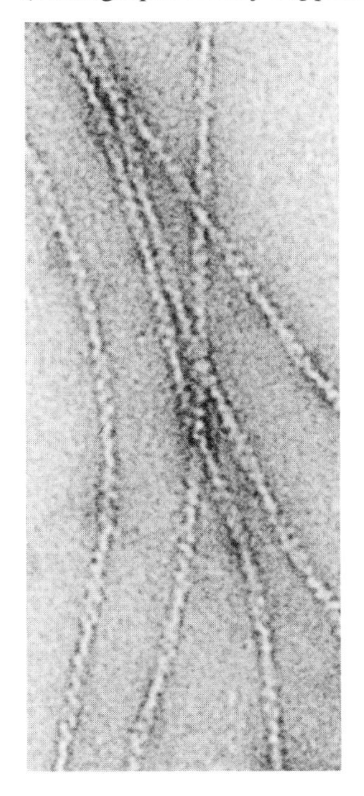

a

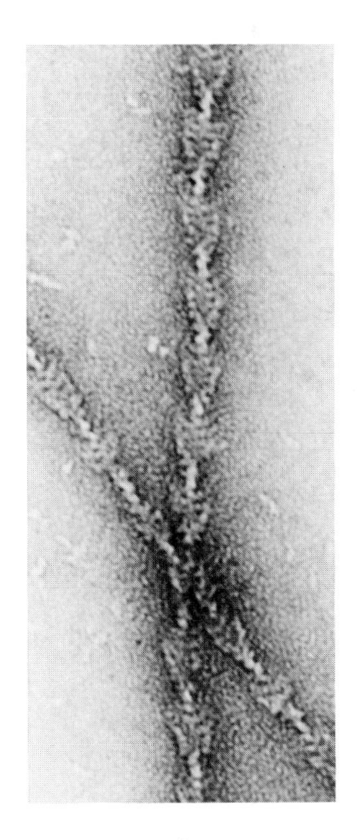

b

with myosin; they suggest that other proteins must be involved in binding these known components of the *M* line together.

The *Z* line (or *Z* disc) of vertebrate muscle has the appearance of a square lattice, when viewed in cross-section. Serial sections show that this is produced by the ends of a series of tetragons formed by connecting filaments ('*Z* filaments') between the *I* filaments on each side of the *Z* line. Each *I* filament is connected by four '*Z* filaments' to four *I* filaments on the other side of the *Z* line, as it shown in fig. 12.33 (Knappeis and Carlsen, 1962; Reedy, 1964).

Fig. 12.31. Three-dimensional reconstruction of the arrowhead structure of a thin filament decorated with myosin S1 heads. Three actin subunits are identified by the letter A. (From Craig, 1986, after Vibert and Craig, 1982.)

The most clearly defined constituent of the *Z* line is α-*actinin*, a protein of molecular weight 180 000 which is not found elsewhere in the myofibril (Masaki *et al.*, 1967). It is probable that other proteins are also present since there seems to be insufficient α-actinin to account for the mass of the *Z* line.

Desmin is a filamentous protein which was first isolated from smooth muscle. It forms a collar round each myofibril at the *Z* line and connects adjacent *Z* discs together (Lazarides and Hubbard, 1976; Granger and Lazarides, 1978). This produces a framework whereby the separate myofibrils are aligned with one another so that their striations are in register across the width of the muscle fibre. These desmin filaments form a subdivision of a group of subcellular structures called

Fig. 12.32. Part of the *M* line showing thick filaments (TF), *M*-filaments (MF), *M*-bridges (M1 and M4) and the postulated secondary *M*-bridges (M3). The structure is symmetrical about the M1 plane. (From Luther and Squire, 1978.)

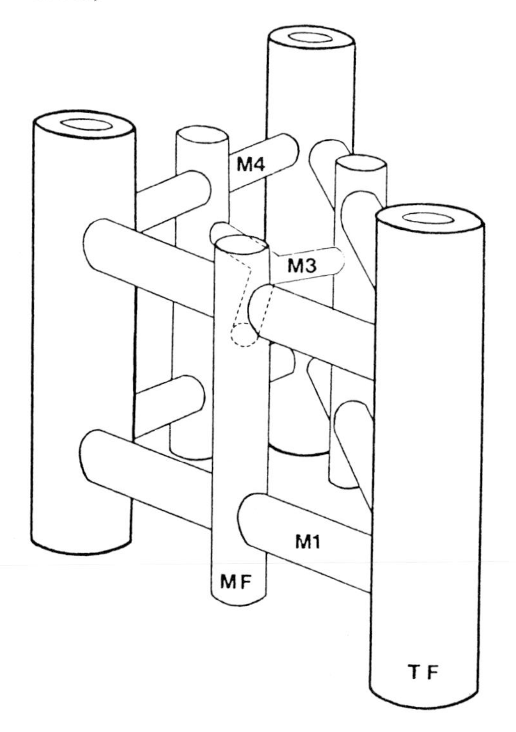

'intermediate filaments' (Lazarides, 1980; Steinert and Perry, 1985).

Maruyama and his colleagues have described a regulatory protein, called β-*actinin*, which affects the polymerization and aggregation of F-actin filaments (Maruyama, 1965; Maruyama *et al.*, 1977*a*). Fluorescent antibody staining shows that it is localized in the middle of the *A* band, but it is not removed by extraction of myosin. Hence it probably binds to the ends of the actin filaments and serves to prevent their further growth or breakdown.

Some electron micrographs of muscle show transverse bands in the *I* band (Page, 1968; Franzini-Armstrong, 1970). These may be similar to the bands which were described by nineteenth century light microscopists and have been called *N* bands. Page found that the distance between the *N* band and the *Z* line was increased in stretched sarcomeres and reduced in shortened ones, implying that the *N* band material is not firmly linked to the *I* filaments. It is possible that the protein *nebulin* (molecular weight about 500 000) is involved (Wang, 1984).

Fig. 12.33. Possible structure of the *Z* line. Four *Z*-filaments (white and shaded rods) are connected to each thin filament (black rod pairs). (From Squire, 1981.)

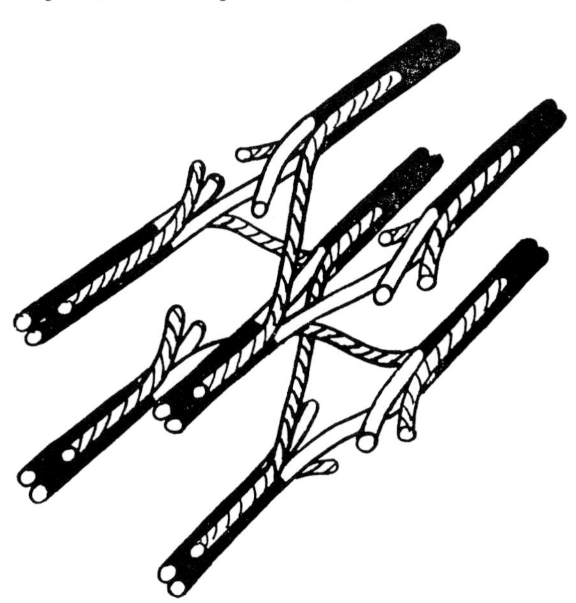

There is a rather puzzling feature of the myosin-extracted myofibrils shown in fig. 12.4*b*. Hanson and Huxley (1955) found that if such a fibre was stretched it would return to its original length on release. After further extraction of actin (fig. 12.4*c*) there is still clearly something holding the *Z* discs in position. Thus it looked as though there should be some longitudinal structures in the myofibril in addition to the thick and thin filaments. The explanation of this may lie with the protein *titin* (also called *connectin*), discovered by Maruyama *et al.* (1977*b*) and Wang *et al.* (1979). Titin is a very large molecule, with a molecular weight of about 1 000 000, which occupies about 9% of the myofibrillar mass. It can be separated by using SDS–polyacrylamide gels with very large pores. Its molecular structure may be in the form of a random coil, which would make it very elastic (Trinick *et al.*, 1984).

Titin and nebulin are both unusually large molecules. This fact has been utilized by Horowits and his colleagues (1986) in a most interesting experiment which suggests that they play an important role in the maintenance of myofibrillar structure. The experiments were done on chemically skinned fibres from rabbit psoas; such fibres have their cell membranes disrupted and made permeable by treatment with solutions containing the calcium-chelating agent EGTA, so that the interior of the fibre is accessible to substances which the experimenter may wish to apply, such as ATP or calcium ions (see Wood *et al.*, 1975, and Eastwood *et al.*, 1979).

Ionizing radiation produces damage to organic molecules, and the proportion inactivated is proportional to their molecular size, since large molecules offer larger targets than do small ones. Hence exposure of muscle fibres to ionizing radiation should result in more breakages in the titin and nebulin molecules than in the other constituents. Horowits and his colleagues showed by SDS–polyacrylamide gel electrophoresis that this was indeed so. Most interesting, however, were the effects on the structure of the sarcomeres. If an untreated fibre was stretched while relaxed (i.e. in the presence of ATP and with no calcium ions) and then activated by irrigation with a solution containing calcium, then there was some slight disorder in the *A* band, produced by a partial misalignment of the thick filaments. But after irradiation there was appreciable disorder in the sarcomere even in the relaxed state, and this was greatly amplified after activation by calcium ions.

This result could be explained if the thick filaments are normally attached at their ends to the Z discs by elastic filaments made of titin and perhaps nebulin, so that they are held in the middle of the sarcomere when the fibre is stretched or activated. When such filaments are broken, the position of the thick filaments becomes inherently unstable, since once one moves more to one end than the other there will be more overlap with thin filaments and so more cross-bridges formed at that end (fig. 12.34). Such elastic filaments would also contribute to the passive elasticity of stretched muscle; Horowits *et al.* found that passive tension was indeed reduced in proportion to the amount of radiation received.

The nature of cross-bridge action

We have seen that the sliding filament theory can provide an excellent explanation for the structural changes that occur when striated muscles contract. When H. E. Huxley (1957) discovered the cross-bridges linking the thick and thin filaments (fig. 12.3), he realized immediately that if the cross-bridges do indeed provide the active force for the sliding of the filaments, then cross-bridge action must be a cyclic process, since the cross-bridges are simply too short to account for the extensive length changes that can occur in contracting muscle. We would therefore expect each cross-bridge to undergo a series of cycles of attachment, pulling and detachment. The energy for each cycle, we would expect, should come from the splitting of ATP.

Can such a concept of cyclic cross-bridge activity account for the properties of contracting muscles as described in the previous chapter? And what actually happens to each cross-bridge as it undergoes a cycle of activity? Partial answers to those questions are now available. Let us begin by looking at a theoretical approach produced by A. F. Huxley soon after the sliding filament theory had been formulated.

A. F. Huxley's 1957 theory

The theory produced by A. F. Huxley in 1957 was a first attempt to devise a precise model based firmly on the sliding filament theory. It has proved to be most successful in the sense that it has suggested a number of experiments and provided the initial inspiration for several daughter theories. Although Huxley's suggestions regarding the particular mechanism whereby force is generated between the myosin and actin filaments may not be correct, some of the conclusions which he reached are of more general application.

Fig. 12.34. Possible role of elastic filaments in the sarcomere, as suggested by experiments with irradiated fibres. In the model each thick filament is connected to the nearest Z disc by elastic filaments made of titin and/or nebulin. The experiments were done on skinned rabbit psoas fibres in the presence of ATP. *a* shows the effects of stretch followed by activation by calcium ions in normal fibres. In *b* the fibre had been subjected to ionizing radiation; breakages produced in the elastic filaments lead to thick-filament misalignment on stretch and activation. (From Horowits *et al.*, 1986.)

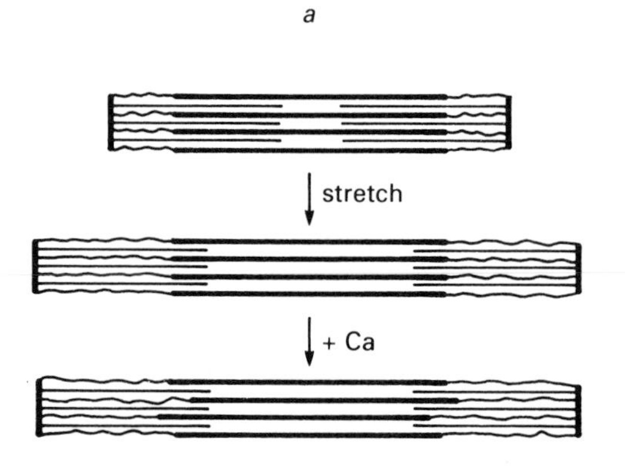

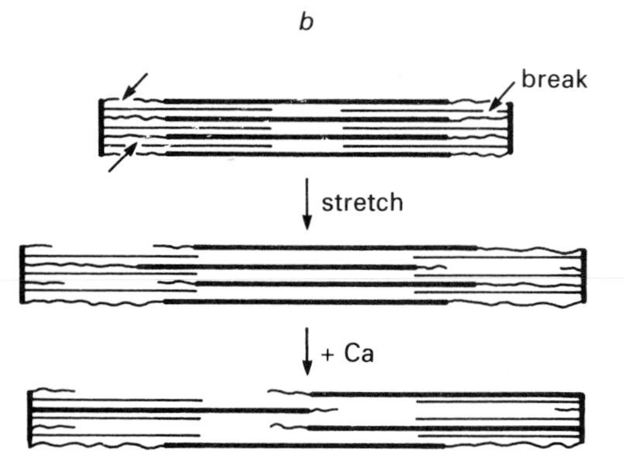

He begins by assuming that shortening and the development of tension are produced by independent tension generators (the cross-bridges) which can be effective only in the region of overlap of the thick and thin filaments. Within each contraction, each tension generator undergoes cycles of activity: attachment, pulling, and detachment, followed by reattachment (perhaps at a different site) and a new cycle. An essential point in what follows is the postulate that the probabilities of attachment and detachment of the cross-bridges are determined by their position.

The hypothetical tension-generating mechanism is shown in fig. 12.35. An active site M on the myosin filament oscillates by thermal agitation backwards and forwards along the length of the filament, but is restrained by elastic elements on each side (the 'springs' in the diagram), its equilibrium position being denoted by O. On the actin filament is an active site A, at a longitudinal distance x from O. The A and M sites can become attached to each other (forming a cross-bridge), and this reaction has a rate constant f:

$$A + M \xrightarrow{f} AM. \qquad (1)$$

This AM link can be broken by combination of the A site with a high-energy phosphate compound XP,* the reaction proceeding with a rate constant g:

$$AM + XP \xrightarrow{g} AXP + M. \qquad (2)$$

* Biochemical evidence now suggests that high-energy phosphate (ATP) combines with myosin rather than with actin, but this detail does not affect the development and conclusions of Huxley's theory.

Fig. 12.35. The tension-generating mechanism in A. F. Huxley's 1957 model. The part of a fibril which is shown is in the right-hand half of a sarcomere. Details in the text. (From Huxley, 1957, by permission of Pergamon Press, Ltd.)

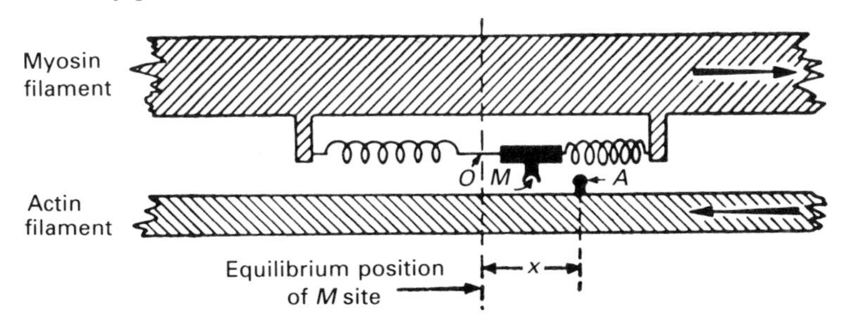

Finally the system is reset so that reaction (1) can occur again by dissociation of AXP and splitting of the high-energy phosphate bond (which supplies the energy needed to bring about this dissociation):

$$AXP \rightarrow A + X + \text{phosphate}. \qquad (3)$$

The rate constants f and g are assumed to be dependent upon x, the position of the A site with respect of O (fig. 12.36). f is zero when x is negative (A to the left of O), and increases linearly with increasing x up to the point h, beyond which it is again zero. g is very high (and constant) when x is negative, but zero at O, and small when x is positive, increasing slowly with increasing x.

During shortening, a single cycle of reactions (1) to (3) occurs as follows. An A site approaches an oscillating M site from the right; reaction (1) cannot occur until it passes the point h. When x is less than h, there is a fairly high probability that

Fig. 12.36. The dependence of the rate constants f and g on x in A. F. Huxley's 1957 model. The unit of the ordinate scale is the value of $(f+g)$ when $x = h$. (From Huxley, 1957, by permission of Pergamon Press, Ltd.)

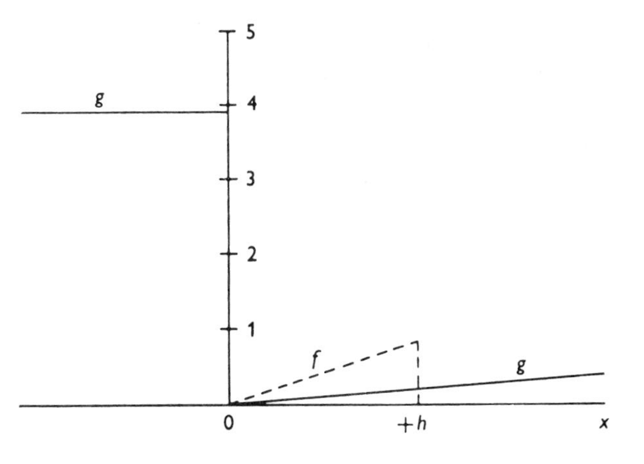

reaction (1) will occur; assume that, in this case, it does. M is now drawn towards O by the elastic force in the left-hand 'spring'. During this time there is a low probability that reaction (2) will occur, but as soon as the link passes O (i.e. as soon as x becomes negative) g becomes very high, and thus the probability of reaction (2) occurring increases enormously, and so the link is broken. Then reaction (3) occurs, and the A site is therefore ready to interact with the next M site that it meets. It is assumed that similar reactions occur asynchronously along the length of the sarcomere. This means that there will always be some links formed at any one time, so that the filaments will slide smoothly past each other.

The tension generated at any one contraction site is the tension in the left-hand 'spring' when the AM link is in existence. The tension produced by the whole muscle is the sum of the tensions at all contraction sites in a length of muscle equal to half a sarcomere (the forces generated in the two halves of a sarcomere, and in other sarcomeres along the length of the muscle, are in series with each other, and are therefore not additive). In an isometric contraction, some sliding occurs until the full isometric tension is reached and the series elastic component is fully stretched. All the AM links formed will then be to the right of O. There will be continual breakage of these links (reaction (2)) since g, though small, is not zero. Thus each M site will be continually going through the cycle of reactions from (1) to (3), and hence the high-energy phosphate compound XP will be continually broken down, with the release of energy. This accounts for the activation heat and the consumption of ATP in an isometric contraction.

When the muscle shortens, the average value of g will rise, since the A sites are continually moving to the left of O. Thus the rate at which the cycle of reactions occurs will also increase. This accounts for the extra energy release during shortening.

Another consequence of shortening of the muscle is that any one A site will only be in a position to react with the corresponding M site for a limited time; this time will decrease as the velocity of shortening increases. Hence, since the rate constant of link formation (f) is finite, the probability of a link being formed between any one pair of sites will decrease as the velocity of shortening increases. This means that, as the shortening velocity increases, the total number of links formed at any one time will decrease, and therefore the tension of the whole muscle will also decrease. Furthermore, as the velocity increases, more and more links will remain in existence when x is negative (since g is also finite); in these cases, the tension in the right-hand 'spring' will pull against the shortening generated by other links where x is positive, so further reducing the tension developed by the whole muscle. These two effects account for the force–velocity relation.

The hypothesis also provides a partial explanation for the transient velocity changes which appear to follow the changes in tension during an isotonic release (see Aidley, 1965*b*, and Civan and Podolosky, 1966). We have seen that more links exist in the isometric condition than when the muscle is shortening at a constant velocity. Immediately after the reduction in tension, these 'extra' links will still be in existence, and so the muscle will be able to shorten more rapidly than usual. However, detailed examination of the transients (see below) shows that they are more complex than the 1957 theory predicts.

Huxley's hypothesis is mainly concerned with the mechanism of contraction. However, it is desirable that it should be related to the properties of muscle under other conditions. In the resting muscle, we must assume that either reaction (1) or reaction (3) cannot occur; Huxley does not specify the activation mechanism which is necessary for contraction (we may assume that calcium ions are involved), but this is not an essential point in relation to the characteristics of contraction. At the end of a contraction, the reaction sequence ceases and the tension falls as reaction (2) proceeds to completion. When the muscle is in rigor, reaction (2) cannot take place, so that the AM links cannot be broken, which accounts for the inextensibility of the muscle in this condition.

H. E. Huxley's 1969 model

Early ideas of cross-bridge movement often invoked binding to actin at the tip of the cross-bridge and motive power supplied by bending or pulling at its base. This view was supplanted by a rather convincing picture produced by H. E. Huxley in 1969 (fig. 12.37). He suggested that that the myosin molecule possesses flexible linkages at two points in its structure: at the link between LMM and S2, the rod-like portion of HMM, and within HMM at the link between S2 and the globular S1 head. The primary source of movement in this model is a rotation of the S1 head about its attachment to the actin filament. The S2 rod serves to transmit the force produced by this rotation to the backbone of the myosin filament.

This model explains the appearance of insect flight muscles in rigor, agrees well with what we know of the structure of the myosin molecule, and is compatible with the results of X-ray diffraction studies. It also enables us to understand how the force per cross-bridge can be unaffected by the change in the transverse distance between the thick and thin filaments at different lengths, since the S2 link allows the S1 head to move out from the backbone of the myosin filament just as far as is necessary to contact the actin filament.

A. F. Huxley and Simmons's theory

This theory differs from A. F. Huxley's 1957 theory in its model of the tension-generating site. It is based on careful measurements of the mechanical transient responses of single muscle fibres and takes into account H. E. Huxley's 1969 model of cross-bridge structure. The apparatus was developed from the spot follower system used by Gordon, Huxley and Julian (p. 259), but had significant improvements in its time resolution. The results and the theory which arose from them were first reported by Huxley and Simmons (1971, 1973) and are discussed by Huxley (1974, 1980). More detailed descriptions and further experiments are given in a series of papers by Ford, Huxley and Simmons (1977, 1981, 1985 and 1986).

Fig. 12.38 shows the two types of transient obtainable from single muscle fibres: the length changes which follow a step change in tension (the velocity transient) and the tension changes which follow a step change in length (the tension transient). The events in the tension transient fall into four fairly distinct phases. First there is a sudden drop in tension during the length change (phase 1). Next, immediately after the length change, there is a rapid rise in tension (phase 2). This is followed by a period of a few milliseconds during which the recovery of tension is greatly slowed or even reversed (phase 3). Finally the tension gradually climbs back to the isometric tension appropriate to its length (phase 4). A related set of four phases can be seen in velocity transients; table 12.2 summarizes these events.

Huxley and Simmons investigated the nature of the first two phases by making a series of length

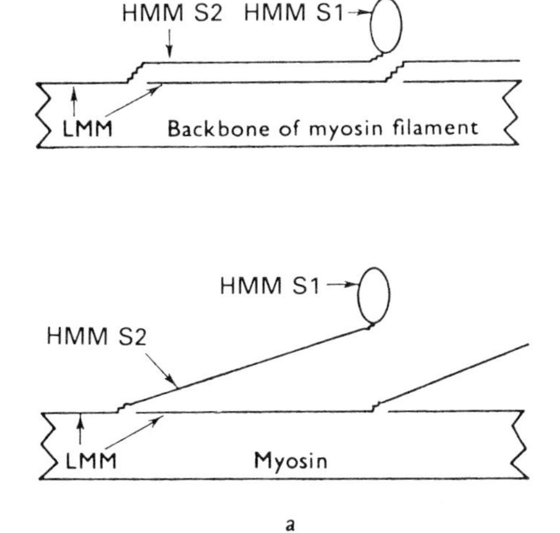

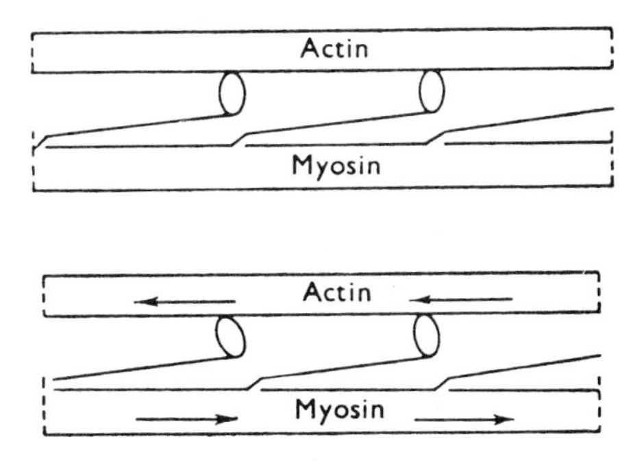

Fig. 12.37. A possible structural basis for cross-bridge action. *a* shows how the cross-bridge could extend different distances from the 'backbone' of the myosin filament. Each myosin molecule is composed of light meromyosin (LLM) and the globular (S1) and rod-like (S2) portions of heavy meromyosin (HMM). *b* shows how a rotational movement at the link between HMM S1 and actin could produce sliding of the filaments past each other. (From Huxley, 1969.)

Table 12.2. *Phases of the transient response to sudden reduction of length ('tension transient') or of load ('velocity transient')*

Phase	Time of occurrence	Events in 'tension transient'	Events in 'velocity transient'
1	During applied step change	Simultaneous drop of tension	Simultaneous shortening
2	Next 1–2 ms	Rapid early tension recovery	Rapid early shortening
3	Next 5–20 ms	Extreme reduction or even reversal of rate of tension recovery	Extreme reduction or even reversal of shortening speed
4	Remainder of the response	Gradual recovery of tension, with asymptotic approach to isometric tension	Shortening at steady speed sometimes with superposed damped oscillation

The stated times are appropriate for frog muscle at *ca.* 5 °C.
(From A. F. Huxley, 1974.)

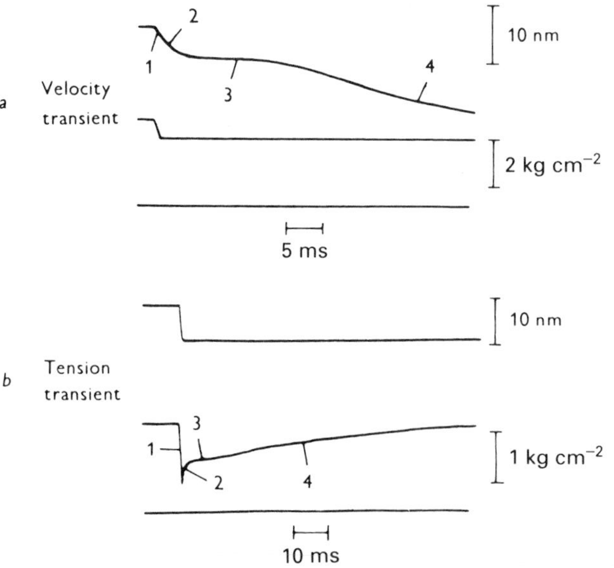

Fig. 12.38. Mechanical transient responses in frog muscle during tetanic stimulation. *a*, the length change following a sudden change in tension; *b*, the tension change following a sudden change in length. In each case the upper trace shows length (shortening downwards), the middle trace tension and the bottom trace the tension zero baseline. The numbers 1 to 4 indicate corresponding phases in the two types of transient response, described in table 12.2. (From A. F. Huxley, 1974.)

steps of different sizes and observing the height of the consequent tension changes. Their results are shown in fig. 12.39, where T_1 represents the tension at the end of the first phase and T_2 that at the end of the second. The T_1 curve is very nearly a straight line and this represents the behaviour of a passive elastic element. But the T_2 curve is quite different from this: tension falls very little for small length changes, after which the curve is roughly parallel to the T_1 curve with a displacement of about 6 nm per half sarcomere. This suggests that T_2 represents the properties of the tension-generators in the cross-bridges, each of which is capable of moving through about 6 nm while still exerting nearly maximum tension. To put it another way, each of the tension generators seems capable of taking up about 6 nm of slack.

What structures contain the passive elasticity represented in the T_1 curve? The experiments were done with single fibres tied to the apparatus at points very close to their ends, and so the compliance of the tendons can be ruled out. This leaves either the thick and thin filaments or the cross-bridges between them. In order to distinguish between these possibilities, Huxley and Simmons measured the T_1 and T_2 curves at different degrees of filament overlap. They found that both curves were scaled down in proportion to the decrease in overlap (fig. 12.40). Thus the stiffness of the elastic element whose properties are described by the T_1 curve is proportional to the number of cross-bridges that can be formed and hence it is very

Fig. 12.39. Tension levels in the early stages of tension transients in a frog muscle fibre. The points show the tension at the end of phase 1 (the T_1 curve) and phase 2 (T_2), in tension transients produced by different length changes. The T_1 curve represents an elastic component of the cross-bridges, and the horizontal difference between the T_1 and T_2 curves represents their capacity to shorten actively. (From A. F. Huxley, 1980, after Ford *et al.*, 1977.)

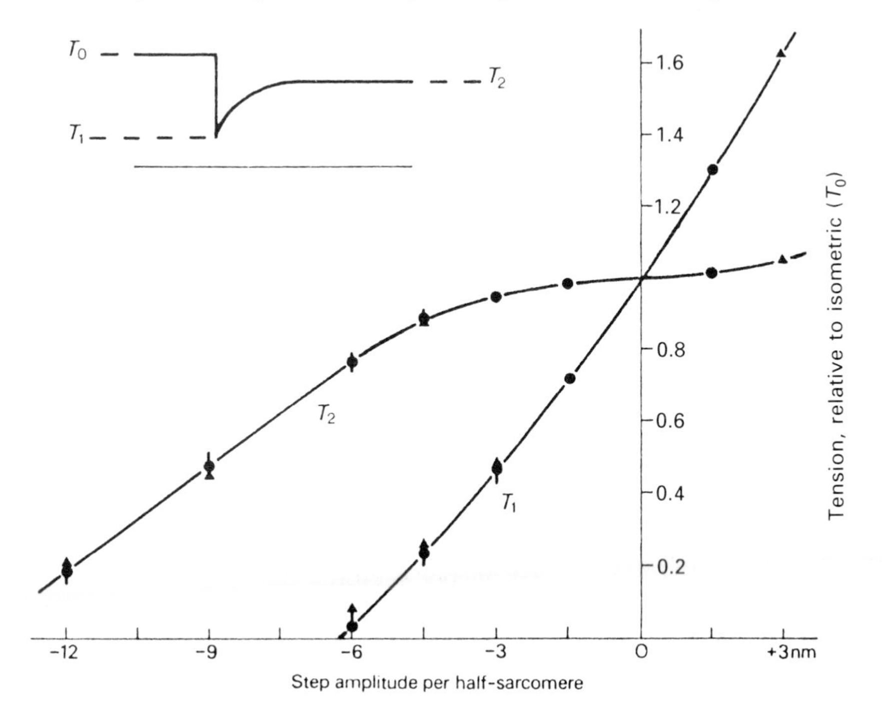

Fig. 12.40. T_1 and T_2 curves from the same frog muscle fibre at two different lengths. The continuous curves are those shown in fig. 12.39, obtained when the sarcomere length was 2.2 μm, at which all the cross-bridges would be overlapped by thin filaments. The crosses show T_1 and T_2 curves from the same fibre when stretched to give a sarcomere length of 3.1 μm, at which the overlap would be reduced to about 39%. The interrupted curves are simply the continuous curves scaled down to 39%. (From A. F. Huxley, 1974.)

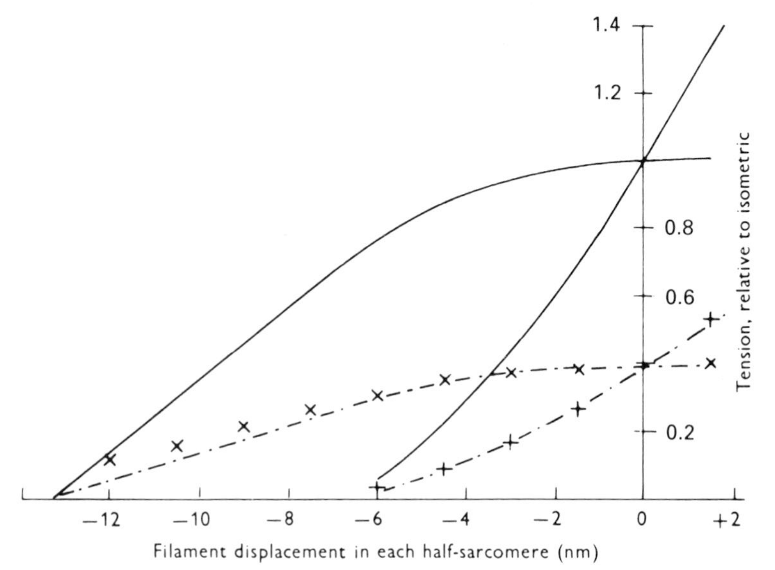

reasonable to suggest that this elasticity actually resides in the cross-bridges.

It is now possible to link this analytical model of a cross-bridge with the structure of the myosin molecule. The tension generator is the S1 subfragment, which can attach to a site on the actin filament and rotate about it. In doing so it pulls on the elastic element which is the S2 subfragment that connects the S1 head to the backbone of the myosin filament.

The time course of the early recovery of tension (phase 2) is much more rapid for large releases than for small ones, and even slower for stretches. Huxley and Simmons showed that this phenomenon could be accounted for if the movement of the tension generator took place as a small series of steps, so that the S1 head could be in two, three or four different positions while bound to the actin filament. The model shown in fig. 12.41 has three such positions.

Fig. 12.41. Huxley and Simmons' model of the cross-bridge, incorporating an elastic element and a stepwise shortening element. Here the elastic element is equated with the S2 portion of the myosin molecule and the stepwise shortening element with the S1 portion and its combination with actin. (From A. F. Huxley, 1974.)

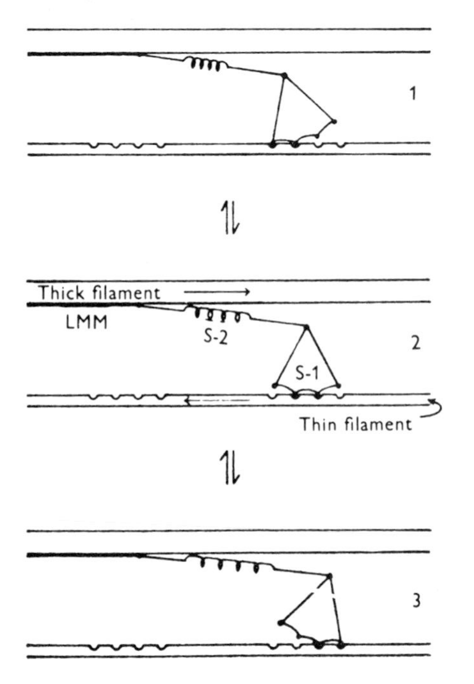

The discovery that some of the series elasticity resides in the cross-bridge enables us to see why A. V. Hill's two-component model cannot accurately describe the mechanical properties of muscle. The more cross-links that are in existence, the stiffer the series elasticity will be, and so it cannot be regarded as an independent element with an invariant force–extension curve.

Biochemical events in the cross-bridge cycle

The cycle of cross-bridge activity – attachment, movement, detachment and resetting ready for reattachment – must somehow be related to the splitting of ATP, which provides the energy for the whole process. We may ask how the reactions between the contractile proteins and ATP in test-tubes relate to the situation in the intact myofibril, and, particularly, at what stage in the cross-bridge cycle is the ATP split?

Lymn and Taylor (1970, 1971) used rapid reaction techniques to investigate the reaction kinetics of ATP hydrolysis by a mixture of actin and heavy meromyosin (HMM). Assay of the products of ATP splitting (ADP and inorganic phophate) by enzymic methods measures their presence in free solution, but assay using strong acid can detect ATP splitting even if the products are still bound to myosin. In this way it was found that the splitting of ATP was much faster than release of its products. The initial rate of hydrolysis of ATP (measured with strong acid quenching) was much the same for both HMM alone and acto-HMM, although the steady state rate for HMM alone was much reduced. This suggests that the first molecule of ATP which an HMM molecule meets is readily split, whereas later ones are not unless actin is present. This situation could arise if actin promotes the dissociation of the products of splitting from myosin.

Further work has led to modification of the Lymn and Taylor scheme (see Tregear and Marston, 1979, Webb and Trentham, 1983, Eisenberg and Hill, 1985, and Goldman, 1987). Thus Stein *et al.* (1979) found that the actin concentration required for 50% maximal ATPase activity was only a quarter of that required for 50% binding of S1. This implies that ATP splitting and actomyosin dissociation are not absolutely linked to one another. From this conclusion arose the concept that there might be two types of actomyosin states, weakly bound and strongly bound, the weakly bound state being converted to the strongly bound state by the release of inorganic phosphate.

Eisenberg and his colleagues have produced a model for the cross-bridge cycle based partly on these later results (Eisenberg and Greene, 1980; Eisenberg and Hill, 1978, 1985). It assumes that the cross-bridge can assume two distinct states, the 45° state during and at the end of the power stroke, during which actin–myosin binding is strong, and the 90° state at other times, when actin–myosin binding is weak. It also assumes that there is a rate-limiting change in the myosin–products complex between ATP splitting and inorganic phosphate release, and that there is a rapid equilibrium between myosin and actomyosin in the 90° state (fig. 12.42).

In the Eisenberg–Greene–Hill model the free energy of the cross-bridge in its various states varies with its position; so also do the rate constants for the various reactions involved. Both these ideas were present in A. F. Huxley's 1957 model and were further developed in a theoretical treatment by Hill (1974, 1976). Fig. 12.43 shows the free-energy profile of the cross-bridge: there are two main curves, corresponding to the 90° and 45° states. Let us see how the model works.

Consider an actin site entering the vicinity of a free myosin site from the right in fig. 12.43. Once the site is within range (i.e. once x is less than about 120 Å) the actin–myosin link can be formed in the 90° (weakly bound) state. The link is swept along to the left, largely by the action of other cross-bridges, until the 90° curve crosses the 45° curve; soon after this there is a transition to the strongly bound 45° state, accompanied by the release of inorganic phosphate. Equilibrium for this state is at $x = 0$, where the 45° curve is at a minimum, and so the link moves towards this position, producing the power stroke. Soon after $x = 0$ the ADP is replaced by ATP and the link is broken. During the whole cycle the actin filament has moved about 120 Å, the free energy of the system has fallen by about 8×10^{-20} J (i.e. 48 kJ per mole) and one ATP molecule has been split.

The Eisenberg–Greene–Hill theory is successful in combining biochemical reaction schemes with models of cross-bridge movement in a plausible manner. A similar scheme was produced by Tregear and Marston (1979). But it is quite clear from such models that a cross-bridge working in the filament array is in a situation very different from that of HMM or S1 molecules in solution. In particular, there is no analogue of muscle force in solution and so no reason why cross-bridge states should be held away from their equilibrium positions. Is it possible, therefore, to investigate the biochemistry of cross-bridges *in situ*?

Fig. 12.42. The Eisenberg–Greene–Hill model of the cross-bridge cycle. (From Eisenberg and Hill, 1985.)

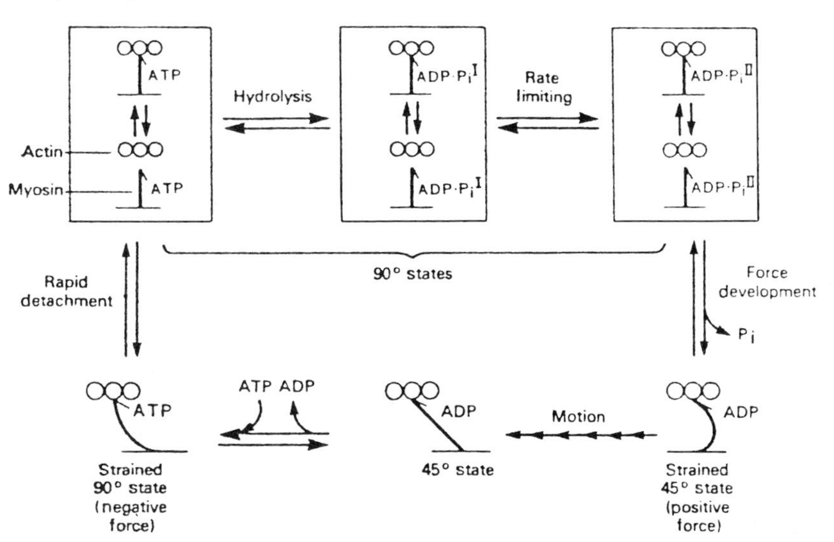

One useful technique which has been used to solve this problem enables the concentration of ATP in the fibre to be suddenly raised from zero to a useful level (Goldman *et al.*, 1984*a, b*). The compound P^3-1-(2-nitro)phenylethylATP is an ATP complex which is itself biologically inert but which can be broken down by an intense flash of light to release ATP; it is known as 'caged ATP'. Such a rapid release of ATP produced very rapid relaxation of muscle fibres in rigor, followed (in the presence of calcium ions) by a redevelopment of tension as the cross-bridges proceeded through the tension-generating steps of their cycle (fig. 12.44). Both these events are rapid, suggesting that the rate-limiting steps in the cycle occur when the

Fig. 12.44. Tension transients initiated by flash photolysis of caged ATP within a single glycerinated muscle fibre. At the start of each trace the fibre was in rigor with 10 mM caged ATP and 100 μM Ca^{2+}. At the time zero a 50 ns flash from a frequency-doubled ruby laser released ATP at a concentration of 700 μM. For trace *i* the fibre was held isometrically at a sarcomere length of 2.6 μm. For *s* the fibre was stretched slightly just before the laser pulse so as to increase the strain on the cross-bridges. The traces show the fall in tension as the rigor cross-bridges are dissociated by ATP, followed by the rise in tension as the active cycle proceeds. (From Goldman, 1987.)

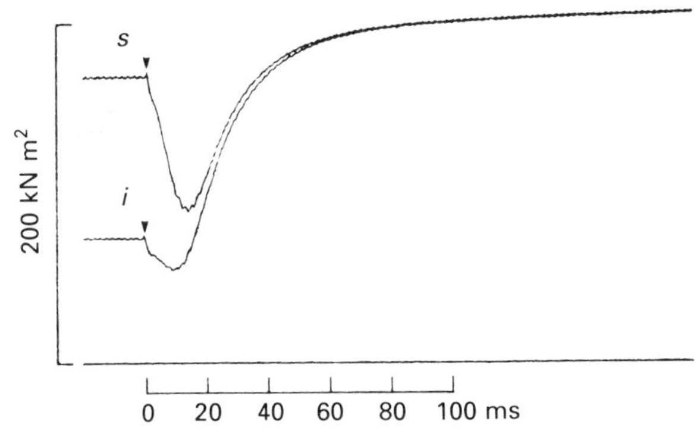

Fig. 12.43. Simplified version of the free energy profile of the Eisenberg–Greene–Hill cross-bridge cycle model. The free energy of an actin-myosin link varies with its position (x) in each of the two states, as shown by the 90° and 45° curves. The 90° curve has a minimum at about $x = 80$ Å whereas the 45° curve has its minimum at $x = 0$. (Simplified after Eisenberg and Greene, 1980.)

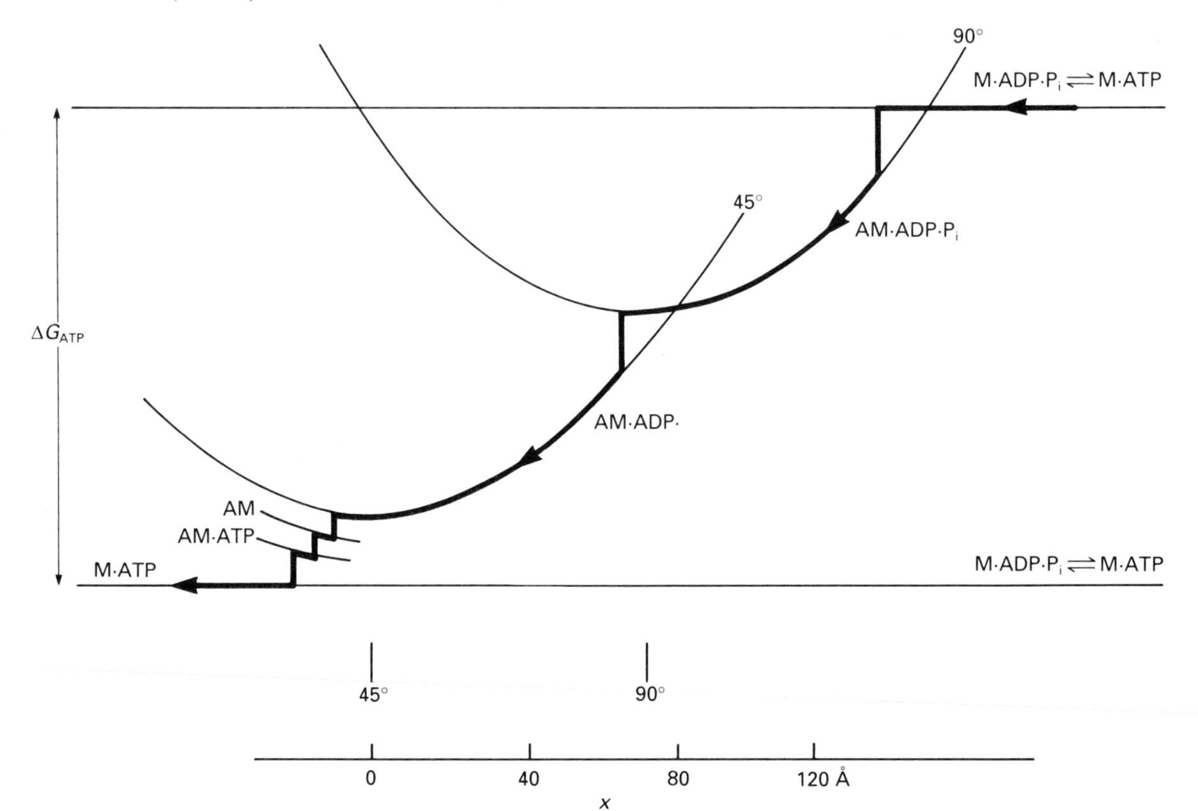

cross-bridges are attached and exerting tension. We can compare this conclusion with A. F. Huxley's 1957 model, in which breakage of the actin–myosin link will not normally occur until it has moved beyond the position $x = 0$.

How do the cross-bridges move?

H. E. Huxley's 1969 model suggests that the S1 heads rotate about their points of attachment to the actin filament. Attempts have been made to find evidence for this view by the use of spectrocopic probes for the detection of molecular motion (see Thomas, 1987). One method which has proved useful is electron paramagnetic resonance (EPR) spectroscopy. This involves measuring the absorption of energy by paramagnetic substances in a radiofrequency oscillating magnetic field. The paramagnetism arises from the presence of unpaired electrons, so a synthetic probe which can be bound to the muscle protein has to be used. Suitable probes include nitroxide derivatives of iodoacetamide or maleimide, which will bind to one of the sulphydryl groups in the S1 head.

The results of EPR spectroscopy show that probes attached to the S1 heads in glycerinated muscle fibres in rigor are uniformly and rigidly oriented (Thomas and Cooke, 1980). Addition of ATP in the absence of calcium (which produces relaxation from the rigor state as the actin–myosin links dissociate) produces a random distribution of probe orientations with rotational movement in the 10 μs range. Similar rotational mobility is seen in isolated myosin filaments and in stretched fibres in rigor; in the latter the proportion of fixed myosin heads is proportional to the degree of filament overlap (Barnett and Thomas, 1984). Thus the rotational mobility of free myosin heads seems to be very high, but heads bound to actin filaments in rigor are held rigidly at a constant orientation. EPR spectra of these two situations are shown in fig. 12.45*a* and *b*.

Fig. 12.45. Electron paramagnetic resonance spectra of spin labels attached to the myosin heads of glycerinated rabbit psoas fibres. The derivative of absorption is plotted against the magnetic field, with a baseline of 100 gauss. Trace *a* is from fibres in rigor (no ATP present), trace *b* from relaxed fibres (ATP present but no calcium ions), and trace *c* from contracting fibres (both ATP and calcium ions present). (From Cooke *et al.*, 1984.)

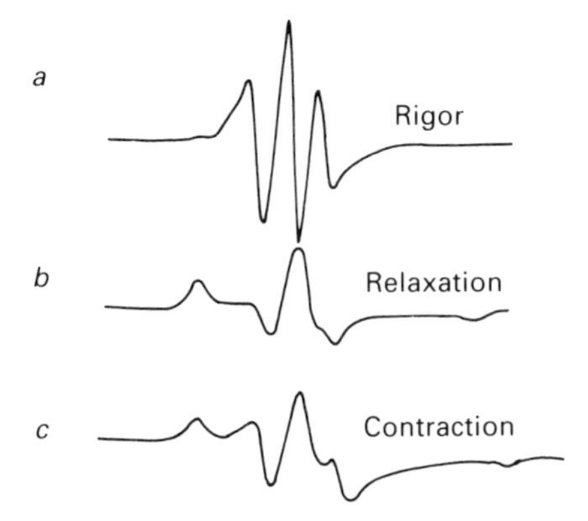

In the presence of both ATP and calcium ions, glycerinated fibres split ATP and contract. The EPR spectrum in isometric tension produced by such conditions is shown in fig. 12.45*c*. It is a combination of the spectra obtained from fibres in rigor and in relaxation: when 81% of the relaxation spectrum is subtracted from contraction spectrum, a spectrum identical to the rigor spectrum is produced (Cooke *et al.*, 1984). This suggests that during an isometric contraction about 20% of the myosin heads have probes oriented at the same angle as in rigor, whereas the remainder of the heads are disoriented.

These results suggest that the myosin heads do not rotate about their point of attachment to the actin filament during the power stroke, contrary to the model shown in fig. 12.37. Two alternatives have been proposed: either the S1 head bends in some way while its anchorage to the actin filament remains firm, or the S2 rod shortens (fig. 12.46).

Harrington and his colleagues have suggested that such shortening of the S2 rod could be produced in the 'hinge' region (the 20 kDa portion that connects to LMM at its C-terminal end) so that it changes from being a coiled-coil α-helical structure to a random coil (Harrington, 1979; Harrington and Rodgers, 1984; Ueno and Harrington, 1986).

More direct evidence on the localization of the motive power in the myosin molecule has been obtained by using some elegant microscopic assays of movement. Thus Hynes *et al.* (1987) induced formalin-fixed bacterial cells coated with myosin or HMM to move along cables of actin from the giant cells of the green alga *Nitella*. Antibodies to

Fig. 12.46. Three models of cross-bridge action. The power stroke may involve rotation of the S1 head about its point of attachment to the actin filament (*a*), a change in shape of the S1 head so as to produce an appreciable movement without rotaion of the area immediately adjacent to the actin filament (*b*), or a shortening of the 'hinge' region of the S2 rod (*c*). Model *b* seems the most likely. (From Pollard, 1987.)

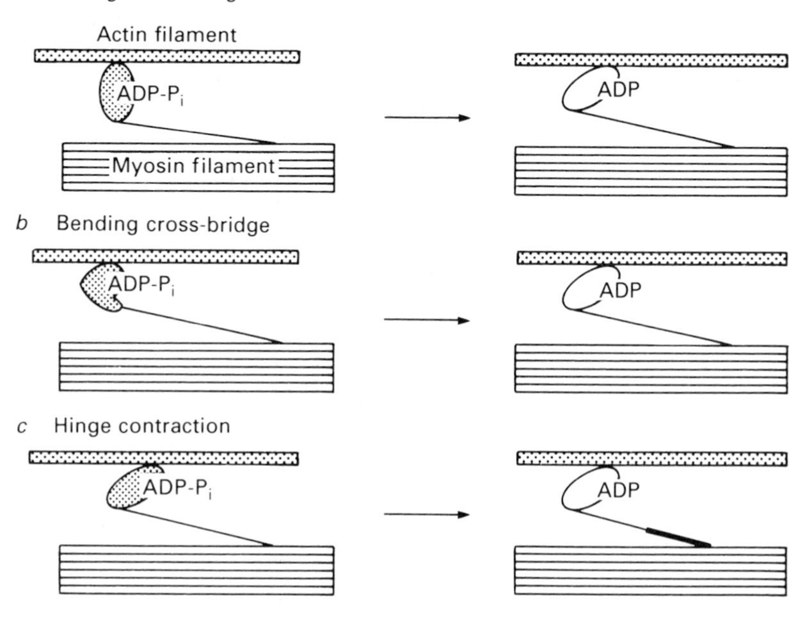

Fig. 12.47. Motile capability of HMM. A formalin-fixed *Staphylococcus aureus* cell has protein A on its surface. This binds to an antibody directed against the rod region of myosin or HMM, so that the myosin heads can interact with actin cables from Nitella. In the presence of ATP, cells coated with myosin, long HMM or short HMM move along the cable. (From Hynes *et al.*, 1987.)

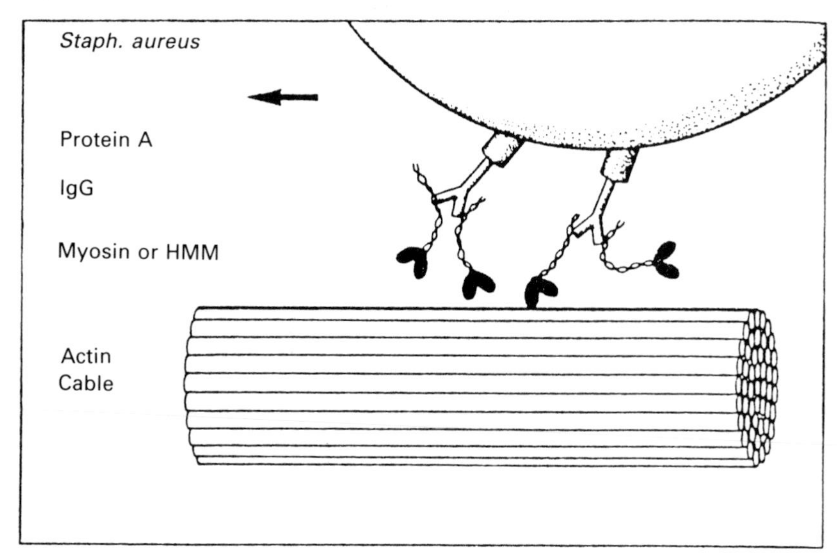

the rod region of myosin or HMM were used to bind the myosin to the bacterial cells (fig. 12.47). In the presence of ATP the coated cells moved along the actin cable at speeds up to $2\,\mu m\,s^{-1}$. Using proteolytic enzymes it was possible to prepare a short form of HMM in which the C-terminal 20 kDa section (the 'hinge' region) was missing. Cells coated with this short HMM moved at a mean speed of $1.2\,\mu m\,s^{-1}$, indicating that the motive power does not arise from the hinge region.

Another set of experiments from the same laboratory involved measurement of the movement of actin filaments on a bed of myosin molecule fragments (Toyoshima *et al.*, 1987). HMM or S1 fragments were bound to a layer of nitrocellulose on a coverslip. The actin filaments were labelled with fluorescein and so could be made visible under the microscope by illumination with ultraviolet light. With HMM the rate of movement of the filaments was about $7\,\mu m\,s^{-1}$, with S1 it was $1\text{–}2\,\mu m\,s^{-1}$. Hence S1 on its own can produce movement, which is in agreement with the 'bending cross-bridge' model shown in fig. 12.45*b*. Perhaps the higher speed of movement with HMM arises because the extra flexibility introduced by the S2 links allows more of the S1 heads to be correctly oriented for binding to actin.

The EPR spectroscopy results suggest that during isometric contraction only about 20% of the S1 heads are firmly fixed to the actin filament in the rigor position. On the other hand, X-ray diffraction results indicate that about 90% of the myosin heads move into the vicinity of the actin filaments in contracting muscle. Since this change does not occur in the absence of filament overlap, it seems likely that the myosin heads are actually bound to the actin filaments for much of the cycle (Haselgrove and Huxley, 1973; Huxley, 1979; Yagi and Matsubara, 1980).

Huxley and Kress (1985) have attempted to reconcile these two sets of observations by suggesting that the cross-bridge movement involves two stages. First, the S1 head binds loosely to the actin filament so that it exerts little tension and can still undergo a range of movements. It will therefore appear mobile on EPR spectroscopy. Secondly, the head becomes much more rigidly bound and produces a power stroke of about 40 Å. Thus the cross-bridge rotates in the first part of its movement and bends in the second part. This idea seems to be compatible with the two-stage models produced by Eisenberg and his colleagues and by Tregear and Marston.

To conclude, then, we have very good evidence that muscular contraction involves the sliding of the thick and thin filaments past each other, and that this movement is brought about by the cyclical action of cross-bridges acting as independent force generators. That much is now well established; indeed Tregear and Marston (1979) consider it more informative to use the term 'cross-bridge theory' in place of the term 'sliding filament theory'. Furthermore we have a great deal of information about muscle struture and function that is relevant to how the cross-bridges work. But there are still many questions to be answered. If it were possible to crystallize the myosin S1 heads for high-resolution X-ray diffraction studies of the structure of the molecule, some of those answers might be a little nearer.

13
The activation of muscular contraction

The normal stimulus for the contraction of a skeletal muscle fibre in a living animal is an impulse in the motor nerve that innervates it. In the twitch muscles of vertebrates with which we are concerned in this chapter, this nerve impulse leads to a propagated action potential in the muscle fibre, which is then followed by a twitch contraction. The time relations of muscle action potential and twitch are shown in fig. 13.1, following electrical stimulation of a single fibre.

The sequence of these events is shown schematically in fig. 11.1. We have examined stages 1 to 4 of this sequence (the excitation processes) in earlier chapters of this book, and stage 6 (contraction) in the two preceding ones. In this chapter we consider how excitation of the muscle fibre membrane initiates contraction of the myofibrils in the interior of the fibre. This constitutes stage 5 of fig. 11.1, the excitation–contraction coupling process. In terms of fig. 13.1, then, how does the action potential produce the contraction?

Fig. 13.1. Electrical and mechanical responses of a single frog twitch muscle fibre to an electrical stimulus. The upper trace shows the action potential, recorded with an intracellular microelectrode, and the lower trace the isometric tension, recorded with a sensitive force transducer. Temperature 20 °C. (From Hodgkin and Horowicz, 1957.)

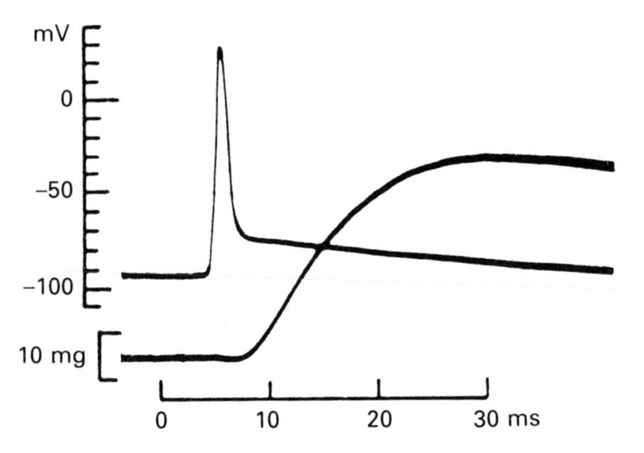

The excitation–contraction coupling process

The importance of depolarization of the cell membrane

When muscle fibres are immersed in solutions containing a high concentration of potassium ions, they undergo a relatively prolonged contraction known as a *contracture*. Contractures can also be produced by various drugs, such as acetylcholine, veratridine and others. In 1946, Kuffler showed that many of these substances produce depolarization of the cell membrane; furthermore, if the substance was applied locally the resulting contracture was limited to that part of the muscle fibre where depolarization occurred. The relation between membrane potential and tension in potassium contracture was determined quantitatively by Hodgkin and Horowicz (1960). They found that no contracture occurred when the membrane potential was more negative than about −55 to −50 mV; above this threshold value, tension increased rapidly with increasing depolarization, reaching a maximum above about −40 mV (fig. 13.2). These results show quite clearly that contracture tension is related to the degree of depolarization of the cell membrane.

What happens when the period of depolarization is very brief, as occurs during the action potential that elicits a twitch contraction? Adrian, Chandler and Hodgkin (1969), and later Costantin (1974) and Gilly and Hui (1980) investigated this question. They used voltage-clamp techniques to produce short depolarizations, tetrodotoxin to prevent action potential production, and visual observation to detect the mechanical threshold. The results showed that the mechanical threshold is commonly around −50 mV at pulse lengths of 20 ms or more, but is progressively higher at shorter pulse lengths, giving a strength–duration curve as is shown in fig. 13.3. Adrian and his colleagues interpreted this curve in terms of the release of an internal activator, with rate constants for the process dependent upon membrane potential.

We must digress a little here to examine a rather more subtle influence of membrane potential on the coupling process, which is seen in certain types of muscle. When a frog rectus abdominis muscle (a 'tonic' muscle) is depolarized by placing it in a solution containing a high potassium ion concentration, the resulting contracture lasts for several minutes. On the other hand, a frog sartorius muscle (a 'twitch' muscle) subjected to the same treatment contracts to maintain a steady tension for a few seconds and then relaxes; this relaxation is not accompanied by any change in membrane potential. A second contracture can only be obtained if the potassium ion concentration is sufficiently reduced for a short time. There is thus some kind of restitution or 'priming' process occurring. Hodgkin and Horowicz showed that the extent of this restitution is inversely related to the potassium ion concentration (and therefore to the membrane potential) in the intervening period between two exposures to a high potassium ion concentration (fig. 13.4). This 'priming' process does not occur in those muscles which show maintained potassium contractures.

The importance of calcium ions

It is clear that depolarization of the cell membrane cannot itself be the ultimate trigger for the contraction process. Apart from the difficulty of seeing how this could work, we know that glycerinated fibres will contract even though there is no membrane present. Now glycerol-extracted muscle fibres bathed in a solution containing ATP and magnesium ions are extremely sensitive to the calcium ion concentration; concentrations as low as 10^{-6} M are sufficient to cause some contraction and ATP splitting. Thus we might expect that depolarization of the cell membrane of an intact muscle fibre would cause an increase in the internal calcium ion concentration, so leading to contraction. There is considerable evidence, which we

Fig. 13.2. The relation between peak contracture tension and potassium ion concentration or membrane potential in single frog muscle fibres. (From Hodgkin and Horowicz, 1960.)

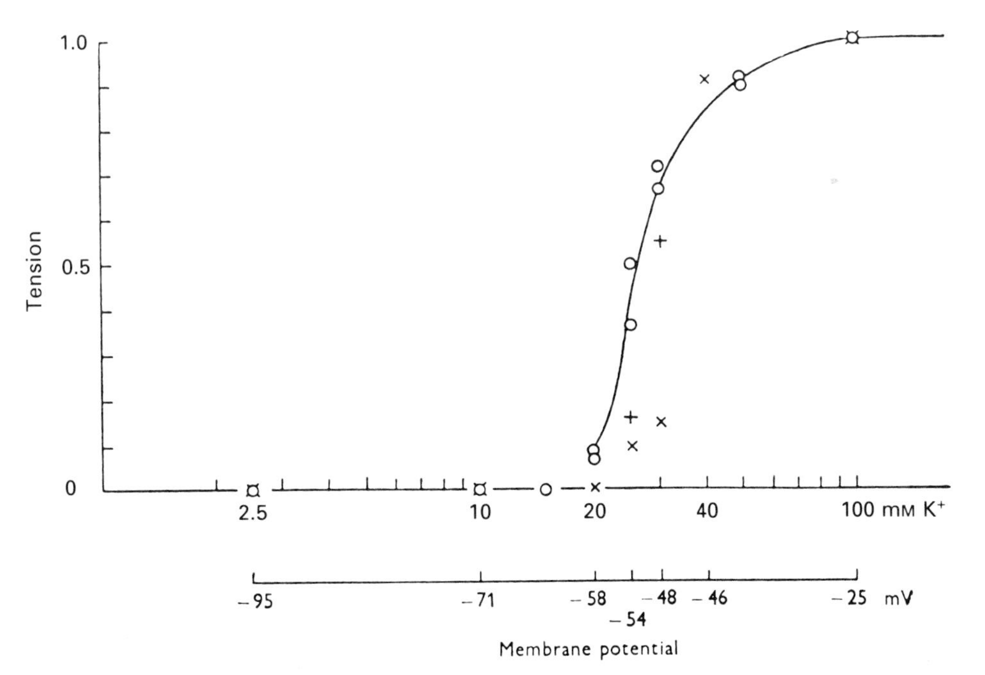

must now consider, that this is so (see Rüegg, 1986).

A number of investigations have been made on the effect of calcium ions on the potassium contractures of various muscles. Contractures eventually fail in the absence of calcium, and low calcium ion concentrations reduce the contracture tension (Niedergerke, 1956; Frank, 1960; Edman and Schild, 1962). An example of this effect in an insect muscle is shown in fig. 13.5; after treatment with a calcium-free solution containing the chelating agent ethylenediamine tetraacetate, the muscle did not contract on immersion in potassium chloride solution, but addition of calcium ions to the depolarized muscle was immediately followed by contraction (Aidley, 1965*a*).

Caldwell and his colleagues developed a technique whereby it is possible to inject solutions into the interior of the large muscle fibres of the spider crab *Maia*. Injection of potassium, sodium or magnesium ions or ATP did not produce any contraction, but the injection of calcium ions did (Caldwell and Walster, 1963). In further experiments (Portzehl *et al.*, 1964), calcium ion buffer solutions, prepared by mixing known quantities of calcium chloride and the chelating agent EGTA,* were injected, so that the free calcium ion concentration inside the fibre could be stabilized at a predetermined value. It was found that the threshold calcium ion concentration necessary for contraction was about 10^{-6} M. This is significantly close to the calcium ion concentration necessary to activate the ATPase system of isolated myofibrils (Weber and Herz, 1963).

The relation between calcium ion concentration and tension has been further investigated with experiments on 'skinned' muscle fibres (Hellam and Podolsky, 1969; Ashley and Moisescu, 1977; Stephenson and Williams, 1981, 1982). A muscle fibre is held under paraffin or silicone oil and its sarcolemma and surrounding connective tissues is removed with a fine needle, so leaving the interior of the fibre directly accessible to aqueous solutions. A sensitive force-transducer measures the tension produced by the fibre when solutions with different concentrations of calcium ions (using calcium–EGTA buffers) are applied.

* Ethylene glycol *bis* (β-aminoethyl ether)-N,N′-tetraacetate.

Fig. 13.3. Strength–duration relation for the mechanical activation of frog twitch muscle fibres, at 9 °C (circles) and at 20 °C (squares). The points are average values of the voltage-clamped membrane potential needed to produce a mechanical response which was just visible under the microscope. (From Gilly and Hui, 1980.)

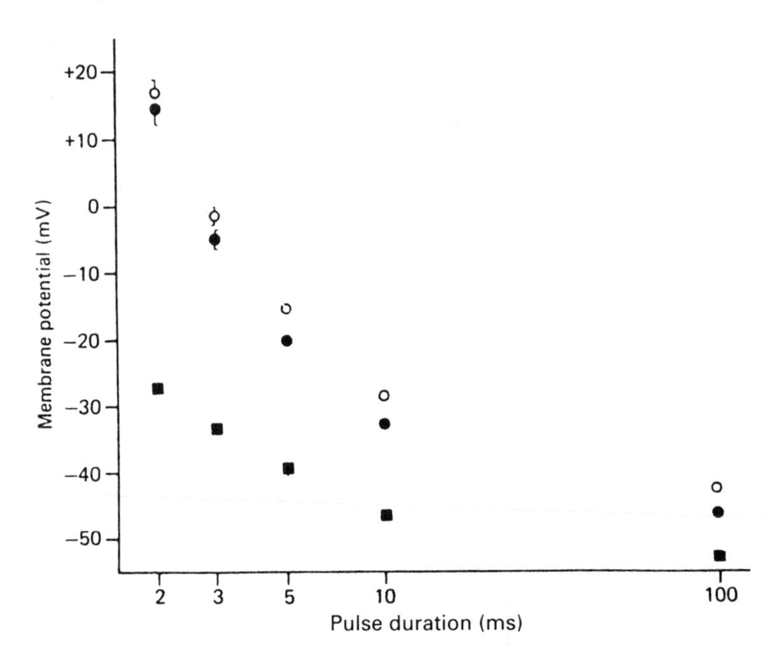

Fig. 13.4. Experiment to show the 'priming' process in a single frog muscle fibre. After a contracture induced by 190 mM K^+, the fibre was allowed to recover for 1 min in x mM K^+ before retesting with 190 mM K^+. (From Hodgkin and Horowicz, 1960.)

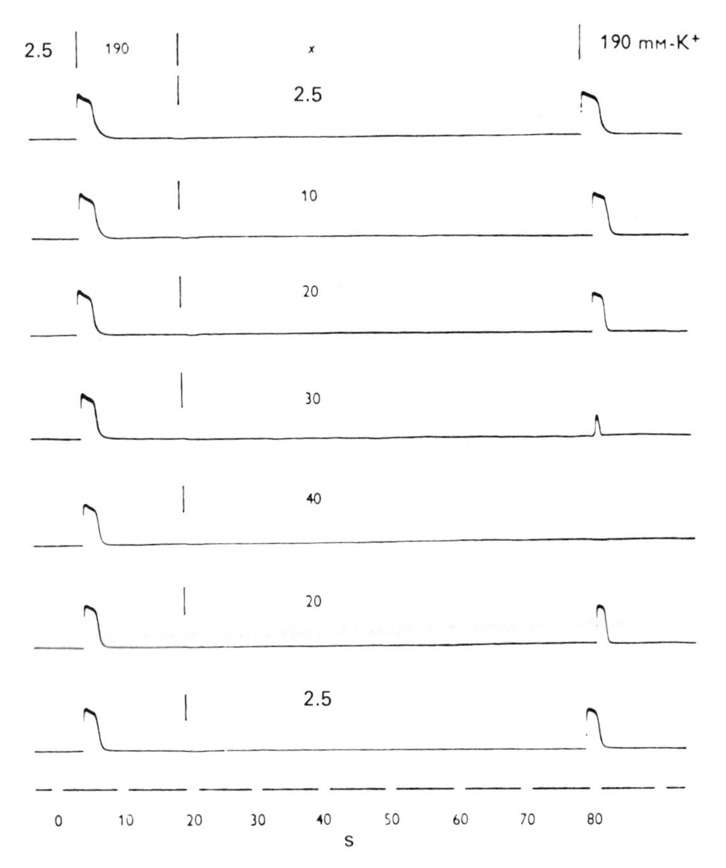

Fig. 13.5. The effect of calcium ion concentration on potassium contracture in a locust leg muscle. The thick horizontal lines under each record show the times of exposure to the high potassium solution. Pretreatment as follows: *a*, Ringer solution (4 mM Ca^{2+}); *b*, calcium-free solution containing a chelating agent; *c*, Ringer solution again. (From Aidley, 1965*a*.)

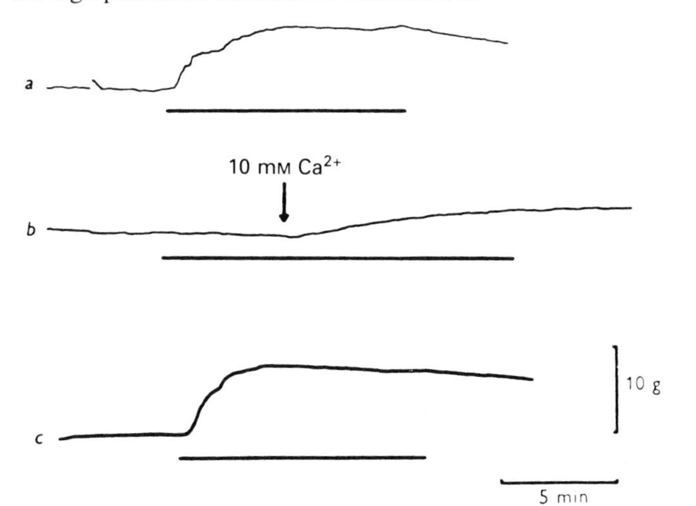

Fig. 13.6 shows the results of an experiment of this type done with a rat fast-twitch muscle fibre. The threshold for contraction is just below 10^{-6} M calcium ions and full contraction is reached by 10^{-5} M. Various factors affect the position of such curves. Increasing magnesium ion concentration and higher acidity (lower pH) push them to the right, i.e. the calcium-sensitivity of the system is reduced. Stretching the muscle makes the system slightly more sensitive to calcium, as is shown in fig. 13.6. There may also be differences between different muscles.

A very direct demonstration that the calcium ion concentration in the sarcoplasm rises immediately after stimulation was made by Ashley and Ridgeway (1968, 1970). Their experiments made use of the protein aequorin, isolated from a bioluminescent jellyfish, which emits light in the presence of calcium ions. Solutions of aequorin were injected into the large muscle fibres of the barnacle *Balanus nubilus*. When such a fibre was stimulated electrically it produced a faint glow of light, indicating the presence of calcium ions in its interior. Relative measurements of the internal calcium ion concentration at different times and under different conditions can be made by measuring the light emission with a photomultiplier tube; the time course of light emission after a stimulus can be called a 'calcium transient'.

Fig. 13.7 gives an example of the results to be got from this preparation, showing the effects of increasing stimulus strength (and therefore increasing depolarization) on the amount of calcium released and the subsequent tension development. Examination of this experiment reveals a number of interesting features. The size of the calcium transient increases with the degree of depolarization, beyond a certain threshold value. The degree of tension development rises with increasing size of the calcium transient. The calcium transient starts after a short latent period and begins to fall soon after the end of the stimulus pulse. But the tension keeps rising until the calcium transient is long past its peak and almost finished; the time course of the calcium transient is in fact very similar to the rate of change of tension during the rising phase of the tension change.

Calcium transients following stimulation were first demonstrated by means of a method which is only slightly less direct than that described above. Jöbsis and O'Connor (1966) used the calcium indicator substance murexide, whose absorption spectrum alters when it combines with calcium. Murexide was taken up by fibres of the sartorius muscle of toads after the substance had been injected into the whole animal some time previously. Then, by shining a monochromatic light beam through the muscle onto a photomultiplier tube, changes in calcium ion concentration could be detected after stimulation of the muscle fibre. The results were very similar to those obtained with the aequorin method from barnacle muscle fibres.

Fig. 13.6. The relation between calcium ion concentration (pCa) and force in skinned muscle fibres. Measurements were made at two different sarcomere lengths; the left-hand curve shows results from the longer length. pCa is given by $-\log_{10}[Ca^{2+}]$, where $[Ca^{2+}]$ is the molar concentration of calcium ions. (From Stephenson and Williams, 1982.)

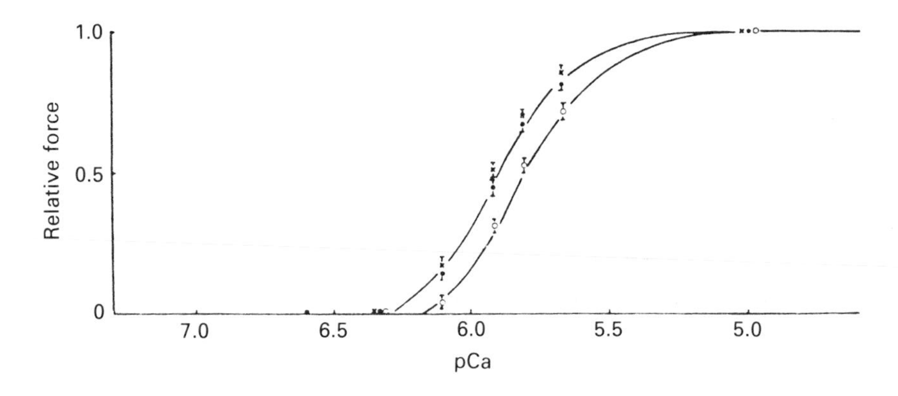

More recently, other metallochromic indicator dyes have been used to follow calcium ion concentrations in muscle cells. Examples are arsenazo III and antipyrylazo III, both useful because they are sensitive to lower calcium ion concentrations than is murexide (Scarpa, 1979; Blinks *et al.*, 1982). Thus Miledi, Parker and Zhu (1982) found that the calcium ion concentration in frog twitch muscles fibres rose to 8 μM soon after an action potential.

In general, it seems clear that the free calcium ion concentrations in skeletal muscle cells rise from a low level of a little over 0.1 μM at rest to up to 10 μM during activity.

The T *system*

The experiments that have been so far described lead to the suggestion that depolarization of the cell membrane causes an increase in the calcium ion concentration in the interior of the muscle fibre. A possible hypothesis as to how this is done is that depolarization releases calcium from the cell membrane or allows calcium ions to enter the muscle fibre from outside; these calcium ions would then diffuse into the interior of the fibre and activate the contractile system. Evidence compatible with this idea was obtained from radioactivity measurements on calcium fluxes; both the influx and efflux of calcium ions are increased on stimulation. There are, however, considerable objections to the idea. First, it seems that the amount of calcium entering the fibre during a twitch is too little to account for the necessary increase in calcium ion concentration inside the fibre. Bianchi and Shanes (1959) showed that the calcium influx during a twitch of frog sartorius muscle is 0.2 pmol cm^{-2}. Sandow (1965) calculated that such an influx would given an internal calcium ion concentration of 8×10^{-8} M (assuming the fibre diameter to be 50 μm), which, as he said, 'is too small by a factor of 10 to cause even threshold effects, and it would be too small by a factor of 100 to elicit maximum activation'. The second objection arises from a calculation by Hill (1948, 1949). He showed that the time taken for a substance released from the membrane to diffuse into the interior of the fibre would

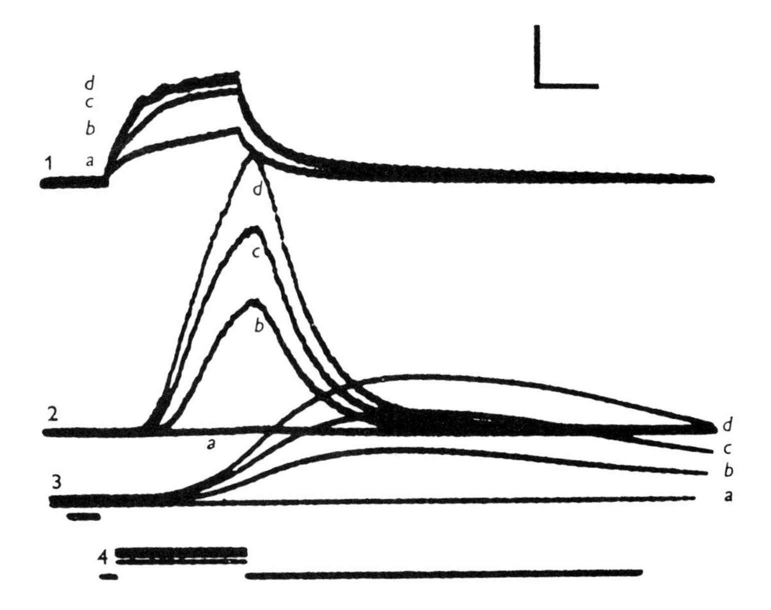

Fig. 13.7. Calcium transients in *Balanus* muscle fibres as measured by the aequorin technique described in the text. Trace 4 shows the stimulus pulse, which is applied via an intracellular silver wire electrode. Trace 1 shows the resulting depolarization, trace 2 the photomultiplier output (the calcium transient) and trace 3 the tension. *a*, *b*, *c* and *d* are the results of applying four different current pulses of increasing intensity. The calibration marks are 100 ms (horizontal), 20 mV (for trace 1) and 5 g (for trace 3). (From Ashley and Ridgeway, 1968.)

be much too long to account for the speed with which activation becomes maximal after the stimulus.

How then does excitation at the cell surface cause release of calcium inside the fibre? The first step in the solution of this problem was provided by the demonstration by A. F. Huxley and Taylor (1955, 1958) that there is a specific inward-conducting mechanism located (in frog sartorius muscles) at the *Z* line. In these experiments, the fibres were viewed by polarized light microscopy (so as to make the striation pattern visible) and stimulated by passing current through an external microelectrode, of diameter 1 to 2 μm, applied to the fibre surface. Hyperpolarizing currents did not produce contraction. Depolarizing currents did produce contraction, but only when the electrode was positioned at certain 'active spots' located at intervals along the *Z* line (fig. 13.8). In these cases the *A* bands adjacent to the *I* bands opposite the electrode were drawn together, and the extent to which this contraction passed into the interior of the fibre was proportional to the strength of the applied current.

At first it was thought that the inward-conducting mechanism was the *Z* line itself, but on repeating the experiments with crab muscle fibres (Huxley and Straub, 1958), it was found that the 'active spots' were localized not at the *Z* line but near the boundary between the *A* and *I* bands. This suggests that there is some transverse structure located at the *Z* lines in frog muscles and at the *A–I* boundary in crab muscles.

Such a structure was found by Porter and Palade (1957) in their study, by electron microscopy, of the endoplasmic (or 'sarcoplasmic') reticulum in various vertebrate skeletal muscles. The sarcoplasmic reticulum consists of a network of vesicular elements surrounding the myofibrils (fig. 13.9). At the *Z* lines in frog muscle, and at the *A–I* boundaries in most other striated muscles (including crab muscle), are structures known as 'triads', in which a central tubular element is situated between two vesicular elements. These central elements of the triads are in fact tubules which run transversely across the fibre and are known as the transverse tubular system or *T system*. There is no communication between the lumina of the *T* system and those of the sarcoplasmic reticulum vesicles, although their respective membranes are in close contact.

Fig. 13.8. The effect of local depolarizations on an isolated frog muscle fibre. Diagrams on the left show the resting condition, those on the right show the condition during passage of a depolarizing current. *a*, With the electrode opposite the *A* band; *b*, with the electrode opposite the *I* band (including the *Z* line). (Based on Huxley and Taylor, 1958.)

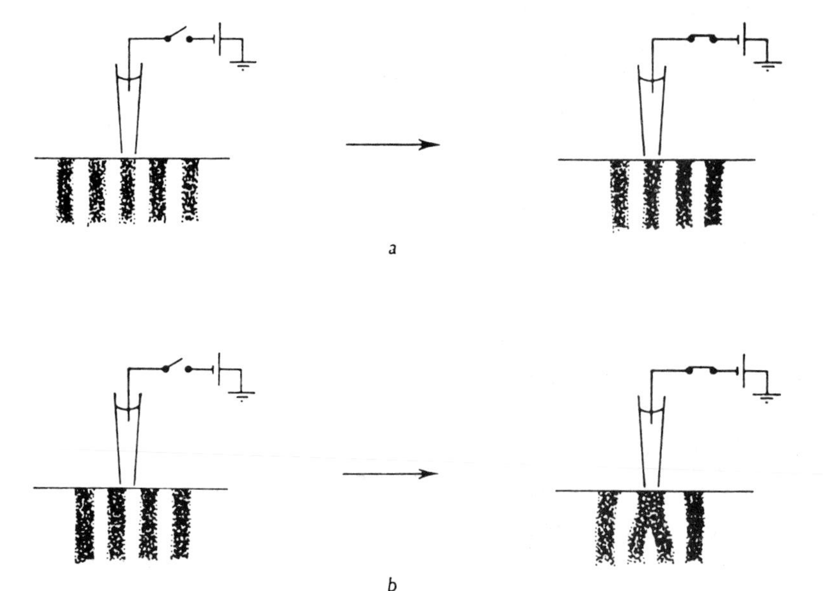

It is clear that the *T* system tubules are in just the right position to account for the 'inward-conducting mechanism' suggested by the experiments of Huxley and Taylor. Hence it is important to know whether or not the *T* system is connected to the cell membrane. In material fixed with osmium tetroxide, as in the earlier studies, the *T* system appears as a row of elongated vesicles, but in material fixed with glutaraldehyde it is clear that the system really is tubular, and in favourable preparations it is possible to see that the tubules are invaginations of the cell membrane (e.g. Franzini-Armstrong and Porter, 1964). Further convincing evidence that this is so was provided by H. E. Huxley (1964), from electron microscopy of muscle fibres which had been soaked in ferritin solutions. Ferritin is an iron-storage protein found in the spleen; because of its high iron content it is electron-dense, and the individual molecules can be seen by electron microscopy. In Huxley's experiments, ferritin appeared in the *T* system tubules, indicating that they must be in contact with the external medium. No ferritin appeared in the sarcoplasmic reticulum or in the rest of the sarcoplasm.

How does the electrical signal at the cell surface membrane travel down the *T* tubules into the interior of the fibre? There are two possibilities to be considered: either it is a passive process of electrotonic spread, or the *T* tubule membranes are electrically excitable and can conduct action

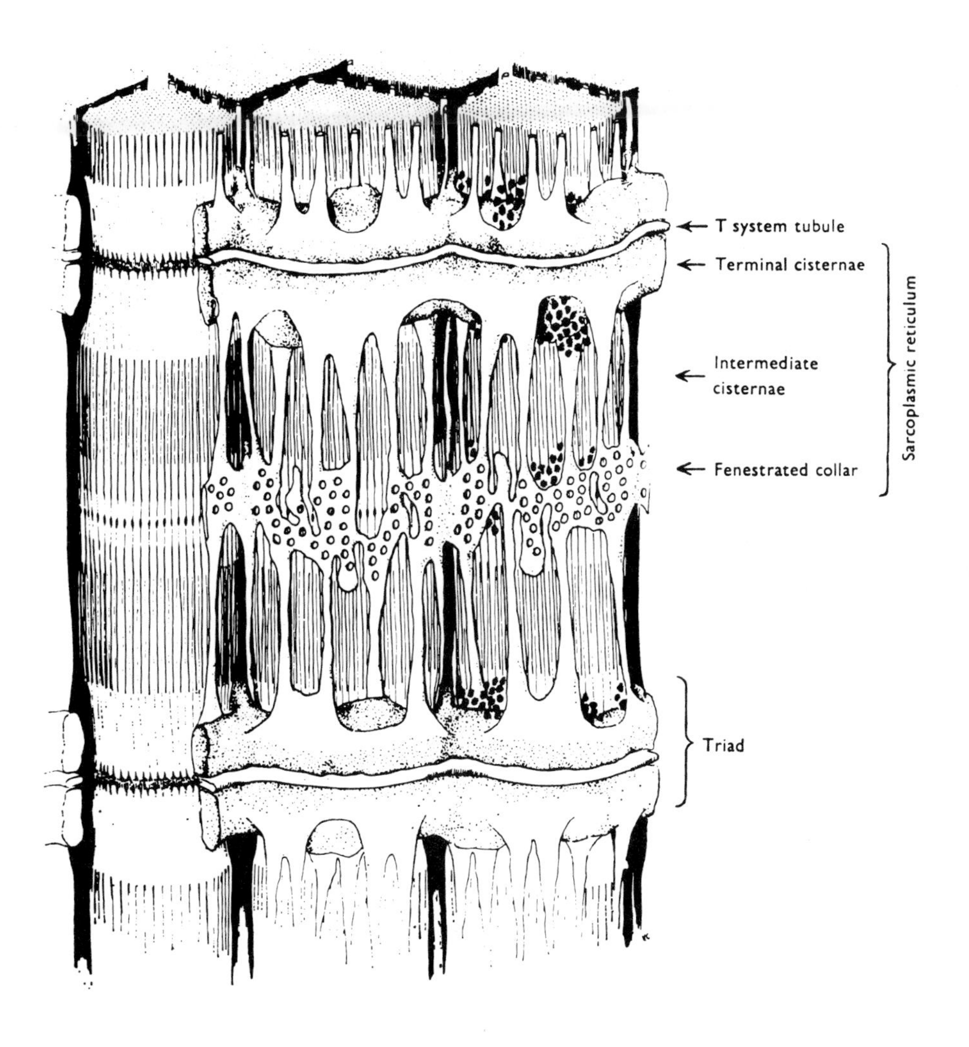

Fig. 13.9. The internal membrane systems of a frog sartorius muscle fibre. (From Peachey, 1965.)

potentials. The T tubules are too small for their membrane potentials to be measured directly, and hence it is necessary to use contraction of the myofibrils in different parts of the muscle fibre cross-section as an indicator of the inward spread of activity. Using voltage-clamped fibres, Costantin (1970) found that the inward spread of activity was reduced in fibres treated with tetrodotoxin or a low-sodium solution. This suggests that inward propagation is an electrically excited process with a propagated action potential involving sodium ion flow, just as at the cell surface membrane.

The speed of inward propagation has been measured in a most ingenious way by Gonzalez-Serratos (1971). He placed single muscle fibres in a Ringer solution containing gelatine, which was then allowed to cool and set. He then compressed the gelatine block so as to shorten the muscle fibre without bending it. This made the myofibrils wavy. But when the myofibrils were activated after stimulation of the fibre, they shortened and so became straight. Using a high-speed movie camera, it was possible to measure the time at which straightening of the myofibrils occurred at different points in the fibre cross-section, and so to determine the speed of propagation of the activating signal. This was about 7 cm s^{-1} at 20 °C, with a Q_{10} of 2. Thus in a fibre 100 μm in diameter the wave of activation would take about 0.7 ms to propagate from the surface to the centre.

The sarcoplasmic reticulum

If a skeletal muscle is homogenized, the myofibrils in the homogenate do not contract on the addition of ATP, and the rate of ATP splitting is very low. However, if the myofibrils are isolated from the rest of the homogenate by low speed centrifugation and subsequent washing, then they do contract in the presence of ATP and have a high ATPase activity. Marsh (1952) suggested that the ATPase activity of the myofibrils was inhibited by a 'relaxing factor' present in the muscle homogenate. Bendall (1953) showed that this factor prevented the contraction of glycerol-extracted muscle fibres. A crucial observation on the nature of relaxing factor was made by Portzehl (1957). She found that if a relaxing factor extract is centrifuged for some time at a high speed, the relaxing factor activity is confined to the resulting precipitate, which is of course particulate; the supernatant is without relaxing effect. Electron microscopy of this precipitate by Nagai, Makinose and Hasselbach (1960) and others showed that it consists of small vesicles, often ellipsoidal or tubular in shape. It would seem that these vesicles are derived from the sarcoplasmic reticulum.

The vesicular fraction is able to accumulate calcium from solutions containing ATP, magnesium ions and a small amount of calcium ions (see Hasselbach, 1964; Hasselbach and Oetliker, 1983; Martonosi and Beeler, 1983). This calcium uptake is associated with ATP splitting; Hasselbach and Makinose (1963) calculate that two calcium ions are taken up for each ATP molecule split at calcium ion concentrations above 10^{-7} M, this ratio falling to one in the range 10^{-7} to 10^{-8} M. When the calcium ion concentration is below 10^{-8} M, the rate of ATP splitting is very low, and no further accumulation of calcium occurs. These experiments show that the vesicles of the sarcoplasmic reticulum can reduce the calcium ion concentration to below that necessary for contraction, by means of an ATP-driven 'calcium pump' in the vesicular membrane. The action of the pump is perhaps aided by binding of calcium inside the sarcoplasmic reticulum by the protein calsequestrin (MacLennan and Wong, 1971).

The pump itself is a calcium–magnesium-activated ATPase of molecular weight about 110 kDa, which is firmly bound in the sarcoplasmic reticulum membrane. Freeze-fracture electron micrographs show numbers of particles densely packed on the cytoplasmic side (see Franzini-Armstrong, 1986). Comparison of the number of particles with the ATPase content suggests that they are oligomers of around three or four molecules each (Inesi and Scales, 1976).

The amino acid sequence of the calcium–ATPase molecule has been determined by cDNA sequencing (MacLennan *et al.*, 1985). The molecule has a globular region containing the ATPase site in the cytoplasm, connected via a stalk to a series of ten hydrophobic α-helices anchored in the membrane.

Pumping of calcium ions from the myofibrils into the sarcoplasmic reticulum appears to be the means whereby relaxation is brought about. We have seen that activation of contraction is effected by a sudden increase in myofibrillar calcium ion concentration, and so it seems very likely that this calcium is released from the sarcoplasmic reticulum.

Somlyo and her colleagues have produced some direct evidence for this view in their studies of the distribution of calcium in muscle cells using electron probe microanalysis (Somylo *et al.*, 1977, 1981). Bombardment of atoms by electrons leads to emission of X-rays, and the frequency of the X-rays is different for the atoms of different elements. Thus one can place a section in the electron microscope, focus the electron beam on a particular part of it, and, with suitable equipment, measure the X-ray spectrum that ensures. The spectrum contains a series of peaks corresponding to the elements present, and the height of each peak is proportional to the quantity of the element. The experiments were done on very thin sections of rapidly frozen frog muscles.

The results showed that 60–70% of the total calcium in relaxed muscle fibres was in the terminal cisternae of the sarcoplasmic reticulum. During a 1.2 s tetanus, 59% of this was released into the cytoplasm, enough to raise the cytoplasmic total calcium concentration by about 1 mM and equivalent to the amount of calcium-binding proteins (troponin and parvalbumin) in frog muscle.

How does the T *system signal to the sarcoplasmic reticulum?*

The nature of the link between depolarization of the *T* system tubule and release of calcium ions from the sarcoplasmic reticulum has long been a major problem. Clues to its solution have emerged from investigations on charge transfer in the *T* tubules and on the fine structure of the triad.

Since the activation of contraction is voltage-dependent, it may be that activation depends upon the movement of charged particles in the potential gradient of the *T* tubule membrane. The problem was investigated by Schneider and Chandler (1973), using techniques similar to those used in investigations of the gating currents in nerve axons. They found that asymmetrical displacement currents that appear to be caused by the movement of membrane charge can indeed be detected. Charge flow was not evident at membrane potentials more negative than −80 mV and saturated above about −10 mV. The time course of the currents was much slower than that of nerve gating currents, by 20 to 100 times. Evidence that these charge movements are related to excitation–contraction coupling comes from a comparison with the mechanical 'priming' process discovered by Hodgkin and Horowicz. After a maintained depolarization followed by a return to the normal resting potential, it takes some seconds for the membrane charge movements to regain their full size; and the time course of their recovery is closely paralleled by the recovery of the ability to contract (Adrian *et al.*, 1976).

How is this intramembrane charge movement related to calcium ion release? The question has been investigated by Melzer and his colleagues (1986*a*, *b*). They used cut segments of single frog muscle fibres, with the ends isolated from the middle by vaseline seals. Solutions in the end pools were in contact with the interior of the fibre, and contained the calcium-sensitive dye antipyrylazo III. The solution in the middle pool was in contact with the outside of the fibre and contained tetrodotoxin and TEA to stop sodium and potassium currents; this allowed the asymmetric currents representing intramembrane charge transfer to be measured. The fibres had to be stretched to the point where there was no filament overlap so as to avoid movement artefacts in the measurement of the calcium transients.

Intramembrane charge movements and calcium transients were determined by voltage clamping a fibre at different membrane potentials. The calcium transient, of course, shows the quantity of calcium which is bound to the dye, and it is necessary to make some assumptions about calcium movements in order to calculate the rate of release of calcium from the sarcoplasmic reticulum. The results of such a calculation are shown in fig. 13.10, which also shows an example of the charge movement.

Fig. 13.11 shows the relations between membrane potential, charge movement and peak calcium release rate as determined in these experiments. The close relation between the three quantities suggests strongly that there is a causal relation between them: that depolarization determines charge movement, which itself produces calcium release. The results are fitted by a three-state model in which the gating units undergo transition from A to B to C, with charge transfer accompanying both transitions but only state C allowing release of calcium ions.

Studies on the charge movements have not produced clear evidence as to just how they are able to produce release of calcium ions. Chandler and his colleagues (1976) suggested that each charge forms part of a long molecule which extends from the *T* tubule membrane to the sarcoplasmic reticulum membrane, and that movement of the charge might unplug channels in the sarcoplasmic reticulum membrane, so allowing calcium ions to escape.

The other clue to the problem comes from studies on the structure of the triad by Franzini-Armstrong and her colleagues (1970*a*, 1975, 1986). She found that the *T* tubule and its adjacent sarcoplasmic reticulum sac are connected by an array of electron-dense structures called 'feet' (fig. 13.12). Examination of the surface structure of isolated sarcoplasmic reticulum vesicles shows that they are relatively large structures each consisting

Fig. 13.11. How membrane potential affects intramembrane charge movement (circles) and peak rate of calcium release (squares) in frog twitch muscle fibres. The points were determined as shown in fig. 13.10. (From Melzer *et al.*, 1986*b*.)

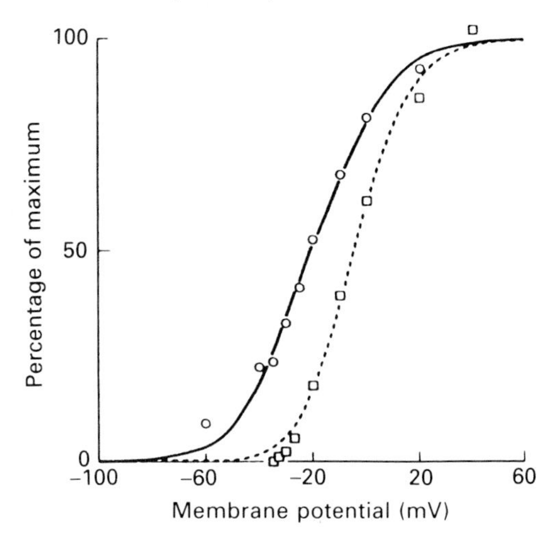

Fig. 13.10. Calcium release and intramembrane charge movement in frog muscle fibres. Calcium transients were measured from the change in light absorbance at a wavelength of 700 nm in a muscle fibre containing antipyrylazo III: *a* shows calcium transients during clamped depolarizations to +40 mV; *b* shows the rate of calcium release from the sarcoplasmic reticulum, calculated from these transients; *c* shows the intramembrane charge movements associated with a similar depolarization. (From Melzer *et al.*, 1986*b*.)

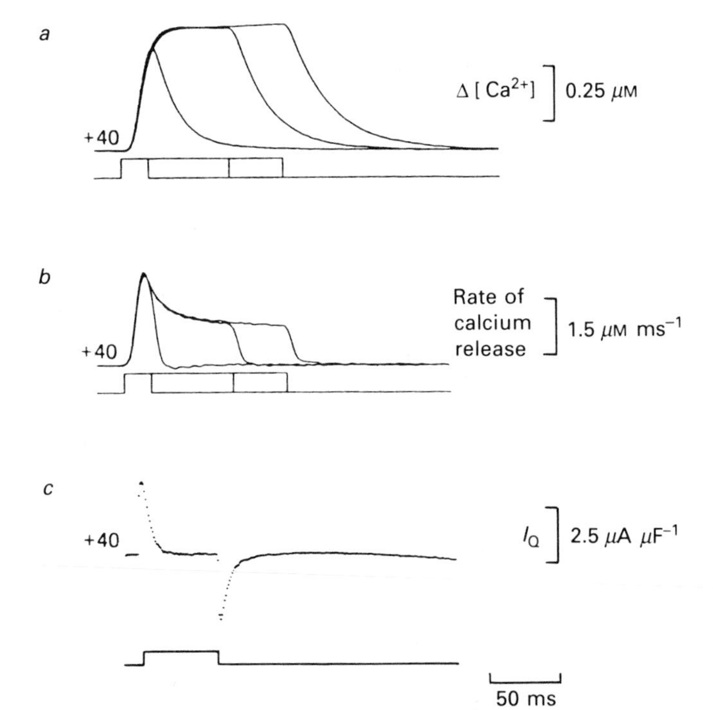

of four subunits (Ferguson *et al.*, 1984). It is tempting to see these feet as possible channels of information transfer.

Further evidence of this view comes from experiments on isolated ryanodine receptors by Lai *et al.* (1988). Ryanodine is a plant alkaloid which locks the calcium-release channels of the sarcoplasmic reticulum open, hence radioactive ryanodine can act as a marker for these channels during their isolation. The receptor is a large protein (molecular weight 400 kDa) which aggregates, apparently in groups of four, to form particles with a characteristic clover-leaf appearance very like that of the feet seen at the triad. It can be incorporated into lipid bilayers, when it forms channels with relatively high conductances, up to 595 pS in some conditions. These results suggest that the feet and the channel for the release of calcium ions are different parts of the same molecule, and fit very well with the speculations of Chandler and his colleagues in 1976.

Fig. 13.12. Longitudinal section of toadfish swimbladder muscle, showing the connections ('feet') between the *T* tubules and the sarcoplasmic reticulum. Magnification 180000. (Photograph kindly supplied by Dr C. Franzini-Armstrong.)

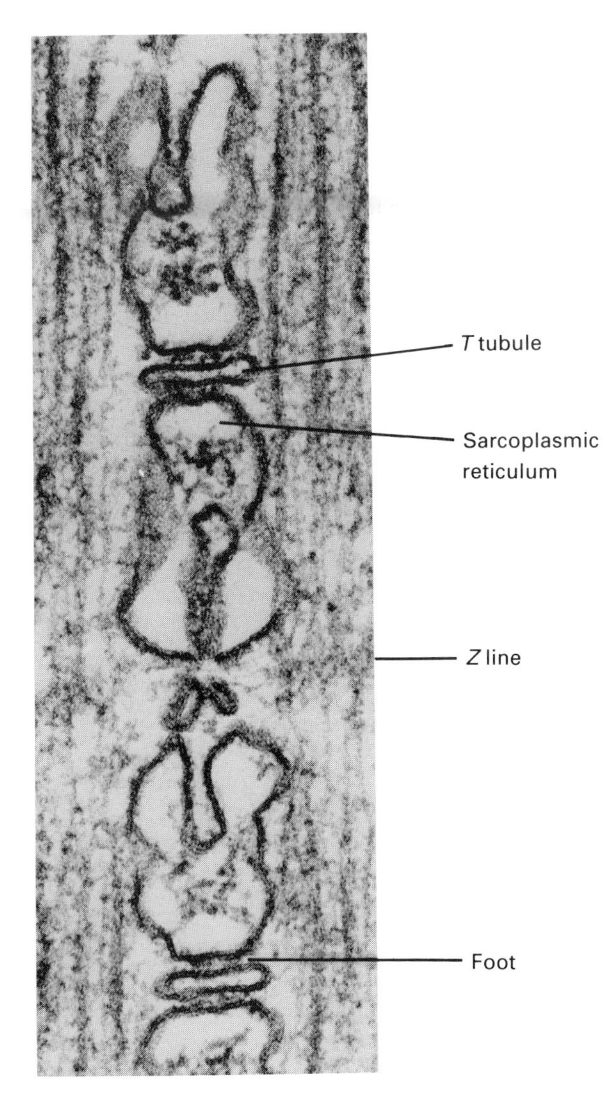

The molecular basis of activation

As mentioned in the previous chapter, a mixture of purified F-actin plus purified myosin shows ATPase activity in the absence of calcium ions, but 'natural' actomyosin (the complex extracted from minced muscle with strong salt solutions) will only split ATP if there is a low concentration of calcium ions present. Ebashi and his colleagues found that this responsiveness to calcium ions is dependent upon the presence of troponin and tropomyosin (Ebashi *et al.*, 1969). So it looked as though an understanding of the mechanism of activation must depend upon knowledge of the behaviour of these two proteins. We have examined some of the properties of tropomyosin already (p. 271), but troponin deserves a further look.

Troponin

Troponin was first isolated by Ebashi and Kodama (1966*a*, *b*). Further work showed that it consists of three components, called troponin-C (18 kDa), troponin-I (21 kDa) and troponin-T (31 kDa) by Greaser and Gergely (1973; see also Perry, 1986).

Troponin-C has four calcium-binding sites in each molecule. It binds firmly to troponin-I and less firmly to troponin-T. Conformational changes occur when calcium is bound. The four sites show homology in their amino acid sequences; each consists of a loop between two α-helices (Collins *et al.*, 1973). There are considerable sequence homologies with other calcium-binding molecules such as parvalbumin, calmodulin and some of the myosin light chains. High-resolution X-ray diffraction studies on troponin-C crystals show that the molecule has a curious dumbbell structure (fig. 13.13), with two distinct calcium-binding domains connected by a long section of α-helix (Herzberg and James, 1985).

The most notable property of troponin-I is that it inhibits actomyosin ATPase, an effect which is much enhanced if tropomyosin is present. This inhibition is relieved by troponin-C if it is binding calcium; thus the presence of calcium ions switches off the inhibition produced by the troponin-I–troponin-C complex.

Troponin-T is an elongated molecule which binds firmly to tropomyosin and troponin-I, and apparently less firmly to troponin-C. Thus it serves to link the whole troponin complex to tropomyosin.

Overall, then, the troponin system forms a complex which binds to tropomyosin and actin and which inhibits actomyosin ATPase unless calcium ions are present. It is not hard to believe that the structural arrangement of this system would be altered by conformational change in one of its members, such as occurs when calcium is bound by troponin-C. Such a rearrangement in the thin filament might well form the basis of the activation process.

Fig. 13.13. The structure of troponin-C, as determined by X-ray diffraction of crystals of turkery troponin-C and heavy metal derivatives of it. The four calcium-binding loops are labelled I to IV; only III and IV were occupied by calcium under the conditions used. (From Herzberg and James, 1985.)

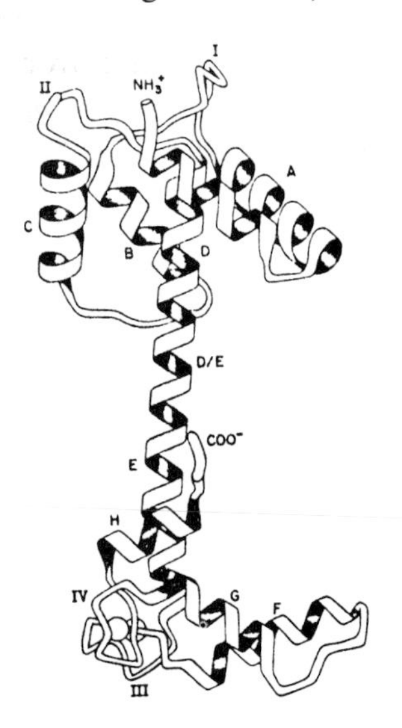

Thin filament activation

Direct evidence that the structure of the thin filaments alters during activation has been provided by X-ray diffraction measurements. The details of the actin pattern alter on contraction; in particular the second layer line at about 180 Å becomes much more evident and the sixth layer line at 59 Å also increases in intensity somewhat (Vibert *et al.*, 1972; Huxley, 1973). Calculations by Parry and Squire (1973), Haselgrove (1973) and Huxley showed that these changes could have been produced by a movement of tropomyosin in the groove between the two long-pitch helices of actin mononers in the thin filament. Fig. 13.14 shows H. E. Huxley's picture of this phenomenon in 1973.

Fig. 13.14. The steric blocking model of thin filament activation as proposed in 1972. The diagram is based on X-ray diffraction studies and three-dimensional reconstruction from electron micrographs. A thin filament is seen in cross-section with actin (A) and tropomyosin (Tm) molecules. Two myosin S1 heads are shown in the position they were thought to occupy in decorated thin filaments. Tropomyosin positions are shown for the muscle at rest (dotted circle) and when activated (solid contours). The probable positions and shapes of the protein molecules have been revised since 1972, but the conclusion that the tropomyosin molecule moves into the 'groove' between the actin monomers remains. The steric blocking model suggests that until this movement takes place the S1 heads are physically unable to attach to the actin monomers. (From Huxley, 1973.)

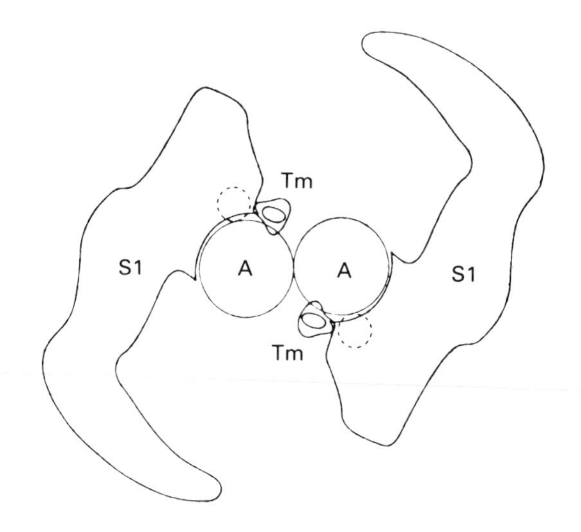

Confirmation for this idea has been sought from electron micrographs of thin filaments, using the three-dimensional reconstruction technique described earlier (Wakabayashi *et al.*, 1975; Taylor and Amos, 1981; Amos, 1985; Toyoshima and Wakabayashi, 1985). By comparing models of actin–tropomyosin–troponin-T–troponin-I filaments with those of actin–tropomyosin–S1(myosin) it is possible to look for changes which might be expected to accompany activation.

But while there is agreement that there really are changes in the structure of the filaments on activation, successive investigations have produced different pictures of the details of thin filament structure. Difficulties arise in ascribing particular regions of the models to the various protein molecules involved; a number of interpretations are shown in fig. 13.15. Perhaps we have to wait for improvements in the resolution of these techniques before we can be sure that we have the details right.

To summarize, then, it seems clear that activation of the thin filament begins with combination of calcium ions with troponin-C. This produces a change in shape of the troponin complex so that the tropomyosin molecule is rolled further into the groove between the actin monomers. Huxley (1973) suggested that such a movement would expose sites on the actin monomers which would allow the binding of the S1 myosin heads, and that such actin–myosin binding could not happen at rest because the tropomyosin was in the way. Each tropomyosin molecule would thus unblock seven actin monomers. This idea is known as the steric blocking hypothesis. It may not be the whole story (Chalovich and Eisenberg, 1982, suggest that the troponin–tropomyosin system affects the kinetics of actomyosin ATPase) but it certainly seems to be a very persuasive suggestion.

The timing of thin filament activation

H. E. Huxley and his colleauges have extended their time-resolved X-ray diffraction measurements using synchroton radiation to a study of the activation process (Kress *et al.*, 1986). They looked at the increase in the intensity of the 179 Å off-meridional reflection (the actin second-layer line) after stimulation; we have seen that this change had been interpreted as a lateral movement of tropomyosin across the thin filament surface.

Fig. 13.15. Successive interpretations of the structure of the actin–tropomyosin–S1 complex, as determined by three-dimensional reconstruction from electron micrographs. The diagrams show cross-sections of decorated thin filaments and the arrowheads point towards us, i.e. we are looking towards the *Z* line. The interpretations are due to: *a*, Moore *et al.*, 1970; *b*, Taylor and Amos, 1981; *c*, Amos *et al.*, 1982; *d*, Toyoshima and Wakabayashi, 1985. S1, myosin heads; Ac, actin; Tm, tropomyosin. (From Toyoshima and Wakabayashi, 1985.)

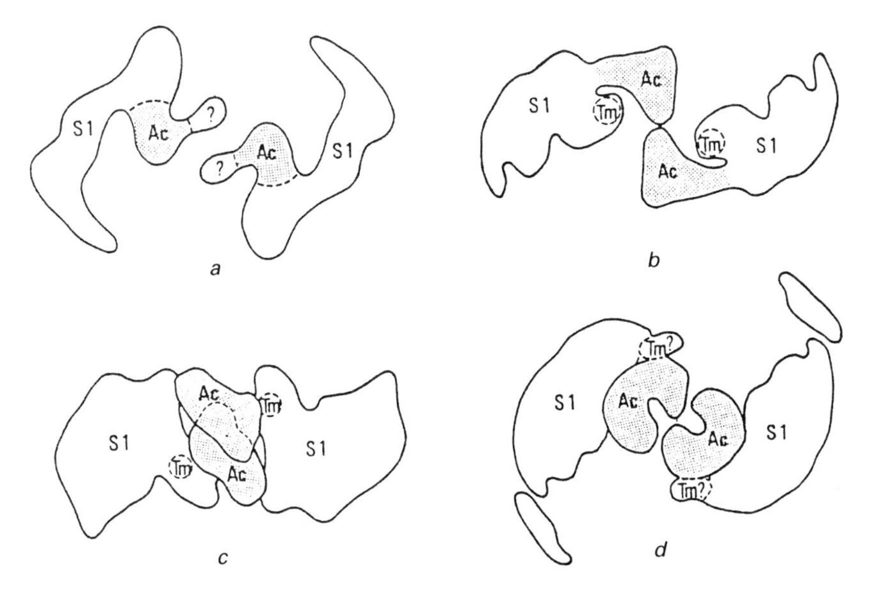

The results of these experiments were clear-cut (fig. 13.16). The increase in the 179 Å reflection intensity was the first indication of structural change in the contractile apparatus, reaching half its final value by about 17 ms after the stimulus. This was considerably quicker than the changes attributed to movement of the myosin heads: the half-times for change of the equatorial pattern and of the 429 Å myosin layer line were at least 25–30 ms (see fig. 12.13). Tension changes were slower still.

When muscles were greatly stretched so that the thick and thin filaments were no longer overlapped, change in the intensity of the 179 Å reflection on stimulation still occurred and followed the same time course. This implies that tropomyosin movement is independent of and precedes cross-bridge attachment. At these lengths the changes on the equator and at 429 Å, attributed to cross-bridge movement, no longer occur.

On relaxation at normal lengths the fall in intensity of the 179 Å reflection is somewhat more rapid than the fall in tension. In greatly stretched muscles, however, the fall in intensity is considerably quicker (fig. 13.17). In one set of experiments, the average time for the intensity to fall to half its

Fig. 13.16. Time-resolved X-ray diffraction measurements on activation, from a frog twitch muscle at 5 °C. The curves show the intensity increases of the second actin layer-line at 179 Å (squares) and of the equatorial (1, 1) reflection (black circles), following an electrical stimulus. These curves indicate the movements of tropomyosin and of the myosin heads respectively. White circles show the development of isometric tension. (From Kress *et al.*, 1986.)

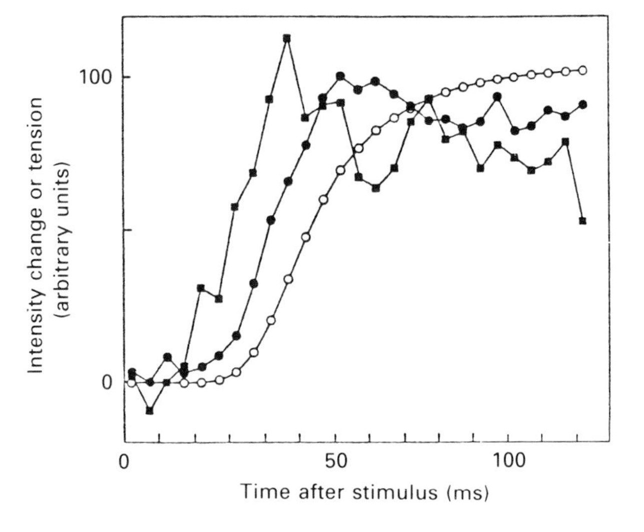

Fig. 13.17. Time-resolved X-ray measurements of the amplitude change of the 179 Å second actin layer-line reflection during relaxation of a frog semitendinosus muscle. Measurements were at normal length (black squares) and in a greatly stretched muscle (white squares). White circles show the tension change. Tetanic stimulation; the time scale is compressed after 300 ms. (From Kress *et al.*, 1986.)

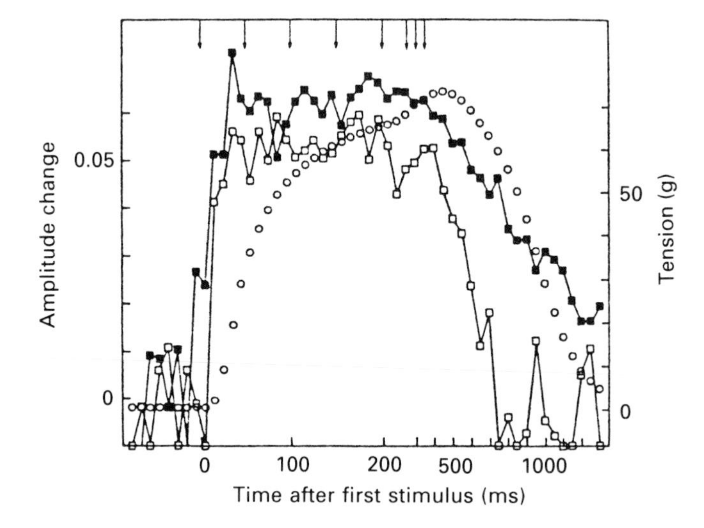

value after the end of stimulation was 407 ms at normal lengths, but only 128 ms in greatly stretched muscles; the half-time for the fall in tension was about 620 ms. This suggests that the presence of attached cross-bridges may interfere with the return of the tropomyosin to its resting position.

These beautiful experiments provide excellent confirmation for the view that movement of tropomyosin further into the groove of the thin filament is the essential prerequisite for the attachment of the cross-bridges. They fit well with the steric blocking hypothesis, although they do not rule out the possibility that tropomyosin movement is associated with allosteric effects on actin structure.

14
The comparative physiology of muscle

In the two previous chapters, we have been very largely concerned with the properties of vertebrate twitch skeletal muscle, exemplified by the sartorious of the frog. But it must be realized that the frog sartorious represents only one of a considerable variety of types of muscles, and a rather specialized one at that. In this chapter we shall attempt to broaden the picture a little by considering some of the variations in the processes of excitation and contraction in different muscles.

Excitation processes and fibre types

Vertebrate skeletal muscles

The skeletal muscle fibres of frogs and toads fall into two distinct classes: *twitch* fibres and *tonic* fibres.* The sartorius is composed entirely of twitch fibres. Tonic fibres are found to varying extents in other muscles; the rectus abdominis is particularly rich in them.

Most twitch fibres have only one motor endplate, although a few may have two. The cell membrane is electrically excitable and, as we have seen in chapter 7, will produce an all-or-none propagated action potential if it is sufficiently depolarized. The end-plate potential is sufficiently large to cause such a depolarization, so that, just as in the nerve axon, excitation consists of 'all-or-nothing' events.

The excitation process in tonic fibres is quite different (Kuffler and Vaughan Williams, 1953*a*, *b*). The fibres are electically inexcitable, so that no propagated action potentials occur. There are a large number of neuro-muscular junctions on each fibre; this condition is known as *multiterminal innervation* (fig. 14.1*b*). Each motor impulse causes the release of only a relatively small amount of transmitter substance, so that the end-plate potentials are small in size. These end-plate potentials show pronounced summation on repetitive stimulation; thus the amount by which the fibre membrane is depolarized is dependent upon the frequency of the nervous input. There is negligible contractile response to a single stimulus; with repetitive stimulation the tension rises with increasing stimulus frequency, reaching a maximum at about 50 impulses per second. It seems probable that the tonic fibres are used for maintained low-tension contractions such as are involved in the maintenance of posture, whereas the twitch fibres are used for rapid movements.

Electron microscopy shows that frog tonic muscle fibres differ from twitch fibres in a number

Fig. 14.1. Varieties of innervation in muscle fibres. *a*: Uniterminal innervation. *b*: Multiterminal innervation. *c*: Multiterminal and polyneuronal innervation.

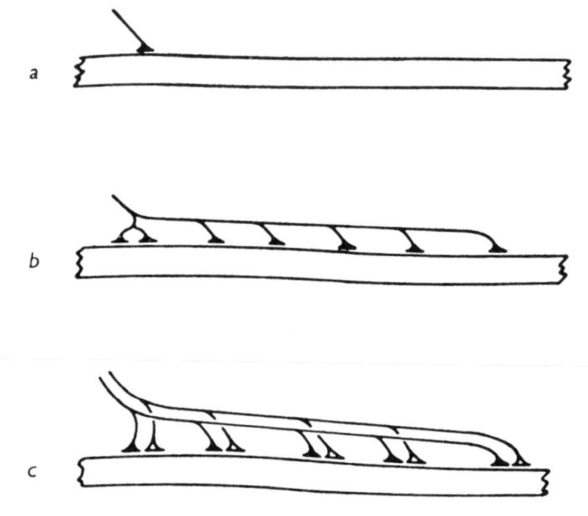

* Twitch and tonic fibres are frequently referred to as 'fast' and 'slow' fibres respectively. Since this nomenclature can lead to some confusion, we shall not adopt it here.

of ways. There are no *M* lines, the *Z* lines are thicker, there is much less sarcoplasmic reticulum and there are fewer triads in tonic fibres (Peachey and Huxley, 1962; Page, 1965).

Lännergen (1979) found a third type of muscle fibre in *Xenopus*. These intermediate fibres had multiterminal innervation and showed maintained potassium contractures just like tonic fibres, but would produce a slow twitch on electrical stimulation.

The muscle fibres of fishes, with some exceptions, fall into two classes which seem to correspond to those of amphibians (Bone, 1978; Johnston, 1980). The tonic fibres are usually reddish since they contain the oxygen-storage pigment myoglobin, and are located peripherally. They differ from the tonic fibres of amphibians in having an *M* line, a well-developed sarcoplasmic reticulum and extensive triads; they are usually well supplied with mitochondria. White (twitch) fibres have rather more sarcoplasmic reticulum but very few mitochondria.

These cytological features are related to the functions of the fibres in the living fish. The red (tonic) fibres are used during continuous cruising at relatively low swimming speeds, and so they will require a continuous supply of ATP produced by oxidative metabolism in the mitochondria. The white (twitch) fibres are responsible for short bursts of high-speed activity, which can be supported by an anaerobic energy supply. Nevertheless it is clear that in many species white muscle fibres are also involved in continuous swimming at faster speeds (Hudson, 1973; Greer Walker and Pull, 1973; Johnston *et al.*, 1977).

Birds also have two types of fibres, with uniterminal and multiterminal innnervation respectively, but those with multiterminal innervation are electrically excitable and may show propagated action potentials (Ginsborg, 1960).

Almost all the skeletal muscles of mammals are composed entirely of twitch fibres. Tonic muscle fibres are found in some of the muscles innervated by the cranial motor nerves, including the extraocular muscles which determine the direction of gaze, the tensor tympani of the middle ear, and some of the muscles of the larynx. The twitch fibres are of two types, named fast-twitch and slow-twitch according to their speeds and duration of contraction (Close, 1972; Buller, 1975). Sometimes whole muscles are composed of the same type of fibre. The soleus, for example, is a slow-twitch muscle while the gastrocnemius (which is larger and acts in parallel with it) contains a majority of fast-twitch fibres. Slow-twitch muscles are usually used in the control of posture, whereas fast-twitch muscles are used for locomotory and other movements.

Fast-twitch fibres have been further divided into two or three types (see Gauthier, 1986). The criteria for this division include the number of mitochondria present and the relative importance of oxidative and glycolytic pathways in respiration (Peter *et al.*, 1972), the ATPase activity at acid or alkaline pH (Brooke and Kaiser, 1970), the contraction time of the fibres (i.e. the time between the onset of a twitch contraction and its peak) and their resistance to fatigue (Burke *et al.*, 1973).

There are thus three main types of twitch fibre in mammals: (1) red (slow), type 1, SO (slow oxidative) or S (slow) fibres, (2) red (fast), type 2A, FOG (fast oxidative-glycolytic) or FR (fast fatigue-resistant) fibres, and (3) white, type 2B, FG (fast glycolytic) or FF (fast fatigue-sensitive) fibres. Table 14.1 summarizes their properties. This classification is simplified in two respects. Firstly, fibres intermediate between types 2A and 2B (called 2C) are sometimes found. Secondly, the subdivisions of the fast-twitch fibres by one criterion may not always agree with those by another scheme; it may not always be correct to equate 2A and 2B fibres precisely with FOG and FG fibres, for example (Green, 1986). Although the motor units in a muscle may show a considerable range of physiological and histochemical properties, the fibres of any one motor unit are always all of one type (Burke *et al.*, 1973; Nemeth *et al.*, 1981).

Arthropod skeletal muscles

All arthropod muscles are multiterminally innervated. In many of them the cell membrane is electrically excitable, so that responses to nervous stimulation may look very similar to those recorded at the neuromuscular junction of a vertebrate twitch fibre (fig. 14.2*a*). However, on closer investigation, it is found that these electrically excited responses are not 'all-or-nothing' propagated action potentials, but graded responses whose size is roughly proportional to the initial

Table 14.1. *Twitch fibre types in mammalian muscles* (simplified after Gauthier, 1986)

Muscle fibre type	1 red (slow) SO S	2A red (fast) FOG FR	2B white FG FF
Mitochondria and SDH	high	high	low
Z-line	wide	wide	narrow
ATPase at pH 10.4	low	high	high
ATPase at pH 4.4	high	low	low
Glycogen content	low	high	high
Fatigue resistance	high	high	low
Contraction time	long	short	short

Abbreviations: SO, slow-oxidative; FOG, fast oxidative-glycolytic; FG, fast glycolytic; S, slow; FR, fast fatigue-resistant; FF, fast fatigue-sensitive; SDH, succinic dehydrogenase.

depolarization (fig. 14.3). In this respect, these graded responses are similar to the subthreshold local responses of nerve axons. It seems reasonable to suggest that they are produced by a similar mechanism, i.e. that the increase in inward current which follows depolarization is too small to counteract the effects of the increase in outward current.

The electrically excited responses of arthropod muscles differ from those of nerve axons and vertebrate twitch muscle fibres in that the inward current is carried very largely by calcium ions. This has been shown, for example, in the muscle fibres of barnacles (Hagiwara and Naka, 1964) and stick insects (Ashcroft, 1981). One of Ashcroft's experiments is shown in fig. 14.4; after treatment with TEA to block the outward potassium current, the membrane potential at the peak of the action potential increased with increasing calcium ion concentration, with a slope of 26 mV per tenfold change. She also found that the action potentials were unaffected by the absence of sodium ions or the presence of tetrodotoxin, showing that sodium channels are not involved.

The motor innvervation of arthropod muscles differs strikingly from that of vertebrate muscles in four respects: (1) many muscles are supplied by an inhibitor axon, stimulation of which causes relaxation if the muscle is excited, (2) each muscle is

Fig. 14.2. Electrical responses of locust leg mucles fibres to stimulation of the 'fast' (*a*), 'slow' (*b*) and inhibitor (*c*) axons. The upper traces show the zero potential level. (From Usherwood, 1967.)

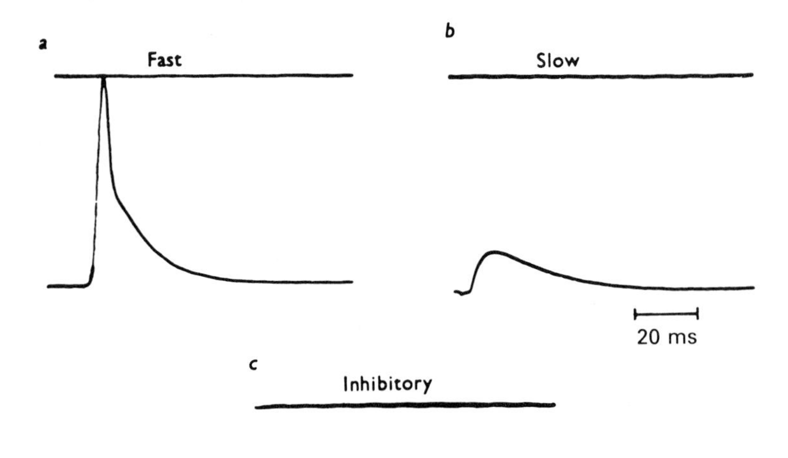

Fig. 14.3. The effect of a neuromuscular blocking agent (tryptamine) on the 'fast' response in a locust leg muscle. The upper traces show the tension developed during a twitch and, initially, zero membrane potential. The lower traces show the electrical response of a muscle fibre. Time signal 500 Hz. Trace 1 shows the normal responses, traces 2 to 4 show responses at 5 s intervals after application of tryptamine. Notice that the electrically excited component of the response is not an all-or-nothing phenomenon. (From Hill and Usherwood, 1961.)

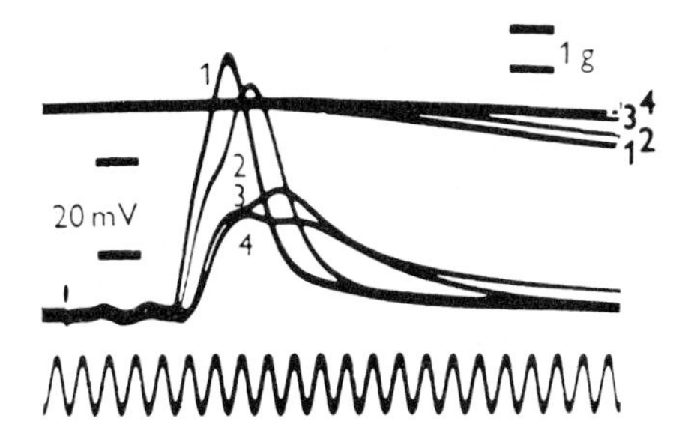

innervated by only a small number of motor axons, (3) *polyneuronal innervation* occurs, in which an individual muscle fibre may be innervated by two or more axons (fig. 14.1*c*), and (4) neuromodulation, involving neural modification of responses without direct synaptic action on the muscle fibres, occurs in some cases.

These points are well illustrated by the jumping muscle (the extensor tibiae) of the hind leg of locusts (Hoyle, 1955*a*, *b*; Usherwood and Grundfest, 1965). This is supplied by three conventional axons: a 'fast' excitor, a 'slow' excitor and an inhibitor. The majority of the fibres are innervated by the 'fast' excitor. These fibres respond to stimulation of the 'fast' axon by means of a postsynaptic potential and a large electrically excited response, similar to that shown in fig. 14.2*a*; the mechanical response of the muscle is a rapid twitch. About a quarter of the fibres in the muscle, including some of those also innervated by the 'fast' excitor, are innervated by the 'slow' excitor. Stimulation of the 'slow' excitor produces postsynaptic potentials which are not large enough to elicit electrically excited responses (fig. 14.2*b*); the mechanical response of the muscle is scarcely detectable after a single stimulus, but repetitive stimulation produces a slow, fairly smooth contraction whose intensity increases with increasing stimulus frequency. Some of the fibres innervated by the 'slow' axon are also innervated by the inhibitor axon. These fibres respond to inhibitory stimulation by

Fig. 14.4. Calcium action potentials from stick insect muscle. The records (*a*) show the response to depolarization in Ringer solutions containing TEA and 5 or 50 mM calcium. The graph (*b*) shows how the size of the overshoot varies with calcium ion concentration. The slope of the line is 26 mV per log unit. (From Ashcroft, 1981.)

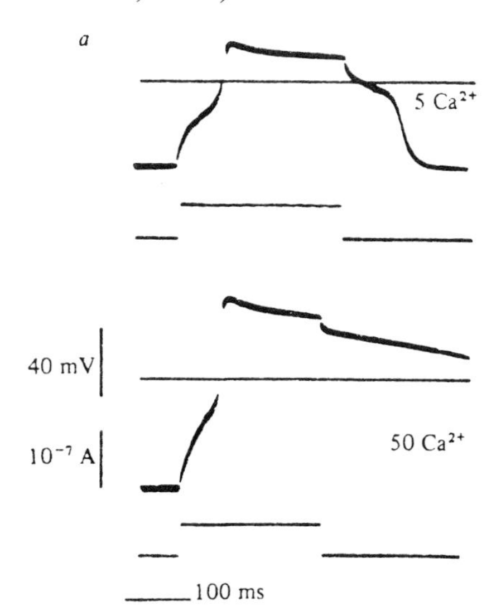

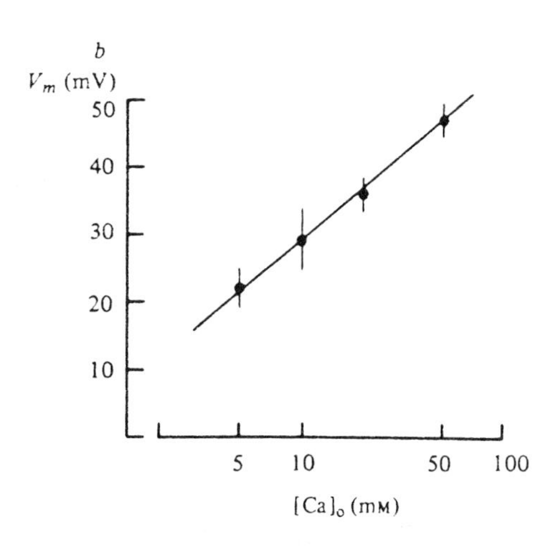

hyperpolarizing potentials (IPSPs) similar to those of vertebrate motoneurons (fig. 14.2*c*); the tension produced by stimulation of the 'slow' excitor axon is reduced if the inhibitor is active at the same time.

The neuromodulator axon for the locust extensor tibiae arises from a dorsal unpaired median (DUM) neuron in the metathoracic ganglion, discovered by Hoyle *et al*. (1974), and called by them the DUMETi neuron. Stimulation of the DUMETi neuron releases octopamine in the region of the muscle, although the neuron does not appear to form neuromuscular synapses of the normal closely apposed type. Such stimulation has no direct effect on the muscle fibre membrane potential, but it does slightly increase the response to slow axon activity and increases the rate of relaxation after both fast and slow axon stimulation (see Evans, 1985). This suggests that there are receptors for octopamine both on the slow axon terminals and on the muscle fibre surface.

Slight differences from this pattern are seen in other arthropod muscles. In the dorsal longitudinal flight muscles of locusts, the fibres are divided into five groups ('motor units'), each innervated by an axon which produces responses of the 'fast' type (Neville, 1963). The fibres of some muscles (such as locust spiracular muscles; Hoyle, 1959) are electrically inexcitable; the 'fast' response is then merely a larger version of the 'slow' response. The EPSPs of crustacean muscles frequently show marked summation and facilitation (fig. 14.5).

Fig. 14.5. Membrane potential changes in a *Panulirus* (rock lobster) muscle fibre during repetitive stimulation of the 'slow' excitor axon. Note the facilitation and summation of the responses. (From Hoyle and Wiersma, 1958.)

The electrical effects of postsynaptic inhibition in arthropod muscle fibres are essentially similar to those in vertebrate motoneurons. The sign and magnitude of the IPSPs can be altered by changing the membrane potential. The reversal potential is usually fairly near to the resting potential; it is practically unaffected by changes in external potassium ion concentration, but very sensitive to changes in external chloride ion concentration (Boistel and Fatt, 1958; Usherwood and Grundfest, 1965). Thus the action of the inhibitory transmitter is to increase the chloride conductance of the membrane.

As mentioned in chapter 10, the excitatory transmitter substance in arthropod muscles is glutamate, which acts by opening channels permeable to cations, and the inhibitory transmitter is γ-aminobutyric acid.

Presynaptic inhibition occurs in certain crustacean muscles, as has been shown very elegantly by Dudel and Kuffler (1961). Using an extracellular microelectrode, they were able to record the currents associated with EPSPs at single nerve terminals on the muscle fibres. It was found that these fluctuated in a discontinuous manner, suggesting corresponding fluctuations in the number of quanta of transmitter released (fig. 14.6*a*). When the inhibitory axon was also stimulated, the number of 'failures' at any one terminal increased, and the EPSP recorded intracellularly (which is caused by transmitter release from all the terminals on the fibre) decreased in size (fig. 14.6*b*). A statistical analysis of the type used by del Castillo and Katz on magnesium-treated frog muscle fibres (see chapter 7) was then carried out. It was found that the average number of quanta released per terminal per impulse was reduced from 2.4 to 0.6 during inhibition, which implies that the inhibitory process must be presynaptic.

Vertebrate heart muscles

The hearts of animals can be divided into two general types according to how their rhythmic contractile activity is initiated. In *myogenic* hearts, found in vertebrates and molluscs, excitation arises spontaneously in the heart muscle fibres themselves. In *neurogenic* hearts, found in many arthropods, the heart muscle fibres only contract in response to nervous stimuli. A full discussion of heart muscle physiology would be out of place here, but it is instructive to examine some of the properties of the excitation process in vertebrate heart muscle fibres.

Intracellular recordings from heart muscle fibres were first made by Draper and Weidmann (1951), using isolated bundles of Purkinje fibres from dogs. In mammalian hearts, the Purkinje fibres form a specialized conducting system which serves to carry excitation through the ventricle. After being isolated for a short time, they begin to produce rhythmic spontaneous action potentials, as is shown in fig. 14.7. The form of the action potentials differs from those of nerve axons and vertebrate twitch muscle fibres in that there is a prolonged 'plateau' between the peak of the spike and the repolarization phase. The action potential is initiated when the slowly rising *pacemaker potential* crosses a threshold level.

A heart would be ineffective if the rhythmic action potentials and their associated contractions in any one fibre were independent of those of its fellows: it is obviously essential that the contractions of the fibres should be synchronized. Vertebrate heart muscle consists of a network of branching fibres which are connected to each other by *intercalated discs* (fig. 14.8). Under the electron microscope these are seen to be concentrations of dense material on each side of two cell membranes. Desmosomes, providing mechanical attachment, are present, and there are gap junctions (p. 179)

Fig. 14.6. Experiment showing presynaptic inhibition in a crayfish claw muscle. Upper traces show intracellular responses to stimulation of the excitor axon at a rate of 1 per second. Lower traces are simultaneous extracellular records showing activity at a single junctional region; the arrows indicate failure of transmission. *a*: Stimulation of the excitor axon alone. *b*: Stimulation of the inhibitor axon 2 ms before each excitatory stimulus. The decrease in the size of the intracellular responses in *b* shows that inhibition is occurring; the increase in the number of failures in transmission recorded by the extracellular electrode indicates that this inhibition must be at least partially presynaptic. (From Dudel and Kuffler, 1961.)

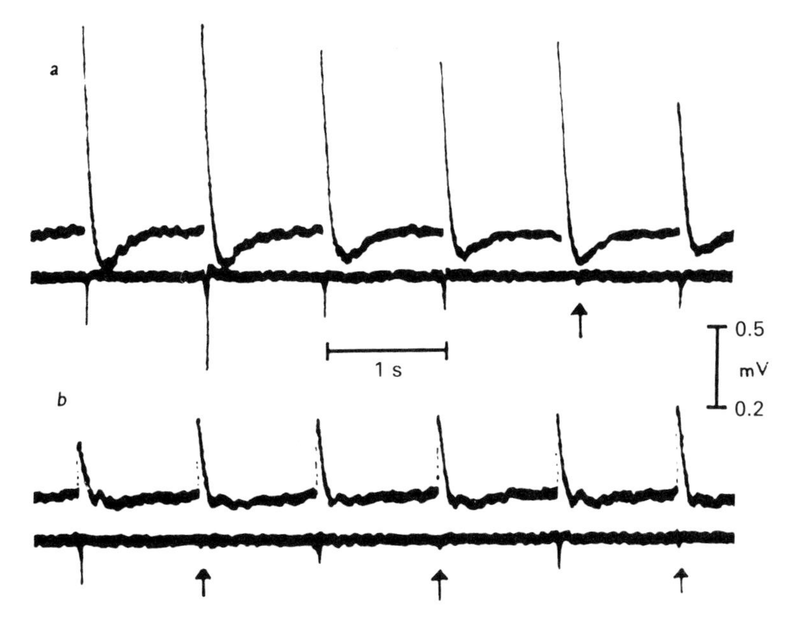

between the cells (Fawcett and McNutt, 1969). These gap junctions offer a low resistance to current flow, so that local circuits set up by an action potential in one cell can cross into the next cell and so excite it. The whole of the heart will then be 'driven' by those cells which have the most rapid spontaneous frequency. In the frog, these *pacemaker* fibres occur in the sinus venosus, and the excitation spreads from there into the auricle and ventricle. In mammals the pacemaker region is the sinuatrial node and excitation is carried into the ventricle along the Purkinje fibres (fig. 14.9). These are specialized for conduction in that they are elongated, relatively large in diameter, and have little contractile material.

It is possible to measure the electrical activity of the human heart simply by attaching leads to the wrists and ankles of the subject. The resulting record is known as an electrocardiogram, or ECG for short. ECGs were first measured by Einthoven, using the string galvanometer which he invented for the purpose (see Einthoven, 1924), and their

Fig. 14.7. Cardiac action potentials. *a* shows an intracellular record from a Purkinje fibre of a dog heart. The microelectrode entered the fibre soon after the start of the trace, and was withdrawn before its end. The interval between successive action potentials was 1.4 s. (From Draper and Weidmann, 1951.) *b* shows the nomenclature for different parts of such action potentials.

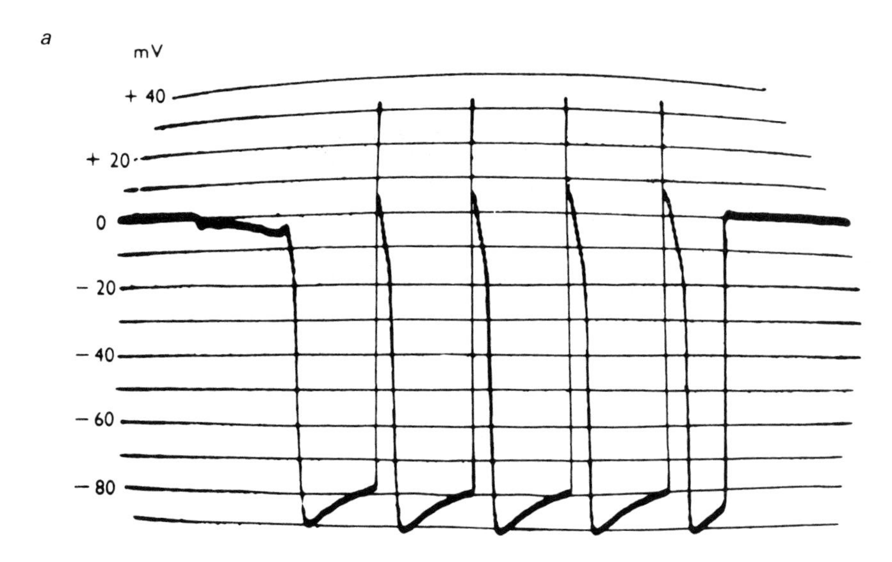

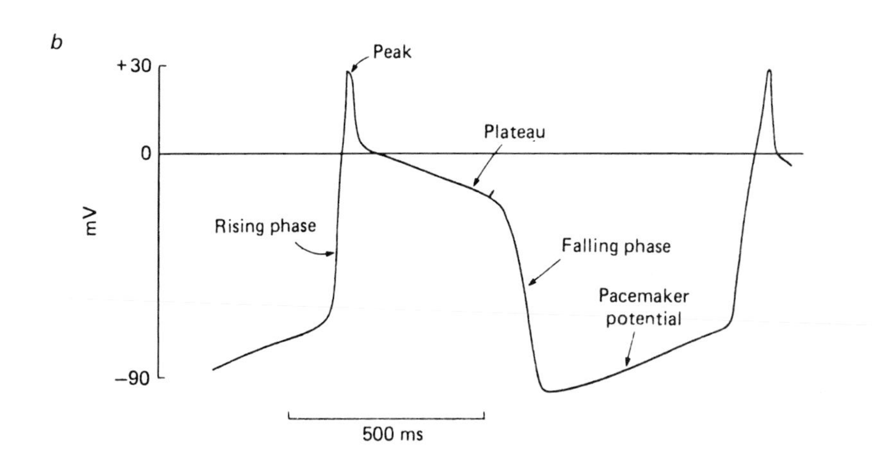

measurement (now usually by means of a hot-wire pen recorder) has long been a standard procedure in medical practice. Fig. 14.10 shows a typical ECG, recorded between the right arm and left leg. The different peaks in the electrical cycle of events were labelled the *P*, *Q*, *R*, *S*, and *T* waves by Einthoven.

The events in the heart cycle to which these electrical waves in the ECG correspond can be worked out by recording with surface electrodes from exposed hearts in experimental animals. The heart beat is initiated by pacemaker activity in the cardiac cells of the sinuatrial node. This excites the adjacent cells of the atria and a wave of depolarization sweeps over the whole of the atria. The currents associated with this are recorded in the ECG as the *P* wave. During the plateau of the atrial action potential there is little current flow and so the level of the ECG is not affected.

The atria and ventricles are separated by a sheet of connective tissue. The only electrically

Fig. 14.8. Schematic diagrams to show the structure of vertebrate heart muscle. Light microscopy (*a*) shows uninucleate muscle cells separated by intercalated discs. Under the electron microscope (*b*) the intercalated discs are seen to be concentrations of dense material on each side of an intercellular boundary.

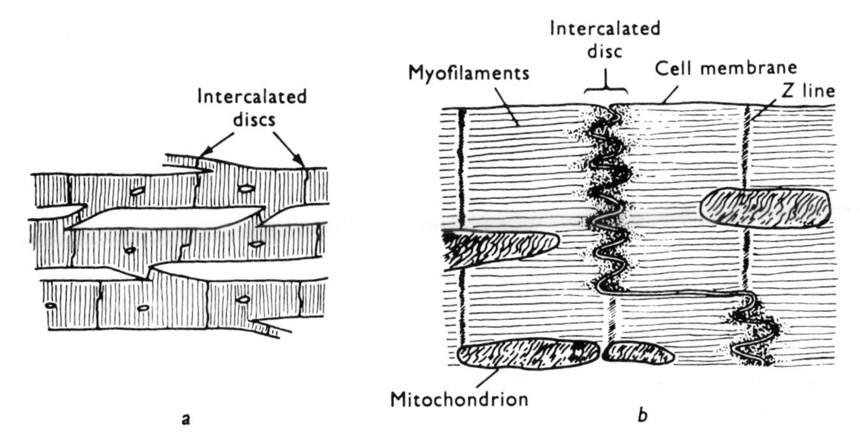

Fig. 14.9. Pacemaker and specialized conductile regions of the mammalian heart. (From Scher, 1965.)

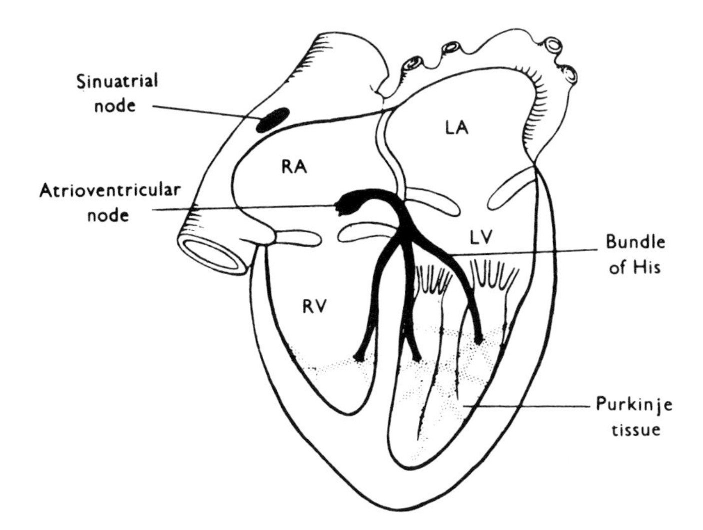

excitable pathway between the two is the atrioventricular bundle, which connects the atrioventricular node at the base of the atrial septum to the bundle of His in the ventricular septum. Conduction in the atrioventricular node is slow, so that there is some delay between the activity in the atria and that in the ventricles.

From the atrioventricular node the wave of depolarization spreads via the atrioventricular bundle to excite the specialized conducting tissue of the ventricles, the Purkinje fibres of the bundles of His. The action potential passes down the left and right branches of this system, but the activity is not evident in the ECG because the number of cells involved is small and hence the current flow is small also. The bundle of His brings excitation to the mass of the ventricular muscle cells, beginning in the septum and then spreading from the apex of the ventricle up to the base. The net currents involved in the depolarization of all these ventricular cells are very large and are seen in the ECG as the *QRS* complex. The precise form of the *QRS* complex is explicable in terms of the detail of this three-dimensional spread of activity in the ventricle, such as is described, for example, by Scher and Spach (1979). After this the whole of the ventricle is depolarized and there is very little electrical current flow (the ventricular muscle is contracting so as to pump blood out along the aorta and pulmonary artery). Then repolarization of the ventricular fibres occurs, at slightly different times in different places, and the current flow associated with this is seen as the *T* wave. After this the heart is electrically at rest except in the pacemaker regions, its muscles are relaxing and it is refilling with blood ready for the next cycle. The pacemaker potentials preceding the next *P* wave are not visible in the ECG.

Recording with intracellular electrodes shows that there is some variation in the form of the action potential in different regions of the heart (fig. 14.11). In the sinuatrial and atrioventricular nodes the action potential is slow to rise and relatively small in amplitude, with no plateau. Elsewhere the action potential rises rapidly and does have a plateau, which is somewhat shorter in the ventricles than in the Purkinje tissue and shorter still in the atria. Cells of the sinuatrial node have pronounced pacemaker potentials; these are not normally found elsewhere, although slow pacemaker potentials may arise after a time in isolated Purkinje fibres.

What is the ionic basis of heart muscle action potentials? Draper and Weidmann showed that the membrane potential at the peak of the early spike was reduced when the external sodium concentration was lowered. This suggests that, as in the action potential of nerve axons, the initial rapid depolarization is brought about by a regenerative increase in the sodium conductance of the membrane. But what happens during the plateau? Weidmann (1951) made the crucial observation that membrane resistance at this time is quite high, being much the same or even higher than it is during the pacemaker potential. So, in contrast to nerve axons, the membrane conductance to potassium cannot begin to rise soon after depolarization; there must be some long delay before this happens. Further evidence for this was produced by Hutter and Noble (1960), who showed that, in the absence

Fig. 14.10. A human electrocardiogram, recorded between electrodes applied to the right wrist and left ankle.

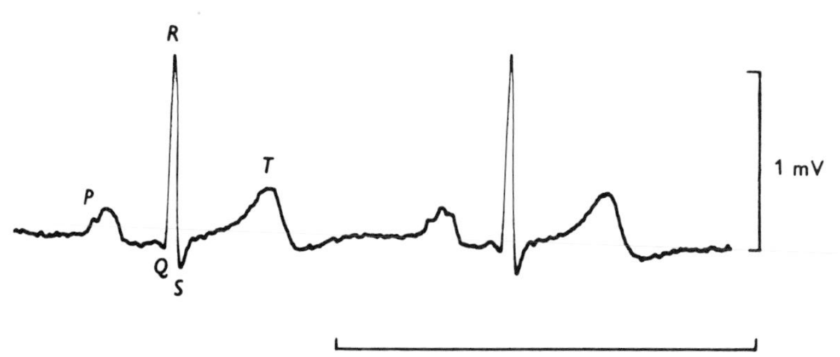

of sodium ions, the cardiac cell membrane shows marked inward rectification: the membrane resistance is higher for depolarizing currents than it is for hyperpolarizing ones.

Direct measurements of membrane ionic currents need an effective voltage-clamp system, but there are difficulties in producing one because of the complex geometry of the heart muscle. The 'sucrose gap' method is illustrated in fig. 14.12. Part of the muscle bundle is enclosed in an isotonic sucrose solution; this penetrates the intercellular spaces between the cells and so renders them incapable of carrying electric current. It is therefore possible to pass current between the two Ringer-filled compartments and know that all of it is passing through the intracellular material; and hence it must all pass out across the cell membrane. Hence we have a measure of membrane current and a means of ensuring that this current flows across the cell membranes that we are interested in. These are those of a cell which is impaled with an intracellular microelectrode to record its membrane potential. The membrane potential is then used to control the current flow through the membrane by means of a feedback loop in the usual manner.

Another method has been used for Purkinje fibres, which are larger than other cardiac cells. Here a short length of fibre is cut out (the injured ends heal over) and two microelectrodes are inserted. Neither method is very good at dealing with large currents.

Isolated cardiac cells have been used increasingly in recent years (see Noble and Powell, 1987). The cells are separated from one another by treatment of the heart with the enzyme collagenase, which breaks down the connective tissue holding them together. Brown, Lee and Powell (1981) used suction electrodes to pass current into the cell and to record its membrane potential in a voltage-clamp system.

One result of the use of the voltage-clamp method is the discovery that depolarization results in an inward flow of calcium ions. Beeler and Reuter (1970) were able to demonstrate this by first depolarizing the fibres to −40 mV so as to inactivate the sodium current; further depolarization then resulted in a slow inward current (fig. 14.13). This slow current was unaffected by tetrodotoxin (which blocks the fast inward current) but was abolished in calcium-free solutions.

Sorting out the full nature of the cardiac action potential has proved to be a complicated and as yet unfinished task (see Noble, 1984, and Reuter, 1984). The number of different ionic currents and channels involved is surprisingly large. The theoretical models developed by Noble and his colleagues have become progressively more complex over the years as they have had to accommodate more and more awkward facts.

The first of these models was produced by

Fig. 14.11. The shape of the action potential in different regions of the mammalian heart. (From Katz, 1977.)

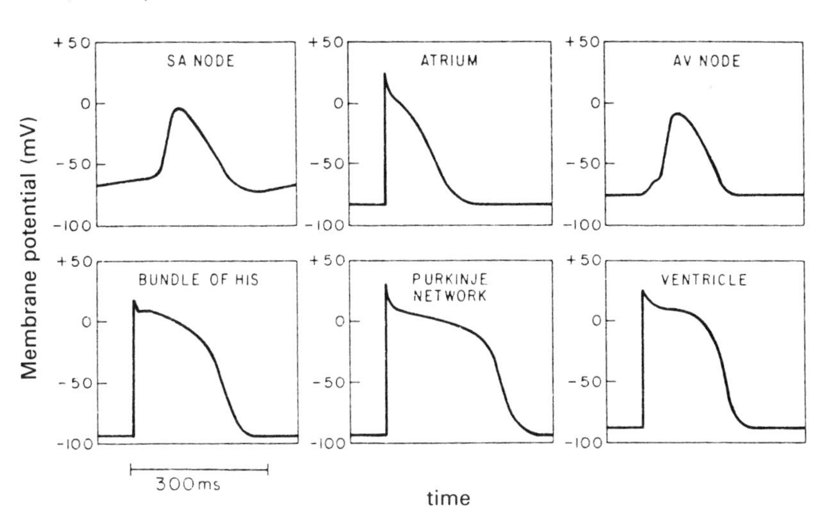

Noble in 1962. It was based on the Hodgkin–Huxley equations for the nerve action potential, with two changes. The leak current was separated into sodium and potassium components, with the latter showing inward rectification, and the potassium current elicited by depolarization was very greatly delayed. Computer simulations mimicked the form of the action potential in what appeared to be a very satisfactory manner.

The next model (McAllister *et al.*, 1975) was produced in response to new observations, especially on the calcium current. In this version the inward calcium current flowed during the whole of the plateau of the action potential. There was also a transient outward current and another outward

Fig. 14.12. Voltage-clamp technique for heart muscle, using the sucrose-gap method.

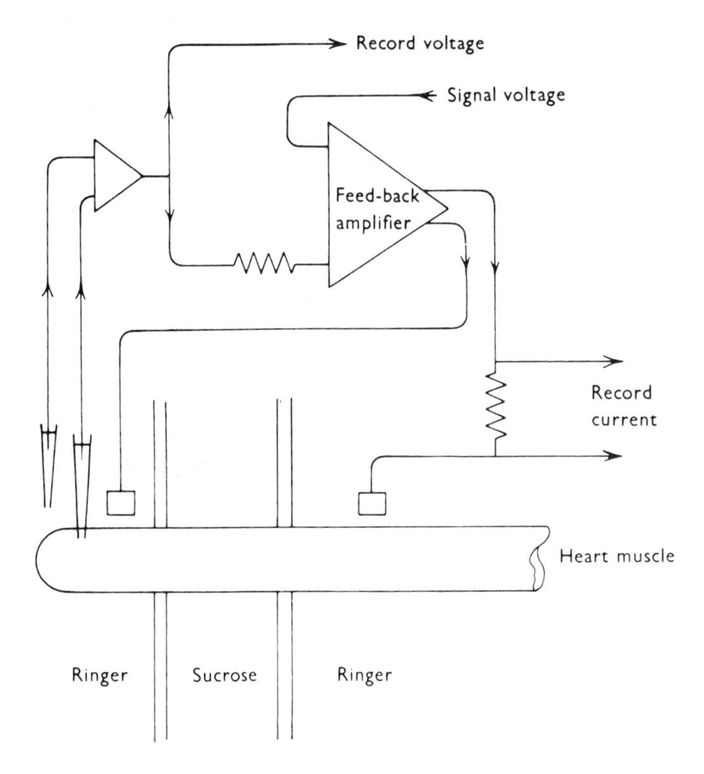

Fig. 14.13. Calcium current in voltage-clamped ventricular muscle from a dog heart. External calcium ion concentration was 1.8 mM in *a* and zero in *b*. The holding potential was −40 mV, which is sufficiently depolarized to inactivate the sodium current. Further depolarization to −25 mV for 570 ms resulted in the slow transient inward current shown in *a*. (From Beeler and Reuter, 1970.)

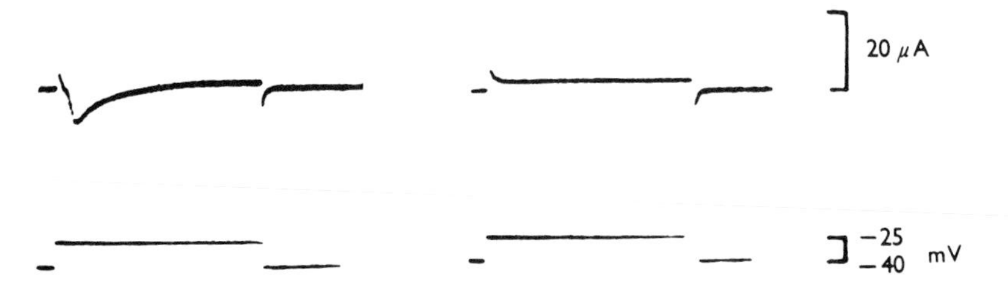

current (called i_{K2} at this time) which was deactivated during the pacemaker potential

It became clear that the 1975 model for Purkinje fibres would have to be replaced when DiFrancesco (1981*a*, *b*) showed that the principal event during the pacemaker phase was not a reduction in outward current (i_{K2}) but an increase in an inward current. A similar inward current had earlier been described in the pacemaker cells of the sinuatrial node (DiFrancesco and Ojeda, 1980). The new current was called i_f. It is time-dependent and activated by hyperpolarization, and is carried largely by sodium ions. The i_f channels are quite different from the fast sodium channels; they are activated by hyperpolarization instead of depolarization and they are insensitive to tetrodotoxin. A new model incorporating i_f has been produced by DiFrancesco and Noble (1985). Other changes in this model include currents activated by internal calcium, currents due to the sodium–potassium pump and the sodium–calcium exchange system, and consideration of the effects of changes in the intracellular and extracellular ionic concentrations as a result of activity.

The sequence of events in the cardiac action potential according to the 1985 model is shown in fig. 14.14. The curves show ionic conductance changes in a computer simulation appropriate to Purkinje fibres. Let us begin at the point in the cycle where the membrane potential is at its most negative, at about 0.4 s on the time trace. It has reached this negative value because the potassium conductance g_K is high. However, the pacemaker conductance g_f has been switched on by the hyperpolarization, and it rises steadily for the next second or so. The slow sodium ion inflow that this permits results in a steady depolarization, the pacemaker potential. After a time the pacemaker potential has depolarized the membrane sufficiently to open the fast-activating sodium channels; these are of the type found in nerve axons, i.e. they are sensitive to tetrodotoxin, and are activated and inactivated rapidly by depolarization. Then follows the familiar runaway relation between

Fig. 14.14. Computer simulation of the mammalian cardiac action potential (based on Purkinje fibres) and the conductance changes which are responsible for it. g_f is the inward current which becomes apparent during the pacemaker potential. The sodium conductance g_{Na} includes both the conductance of the fast sodium channel and the sodium component of g_f. (From DiFrancesco and Noble, 1985.)

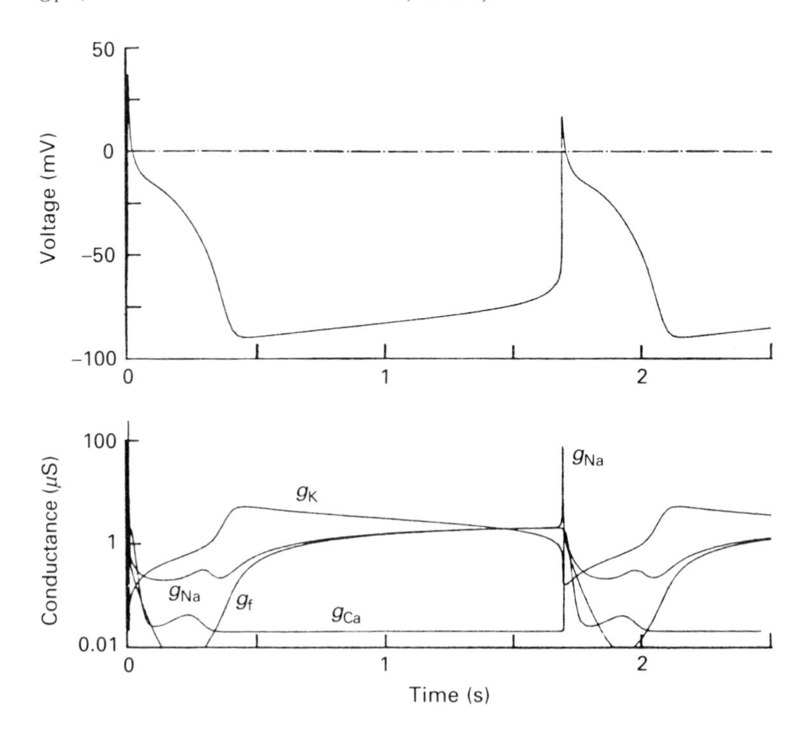

membrane potential and sodium conductance just as in the nerve axon, so that there is a massive inflow of sodium ions and a rapid overshooting depolarization. But now the model departs radically from the situation in nerve axons. The potassium conductance has fallen to a low level and there is an elevated calcium conductance, hence the membrane potential remains near zero for some hundreds of milliseconds. This plateau declines gradually and is brought to an end as a result of the long-delayed increase in potassium conductance, and any calcium and fast sodium channels remaining open are finally closed during the repolarization phase. By the end of the action potential the pacemaker conductance g_f has already begun to rise and so the new cycle continues on its way.

The DiFrancesco–Noble equations can be modified to account for the form of the action potential at the sinuatrial node (Noble and Noble, 1984; Brown *et al.*, 1984). The currents involved in this simulation, which takes account of the results of voltage-clamp experiments, are shown in fig. 14.15. Notice particularly that there is no fast sodium current; sinuatrial node activity is largely unaffected by tetrodotoxin. The pacemaker potential is rather different from that in the Purkinje fibre model: it is produced by (1) a fall in the potassium current i_K and (2) an increase in the slow inward current i_{is}. The latter consists of the voltage-gated calcium current $i_{Ca,\ fast}$ plus the current $i_{Na\ Ca}$ from the electrogenic sodium–calcium exchange system, which is stimulated by the inflow of calcium ions. The current i_f is small and has little effect on the total pacemaker current.

Although the heart-beat is myogenic in origin, its rate and amplitude can be modified by the action of the autonomic nervous system. The parasympathetic system, acting via muscarinic acetylcholine receptors, is inhibitory in that its activity reduces the heart rate and heart output. The

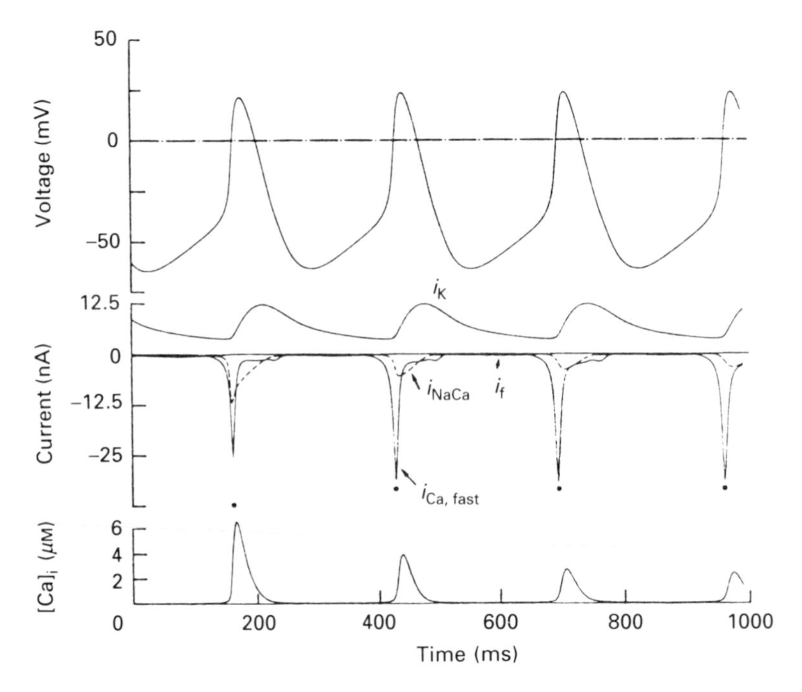

Fig. 14.15. Computer modelling of pacemaker activity in the mammalian sinuatrial node. The upper trace shows computed voltage changes. Middle traces show voltage-gated currents and the current due to the sodium–calcium exchange system. The latter is stimulated by increase in internal calcium ion concentration, which is shown in the lower trace. Black dots show the maximum value of the slow inward current. There is no fast sodium current in these cells. The simulation was started with a higher than usual value of internal calcium, and took about two cycles to reach the steady state. (From Brown *et al.*, 1984.)

sympathetic system and the hormone adrenaline act largely via β-adrenergic receptors to stimulate heart rate and output.

Stimulation of parasympathetic fibres in the vagus nerve, or application of acetylcholine, causes slowing of the heart rate, a decrease in the height and duration of the action potential, and a decrease in the tension during contraction. Strong inhibition may prevent spontaneous activity altogether. These effects are largely confined to the atria and the sinuatrial and atrioventricular nodes; the ventricles are not much affected. The effects on the action potentials are caused by an increased permeability to potassium ions, with the opening of more inward-rectifier potassium channels (Hutter, 1961; Sakmann *et al.*, 1983) and a reduction in the slow inward current (Giles and Noble, 1976).

Stimulation of the sympathetic nervous input to the heart produces acceleration of the heartbeat, shortening of the action potential, increased conduction velocity in the atrioventricular node, and increased contractility in the atria and ventricles. Most of these effects appear to be mediated through β-adrenergic receptors, causing an increase in cyclic AMP levels in the cell, with consequent effects on calcium channels (see Reuter, 1983, 1984). Reuter and Scholz (1977), for example, showed that adrenaline increases the magnitude of the slow inward current, apparently by raising the number of calcium channels that are opened by depolarization. Injection of cyclic AMP into single ventricular cells produces similar effects (Irisawa and Kokobun, 1983). We shall return to the question of increased contractility later in this chapter.

Vertebrate smooth muscle

Smooth muscles – muscles in which there are no oblique or cross striations – form the muscular walls of the viscera and blood vessels in vertebrates. They also occur in the iris, ciliary body and nictitating membrane in the eye, in the tracheae, bronchi and bronchioles of the lungs, and are the small muscles which erect the hairs in mammals. Individual smooth muscle cells are uninucleate, and much smaller than the multinucleate fibres of skeletal muscles; they are usually about 4 μm in diameter and up to 400 μm long. The cells are held together in bundles, 20–200 μm in diameter, by thin sheets of connective tissue, and these bundles themselves are frequently interconnected. Adjacent cells are connected by a small flask-shaped protrusion from one into a pocket of the other, in which the opposing cell membranes are brought close together to form gap junctions; it seems very likely that current can flow from one cell to another through these gap junctions (see Gabella, 1981).

Many vertebrate smooth muscles show a great deal of spontaneous activity. This is particularly so in intestinal muscles, where the spontaneous contractions serve to mix and move the gut contents. The electrical activity consists of slow waves of variable amplitude and all-or-nothing action potentials (fig. 14.16). The fibres are depolarized and the frequency of the action potentials increases if the muscle is stretched (Bülbring, 1955). The spontaneous activity can be modified by the action of extrinsic nerves, by adrenaline, and, in uterine muscle, by hormones of the reproductive cycle. The smooth muscles of the iris, nictitating membrane and vas deferens are not spontaneously active.

Action potentials can be initiated in sheets or strips of smooth muscle by electrical stimulation; they will then propagate along the axes of the muscle cell and bundles and also, more slowly, across them.

The action potentials of smooth muscles are slower than those of nerve axons and vertebrate skeletal muscles; their duration at half maximum amplitude varies from 7 to 20 ms in different muscles. They are insensitive to tetrodotoxin and can be produced in the absence of sodium ions in some muscles, but are prevented by calcium channel blocking agents such as manganese, cobalt or lanthanum ions or the drugs verapamil or nifedipine. This suggests that the principal inward current during the action potential is via calcium channels rather than sodium channels (see Holman and Neild, 1979; Tomita, 1981).

The terminals of the sympathetic nerves supplying smooth muscle can be prepared for light microscopy by exposing freeze-dried tissues to hot formaldehyde vapour; the catecholamines present are then converted to substances which fluoresce strongly under ultraviolet light. The relations of the terminals to the smooth muscles cells can be determined by electron microscopy. These methods show that the fine terminal branches have frequent swellings along their length, known as varicosities, which contain synaptic vesicles. In some muscles each varicosity is intimately apposed to an individual muscle cell (fig. 14.17*b*). In the spontaneously active muscles of the gut, the axon terminals remain in small bundles, associated with Schwann cells, and their varicosities do not form close contacts with individual muscle cells (fig. 14.17*a*). And in some muscles innervation is via both small axon bundles and close-contact varicosities.

Stimulation of the motor nerves innervating smooth muscle results in junction potentials (postsynaptic potentials) of various types (fig. 14.18). Excitatory nerves produce depolarizing potentials whereas inhibitory nerves produce hyperpolarizing ones. In muscles with innervation via close-contact varicosities, these have a relatively rapid time course, whereas in those with innervation via axon bundles only, the junction potentials may last for half a second or more.

The synaptic mechanisms involved in these various junction potentials are correspondingly diverse (Bolton, 1979; Bülbring *et al.*, 1981*a*; Bolton and Large, 1986). Their slowness, in

Fig. 14.16. Intracellular records of membrane potential in smooth muscle fibres of guinea pig intestine, showing various patterns of spontaneous activity. (From Bülbring *et al.*, 1958.)

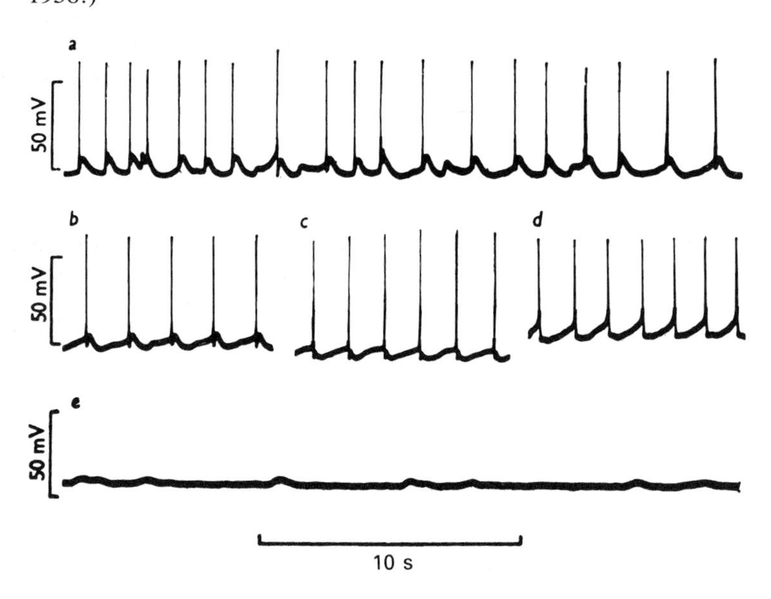

Fig. 14.17. Neuromuscular junctions in vertebrate smooth muscle. In *a* the axons occur in small bundles (*s*); they do not closely contact the muscle cells. In *b* the axons (*N*) occur singly and their varicosities make close synaptic contacts (with a 20 nm synaptic cleft) with the muscle cells. (From Bennett, 1972.)

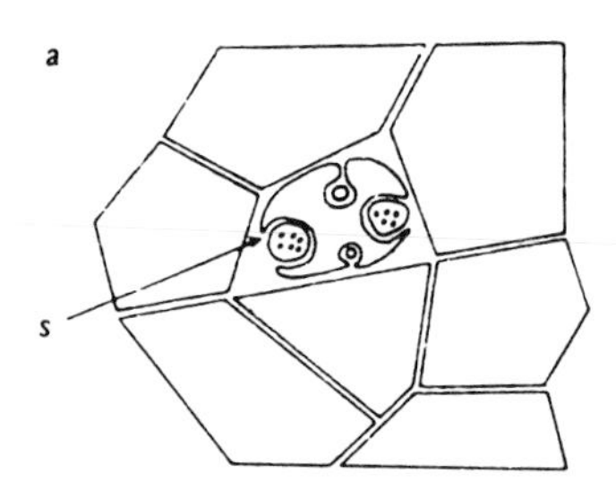

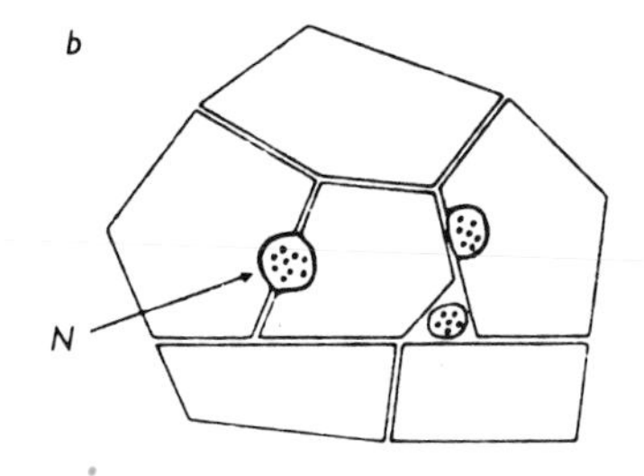

comparison with the end-plate potentials of skeletal muscles, would suggest that the channels involved are activated indirectly via G proteins (fig. 10.2 *b* and *c*) rather than directly as part of a unitary receptor–channel complex such as the nicotinic acetylcholine receptor. This conclusion is in part confirmed by what we know of the postsynaptic receptors involved: both the muscarinic acetylcholine receptor and the β- adrenergic receptor act via G proteins (see chapter 10).

Acetylcholine, released from parasympathetic nerve fibres, produces excitatory responses in a large variety of visceral muscles (Bolton, 1981). Since these responses are blocked by atropine the receptors involved are all muscarinic. Excitatory responses are also produced in some smooth muscles via α-adrenergic receptors, purinergic receptors activated by ATP, or both these acting together. In the guinea pig vas deferens, for example, there is good evidence that cotransmission occurs in the sympathetic nerves, with both ATP and noradrenaline being released on stimulation; the ATP produces a relatively fast response and the noradrenaline a slower one (Sneddon and Westfall, 1984). Substance P may act as a neurotransmitter in guinea pig small intestine (Gintzler and Hyde, 1983).

Many inhibitory responses are produced by stimulation of sympathetic nerves acting via adrenergic receptors. In the guinea pig taenia coli, for example, activation of α-adrenergic receptors produces hyperpolarization and relaxation. Activation of β-adrenergic receptors also produces relaxation but with only slight hyperpolarization, suggesting that they may act directly on the calcium stores of the cell (Bülbring *et al.*, 1981*b*). In some intestinal muscles, nervous stimulation produces strong hyperpolarizations by the action of a transmitter whose identity is as yet unknown; it is neither acetylcholine, noradrenaline nor ATP (Westfall *et al.*, 1982).

Fig. 14.18. Junction potentials in smooth muscle. *a*, An excitatory junction potential in guinea pig taenia coli muscle; *b*, an inhibitory junction potential in guinea pig taenia coli; *c*, an excitatory junction potential in mouse vas deferens. Innervation is by small axon bundles in the taenia coli (*a* and *b*), and by close-contact varicosities in the vas deferens (*c*). Notice that *c* has a much faster time course than *a* and *b*. (From Bennett, 1972.)

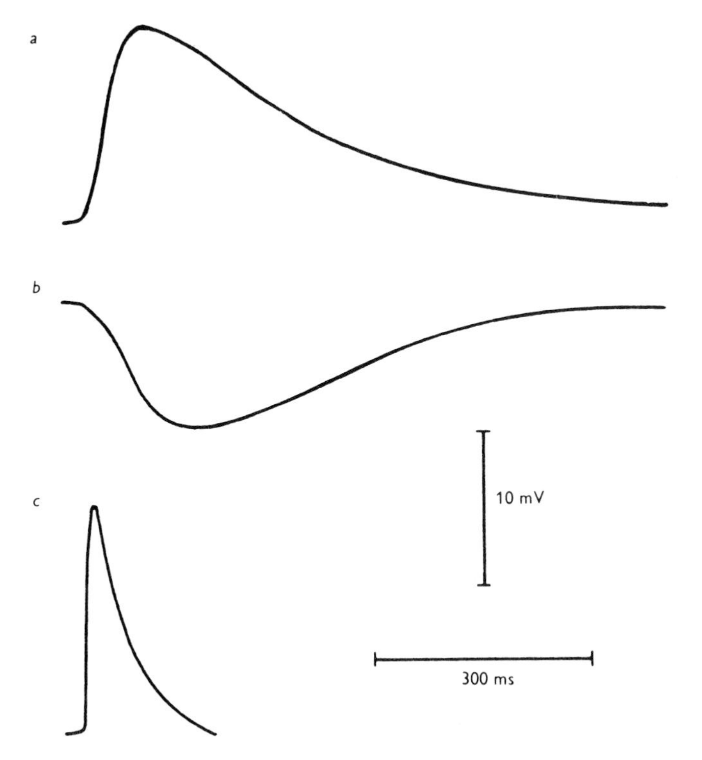

Gradation of contraction

It is clearly necessary that the force exerted by a muscle should be capable of being regulated by its nervous input. In skeletal muscle, there are two main ways in which this can be brought about; either the proportion of the fibres in the muscle that are excited is varied, or the degree to which the individual fibres are excited is controlled.

The first system is the main method in vertebrate skeletal muscles, indeed the only one in those muscles that consist entirely of twitch fibres. Each muscle is innervated by a large number of motor axons, and each axon innervates a group of muscle fibres known as a 'motor unit'. Since an action potential in a motor axon initiates propagated action potentials in each of the muscle fibres in the motor unit that it innervates, the excitation process in any one motor unit is all-or-nothing. The contractile response that follows excitation depends to some extent on the frequency of nervous stimuli (since the twitch tension is less than the tetanus tension), but the range over which this can occur is not great. For an isometric load, we might describe the contractile response as being 'half-or-more-or-nothing'. Thus fine control of tension development has to be brought about by variation in the number of active motor units. The excitation of different motor units in the muscle is usually asynchronous, so that contraction of the whole muscle is a fairly smooth process.

In arthropods, the mechanism of gradation is quite different. Here, as we have seen, the whole muscle is frequently innervated by only two or three motor axons, and the excitation of an individual fibre is definitely not an all-or-nothing process. The size of the electrical response in any fibre is dependent on (1) which axon, 'fast' or 'slow', is active, (2) the frequency of arrival of excitatory nerve impulses, which affects the degree of membrane depolarization by summation and facilitation of the end-plate potentials, (3) the activity of the inhibitor axon, and (4) the activity of the neuromodulator axon, if there is one.

The situation in vertebrate heart muscle is different again. All the muscle fibres are active during the heart beat, and full relaxation must occur before the next action potential arrives, so gradation must depend on changes in the size of the contraction following an action potential in each fibre. We have seen that nervous stimulation may alter the form and duration of the action potential, but there are also changes in the relation between depolarization and coupling, as we shall see in the next section.

In vertebrate smooth muscles we find a number of these various mechanisms working together. The extent of contraction is controlled by the number of excitatory and inhibitory nerve fibres that are active and the frequency at which they fire, but there may also be some modulation of the coupling process by neurotransmitters or hormones.

Excitation–contraction coupling

It seems to be generally true that, in all muscles, contraction is brought about by an increase in the intracellular concentration of calcium ions. Different muscles may show different sensitivity to calcium ions, but typical concentration changes from rest to maximal activity fall within the range from about 10^{-7} M to 10^{-5} M. In most cases this increase in internal calcium ion concentration is brought about by depolarization of the cell membrane, whether by nervous action or spontaneous activity. Within this general pattern, however, there are a number of differences in the details of the coupling process in muscles of different types.

Internal membrane systems

The sarcoplasmic reticulum and the *T* system are well developed in muscle fibres that contract and relax rapidly, such as vertebrate twitch fibres and many arthropod muscles. This is especially evident in many muscles concerned with sound production: large amounts of sarcoplasmic reticulum are found in the sound-producing muscles of fishes (Franzini-Armstrong, 1980), bats (Revel, 1962), insects (see Aidley, 1985) and lobsters (Rosenbluth, 1969). These systems are less evident, though still present, in vertebrate tonic fibres and in cardiac muscle. Thus the degree of development of the sarcoplasmic reticulum seems to be related to the speeds of contraction and relaxation of the muscles.

The *T* system is absent from fibres of small diameter such as occur in annelids and molluscs and in vertebrate smooth muscles, and also from the thin striated fibres of amphioxus (Flood, 1977; see Peachey and Franzini-Armstrong, 1983). The sarcoplasmic reticulum in these muscles is small in quantity and often forms junctions with the cell surface membrane.

What limits the size of the sarcoplasmic reticulum? The muscle cell can be regarded as an arena of competing subcellular interests: the principal contestants are the myofibrils, the mitochondria and the endoplasmic reticulum. To maximize the work output per contraction, the myofibrillar fraction should be maximal. To increase the rate of activation and relaxation, the sarcoplasmic reticulum should be increased in volume. And to increase the level of maintained power output, the mitochondrial mass should be increased.

Fig. 14.19 shows the relative proportions of these three types in various insect muscles. Slowly contracting leg muscles (6 and 7 in fig. 14.19) do not require or possess large quantities of the non-myofibrillar components. Leg muscles which need to be rapidly activated (such as 8 to 11) need to have much more sarcoplasmic reticulum, and hence must make do with a corresponding reduction in myofibril content; but they have no requirement for maintained power output, hence they have few mitochondria. Flight muscles and singing muscles (1 to 5) are repetitively active for long periods, so they need a large mitochondrial component to supply the energy for contraction as well as (in the conventional synchronous muscles, 3 to 5) an appreciable sarcoplasmic reticulum. The asychronous muscles, 1 and 2 in fig. 14.19, have very little sarcoplasmic reticulum; we shall see later that this is related to their ability to produce oscillatory power output while being fully active.

Fig. 14.19. The relative proportions of myofibrils, mitochondria and sarcoplasmic reticulum in various insect muscle cells. The axes show the relative proportions by volume of each of the three components, expressed on a scale in which the volume of all three is 100%; the proportions of nuclei, sarcoplasm etc are thus neglected. 1 and 2 are asynchronous (fibrillar) flight or singing muscles; 3 to 5 are synchronous flight or singing muscles; 6 to 11 are leg muscles. The muscles are: 1, blowfly flight muscle; 2, a cicada asynchronous tymbal muscle; 3, a cicada synchronous tymbal muscle; 4, katydid flight/singing muscle; 5, dragonfly flight muscle; 6, locust jumping muscle, tonic and, 7, phasic fibres; 8, locust toe muscle; 9, flea jumping muscle; 10, wasp extensor tibialis and, 11, toe muscle. (From Aidley, 1985.)

The regulatory proteins

In vertebrate skeletal muscles, as we have seen in chapter 13, contraction is regulated by the calcium-sensitive troponin–tropomyosin system on the thin filaments. Similar thin-filament linked regulatory systems are found in amphioxus and mysid crustaceans and in the fast muscles of decapod crustaceans (Lehman, 1976).

In molluscan muscles, however, Kendrick-Jones, Lehman and Szent-Györgyi (1970) found that there is no troponin present. Molluscan thin filaments will react with rabbit myosin so as to split ATP, in the absence of calcium ions. But there is a calcium-sensitive mechanism on the thick filaments: molluscan myosin will only combine with pure actin and split ATP in the presence of calcium ions. Similar myosin linked regulatory systems are found in echinoderms and in a number of minor groups.

In insects and most other arthropods, and in annelids and nematodes, both types of regulation are present simultaneously.

The mechanism of molluscan myosin-linked regulation has been investigated in the fast striated adductor muscle of the scallop, *Pecten*. Each myosin molecule is similar in general structure to those of vertebrate skeletal muscles, with two heavy chains, two regulatory light chains which can be removed with EDTA, and two essential light chains which contain SH groups. We may assume that there is one of each type of light chain associated with each of the two myosin heads.

After removal of the two regulatory light chains, ATPase activity and tension development will occur in the absence of calcium ions (Szent-Györgyi *et al.*, 1973; Chantler and Szent-Györgyi, 1980; Simmons and Szent-Györgyi, 1985). The regulation does not depend on the presence of actin; the ATPase activity of the myosin alone is 100 times greater in the presence of calcium than in its absence (Wells and Bagshaw, 1985).

Hardwicke and his colleagues have used photo-crosslinking techniques to investigate myosin regulation in the scallop. They found that a substituted photoactivated agent would link the regulatory and SH light chains under conditions when rigor was produced (calcium ions but no ATP) but not under conditions of rest (ATP but no calcium ions). This implies that the distance between the two light chains is greater at rest and suggests that movement of at least one of them occurs during activation by calcium (Hardwicke *et al.*, 1983; Hardwicke and Szent-Györgyi, 1985).

The situation in vertebrate smooth muscle is rather less clear. Purified smooth muscle actomyosin will not split ATP even in the presence of calcium, suggesting that the purification process may have removed an activator of some sort. Calmodulin is a globular protein with a molecular weight of 17 kDa and four calcium-binding sites per molecule (Babu *et al.*, 1985). It combines with calcium in the physiological important concentration range of 10^{-7}–10^{-5} M, and will then combine with a number of target proteins. One of these is myosin light-chain kinase, which phosphorylates the regulatory light chains of myosin. Thus it looks as though calcium combines with calmodulin, which then activates myosin light-chain kinase, which then phosphorylates the regulatory light chains, and this somehow activates the actomyosin ATPase (see Marston, 1982; Kamm and Stull, 1985). This activation mechanism can be modulated by a cyclic-AMP-dependent protein kinase, since phosphorylation of the myosin light-chain kinase reduces its affinity for calmodulin (Conti and Adelstein, 1980).

There is also a thin filament regulatory system present in smooth muscle. The thin filaments contain tropomyosin. There is no troponin, but in its place the protein caldesmon, which will combine with calmodulin in the presence of calcium ions (Sobue *et al.*, 1981). Caldesmon is thus analogous to troponins I and T in skeletal muscle, with calmodulin playing the part of troponin C. The mechanism of regulation must differ from that in skeletal muscle, however, since there is a molecular ratio of 28 : 4 : 1 for actin : tropomyosin : caldesmon (Marston and Smith, 1985).

Vertebrate heart muscle

Our common experience shows that the heart is a remarkably adaptable organ. If we take exercise, both the rate at which it beats and the volume of blood pumped per beat (the stroke volume) increase: we become conscious of the heart pounding away within us. We have seen that these effects are controlled via the autonomic nervous system. The cardio-accelerator action of the adrenergic nerves is called their *chronotropic* action, and their effect on the strength of the individual heart beat is an example of positive *inotropic* action.

The contractility of isolated cardiac muscle is affected by the timing of electrical stimuli. Up to a point higher frequencies lead to stronger contractions; this is called the staircase effect. Thus if the heart rate rises the stroke volume usually rises also. A number of poisonous drugs called cardiac glycosides increase the strength of the heart beat. Examples are ouabain and digitoxin; they inhibit the sodium pump of cell membranes. In heart muscle this may produce a number of indirect effects as a result of the consequent increase in internal sodium ion concentration; one such change may be an increase in the activity of the sodium–calcium exchange system.

Since each heart beat involves only one action potential in every cardiac muscle cell, the large changes in contractile force which can take place must involve changes in the relation between depolarization and contraction. It is now clear that there are two ways in which this can be brought about: either the amount of calcium released within the fibre is changed or the sensitivity of the myofibrils to calcium ion concentration alters.

We have seen in the previous chapter how the calcium-sensitive luminescent protein aequorin can be used to measure the rise and fall of the intracellular calcium ion concentration. Similar measurements have been made on frog and mammalian cardiac fibres (Allen and Blinks, 1978; Wier, 1980; Morgan and Blinks, 1982). Thin strips of muscle are used, with aequorin microinjected into the superficial cells. Each electrical stimulus produces a calcium transient and a twitch contraction. The size of the calcium transient in these cardiac muscles fibres is much less than in skeletal muscle fibres, and would normally provide insufficient calcium to saturate all the troponin C sites in the cell. This allows changes in the amount of calcium released to alter the tension produced in the ensuing contraction.

The amplitude of these calcium transients (and of the consequent twitches) can be increased by a number of agents known to have positive inotropic effects. These include increased external calcium concentration, increased stimulus frequency (the staircase effect), noradrenaline and other catecholamines acting via β-adrenergic receptors, and cardiac glycosides (fig. 14.20). But the inotropic effect of increasing length (see below) is not accompanied by an increase in the size of the calcium transient.

The calcium transient is probably derived from two sources. First there is entry of calcium via the calcium channels which open to allow the slow inward current during the action potential. Secondly there is release of calcium ions from the sarcoplasmic reticulum. There is good evidence

Fig. 14.20. Calcium transients in cat heart muscle in response to electrical stimulation. The noisy traces show the aequorin signal, averaged from a number of successive responses, the smooth traces show force. Pairs of records show the effects of *a*, stimulation frequency; *b*, muscle length; *c*, strophanthidin (a cardiac glycoside); and *d*, noradrenaline. Calibrations of light output and force are the same for each member of a pair of records, but differ between pairs. (From Morgan and Blinks, 1982.)

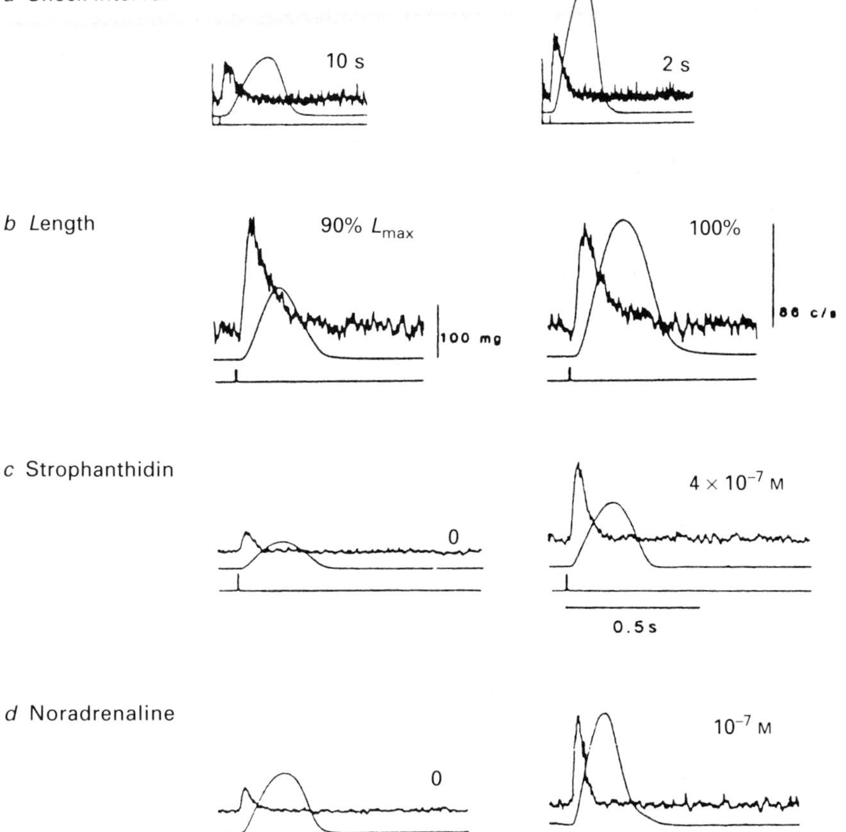

that the second of these processes is triggered by the first, i.e. that calcium release from the sarcoplasmic reticulum is induced by a rise in calcium ion concentration within the cell (Fabiato, 1983, 1985).

It is well known that increasing the volume of the heart immediately before a beat increases the force of the contraction. This relation is known as Starling's law of the heart or alternatively the Frank–Starling law (Frank, 1895; Patterson and Starling, 1914; Allen and Kentish, 1985). The length–tension curve of intact cardiac muscle, at lengths below that at which maximum tension is obtained, is much steeper than the corresponding curve for skeletal muscle, as is shown in fig. 14.21 (Allen *et al.*, 1974; Sonnenblick and Skelton, 1974; Ter Keurs *et al.*, 1980). But if we look at the length–tension curve for skinned cardiac fibres fully activated by relatively high calcium concentrations, this steepness disappears (Fabiato and Fabiato, 1975). This implies that the degree of activation of the contractile apparatus is affected by length.

This conclusion has been confirmed by further experiments on skinned fibres. Thus Hibberd and Jewell (1982) found that the steepness of the length–tension relation was greater at lower calcium concentrations (fig. 14.22*a*) and that the force production was more sensitive to calcium ion concentration at longer lengths (fig. 14.22*b*). The precise mechanism of this length-dependent sensitivity to calcium is not clear. Perhaps the affinity of troponin for calcium is higher at longer sarcomere lengths, but it is not easy to see what might bring this about.

Vertebrate smooth muscle

We have seen that calcium activates smooth muscle via its combination with calmodulin, which itself then activates the myosin light chain kinase. In some smooth muscles the activator calcium probably enters via voltage-dependent calcium channels. In others there is an appreciable amount of sarcoplasmic reticulum from which calcium ions can be released (see Bolton, 1986).

Experiments on smooth muscles from the coronary arteries of the pig show that both acetylcholine (acting via muscarinic receptors) and noradrenaline (acting on α_1-adrenergic receptors) produce inositol trisphosphate as a second messenger, and that this releases calcium from the sarcoplasmic reticulum (Suematsu *et al.*, 1984;

Fig. 14.21. Length–tension relations of cardiac muscle fibres. In each case the broken curve (i) shows the corresponding curve for skeletal muscle (fig. 12.16), and all curves are normalized to 100% at a sarcomere length of 2.2 μm. Curves in *a* show peak tension in twitches of intact cat papillary muscle at high (ii) and low (iii) calcium ion concentrations. Curves in *b* show the contracture tensions in skinned rat ventricular muscle at high (ii) and low (iii and iv) calcium ion concentrations. (From Allen and Kentish, 1985.)

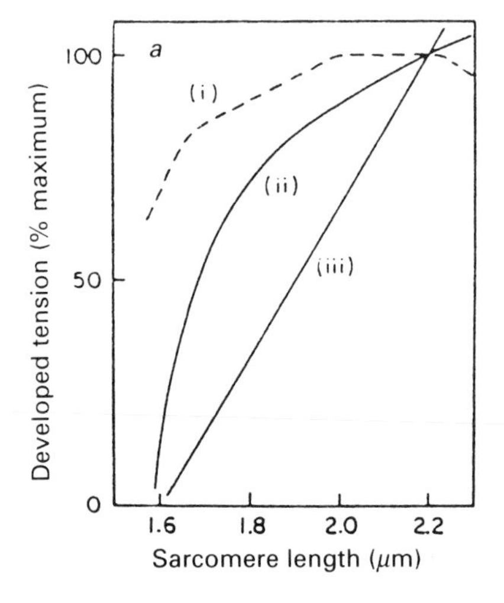

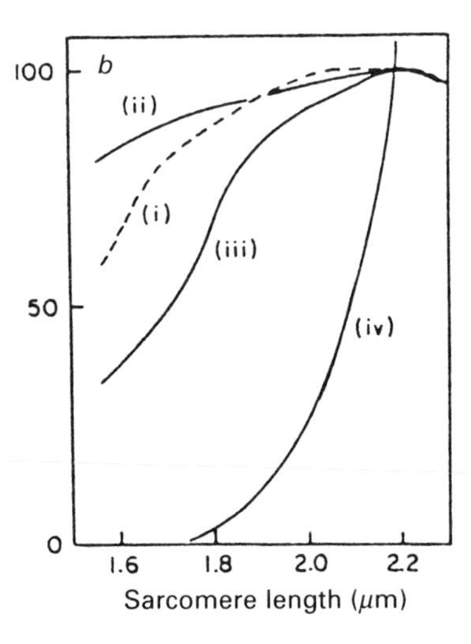

Hashimoto *et al.*, 1986). Many smooth muscles are relaxed by noradrenaline acting via β- adrenergic receptors. The second messenger here is cyclic AMP, which probably causes some reuptake of calcium into the sarcoplasmic reticulum, together with a more direct action which makes the contractile system less sensitive to calcium ions (Morgan and Morgan, 1984; Rüegg and Pfitzer, 1985).

The organization of the contractile apparatus

Striated muscles

The structure of striated muscles has been dealt with in chapter 12. The skeletal and cardiac muscles of vertebrates and arthropods, and the visceral muscles of arthropods, are almost all striated. Striated muscles also occur locally in many other animals; some examples are the swimming muscles of jellyfish, the pedicellariae muscles of echinoderms, and certain muscles in the heads of some annelids.

The sarcomere length, or, to be more precise, the length of the *A* band, is not always the same in different muscles. What effect will the length of the *A* band have on the contraction of the muscle? Let us assume that an individual cross-bridge always produces the same amount of tension in an isometric contraction. Cross-bridges acting between any one pair of actin and myosin filaments act in parallel, more of such cross-bridges can be formed with longer *A* bands, and therefore the maximum tension per unit cross-sectional area is proportional to the length of the *A* band. The forces generated in different half-sarcomeres are in series with each other, and so the total force is unaffected by the number of sarcomeres in a fibre. Hence we would expect muscles with long *A* bands to have a higher maximum tension per unit cross-sectional area than those with short *A* bands. A corollary of this idea is that, other things being equal, the velocity of shortening should decrease with increasing length of the *A* band. However, it seems that other factors are involved in the determination of shortening velocities; the sarcomere lengths of frog twitch and tonic fibres are the same, but their speeds of contraction are very different.

Fig. 14.22. Relations between length, calcium ion concentration and force in chemically skinned rat ventricular muscle. *a* shows force–length curves at different calcium ion concentrations. *b* shows force at different calcium ion concentrations for two different sarcomere lengths: long (2.3–2.5 μm, black circles) and short (1.9–2.0 μm, white circles). Maximum force at the shorter length in *b* was about 80% of that at the longer length; the curves have been normalized to show that the muscle is more sensitive to calcium at the longer length. (From Hibberd and Jewell, 1982.)

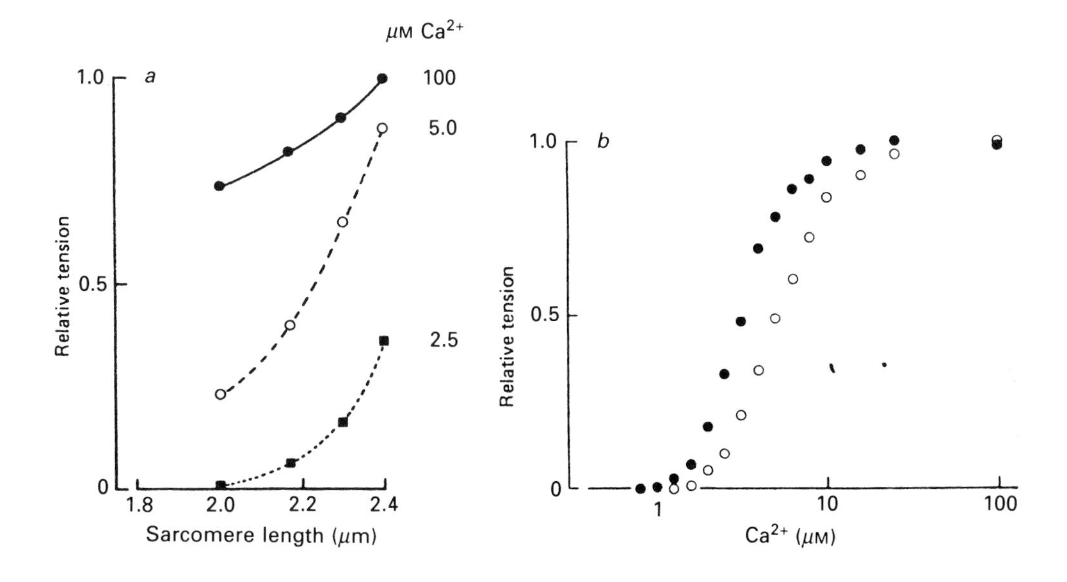

We have seen that, in vertebrate skeletal muscles, the myofilament array is such that each *A* filament is surrounded by six *I* filaments and each *I* filament by three *A* filaments, so that there are two *I* filaments per *A* filament in each half-sarcomere. Different arrangements occur in arthropod muscles. In most insect flight muscles each *A* filament is surrounded by six *I* filaments, but each *I* filament is placed between two *A* filaments so that the *I* to *A* ratio is three to one (fig. 14.23). In thoracic muscles of the cockroach there are eight or nine *I* filaments round each *A* filament and the *I* to *A* ratio is four to one (Hagopian and Spiro, 1968). Finally, in some insect leg and visceral muscles there are twelve *I* filaments round each *A* filament and the *I* to *A* ratio is six to one (Hagopian, 1966; Smith *et al.*, 1966).

What is the significance of these 'extra' *I* filaments? We have seen that, other things being equal, the maximum isometric tension produced by a muscle is proportional to the length of the *A* band. Hence the longer the *A* band, the greater the strain on the myofilaments during contraction. The presence of 'extra' *I* filaments is associated with long *A* bands, so it is possible that the greater total number of *I* filaments is needed to reduce the strain on each one of them.

Insect asynchronous flight muscles contain another contractile protein, paramyosin (Bullard *et al.*, 1973). This is a rod-like protein, molecular weight about 100 000, which aggregates readily to form structures with a repeat periodicity in the region of 720 Å. It was first found in molluscan muscles, where it is a major component of the thick filaments. In insect flight muscles, it may form a core in the thick filaments, and could conceivably connect them to the *Z* lines.

Fig. 14.23. Myofilament arrays in vertebrate and insect striated muscles. The diagrams show the arrangement of myosin (large spots) and actin (small spots) filaments as seen in transverse sections through the overlap region. *a*: Vertebrate muscles. *b*: Insect flight muscles. *c*: Insect leg and visceral muscles. (From Smith *et al.*, 1966.)

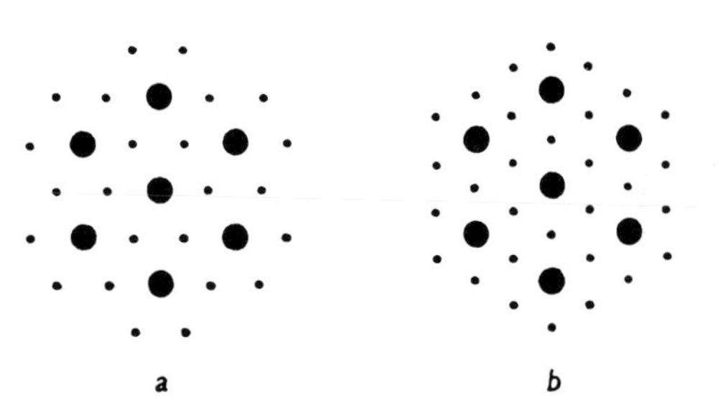

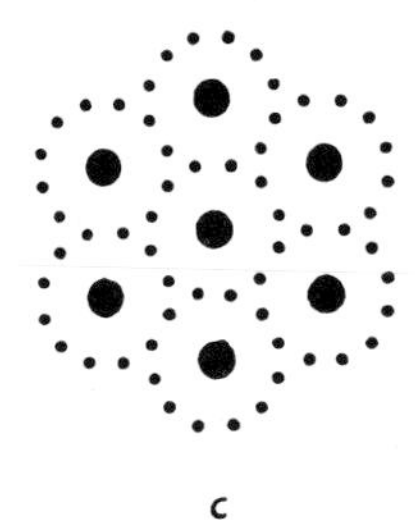

Oblique striated muscles

Muscles of this type possess striations which follow a helical or oblique course within the fibres. They are found in annelids, molluscs, nematodes and some other invertebrates (Rosenbluth, 1968). The fibres contain two types of filament, corresponding to the *A* and *I* filaments of striated muscles. The oblique striations arise from a staggered arrangement of the thick filaments, as can be seen in electron micrographs of tangential longitudinal sections (fig. 14.24). During contraction the thin filaments slide between the thick ones, so shortening the distance between successive striations. Since the muscle cell is a constant volume system, shortening is accompanied by widening, hence the angle of the striations alters: they become less oblique (Lanzavecchia, 1968; Mill and Knapp, 1970).

Paramyosin occurs in the thick filaments of the oblique striated muscles of annelids and molluscs.

Invertebrate smooth muscles

Many muscles in invertebrates show no signs of striation when examined by light microscopy. Electron micrograph sections show two types of filaments, thick and thin, but these are not transversely or obliquely aligned (Hanson and Lowy, 1960).

The 'catch' muscles of molluscs (p. 334) have been particularly investigated (Lowy and Hanson, 1962; Sobiescek, 1973; Elliott and Bennett, 1982). The thin filaments are similar to the *I* filaments of striated muscles and show the same F-actin double helix structure when isolated. But the thick filaments vary in thickness from 150 to 1500 Å, and appear to consist of a number of closely packed ribbon-like elements. The muscles contain a high proportion of paramyosin, and it is this that forms the major part of the thick filaments. Myosin can be extracted from the thick filaments, when an array of roughly triangular pits are seen on their surface (Szent-Györgyi *et al.*, 1971). Neighbouring triangles all point in the same direction, but this reverses in the middle of the filament, so that the two halves have opposite polarities, rather as in the myosin filaments of vertebrate striated muscles. Many of the thin (actin) filaments are attached to dense bodies. They form 'arrowheads' when decorated with heavy meromyosin, and these are of opposite polarity on each side of a dense body, just as are those on each side of a *Z* band in striated muscles.

Vertebrate smooth muscle

Vertebrate smooth muscles contain the major contractile proteins actin and myosin, together with tropomyosin, and the actin occurs in thin filaments, which are readily seen by electron microscopy. The relative proportion of myosin is much less than that in skeletal muscles.

Early studies by electron microscopy were unable to demonstrate the presence of thick filaments. This was rather surprising, not least because of the existence of a 143 Å reflection in the X-ray diffraction pattern (Lowy *et al.*, 1970). Improvements in fixation technique showed that thick filaments are indeed present, but there was then some controversy as to whether the myosin occurs as filaments 120–300 Å in diameter or as ribbons measuring 80 Å by 200–1100 Å in section (see Shoenberg and Needham, 1976). More recently it has been concluded that the ribbons are artefacts, perhaps produced by low temperature, but the episode serves to demonstrate the surprising lability of organization of the myosin in smooth muscle.

It seems likely that the structure of smooth muscle myosin filaments is rather different from those in skeletal muscles. In particular, there is some evidence that the myosin molecules are oriented in opposite directions on the two faces of the filament (Small and Squire, 1972; Small, 1977). Such filaments would have no bare zone in the middle but would have a bare region on one side at each end (see fig. 14.25). Craig and Megermann (1977) have observed this 'side polar' pattern in synthetic filaments formed by aggregation of smooth muscle myosin.

Many thin filaments appear to be attached at one end to structures in the cytoplasm called dense bodies, which are probably analogous to the *Z* lines of striated muscle. Others are attached to dense patches next to the cell membrane. The dense patches of adjacent cells appear to connect with each other so that there is mechanical continuity between the constituent cells of the muscle (see Gabella, 1981, 1984).

It is possible that bundles of filaments in smooth muscle form 'contractile units' analogous in function to a short series of sarcomeres in a

Fig. 14.24. Diagram showing the arrangement of myofilaments in an oblique striated earthworm muscle fibre. The thin filaments are attached to *Z* rods, shown in black here. Alternating with the *Z* rods are tubules of the sarcoplasmic reticulum. The angle of the oblique striations is exaggerated. (From Mill and Knapp, 1970.)

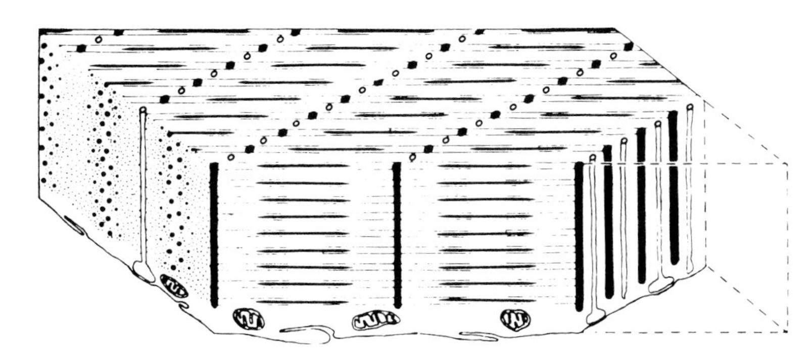

striated muscle cell (Small and Squire, 1972; Small, 1974). According to this idea, contraction occurs as thin filaments are pulled past thick filaments by cross-bridge action just as in striated muscle (fig. 14.26). But because of the side-polar arrangement of the cross-bridges, a long thin filament can completely overlap a short thick filament and can be effectively pulled over the whole of its length by that filament. This allows the muscle to operate at near maximum tension over a very wide range of lengths.

Fig. 14.25. Possible packing models for myosin filaments in vertebrate smooth muscles, showing the 'side-polar' arrangement in which the molecules on one side of the filament are all oriented in one direction while those on the other side are all oriented in the opposite direction. Notice that there is a bare zone, free of myosin heads, at one end on each side of the filament. (From Squire, 1986.)

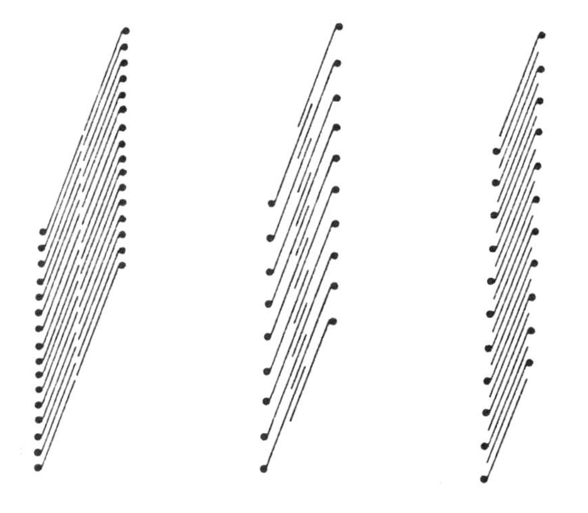

Fig. 14.26. A contractile unit of vertebrate smooth muscle, showing how the filaments could slide past each other during contraction. The model assumes a side-polar arrangement in the thick filaments. (From Squire, 1986, based on Small and Squire, 1972.)

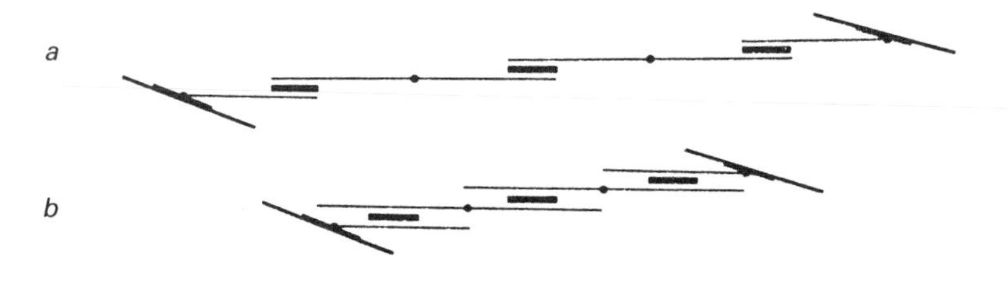

Mechanical properties

The mechanical properties of muscles are related to the functions that they subserve in the body. For example, muscles that are involved in rapid movements have greater speeds of contraction than do those involved in the maintenance of posture. The muscles of small animals tend to contract more rapidly than those of larger animals, since their locomotory movements are relatively more rapid (see Hill, 1950*b*). In some cases, specialization is carried further, and apparently involves the development of new mechanical properties. We shall consider two examples of these specializations: the fibrillar muscles of insects, and the 'catch' muscles of molluscs.

Insect fibrillar muscles

The wing-beat frequency of many insects is very high; the record is held by the midge *Forcipomyia*, in which is it about 1000 beats per second (Sotavalta, 1953). This means that the tension in the flight muscles in *Forcipomyia* must rise and fall within a millisecond, which seems almost impossibly fast. Pringle (1949) measured the electrical responses of the flight muscles of a flying blow-fly (*Calliphora*) and showed that their frequency was very much less than that of the wing-beat. This indicates that the individual contractions of the muscles, which produce the wing-beat, are not 'twitches' analogous to the twitches of other muscles. Pringle suggested that the function of excitation is to bring the contractile apparatus into a state of activity (analogous to the tetanus condition in other muscles) in which rhythmic contractions are possible.

Further work by Roeder (1951) and others (see Pringle, 1981) showed that this asynchronous relation between the neurally excited muscle potentials and the contraction frequency occurs in the flight muscles of flies, beetles, bees and wasps, and certain bugs; it is also found in the tymbal (sound-producing) muscles of some cicadas (Pringle, 1954). These muscles are described as 'fibrillar' (since their fibres contain closely packed myofibrils) or 'asynchronous'. The flight muscles of more primitive insects, such as locusts, dragonflies and moths, are not of the fibrillar type, and show the familiar one to one ratio between the electrical and mechanical responses to nervous stimulation.

A closer investigation of the mechanical properties of fibrillar muscle was carried out by Machin and Pringle (1959), using a flight muscle from the rhinoceros beetle *Oryctes*. The muscle was partly dissected from the insect and connected to a piezo-electric force transducer at one end and to a moving-coil vibrator at the other. The position of the vibrator arm was measured by means of a light beam and a phototransistor, so as to give the length of the muscle. The load on the muscle (produced by the action of the vibrator) was electrically controlled from the output of the phototransistor, allowing regulation of its stiffness, damping and mass (inertia), as is shown in fig. 14.27. Examination of the load control circuit shows that it contains two differentiating units; if the output of the phototransistor is V, then stiffness is proportional to V, damping is proportional to dV/dt, and mass is proportional to d^2V/dt^2. The outputs of the length and tension transducers were connected to the X and Y amplifiers of an oscilloscope, giving a direct display of the length–tension diagram.

In the resting condition, the muscle was very stiff, giving a steep passive length–tension curve. With an isometric load, stimulation of the motor nerve resulted in the development of an extra steady maintained tension, just as in other muscles. But with an inertial load, obtained by a suitable setting of the load parameters, the muscle would undergo an oscillatory contraction, giving a length–tension loop such as is shown in fig. 14.28. The frequency of this oscillatory contraction is the same as the mechanical resonant frequency of the muscle and its load. In a flying insect, the load is determined by the inertia of the wings, the aerodynamic forces on the wings, and the stiffness of the thoracic skeleton by whose movements the wings are moved. Thus the wing-beat frequency of insects with fibrillar flight muscles can be altered by altering the resonant

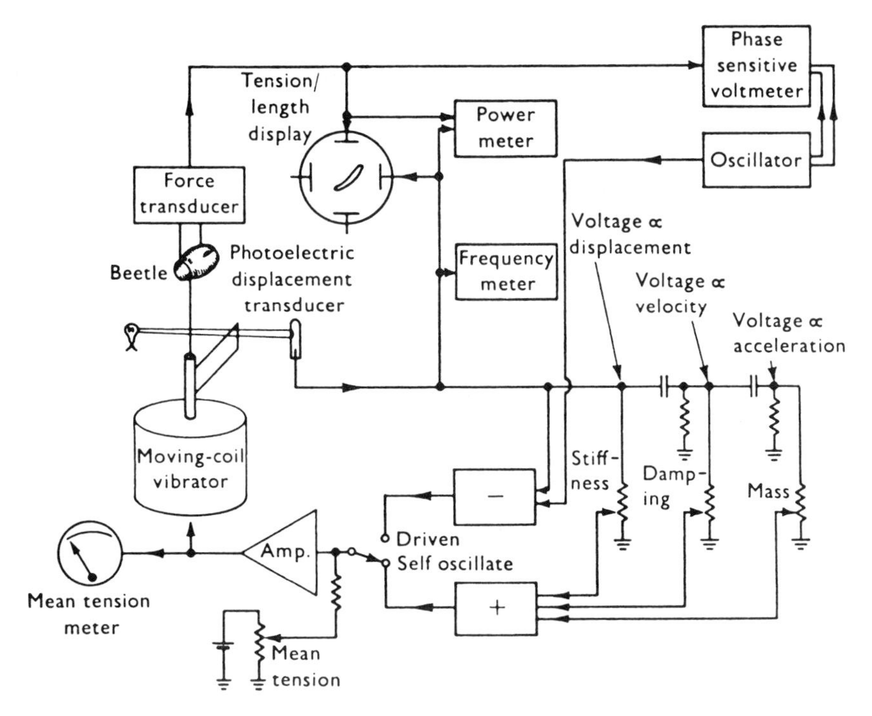

Fig. 14.27. Block diagram of the apparatus used to measure the mechnical properties of rhinoceros beetle flight muscle. (From Machin and Pringle, 1959, by permission of the Royal Society.)

frequency of the wing–thorax system, such as by cutting off the tips of the wings (when the frequency increases) or by loading them with small amounts of wax (the frequency decreases).

Much information about the properties of fibrillar muscles has been obtained by using the method of sinusoidal analysis. The muscle, or a bundle of glycerol-extracted fibres, is subjected to a sinusoidal length change of small amplitude, and the resulting sinusoidal change in tension is measured. The process is then repeated with length changes of the same amplitude but of different frequencies. In experiments on intact muscles, the apparatus shown in fig. 14.27 can be used if the switch is placed in the 'driven' position.

The results obtained from this type of experiment are usefully expressed in a diagram known as the *Nyquist plot*, which is constructed as follows. Referring to fig. 14.29*a*, we see that the tension output is a sinusoidal waveform of the same frequency ($1/b$) as the length input, but that it suffers a phase shift a. The magnitude of this phase shift can be expressed either as a lead or lag of so many units of time, or as the *phase angle*, which is given by a/b revolutions, $360a/b$ degrees or $2\pi a/b$ radians. We then construct a graph whose axes are both in units of tension and to the same scale, the abscissa being labelled 'in-phase component' and the ordinate 'quadrature component'. A point is plotted on this graph at a distance from the origin equal to the amplitude of the tension trace and at an angle to the abscissa equal to the phase angle (fig. 14.29*b*). Similar measurements are then made with length inputs of the same amplitude but different frequencies, to give a complete Nyquist plot as is shown in fig. 14.29*c*.

Fig. 14.28. The oscillatory contraction of a rhinoceros beetle basalar muscle. The lower sloping line shows the length–tension relation of the unstimulated muscle, the upper sloping line shows the length–tension relation when the muscle is stimulated under isometric conditions. The loop shows the oscillatory contraction produced by stimulation when the load possesses suitable inertial and damping components; it is traced out in an anticlockwise direction and at a frequency in the region of 25 Hz. (From Machin and Pringle, 1959, by permission of the Royal Society.)

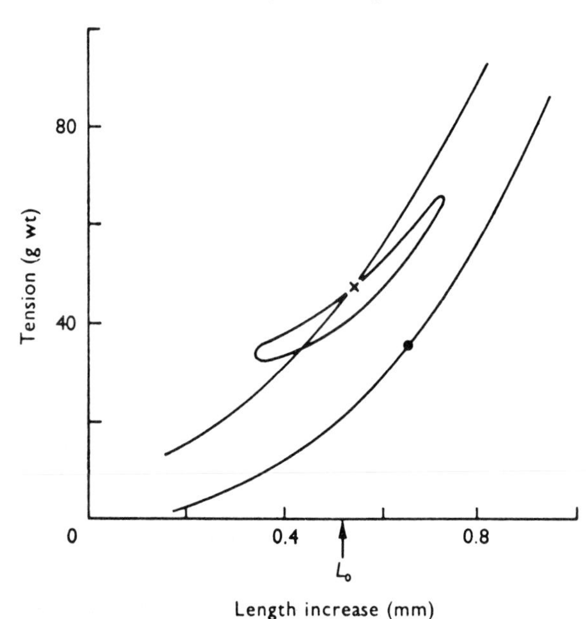

It is essential to use length changes of small amplitude in these experiments, since the analysis depends on the tension change being almost perfectly sinusoidal; this will not be so for large length changes. The process of analysis can be made easier by using a resolved components indicator, an electronic device which gives a direct measure of the in-phase and quadrature components of the tension change. Results from different muscles can be compared by converting the in-phase and quadrature components of tension into elastic and viscous moduli respectively, by means of the formulae

$$E_e = \frac{\Delta F_e \cdot L}{A \cdot \Delta L}$$

and

$$E_v = \frac{\Delta F_v \cdot L}{A \cdot \Delta L}$$

where E_e and E_v are the elastic and viscous moduli, ΔF_e and ΔF_v are the in-phase and quadrature components of the tension change, L is the length of the muscle, ΔL is the amplitude of the length change, and A is the cross-sectional area of the muscle (Machin and Pringle, 1960). E_e and E_v are then conveniently expressed in kilograms per square centimetre, and the Nyquist plot has now been converted into a *vector modulus plot*.

Vector modulus plots for a beetle flight muscle in the resting condition and when stimu-

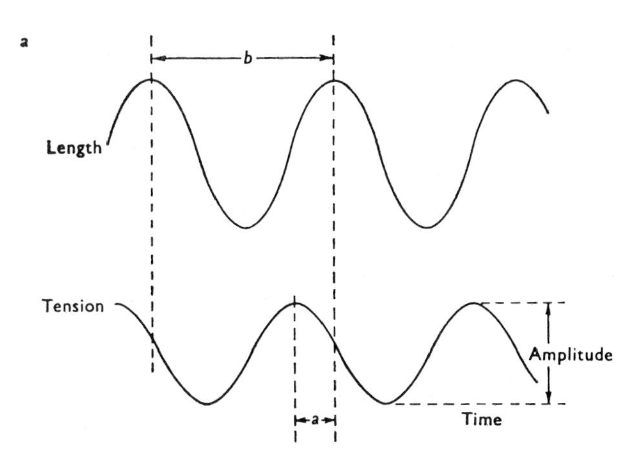

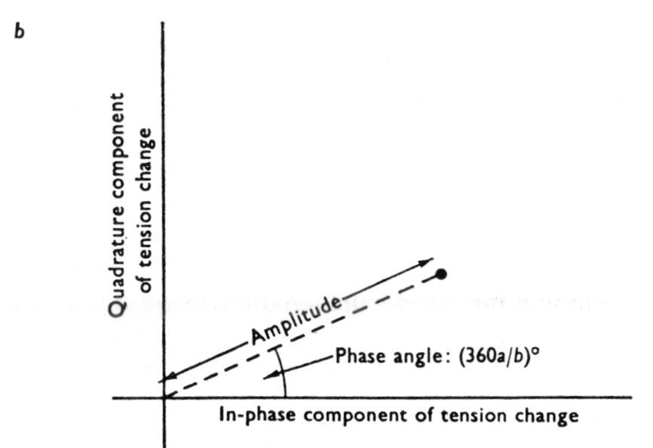

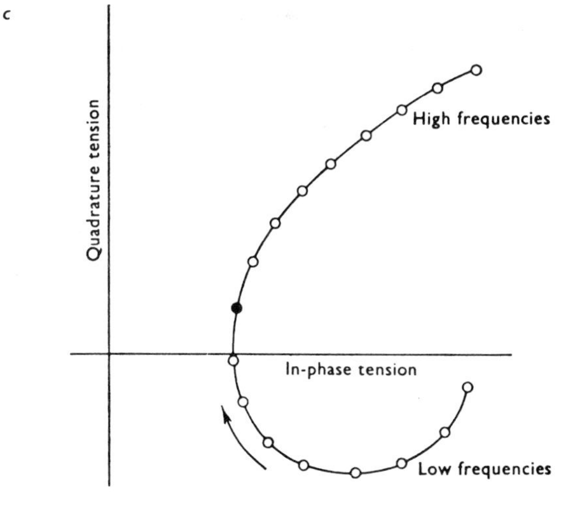

Fig. 14.29. How a Nyquist plot is constructed. *a* shows the imposed length change at one particular oscillation frequency, and the consequent tension change. The amplitude of this tension change and its phase angle with respect to the length change are then plotted on the Nyquist graph as shown in *b*. Similar measurements are made for other frequencies, to give the complete Nyquist plot shown in *c* (the curve shown is such as is obtained from an active fibrillar muscle). Further explanation in text.

lated are shown in fig. 14.30*a*. If we compare these with that obtained from the mechanical visco-elastic system shown in fig. 14.30*b*, it is clear that the resting muscle behaves very similarly to the visco-elastic model. The active muscle, however, differs from the model in that its viscous modulus is negative at low frequencies; this indicates that tension changes are lagging behind length changes (so giving the negative phase angle) and that the muscle is therefore doing work on the apparatus. Nyquist and vector modulus plots of this type are characteristic of insect fibrillar muscle; driven oscillations at negative phase angles are equivalent to free oscillations against a suitable load.

The work output per cycle of an oscillating fibrillar muscle is greater when the frequency is such that the viscous modulus is at its maximum negative value. Maximal power output occurs at a slightly higher frequency, when the product of the frequency and the viscous modulus is at its maximum negative value. Since the wing-beat frequency of a flying insect is determined by the mechanical resonant frequency of the wing–thorax system, we would expect the Nyquist plot of the flight muscles to be such that maximum power output occurs at or near this frequency. If measurements are made at the temperature of the flight muscles in the flying insect, this is found to be so (Machin *et al.*, 1962).

Jewell and Rüegg (1966) showed that glycerinated fibres from the wing muscles of giant water bugs (Belostomatidae) showed the typical oscillatory phenomena in the presence of suitable concentrations of ATP and calcium ions. This indicates that the oscillation really is a property of the contractile apparatus, not of the excitation or coupling processes, which merely serve to keep the muscle in the active condition. Fig. 14.31 shows, in the form of vector modulus plots, the effects of ATP and calcium ion concentration on the response of these glycerinated fibres to imposed oscillations. A feature of these results is that the frequency at which the viscous modulus is maximally negative

(2–5 Hz) is much lower than the wing-beat frequency of the intact insects (20–30 Hz); the discrepancy can be explained partly in terms of temperature effects.

The essential property of fibrillar muscle is called activation by stretch: an increase in length is followed, after a short delay, by a rise in tension. There is a corresponding delayed fall in tension after shortening, and the two effects together produce the oscillatory capabilities of the muscle. The question that arises is, therefore, what is the mechanism of this activation by stretch?

The ATPase activity of glycerol-extracted water bug flight muscle fibres in the presence of calcium ions was investigated by Rüegg and Tregear (1966). They found that more ATP is split when the fibres are subjected to oscillatory length changes, and that this 'extra' ATPase activity is greatest when the frequency is that at which maximum power output occurs. This suggests that oscillatory activity is associated with an increased cycling of cross-bridges, a conclusion which led Thorson and White (1969, 1983) to propose that stretching the muscle alters the rate constants for cross-bridge cycling (corresponding to f and g in A. F. Huxley's 1957 model).

Thorson and White's model of activation by stretch is related to one of the striking features of fibrillar muscle, its very great stiffness both at rest and during contraction. White (1983) has demonstrated that this is accounted for by the existence of '*C* filaments' which connect the ends of the *A* filaments to the *Z* line; stretching the muscle as a whole will thus also stretch the individual *A* filaments. Thorson and White suggest that such an alteration in *A* filament length could alter the rate constants for cross-bridge action.

Wray (1979) has produced an alternative suggestion for the origin of stretch activation, arising from structural considerations. The *A* filaments of insect flight muscles are different from those of vertebrate skeletal muscles; there are four myosin heads per crown, emerging every 145 Å along the filament, but the rotation of successive crowns gives an overall pitch of 4 × 385 Å (Reedy *et al.*, 1981; see Tregear, 1983). Thus the pitch of the thick filaments matches almost precisely that of the thin filaments, a very unusual situation. Hence the

Fig. 14.30. Vector modulus plots of a beetle flight muscle (*a*) and a visco-elastic model (*b*). In *a*, the continuous curve shows the results from a stimulated muscle and the dotted curve shows those from an unstimulated one; the points are shown at frequency increments of 10 Hz, starting with 5 Hz. (From Machin and Pringle, 1960, by permission of the Royal Society.)

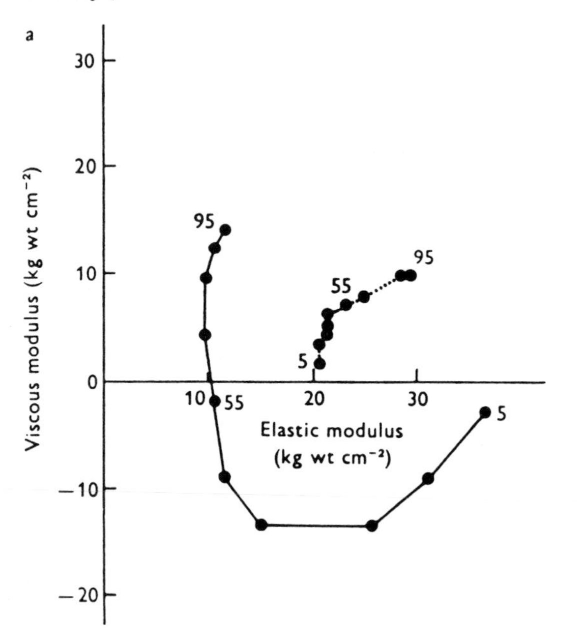

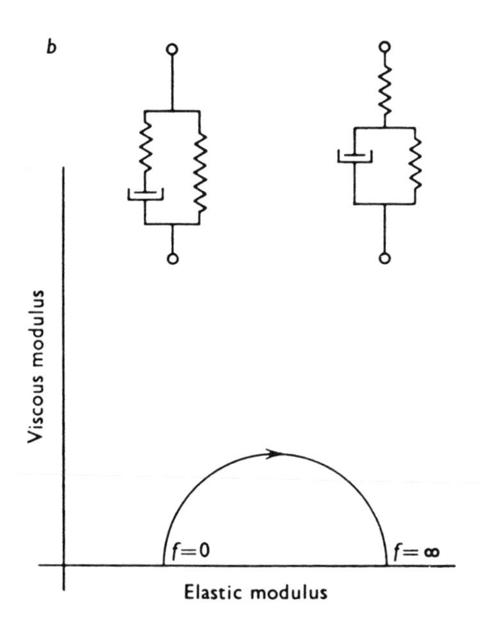

number of cross-bridges that can be formed must depend critically on the relative axial alignment of the two sets of filaments. Wray suggested that stretch brings more myosin heads into positions where they can readily form links with actin molecules (fig. 14.32).

The high degree of geometrical regularity which is a characteristic of fibrillar muscles could be related to this mechanism. Their well-developed *M* bands and *Z* lines might be needed to maintain the lateral alignment of the *A* and *I* filament arrays. A prime function of the *C* filaments could be to ensure that the *A* filaments are positioned precisely midway between the two ends of the sarcomere so that the relative positions of the two sets of filaments are the same in each half-sarcomere and at different positions along the length of the muscle.

From an evolutionary of view, it is interesting to speculate on whether the oscillatory characteristics of fibrillar muscle are dependent on some basically new properties of the muscle, or whether they arose by enhancement of properties already present to some extent in other muscles. Some evidence in favour of the second possibility is provided by the properties of the sound-producing muscles in the cicada *Fidicina*: although these muscles are neurogenically activated, with the conventional one-to-one relation between nerve impulse and muscle contraction, their glycerinated fibres show all the mechanical properties of glycerinated water bug flight muscles (Aidley and White, 1969). This suggests that the work on insect fibrillar muscles (reviewed by Pringle, 1981, and Tregear, 1983) may help in understanding how more conventional muscles work.

Fig. 14.31. Vector modulus plots of a bundle of glycerinated fibres from the flight muscles of a giant water bug, *Lethocerus*, in response to oscillatory length changes of constant amplitude. Curve *a* shows the response with the fibre in a buffered salt solution containing neither ATP nor calcium ions (the rigor state). Curve *b* shows the response in the presence of ATP (5 mM) but the virtual absence of calcium ions. Curves *c* to *e* show responses in the presence of both ATP and calcium ions, the calcium ion concentrations being: *c*, 5.2×10^{-8} M; *d*, 1.4×10^{-7} M; and *e*, 2.1×10^{-7} M. The frequencies for curve *c* and *d* were the same as for curve *e*. (From Jewell and Rüegg, 1966, by permission of the Royal Society.)

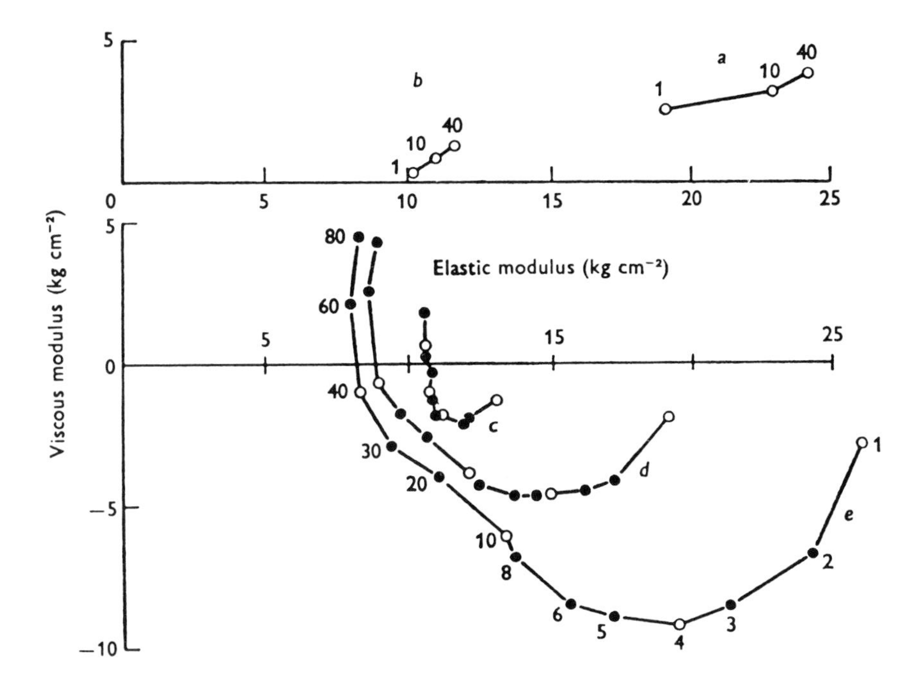

Molluscan 'catch' muscles

Certain muscles of molluscs, of which the most closely investigated is the anterior byssus retractor of the mussel *Mytilus*, are remarkable in that their rate of relaxation is dependent upon the nature of the excitatory stimulus. In 'phasic' responses, the relaxation following an isometric contraction is complete within a few seconds, whereas in 'tonic' or 'catch' responses, the relaxation phase lasts for several minutes or sometimes hours. Phasic responses are usually produced by electrical stimulation using repetitive brief pulses or alternating current. Tonic responses are usually produced by continuous direct current stimulation or by treatment with acetylcholine. Tonic responses are abolished by 5-hydroxytryptamine.

The mechanical responses of the *Mytilus* anterior byssus retractor muscles were investigated by Jewell (1959) and Lowy and Millman (1963), among others. Jewell found that the behaviour of the muscle during stimulation was not greatly different from that of frog leg muscles, except that the time course of contraction was much slower. During the tonic response, however, the ability to shorten and to redevelop tension after a quick release was very much reduced (fig. 14.33*c*). If the muscle was restretched after a quick release during this time, the full 'catch' tension was re-established.

The control mechanism for the catch system has been illuminated by Cornelius (1982), in experiments on *Mytilus* anterior byssus retractors muscles which had been chemically 'skinned' with saponin. He found that maintained contractions, with the ability to shorten, were produced in the presence of ATP by calcium ion concentrations over 10^{-7} M. Reduction of the calcium ion concentration to below this value produced a very slow relaxation during which the muscle was in the catch state, i.e. tension was maintained but shortening could not occur. The duration of the catch was greatly reduced in the presence of cyclic AMP or 5-hydroxytryptamine. Cyclic AMP had no effect on the response in the presence of calcium.

5-Hydroxytryptamine activates adenylate cyclase and so leads to cyclic AMP production in catch muscles. There is some evidence that catch is associated either with the phosphorylation of

Fig. 14.32. Wray's suggestion about the geometry of actin–myosin interaction in fibrillar (asynchronous) insect flight muscle. Two arrays are shown: the myosin heads on a single thick filament (circles), and the sterically accessible regions of the six thin filaments which surround it (bars). Since the pitch (385 Å) of the two helices is the same, the two arrays may be in or out of register with each other, so allowing more or fewer cross-bridges to be formed. Wray suggests that the two arrays are largely out of register at rest length (*a*) and are brought into register on stretching (*b*), so accounting for the activation by stretch that occurs in these muscles. (From Wray, 1979.)

Rest length

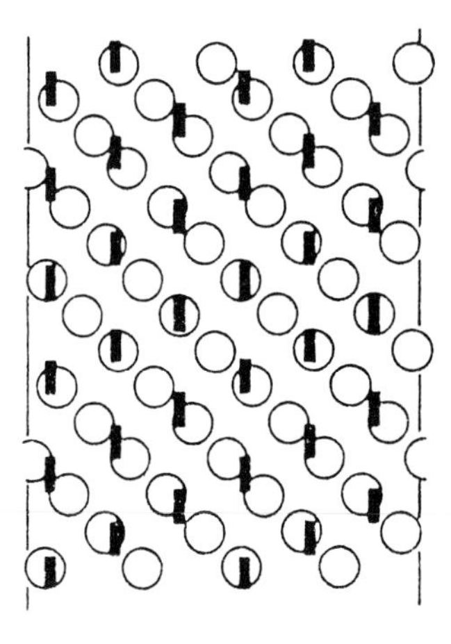

Stretched

paramyosin (Cooley *et al.*, 1979; Achazi, 1982), or with the dephosphorylation of myosin (Castellani and Cohen, 1987).

Myosin forms only 20% of the content of the thick filaments of catch muscles, the rest being paramyosin. It would not be surprising, therefore, it paramyosin were crucially involved in the mechanism of the catch. At one time it was thought that separate links between the paramyosin molecules, in parallel with the normal myosin–actin cross-bridges, might be involved in the catch system, but no such links have been seen in electron micro graphs. It seems more likely that changes in the paramyosin might affect the rate of breakage of the myosin–actin links (Szent-Györgi *et al.*, 1971). The structure of the thick filaments, in which myosin molecules form a layer just one molecule thick over the surface of the paramyosin core, would seem to permit some such interaction. Perhaps, as Cohen (1982) suggests, catch occurs when phosphorylation of the paramyosin somehow restricts the mobility of the myosin cross-bridges.

Fig. 14.33. Experiments showing some of the mechanical properties of a *Mytilus* anterior byssus retractor muscle. *a* and *b*, Phasic responses, induced by repetitive stimulation; *c*, tonic response, produced after washing the muscle for a short time in an acetylcholine solution. The left record in *a* shows the isometric tension developed during the phasic response, the right record shows the shortening produced during an after-loaded isotonic contraction against a 30 g load. Record *b* shows an isotonic release during phasic stimulation. In record *c*, the muscle first undergoes an isotonic release and is then restretched, and the process is then repeated twice; notice the maintenance of tension but absence of the ability to shorten during the tonic response. (From Jewell, 1959.)

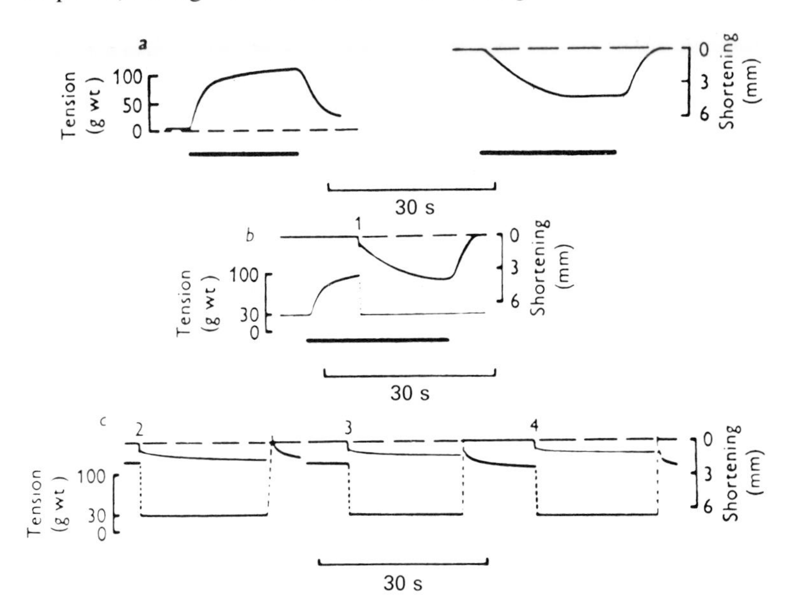

15
The electric organs of fishes

A number of different types of fish are capable of producing appreciable electric currents in the water surrounding them. These currents are produced by special organs known as *electric organs*. The position of the electric organs in various electric fish is shown in fig. 15.1.

Electric organs are composed of columns of cells called *electroplaques* (sometimes called electroplates, electroplaxes or electrocytes), each of which is innervated by an excitor nerve. These electroplaques (with one exception – see p. 343) appear to be modified muscle cells which have lost their contractile function. In this chapter we shall examine the means whereby the electroplaques produce the electric currents constituting the discharge of the electric organs.

The electroplaques of the electric eel

The electric eel, *Electrophorus electricus*, is capable of producing an electric discharge of over 600 V (the record seems to be 866 V), consisting of about half a dozen pulses each lasting 2–3 ms. The gross structure of the electric organ is shown in fig. 15.2. Each electroplaque is about 100 μm thick (longitudinally), 1 mm wide (vertically) and 10–30 mm long (radially). The nerve endings are restricted to the posterior face. The anterior (non-innervated) faces are much folded, giving rise to numerous papillae. The high-voltage discharges are produced by the main organ; Sachs' organ gives much smaller discharges. In the main organ, up to 6000 rows of electroplaques are arranged in series with each other. This series arrangement of the electroplaques leads to addition of the voltages produced by each one of them; for example, if a discharge of 600 V were produced by an organ containing 4000 rows of electroplaques, each electroplaque would have to produce a potential of 150 mV across it.

The electrical properties of the electroplaque were investigated by Keynes and Martins-Ferreira (1953), using intracellular electrodes. They used the electroplaques from the organ of Sachs for their experiments, since these electroplaques can be more easily isolated than can those of the main organ. In the resting condition, there is a resting potential of about −80 to −90 mV. Electrical stimulation of the electroplaque results in an all-or-nothing action potential appearing across the innervated (posterior) face, but there is no potential change across the non-innervated (anterior) face. The experimental evidence for this statement is as follows. In fig. 15.3, trace *a* is recorded from two microelectrodes placed just outside the innervated face; there is, of course, no potential difference between them either at rest or during activity. For trace *b*, one of the microelectrodes was inserted into the cell; this then records the resting potential, and, on stimulation, an 'overshooting' action potential of about 150 mV is observed. This means that the innervated face is electrically excitable, just as the cell membranes of nerve axons and vertebrate twitch muscle fibres are. In trace *c*, one of the electrodes is pushed right through the electroplaque so that the potential across the whole electroplaque is recorded. There is no steady potential at rest, indicating that the resting potential across the non-innervated face is equal to that across the innervated face. On stimulation, an action potential of the same size and shape as that across the innervated face appears across the whole electroplaque. This suggests that the non-innervated face is electrically inexcitable, and that the whole of the discharge is accounted for by the activity of the innervated face.

Confirmation of this view is provided by the experiment shown in fig. 15.4. Here, the potential across the non-innervated face (trace *b*) is not affected by stimulation (except for some small electrotonic changes), whereas the potential across the whole electroplaque (trace *c*) shows the familiar action potential.

Thus the voltage produced by a stimulated electroplaque is caused by the asymmetry of the responses of its two faces, as is shown diagrammatically in fig. 15.5. In the complete electric organ, the electroplaques are arranged in series so that, as we have seen, their voltages are additive. Notice that the innervated face becomes negative to the non-innervated face during the discharge; this situation occurs in many other electric fish, and is sometimes known as 'Pacini's Law' – but by no

Fig. 15.1. Diagrams showing the positions of the electric organs in some electric fish. Small arrows point to the electric organs. Large arrows show the direction of current flow through the electric organs during their discharge; for the three smaller species (*Gnathonemus*, *Gymnotus* and *Sternarchus*) the two or three arrows indicate successive phases of the discharge. (After Bennett, 1968, redrawn.)

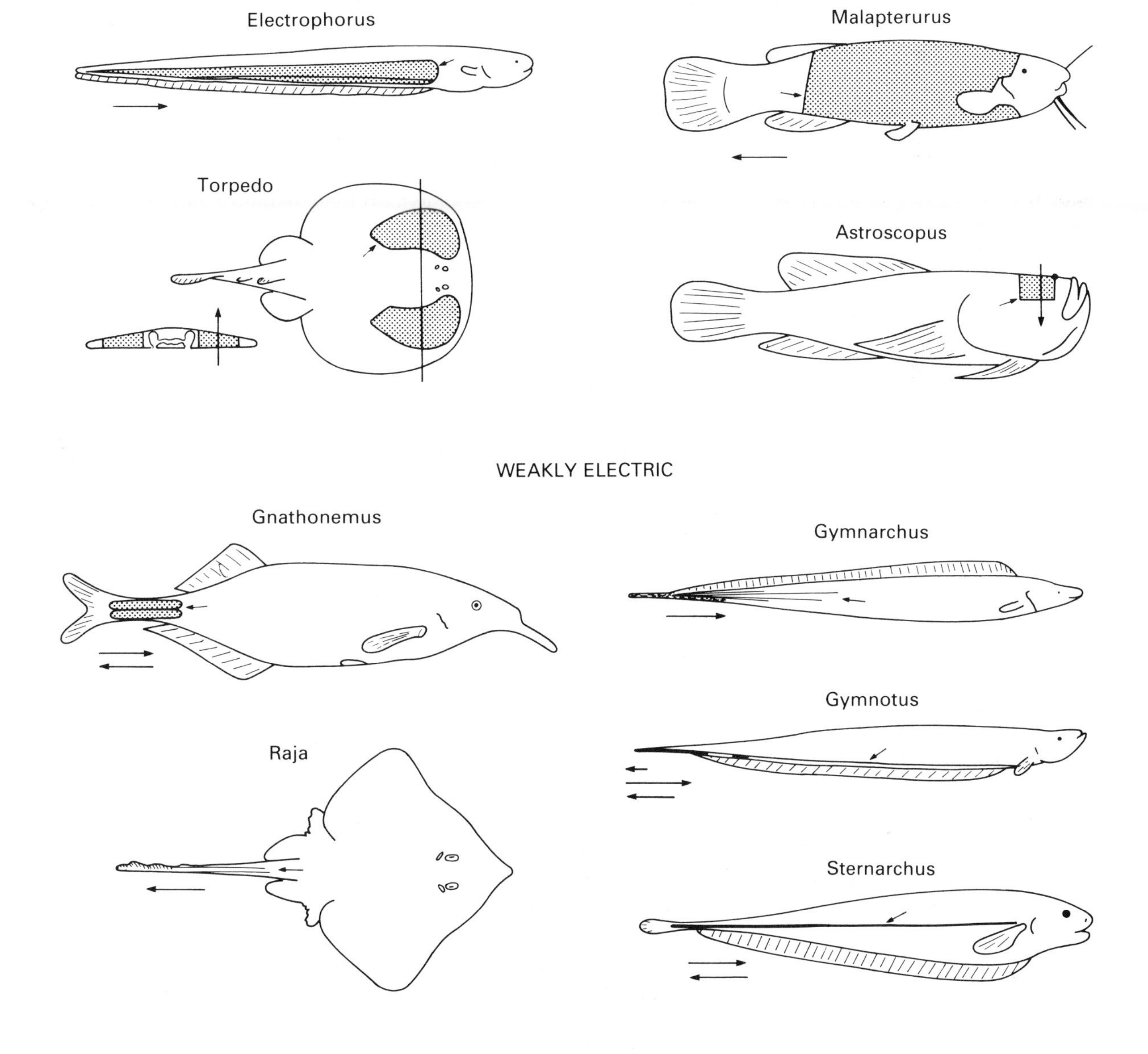

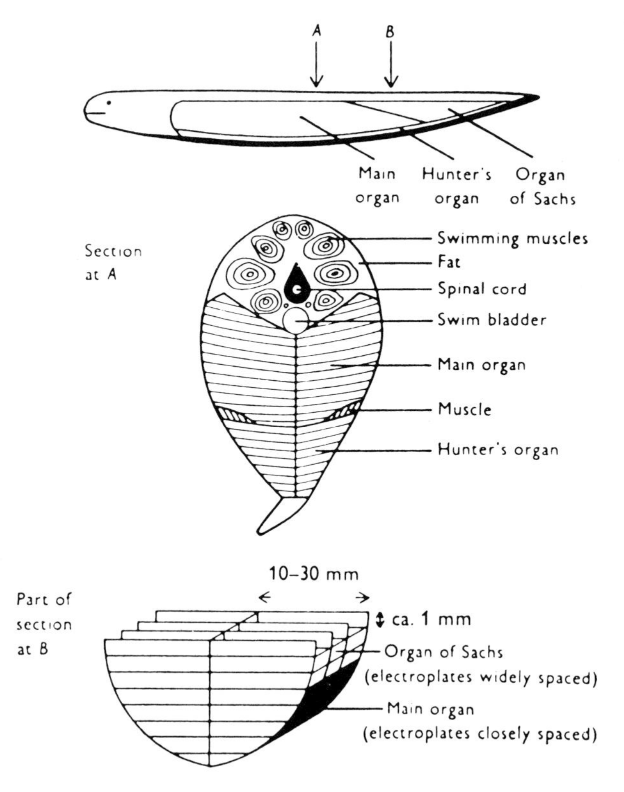

Fig. 15.2. The gross structure of the electric organs in the electric eel. (From Keynes and Martins-Ferreira, 1953.)

means all electric fish obey this rule. In *Electrophorus*, since it is the posterior face of the electroplaques that is innervated, the head end becomes positive to the tail during the electric discharge.

The ionic basis of the electroplaque action potential is much the same as that of the action potentials of nerve axons and twitch muscle fibres. Keynes and Martins-Ferreira showed that its size is dependent on the external sodium ion concentration, and Schoffeniels (1959), from radioactive trace measurements on isolated electroplaques, found that the sodium inflow across the innervated membrane increases greatly during activity.

Tetrodotoxin combines with the sodium channel protein and renders the electroplaques inexcitable. Antibodies to the tetrodotoxin-binding protein can be prepared. When these are allowed access to thin tissue sections, they bind entirely to the innervated faces of the electroplaques and not at all to the non-innervated faces (Ellisman and Levinson, 1982).

We have seen in chapter 6 that *Electrophorus* electric organ is one of the richest sources of sodium channel protein, from which it can be isolated (Agnew *et al.*, 1978, 1983) and sequenced by isolating its messenger RNA followed by complementary DNA cloning (Noda *et al.*, 1984).

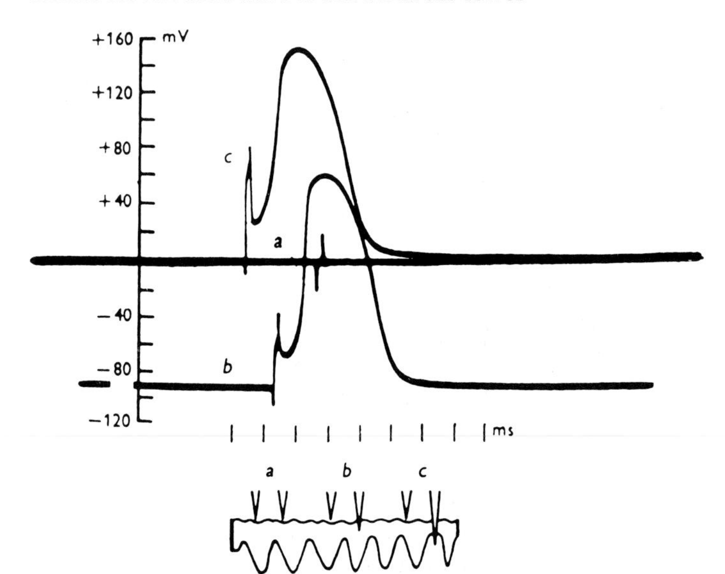

Fig. 15.3. Responses of a Sachs' organ electroplaque of *Electrophorus* to electrical stimulation. The position of the recording electrodes for each trace is shown in the lower diagram, in which the innervated face of the electroplaque is uppermost. (From Keynes and Martins-Ferreira, 1953.)

Fig. 15.4. Results of an experiment similar to that of fig. 15.3, but this time the non-innervated face of the electroplaque is uppermost. (From Keynes and Martins-Ferreira, 1953.)

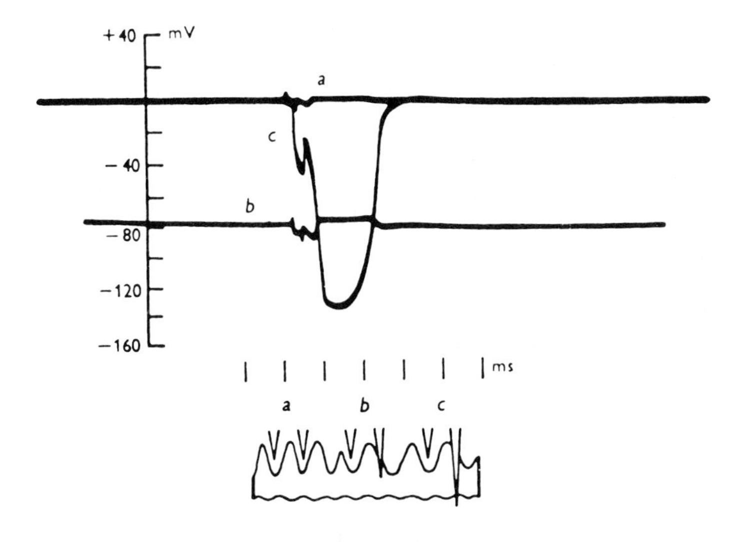

Fig. 15.5. Diagram to show membrane potentials across the innervated (uppermost in the diagram) and non-innervated faces of electroplaques in the Sachs organ of the electric eel, at rest (*a*) and at the peak of the discharge (*b*). (From Keynes and Martins-Ferreira, 1953.)

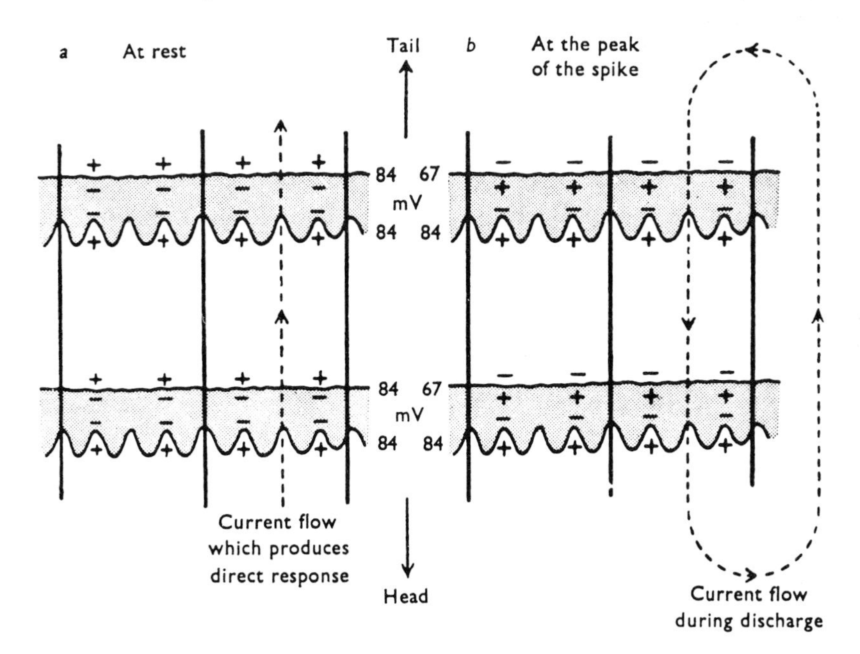

In the living animal, discharge of an electroplaque is not, of course, initiated by direct electrical stimulation, but by excitation via the efferent nerves innervating it. Altamarino, Coates and Grundfest (1955) found that stimulation of the motor nerves elicits an excitatory post-synaptic potential across the innervated surface of the electroplaque, which then induces the action potential (fig. 15.6). This system is biophysically similar to that whereby vertebrate twitch muscle fibres are excited, except that each electroplaque is innervated by a number of nerve fibres; if only a small proportion of these fibres is stimulated (as in fig. 15.6*c*), the resulting postsynaptic potential is not large enough to elicit an action potential.

The electroplaques of some other electric fish

Electric organs occur in a variety of different fish. They have probably arisen independently six times in evolution: in (1) the rajoid and (2) torpedoid rays among the cartilaginous fish, in (3) the electric catfish *Malapterurus* and (4) the stargazer *Astroscopus*, and in two large freshwater groups separated by the Atlantic ocean, (5) the six gymnotiform families of Central and South America and (6) *Gymnarchus* and the mormyrids of tropical Africa. Much detail is now known about the nature of their various electric discharges (see Bennett, 1971, and Bass, 1986); we shall here look at a few examples.

Fig. 15.6. Comparison of electroplaque responses to direct and nervous stimulation in the electric eel. *a*: Direct stimulation at moderate intensity. *b*: Direct stimulation at high intensity; note the extremely low latency of the response. *c*: Nervous stimulation at low intensity, producing a postsynaptic potential which is not large enough to elicit an action potential. *d*: Nervous stimulation at high intensity; note the appreciable latency of the response. Upper traces show zero potential levels and, in *c*, a 100 mV peak-to-peak 1000 Hz calibration wave. (From Altamarino *et al.*, 1955.)

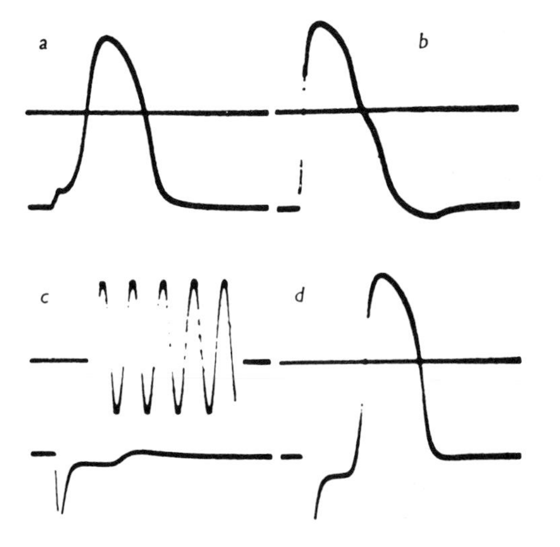

Marine electric fish

This group includes the electric rays *Torpedo, Narcine* and *Raia*, and the 'stargazer', *Astroscopus*. In all these fish, the innervated faces of the electroplaques are not electrically excitable, and the response to nervous stimulation consists solely of a depolarizing postsynaptic potential (Bennett and Grundfest, 1961*a*, *b*; Bennett *et al.*, 1961; Brock *et al.*, 1953), as is shown for *Raia* in fig. 15.7. Synaptic transmission is cholinergic, with large numbers of nerve endings and high densities of nicotinic acetylcholine receptors on the innervated face of the electroplaque. Ionic current flow across the innervated face occurs almost entirely via the channels of the acetylcholine receptors. Hence the maximum potential across the active electroplaque cannot be as large as it is in *Electrophorus*; it is usually less than 70 mV.

The electric organ of *Torpedo* has been much used as a rich source of acetylcholine vesicles (Whittaker, 1984) and acetylcholine receptors and their messenger RNA (Raftery *et al.*, 1976; Noda *et al.*, 1982), with the results that we have seen in chapters 7 and 8. Volta's invention of the voltaic pile in 1800, a crucial step in the science of electricity, was described by him as 'an artificial electric organ' and may well have been inspired by the structure of the *Torpedo* electric organ (Wu, 1984).

Now why should the electroplaques of marine electric fishes be electrically inexcitable? If it were not for the existence of *Astroscopus*, we could assume that the situation was of no functional significance, since the marine rays and skates are phylogenetically distinct from the fresh-water teleosts; but *Astroscopus* is much more closely related to the fresh-water forms than to the rays, yet it too has electrically inexcitable electroplaques. There is no generally agreed answer to this question, but a possible explanation is as follows. We can regard the excited electroplaque membrane of a fresh-water fish as consisting of a battery V_{Na} in series with a resistance R_{Na}. The current produced by the discharge flows through the resistance R_{ex}, which consists mainly of the resistance of the external medium in which the fish swims; the voltage produced by the membrane is V. This rather simplified scheme is shown in fig. 15.8*a*. In this circuit, it is evident that

$$V = V_{Na}\left(\frac{R_{ex}}{R_{Na} + R_{ex}}\right).$$

Thus if R_{ex} is reduced, V also falls. But we know that, in an electrically excitable system such as this one is, R_{Na} rises as V falls, which will itself produce a further fall in V, and so on. Hence it follows that an electrically excitable system, based essentially on a positive feedback relation between membrane potential and sodium conductance, is ineffective if the external resistance is sufficiently low. An analogous circuit for an electrically inexcitable electroplaque is shown in fig. 15.8*b*. In this system the resistance of the excited synaptic membrane (R_s) is independent of the membrane potential, hence the membrane potential during excitation is not disproportionately lowered if the external resistance is low. Since the external resistance is very much lower for fishes living in sea water, we might expect such fish to rely on

Fig. 15.7. Responses of *Raia* electroplaques, as recorded by electrodes in various positions. (From Brock *et al.*, 1953.)

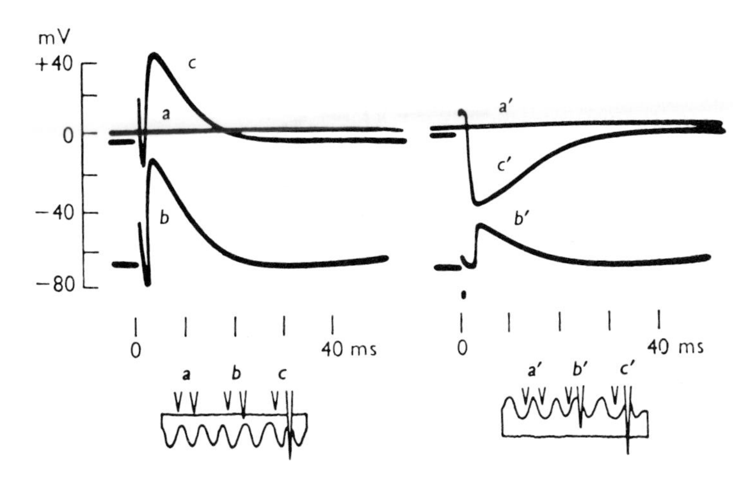

Fig. 15.8. Simplified equivalent circuits of the electric organs in fishes whose electroplaques are (*a*) and are not (*b*) electrically excitable.

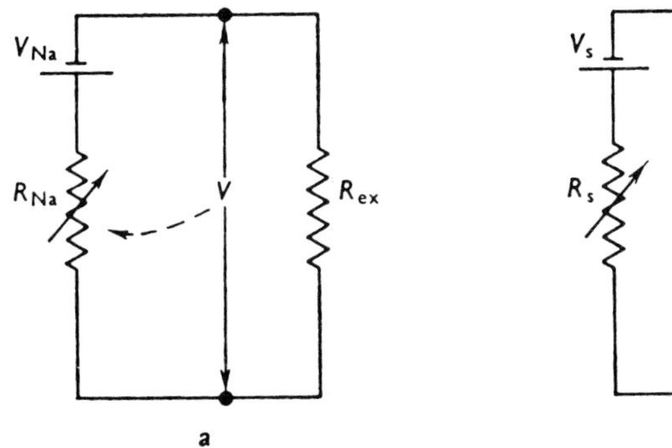

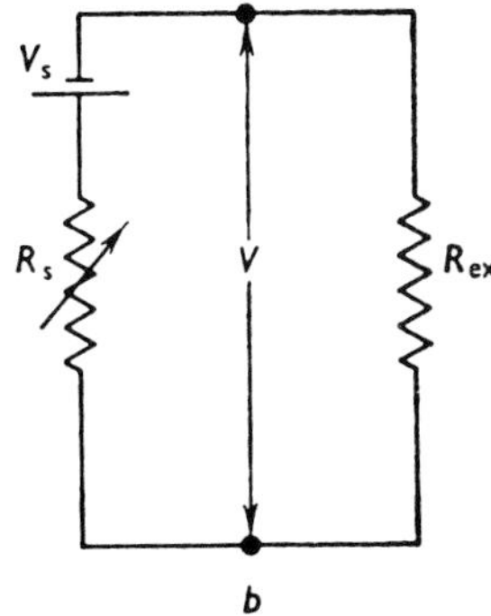

purely synaptic excitatory mechanisms for the discharge of the electroplaques, as indeed they do.

Gymnotus

This is a South American fresh-water fish which produces weak electric discharges. The electroplaques lie in eight tubes, four on each side of the animal. Those in the dorsal pair of tubes are innervated on the anterior surface, whereas those in the other tubes are innervated on the posterior surface. The physiology of the discharge was investigated by Bennett and Grundfest (1959). They found that both faces of the electroplaques are electrically excitable, but, when the motor nerves are stimulated, the innervated surfaces are excited before the non-innervated surfaces, so that the potential change across the whole electroplaque is diphasic. During the natural discharge, the upper tube on each side (tube 1) fires first, and then the lower tubes (2 to 4) fire together. The result of this is that the potential change over the whole fish is triphasic (fig. 15.9), the sequence of events being: (*a*) excitation of the innervated faces in tube 1; (*b*) excitation of the non-innervated faces in tube 1 and of the innervated faces in tubes 2 to 4; and (*c*) excitation of the non-innervated faces in tubes 2 to 4.

Malapterurus

The strong electric discharge of *Malapterurus electricus* (the electric catfish) is of considerable interest in that it does not obey 'Pacini's law': the electroplaques are innervated posteriorly, yet the head becomes negative to the tail during activity. At one time it was thought that the electroplaques in this fish were derived from glandular tissue rather than muscle, and, in the absence of physiological observations, it was suggested that the innervated face became hyperpolarized during activity. Later observations showed both these ideas to be wrong.

The electroplaques of *Malapterurus* are rather unusual in that innervation occurs at the end of a stalk which is produced from the centre of the posterior face, as is indicated in fig. 15.11 (a somewhat similar arrangement occurs in mormyrids; see Bennett and Grundfest, 1961*c*). The sequence of events during discharge of an electroplaque has been determined by Keynes, Bennett and Grundfest (1961), using intracellular microelectrodes. Excitation arises in the stalk and then passes into the electroplaque disc, where both faces are excited simultaneously. However, the action potential across the anterior ('non-innervated') face lasts only for about 0.3 ms (figs. 15.10 and 15.11). Hence, during the later stages of the response, only the anterior face is active, so that there is a net potential difference across the whole electroplaque.

Fig. 15.9. Waveforms of a single pulse from the repetitive discharge of *Gymnotus carapo*, recorded at different positions near the body surface. Notice (*a*) that the potential is triphasic near the tail, rather more complex elsewhere, and (*b*) that the potentials are maximal near the tail and the head, suggesting that the electric organ acts as a dipole generator. (From Bennett and Grundfest, 1959.)

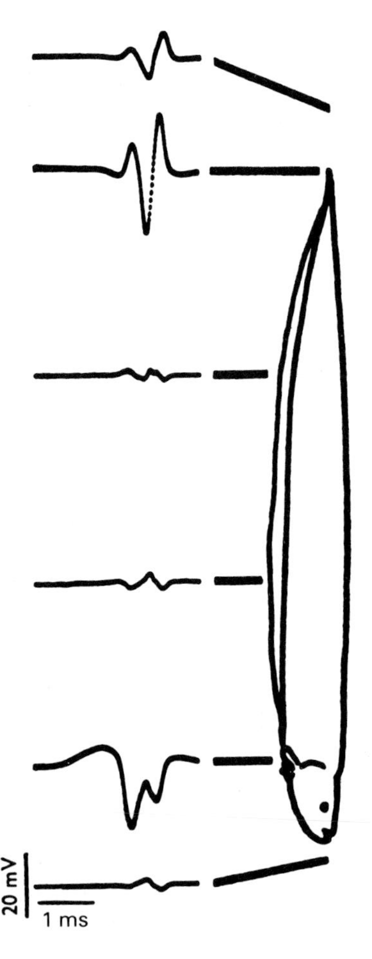

Fig. 15.10. Responses of a *Malapterurus* electroplaque to nervous stimulation, recorded across the whole electroplaque (V_1), across the anterior face (V_2) and across the posterior face (V_3). (From Keynes *et al.*, 1961.)

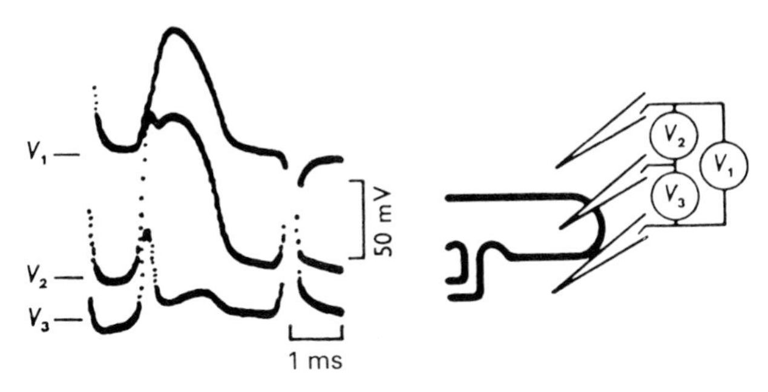

Fig. 15.11. Schematic diagram showing the sequence of potential changes during the discharge of a *Malapterurus* electroplaque. (From Keynes *et al.*, 1961.)

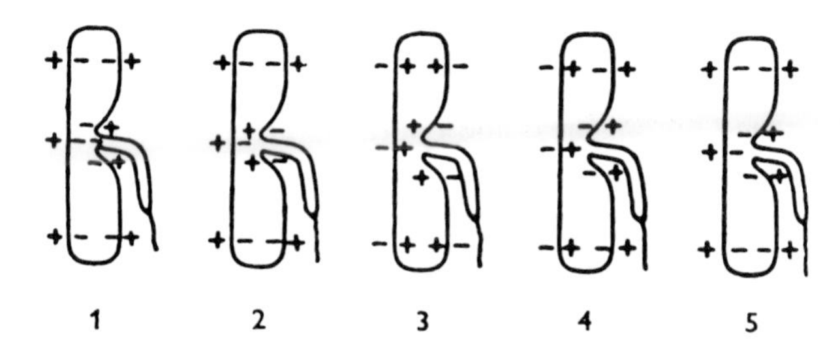

Sternarchus

The electric organs of *Sternarchus* and its immediate relatives are unique in that they are composed of tissue derived from nerve axons rather than from muscle cells (Couceiro and de Almeida, 1961). The electromotor axons leaving the spinal cord are swollen distally to form 'electrotubes' which are analogous in function to the electroplaques of other electric fishes. Fig. 15.12 shows the structure of these swollen axons. After entering the electric organ each axon passes forwards for several millimetres, and then turns back on itself and runs backwards for an approximately equal distance. The diameter of the axon is about 100 μm in the swollen portions but only about 20 μm elsewhere. Each arm of the loop has three very large nodes (about 50 μm long) in the region just distal to the swollen portion (Waxman *et al.*, 1972).

Waxman and his colleagues investigated the functioning of these remarkable axons (fig. 15.13). The greatly enlarged nodes on the distal side of each swollen portion are electrically inexcitable, whereas those on the proximal side behave in the normal manner. Thus inward current flow at the proximal nodes will pass outwards at the enlarged distal ones. The net external current produced by

Fig. 15.12. The anatomy of the electric organ of *Sternarchus*. The middle diagram shows the course of one of the modified nerve fibres which constitute the organ and the lower one shows its structure. (From Bennett, 1971.)

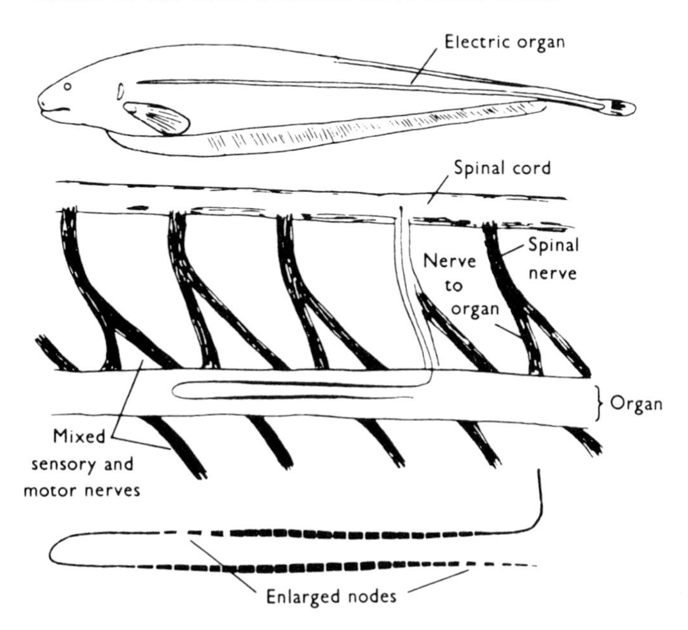

the whole axon reverses in direction as the second arm of the loop is excited. The swollen portions will allow the internal current to pass along to the enlarged distal nodes without much attenuation, and the enlarged area of the distal nodes will facilitate its outward flow.

The functions and evolution of electric organs

The powerful electric organs of *Electrophorus* and *Torpedo* are used both offensively, in temporarily paralysing smaller fishes which can then be eaten, and defensively, in deterring predators. The discharges of *Astroscopus* and *Malapterurus* are also able to stun smaller fishes, and are probably used for offensive purposes.

However, it is clear that the low-voltage discharges of the mormyrids, *Gymnarchus* and the gymnotiformes cannot act as offensive or defensive weapon systems. Lissmann (1951) showed that *Gymnarchus* could detect the presence of electrical conductors and non-conductors in its environment, and was sensitive to weak electric currents. He suggested that the electric organs functioned as the 'transmitter' of a direction-finding system, the 'receiver' being electro-sensitive sense organs capable of detecting changes in the pattern of current flow around the fish. We shall examine these electro-receptors in a little more detail in chapter 17; suffice it to say, here, that Lissmann's hypothesis is now well substantiated (Lissmann and Machin, 1958).

Electric signals are also used as methods of communication between individuals of the same species. The signals are species-specific, and in at least some species they may be involved in species identification, sex recognition, territorial or aggregational behaviour, the establishment of dominance hierarchies, and courtship (Hopkins, 1974;

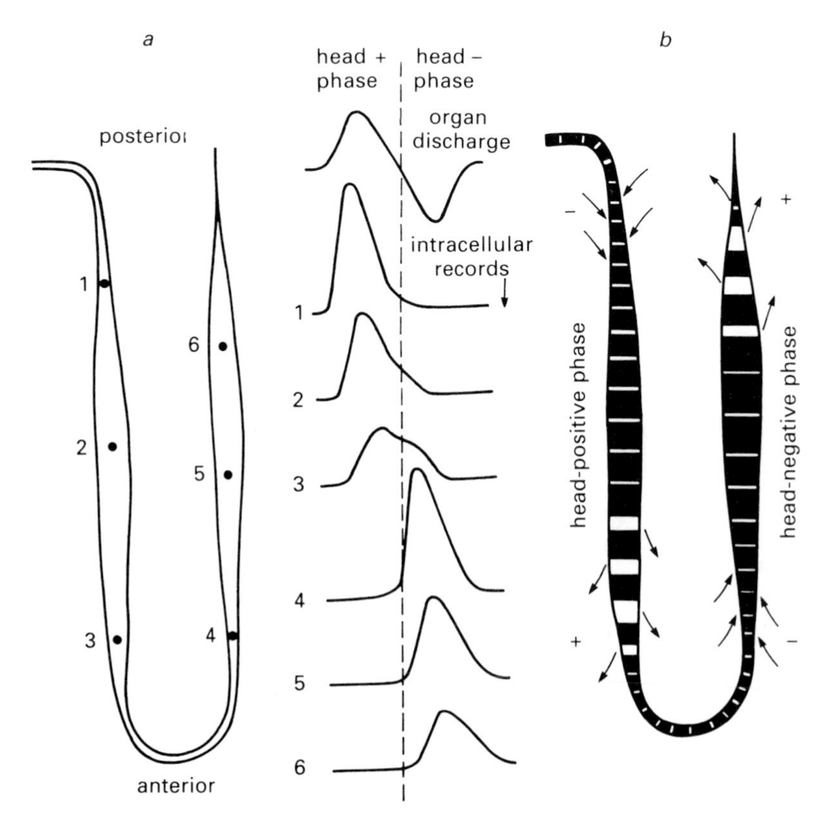

Fig. 15.13. The mode of action of *Sternarchus* electrocytes. *a* shows intracellular records from different sites during the discharge; notice that the potentials in the regions with large nodes are smaller than elsewhere, suggesting that these nodes are not electrically excitable. *b* shows the current flow deduced from these observations. (From Waxman *et al.*, 1972.)

Hopkins and Bass, 1981; Hagedorn, 1986). In the gymnotiform *Hypopomus*, the males have larger electroplaques than the females, but the difference disappears in females injected with male sex hormone (Hagedorn and Carr, 1985).

In discussing various objections to his theory of evolution by natural selection, Darwin (1859) mentioned the problem of the electric organs of fishes. At that time only the powerful electric organs (such as those of *Electrophorus* and *Torpedo*) were known, and it was impossible to see how these could have gradually developed through ancestors with weaker discharges, since such weak discharges would not be effective as weapons of offence or defence and would therefore give no selective advantage to a fish possessing them. Lissmann (1958) suggested that the offensive–defensive functions of strong electric organs are a development of the direction-finding function of weak electric organs. The first stage in the evolution of the direction-finding system must have been the development of electroreceptors; these may have been capable of detecting muscular action potentials, and the sensitivity and accuracy of the system could then have been increased by modification of the muscle cells to form electric organs.

16
The organization of sensory receptors

All animals are sensitive to some extent to changes in their environment. In all except the most lowly animals, special parts of the body are developed that are responsive to some of these changes and that feed information concerning them into the central nervous system. Information about the workings of the animal's own body, or about communication signals from other members of its species, may be acquired in a similar fashion. These specially sensitive structures are known as sense organs or sensory receptors.

Since there is a very great variety of different types of receptor, the following pages must necessarily be no more than a rather brief introduction to their physiology. This chapter attempts to give a general outline of the properties of sensory receptors, chapter 17 surveys some selected different types, and chapter 18 deals with a particular sense organ, the vertebrate eye, in a little more detail. The books edited by Barlow and Mollon (1982), Dawson and Enoch (1984), Bolis, Keynes and Maddrell (1984) and Darian-Smith (1984) are some of the many useful sources of further information.

The methods used for investigating receptors fall into two main categories. Firstly, there is the behavioural, or psychophysical, approach, where the receptor is investigated indirectly by observation of the response of the animal to a sensory stimulus. For example, suppose we are interested in the ability of an animal to discriminate colours. In the case of man, the experiments are not too difficult, since we can ask the subject if two colours look different to him. With other animals, however, we have to devise experiments in which the behaviour of the animal depends upon the ability to discriminate colour; for example, we can train a fish to feed from a red container and then see if it always chooses the red container from among a series of greys. Secondly, there is the electrophysiological method, in which the receptor is investigated by means of electrodes placed in its vicinity or on the sensory nerves leading from it. This type of study gives precise, quantitative information about the properties of the receptor, but it does not always enable us to decide which of these properties are relevant to its normal function. For example, we may be able to show by electrophysiological means that a receptor can be excited both by mechanical stimuli and by changes in temperature, but we cannot from these experiments alone determine how the animal interprets this information. Hence a thorough investigation of a sensory system must involve both behavioural and electrophysiological studies.

Now let us consider how sensory systems can be classified. A moment's thought will show that Aristotle's list of 'the five senses' – sight, hearing, touch, taste and smell – is incomplete. We are also sensitive to changes in temperature, to our position with respect to gravity, to the movements of our bodies, and so on. These various categories of sensation are called sensory *modalities*. Some of these modalities can further be divided into *qualities*: for example, we can determine the pitch as well as the loudness of a sound, the colour as well as the brightness of a light source.

The sense organs themselves can be classified according to the type of stimulus which normally excites them. Thus *mechanoreceptors* are excited by mechanical stimuli, *photoreceptors* are sensitive to light, *thermoreceptors* are temperature-sensitive, *chemoreceptors* are sensitive to the chemical composition of the surrounding medium, and *electroreceptors* are sensitive to very weak electric currents.

It is important to realize that the specificity of receptors is not absolute; most receptors, for instance, can be excited by strong electric currents. However, as was first pointed out in 1826 by

Table 16.1. *Types of sensory receptors*

Type of receptor	Exteroceptors		Interoceptors
	Contact	Distance	
Mechanoreceptors	Touch	Hearing (phonoreceptors) Lateral line organs of fishes	All proprioceptors All equilibrium receptors Some visceroceptors, e.g. baroreceptors of carotid body (sensitive to blood pressure)
Photoreceptors	—	Sight. May or may not be image-forming and/or colour sensitive	—
Chemoreceptors	Gustatory (taste)	Olfactory (smell)	E.g. chemoreceptors of carotid body
Thermoreceptors	Most	Sensitivity to radiant heat. E.g. facial pit of crotaline snakes	E.g. hypothalamic thermoreceptors
Electroreceptors	—	Found in electric and some other fishes	—

Müller in his 'law of specific nerve energies', the sensation produced by stimulation of sensory receptors for a particular modality is the same whatever the nature of the stimulus. One can easily demonstrate this by closing an eye and pressing it gently: the mechanical stimulation of the retina produces a visual sensation.

An alternative classification is based on the position of the receptors and the stimulus. Thus *exteroceptors* are sensitive to stimuli originating outside the body, and *interoceptors* are excited by stimuli inside the body. Exteroceptors can be divided into *distance receptors* (such as those involved in sight and hearing), which are able to detect phenomena at a distance, and *contact receptors* (such as those involved in touch and taste), which can only be stimulated by contact with the stimulus. Interoceptors are divided into *equilibrium receptors*, which give information about the movement and position of the whole body, *proprioceptors*, which give information about the relative positions and movements of the muscles and skeleton, and *visceroceptors*, which monitor the conditions in the rest of the body.

Examples of these various categories are shown in table 16.1. In addition to these, there is a rather ill-defined sensation known as pain. Since pain is a subjective phenomenon, it is very difficult to study it in animals other than man. Stimuli that produce reactions which appear to involve the sensation of pain in animals are sometimes called *nociceptive*.

We should also mention here the question of sensitivity to magnetism. There is excellent evidence that the dances of bees are affected by magnetic fields (Lindauer and Martin, 1972) and good evidence that migratory birds and homing pigeons can use the Earth's magnetic field for orientation (see Keeton, 1981, for example). But how this magnetic sense works is still a mystery; it is too soon for us to add a further category, of 'magnetoreceptors', to table 16.1.

The anatomy of sense organs shows a great deal of variety. All sense organs are innervated by sensory neurons. In the simplest type, there is but a single sensory neuron whose peripheral end is excited by the sensory stimulus. In the Pacinian corpuscle (fig. 16.1) for example, there is a single axon whose terminal is unmyelinated and surrounded by a series of lamellae. The mechanoreceptive hairs of arthropods (fig. 16.2) are also innervated by a single sensory hair, although in this case the neuron soma is placed distally instead of in the central nervous system. Notice that, in each case, the neuron ending is associated with *accessory structures* (the lamellae in the Pacinian corpuscle and the cuticular hair in the arthropod touch receptor). In some cases, no specialized accessory structures can be seen, and the nerve endings are 'free' in the tissues.

A more complex type of organization occurs when the sensory neuron is connected to a specialized *receptor cell* (or 'secondary sense cell'), so that excitation of the sensory neuron is preceded by excitation of the receptor cell. In the taste buds of the mammalian tongue, for example, there are a number of receptor cells, each connected, either alone or with others, to a sensory neuron. In the very complex sense organs, such as the vertebrate eye and ear, there are large numbers of receptor cells and sensory neurons, and the accessory structures (such as the lens and the iris in the eye) may reach a high degree of sophistication.

The coding of sensory information

We must now consider how sensory information is passed from the receptor to the central nervous system. It is very easy to show that the information pathway involves the sensory nerves, since, after cutting these nerves, an animal no longer responds to stimulation of the receptor which they innervated. The first records of electrical activity in sensory nerves in response to sensory stimulation were made by Adrian in 1926. Using frog skin and muscle nerves, he showed that sensory stimulation was followed by action potentials in the sensory nerves. A more delicate

Fig. 16.2. Diagram showing how the sense cell is attached to the base of a mechano-receptive hair in the hair-plate organ in a honey-bee's neck. (From Thurm, 1965.)

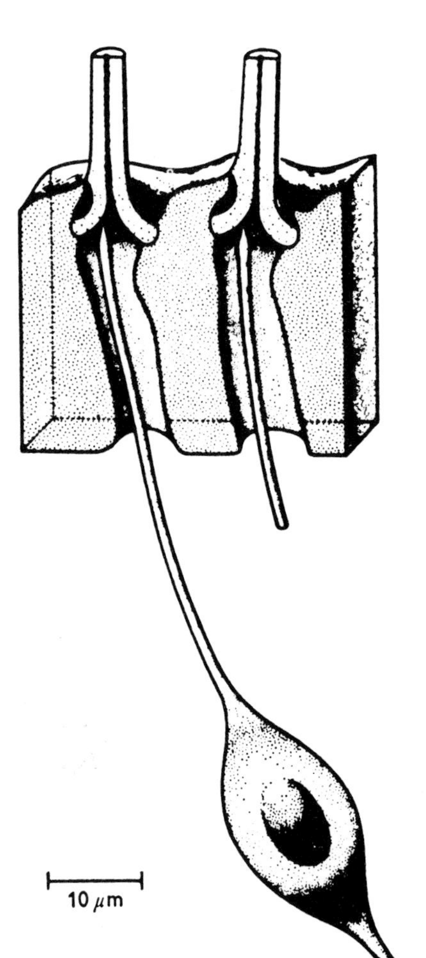

Fig. 16.1. Diagrammatic section through a Pacinian corpuscle. (From Quilliam and Sato, 1955.)

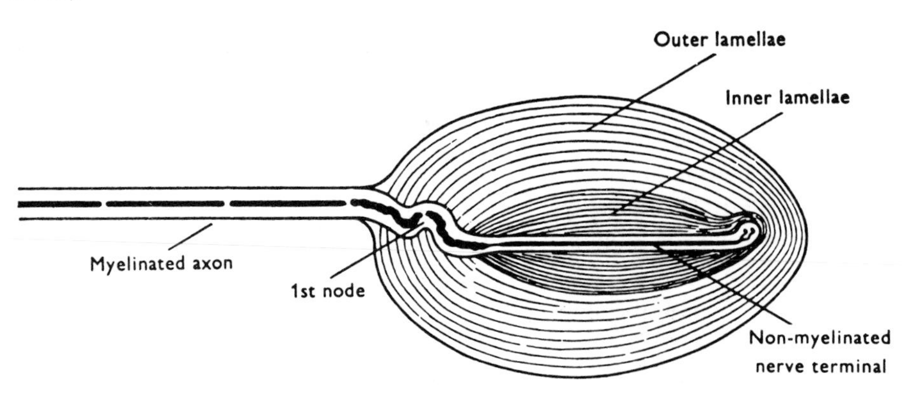

analysis followed the recording of the electrical activity in a single sensory nerve fibre by Adrian and Zotterman (1926), whose experiments we shall now consider.

The skeletal muscles of tetrapod vertebrates contain sense organs known as muscle spindles, which are responsive to stretch. Each spindle (fig. 16.3) consists of a bundle of modified muscle fibres whose central region is innervated by a sensory nerve fibre. The sternocutaneous muscle of the frog, a small muscle which is attached at one end to the sternum and at the other to the skin, contains three or four of these muscle spindles. Adrian and Zotterman cut the nerve supplying the muscle so that it was connected to the muscle only, and placed it across their recording electrodes. The skin was cut so as to leave a small piece attached to the muscle, which could then be stretched by means of small weights hung from a thread tied to this piece of skin. They found that stretching the muscle produced a series of action potentials in the nerve. By progressively removing strips of muscle, they were able to reach a situation in which only one muscle spindle remained connected to the sensory nerve. Under these conditions the action potentials produced by stretch appeared at fairly regular intervals and were always the same size, thus showing that the all-or-nothing law applied to sensory nerve impulses.

The next step was to see how this rhythmical discharge was affected by the intensity of the stimulus. Adrian and Zotterman discovered that the frequency of impulse discharge increased with increasing stretch of the muscle (fig. 16.4). This was a most important result, which has since been verified for a wide variety of different receptors.

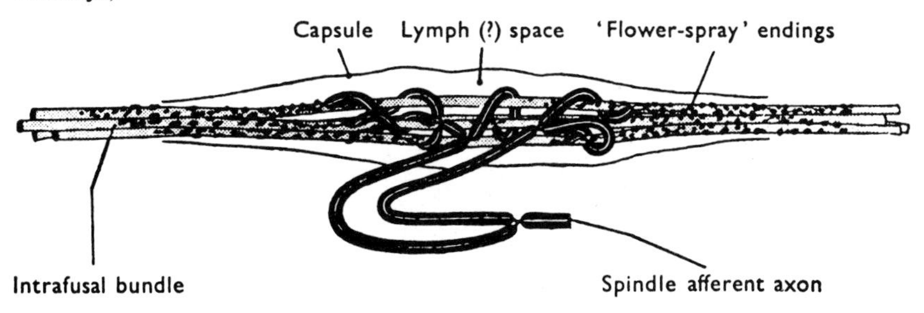

Fig. 16.3. Capsular portion and its sensory nerve ending in a frog muscle spindle. (From Gray, 1957, by permission of the Royal Society.)

We can therefore say that, in a single sensory nerve fibre, information as to the intensity of the sensory stimulus is carried as a *frequency code* – the frequency of the all-or-nothing action potentials in the fibre.

With many sensory stimuli, a stimulus of constant intensity produces a sensation which declines with time. This phenomenon, which is known as *sensory adaptation*, can readily be seen by looking at a bright light or by resting light weights on the back of the hand. It is obviously of some interest to see if adaptation is accompanied by any change in the nervous output of the sensory organ. Using their frog muscle spindle preparation, Adrian and Zotterman found that the impulse frequency in the sensory nerve fibre declined during a constant stretch, as is shown in fig. 16.5. This observation provides a partial explanation for the existence of adaptation, and is further evidence for the frequency code hypothesis.

Adrian and Zotterman summarized their observations on the frog muscle spindle in the following words: 'The impulses set up by a single end-organ occur with a regular rhythm at a frequency which increases with the load on the muscle and decreases with the length of time for which the load has been applied.' It is a remarkable tribute to the quality of their investigation that much of the work done in sensory physiology since that time has been concerned with quantifying this statement for a wide variety of different sense organs.

The initiation of sensory nerve impulses

Having seen that sensory information is conveyed to the central nervous system by means of impulses in the sensory nerve fibres, the next problem for us to consider is how these impulses arise. Using the frog muscle spindle preparation with an external electrode placed as near the sensory nerve ending as possible, Katz (1950) found

Fig. 16.4. The relation between the load applied to a frog muscle and the impulse frequency in the sensory nerve fibre from one of its muscle spindles. Results from six different experiments, denoted by different symbols; in each case the weight had been applied to the muscle for 10 s before the measurement of nerve impulse frequency was made. (From Adrian and Zotterman, 1926.)

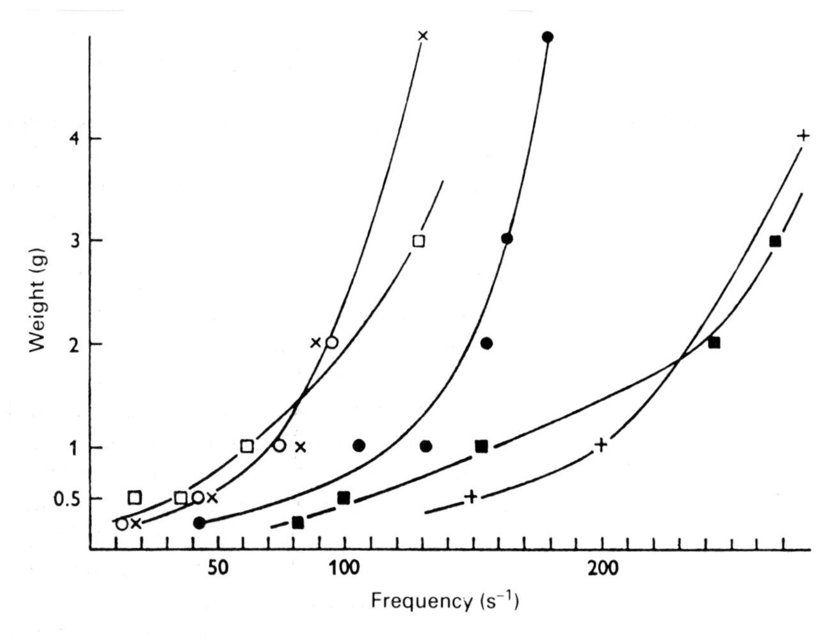

Fig. 16.5. Adaptation in a frog muscle spindle. (From Adrian and Zotterman, 1926.)

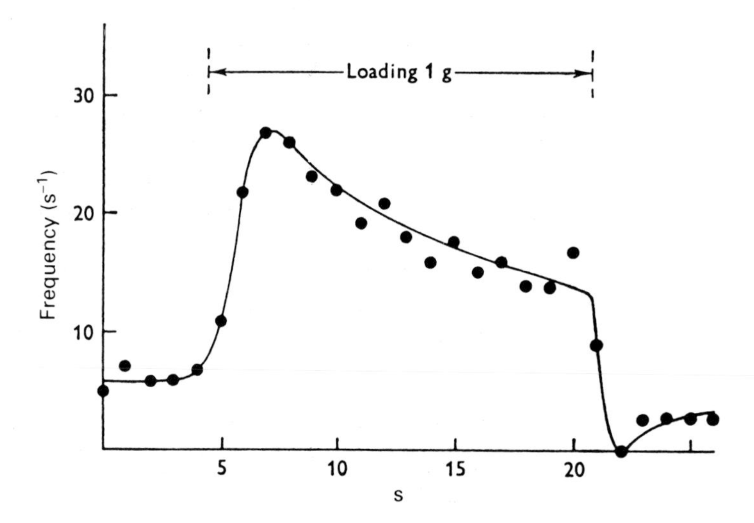

that a local depolarization of the terminals appeared when the muscle was stretched, as well as the impulses in the sensory nerve axon (fig. 16.6). Treatment of the preparation with procaine prevented the production of impulses but did not affect the maintained depolarization. Finally, the degree of depolarization increased with increasing stretch, and the impulse frequency in the sensory nerve was linearly proportional to the extent of the local depolarization (fig. 16.7). Katz concluded from these results that the local depolarization is an intermediate link between the stimulus and the action potentials in the sensory nerve fibre.

Fig. 16.6. The receptor/generator potential recorded extracellularly from a frog muscle spindle in response to various degrees of stretch. The upper trace in each record indicates the change in length of the muscle and the lower trace records the activity of the sensory axon. Record 1 is from an intact preparation, records 2 to 6 from the same preparation after treatment with procaine, and record 7, obtained after crushing the sensory axon, shows that the potential changes in 1 to 6 are not artefacts. (From Katz, 1950.)

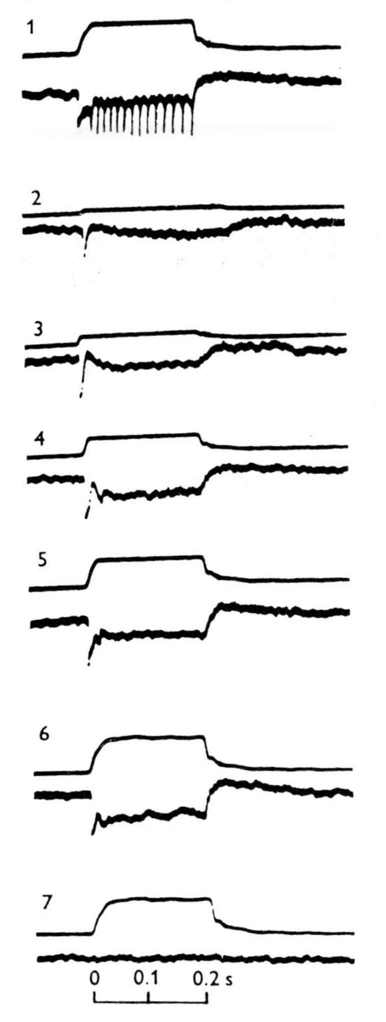

We may call this local depolarization the *generator potential*, since it generates nerve impulses. In the receptor neurons of crustacean stretch receptors (fig. 16.8), Eyzaguirre and Kuffler (1955*a*) were able to record generator potentials with intracellular electrodes; their results, shown in fig. 16.9, are in full agreement with those of Katz. Generator potentials have since been recorded from a wide variety of sensory nerve endings. The size of the generator potential is always roughly proportional to the intensity of the stimulus; in other words, the generator potential is a graded potential change, not an all-or-nothing one.

We shall define a *receptor potential* as a potential change, in the receptor cell or (in those cases where the sensory stimulus excites the sensory nerve terminal directly) in the sensory nerve terminal, induced by the action of the stimulus.* In the case of the stretch receptors of frogs and crayfish, it is clear that the receptor potential is the same as the generator potential, since the stimulus produces a potential change which then generates impulses. However, where there is a separate receptor cell, it is evident that the generator potential is confined to the nerve terminal, and that any potential change that occurs in the receptor cell must be a receptor potential. It seems clear that the microphonic potentials of the ear (see p. 382) are receptor potentials in this sense.

The mechanism whereby the energy of the stimulus produces excitation of the receptor cell or sensory neuron terminal is known as the *transducer mechanism*. In photoreceptors, we know that the primary stage in this process is the capture of light quanta by, and the consequent photochemical changes in, the visual pigment molecules (see chapter 18). In other cases, however, we have very little idea what the nature of the transducer mechanism is. Where receptor potentials have

* There have been some differences in the use of the terms 'generator potential' and 'receptor potential' by different authors. The terminology adopted her is that suggested by Davis (1961).

been detected, such as in many mechanoreceptors and some chemoreceptors, it seems reasonable to suggest that the stimulus causes an increase in the ionic permeability of the receptor membrane, probably by mechanical distortion of the membrane so as to open new ionic channels. In many cases, the transduction process acts as an amplification stage, in that the energy change comprising the receptor potential is much greater than the incident energy which produces it. A system whereby the incident energy alters the ionic permeability of the receptor membrane is, of course, ideally suited to provide such amplification.

In those systems where there are separate receptor cells, there must be some mechanism whereby excitation of the receptor cell can produce the generator potential in the sensory neuron terminal. In other words, there is a synapse between the receptor cell and the neuron terminal, and our problem is to decide what the mechanism of transmission at this synapse is. There is some evidence that a chemical transmission process is involved. Vesicles, which appear to be very similar to synaptic vesicles, can be seen in the presynaptic regions of the rods and cones of the vertebrate eye (de Robertis and Franchi, 1956) and in the mechanoreceptive hair cells of the vertebrate labyrinth; vesicles of a slightly different type are found in the receptor cells of mammalian taste buds

Fig. 16.7. The relation between the size of the receptor/generator potential and the frequency of impulse discharge in a frog muscle spindle. (From Katz, 1950.)

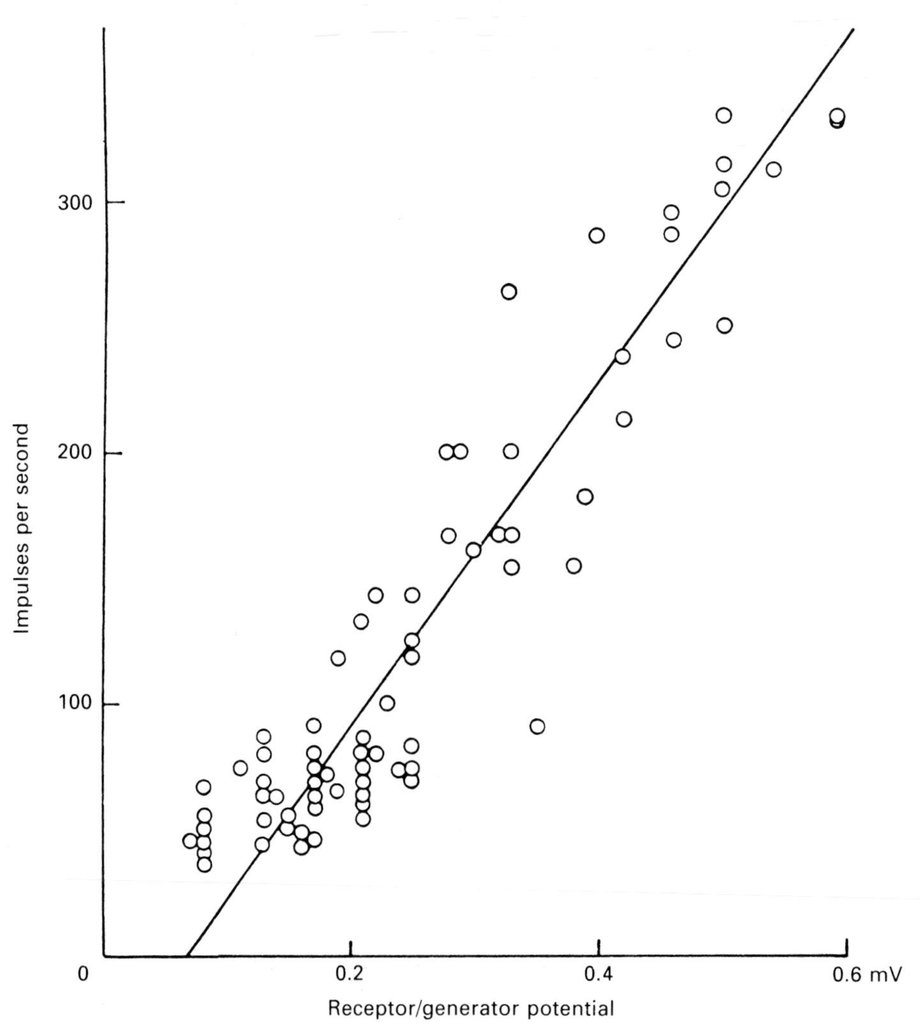

(de Lorenzo, 1963). Fuortes (1959) showed that the generator potential in the eccentric cell of a *Limulus* ommatidium is accompanied by an increase in membrane conductance; this suggests that transmission between the retinula cells (the receptor cells) and the eccentric cell (the sensory neuron) is chemical rather than electrical.

The general theme of this section – that there is a definite control sequence of events in the excitation of receptors – is illustrated diagrammatically in fig. 16.10. We must now take a closer look at the final stages in this sequence, the generator potential and the initiation of action potentials.

Fig. 16.8. Diagram to show the structure of the abdominal muscle receptor organs in the crayfish. RM_1, muscle bundle of the tonic receptor neuron, SN_1. RM_2, muscle bundle of the phasic receptor neuron, SN_2. MO_1 and MO_2, motor fibres of, respectively, the tonic and phasic muscle bundles. *I*, an inhibitory axon. (From Burkhardt, 1958.)

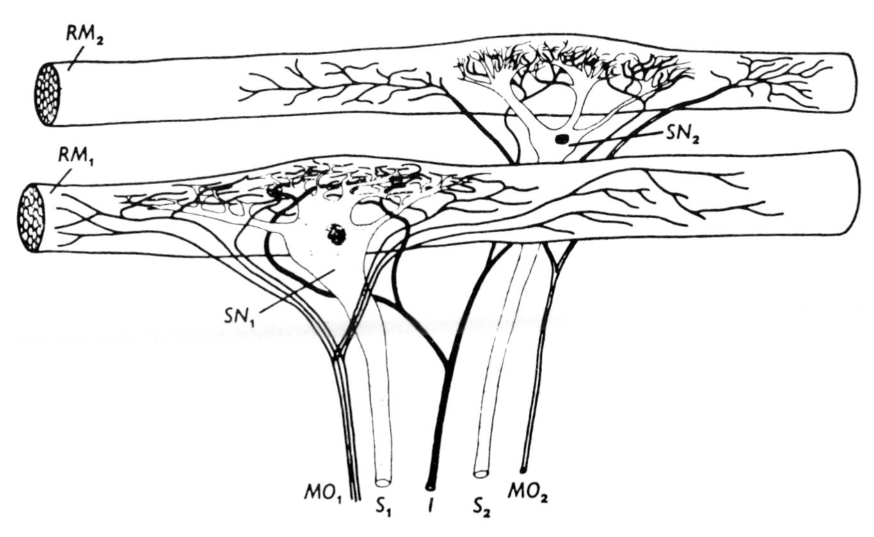

Fig. 16.9. Intracellular records from a crayfish stretch receptor cell. The receptor muscle was stretched between the times marked by the arrows. (From Eyzaguirre and Kuffler, 1955*a*.)

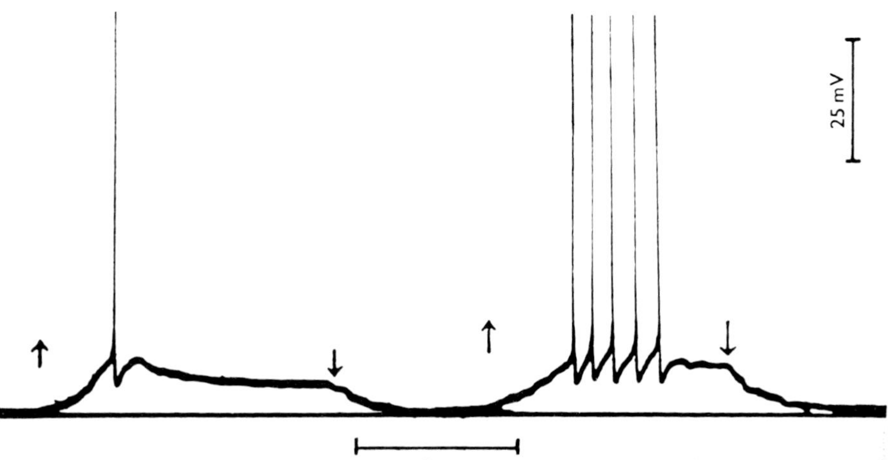

There is some evidence that generator potentials are produced, in part, by an increase in the sodium conductance of the membrane. Diamond, Gray and Inman (1958) investigated the effects of sodium ion concentration on the response of Pacinian corpuscles to mechanical stimulation. Since the receptor region – the sensory neuron terminal – is closely surrounded by the lamellae of the corpuscle, the experimental solutions had to be perfused through the arterial supply of the corpuscle. They found that the production of action potentials was quite rapidly abolished on perfusion with sodium-free solutions, and that the generator potential fell to about 10% of its initial value. With intermediate sodium ion concentrations, the size of the generator potential increased with increasing sodium ion concentration. These results suggest that the generator potential normally results from an inflow of sodium ions, but that other ions are probably involved as well, since the response is not completely abolished in sodium-free solutions. Similar results were obtained by Ottoson (1964), on the generator potential of the frog muscle spindle.

We have seen that the generator potential carries sensory information in the form of an 'amplitude code' in which the degree of depolarization is related to the intensity of the stimulus. We must now consider how this amplitude code is converted into the frequency code of action potentials in the sensory nerve fibres. The precise details of his process are not fully agreed upon (see Hodgkin, 1948; Fuortes and Mantegazzini, 1962), but the following explanation is probably somewhere near the truth.

Consider a sensory nerve terminal which can be functionally divided into two regions (fig. 16.11): a receptor region, which is particularly sensitive to the stimulus and responds to it by means of a graded receptor/generator potential, and a conductile region whose activity consists of all-or-nothing action potentials; at the boundary between these two regions is the impulse initiation site. We assume that the stimulus causes an increase in the ionic permeability of the membrane so that there is an inward flow of current across the cell membrane in the receptor region; this will produce an outward flow of current at the impulse initiation site. We further assume that the stimulus is maintained and that the current it produces is constant. The electrical response at the impulse initiation site to this constant generator current is shown in fig. 16.12. The initial depolarization, from *a* to *b*, is a purely electronic potential. Then, from *b* to *c*, the depolarization is sufficient to cause a local response of the membrane, i.e. a further

Fig. 16.10. Sequences of events in the excitation of sensory nerve endings. The sequence on the left applies where there is no specialized receptor cell, that on the right applies where there is such a cell.

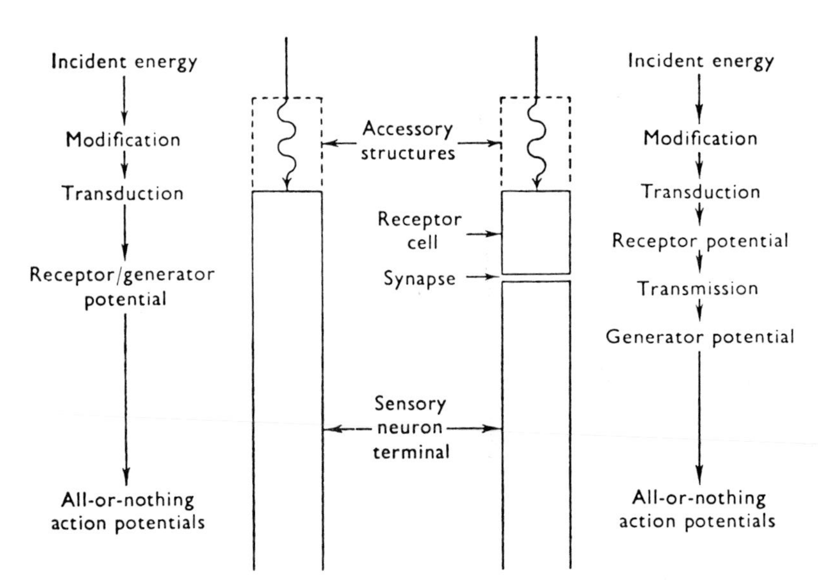

depolarization caused by an increase in sodium conductance. At *c*, the membrane potential crosses the threshold, the sodium conductance increases regeneratively, and an action potential results. After the action potential, there is a positive phase (*d*) produced by the high potassium conductance of the membrane. From *d* to *e*, the combined effects of the decrease in potassium conductance and the generator current produce depolarization again; at *e* the sodium conductance begins to increase, and the process becomes regenerative, producing another action potential, at *f*. After this action potential, the sequence repeats itself for as long as the generator current persists.

If the intensity of the generator current is increased, the depolarization between *d* and *f* will occur more rapidly. This means that the time interval between successive impulses will be less, so that the impulse frequency is higher. At relatively high impulse frequencies, the threshold for production of an action potential, which is crossed at *f*, will be raised, since the second action potential will fall within the refractory period of the first. This effect sets an upper limit to the discharge frequency.

Ripley, Bush and Roberts (1968) described an unusual phenomenon in the coxal muscle receptor organs of crabs. Here the sensory nerve fibres do not conduct action potentials. Instead the receptor potential spreads electrotonically along the

Fig. 16.11. Local circuit currents producing impulse initiation in a sensory nerve terminal.

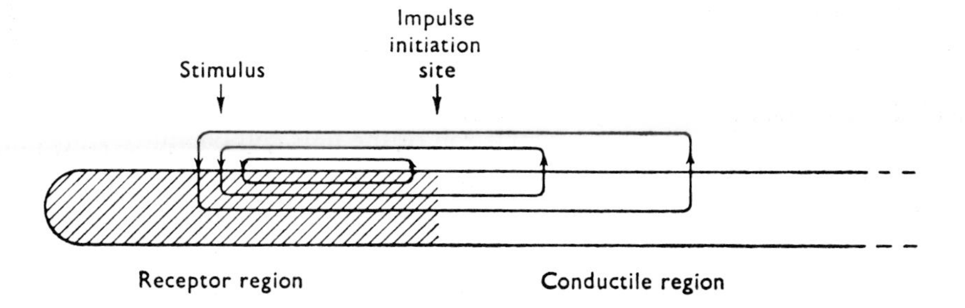

Fig. 16.12. Generation of the frequency code in a sensory nerve terminal. Explanation in text.

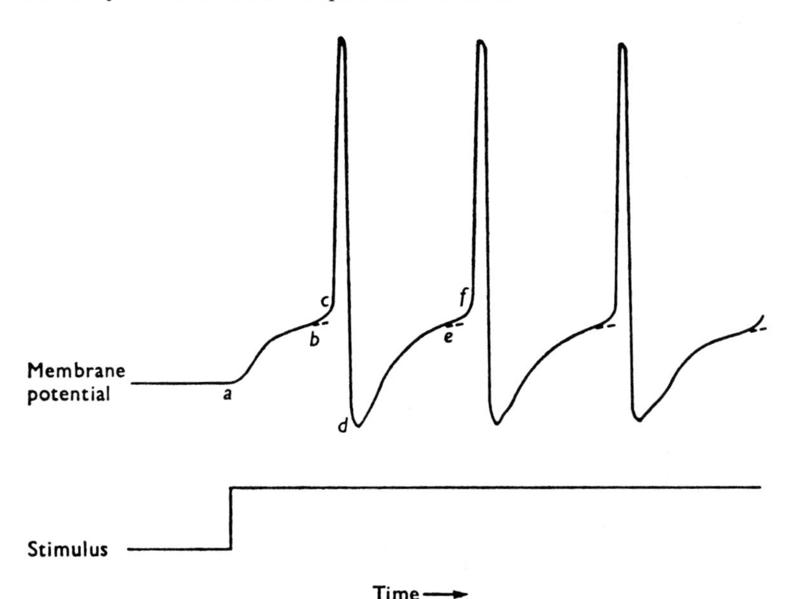

nerve fibre (which is a few millimetres long) into the thoracic ganglion.

Further aspects of sensory coding

Adaptation

Different receptors show different adaptation rates. The rate and extent of adaptation is related to the function of the receptor, depending upon whether the animal needs information about the steady level of the stimulus or about changes that occur in the stimulus intensity. We find that the response of many receptors consists of two parts, a tonic response, which is proportional to the intensity of the stimulus, and a phasic response, which is produced by changes in stimulus intensity. An example of a case in which the response is predominantly tonic is in receptors indicating the position of the animal with respect to gravity. In touch receptors, on the other hand, the tonic response is very small and the phasic response is rapidly adapting; a circumstance which probably accounts in part for our not feeling the presence of the clothes we are wearing. Muscle spindles provide an example of a receptor which has both tonic and phasic responses.

The presence of adaptation probably has a number of different advantages. In many cases, it is much more important for an animal to be able to detect changes in its environment than to measure accurately the static properties of the environment. For example, objects moving in the visual field are likely to be associated with danger or (when the animal is a carnivore) food. Adaptation thus serves to reduce the amount of unimportant information which reaches the central nervous system. Adaptation also enables a receptor to supply information about the rate of change of a stimulus, and can thus make a receptor much more sensitive to small changes in stimulus intensity without sacrificing the ability to respond over a wide range of stimulus intensities.

An example of a receptor system in which adaptation is extremely rapid occurs in the pad of the cat's foot (Armett *et al.*, 1962). Here, the receptors sensitive to pressure on the skin only respond if the displacement of the skin reaches the threshold amplitude in less than 0.7 ms, and the response of a receptor consists of a single nerve impulse. It is obvious that such a system cannot use the normal frequency code for measuring the intensity of the stimulus. Instead, stimulus intensity seems to be signalled by means of the number of receptors excited.

Adaptation can occur at a number of stages in the control sequence of sensory excitation. In the Pacinian corpuscle, the accessory structures (the lamellae) are involved in the process; when pressure is applied to the outside of the corpuscle, the lamellae are temporarily deformed, so deforming the nerve terminal, but the inner lamellae rapidly move back to their original positions, so that excitation of the nerve terminal is not maintained (Hubbard, 1958). In other cases, such as in many photoreceptors, the receptor potential shows adaptation. Adaptation at the impulse initiation site, by a mechanism similar to that of accommodation in nerve fibres, occurs in the tactile spine on the femur of cockroaches (French, 1984) and probably also in Pacinian corpuscles (Gray and Matthews, 1951). Finally, the generator potentials in the afferent nerve terminals innervating the hair cells of the goldfish inner ear show adaptation as a result of exhaustion of the hair cell chemical transmitter (Furukawa and Matsuura, 1978).

The sensory threshold

Suppose we perform an experiment in which a human subject listens to a series of sound pulses whose intensity is varied from quite loud to very soft, and is asked to state whether or not he can hear each pulse. At very low sound intensities, he will not be able to detect the pulses, whereas at higher intensities he will. In an intermediate range, he will be able to detect the pulses in a fraction of the times each pulse is presented, and this fraction will increase with increasing intensity. We may call the mid-point of this range, i.e. the sound intensity at which 50% of the pulses are detected, the sensory threshold. Obviously, this is a rather arbitrary concept: we could equally well call the sensory threshold the intensity at which there is a 10% or 75% probability of detection. But the important point here is that, with a stimulus whose intensity is variable over a continuous range (this proviso excludes vision and olfaction, where the stimulus is made up of discrete quanta or molecules), the probability of detection is a continuous function of the stimulus intensity; there are no discontinuities in the relation (fig. 16.13).

Some sense organs are amazingly sensitive.

The rods of the human eye can respond to the absorption of a single photon, the ear can detect sounds which produce movements of the basilar membrane less than 1 Å in amplitude, and even the poor old human nose can detect as few as 50 odorant molecules under optimum conditions (see Bialek, 1987).

Stimulus–sensation relations

Let us now consider the question, 'what is the relation between the intensity of the stimulus and the intensity of the sensation produced by it?' Weber, in 1846, found that human subjects could distinguish two weights, as long as the difference between them was about 2½%, irrespective of the absolute magnitude of the weights (this value – the minimum difference that can be detected – is known as the *increment threshold* or difference threshold). This conclusion was apparently confirmed by Fechner (1862), who described the results in the form of an equation:

$$\Delta I/I = k\Delta S \qquad (16.1)$$

where I is the stimulus intensity, ΔI is the increment threshold, k is a constant, and ΔS is a 'unit of sensation'. Fechner then suggested that this equation could be converted into a differential equation by making $\Delta I = \mathrm{d}I$ and $\Delta S = \mathrm{d}S$, so that

Fig. 16.13. Diagram to show how the probability of detection of a stimulus rises with increasing stimulus intensity. The threshold is here arbitrarily defined as that stimulus intensity which is detected 50% of the times it is presented.

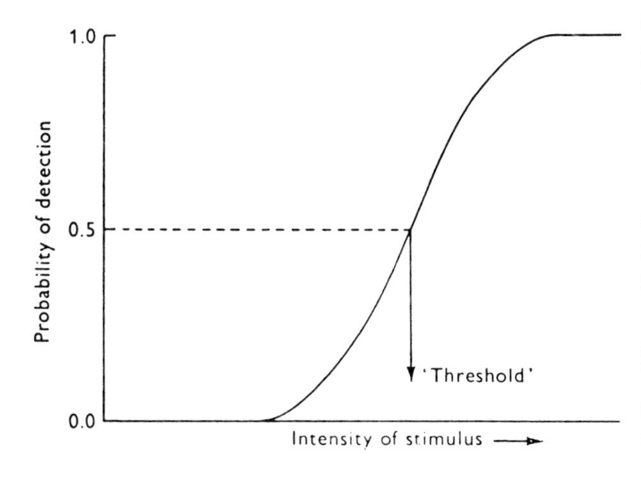

$$\frac{\mathrm{d}S}{\mathrm{d}I} = \frac{1}{kI}$$

which, when integrated, gives

$$S = a \log I + b, \qquad (16.2)$$

where a and b are constants. Equation (16.2) is known as the 'Weber–Fechner law', and states that the intensity of the sensation is proportional to the logarithm of the intensity of the stimulus.

In fact, it is found that the Weber–Fechner law is not universally applicable. In the human eye, for example, while it is approximately true for the medium range of light intensities, it does not apply at very low or very high light intensities.

An alternative formulation of the stimulus–sensation relation has been produced by Stevens (1961*a, b*). For a number of different modalities, he found that the magnitude of the sensation, ψ (psi), equivalent to S in (16.2), was given by

$$\psi = k\phi^n \qquad (16.3)$$

where ϕ (phi) is the intensity of the stimulus and k and n are constants. This relation has been called 'the power law' or 'the psychophysical law'.

Taking logarithms, (16.3) becomes

$$\log \psi = n \log \phi + \log k \qquad (16.4)$$

so a plot of log ψ against log ϕ will be straight line with a slope equal to the exponent n. Fig. 16.14 shows three examples.

Stevens later produced evidence that the power law can be demonstrated in physiological events as well as in psychological measurements. Thus the frequency of action potentials in a *Limulus* optic nerve fibre was proportional to the light intensity to the power 0.3, and the mammalian cochlear nerve compound action potential was proportional to sound pressure to the power 0.4 (Stevens, 1970).

Spontaneous activity

A number of receptors are spontaneously active: they produce nerve impulses in the sensory nerve in the absence of stimulation. An example of this is in the lateral line organs of fishes (Sand, 1937). In this case, it is clear that the spontaneous activity enables the mechanoreceptors in the lateral line canals to be sensitive to movements of the canal fluid in both directions. In fig. 16.15, headward perfusion of the canal causes an increase in the discharge frequency, and tailward perfusion causes a decrease; thus the receptor is analogous to a centre-reading voltmeter, which can register both positive and negative voltages, as opposed to a voltmeter in which the zero position is at one end of the scale.

An alternative function of spontaneous activity may be to increase the sensitivity of a receptor. If the receptor is spontaneously active, then there is no such thing as a subthreshold stimulus, since any increase in stimulus intensity will change the frequency of impulse discharge. The problem for the animal is then the detection of this change in sensory nerve impulse frequency. The facial pit

Fig. 16.14. Scales of apparent magnitude for three different sensory qualities, plotted on log–log coordinates, to illustrate Steven's power law. The lines follow (16.4), with different values of the exponent *n*. (From Stevens, 1976*b*.)

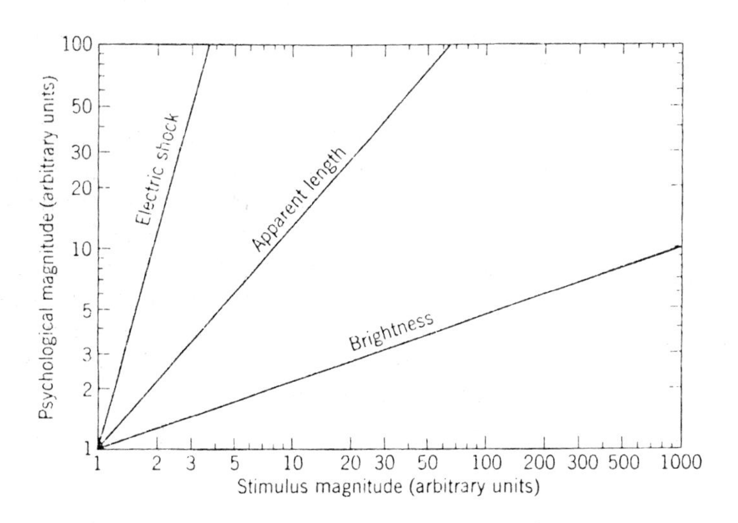

Fig. 16.15. Impulse frequency in a single lateral line fibre of a ray in response to perfusion of the hyomandibular canal. Black circles, headward perfusion (between 0 and 10 s); white circles, tailward perfusion. Graph *b* is a continuation of *a*. (From Sand, 1937, by permission of the Royal Society.)

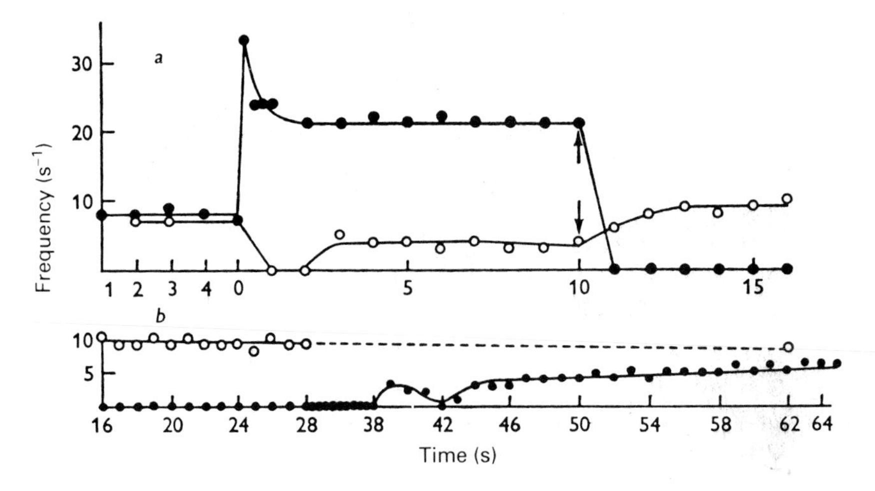

receptors of crotaline snakes show spontaneous activity, which is modified by very small changes in the radiant heat impinging upon them; it seems probable that the existence of this spontaneous activity contributes to the very great sensitivity of these organs (Bullock and Dieke, 1956).

Finally, there is considerable evidence that the activity of sense organs, spontaneous or induced, plays an important part in determining levels of activity in the central nervous system.

The use of multiple channels

A large number of sense organs send sensory information into the central nervous system via a number of nerve fibres. Such an increase in the number of transmission lines will obviously increase the amount of information that can be carried by the system. It is useful at this point to introduce the concept of a *receptor unit* (Gray, 1962); this consists of a single sensory nerve fibre together with its branches and any receptor cells which may be connected with it. A multiple channel system is thus a sense organ, or a group of closely associated receptors, consisting of more than one receptor unit.

A multiple channel system can increase the sensitivity of a sense organ, as the following theoretical argument shows. Consider a sense organ consisting of n identical receptor units excited independently of one another, and assume that each unit has a probability x of firing in response to a small constant stimulus. Then the probability of any unit not firing is $(1-x)$, and the probability of no unit firing is $(1-x)^n$. Hence the probability (p) of one or more units firing (which, we shall assume, constitutes detection of the stimulus) is given by

$$p = 1 - (1-x)^n. \quad (16.5)$$

Fig. 16.16 shows how p varies with x for various values of n. It is evident that an increase in the number of channels can produce (a) a considerable increase in sensitivity (just how much will depend on how x varies with the size of the stimulus), and (b) an increase in the sharpness of the threshold, since the slopes of the curves increase with increasing values of n. For the particular case where $n=2$, (16.5) becomes

$$p = 2x - x^2. \quad (16.6)$$

Fig. 16.16. Graphical solutions of (16.5) for various values of n.

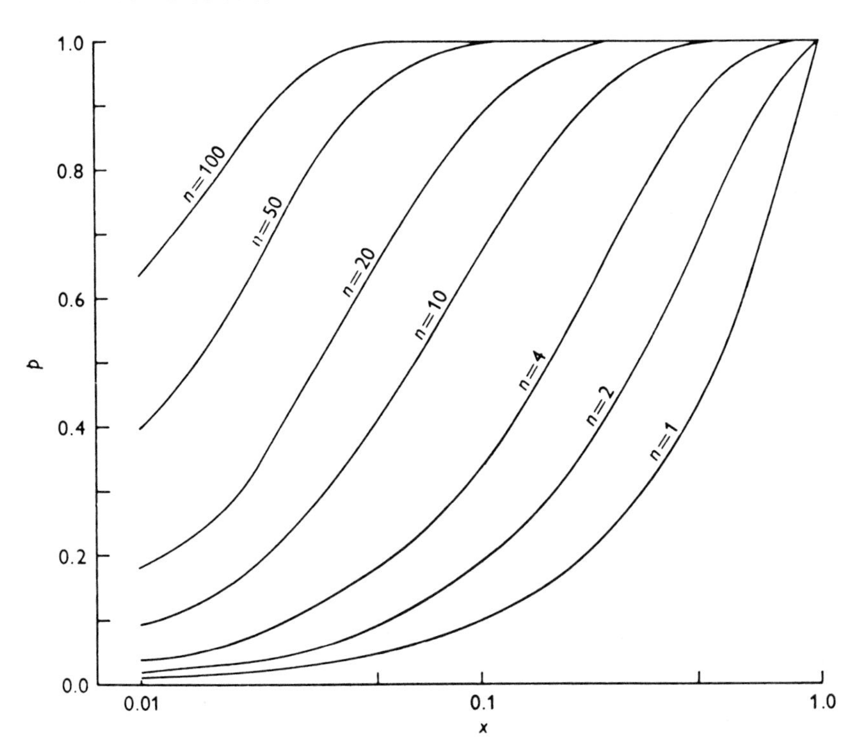

This equation is fairly easy to test by psychophysical methods, since a number of sense organs are paired. Thus Pirenne (1943) found that the threshold for binocular vision in man is lower than that for monocular vision, and Dethier (1953) showed that the threshold of the tarsal taste receptors in flies followed a similar relation (fig. 16.17).

Most information channels are 'noisy' to some extent, i.e. there is some random activity (noise) in addition to the signal. In a sensory nerve fibre, this noise will be seen as random fluctuations in the interval between successive nerve impulses. How, then, is the animal to detect a small change in nerve impulse frequency against this noisy background? One way of doing this would be to measure the average frequency over a sufficiently long time, both before and during the stimulus. Use of this system must necessarily mean that the latency of detection will be appreciable, and that there will be no detection of very brief stimuli. However, by using a number of channels the same information can be obtained in a fraction of the time. Suppose that, in a single noisy sensory channel, the time needed to determine whether or not a new average frequency is significantly higher than the original one is t; then the equivalent time for n identical channels is t/n. Hence an increase in the number of sensory channels can reduce the response time. Another way of looking at this problem of detection is to consider the signal-to-noise ratio of the system. If the signal-to-noise ratio of a single channel is q, then the signal-to-noise ratio of n identical channels is $q\sqrt{n}$. Hence increase in the number of channels increases the sensitivity of a receptor to changes in stimulus intensity. It seems very probable that considerations of this type apply to the large number of spontaneously active sensory channels found in the facial pit sense organs of crotaline snakes.

A further use of multiple channel systems is

Fig. 16.17. The greater sensitivity of flies with two legs in contact with a sucrose solution as compared with those having only one leg so applied. The broken line shows the expected relation between surcrose concentration and the probability of response for two-legged flies, as calculated from application of (16.6) to the results for one-legged flies. The units (probits) on the ordinate are equal to one standard deviation; 5 probits correspond to 50% acceptance. (From Dethier, 1963.)

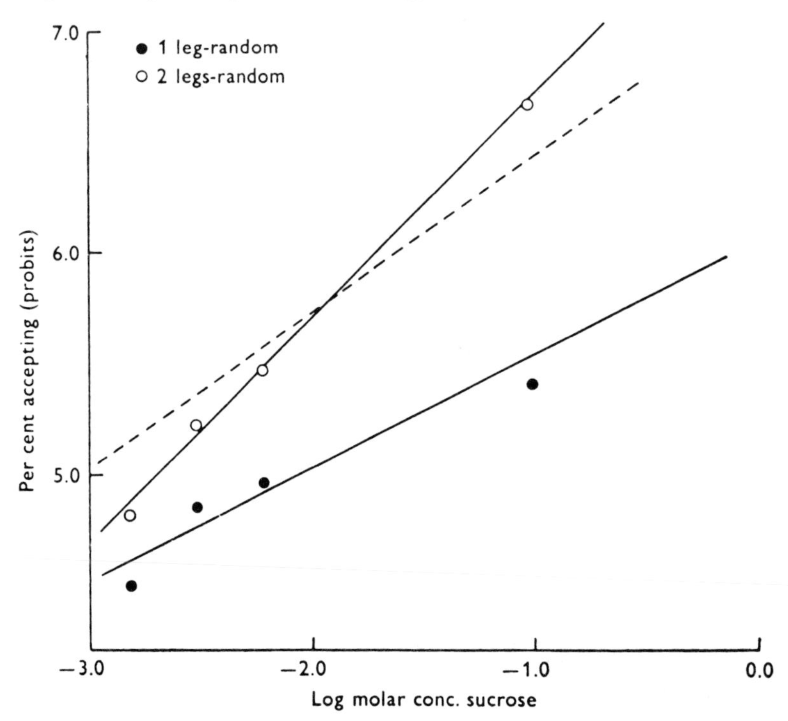

to ensure that the sense organ is sensitive to the direction or location of the stimulus. The most obvious example of this occurs in image-forming eyes, in which different photoreceptors respond when the stimulus (a light source, for example) is in different parts of the visual field.

We have so far considered the advantages of multiple channel systems only for cases in which the channels have identical receptor properties. A considerable increase in the information derived from a sense organ can be obtained by using a multiple channel system whose receptor units differ, quantitatively or qualitatively, in their sensitivity. A common type of organization is what may be called *range fractionation* (Cohen, 1964), in which different receptor units respond to different parts of the range of stimulus intensities or qualities covered by the whole sense organ. This leads to an increase in the sensitivity of the sense organ, since a small change in the stimulus can produce a large change in the response of a particular receptor unit, whereas if that receptor unit had to be responsive over the whole of the stimulus range, its sensitivity over a fraction of the range would be correspondingly reduced. An example of this type of organization is shown in fig. 16.18, where the different receptor units of the Golgi organs in the ligaments of a cat's knee have different restricted sensitivity ranges, but between them can provide information about the position of the knee joint over the full range from full flexion to full extension. Another example is in the vertebrate eye, where the rods and cones deal with low and high light intensities respectively.

Receptor units in a system which utilizes range fractionation frequently show bell-shaped response curves (figs. 16.18 and 16.19). It follows that there must be some ambiguity in the information supplied by any one receptor unit; in the unit whose response is shown in curve *a* of fig. 16.19, for example, an impulse frequency of 7 per second is produced at both 24° and 35°. This ambiguity can be resolved by comparison with another receptor unit with a different, but overlapping, response range; thus, in fig. 16.19, the unit whose response is shown by curve *b* has an impulse frequency of 6 per second at 24° and 9 per second at 35°.

A similar type of argument can be applied to the fractionation of receptor units with respect to the qualitative aspects of the stimulus. Cohen, Hagiwara and Zotterman (1955) measured the responsiveness of different gustatory receptor units in the afferent nerves of the cat's tongue. They found that many fibres were responsive to more than one of the modalities which psychophysical measurements suggest can be distinguished, and concluded that the different sensations arise from comparisons of the activity in a number of different

Fig. 16.18. The relations between the impulse discharge frequency in a number of cat knee joint receptors and the angle of the knee joint. (From Skoglund, 1956.)

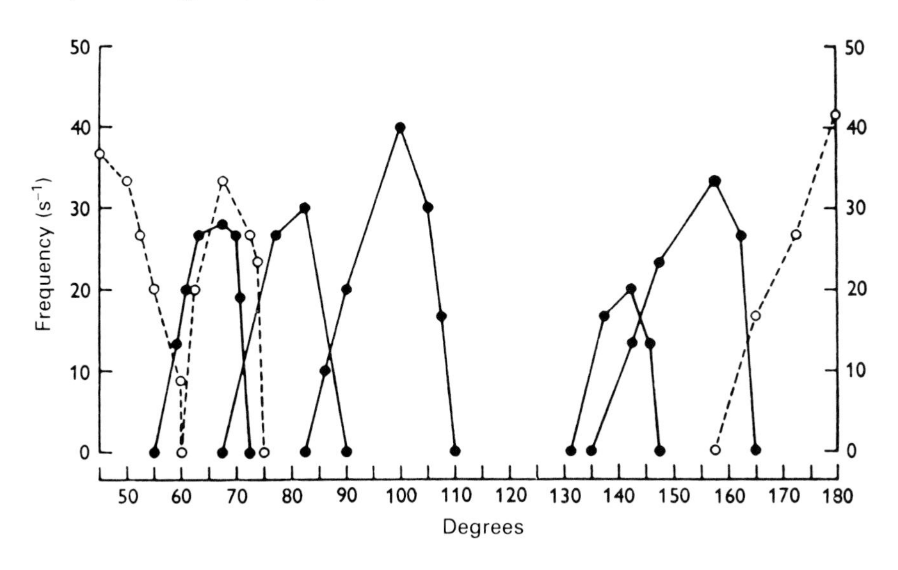

receptor units. For example, they suggested that the 'sour' sensation is produced by simultaneous activity in the 'water', 'salt' and 'acid' fibres; if the 'water' and 'acid' fibres cease their activity, the sensation is then one of 'salt'. Thus the specificity of the response is sharpened by utilizing the multiple channel system. The trichromatic colour vision system in man is another example of such sharpening of the specificity of sensation. An extreme example of this type of system is found in the olfactory receptors in the antennae of the silkmoth (Schneider *et al.*, 1964). In over 50 cells whose response to various organic compounds was tested, no two cells showed the same response spectrum (fig. 17.37).

Interaction between receptors: lateral inhibition

A very interesting phenomenon occurs in the compound eye of *Limulus*, known as lateral inhibition (Hartline *et al.*, 1956; see also Hartline *et al.*, 1961). Hartline and his colleagues discovered that the discharge rate of a single eccentric cell decreases if the adjacent ommatidium is illuminated, indicating that adjacent ommatidia are mutually inhibitory (fig. 16.20). This phenomenon leads to two interesting features in the pattern of information produced by the eye.

Fig. 16.19. Steady impulse discharge frequencies of five 'cold' fibres in a cat's tongue at different temperatures. Curves *a* and *b* are referred to in the text. (From Hensel and Zotterman, 1951.)

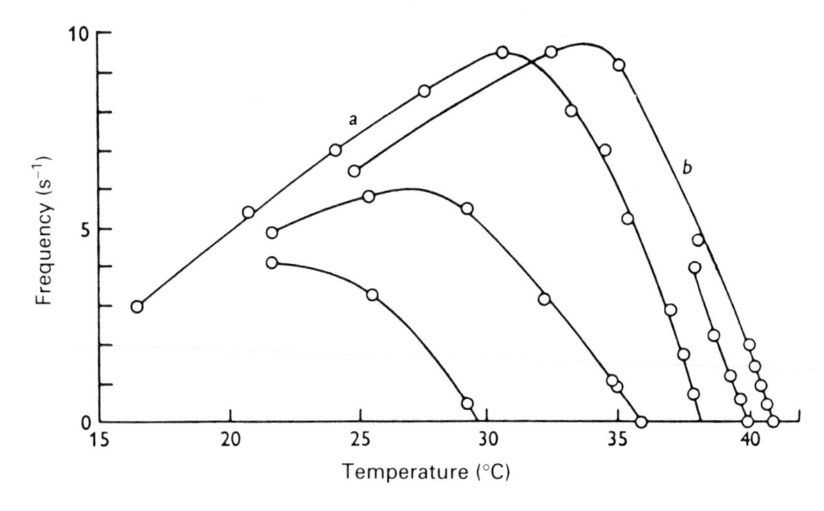

Firstly, the eye must become much more sensitive to edges. Consider the model system shown in fig. 16.21, where we have a row of ommatidia *A* to *J* of which *A* to *E* are exposed to a relatively high light intensity, and *F* to *J* are partially shaded. The extent of the inhibition produced by each cell on its immediate neighbours is then indicated by the diagonal arrows in the diagram, full arrows indicating strong inhibition and broken arrows light inhibition. Cells *A* to *D* are mutually inhibited to some extent, and hence their discharge frequencies are considerably less than they would be if there were no lateral inhibition. But cell *E* is only slightly inhibited by cell *F*, hence its discharge rate is higher than those of cells *A* to *D*. Similarly, cell *F* is more strongly inhibited than are cells *G* to *J*, hence its discharge rate is lower than theirs are. The net effect of these interactions is to emphasize the presence of the edge. A similar process appears to form the basis of edge perception in our own eyes; it may also account for the well-known optical illusion shown in fig. 16.22 – we may suggest that the grey spots seen at the junctions of the white lines are produced by lateral inhibition from the surrounding receptors produced by the image of the white lines.

The second feature of lateral inhibition is that it can increase the resolving power of the eye. The 'fields of view' of adjacent ommatidia show some considerable overlap. Lateral inhibition has the effect of cutting down this overlap, so that each optic nerve fibre only responds to the illumination of a small part of the visual field (Reichardt, 1962).

Transmitter–receiver systems

The energy for the activation of most distance exteroceptors is derived from the environment; a crotaline viper, for example, perceives the radiant heat emitted by its prey, and the prey (if it is lucky enough) hears the sounds produced by the approaching viper. However, there exist some specialized sensory systems in which the energy that stimulates the receptors is produced by the animal itself, being reflected

Fig. 16.20. Simultaneous records of the sensory discharges in two *Limulus* optic nerve fibres serving adjacent ommatidia (*a* and *b*), in response to illumination of either *a* or *b* alone or both together. The records last 1.5 s in each case, and the figures to the right of them show the number of action potentials occurring in that time. (From Hartline and Ratliff, 1957.)

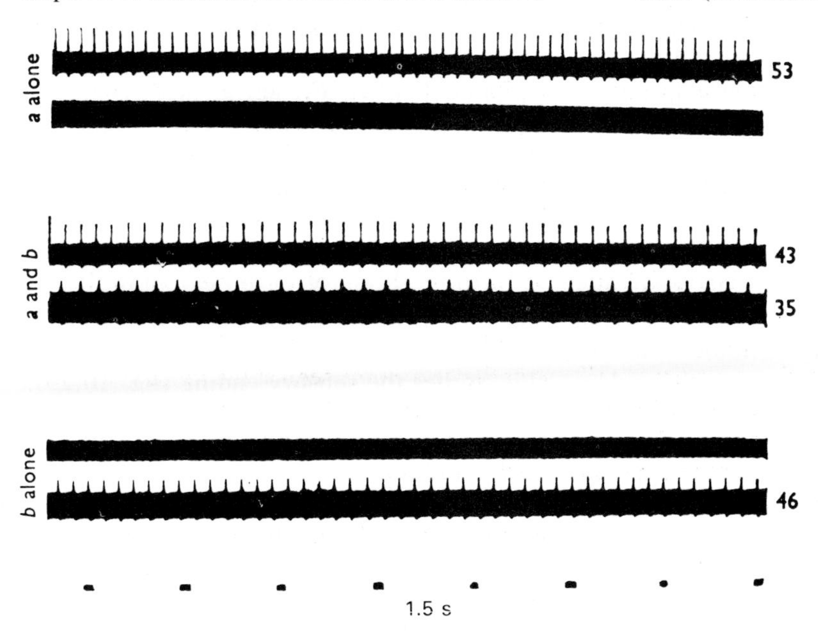

Fig. 16.21. A model system to show lateral inhibition. Explanation in text.

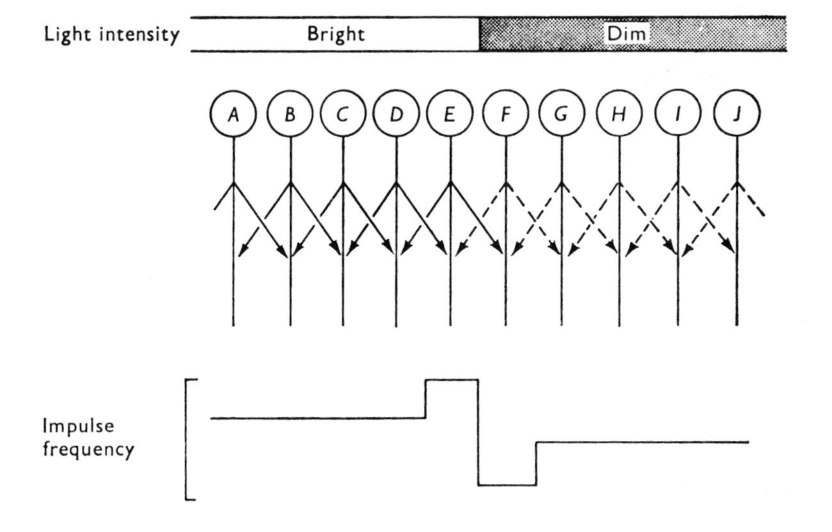

back from or modified by objects in the environment.

One of the most well-known of such 'transmitter–receiver systems' is the echolocation system used by bats (see Griffin, 1958). It has long been known that blinded bats can avoid obstacles and catch insects in flight just as well as normal bats. Pierce and Griffin (1938) discovered that flying bats emit a continuous series of ultrasonic cries, and Griffin later showed that the frequency with which these cries are made is much increased when a bat approaches an obstacle. Griffin and Galambos (1941) measured the ability of bats to fly through a barrier of metal wires spaced just wide enough apart to let them through. They found that deaf or dumb bats frequently hit the wires whereas normal or blind bats rarely did so. The conclusion from these experiments was that bats detect obstacles by emitting ultrasonic sound pulses and hearing the sound waves reflected back to them; the position of an object is determined by localizing the source of the reflected sound. Echolocation systems which are similar in principle to those used by bats are also found in porpoises, certain cave-dwelling birds and possibly some other animals.

A different type of transmitter–receiver system is seen in the object location system used by some electric fish. Here the fish produces an electric field in the water by means of its electric organs, and can detect changes in this field with its electoreceptors. We shall examine these electroreceptors in the following chapter.

Finally, there are a number of deep-sea fishes which are bioluminescent. Since they have well-developed eyes, it seems probable that one of the functions of their luminescence is to produce light by which the fish can see.

Fig. 16.22. An optical illusion which may be caused by lateral inhibition: there is an appearance of grey spots at the intersections of the white lines.

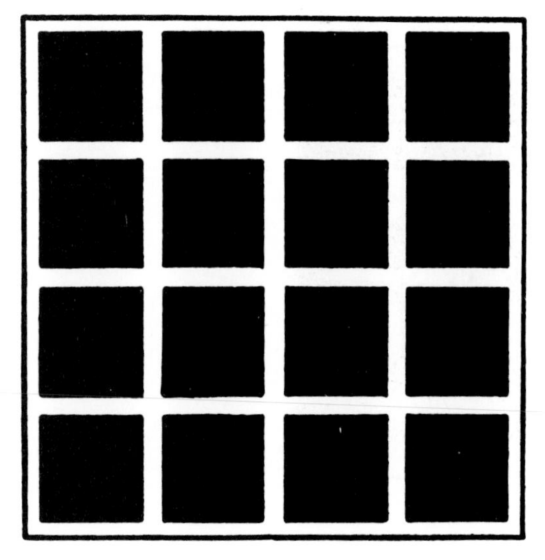

Central control of receptors

The efferent nerve fibres innervating mammalian limb muscles are of two types: the large, rapidly conducting α fibres, and the small, more slowly conducting γ fibres. The γ fibres innervate the muscle fibres (intrafusal fibres) in the muscle spindles, whereas the α fibres innervate the extrafusal fibres and so are responsible for the contraction of the muscle (Leksell, 1945; Kuffler *et al.*, 1951). Stimulation of the γ fibres causes contraction of the intrafusal fibres so that the middle regions (which are not contractile) are stretched; this excites the sensory ending (fig. 16.23). Thus the output of the sensory fibres can be modified by the action of the γ motor fibres; in other words, it is to some extent under the control of the central nervous system.

Such central nervous control of receptor output via efferent nerve fibres is not uncommon. The efferent fibres may act at various points in the sensory excitation control sequence. In the muscle spindles they act on the accessory structures; other examples of this type of action are the reflex control of pupil diameter in the vertebrate eye, and the action of the stapedius and tensor tympani muscles in the mammalian middle ear, which decrease the sensitivity of the ear to loud sounds. Direct efferent control of the size of the generator potential occurs in crustacean muscle receptor organs, where the sensory neuron is inhibited by an inhibitor axon (fig. 16.8) which, when stimulated, depresses the generator potential and prevents the initiation of sensory impulses (Eyzaguirre and Kuffler, 1955*b*).

Loewenstein (1956) showed that the sensitivity of mechanoreceptors in frog skin is increased if the sympathetic nerve supply to the skin is stimulated. Similar effects are produced by application of the sympathetic neurotransmitters, adrenaline and noradrenaline, to the skin. Livingston (1959) suggests that such sympathetic actions on sense organs are likely to be widespread.

Efferent nerve fibres occur in the auditory nerves of mammals (Rasmussen, 1953), and stimulation of them suppresses the afferent response in the auditory nerve to sound clicks. Electron microscopy shows that the receptor cells (hair cells) of the cochlea are innervated by two types of nerve ending, one of which contains synaptic vesicles and is therefore probably efferent (Engström and Wersäll, 1958). Similar apparently efferent endings are found on the homologous receptor cells in other parts of the ear and in the lateral line organs of fishes and amphibians. Russell (1971) found that the lateral line organs of *Xenopus* (which are stimulated by water currents) are 'switched off' by the efferent nerves during voluntary movements of the animal.

Perception

Many questions about sensory processes can be considered at a number of different levels. How do we see things, for example? At one level we can look at the optics of the eye or investigate what happens in a photoreceptor cell when its visual pigment absorbs photons of light. At a higher level we can look at how the photoreceptor cells interact with each other and with other cells in the retina. Or we can proceed up the visual pathway into the brain and look at the properties of the nerve cells there that receive inputs from the retina. We would find much complexity there, and an increasing abstraction of the particular properties of the image falling on the retina. Just what these abstractions imply is still a matter for discussion.

Such investigations, all of them falling within the general field of sensory physiology, still do not answer the question 'how do we see things?' Our conscious perceptions are not the sensory messages themselves, they are interpretations of them (see, for example, Hochberg, 1984, and Gregory, 1970, 1986). In vision, for example, we are able to understand and make sense of what we are looking at; it is very difficult to make a machine that will do this. There is a strong element of recognition in many perceptual processes, hence they are much affected by our previous experience of the world we live in. But questions of this type, by transgressing the cellular level of animal organization, would take us outside the bounds of this book.

Fig. 16.23. The effect of γ motor stimulation on the afferent discharge from a cat muscle spindle. Upward deflections on the records are sensory nerve impulses; downward deflections are stimulus artefacts produced by stimulating a single γ fibre at 200 per second. Columns show different loads on the muscle. Rows show effects of stimulation: *a*, no stimulation; *b*, 4 to 6 stimuli; *c*, 9 to 11 stimuli; and *d*, 14 to 16 stimuli. (From Kuffler *et al.*, 1951.)

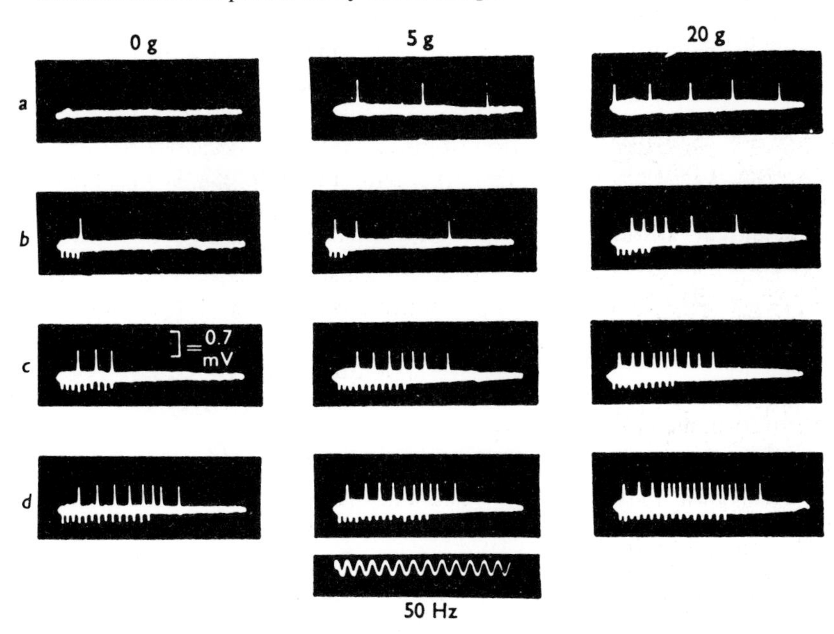

17
A variety of sense organs

Having considered some general properties of sensory receptor processes, we shall now examine the physiology of some particular sense organs in a little more detail. It is not possible in a book of this nature to deal with more than a small fraction of the different types of sense organ that have been described, hence this chapter merely provides a brief look at a vast and complex subject.

The acoustico–lateralis system of vertebrates

In vertebrates, the receptors involved in hearing, equilibrium reception, and the detection of water movements are all of one basic type and are clearly of common evolutionary origin. In each case the receptor cells, or hair cells (fig. 17.1) possess fine processes or 'hairs' whose movement leads to modification of the sensory output. The receptor cells are all connected to sensory neurons which the enter the brain via the cranial nerves; neurons concerned with hearing and balance enter via the eighth nerve and those concerned with the detection of water currents enter via the various lateralis branches.

It is a fascinating feature of the acoustico–lateralis system that we can see how a particular type of receptor cell can be used to respond to a whole variety of different mechanical stimuli, largely by elaboration of the accessory structures in the various sense organs (see fig. 17.2).

The lateral line organs of fishes

The sensory cells of the 'ordinary' lateral line system* are grouped together in organs called *neuoromasts*. Each neuromast (fig. 17.2*a*) consists of a number of receptor cells and supporting cells situated in the epidermis, and a gelatinous projection, the *cupula*, into which the sensory hairs of

Fig. 17.1. The structure and innervation of a typical receptor cell in the vertebrate acoustico–lateralis system. (From Flock, 1971.)

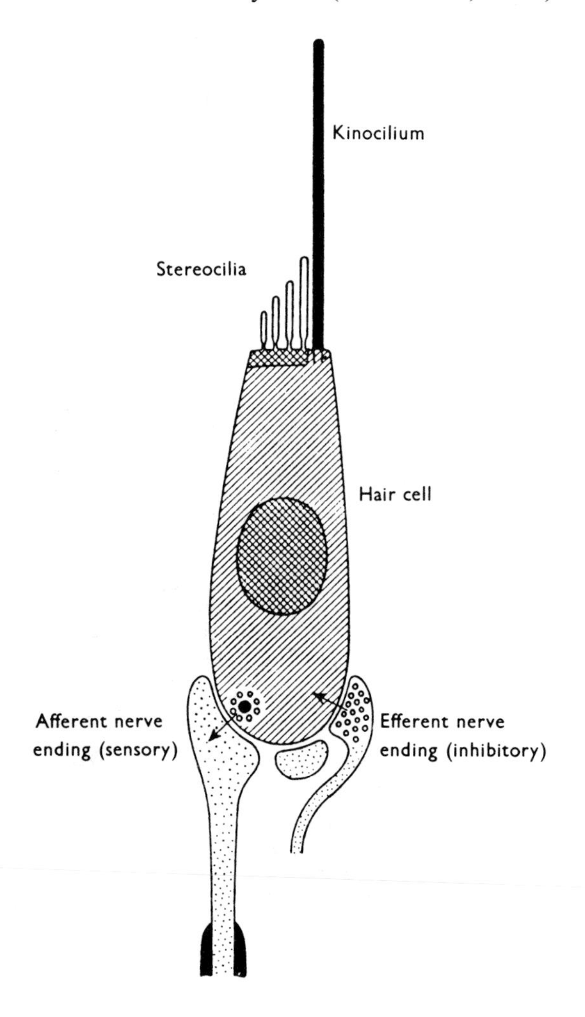

* The 'ordinary' lateral line system (Dijkgraaf, 1963) is that part of the system which contains neuromasts and is primarily responsive to water movements. Some fishes possess, in addition, numbers of 'ampullary' organs associated with the lateral line system, which appear to be electroreceptors (p. 298).

the receptor cells project. Neuromasts may be situated either freely on the outer surface of the animal or, typically, in canals which open to the surface at intervals through small pores. The distribution of these canals in a typical fish is shown in fig. 17.3.

If a water current impinges on the cupula of a neuromast it is moved to one side, and it is this movement ('detected' by the sensory hairs of the receptor cells) which acts as the effective stimulus.

A fish whose lateral line system has been partially denervated no longer responds to water currents applied to the denervated region. This sensitivity to water movements enables the fish to detect moving objects in its immediate vicinity by means of the water currents which they produce (Dijkgraaf, 1934); fig. 17.4 shows how an approaching object produces movements of neuromast cupulae.

The electrical responses of the lateral line nerves to stimulation of the neuromasts by water currents of controlled velocity were investigated by Sand (1937), using a perfusion technique on the hyomandibular canal of the ray. He found that there was a considerable discharge in the nerve in the resting state, which was much increased by water movements in either direction. After dissect-

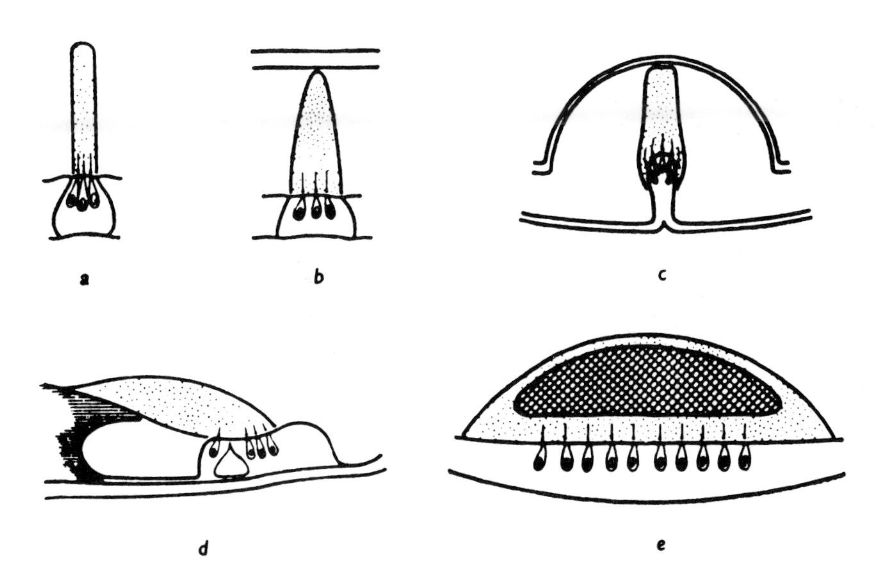

Fig. 17.2. Different accessory structures associated with receptor cells of the vertebrate acoustico–lateralis system. *a*: Free neuromast of the lateral line system. *b*: Neuromast in a lateral line canal. *c*: Ampullary sense organ of a semicircular canal. *d*: Organ of Corti in the mammalian cochlea. *e*: Otolith organ. (From Dijkgraaf, 1963.)

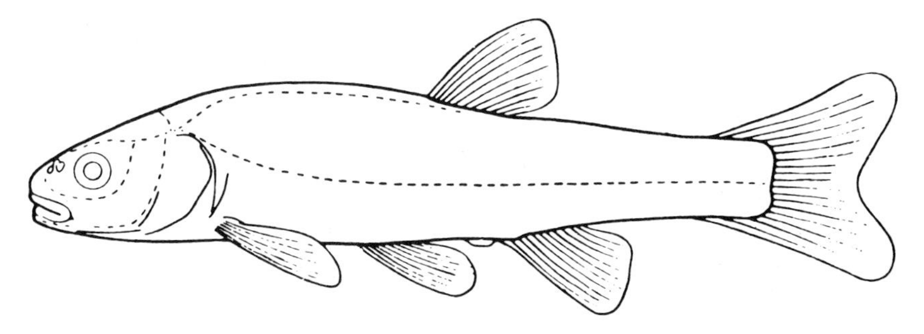

Fig. 17.3. Distribution of the lateral line canals on the body surface of a fish. (From Dijkgraaf, 1952.)

ing the nerve so that only a single active unit remained, the spontaneous discharge in this unit was increased by perfusion in one direction but decreased by perfusion in the other direction (fig. 16.15). Some units were excited by headward perfusion of water in the canal, others by tailward perfusion.

The semicircular canals

The inner ear, or labyrinth, of vertebrates is a complex organ concerned with equilibrium reception and hearing. It consists of (i) the membranous labyrinth, which is a series of interconnected sacs and tubes containing the receptor cells and filled with an aqueous fluid called the *endolymph*, and (ii) the osseous labyrinth, which is a cavity in the bone of the skull in which the membranous labyrinth is situated. The inner wall of the osseus labyrinth and the outer surface of the membranous labyrinth are separated by a thin layer of another fluid, the *perilymph*. The membranous labyrinth arises embryologically by an invagination of the ectoderm, and its receptor cells are very similar to those of the lateral line neuromasts.

The structure of the membranous labyrinth is shown in fig. 17.5. There are three semicircular canals, set in planes approximately at right angles to one another. The horizontal canals of the two labyrinths in the head lie in the same plane; the anterior vertical canal on one side is in approximately the same plane as the posterior vertical canal on the other, and *vice versa*. The canals are connected at each end to a sac called the utriculus, so that endolymph can move between canal and utriculus. At one end of each canal is a swelling, the ampulla, in which a group of receptor cells and supporting cells, the crista, is found (fig. 17.2*c*). Sensory nerve fibres leave the crista to join the auditory nerve.

The sensory hairs of the crista are inserted into a gelatinous cupula. In sections of preserved material the cupula appears as a rather small structure, but in the living animal it extends right across the ampulla to form a fairly close-fitting seal with the opposite wall. This fact, which has important connotations for the mode of action of the semicircular canals, was discovered by Steinhausen (1931) by observing the appearance of the horizontal semicircular canal in a pike after indian ink had been injected into the endolymph.

It has been known for many years that the semicircular canals are concerned with the perception of rotational movements of the head. Such movements produce compensatory reflexes, especially in the extraocular muscles of the eyes, which are abolished by bilateral extirpation or denervation of the canals.

When a vertebrate is rotated at a constant angular velocity in the horizontal plane, its eyes move so as to compensate for this movement. If the head is passively turned to the left, for example, the eyes move to the right, so keeping the visual image fixed in position on the retina. As the rotation continues, the eyes flick rapidly to the left, and then revert to moving to the right again. This pattern of movements, which is known as *nystagmus*, may be repeated for a time. With prolonged rotation however, the nystagmus dies away and the eyes remain in their normal position with respect to the head. On cessation of a prologed rotation, the eyes

Fig. 17.4. The movements of neuromast cupulae produced by the water currents (small arrows) set up by an approaching object. (From Dijkgraaf, 1952.)

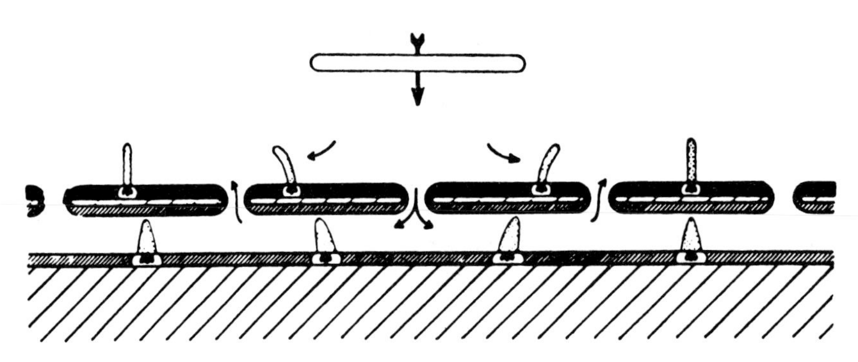

show *after-nystagmus* for several seconds, i.e. they behave as though the head were being rotated in the opposite direction.

Can we relate these eye movements to the mechanics of the semicircular canals? At one time it was thought that the action of the semicircular canals was as follows. When the head was rotated, endolymph would flow freely through the canals by inertial forces, and the cupula, extending only part-way across the ampulla, would respond to the velocity of this stream in the same way that the free neuromasts on the surface of a fish respond to water currents. This scheme (which we may call the 'flow' theory) could not account for the duration of nystagmus and after-nystagmus since the friction between the endolymph and the walls of the canal would bring the endolymph to rest with respect to the canal within a fraction of a second of the change in rotational velocity of the canal.

Steinhausen's observation that the cupula extends across the whole width of the ampulla in the living animal is incompatible with the 'flow' theory. In a further series of experiments, Steinhausen (1933) was able to show that when the cupula is displaced by a sudden change in the velocity of rotation, it takes several seconds to return to its resting position. As a result of these observations, the 'flow' theory was replaced by the 'displacement' theory, according to which the displacement of the cupula is strictly linked to the displacement of the endolymph, not to its velocity of flow. An elegant piece of evidence for this theory was provided by Dohlman (1935) who, using Steinhausen's technique for observation of the movements of the cupula, was able to inject a small oil droplet into a semi-circular canal of a cod. He found that the droplet only moved a short way round the canal during full displacements of the cupula (fig. 17.6); hence there can be very little flow of endolymph past the cupula.

Fig. 17.5. Membranous labyrinths of a fish (*a*), a reptile (*b*), a bird (*c*) and a mammal (*d*). *U*., Utriculus; *S*., sacculus; *L*., lagena; *P.B*., papilla basilaris; *B.M*., basilar membrane; *C*., cochlea. (From von Frisch, 1936.)

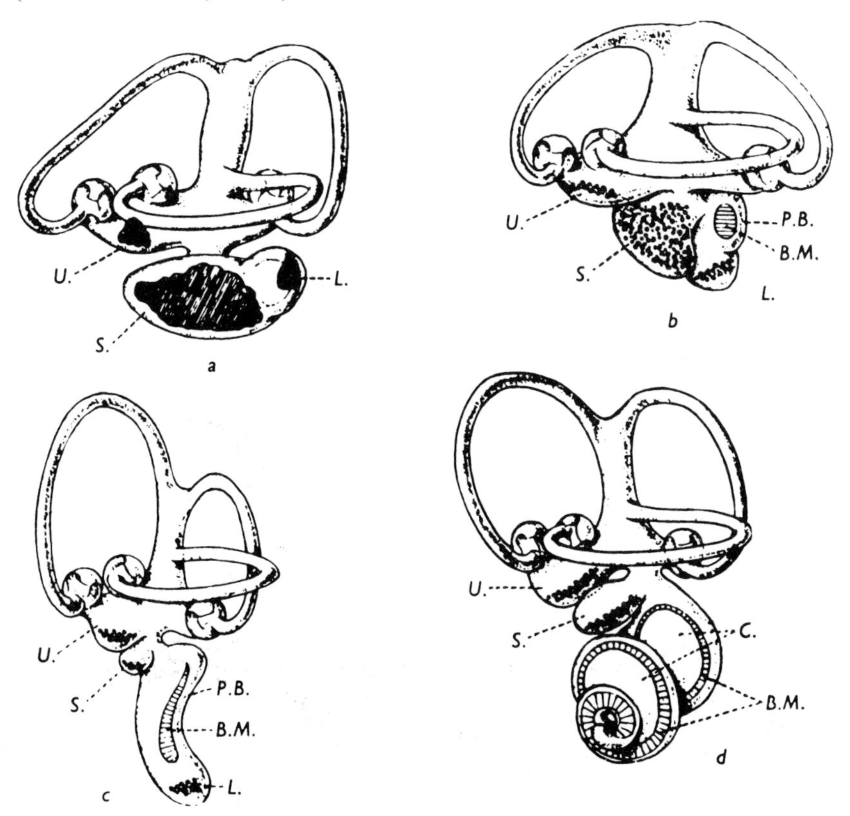

Let us now see how, according to the displacement theory, the semicircular canal works. The essential stimulus for the hair cells of the crista is the angular deviation of the cupula from its normal position in the canal. This deviation is produced by the combination of three forces: (i) an inertial force, proportional to the moment of inertia of the endolymph and the angular acceleration of the canal, (ii) a viscous (damping) force, proportional to the viscosity of the endolymph, the dimensions of the canal and the velocity of movement of the endolymph through it, and (iii) an elastic restoring force in the cupula, proportional to its angular deviation from the resting position.

Now consider what happens if the canal is subjected to a period of rotation at a constant angular acceleration. In this case the inertial force is constant and the viscous and elastic forces decrease and increase respectively as the cupula moves smoothly to a position determined by the acceleration. The behaviour of the system is more complex when it is subject to a long period of rotation at a constant angular velocity (fig. 17.7). At the start of this rotation the acceleration is very high and so the inertial force produces a rapid deflection of the cupula. Then, since there is no more acceleration, the inertial force falls and the cupula begins to return to its resting position under the influence of the elastic force. In order to do this it has to act against the viscous force by pushing the endolymph back through the canal, so that it is some time (about 30 s) before the cupula reaches its resting position. If the rotation is now stopped, the endolymph tends to continue on its way, so that there is for a short time a large inertial force which pushes the cupula in the opposite direction to that in which it was previously deflected; it is this 'overswing' that produces after-nystagmus. Finally, over the next 20–30 s, the cupula returns to its resting position, again by pushing the endolymph through the canal against the viscous force.

Fig. 17.6. The appearance of part of a fish semicircular canal after injection of indian ink into the endolymph followed by a further injection of an oil droplet into the canal. The upper diagram shows the system at rest, the lower diagram shows the displacement of the cupula and oil droplet after rotation. (Drawn from photographs by Dohlman, 1935.)

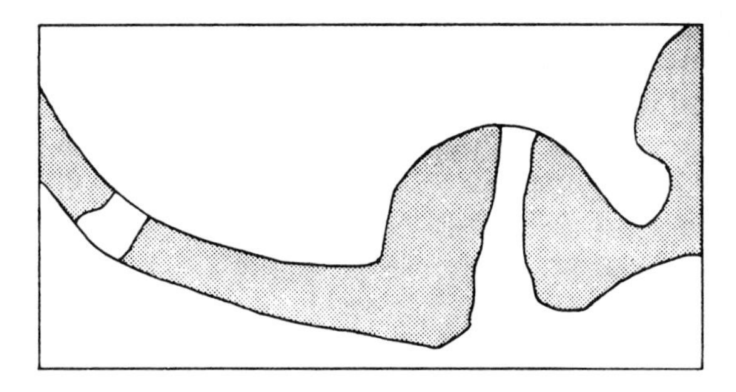

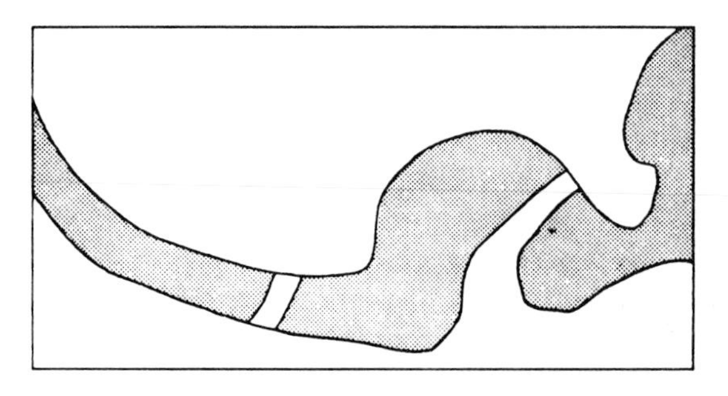

Excellent further evidence for Steinhausen's 'displacement' theory has been provided from two sources: by psychophysical measurements on the sense of rotation in man, and by electrophysiological measurements on the nervous output from the hair cells of the crista in cartilaginous fish. The analysis of the cupula movements in terms of three forces, given above, can be expressed in terms of a differential equation as follows:

$$A\frac{\mathrm{d}^2x}{\mathrm{d}t^2} + B\frac{\mathrm{d}x}{\mathrm{d}t} + Cx = 0,$$

where x is angular displacement, and A, B and C are constants. Van Egmond, Groen and Jongkees (1949) made a series of measurements on the sensation of rotation in human subjects for different values of x and its derivatives, and showed that the results were consistent with this equation.

Lowenstein and Sand (1936, 1940*a*) measured the nervous discharge from the horizontal semicircular canals of the dogfish and the ray in response to rotations in the horizontal plane. They found that the changes in the nerve impulse frequency in the sensory fibres followed the same time course as the displacement of the cupula as measured by Steinhausen. These sensory fibres showed a steady spontaneous discharge when the cupula was in its resting position, which increased when the cupula was deflected towards the utriculus and decreased when it was deflected away from the utriculus. Similar features were seen in the responses of the vertical canals (Lowenstein and Sand, 1940*b*), except that here excitation occurred when the head was rotated so as to deflect the cupula away from the utriculus.

The effective stimulus for the canals is angular acceleration produced by a rotation in the plane of the canal. Linear accelerations will have no effect since the inertial forces produced in all parts of the endolymph will be of equal magnitude and in the same direction, hence there will be no tendency for the endolymph to rotate.

So far we have considered the responses of a semicircular canal to rotation in the plane in which it lies, i.e. about an axis perpendicular to that plane. What happens if the axis of rotation is not perpendicular to the plane of the canal? This question was theoretically investigated by Summers, Morgan and Reimann (1943); the following analysis is partially based on their account. An angular acceleration can be represented by a vector parallel to the axis of rotation whose length represents the magnitude of the acceleration and whose direction indicates the direction of rotation (it is conventional to draw this so that the rotation is clockwise when seen along the direction of the arrow). Now consider fig. 17.8, in which the line B is perpendicular to the plane in which the semicircular canal lies. The canal is subjected to an angular acceleration about an axis parallel to the line A, which lies at an angle θ with B. This angular acceleration is represented by the vector $\boldsymbol{\alpha}$. The canal is then effectively subjected to an angular acceleration about B represented by the vector $\boldsymbol{\beta}$, which is obtained by dropping a perpendicular from the tip of the vector $\boldsymbol{\alpha}$ onto B. It is clear that

$$\boldsymbol{\beta} = \boldsymbol{\alpha} \cos \theta.$$

Hence if θ is 90°, i.e. if A lies in the plane of the semicircular canal, there can be no effective stimulation of the canal. This is the only set of positions of A for which this will occur; for all other positions there will be some response, and this will be maximal for a given angular acceleration when A coincides with B. Thus a single canal is subjected

Fig. 17.7. Diagram to show the deflection of the cupula of a semicircular canal in response to rotations at constant velocity for various periods of time. The resting position is indicated by the horizontal line. The curves apply equally well to the changes in sensory nerve discharge produced by constant velocity rotations. (From Lowenstein and Sand, 1940*a*, by permission of the Royal Society.)

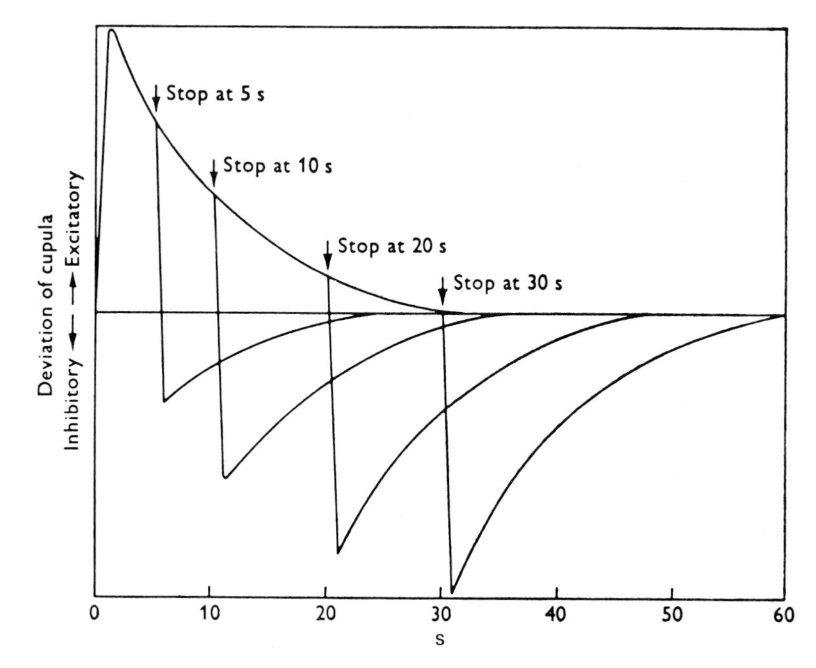

to the same effective stimulus when rotated about A with an angular acceleration α as it is when rotated about B with an angular acceleration $\alpha \cos \theta$. A single canal is therefore incapable of providing precise information as to the nature of the angular acceleration to which it is subjected.

In order to provide such precise information, it is necessary to have three canals set in three planes which intersect in lines that are not parallel to one another, since there are three degrees of freedom in the orientation of an axis in space. Any particular angular acceleration can then be resolved as is shown in fig. 17.9, by dropping perpendiculars from the tip of its vector onto lines perpendicular to the planes in which the semicircular canals lie. It is in accordance with this that there are three semicircular canals in the labyrinths of all jawed vertebrates. It is not essential that the planes of these canals should be mutually at right angles to each other, but the sensitivity of measurement is maximal if they are.

Further aspects of the geometry of the semicircular canals were discussed in earlier editions of this book.

The otolith organs

In addition to the semicircular canals, the labyrinth contains three sacs known as the utriculus, the sacculus and the lagena (fig. 17.5). The receptor cells of these sacs are arranged in groups to produce areas of sensory epithelia called *maculae*. The gelatinous material covering each macula is filled with calcareous granules (*otoconia*); in bony fishes the otoconia are fused to form *otoliths*

Fig. 17.9. How the semicircular canals resolve an angular acceleration about any axis in space into three components. The lines H, P and A represent perpendiculars to the planes of the horizontal, posterior vertical and anterior vertical canals respectively, and are therefore very approximately at right angles to one another. The black arrow is a vector indicating the imposed angular acceleration. Perpendiculars dropped from the tip of this arrow onto H, P and A give the white arrows, which are vectors showing the components of angular acceleration about these axes.

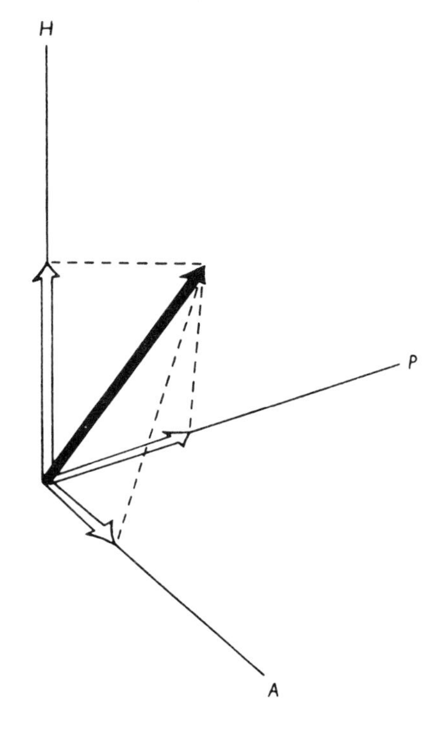

Fig. 17.8. Rotation of a semicircular canal about an axis (A) which is not perpendicular to the plane in which the canal lies. The vector α represents the magnitude and direction of the angular acceleration, and β is the projection of α onto the line B, which is perpendicular to the plane of the canal. β is then the effective acceleration to which the canal responds.

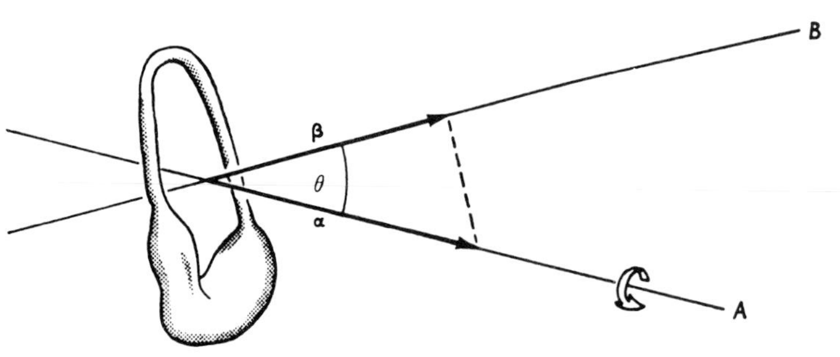

(fig. 17.2*e*). There are typically three otolith organs, the macula utriculi, the macula sacculi and the macula lagenae, but the latter is absent in most mammals.

It seems that the essential stimulus for the receptor cells is a displacement of their sensory hairs, just as in the neuromasts of the lateral line and the cristae of the semicircular canals. In this case such displacement is brought about by movements of the otoconial mass under the influence of gravity or linear accelerations. Consider a flat macula which is horizontal when the head is in its normal position. The otoconial mass exerts a downward force on the cilia of the receptor cells; in the normal position these are upright and there is therefore no lateral force applied to them. If now the head is tilted through an angle of α, then the cilia will be subject to a lateral force proportional to $\sin \alpha$ (fig. 17.10). A system such as this will give a sinusoidal relationship between the angle of tilt and the degree of sensory excitation whatever the initial orientation of the receptor cell is. Now let us see how this model system corresponds with actuality.

The equilibrium function of the otolith organs was investigated electrophysiologically by Lowenstein and Roberts (1950), using single-fibre preparations from the ray, *Raia*. In almost all cases the sensory fibres showed a resting discharge in the normal position, which was increased or decreased by tilting the head. Many units appeared to act as true 'position receptors', with a more-or-less sinusoidal relation between the angle of rotation about one of the horizontal axes and the frequency of the sensory nerve impulses (fig. 17.11). In general, these units were of two types, those which had their positions of maximum activity in the 'side-up' and 'nose-up' positions, and those in which these positions were 'side-up' and 'nose-down'. In addition to these 'static' receptors, some units responded to changes in position irrespective of the direction of the change.

Lowenstein and Roberts (1951) later showed that fibres from certain parts of the otolith maculae in the ray are very sensitive to vibration. The otoliths are thus potentially hearing organs, and the sacculus has been so developed in bony fishes and amphibians. This function is taken over by the cochlea in birds and mammals.

Orientation of the hair cells

The receptor cells of the different components of the acoustico–lateralis system are remarkably uniform in structure. They are all set in an epithelial layer, have ciliary processes projecting from their distal surfaces, and are innervated proximally (fig. 17.1). Each cell (except those of the organ of Corti in the cochlea) possesses two types of ciliary processes: a group of *stereocilia*, which are short and densely packed with actin filaments inside, and a *kinocilium*, which is longer and contains nine peripheral microtubules and two central ones along its length just as in motile cilia. The kinocilium is absent from the mature hair cells

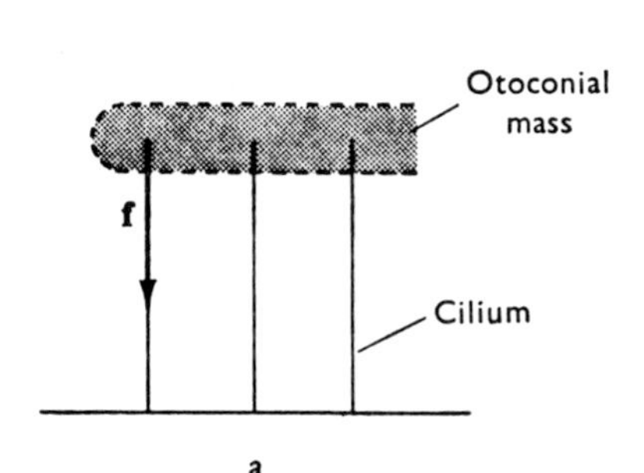

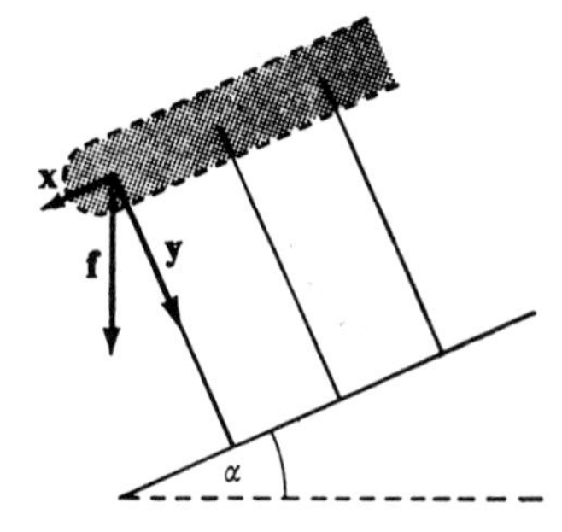

Fig. 17.10. The forces on the sensory cilia of a flat macula when it is horizontal (*a*), and when it is tilted through an angle α (*b*). The vertical force **f** represents the weight of the otoconial mass that is supported by one of the cilia. In *b* this force can be resolved into a component acting along the axis of the cilium (**y**) and a force at right angles to this (**x**). It is clear that $\mathbf{x} = \mathbf{f} \sin \alpha$.

of the organ of Corti, but a basal region (the 'centriole') is still present.

The kinocilium is always placed on one side of the group of stereocilia, and possesses a 'basal foot' which projects from its basal body on the side away from the stereocilia; the sensory hair bundle is thus polarized in a particular direction. From electron microscopy of the sensory cells in fish labyrinths, Lowenstein and his colleagues were able to show that this polarization determines the direction in which the sensory hair bundle must be bent in order to produce excitation. Fig. 17.12 shows the orientation of the hair cells in the labyrinth of the burbot (Wersäll *et al.*, 1965); similar observations were made on the ray (Lowenstein and Wersäll, 1959; Lowenstein *et al.*, 1964). The cells in the crista of any particular semicircular canal are all oriented in the same direction. As mentioned earlier, Lowenstein and Roberts had found that the receptor cells of the vertical canals are excited by movements of the cupula away from the utriculus whereas those of the horizontal canal are excited by movements towards it. In the vertical canals, the kinocilia are positioned on the side of the sensory hair bundle away from the utriculus, but those of the horizontal canal are on the side near to it. This means that the cells are excited when the sensory hair bundle is bent towards the kinocilium and inhibited when it is bent away from it (fig. 17.13).

Fig. 17.11. Discharge frequencies in a single sensory fibre from the utriculus of *Raia* during slow rotation about the longitudinal axis. The continuous curve should be read from left to right, and the dotted curve from right to left. (From Lowenstein and Roberts, 1950.)

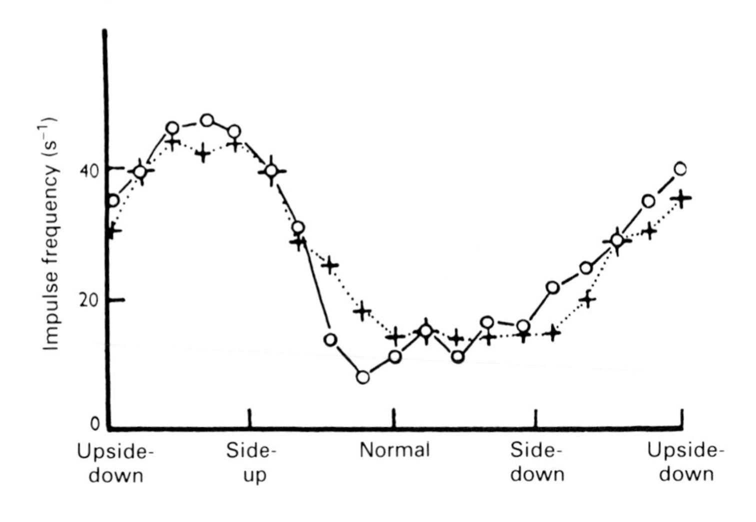

The polarization of the sensory cell in the macula utriculi of *Lota* is shown in fig. 17.12*b*. Notice that there are two principle directions of polarization, set approximately at right angles to each other. This type of arrangement is just what is required to give information about the direction of tilting. It is instructive to compare this arrangement with the findings of Lowenstein and Roberts on the sensitivity of nerve fibres from the utricular macula of the ray. If the receptor cells are essentially unidirectional, one which is excited by both 'side-up' and 'nose-up' deflections will be directed contralaterally backwards at about 45° to the body axis, as are the cells in the anterior peripheral region of the macula utriculi in *Lota*. One which is excited by 'side-up' and 'nose-down' deflections will be directed contralaterally forwards, as are the cells in the posterior peripheral region in *Lota*. We would expect the majority of the cells in *Lota* to be excited by 'side-down' deflections (Flock, 1964) since they are ipsilaterally directed, but this question has yet to be investigated.

In the lateral line, the discharge frequency in any single sensory nerve fibre is increased by movement of the cupula in one direction and decreased by such movement in the opposite direction; each neuromast contains some fibres responsive to movements in one direction and some to movements in the other (p. 368). Flock (1965)

Fig. 17.12. *a*: The orientation of the hair cells in the right labyrinth of the fish *Lota vulgaris*: the labyrinth is viewed from above. The arrows point towards the side of the ciliary bundle on which the kinocilium is situated. *b*: Hair cell orientation in the macula utriculi, shown in more detail. (From Wersäll *et al.*, 1965.)

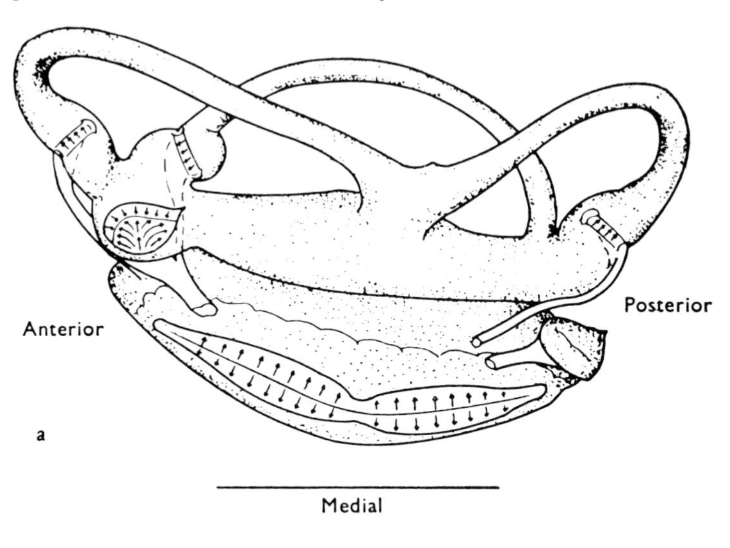

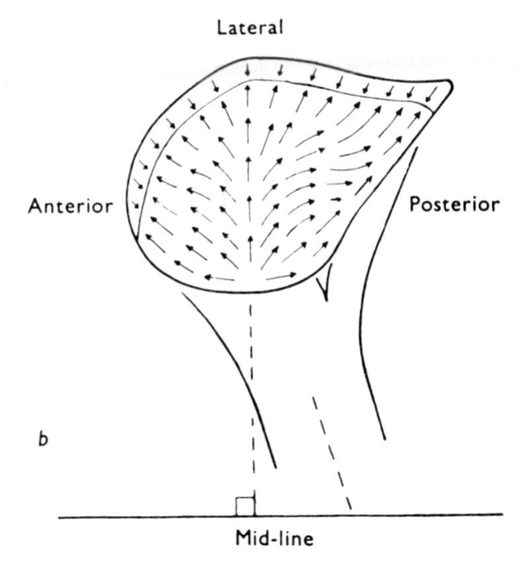

Fig. 17.13. Schematic section through a hair bundle showing an array of stereocilia and a single kinocilium; the position of the basal foot of the kinocilium is also shown. Deflection towards the kinocilium excites the hair cell, deflection in the opposite direction inhibits it. (From Lowenstein *et al.*, 1964.)

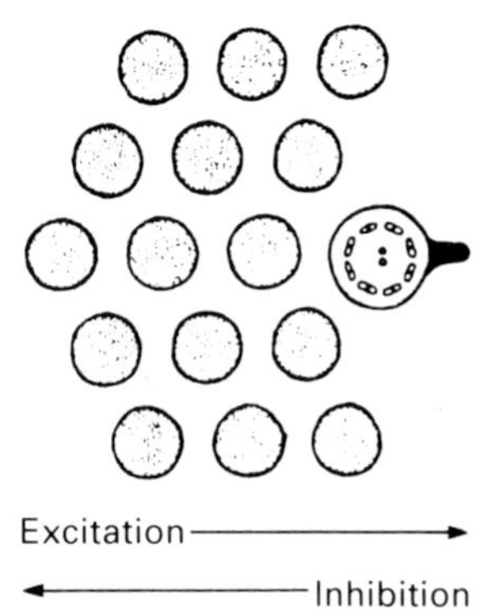

demonstrated the structural counter-part of these observations; adjacent hair cells in the neuromast are oriented with their kinocilia pointed in oppposite directions, either up or down the canal. A similar situation occurs in free neuromasts (Dijkgraaf, 1963).

The centriole in the hair cells of the organ of Corti is always on the outer side of the ciliary bundle.

The mammalian cochlea

The cochlea contains the receptor cells responsible for the perception of sound in mammals. It consists of a tubular outgrowth of the membranous labyrinth, the *scala media*, surrounded by two tubular outgrowths of the osseous labyrinth, the *scala vestibuli* and the *scala tympani*. The structure of this complex, as seen in cross-section, is shown in fig. 17.14. The receptor cells, or 'hair cells' are mounted on the basilar membrane and their sensory hairs are inserted into the tectorial membrane (the whole complex forms the *organ of Corti*); hence we would expect these cells to be excited by movements of the basilar and tectorial membranes relative to each other.

Fig. 17.15 shows how the displacements produced by sound waves pass through the ear. Sound waves entering the external ear cause vibrations of the ear drum, which are transmitted through the air-filled middle ear by a chain of small bones (the auditory ossicles) called the *malleus, incus* and *stapes*. These ossicles form an 'impedance matching' device whereby the forces over the relatively large area of the ear drum (50–90 mm^2 in man) are concentrated on the relatively small (3.2 mm^2) area of the foot-plate of the stapes. This concentration of forces produces an increase in pressure sufficient to ensure that the sound vibrations can be successfully transmitted from the air outside to the liquid perilymph of the inner ear. if there were no such mechanism, and the ear drum were directly backed by liquid, only a very small proportion of the sound energy would cross the air/liquid interface – the rest would be reflected.

The foot-plate of the stapes rests on a small membrane, the *oval window*, which covers a gap in

Fig. 17.14. Semi-diagrammatic transverse section of the organ of Corti in a guinea pig. (From Davis *et al.*, 1953.)

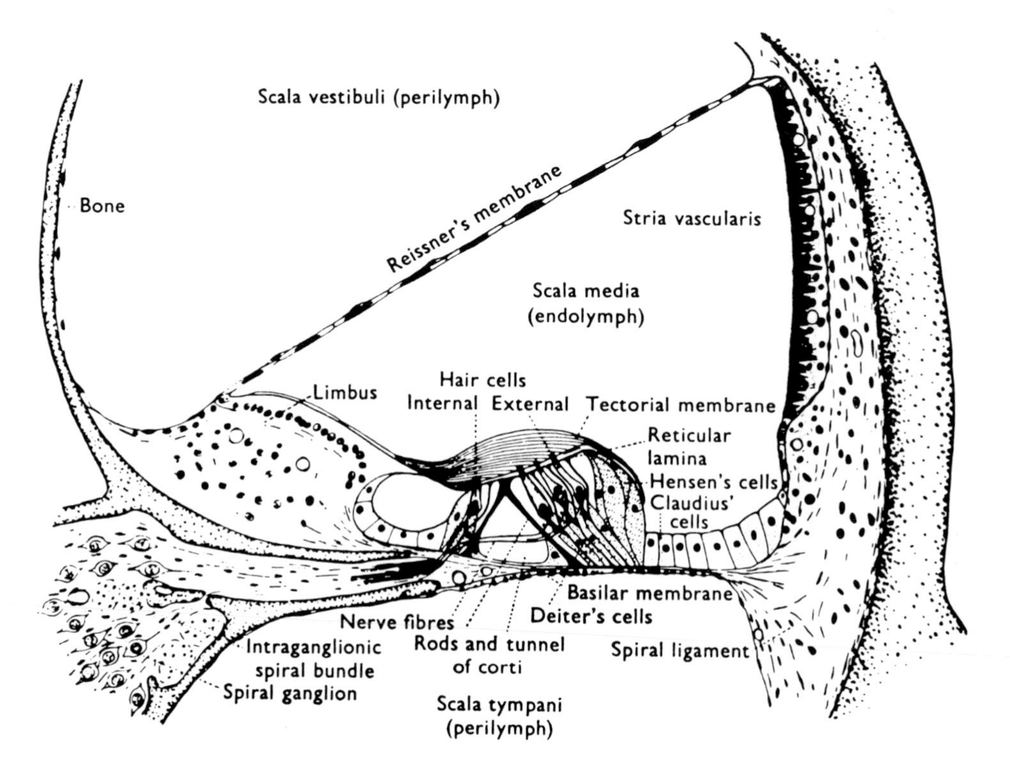

the bone in which the inner ear is set. Behind this is the perilymph of the scala vestibuli, so that an inward movement of the stapes causes a displacement of fluid along the scala vestibuli. This displacement is accompanied by movement of the basilar membrane and a corresponding displacement of fluid in the scala tympani, which results in a bulging of the *round window*, a membrane between the scala tympani and the middle ear. The tectorial membrane moves up and down with the basilar membrane, but there is a lateral shearing action between them which displaces the ciliary processes of the hair cells and so provides the immediate stimulus for their excitation.

The mechanical properties of this system are complex and difficult to investigate. The general problem in cochlear function is how frequency discrimination arises. At moderate sound intensities, for example, we can tell the difference between pure tones of 1000 Hz and those of 1003 Hz (Moore, 1982). In a total frequency range of 20–17 000 Hz this is a high degree of sensitivity. Is such sensitivity in frequency discrimination built into the 32 mm length of the human cochlea? And if so how is it done? Or is frequency discrimination partly or entirely carried out in the higher centres of the auditory processing system in the brain?

These questions began to be asked in the nineteenth century, and the most influential answer was that suggested by Helmholtz in 1863. He assumed that the basilar membrane consisted of transverse fibres under tension, rather like the strings of a musical instrument. Since the width of the basilar membrane varies along its length these fibres would resonate at different frequencies. The cochlea is supplied along its length by a large number of nerve fibres, so it seemed reasonable to suppose that each nerve fibre was excited by the

Fig. 17.15. Schematic diagram of the middle and inner ear in a mammal. The cochlea is shown uncoiled. The arrows show the displacements produced by an inward movement of the ear drum. (From von Békésy, 1962.)

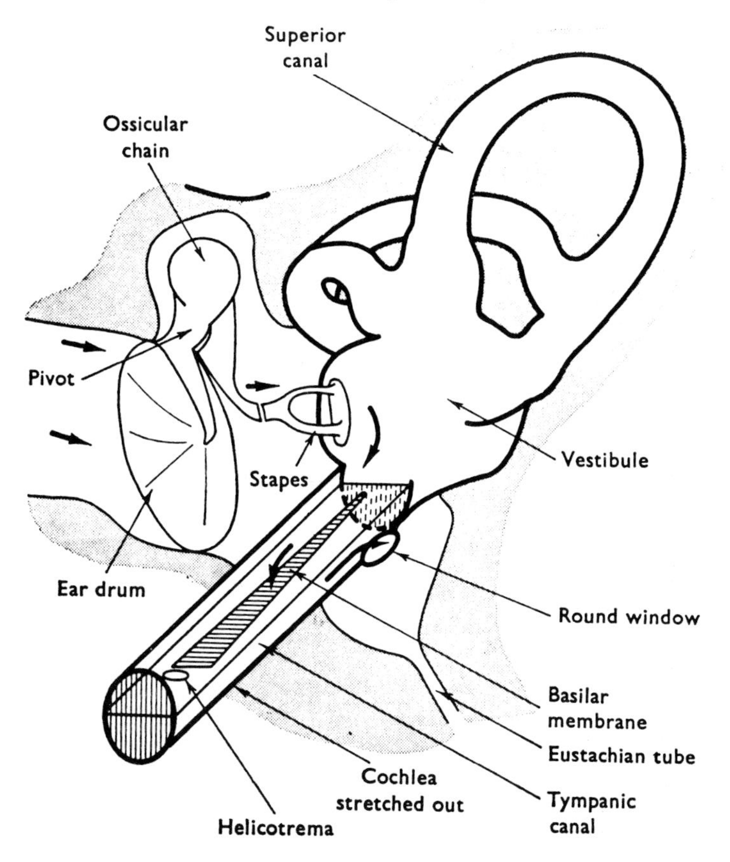

vibration of a particular part of the basilar membrane.

The first measurements of how the basilar membrane actually behaves were made by von Békésy, in an extensive and beautiful series of experiments on the physics of hearing, for which he was awarded the Nobel prize in 1961 (see von Békésy, 1960, 1964). He used cochleas derived from human cadavers and applied vibration at the oval window. He was then able to measure the displacement of different parts of the basilar membrane by placing silver grains on it and observing their movement with the aid of a microscope and a stroboflash. He found that a travelling wave of the same frequency as the sounds moves along the basilar membrane from base to apex, its amplitude varying with the distance along the membrane (fig. 17.16). The position at which the vibration of the basilar membrane reached its maximum varied with sound frequency; it moved nearer to the stapes as the frequency was raised.

Von Békésy's results showed clearly that the basilar membrane responds differentially to different frequencies, and so it must be at least partially responsible for the process of frequency discrimination. However, they also seemed to show that the basilar membrane was not sharply tuned in the way that Helmholtz had suggested; any one point would vibrate in response to a number of different sound frequencies. This implied that frequency discrimination must involve central analysis of the differential excitation of relatively large numbers of receptors.

The first results on the frequency responses of primary auditory nerve fibres were obtained by Tasaki (1954) from the guinea pig cochlea. He found that individual fibres were most sensitive at a particular frequency but that they would respond over a considerable frequency range, in accordance with von Békésy's observations on cochlear mechanics. Much sharper tuning curves were later obtained by Kiang (1965) in cats and Evans (1972) in guinea pigs. They found that any particular fibre is most sensitive at one particular frequency (the 'characteristic frequency'): the threshold rises very rapidly at higher frequencies and rather less rapidly at lower frequencies. Different fibres have different characteristic frequencies (fig. 17.17 shows a selection), and are connected to inner hair cells at different places on the basilar membrane (Liberman, 1982).

The discrepancy between these sharp neural tuning curves and the broad curves produced by von Békésy, which is illustrated in fig. 17.17, was a considerable puzzle. For a time it seemed likely that there was a 'second filter' that in some way could sharpen up the tuning between the basilar membrane and the hair cells (Evans and Wilson, 1975). But further investigations on cochlea mechanics have since provided much of the answer to this problem.

Fig. 17.16. Von Békésy's demonstration of the travelling wave in the cochlea. The full line shows the position of the basilar membrane at one instant during the response to a loud 200 Hz tone, and the broken line shows its position one quarter of a cycle later. The outer broken curves indicate the envelope of the movement. (From von Békésy, 1960.)

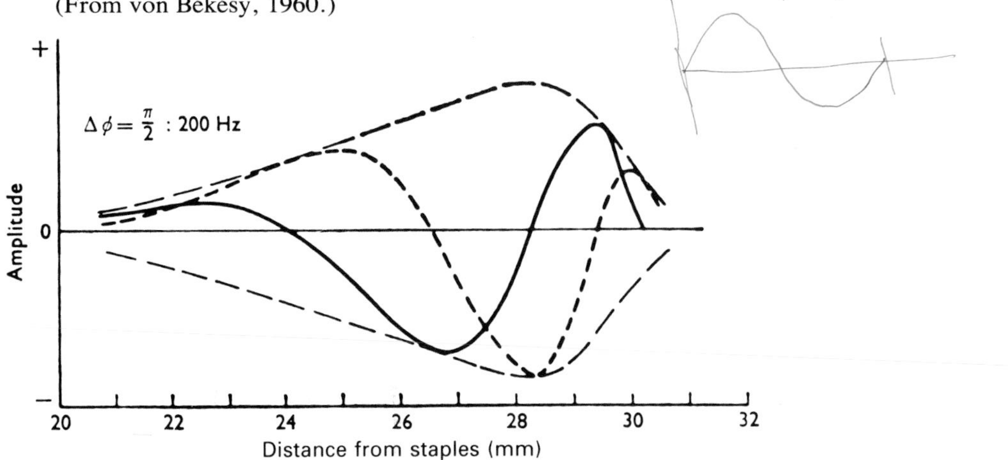

Von Békésy had to use stimuli equivalent to very high sound pressures in order to produce visible movements of the basilar membrane, and could not make measurements at high frequencies. More sensitive techniques have since been developed, and there has been much more emphasis on the use of living cochleas in good condition. Some very fruitful experiments have been performed by Johnstone and his colleagues at the University of Western Australia, utilizing the Mössbauer effect (Johnstone and Boyle, 1967; Johnstone *et al.*, 1970; Sellick *et al.*, 1982).

The principle of movement measurement by the Mössbauer technique is shown schematically in fig. 17.18. A small piece of foil impregnated with ^{57}Co, which produces gamma rays by radioactive emission, is place on the basilar membrane. The gamma rays are measured by a proportional counter, but before meeting the counter they have to pass through a piece of stainless steel foil which is enriched with ^{57}Fe. The energy of the gamma rays is precisely that which is absorbed by the ^{57}Fe in the crystalline structure of the foil, and so only a proportion of them pass through the foil to be recorded by the counter. However, if the source is moving toward the absorbant, the frequency of the gamma rays 'seen' by the absorbant is raised; when it moves away, the frequency is lowered. (This is an example of the Doppler effect.) These changes in frequency mean that the energies of the gamma rays are no longer so precisely matched to the ^{57}Fe atoms, and so fewer of them are absorbed. Thus over a limited range (up to about $\pm 1\ \mathrm{mm\,s^{-1}}$) the output of the proportional counter increases with increasing velocity; and the velocity of the vibration is proportional to amplitude.

First results with the Mössbauer technique were largely compatible with von Békésy's measurements. But then doubts arose with the discovery of nonlinearities in the behaviour of the basilar membrane at the point of maximum displacement: its movement was relatively greater, in relation to the movement of the malleus, at lower sound pressure levels (Rhode and Robles, 1974). These nonlinearities disappeared after death, an effect paralleled in the neural tuning curve, which

Fig. 17.17. Sharp tuning of neural responses from the cochlea. The upper set of curves show how the threshold varies with sound frequency for eight different primary afferent fibres in the cochlear nerves of guinea pigs. There is a characteristic frequency for each fibre at which it is most sensitive, with the threshold intensity rising sharply at lower and higher frequencies. The dotted curves show tuning curves for the basilar membrane as they appeared to be in 1972, when this diagram was first published. More recent mechanical response curves are shown in fig. 17.19. (From Evans, 1972.)

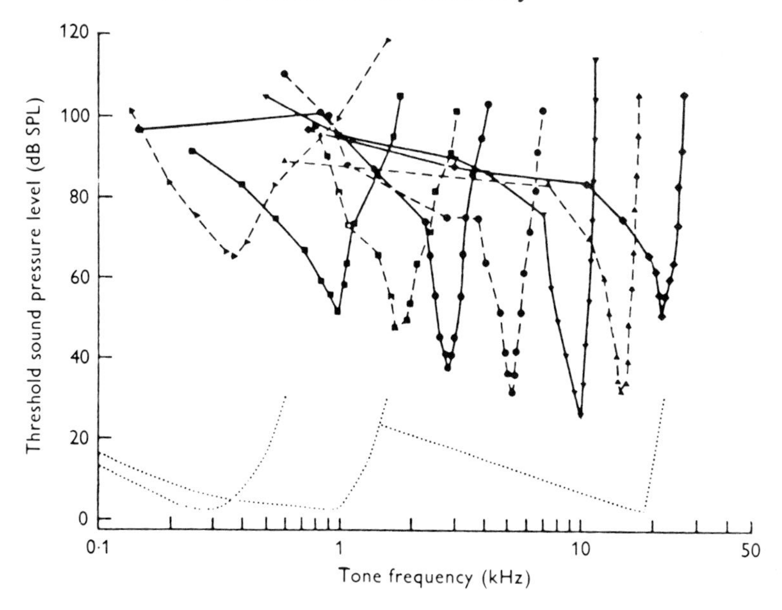

becomes much less sharp and loses its sensitive 'tip' when the oxygen supply is reduced (Evans, 1974).

Definitive results with the Mössbauer technique have been produced by Sellick *et al.* (1982). They showed that the basilar membrane is indeed finely tuned in the living cochlea, and the mechanical tuning curve is similar in shape to that of individual auditory nerve fibres (fig. 17.19). This implies that, for soft sounds, the travelling wave on the basilar membrane shows a sharp peak in amplitude at a particular place determined by the frequency of the sound.

A remarkable property of the cochlea was discovered by Kemp (1978). He found that application of a brief click is followed by sound output from the cochlea itself for several milliseconds afterwards. The frequency spectrum of the emitted sounds is very similar to that of the stimulus, suggesting that the sounds are generated at sites along the cochlea which correspond to their frequencies (Norton and Neely, 1987). It may be that the phenomenon is connected with the sharpness of tuning of the basilar membrane; one possibility is that these active responses are in some way produced by the outer hair cells.

There are about 25 000 hair cells in the human cochlea; in the cat there are about half this

Fig. 17.18. How the Mössbauer effect can be used to measure velocities. The ^{57}Co atoms which form the γ-ray source are part of the crystal lattice of a small piece of steel foil. The γ-rays emitted are then all of nearly the same frequency (this is what Mössbauer discovered). The absorber is another piece of steel foil which contains ^{57}Fe atoms in its crystal lattice; these absorb γ-rays over precisely the same narrow frequency band. If the γ-ray source is moving away from or towards the absorber, the frequency of the γ-rays hitting it will be lowered or raised, and so less absorption will occur. A γ-ray counter on the far side of the resonance absorber will then produce a higher count rate than when the source was not moving. *a* shows how the components of the apparatus are arranged; *b* shows how the count rate varies with the velocity of the source.

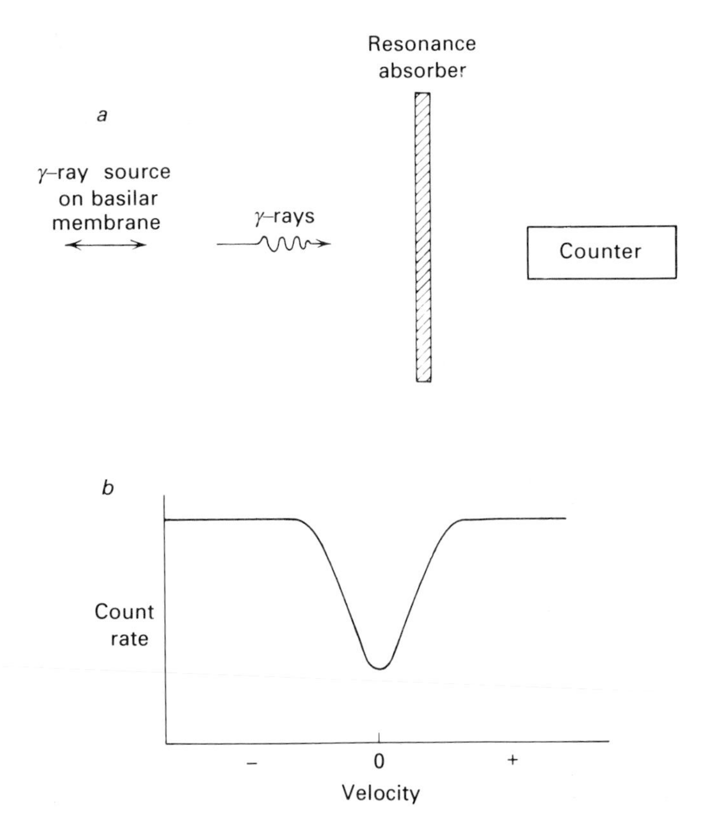

number, supplied by about 50 000 nerve fibres. Much new information about the innervation pattern was provided by Spoendlin in a study of the cat cochlea by electron microscopy (see Spoendlin, 1975, 1984). He was able to distinguish the afferent and efferent nerve fibres by sectioning part of the auditory nerve; the efferent fibres in the cochlea, cut off from their cell nuclei, would degenerate and disappear after two weeks or so, leaving the afferent innervation intact.

Spoendlin's conclusions, illustrated in fig. 17.20, were surprising and very interesting. Although there are three times as many outer hair cells as there are inner hair cells, 95% of the afferent nerve fibres are connected to inner hair cells. Each inner hair cell is connected to about 20 unbranched afferent neurons. Outer hair cell afferents branch to contact about 10 separate outer hair cells, and each outer hair cell is contacted by about four afferents.

This disparity in the afferent innervation of the inner and outer hair cells is reflected in recordings of single fibre responses to sound. Thus

Fig. 17.19. Sharp tuning in the basilar membrane, as determined by the Mössbauer technique. The continuous lines show the sound intensities required to produce standard movements (a displacement of 3.5 Å or a velocity of 40 $\mu m\,s^{-1}$) of a particular point on the basilar membrane of a guinea pig at various frequencies. The dotted line shows a similar fine tuning curve for the response of a nerve fibre with a similar characteristic frequency. (From Sellick *et al.*, 1982.)

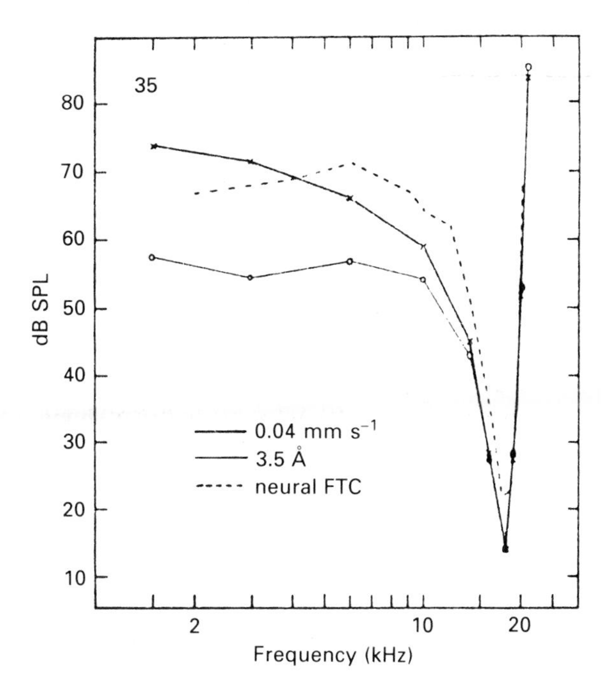

Fig. 17.20. Innervation of the cat organ of Corti. The afferent neurons are shown by full lines, originating on the left of the diagram. Efferent neurons are shown by broken lines on the right. Most of the efferent neurons arise in the opposite (contralateral) side of the brain. (From Spoendlin, 1975.)

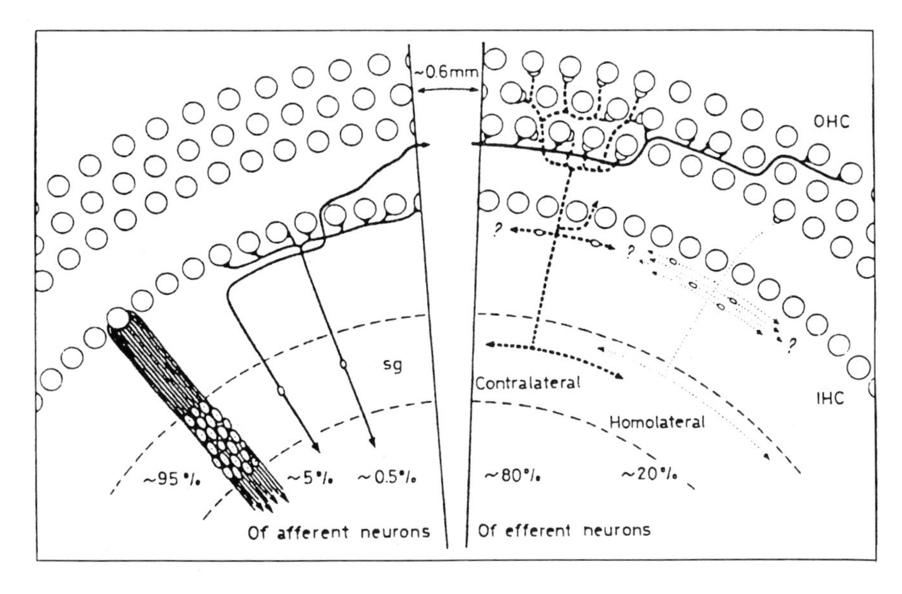

Liberman (1982) recorded from 56 different primary afferents with an intracellular electrode, and in each case he then injected horseradish peroxidase through the electrode to label the fibre for microscopic analysis. All 56 fibres innervated inner hair cells. This suggests that all the tuning curves obtained by Kiang and by Evans (fig. 17.17) were similarly from afferent fibres connected to inner hair cells. Nothing is known about the activity or functions of the afferent neurons that innervate the outer hair cells (Kiang *et al.*, 1984).

Spoendlin's degeneration experiments allowed him to look again at the normal cochlea and so see which neurons were efferent. He found that efferent neurons branch to supply a number of outer hair cells, each of which is contacted by a number of efferent nerve terminals. Separate neurons innervate the afferent dendrites from the inner hair cells (fig. 17.21).

Spoendlin's results led to some hard thinking about cochlear function. It seemed particularly odd that the outer hair cells should be connected to such a small proportion of the afferents and yet should be so well supplied with efferents. Stimulation of the crossed olivocochlear bundle (the efferent pathway running from the brainstem to the cochlea, whose neurons end on the outer hair cells) causes tuning curves to become less sensitive at their tips (Kiang *et al.*, 1970; Brown *et al.*, 1983). Hence it may be that the outer hair cells are primarily

Fig. 17.21. Innervation of cochlea hair cells. (From Spoendlin, 1984.)

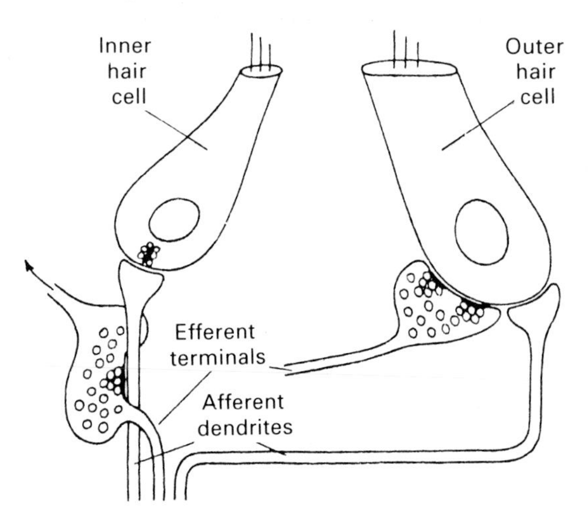

important in connection with the fine tuning of the basilar membrane, perhaps by some form of active response to sound (see Pickles, 1985; Russell *et al.*, 1986).

In accordance with this idea, it has been found that isolated outer hair cells change their length when current is passed through them (Brownell *et al.*, 1985; Ashmore and Brownell, 1986). Just how this surprising effect is involved in cochlear mechanics has yet to be worked out.

How do the hair cells work?

We have seen that deflection of the sensory hair bundle towards its kinocilium side increases the frequency of action potentials in the afferent nerve fibres which contact the hair cell, and movement in the opposite direction reduces it (fig. 17.13). Corresponding excitatory and inhibitory changes are seen on intracellular recording from hair cells. Thus movement of the hair bundle towards the kinocilium produces depolarization and movement away from it produces hyperpolarization, both in lateral line cells (Flock, 1971) and in bullfrog sacculus (Hudspeth and Corey, 1977), as is shown in fig. 17.22. It seems reasonable to suppose that these receptor potentials control the release of chemical transmitter from the hair cells at their synapses with the afferent nerve fibres (see Furukawa *et al.*, 1978).

An electrode placed in the mammalian cochlea will record an oscillating potential of the same frequency as the sound stimulus when the ear is stimulated by a pure tone in the audible range. This is called the cochlear microphonic potential. Since it is largest in the immediate vicinity of the hair cells, it is very probable that it is produced by them and is therefore a receptor potential (Tasaki *et al.*, 1952, 1954). Intracellular measurements from inner hair cells by Russell and Sellick (1978) show receptor potentials with an a.c. component corresponding to the cochlear microphonic superimposed on a maintained depolarization. Tuning curves for these responses were sharp for soft sounds (i.e. depolarization only occurred at sound frequencies very near to the characteristic frequency of the cell) but broader for loud ones.

A crucial question about hair cells is the nature of the transducer mechanism (Hudspeth, 1983, 1985): what is the causal link between deflection of the sensory hair bundle and the production of the receptor potential? Using microprobe stimulation of bullfrog saccular hair cells, Hudspeth and Jacobs (1979) found that movement of the kinocilium alone produced no receptor potential, whereas movement of the stereocilia alone (the kinocilium was held down flat by a microneedle) produced normal receptor potentials. So it seemed likely that the transducer mechanism is associated in some way with the stereocilia.

The stereocilia are rooted in an electron-dense structure called the cuticular plate, from which they emerge in a hexagonal array. They are more or less constant in diameter for most of their length but taper sharply at the base. They are packed with longitudinal actin filaments, all oriented in the same direction and cross-linked (Flock and Cheung, 1977; Tilney *et al.*, 1980). This means that they are stiff structures which move essentially as rigid rods with basal pivots.

There is some variation in the numbers and dimensions of the stereocilia in different hair cells. In the chick cochlea, for example, the hair cells at the distal end have about 50 stereocilia, the longest of which are 5.5 μm long and 0.12 μm thick, whereas those at the proximal end have about 300 stereocilia, with the longest 1.5 μm by 0.2 μm (Tilney *et al.*, 1983). Corresponding differences occur between hair cells in different organs and different species.

The stereocilia are grouped in ranks or rows across the hair bundle, with each rank containing stereocilia of a different length. The shortest are in the rank on the side away from kinocilium (or the centriole in the mammalian cochlea), those in the next row are longer, and so on until we reach the tallest row next to the kinocilium. There are filamentous cross-links between adjacent stereocilia just above the basal tapered region; perhaps these account for the fact that the hair bundle moves as a whole when it is pushed by a microprobe (Flock, 1977). There is also a fine filament rising upwards from the tip of each sterocilium to the side of its taller neighbour (Pickles *et al.*, 1984; Furness and Hackney, 1985). The ranked arrangement of the stereocilia means

Fig. 17.22. Receptor potentials recorded from bullfrog saccular hair cells with an intracellular microelectrode. The hair bundles were exposed by removal of the otolith and stimulated by a vibrating probe. The three upper traces in *a* show the responses to the 10 Hz triangular wave stimulus monitored in the lower trace. The topmost trace, produced by a low-amplitude stimulus, follows the waveform well, but the higher-amplitude stimuli produce receptor potentials which show saturation effects, as seen in the next two traces. The graph *b* shows how the amplitude of the responses varies with that of the stimulus. (From Hudspeth and Corey, 1977.)

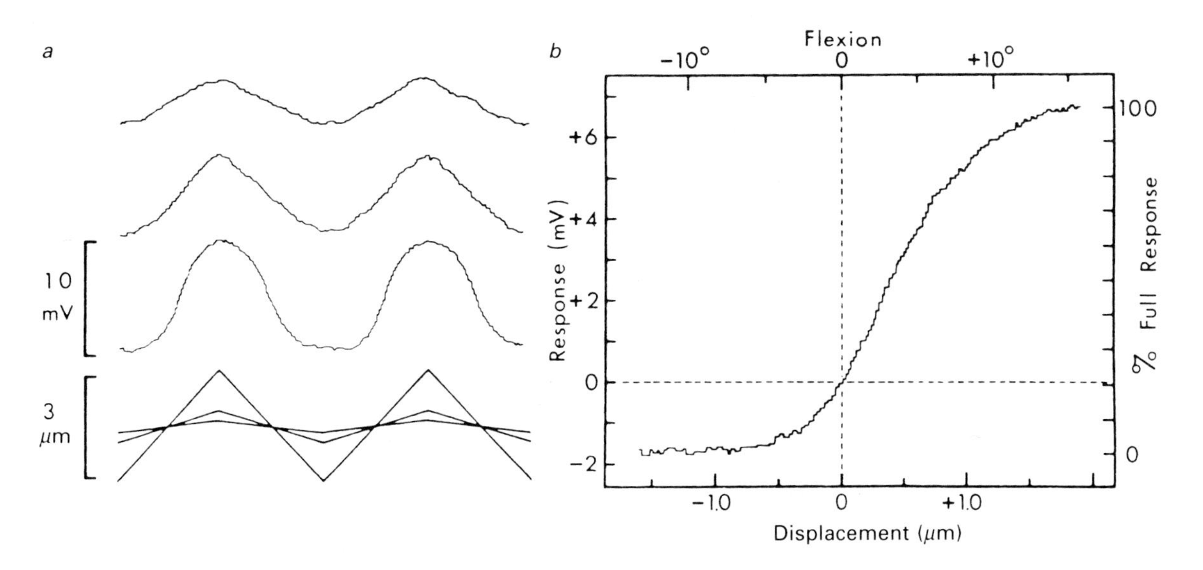

that these upward-pointing filaments are aligned with the axis of directional sensitivity of the cell, hence they will be stretched when the bundle is moved in the excitatory direction (fig. 17.23).

We might expect that the receptor potential is produced by the opening of ionic channels somewhere in the hair cell membrane as a result of movement of the stereocilia. Hudspeth (1982) used an extracellular electrode to map the electrical field around bullfrog sacculus hair cells. Rather surprisingly, he found that current flow is strongest near the tip of the hair bundle, suggesting that the transducer channels are located at or near the tips of the stereocilia. This, together with the observations of the upward-pointing filaments, led to a model for transduction in hair cells that is illustrated in fig. 17.24 (Hudspeth, 1985). The idea is that the transducer channels in the tips of the shorter stereocilia are more likely to be open if the upward-pointing filament is under tension, which will occur when the hair bundle is moved towards the taller stereocilia.

The characteristics of the transducer channels have been investigated by Holton and Hudspeth (1986), using the ensemble-variance method of noise analysis which we looked at in chapter 6. They subjected a bullfrog saccular hair cell to whole-cell voltage clamp by using a gigaohm-seal patch-clamp electrode, and stimulated it by applying a vibrating probe to the hair bundle (fig. 17.25). The vibration was sinusoidal and its amplitude (1 μm peak to peak) was large enough to produce an inward current which alternated between zero and a saturating maximum (fig. 17.26*a*).

The variance of this current, measured from a large number of successive cycles, was largest when the current was about half-maximal (fig. 17.26*b*). This is not too surprising since we would expect the variance to be low when all the channels are closed (giving zero current) and also when all of them are open (maximal current). Fig. 17.26*c* plots the variance against the current for the two halves of the cycle; the results are in good agreement with the parabolic relation predicted by equation (6.2), with the number of channels (N) equal to 160 and the individual channel current (i) equal to 1.0 pA. The single channel conductance was estimated as 11.8 pS.

Fig. 17.23. Schematic sections through the hair bundles of cochlear hair cells to show the position of the upward-pointing links, here called transduction links to indicate their suggested function. (From Pickles *et al.*, 1984.)

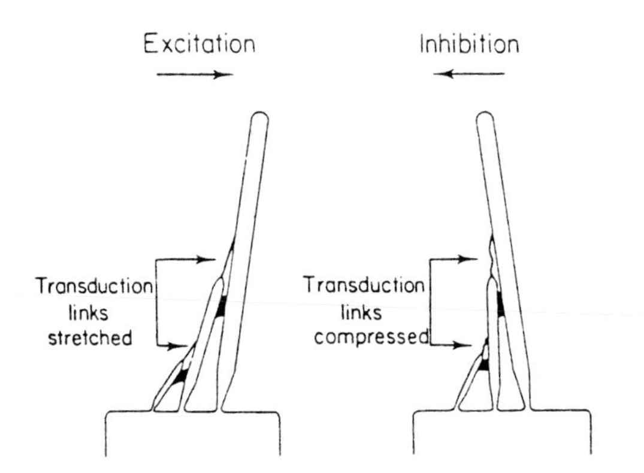

Fig. 17.24. A model for mechanoelectrical transduction in hair cells. At any instant each transduction channel at the tip of a stereocilium may be either closed (*a*) or open (*b*). The rate constants for channel opening and closing are k_{12} and k_{21} respectively, so the fraction of open channels is $k_{12}/(k_{12} + k_{21})$. When the hair bundle is deflected by an excitatory stimulus (*c* and *d*), the rate constant for opening increases and that for closing decreases, so the fraction of channels open increases, and the cell is depolarized. Deflection in the opposite direction has the opposite effect on the rate constants, so the fraction of open channels is reduced and the cell becomes hyperpolarized. (From Hudspeth, 1985.)

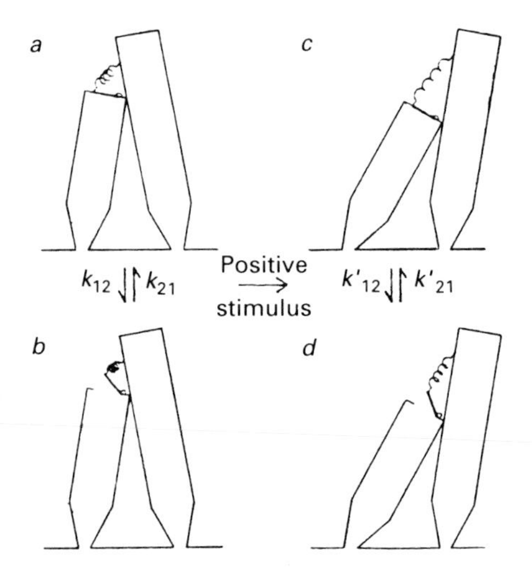

Data from different cells in these experiments gave values of N between 30 and 160 (the lower values may have been a result of damage during the preparation), but values of γ were all close to 12 pS. Thus, with about 40 stereocilia linked to their taller neighbours by upward-pointing filaments, it looks as though there are about four channels located near the tip of each stereocilium.

How sensitive are the hair cells? We can attempt to answer this question by estimating the movement of the stereocilia at the behavioural threshold. The first calculations in this area were made on the cochlea, using linear extrapolation from von Békésy's measurements. At a sound intensity of 1000 dyn cm^{-2} (a very loud sound, 134 dB above threshold) the amplitude of the basilar membrane vibration was about 10^{-4} cm or 10 000 Å. At the standard threshold level of 0.0002 dyn cm^{-2}, therefore, we might expect the amplitude to be $10\,000 \times 0.0002/1000$ Å, i.e. 0.002 Å. If this calculation is correct, the ear should be able to detect sound when the basilar membrane moves by as little as 1/300th the diameter of a hydrogen atom. This 'astonishing and baffling conclusion', as von Békésy and Rosenblith (1951) described it, was current for some years.

We now know, however, thanks to the Mössbauer technique, that the amplitude of the basilar membrane movement is not a linear function of the sound pressure level, a conclusion which invalidates the calculation given above. These later results suggest that the absolute threshold corresponds to a basilar membrane movement of about 1 Å or a little less (Sellick *et al.*, 1982; Bialek, 1987). This is still a very small distance, but one which is in the atomic rather than the subatomic range. The amount of shear between the hair cells and the tectorial membrane is probably much the same as the displacement of the basilar membrane. In the basal turn of the cochlea the tallest stereocilia are about 3 μm in length, so a 1 Å movement corresponds to an angular deflection of about 2×10^{-3} degrees.

The threshold for detection of angular acceleration in man is about 0.1 degree per square second. Hudspeth calculates that this corresponds to an angular deflection of the semicircular canal stereocilia by $2–7 \times 10^{-3}$ degrees, a similar figure to that obtained for the cochlea.

Fig. 17.25. Arrangement for fluctuation analysis in a bullfrog saccular hair cell. The vibrating stimulus probe is applied to the side of the hair bundle; it is shown in contact with a swelling at the top of the kinocilium. The recording patch electrode is applied to the apical surface of the cell. Suction forms a gigaohm seal round the edge of the electrode and then breaks the cell membrane within the seal, so that current flow across the whole of the rest of the cell surface can be measured. The electrode forms part of a voltage-clamp circuit which holds the membrane potential constant. (From Holton and Hudspeth, 1986.)

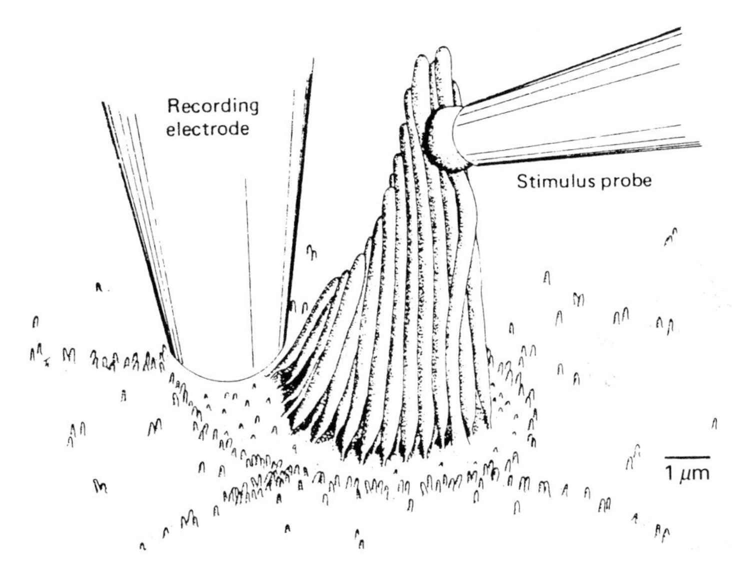

Finally, we may ask what the kinocilium is for. It may well be vitally important in the morphogenesis of the hair cell, since its position and orientation indicates precisely the directional sensitivity of the cell. In some cells the kinocilium may serve to help couple the hair bundle to the surrounding accessory structures.

Mammalian muscle spindles

The afferent nerve fibres from mammalian limb muscles are of a number of different types (table 4.2), distinguishable according to their diameters and the nature of their sensory endings. There are two main types of receptor in the muscles: (i) the muscle spindles, which consist of small modified muscle fibres and are innervated by group Ia (primary) and group II (secondary) fibres and by motor fibres of the γ system, and (ii) the Golgi organs in the tendons, which are innervated by group Ib fibres. In addition there are a number of small diameter fibres (groups III and IV) which have free or encapsulated endings and may be responsive to pressure or pain.

This pattern of innervation can be illustrated by reference to the soleus muscle of the cat; Matthews (1964) combines the results of several workers to give the following figures. There are about 50 spindles, with the same number of primary

Fig. 17.26. Fluctuation analysis of bullfrog sacculus hair cell responses. *a* shows the mean receptor current during one cycle of a relatively large vibratory stimulus, averaged over a large number of cycles. The dashed line shows zero current. *b* shows the variance of the current over the same period of time. *c* shows the variance-mean plot derived from *a* and *b*; open symbols are for the first half of the cycle and closed symbols for the second half. (From Holton and Hudspeth, 1986.)

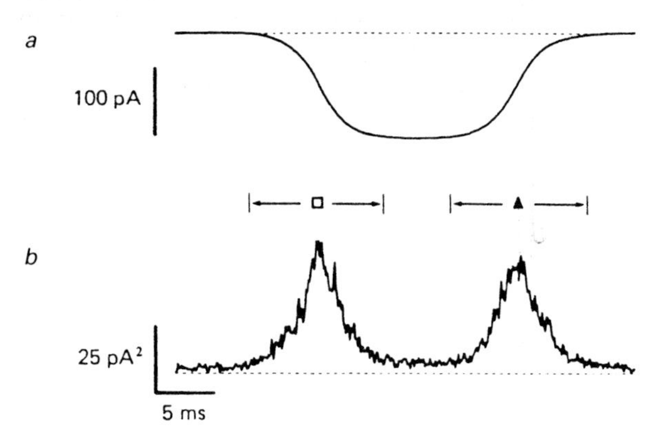

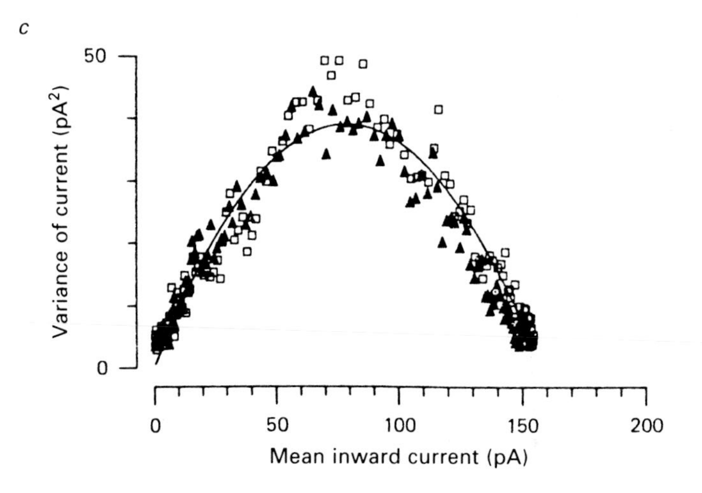

endings and 50–70 secondary endings. There are also about 45 tendon organs, but individual group Ib fibres may innervate more than one tendon organ. A few group III and group IV afferents occur. The spindles contain a total of about 300 intrafusal muscle fibres, supplied by about 100 fusimotor nerve fibres of the γ system. Finally, there are about 25 000 extrafusal muscle fibres innervated by about 150 α motor nerve fibres.

Before examining the details of muscle spindle action, it would be as well to establish some general ideas as to how we might expect them to act under various conditions. Fig. 17.27 shows, in a schematic and much simplified fashion, the mechanical relations between the extrafusal muscle fibres, the muscle spindles and the Golgi tendon organs. Notice that the tendon organs are in series with the extrafusal muscle fibres, whereas the muscle spindles are in parallel with them: these positions are just right for the tendon organs to act as tension receptors and the spindles to act as length receptors.

Our knowledge of the workings of mammalian muscle receptors begins with the work of B. H. C. Matthews (1933), and it is instructive to relate his results (some of which are shown in fig. 17.28) to the scheme shown in fig. 17.27. If the resting muscle is passively stretched, then both the muscle spindles and the tendon organs will be stimulated (figs. 17.28*a* and *b*), the former by the stretch itself and the latter by the consequent increase in passive tension. If the α fibres are stimulated so that the muscle contracts, the tendon organs will be stimulated by the increase in tension (fig. 17.28*c*) and the muscle spindle will shorten (even in an 'isometric' contraction there will be some shortening because of the compliance of the tendons) and so cease to fire (fig. 17.28*d*). Stimulation of the γ fibres in an otherwise resting muscle will stretch the equatorial regions of the intrafusal muscle fibres and so produce excitation of the muscle spindle afferents (fig. 16.23).

So much for the basic outlines of muscle spindle physiology. Detailed investigation has shown that muscle spindles are remarkably complex organs, whose study is no easy matter. In the sixties there were divergent views about the structure of spindles and how it related to their physiology. More recently, the conflicts have been largely

Fig. 17.27. Simplified diagram of the innervation of a mammalian muscle and muscle spindle. Contractile regions are shaded.

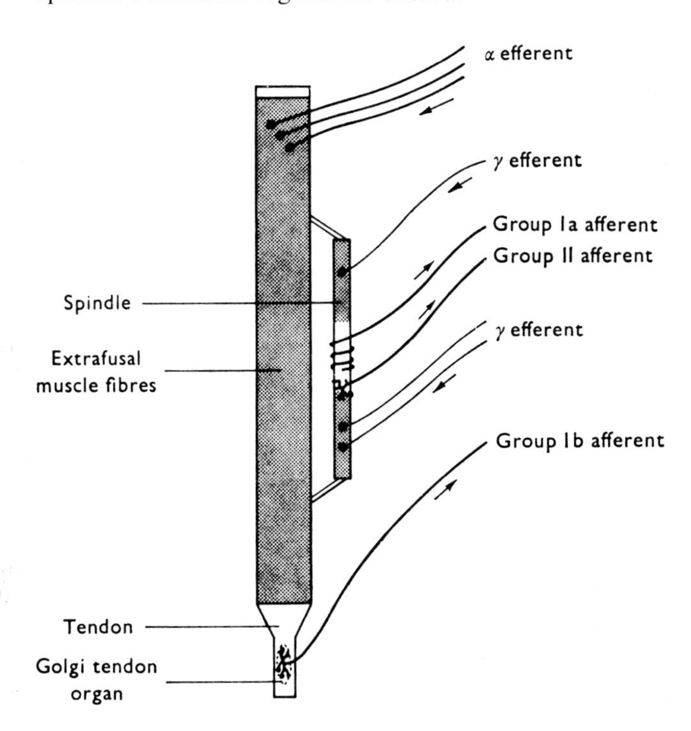

resolved and the participants have been able to look back and see that each was partly right (Boyd, 1981; Barker and Banks, 1986; also Matthews, 1981). Let us look at some of the details.

Structure

The most comprehensive studies on muscle spindles are of those in cat hind-limb muscles. What follows relates principally to these, but the spindles of other mammals, including man, are not very different.

Each spindle contains five to ten intrafusal muscle fibres, held together centrally in a fluid-filled capsule. Two different types of muscle fibre were distinguished in the fifties (Cooper and Daniel, 1956; Boyd, 1962). The *nuclear bag* fibres are about 8 mm long and their equatorial regions are swollen and contain numbers of nuclei clustered together. The *nuclear chain* fibres are narrower than the nuclear bag fibres and about half their length; their equatorial regions are not swollen but contain a single string of nuclei. Myofibrils are almost absent from the equatorial regions in both cases, specially in the nuclear bag fibres. Later it became clear that there are two types of nuclear bag fibres, one of them ('bag_1') shorter and thinner than the other ('bag_2'); there are also histochemical differences between the two types (Ovalle and Smith, 1972; Banks *et al.*, 1977). Most spindles contain two nuclear bag fibres, one of each type, and four or five nuclear chain fibres.

Can we relate these intrafusal fibre types to the different types of extrafusal fibre? Rowlerson *et al.* (1985) have used immunohistochemistry to investigate the myosin isoforms of intrafusal fibres. They found that bag_1 fibres possessed myosin very like that of tonic fibres, and that the myosin of chain fibres was similar to that of embryonic and

Fig. 17.28. Responses of sensory fibres in cat leg muscles to passive stretch (*a* and *b*) and to twitch contractions of the muscle (*c* and *d*). *a* and *c* show responses of tendon organ endings; *b* and *d* show responses of spindle endings. The thick lines show nervous activity, the thin traces show tension changes in the muscle. Time scales 0.1 s. (From Matthews, 1933, redrawn.)

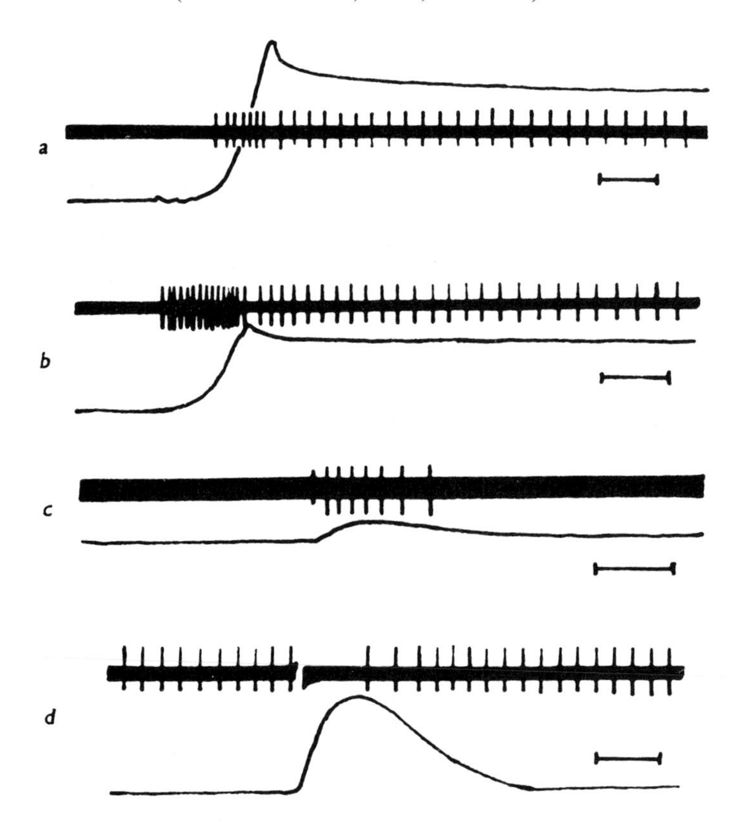

neonatal fast-twitch muscles. Myosin in bag_2 fibres seemed to be a distinct type, probably with some similarities to that of slow-twitch muscles. Bag_1 fibres show other similarities to extrafusal tonic fibres: the *M* line is absent and the sarcoplasmic reticulum is relatively sparse.

It was shown by Ruffini in 1898 that muscle spindles possess two types of sensory endings; later work has provided more detail but no cause to revise this conclusion (Banks *et al.*, 1982). Each spindle receives one group Ia afferent nerve fibre and from none to five group II fibres. The group Ia fibres end at the *primary endings*, where they wind round the equatorial regions of the intrafusal fibres to form 'annulospiral' terminations. All nuclear bag fibres have primary endings on them, and almost all nuclear chain fibres do. *Secondary endings*, derived from group II fibres, are found on either side of the primary endings on nuclear chain fibres; occasionally they also occur in a similar position on nuclear bag fibres. Most of the secondary endings in the cat are of the 'annulospiral' type, but there are some 'flower-spray' endings, in which the fibres split up into a number of branches which have slight thickenings at their ends.

The efferent nerve fibres supplying the muscle spindle are called fusimotor fibres. A number of these innervate extrafusal as well as intrafusal muscle fibres; they are known as β fibres and have terminals described as 'plate (p_1)' endings on the intrafusal muscle fibres. The γ fibres, which are exclusively fusimotor, have two types of terminal: 'plate (p_2)' and 'trail' endings. Plate endings form distinct end-plates, the p_2 endings being larger than the p_1 ones. Trail endings are more diffuse endings in which the fibre splits up into a number of terminals. γ-Plate axons innervate bag_1 fibres and γ-trail axons innervate bag_2 or chain fibres, or both. β axons replace γ axons on either bag_1 or long-chain fibres in some spindles. Fig. 17.29 summarizes the situation.

Physiological properties

The properties of the primary and secondary endings were investigated by Cooper (1961) and Bessou and Laporte (1962), among others, by recording the responses of single afferent fibres from the limb muscles of anaesthetized cats. The group Ia (primary ending) and group II (secondary ending) afferents could be distinguished by their

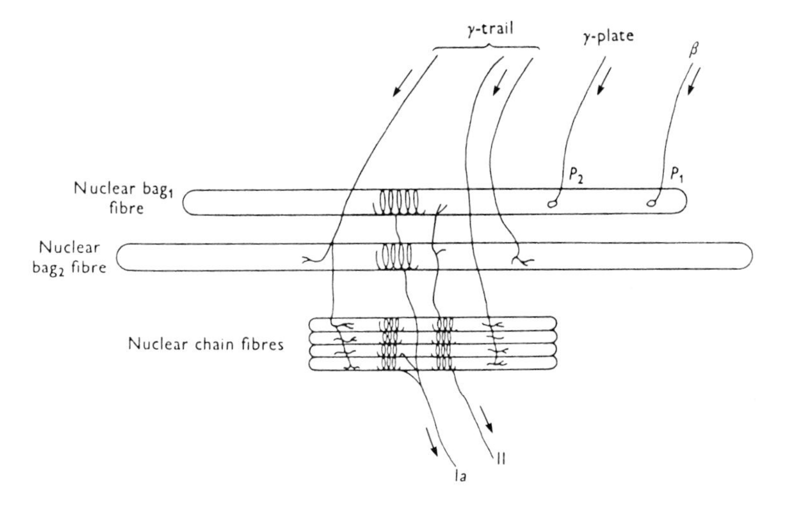

Fig. 17.29. Schematic diagram showing the typical innervation pattern of a mammalian muscle spindle. Sensory nerve axons (a group Ia fibre from the primary ending and a group II fibre from the secondary ending) are shown emerging below; the number of secondary endings per spindle varies from nought to five. Fusimotor axons are shown above; axons with γ-plate terminals are now often called γ-dynamic axons and those with γ-trail endings are called γ-static axons. Similarly, the β axon as shown may be called a β-dyamic axon; sometimes β-static axons occur, innervating the nuclear chain fibres. Based largely on Boyd *et al.* (1977), and Boyd and Gladden (1986).

different conduction velocities. Both types of ending produced more action potentials at longer lengths, but only the primary ending gave a high-frequency burst of impulses when the muscle was being stretched (fig. 17.30). The primary ending is very sensitive at the start of a movement, less so as it proceeds; this nonlinearity means that even quite small movements are readily detected.

Hunt and Otterson (1975) recorded receptor potentials of primary and secondary endings in isolated cat muscle spindles. They used extracellular electrodes and blocked the action potentials with tetrodotoxin; thus their experiments were similar in principle to those by Katz on frog muscle spindles, which we examined in the previous chapter. The receptor potentials of primary endings showed a marked dynamic component: depolarization was greater during a stretch than after its completion. This effect was especially evident at the beginning of a stretch, in line with the high sensitivity. In secondary endings the dynamic component of the receptor potential was much smaller (fig. 17.31). Thus the nervous output of the two types of sensory ending is largely predictable from their receptor potentials.

Why should the receptor potentials of the two endings differ in their dynamic responsiveness? We might look for an answer in the mechanical properties of the intrafusal muscle fibres. This is likely to be more complicated than in extrafusal fibres, since the quantities of myofibrils and of elastic material vary along their length. Furthermore, there is evidence for some unexpected effects of stretching the bag fibre. Thus Poppele and Quick (1981), using cinematography to look at isolated spindles, found that the sarcomere length in the region next to the sensory endings would actually decrease during a stretch. The mechanism of this stretch activation is probably rather different

Fig. 17.30. Diagram summarizing the responses of primary and secondary endings to various length changes. (From Matthews, 1964.)

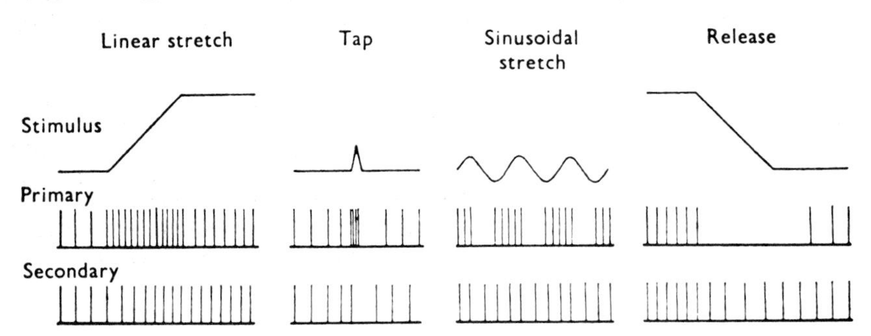

Fig. 17.31. Receptor potentials from an isolated cat muscle spindle in response to stretching at three different velocities. Records were made with extracellular electrodes from the afferent fibres close to their sensory endings, with tetrodotoxin to prevent action potential production. The upper trace shows the response of the primary ending, the middle trace that of a secondary ending, and the lower trace monitors the length change applied to the spindle. (From Hunt and Ottoson, 1975.)

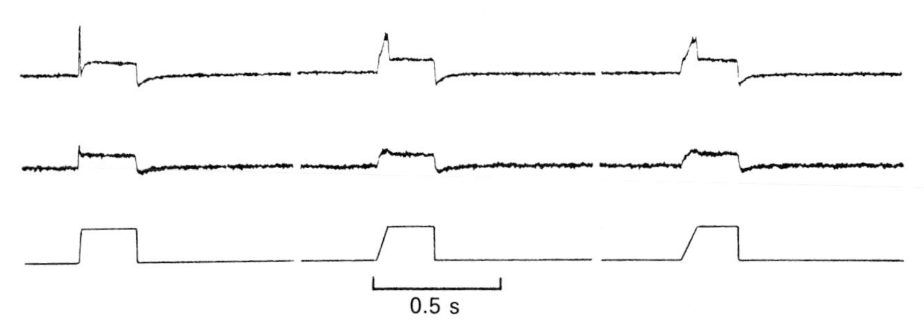

from that in insect fibrillar muscles since it occurs in apparently resting muscle fibres.

Stretch activation has been evoked to explain some of the after-effects of stretching and fusimotor stimulation (Emonet-Dénand *et al.*, 1985). An alternative view is that these effects can be ascribed to the formation of a number of cross-bridges in the resting state (Morgan *et al.*, 1984). There is clearly some way to go before the mechanical properties of muscle spindles are fully understood.

P. B. C. Matthews and his colleagues (Matthews, 1962; Crowe and Matthews, 1964; Brown *et al.*, 1965) showed that the γ efferent fibres in the cat are of two functionally distinct types, distinguished by their effects on the velocity-sensitive responses of the primary endings during extension. They defined the 'dynamic index' of a response to stretching at a constant velocity as the difference between the frequency of firing of an afferent fibre just before the end of the period during which the muscle is being extended and that occurring at the final length half a second later. The two types of fusimotor fibres are then *dynamic fibres*, which increase the dynamic index of the primary afferent fibres, and the *static fibres*, which reduce it. In other words, dynamic fibres make the primary endings relatively more sensitive to the velocity of stretching, and the static fibres make them less sensitive. These features are shown in fig. 17.32.

Is it possible to find any anatomical basis for this physiological division of fusimotor fibres into two types? An ingenious technique was applied to the problem (Brown and Butler, 1973; Barker *et al.*, 1976). If a fusimotor axon is stimulated repetitively for some time, all the intrafusal fibres which it innervates will contract and so deplete their glycogen reserves, and this depletion can be detected by suitable histological staining. In this way it has been found that static γ axons innervate both bag and chain muscle fibres, whereas dynamic γ axons innervate predominantly bag_1 fibres. It looks as though static γ axons correspond to γ-trail axons and dynamic γ axons correspond to γ-plate axons. To put it another way, fusimotor axons innervating bag_1 fibres are dynamic and those which innervate bag_2 or chain fibres are static. In accordance with this, β axons have dynamic actions if they end on bag_1 fibres, static actions if they do

Fig. 17.32. Effect of two types of fusimotor stimulation on the response of a single primary ending, in a cat soleus muscle, to stretching the muscle at different velocities. Each action potential is shown as a dot whose vertical position is proportional to the instantaneous frequency (i.e. the reciprocal of the time since the preceding action potential). The time scale at the bottom right (1 s) does not apply while the length is changing. *a–c*: No stimulation of fusimotor fibres. *d–f*: Stimulation of static fibre at 70 per second. *g–i*: Stimulation of dynamic fibre at 70 per second. (From Crowe and Matthews, 1964.)

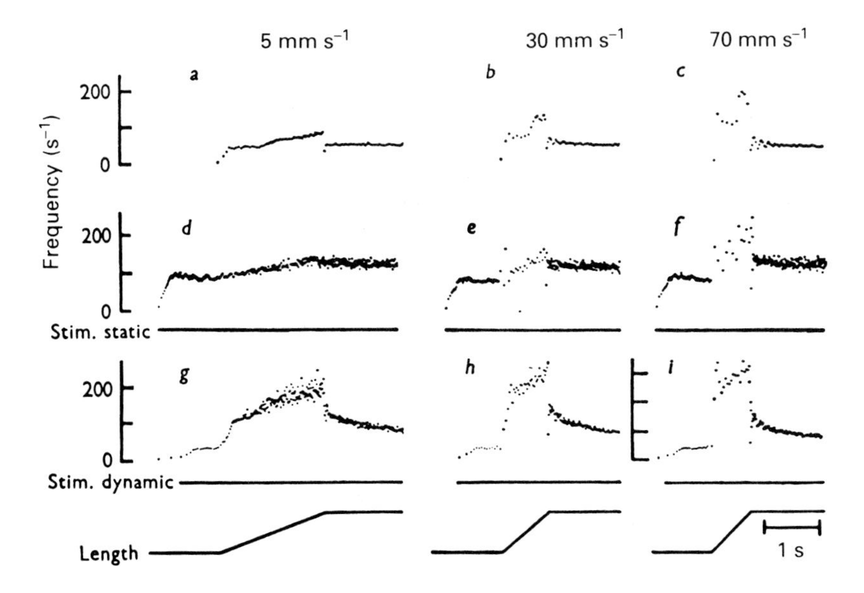

not (Barker *et al.*, 1977; Jami *et al.*, 1982; Banks *et al.*, 1985).

This idea that the differences between the effects of static and dynamic γ axons arises from differences in the mechanical properties of the intrafusal muscle fibres which they innervate has an interesting history. Thus Matthews (1964) suggested that dynamic γ axons innervated bag fibres while static γ axons innervated chain fibres; this suggestion fitted well with Boyd's anatomical views at the time, but not with Barker's. The ensuing controversy was not resolved until the existence of the two types of nuclear bag fibres was established.

There is still a lot to be discovered about muscle spindles. In particular, it would be very pleasant to understand how the contractions of the intrafusal muscle fibres produce their static or dynamic effects on the primary afferent discharge.

Chemoreception in insects

Chemical stimuli play an important part in the lives of insects, as they do in many other animals, especially in the identification of food substances and the location of mating partners. One of the great advantages of insects as material for the study of chemoreception is that single sense organs (sensilla) can be stimulated and recorded from individually in isolation from their neighbours.

The variety of form in these chemoreceptive sensilla is indicated in fig. 17.33. Each consists of a modified region of cuticle through which the dendrites of one or more bipolar nerve cells make contact with the exterior. The cuticular part of the sensillum may be in the form of a hair (sensilla trichodea), a projecting peg (sensilla basiconica), a flat plate (sensilla placodea), or a short peg set in a small pit (sensilla coelonconica). A typical olfactory sensillum trichodeum will have many thousands of fine pores along its length, whereas a similar gustatory hair will have a single pore at its tip. Olfactory sensilla sunk in pits are adapted for use with high concentrations of odorant substances, those in the form of protruding pegs or hairs are for lower ones.

Fig. 17.34 shows the structure of a typical olfactory sensillum. The sense cell is a neuron whose cell body sits among some accessory cells in the epidermis. On its inner side its axon connects it to the central nervous system. On its outer side is the sensory dendrite, divided into inner and outer segments by a short ciliary section. The outer segment contains longitudinal microtubules; it is in contact with a fluid called the sensillum liquor or lymph. The ciliary section contains a ring of nine microtubular doublets, as in motile cilia, but there is usually no central pair. The inner segment contains mitochondria and other components of cytoplasm as in the cell body with which it connects.

There are usually four sheath cells associated with each sensillum. They are connected to each other, to the sense cells and to the surrounding epidermal cells by tight junctions. This ensures

Fig. 17.33. Various types of olfactory sensilla. *a*, sensillum trichodeum; *b*, sensillum basiconicum; *c*, sensillum coelonconicum; *d*, sensillum ampulaceum; *e*, sensillum placodeum. The numbers of sensory cells in each sensillum are one in *d*, two in *a*, three in *b* and *c*, and 18 in *e*. *a* and *b* from a moth, *c* from a locust and *d* and *e* from a bee. (From Kaissling, 1971.)

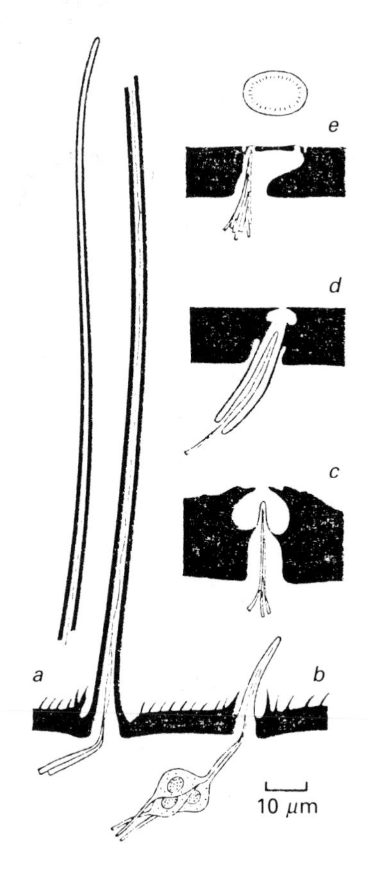

that the sensillum liquor is kept separate from the haemolymph, which is the main body fluid. The sensillum liquor is probably secreted by one of the sheath cells; it has a much higher potassium concentration than the haemolymph.

Behavioural experiments on taste

A remarkably large amount of information has been obtained from experiments involving measurement of the concentration of a substance required to elicit or inhibit a feeding response (Dethier, 1955, 1976). Suppose, for example, that we wish to measure the minimum concentration of sucrose which is detectable by a blowfly such as *Phormia* or *Calliphora.* The fly is suspended from a glass rod, starved, and then allowed to drink as much water as it wants. A drop of sucrose solution of known concentration is then brought into contact with one of the tarsi; if the sugar concentration is sufficient, the fly extends its proboscis, but there is no response if it is not. In this way it has been shown that the tarsal receptors of blowflies are most sensitive to sucrose and maltose, less sensitive to glucose, and insensitive to lactose.

The relative sensitivity to compounds which do not evoke the feeding response can be found by measuring the 'rejection thresholds' of the various compounds. In this technique, the concentration of the compound which is just necessary to prevent the feeding response to a sucrose solution is determined. Dethier and his colleagues measured the

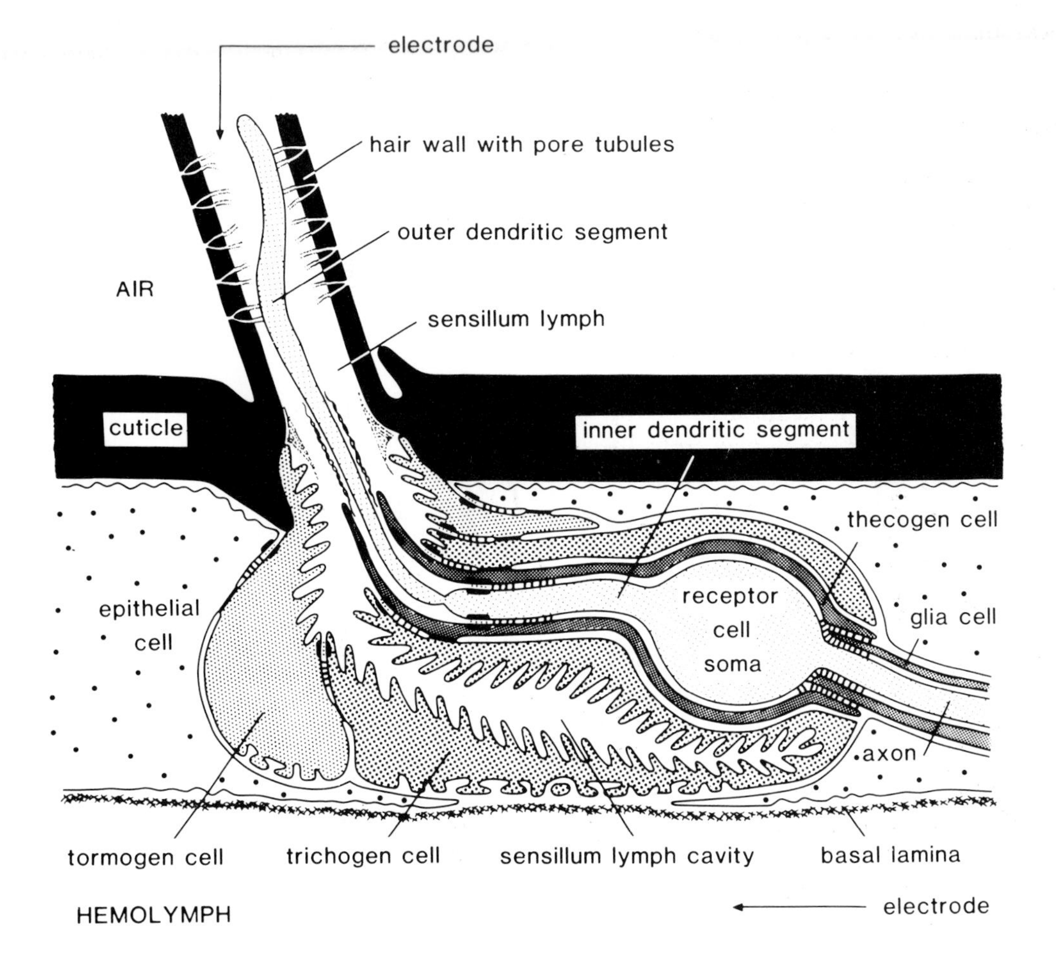

Fig. 17.34. Schematic diagram of an insect olfactory sensillum trichodeum with one receptor cell and three auxiliary cells. The ciliary portion of the dendrite occurs at the narrow connection between its inner and outer segments. (From Kaissling, 1986.)

chemoreceptor sensitivity of blowflies to over 2000 aliphatic organic compounds by this technique. A number of correlations between activity and molecular structure appeared, of which perhaps the most clear-cut was that the rejection thresholds were inversely related to the water solubility of the compounds involved. The simplest explanation of this phenomenon is that a response should depend on the stimulating substance becoming incorporated into the lipid membrane of the receptor cell.

Electrophysiological experiments on taste

Owing to the small diameter of the axons of the chemoreceptor sense cells, their responses to stimulation cannot readily be recorded in the usual manner; two novel techniques had to be developed before electrophysiological investigations could be carried out. In the first of these, shown in fig. 17.35*a*, a pipette containing the stimulating solution is placed over the tip of the sensory hair, and also acts as the recording electrode (Hodgson *et al.*, 1955). This method can only be used with electrolytes in the stimulating solution. The second method, sometimes known as 'side wall cracking' (fig. 17.35*b*), is not subject to this limitation, since the recording electrode is applied to the base of the hair and gains contact with its interior by fracturing the cuticle; stimulating solutions are applied, as before, by means of a pipette applied to the tip of the hair (Morita and Yamashita, 1959).

Some records obtained with the side-wall cracking technique, with glucose as the stimulant, are shown in fig. 17.36. The response consists of a negative receptor potential upon which are superimposed a number of small, largely positive-going action potentials. Notice that the size of the receptor potential and the frequency of the action potentials both increase with increasing glucose concentration, that the response shows adaptation, and that at least two sizes of action potential are evident, indicating that at least two sense cells are stimulated.

How do these responses arise? We may assume that the dendrite outer segment contains specific receptor sites, and that combination of the stimulant substance with these receptors opens ionic channels in the outer segment membrane.

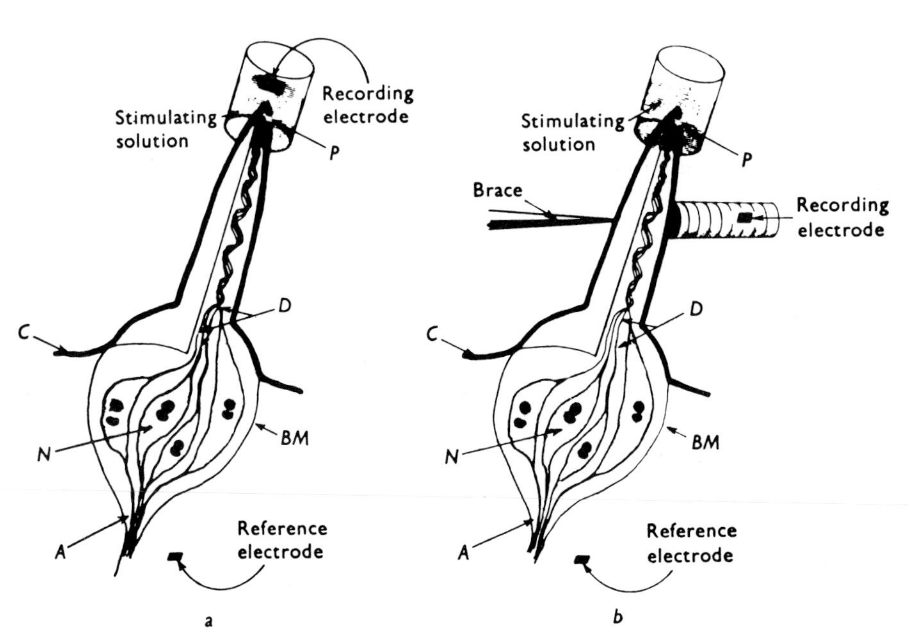

Fig. 17.35. Methods of stimulating and recording from individual taste receptor hairs in insects. *a* shows the technique in which the stimulating solution is held in the recording electrode, *b* shows the 'side-wall cracking' technique. *A*, axons; *BM*, basement membrane; *C*, cuticle; *D*, dendrites; *N*, sensory nerve cell; *P*, pore in cuticle at tip of hair. (From Wolbarsht, 1965.)

Movement of ions through the channels would then depolarize the membrane to produce the receptor potential. This depolarization, we assume, spreads down the dendrite to reach an impulse initiation site near the cell body. Action potentials then propagate down the sensory axon. The local circuits associated with these action potentials spread back up the dendrite and are recorded by the extracellular electrode next to the outer segment. Precise details of this mechanism are still under discussion (Morita and Shiraishi, 1985).

Electron micrographs show that there are four, or sometimes five, receptor cells associated with each sensillum trichodeum in *Phormia*. In accordance with this, four physiologically distinct receptors have been found by electrophysiological methods. These are the 'salt' and 'sugar' receptors (Hodgson and Roeder, 1956), which respond to monovalent salts and certain sugars respectively, the 'water' receptor (Evans and Mellon, 1962), and a mechanoreceptor whose dendrite is attached to the base of the hair (Wolbarsht and Dethier, 1958). The function of the fifth fibre may be connected with sensitivity to anions (Dethier and Hansen, 1968).

Fig. 17.36. Records obtained from a single sensillum on the labellum of a fleshfly by the side-wall cracking method. The stimulus was glucose solutions of different concentrations. (From Morita and Shiraishi, 1985.)

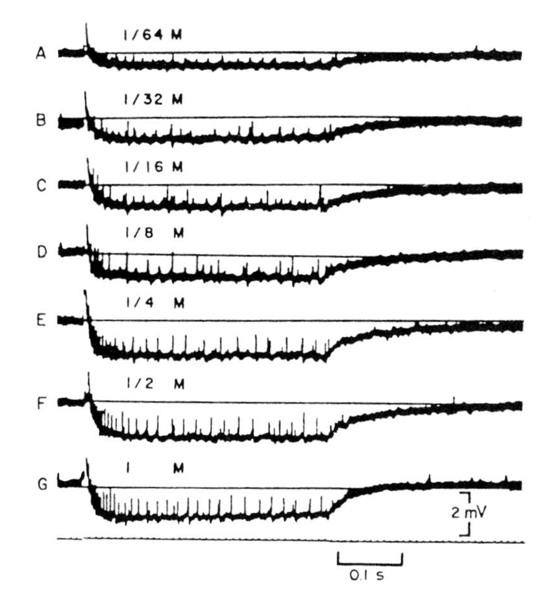

Some very specific contact chemoreceptors exist. The beetle *Chrysolina brunsvicensis* feeds only on plants of the genus *Hypericum* (Saint John's wort). These plants contain the compound hypericin, which is found nowhere else. Rees (1969) found that the tarsal chemoreceptors of the beetle are stimulated by hypercin, whereas those of related insects are not. Another example is provided by the larvae of the cabbage white butterfly, *Pieris brassicae.* Ma (1972) found that sensilla on the mouthparts are stimulated by the mustard oil glucosides, and that this leads to feeding; thus compounds which act as a feeding deterrent for most other insects are here used by the larva to keep it eating the right food plant. Further examples are given by Städler (1984).

Olfaction

Insects use the sense of smell to detect odours from such sources as plants, animals, carrion or dung, so as to find their food, oviposition sites and so on. Many insects also communicate by chemical means: one individual releases a substance which is detected by the olfactory sensilla of other members of the species. Such substances are called pheromones. They include sex attractants, aggregation pheromones and alarm substances.

Much of our knowledge of insect olfaction is derived from work by Schneider and his colleagues at the Max Planck Institute at Seewiesen (Boeckh *et al.*, 1965; Kaissling, 1971, 1986; Schneider, 1965, 1984). They used two methods of measuring the receptor responses: (*a*) the mass response from the whole antenna (the electroantennogram), obtained with recording electrodes placed at the tip and base of an antenna, and (*b*) single unit responses recorded by means of a microelectrode inserted through the thin cuticle of a sensillum. Solutions of odoriferous substances were placed on small pieces of filter paper placed in the orifice of an air jet which was then directed at the antenna.

By measuring the responsiveness or otherwise of a large number of sensilla on the antennae of moths and honey-bees, it was found that the receptor cells fall into two main classes, called 'generalists' and 'specialists'. Odour 'generalists' are cells which respond to a number of different substances, some causing an increase in discharge frequency and others inhibiting it. Schneider describes the pattern of responsiveness of any particular receptor cell to a particular set of compounds as its 'reaction spectrum'. A number of such reaction spectra for the sensilla basiconica of a moth are shown in fig. 17.37. What is particularly interesting about these results is that no two cells have the same reaction spectrum. It seems probable that odours are identified by the patterns of responsiveness which they elicit in a large number of receptors. The system is an interesting example of the way in which a multi-channel sensory system can be used to sharpen the specificity of perception.

Odour 'specialists', the second type of receptor cell in Schneider's classification, are olfactory receptor cells which respond to a very limited number of compounds and which are usually present in large numbers, each cell having the same action spectrum. The antennae of male silkmoths (*Bombyx mori*), for example, contain a large number of sensilla trichodea which are specifically responsive to the sex attractant pheromone which is produced by the females (fig. 17.38). The main component of the *Bombyx* pheromone was eventually isolated (from half a million female moths!) and synthesized by Butenandt and his colleagues in 1961 (see Hecker and Butenandt, 1984). It is a long-chain unsaturated alcohol (hexadeca-dien-10-*trans*-12-*cis*-1-ol), which was given the name bombykol.

Fig. 17.37. Olfactory responses of 27 sensilla basiconica in the moth *Antherea pernyi* to 10 different compounds. Filled circles indicate excitation, open circles indicate inhibition; the sizes of these circles indicate the extent of the change in nerve impulse frequency. Small horizontal lines: no effect. (From Boeckh *et al.*, 1965.)

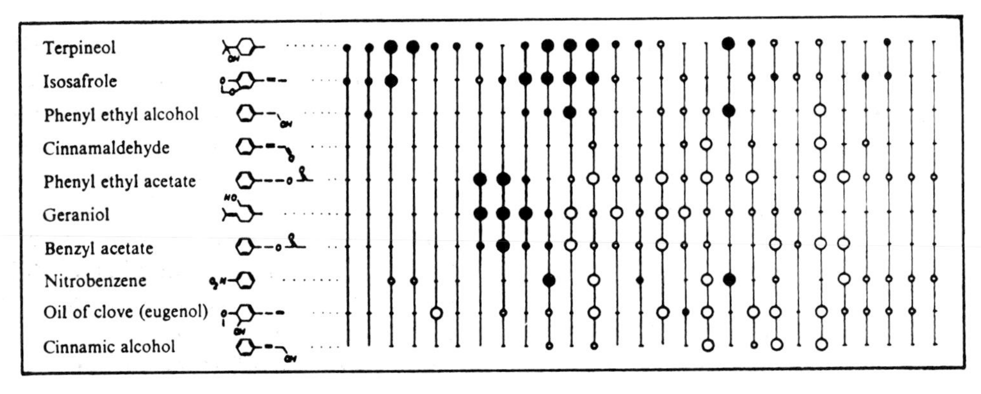

The sensitivity of the bombykol response was investigated by Kaissling and Priesner (1970; see also Kaissling, 1971). For the absolute threshold the relevant questions are how many molecules are required (*a*) to excite one sensory cell, and (*b*) to produce a behavioural response in the male. The response of an individual sensillum to puffs of air containing different concentrations of bombykol is shown in fig. 17.39. With an odour source of 10^{-4} μg, the stimulus sometimes elicits an action potential, sometimes not; the mean response was 0.34 action potentials per puff. The average number of molecules reaching the sensillum at this concentration is also less than one, hence we can say that one molecule of bombykol is sufficient to excite one sensory cell.

Males respond to a puff of air containing bombykol by fluttering their wings. This reaction occurred in some moths with an odour source of 3×10^{-6} μg. Kaissling and Priesner calculated that under these conditions about 310 molecules would contact sensilla, out of a total number of about 30 000 sensilla in each antenna. The moth thus responds when less than 1% of its pheromone detectors are activated.

The cuticular wall of the hair sensillum is perforated by large numbers of pores (fig. 17.38). Each one is connected to a number of tubules which open into the sensillum liquor space surrounding the sensory cell dendrite. Experiments with radioactive pheromone in *Antheraea polyphemus* show that diffusion occurs rapidly from the outer surface of the sensillum through the pores to the sensillum liquor; 40% of the radioactivity appeared in sensillum liquor extruded two minutes after exposure (Kanaujia and Kaissling, 1985).

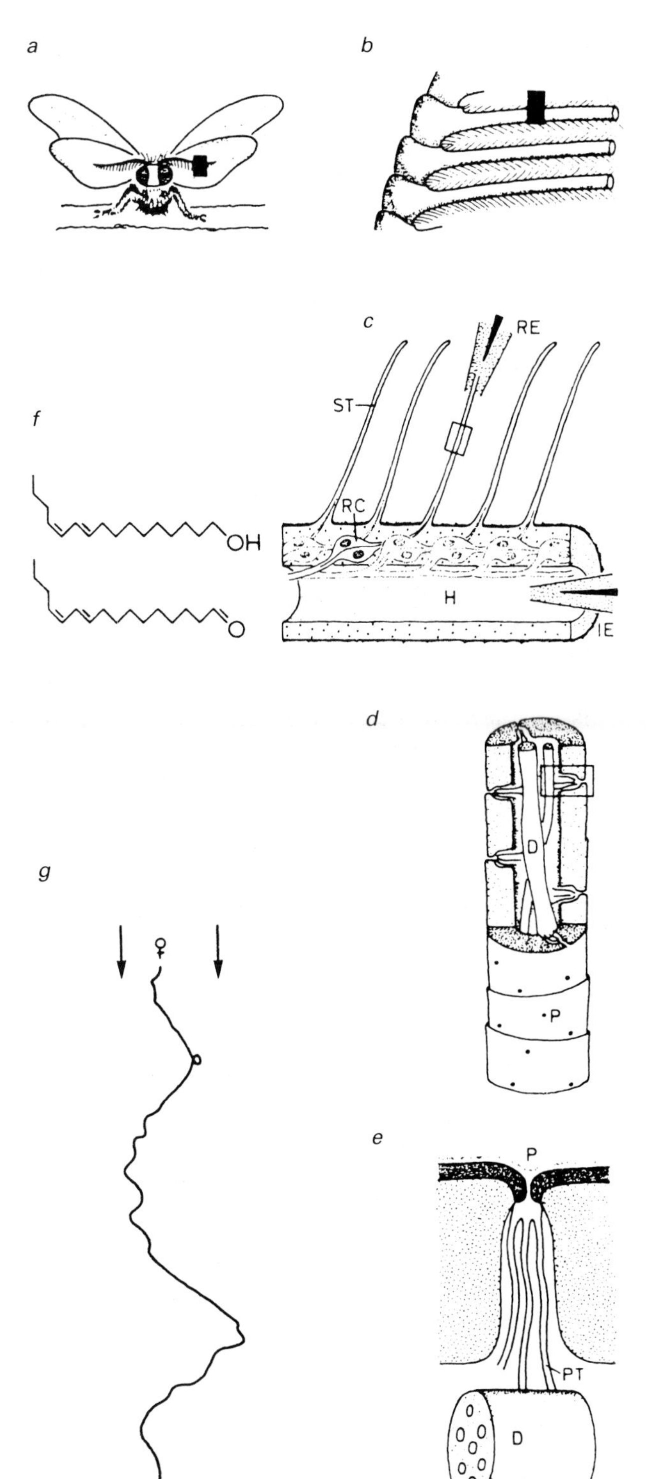

Fig. 17.38. Pheromone reception in *Bombyx mori*. The male moth is shown in *a*, and parts of its antennae are shown at increasing magnifications in *b* to *e*. In *c* the tip of a sensillum ST is cut off and the recording electrode RE placed over it; the indifferent electrode IE is placed in the haemolymph H. Each sensillum contains the dendrites from two sensory receptor cells RC. *d* shows part of the two dendrites D; the outer surface of the sensillum is covered with pores P. A section through one of the pores is shown in *e*; pore tubules PT connect the pore to the sensory liquor and possibly to the dendrite surface. The formulae of the two components of the pheromone, bombykol (above) and bombykal, are shown in *f*. A male moth reaches a female by flying along the odour plume containing her pheromone as is indicated in *g*. (From Steinbrecht and Schneider, 1980.)

What happens to the pheromone after it has contacted the sensory dendrite? The moth needs to respond rapidly to a fall in pheromone concentration, otherwise it would not know when it had left the odour plume downwind of the female. Prolonged after-effects of pheromone exposure would interfere with this. Vogt and Riddiford (1981) found that the sensillum liquor in *Antheraea* contains a very active esterase which will hydrolyse the pheromone (which in this case is an acetate ester) and so inactivate it. There is also a pheromone-binding protein present, and Vogt, Riddiford and Prestwich (1985) suggest that this serves to protect the pheromone from the enzyme during its passage from the pore tubules to the receptors on the dendrite membrane.

In the early years of pheromone research it was assumed that each sex attractant pheromone consisted of a single substance only. Then Meijer *et al.* (1972) found that the pheromone of the summer fruit tortrix moth *Adoxophyes orana* contains two components, *cis*-9- and *cis*-11-tetradecen-1-ol acetate. Traps containing either component alone were ineffective in capturing males, but those containing both components, in the ratio 3 to 1 of *cis*-9 to *cis*-11, were highly effective. Den Otter (1977) showed that each antennal sensillum contains two sensory cells, one responsive to the *cis*-9 isomer only and the other much more responsive to the *cis*-11. Thus the sensitivity to the ratios of the two components resides in the central nervous system of the moth, not in its individual sensory cells.

Further work showed that it is the normal situation for sex-attractant pheromones to have two or more components. In accordance with this, the aldehyde of bombykol, called bombykal, was discovered as a minor component in the pheromone of *Bombyx mori* (Kaissling *et al.*, 1978). The sensitivity to the component ratio means that male moths can readily distinguish their own species pheromone from that of a related species which uses the same components in a different ratio (Priesner, 1979; Steinbrecht and Schneider, 1980).

Electroreceptors

Most sense cells are responsive to electric currents of sufficient magnitude, but this does not justify description of them as 'electroreceptors'. As Machin (1962) points out, a sense organ can only be described as an electroreceptor if it responds to electrical stimuli present in the environment and if the organism responds in a way appropriate to the detection of these electrical stimuli.

Fig. 17.39. Action potentials from a bombykol receptor cell in a male *Bombyx mori*. The quantities of bombykol on the odour source are shown at the left, and the bar indicates the duration of the puff of air containing it. (From Kaissling and Priesner, 1970.)

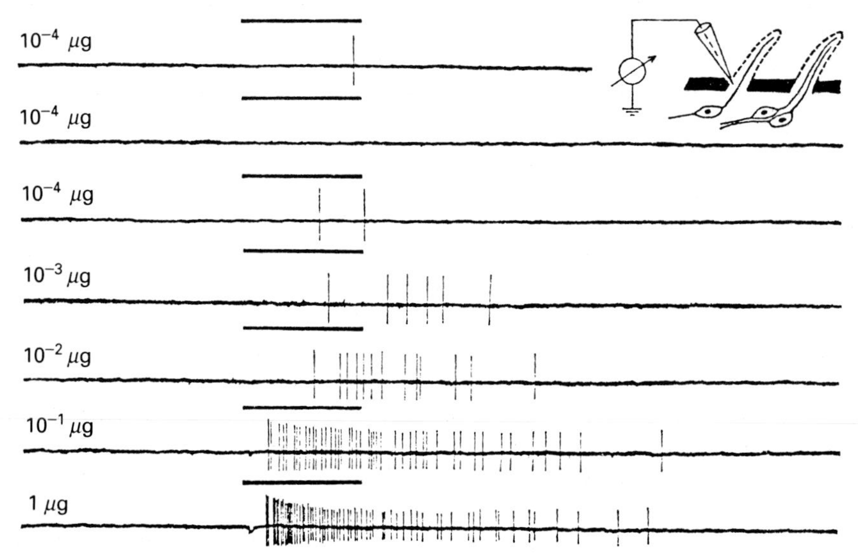

As mentioned in chapter 15, Lissmann (1951) suggested that fish with weak electric organs possess an object-location system. According to this theory, the activity of the electric organ sets up an electric field in the water, and the form of this field will be distorted if objects of different conductivity are placed in it, as is shown in fig. 17.40. The disturbance could be detected by the fish if it is able to measure the currents flowing across the skin at

Fig. 17.40. The electric field produced by an electric fish in the presence of objects of low (*a*) and high (*b*) conductivity. (From Lissmann and Machin, 1958.)

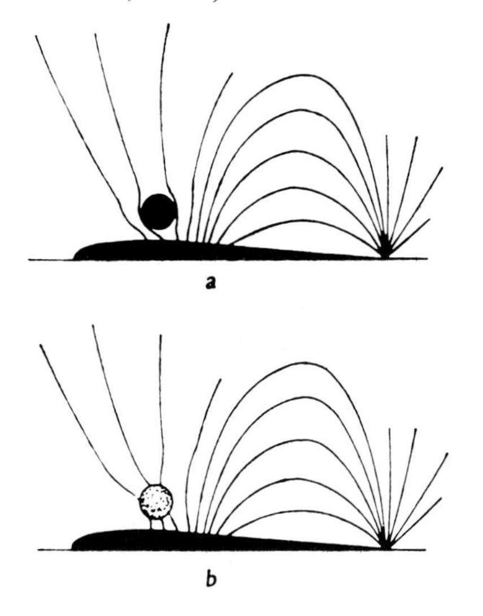

different points on its body surface, for which the presence of an array of electroreceptors must be necessary.

A number of experiments were adduced by Lissmann (1958) as evidence for sensitivity to weak electric fields in electric fish. Specimens of *Gymnarchus niloticus* responded to the closing of a switch between two wires dipped into the aquarium tank, and to movements of magnets or electrostatic charges outside the tank. He was able to show that *Gymnotus carapo* can be trained to differentiate between metal and plastic discs in total darkness, or between presence and absence of a magnet placed outside the tank and out of sight of the fish.

These experiments clearly establish that electric fish are sensitive to electrical changes in the environment, but they do not necessarily imply that the fish can detect objects by means of the distortion of the electric field produced by the electric organs, since metals immersed in water can give rise to small voltages and it might be these that the fish is responding to. The question was further investigated by Lissmann and Machin (1958) in some elegant experiments on *Gymnarchus*. They used as test objects porous pots which were filled with materials of different conductivities; porous pot is opaque and is an effective diffusion barrier for short-term experiments, but does not offer any considerable resistance to electric current when it is immersed in a conducting solution. The fish could easily be trained to distinguish between pots filled with aquarium water and pots filled with distilled water. Such a fish would then respond differently to a pot filled with aquarium water according to whether or not it contained a glass rod more than 1.5 mm in diameter; the presence of the glass rod would, of course, alter the overall conductance of the pot.

In discussing the evolution of electric fishes, Lissmann (1958) argued that the possession of an electric sense must be a prerequisite for the evolution of an electrolocation system involving electric organs, and thus that we might expect to find electroreceptors in some non-electric fish. He suggested that certain sense organs known to exist in the skin of fishes, often connected to the surface by jelly-filled pores, would in fact be electroreceptors. Such structures included the ampullae of Lorenzini in elasmobranchs, the pit organs of catfish, and the mormyromasts of mormyrids.

Confirmation of Lissmann's predictions was rapidly forthcoming. Murray (1962) showed that the ampullae of Lorenzini were extremely sensitive to electric currents. Dijkgraaf and Kalmijn (1963) found that dogfish and skate show behavioural responses to weak electric currents, and that these are abolished by cutting the nerves from the ampullae. Action potentials in the nerves supplying the electroreceptors of weak electric fish were recorded by Bullock *et al.* (1961) in gymnotiforms and by Fessard and Szabo (1961) in mormyrids. Since then electroreceptors have been described in catfish, African knifefish, most non-teleost fish and some amphibians (see Bullock and Heiligenberg, 1986), and even in the duck-billed platypus (Scheich *et al.*, 1986).

The anatomy of the electroreceptors of various fish was investigated by Szabo (1965, 1974). He distinguished two general types, called the ampullary organs and the tuberous organs (fig. 17.41). An ampullary organ consists of a group of sensory cells at the bottom of a jelly-filled canal which opens to the exterior. A tuberous organ consists of a small group of sensory cells in a capsule which projects into the epidermis from below. The capsule is connected to the outer surface of the fish by a canal which contains loosely packed epidermal cells with large extracellular spaces. The receptor cells in each organ synapse with the branches of a single afferent axon, and sometimes a number of organs are connected to the same axon. Szabo found that a small specimen of *Sternarchus* possessed about 130 ampullary organs and over 2000 tuberous organs.

Ampullary organs are found in most electrosensitive fish. Tuberous organs are found only in the Gymnotiformes and Mormyriformes, both of them teleost orders whose members possess electric organs. In line with this distribution we find that the two types have different functions: ampullary organs are used primarily for detecting electric fields arising in the environment, tuberous organs are used by electric fish for electrolocation and interspecific communication.

Electric fish can be roughly divided into pulse species, which produce brief discharges at various intervals, and wave species, which produce high-frequency discharges more or less continuously. The coding methods of their tuberous electroreceptors, in their responses to the electric organ discharge are affected by these differences. Fig. 17.42*a*, for example, shows responses of two types of unit in the pulse species *Hypopomus*; M units respond with a single action potential, B units give a burst of impulses whose number and frequency are proportional to the strength of the stimulating current. In *Eigenmannia*, a wave species, there are again two types of unit (fig. 17.42*b*); T units always respond once per cycle when the electric organ discharge intensity is above threshold, while P units increase their probability of firing in any one cycle (and therefore their overall firing frequency) as the intensity of the stimulating current increases.

Units analogous to the M and B units of gymnotiform pulse species occur in mormyrids; they correspond to two different morphological types, known as knollenorgans and mormyromasts (Bennett, 1965). In both gymnotiforms and mormyrids, the units that simply respond to the presence or absence of the electric organ discharge (M and T units, knollenorgans) may serve to detect the presence of other discharging individuals. On the other hand, the units which code for the intensity of the electric organ discharge (B and P units, mormyromasts) are probably concerned with electrolocation (Heiligenberg, 1977).

Fig. 17.42. Coding in tuberous electroreceptors of gymnotiforms. *a* shows responses of an M unit and a B unit in response to a single discharge of the electric organ (EOD) in a pulse species, *Hypopomus*. *b* shows corresponding responses of a T unit and a P unit in the wave species *Eigenmannia*. (Data of J. Bastian from Heiligenberg, 1977).

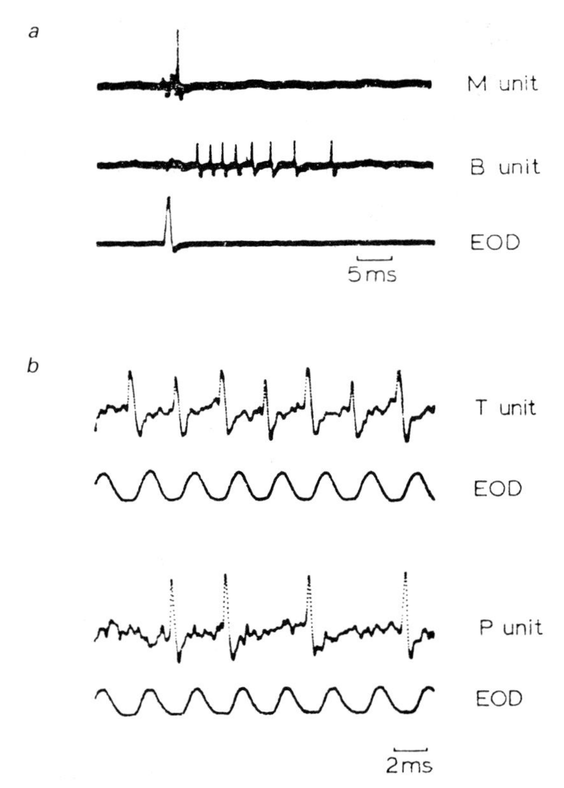

Fig. 17.41. Gymnotid electroreceptors. A diagrammatic section through the skin of a gymnotid fish showing an ampullary organ (*a*) and a tuberous organ (*b*). (Redrawn after Szabo, 1965.)

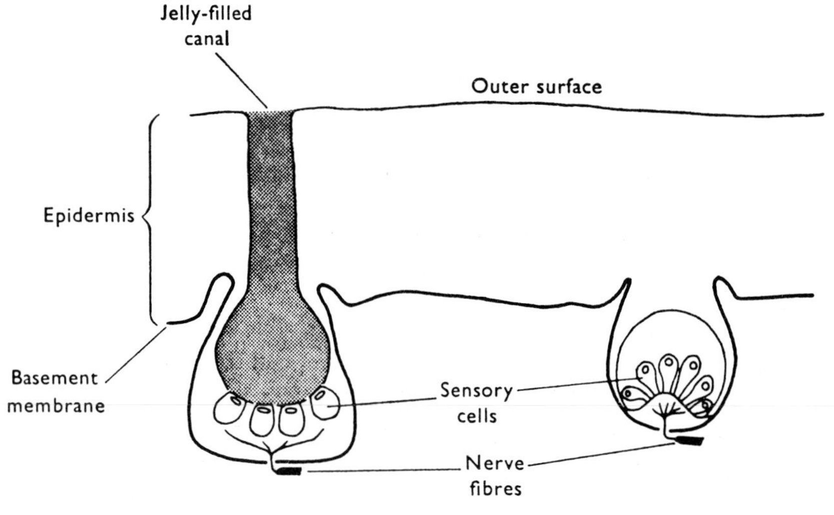

Fig. 17.43 shows how a single P unit in the gymnotiform wave species *Sternarchus* can respond to the position of objects of different conductivity in its immediate vicinity, as determined by Hagiwara and his colleagues (1965). The normal firing rate of the unit is about 130 per second; at an electric organ discharge frequency of 800 per second this corresponds to a probability of firing of 0.16 per cycle. When a metal plate is brought towards the electroreceptor from the tail, the firing rate increases to reach a maximum of over 400 per second, corresponding to a firing probability of about 0.52 per cycle. As the plate passes over the electroreceptor, the firing rate drops markedly to a very low value (probability about 0.01) and then recovers as the plate is taken away towards the head. An inverse set of responses occurs when a plastic plate is moved over the same track. Clearly an array of a large number of electroreceptors of this type will provide the information for a highly sensitive location system.

The electric sense of elasmobranchs was investigated by Kalmijn (1971). He found that plaice (flatfish upon which dogfish and skates feed) produce an appreciable electric field, up to 1000 μV cm^{-1}, from the action potentials of muscles used in respiration or movement. This compares with Murray's value of 1–10 μV cm^{-1} for the threshold field detectable by a single ampulla of Lorenzini. A dogfish can find a plaice even when it is held in an agar chamber under sand, whereas it cannot find pieces of fish under the same conditions. So it seems as though these elasmobranchs are using their electroreceptors as highly sensitive detectors of live food buried in the sand.

It is also possible that elasmobranchs are sensitive to the induced currents set up in moving water currents or a moving fish by the earth's magnetic field (Kalmijn, 1982, 1984). Kalmijn was able to train dogfish and rays to respond to fields as low as 5 nV cm^{-1}. Fields of higher strengths than this occur in ocean currents, and would be induced in a fish swimming at the low speed of just a few centimetres per second. The fact that the behavioural threshold is below the apparent threshold for a single ampulla (which is detected by the experimenter as an increase or decrease in the spontaneous level of neural firing) can probably be explained by the multiple channel effects on the signal to noise ratio that were discussed in chapter 16.

Fig. 17.43. The relation between nerve impulse frequency in an electroreceptor nerve fibre of *Sternarchus* and the position of metal and plastic plates placed close to the fish. The arrow indicates the location of the electroreceptor ending. The plates (the metal plate is shown cross-hatched and the plastic plate plain) are drawn to scale in the positions where they elicted the maximal response. (From Hagiwara *et al.*, 1965.)

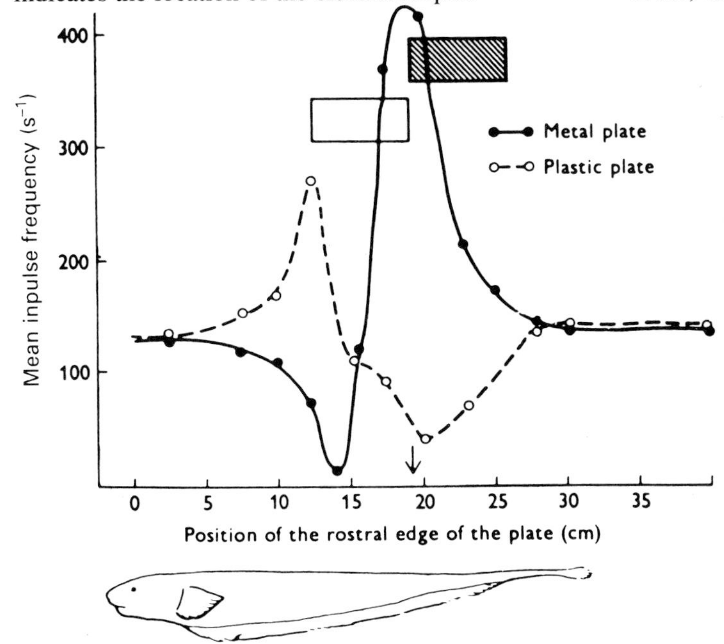

18
Vertebrate photoreceptors

In this chapter we shall examine some aspects of the sensory receptor cells in our most complex sense organ, the eye. The optical apparatus of the eye focuses an image of the visual field on the retina. The retina contains, in man, about 100 million receptor cells, which are connected in a rather complicated fashion to about a million fibres in the optic nerve. When light falls upon the receptor cells they are excited, and their excitation eventually leads to the production of action potentials in the optic nerve fibres.

The light sensitivity of the eye is primarily caused by the existence in the receptor cells of a *visual pigment*, whose function is to absorb light and, in so doing, to change in some way so as to start the chain of events leading to excitation of the optic nerve fibres. As a reflection of this photochemical change in the pigment, we find that the pigment molecules are bleached by illumination, and have to be regenerated before they regain their photosensitivity.

The range of sensitivity of the eye is enormous: the brightest light in which we can see is about 10^{10} times the intensity of the dimmest. There are a number of mechanisms which enable this wide range to be perceived, which constitute the phenomena of visual adaptation. *Dark adaptation* is the increase in sensitivity which occurs when we pass from brightly lit to dim surroundings, and *light adaptation* is the reverse of this.

Visible light consists of electromagnetic radiation within a limited range of wavelengths. The visible spectrum, i.e. that range of electromagnetic radiation that can pass through the eye and cause a photochemical change in the visual pigments of the retina, covers wavelengths from about 400 nm (violet) to about 700 nm (deep red); within this range a number of different colours can be seen, as is shown in table 18.1. The colours, of course, are essentially properties of the retina, not of the light (Young, 1802; Wright, 1967).

Table 18.1. *The colours of the visible spectrum. The wavelength ranges given in this table are only an approximation, since the naming of colours is a subjective phenomenon and differs for different observers*

Colour	Wavelength (nm)
Red	Above 620
Orange	590–620
Yellow	570–590
Green	500–570
Blue	440–500
Violet	Below 440

Light energy is emitted and absorbed in discrete packets known as quanta or photons; there is no such thing as half a quantum. In a photochemical reaction, one molecule of pigment absorbs one quantum of light. The energy of a quantum varies with the wavelength of the light, and is given by $h\nu$, where ν is the frequency (the velocity of light divided by the wavelength) and h is Planck's constant, 6.63×10^{-27} erg s.

The structure of the eye

The photoreceptor cells of the eye, and the nerve cells that they innervate, are found in a thin layer, the retina, which coats the inner surface of the eye. The rest of the eye consists of accessory structures concerned, directly or indirectly, with assisting the retina in the perception of the visual field.

The accessory structures

The gross structure of the human eye is shown in fig. 18.1. The outer coat of the eye is protective in function; it is called the *sclera*. The *cornea* is the transparent region of the sclera. Inside the sclera is a vascular layer, the *choroid*, which is usually pigmented. In some animals the choroid contains a reflecting layer, the *tapetum lucidum*, which serves to increase the effectiveness of the retina in catching light, but must necessarily lead to some loss of definition in the visual image. It is this layer which accounts for the reflecting properties of cat's eyes at night. In the bush-baby, the choroid contains a pigment which fluoresces in ultraviolet light and so enables ultraviolet light to be 'seen' by the retinal receptors (Dartnall *et al.*, 1965).

The interior of the eye contains the *lens*, the *aqueous humour* in front of it, and the *vitreous humour* behind it. The curved surface of the cornea acts as a lens, so that a parallel beam of light entering a human lensless eye is brought to a focus at about 31.2 mm behind the front face of the cornea. Since the length of the eye is about 24.4 mm, the cornea, aqueous and vitreous alone cannot produce a sharp image on the retina. The necessary extra focusing power is provided by the lens, which is of higher refractive index than the aqueous and vitreous humours. The power of the lens can be increased by contraction of the ciliary muscle; this process, which is known as *accommodation*, allows the images of near objects to be focused on the retina. In many of the lower vertebrates (fishes, amphibia and snakes – see Walls, 1942), accommodation is produced not by altering the power of the lens but by moving it backwards and forwards.

In front of the lens is a diaphragm, the *iris*, the diameter of whose aperture (the *pupil*) can be varied by contraction of the iris muscles. Thus the activity of the iris can regulate the amount of light entering the eye; it is, in fact, reflexly controlled by the light intensity.

The direction in which the eye looks is controlled by the extraocular muscles, of which there are six.

The retina

The nervous structure of the retina is much more complex than that of most other peripheral sense organs. The developmental reason for this is that it is produced by an outgrowth of the brain in the embryo. The photoreceptors themselves are probably derived from flagellated cells forming the ependymal lining of the cavities of the brain, so that they are on the internal side of the mature

Fig. 18.1. Diagrammatic horizontal section of the human right eye.

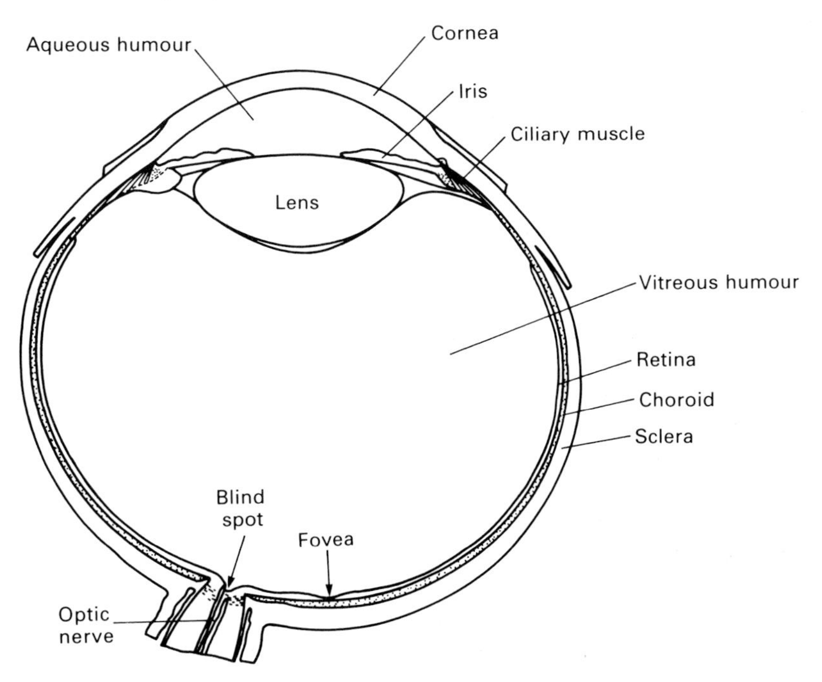

retina, and light must first pass through the rest of the retina before reaching them.

The structure of the retina, as determined by light microscopy, is shown in fig. 18.2. Using a conventional non-selective stain, a number of layers can be distinguished. The interrelations of the cells in these layers can be observed by using a selective stain such as the Golgi silver impregnation method (see Polyak, 1941). The photoreceptor cells are of two types, described, from their shapes, as *rods* and *cones*. These synapse with small interneurons, the *bipolar cells*, which themselves synapse with the *ganglion cells*. The axons of the ganglion cells form the optic nerve, and carry visual information from the retina into the brain. In addition to this sequential system, there are two lateral systems of neurons: the *horizontal cells*, which form interconnections between the receptor cells, and the *amacrine cells*, which synapse with each other, with the ganglion cells, and with the proximal ends of the bipolars. Filling the spaces between these various neurons are the *Müller fibres*, which are elongated glial cells.

These retinal cells are arranged in the retinal layers as follows. Inside the *pigment epithelium* is the *receptor layer*, which consists of the inner and outer segments of the rods and cones. Beneath the receptor layer is the *outer nuclear layer*, which contains the nuclei of the receptor cells. Between the receptor layer and the outer nuclear layer is the 'external limiting membrane'; electron micrographs show that this is not a true membrane but a level at which the receptor cells are closely attached to each other via thickenings of the cell membrane. The next layer is the *outer plexiform layer*, which contains the dendrites and synapses of the rods and cones, the bipolars and the horizontal cells. The *inner nuclear layer* contains the nuclei of the bipolars, horizontal cells and Müller fibres. The *inner plexiform layer* contains the synapses and dendritic processes of the bipolar cells, amacrine cells and ganglion cells. The *ganglion cell layer* contains the nuclei of the ganglion cells and, finally, the *nervous layer* contains the axons of the ganglion cells.

The rods and cones are not distributed evenly over the surface of the retina. In man (and in a number of other vertebrates, especially birds) there is a more or less central region which is specially modified for high visual acuity, known as the *fovea*. The fovea in man contains cones only; it is surrounded by a region in which some rods occur, the *parafovea*. In the extrafoveal retina the proportion of cones (in man) is very small.

There are about 100 million rods and about six million cones in the human eye. Since there are

Fig. 18.2. Cells of the vertebrate retina, based on observations on *Necturus*. The top of the section in this diagram is next to the pigment cell layer on the outer surface of the retina; light arrives from the bottom. A, amacrine cell; B, bipolar cell; G, ganglion cell; H, horizontal cell; M, glial cell or Müller fibre; R, receptor cell. (From Dowling, 1970.)

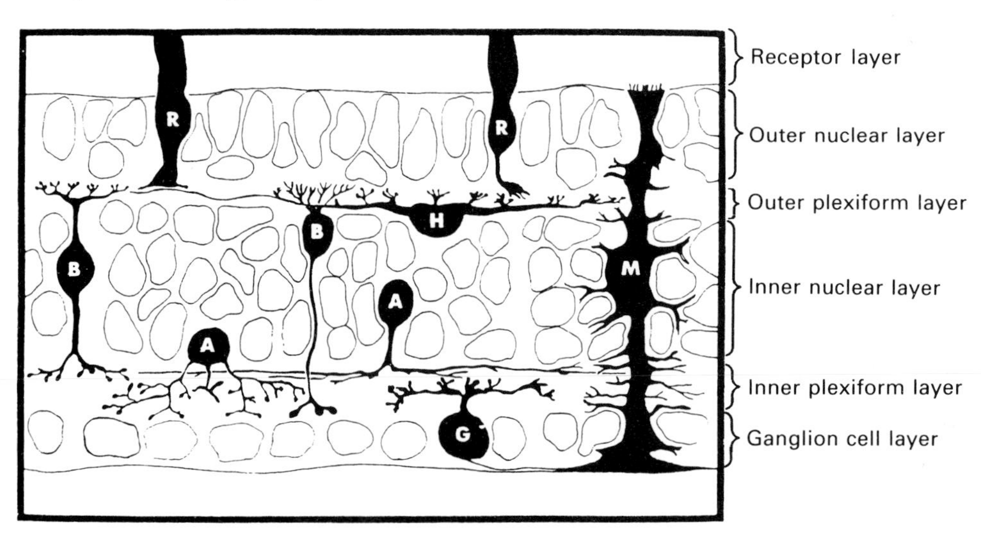

only about one million ganglion cells and optic nerve fibres, it follows that there must be some considerable convergence of the photoreceptors onto the ganglion cells. The anatomical basis of this convergence can be found in the synaptic contacts of the retinal cells. Fig. 18.3. summarizes the results of a number of investigations by electron microscopy (see, for example, Dowling, 1968, 1970, 1987; Dowling and Dubin, 1984). The presynaptic membrane of each receptor cell terminal is invaginated to form a pocket into which processes from bipolar and horizontal cells fit. These are called *ribbon synapses* since there is a dense ribbon or bar in the presynaptic cytoplasm, surrounded by an array of synaptic vesicles. Synapses of a more conventional structure are found between horizontal cells and bipolar cells. Some of the terminals of the bipolar cells also show ribbon synapses, again where there are two types of postsynaptic cell: amacrine and ganglion cells in this case. Many of the synapses between bipolar cells and amacrines are *reciprocal*: presynaptic and postsynaptic areas may occur at different places on each of the membranes bounding the synaptic cleft between the same two cells. Amacrine cells also form conventional synapses with ganglion cells and with other amacrine cells. Some of the latter form *serial synapses* in which a terminal may be postsynaptic to one cell and presynaptic to another. All this structural complexity indicates that there must be an enormous amount of interaction between the various cells of the retina. Thus the information that passes up the optic nerve into the brain is already a highly processed version of the visual image.

Fig. 18.3. Synaptic contacts in the vertebrate retina, based largely on the frog. R, synaptic terminals of the photoreceptor cells. Notice their synaptic ribbons and the invaginations into which processes of the horizontal cells (H) and invaginating bipolars (IB) fit, and their simpler synapses with flat bipolars (FB). A, amacrine cells; notice their reciprocal synapses with bipolars and serial synapses with other amacrines and ganglion cells (G). The ganglion cells may receive input mainly from bipolars (G_1), from both bipolars and amacrines (G_2) or entirely from amacrines (G_3). (From Dowling and Dubin, 1984.)

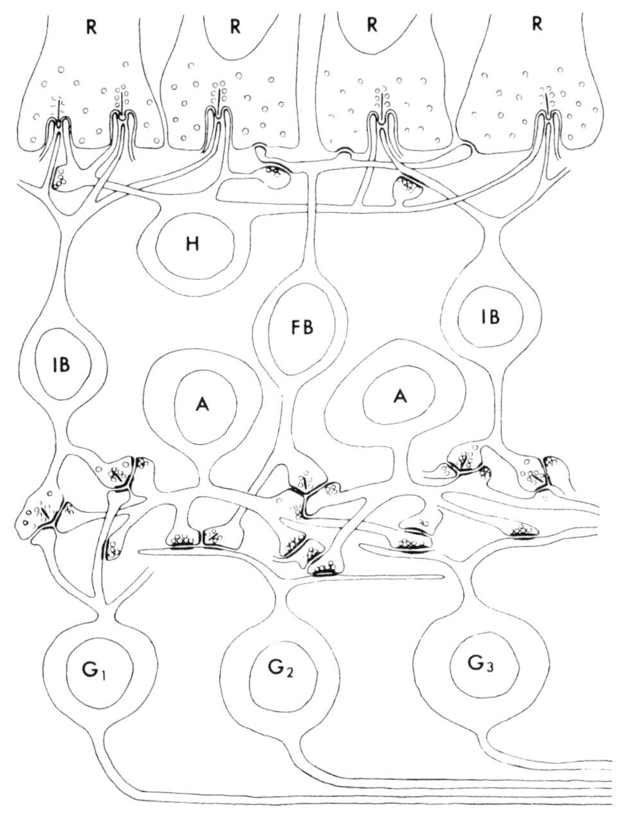

The structure of rods and cones is shown schematically in fig. 18.4, and part of a rod is shown in more detail in fig. 18.5. The outer segment, which contains the visual pigment, contains a stack of membranous discs (in rods) or infoldings of the cell membrane (in cones). The outer segment is connected to the inner segment, which contains numerous mitochondria, via a thin neck whose structure is very like that of a cilium. Below the inner segment is a region which contains the nucleus, and finally ends at the synaptic terminal.

The duplicity theory

The rods and cones have different functions as photoreceptors. The rods are used for vision at low light intensities and are not involved in colour vision. The cones are used at higher light intensities, and for colour vision. Visual acuity is higher for cone vision than for rod vision. These statements constitute the *duplicity theory*.

The duplicity theory was first propounded by Schultze in 1866. It is well known that, in very low light intensities such as occur on a dark night, the fovea is practically blind, and vision depends upon the extrafoveal regions of the retina; colour vision is absent under these conditions. Schultze pointed out that these features tie in with the distribution of the rods: there are no rods in the fovea. He went on to examine the retinae of a variety of different vertebrates, and showed that nocturnal animals tend to have a great preponderance of rods, and diurnal animals have a corresponding preponderance of cones.

The spectral sensitivity of rod (scotopic) vision differs from that of cone (photopic) vision. Scotopic vision is most sensitive to blue–green light and insensitive to red light, whereas photopic vision covers the whole of the visible spectrum, with the greatest sensitivity being in the yellow region. This movement of the region of maximum sensitivity to longer wavelengths, which accompanies the change from scotopic to photopic vision, is known as the 'Purkinje shift'.

A number of psychophysical experiments provide further evidence for the duplicity theory (see Hecht, 1937). A particularly useful experimental method is what is known as the 'fixation and flash' technique for determining visual thresholds. Since the threshold varies somewhat over different parts of the visual field, it is usually necessary in such experiments to ensure that a visual stimulus is always applied to the same part of the retina. This is done as follows. The subject 'fixates' on a small red light, so that its image falls on the fovea. The visual stimulus is then presented at some known angle to the line of fixation (fig. 18.6). In determining the threshold, the light intensity of the test field is adjusted so that the subject can just see it (or, when flashes of short duration are used, when the subject sees the flash in 50% of the times that it is presented).

Fig. 18.4. Rod and cone cells of the mammalian retina. (From Ripps and Weale, 1976.)

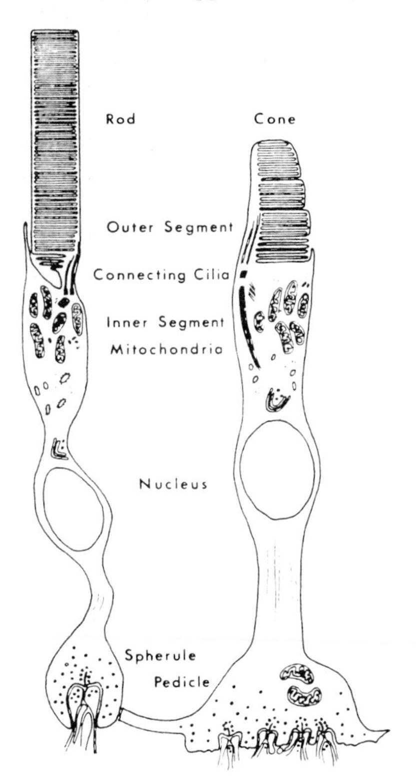

Using this technique, Pirenne (1944) investigated the threshold of the dark-adapted eye to deep red and deep blue test fields of small diameter presented at small angles to the fixation line. His results are shown in fig. 18.7. The threshold for red light increases slightly as the field is presented further away from the fovea. The interpretation of this is that the cone density in the retina decreases correspondingly, and the rods are insensitive to red light. With blue light, however, to which the rods *are* sensitive, the threshold falls markedly with increasing eccentricities. This corresponds very neatly with the distribution of the rods, which are absent in the fovea and appear in increasing numbers at eccentricities greater than 0.5°. Furthermore, the blue flash appears to the observer to be blue when seen by the fovea (cone vision), but appears colourless at the threshold in the extrafoveal region (rod vision), as is indicated by the filled and open circles in fig. 18.7.

During dark adaptation, the threshold falls progressively with time. Fig. 18.8 shows the result of one experiment on this phenomenon, in which the test field was a blue flash placed at an eccentricity of 7° to the fixation line. The curve consists of two branches with a definite 'kink' at their intersection, and the flash appears blue above this kink and colourless below it. The interpretation of these results in terms of the duplicity theory is that the initial section of the curve is due to cone vision, and the later section to rod vision. In accordance with this, the later section is absent if the test field is small in size and viewed by the fovea, or if it is deep red in colour; in each case we would not expect the rods to be excited.

Fig. 18.5. Diagram to show the structure of a mammalian rod, as seen by electron microscopy. *OS*, outer segment; *CC*, connecting cilium; *IS*, inner segment; *C*1 and *C*2, centrioles. Transverse sections through the connecting cilium (*a*) and the centriole (*b*) are shown at the right. *rs*, rod sacs; *cf*, ciliary filaments; *cm*, cell membrane; *mi*, mitochondrion; *er*, endoplasmic reticulum. (From de Robertis, 1960.)

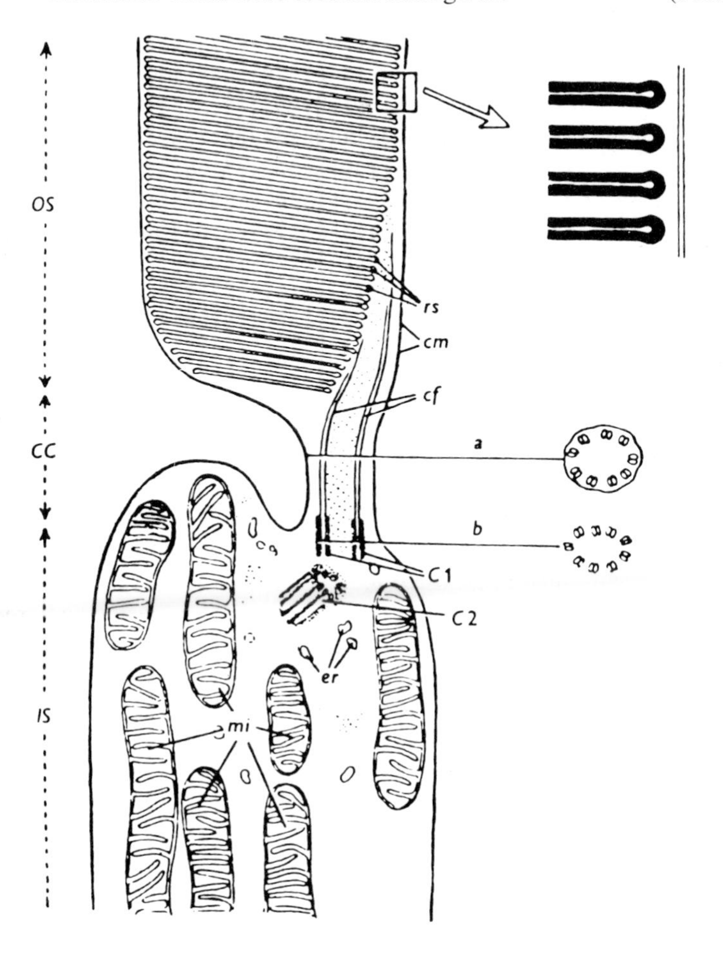

Fig. 18.6. The fixation and flash technique. (After Pirenne, 1962.)

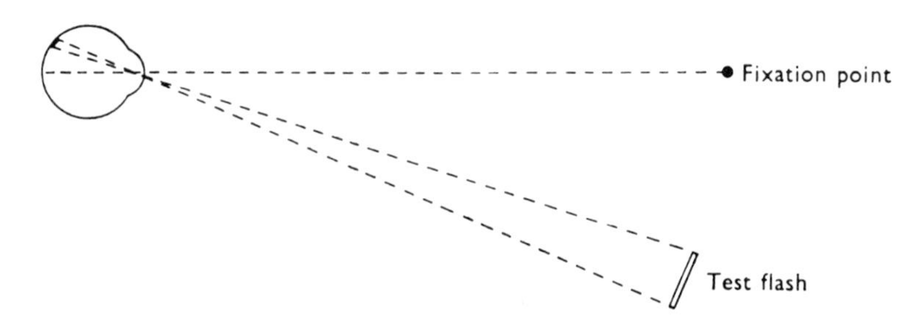

Colour vision

By 'colour vision', we mean the ability to distinguish lights of different wavelengths. A glance at a paint specification chart will immediately show that we are able to recognize an enormous variety of different colours. Although the spectrum is conventionally divided into six or seven different colours (table 18.1), lights of wavelengths as little as 1 or 2 nm apart can be clearly distinguished from each other. What is the physiological basis of this ability?

Colour vision is trichromatic. The meaning of this statement, and the main evidence for it, has been explained by Brindley (1970) as follows: 'Given any four lights, whether spectroscopically pure or not, it is always possible to place two of them in one half of a photometric field and two in the other half, or else three in one half and one in

Fig. 18.7. Thresholds for deep blue (circles) and deep red (triangles) flashes of 10′ diameter at points near to the fovea. Black symbols show points at which the flash appeared coloured, open symbols where it did not. (From Pirenne, 1944.)

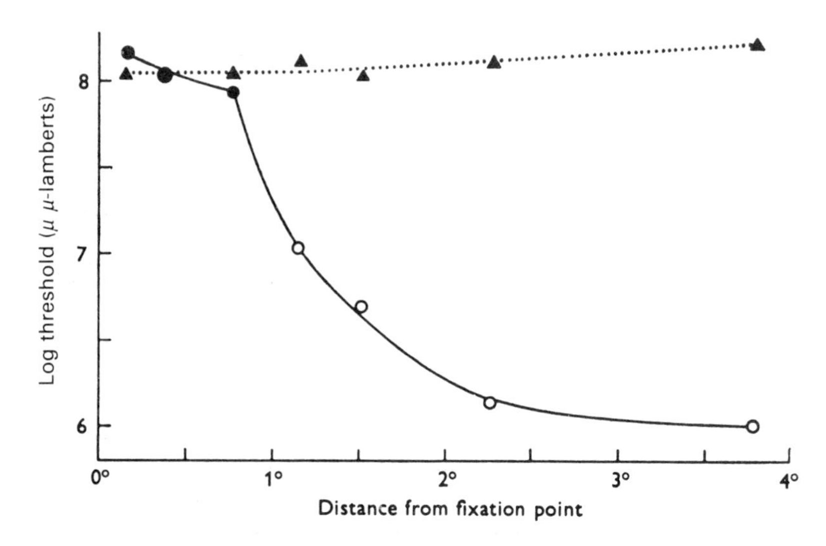

Fig. 18.8. The fall in threshold in the dark following a 3 min exposure to bright light, as tested by deep blue light flashes seen at an eccentricity of 7°. For the first five points (black circles) the flash appeared blue or violet in colour, thereafter (open circles) it was apparently colourless. (From Hecht and Schlaer, 1938.)

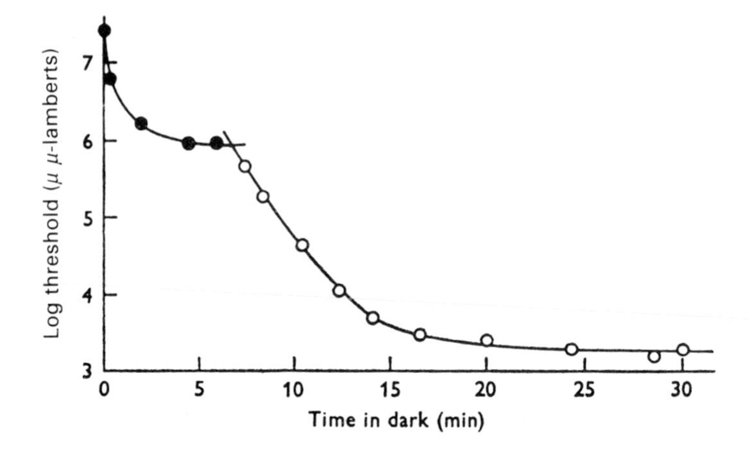

the other, and by adjusting the intensities of three of the four lights to make the two halves of the field appear indistinguishable to the eye. This property of human colour discrimination, whereby the adjustment of three independent continuous controls makes possible an exact match, though two are generally insufficient, is known as trichromacy.'

Trichromacy was first hinted at by Newton, and became generally accepted during the eighteenth centry. But at this time it was thought to be a property of light, and it needed the genius of Thomas Young (1802) to suggest that trichromacy arose from the properties of the eye; that, to put it in modern terms, colour vision is mediated by three sensory channels which have different spectral sensitivities. Young's suggestion was adopted by Helmholtz (1866) in his great work on physiological optics, so that the idea that colour vision is brought about by three mechanisms with differing spectral sensitivities is frequently known as 'the Young–Helmholtz theory'.

We can state the Young–Helmholtz theory in modern terms as follows: colour vision is mediated by three types of cones, each containing a different visual pigment. Two colours can be distinguished if they stimulate at least two of the different cone mechanisms to different extents. Excellent evidence for the existence of three different cone pigments has been provided by the technique of microspectrophotometry, as we shall see later. There are still many problems to be solved in colour vision (see Mollon and Sharpe, 1983, for example), but it is quite clear that Thomas Young got the answer right.

Visual pigments

Visual pigments were discovered in the nineteenth century, initially from observations on the retinas of frogs. Boll (1877) found that the reddish colour of the receptor outer segment layer fades on exposure to light. Kühne (1878) called the pigment sehpurpur or visual purple; it later became known as rhodopsin. He extracted it from the retina with bile salts, measured the spectral sensivity of its bleaching by light, and concluded that the absorption of light by the pigment is the primary event in photoreception. Wald (1933, 1934) showed that vitamin A and a carotenoid which he called retinene were involved in visual function, and concluded that the visual pigment is a conjugated protein with retinene as a prosthetic group (see Wald, 1968).

The properties of the visual pigments can be examined in two situations, either *in vitro*, after extraction from the eye, or *in vivo*, while still in the intact eye. As a half-way house between these two extremes, they can also be examined in isolated retinae or in fragments of the photoreceptor cells. The two methods are complementary to each other: *in vivo* studies must be interpreted in the light of the properties of the pigments, which can be determined much more precisely *in vitro*, and the relevance of *in vitro* studies can only be tested by reference to what happens in the intact eye.

Methods of investigating extracted pigments

In extracting a visual pigment from the retina of an animal, it is essential that the pigment should not be bleached by exposure to bright light. Hence it is usual to use dark-adapted animals, to dissect out the retina under darkroom conditions, using a dim deep red light, and to carry out the extraction procedures from then on in the dark. Visual pigments are not normally soluble in water, and have to be extracted from the retina by aqueous solutions of various substances; digitonin is a commonly used extractant. Much purer solutions are obtained if the pigment is extracted from the outer segments of the rods and cones than if the whole retina is used; in particular, this method avoids contamination by haemoglobin from the retinal blood vessels. The outer segments are obtained by shaking the retina vigorously in 35–45% sucrose solution followed by low-speed centrifugation, which leaves the outer segments suspended in the supernatant fluid.

Visual pigments can be characterized by measurement of the amount of light that they absorb at different wavelengths (see Dartnall, 1957; Knowles and Dartnall, 1977). Light from a monochromator is passed through a cell or cuvette containing an extract of the visual pigment, and measured by means of a photocell. The fraction of light absorbed is called the *absorptance, J*, of the solution. Thus if I_i is the intensity of light incident upon the solution in the cell and I_t is the intensity of light transmitted by it, then

$$J = (I_i - I_t)/I_i. \qquad (18.1)$$

The absorptance of the solution varies with

the wavelength of the light. If we plot the absorptance against the wavelength, the curve obtained is called the *absorptance spectrum* of the solution. Fig. 18.9*a* shows the absorptance spectra of frog visual pigment solutions at different pigment concentrations, and in fig. 18.9*b* these curves have been replotted with their maxima made equal to unity.

Fig. 18.9. Absorptance spectra of solutions containing different concentrations of frog rhodopsin. In *a* the ordinate is the proportion of light absorbed by the pigment (*J* in equation (18.1)). In *b* these curves are replotted so that their maxima coincide; notice that the curves become broader as the concentration of pigment increases. (From Dartnall, 1957, and Knowles and Dartnall, 1977.)

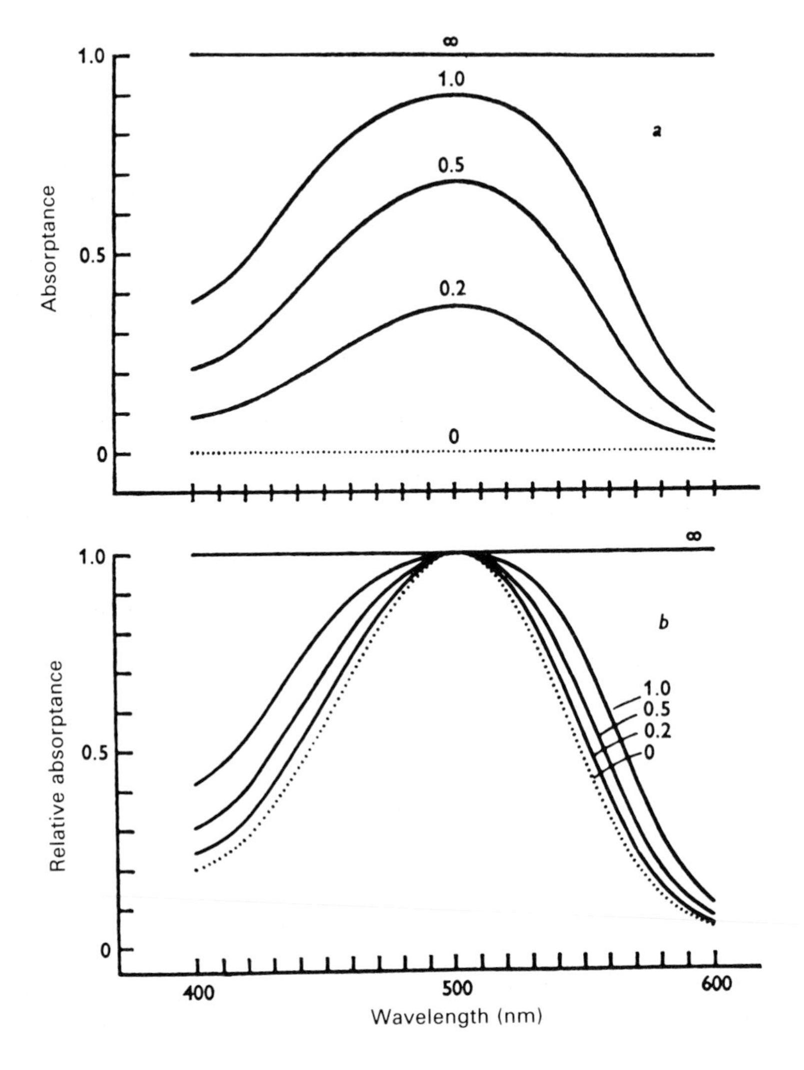

It is evident in fig. 18.9 that the absorptance spectra become broader as the pigment concentration rises. The reason for this can be explained by an example. Suppose that the amount of light transmitted is halved if we double the concentration of the pigment. Then, at a wavelength where the absorptance is 90% of the maximum, doubling the pigment concentration will raise the absorptance to 95%. But, at a wavelength where the absorptance is only 40% of the maximum, doubling the concentration will raise it to 70%.

This dependence of the absorptance spectrum on the pigment concentration makes it difficult to compare results from solutions either of different strengths or in cells of different thicknesses. This difficulty is obviated by the use of absorbance spectra.

Consider a very thin plane of pigment solution, of thickness dl. Of the light I incident on this plane, a portion dI will be absorbed. The fraction dI/I will be proportional to the thickness of the plane, dl, and the concentration of the pigment, c. Thus

$$\frac{dI}{I} = \alpha_\lambda \cdot c \, dl,$$

where α_λ is a constant, the *extinction coefficient*, for any particular pigment at any particular wavelength λ. Integrating this equation between the limits I_i and I_t, we get

$$\alpha_\lambda \cdot c \cdot l = \log_e I_i - \log_e I_t,$$

where l is the width of the cell. That is

$$\log_e (I_i / I_t) = \alpha_\lambda \cdot c \cdot l.$$

When converted into decadic logarithms, this gives

$$A_\lambda = \alpha_\lambda \cdot c \cdot l/2.303. \qquad (18.2)$$

This quantity A_λ is called the *absorbance* of the solution. If we plot A_λ against λ, we obtain the *absorbance spectrum* of the solution, as is shown in fig. 18.10*a*.

When absorbance spectra for different concentrations of the same pigment are plotted as fractions of their maxima, all the curves coincide (fig. 18.10*b*). If there are two or more pigments in a solution, then the total absorbance is the sum of the absorbances of the individual pigments. These two properties make the absorbance spectrum a most useful tool for the characterization and comparison of different pigments. The absorptance spectrum, on the other hand, is normally used when questions involving the concentration of a pigment arise.

(The absorbance is sometimes known as the optical density, and the absorbance spectrum may correspondingly be called the density spectrum. The term 'extinction spectrum' has also been used for the absorbance spectrum. To avoid confusion between absorptance and absorbance, the student may like to remind himself that the longer word refers to the broader spectrum.)

One of the difficulties of measuring the absorbance spectra of visual pigments is that it is frequently very difficult to get rid of impurities in the extract, whose absorbance spectra add to those of the visual pigment. This difficulty can be sidestepped by the measurement of the *difference spectrum*. It is a characteristic feature of visual pigments that they are bleached by exposure to bright lights. Hence it follows that the absorbance spectrum of the products of bleaching differs from that of the unbleached pigment. The difference between the absorbance spectra before and after bleaching constitutes the difference spectrum. Since any impurities which are not visual pigments will have the same absorbance spectrum after bleaching as they did before, they do not contribute to the difference spectrum.

Finally, how are we to distinguish between an extract containing only one visual pigment and one containing two or more? The technique for dealing with this problem is known as partial bleaching, and was developed by Dartnall (1952). The method consists, essentially, in successively bleaching the extract with light of different wavelengths and measuring the difference spectrum after each bleach. If these difference spectra are different (when plotted on a percentage scale with their maxima made equal to 100%), then more than one pigment is present. For example, suppose we have a retinal extract containing two visual pigments, with maximal absorbances at wavelengths (λ_{max}) of 500 nm and 540 nm. If we illuminate this with red light (λ equal to, say, 640 nm), the 540 pigment will absorb much more than the 500 pigment and will therefore be preferentially bleached (the extinction coefficients at this wavelength are likely to be, respectively, about 8% and less than 1% of their maxima). Hence the difference spectrum after this first bleach will have its λ_{max} at about 540 nm. With a long enough exposure, nearly all the 540 pigment can be bleached. If we now illuminate the pigment with white light, or with light at any wavelength between about 560 and 420 nm, the 500 pigment will be bleached, and the difference spectrum for this second bleach will have its λ_{max} at about 500 nm. This shift in the difference spectrum indicates the presence of at least two visual pigments in the extract.

The human scotopic visual pigment: rhodopsin

Rhodopsin (visual purple) is the principal photosensitive pigment of many vertebrate retinas. It is found throughout the human retina except at the fovea, hence it is characteristic of the rods, and absent from the cones. Thus we would expect rhodopsin to be the visual pigment involved in scotopic vision, or, to put it more precisely, the primary photoreceptor event in human scotopic vision should be the capture of light by rhodopsin. The correspondence between these two phenomena has been investigated many times, beginning with Kühne in 1878. Very clear evidence was produced by Crescitelli and Dartnall (1953); let us look at their results.

Crescitelli and Dartnall extracted human rhodopsin from an eye which had to be removed because of the presence of a ciliary body melanoma. The patient was fully dark-adapted before the operation, which was then performed under deep red light. The absorbance spectrum of the rhodopsin extract was measured before and after bleaching, with the results shown in fig. 18.11*a*.

If the extract did not contain any light-absorbing impurities, then the 'unbleached' curve in fig. 18.11*a* would be the absorbance spectrum

Fig. 18.10. Absorbance spectra of solutions containing different concentrations of frog rhodopsin. In *a* the ordinate is the absorbance of the solution (*A* in equation (18.2)). In *b* these curves are replotted so that their maxima coincide; all concentrations now give identical curves. (From Dartnall, 1957, and Knowles and Dartnall, 1977.)

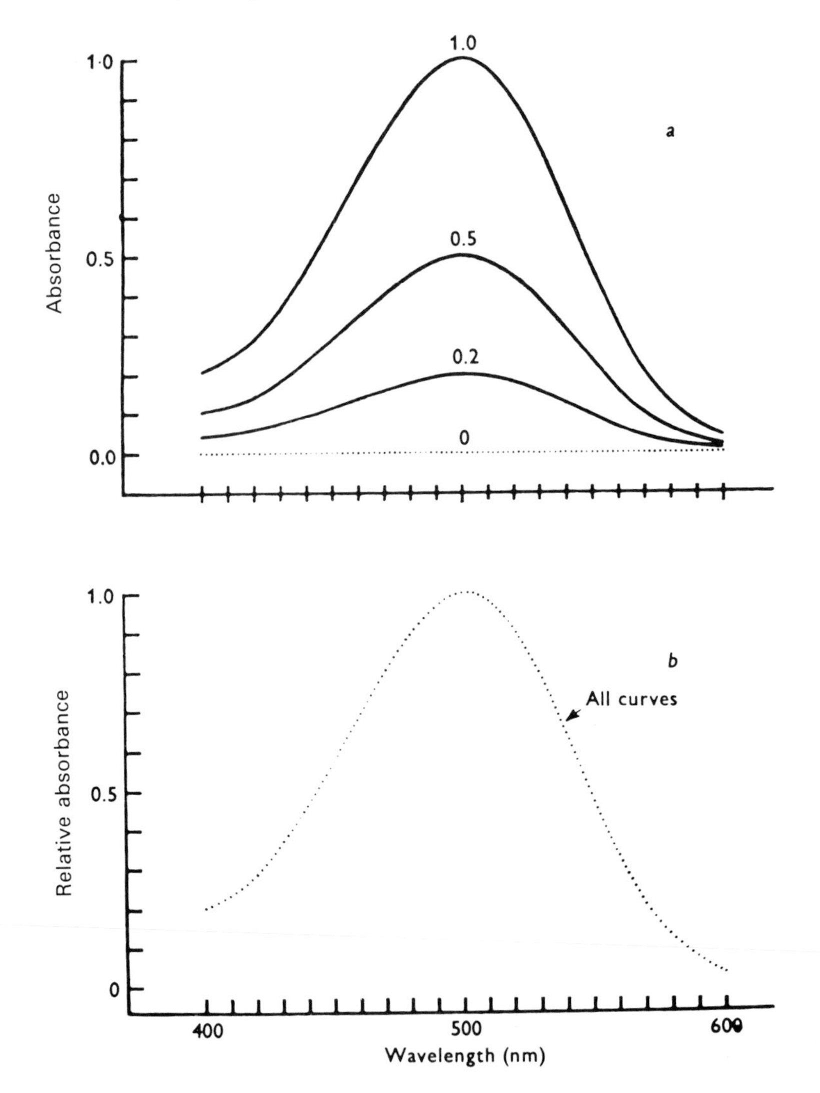

of human rhodopsin. But this is unlikely to be so, and indeed the 'unbleached' curve, by comparison with pure extracts of frog and cattle rhodopsin, looks as though it is derived from a mixture of rhodopsin and some impurities which absorb light increasingly at wavelengths below about 480 nm. Is it possible, then, to make a reasonable calculation as to what the absorbance spectrum of pure human rhodopsin should be? Fig. 18.11*b* shows the difference spectrum of the retinal extract; this is effectively identical with that of rat rhodopsin, and very similar in shape to that of frog rhodopsin, but shifted along the wavelength axis so that its λ_{max} is about 5 nm less. This indicates that the absorbance spectrum of human rhodopsin should be similar to that of frog rhodopsin but, again, with its λ_{max} about 5 nm less.

Now the absorbance spectrum of frog rhodopsin is well established; its λ_{max} is 502 nm, and the full spectrum is shown by the broken line in fig. 18.12. Dartnall (1953) produced a nomogram which could be used to predict the form of the absorbance spectrum of any visual pigment, given the value of λ_{max}: since then this nomogram has been

Fig. 18.11. Absorbance and difference spectra of an extract containing human rhodopsin. Absorbance spectra of the extract before and after bleaching are shown in *a*; readings were taken in sequence from short to long wavelengths (open circles) and back again (filled circles). The differences between the two curves in *a* are plotted in *b* (circles) on a percentage scale to give the difference spectrum; similar spectra for rat and frog rhodopsin are also shown. (From Crescitelli and Dartnall, 1953.)

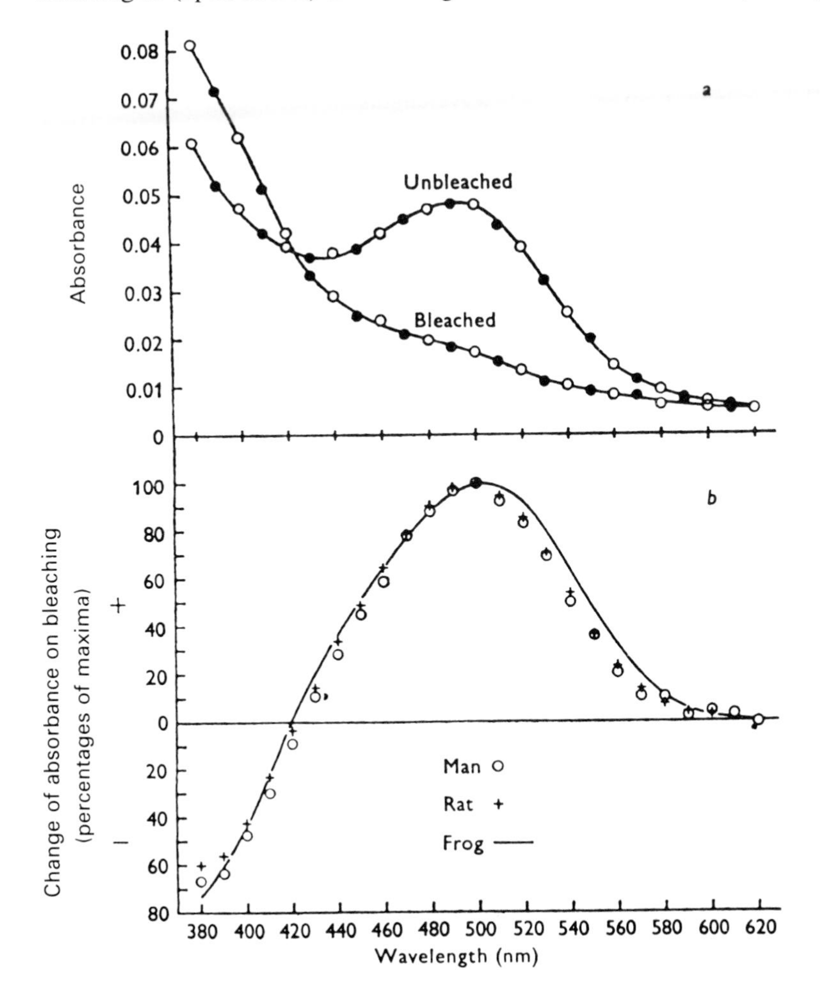

tested against a variety of visual pigments, and its use seems to be fully justified in most cases. Using this nomogram to determine the absorbance spectrum of a pigment whose λ_{max} was equal to 497 nm, Crescitelli and Dartnall obtained the curve shown by the full line of fig. 18.12. This curve, they suggested, represents the true absorbance spectrum of human rhodopsin; it is difficult to believe that this conclusion is not justified.

The next step was to compare this curve with the scotopic sensitivity to light of different wavelengths. The scotopic visibility function was determined by Crawford (1949) from observations on 50 subjects under the age of 30 (this matter of the age of the subjects is rather important, since sensitivity to short wavelengths declines after 30 because of accumulation of yellow pigment in the lens). The subjects were dark-adapted, and looked at a field 20° in diameter which was divided vertically into two halves. One half of the field was illuminated with white light of constant low intensity, and the other half with monochromatic light of different wavelengths, whose intensity could be altered so as to make both halves of the field appear equally bright. The actual intensities were about 15 times threshold, well within the scotopic range.

Crawford's results were expressed as relative sensitivities, derived from the reciprocals of the light intensities needed to produce a certain sensation of brightness. Light intensity is usually measured in units equivalent to a certain amount of energy per unit time, hence Crawford's results

Fig. 18.12. Absorbance spectrum of human rhodopsin and the scotopic visibility function. The broken curve is the absorbance spectrum of frog rhodopsin, as determined experimentally from very pure solutions. The continuous curve is the absorbance spectrum of human rhodopsin, calculated from the peak wavelength of the difference spectrum seen in fig. 18.11 using Dartnall's nomogram (see text). The points represent the sensitivity of the eye at different wavelengths in very dim light, as determined by Crawford, corrected for the absorption of short wavelengths between the cornea and the retina and expressed as an equal quantal content spectrum. (From Crescitelli and Dartnall, 1953.)

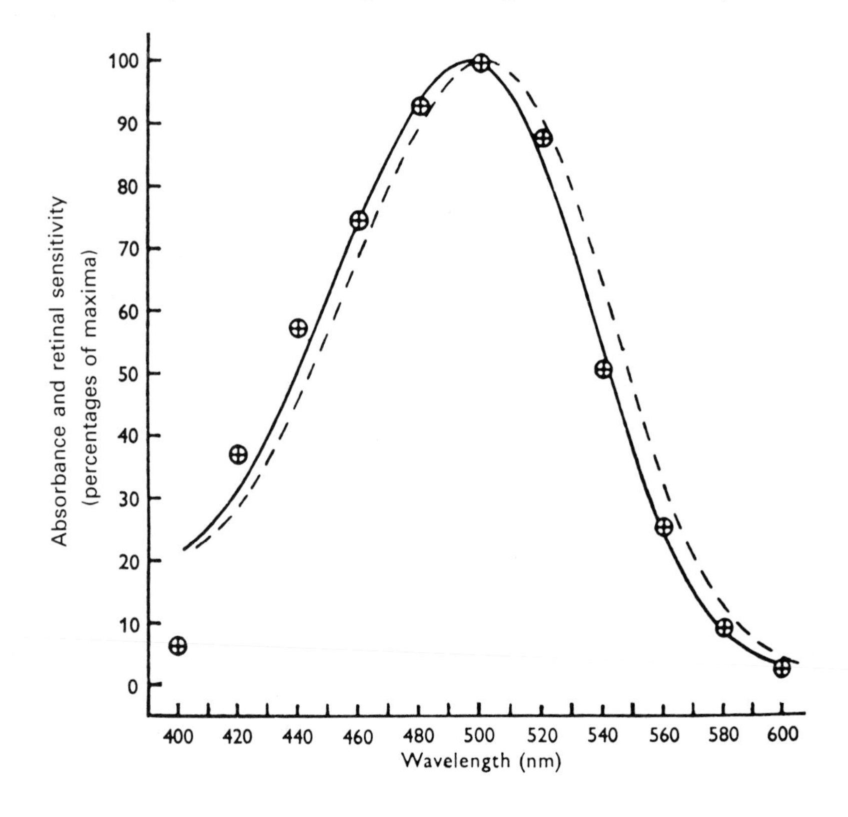

were expressed as an 'equal energy spectrum'. But the absorption of light by a pigment takes place in quanta, whose energy is inversely proportional to the wavelength. Hence, in order for Crawford's results to be compared with the rhodopsin absorbance spectrum, they must be recalculated in terms of an 'equal quantal content spectrum'. A second correction is necessary, to allow for the absorption of light between the cornea and the retina, which is more pronounced at short wavelengths. Values for this preretinal absorption were obtained from the results of Ludwigh and McCarthy (1938). The points shown in fig. 18.12 are Crawford's results as plotted by Crescitelli and Dartnall after these two corrections had been made. It is evident that the agreement between the scotopic sensitivity function and the absorption of light by rhodopsin is very precise, and we can therefore conclude that rhodopsin is the visual pigment used in human scotopic vision.

(There is a further correction that is not taken into account in fig. 18.12. The absorption of light by the retinal rhodopsin is given by its absorptance spectrum, not by its absorbance spectrum; the two are only equal when the net percentage of light absorbed is very low. Crescitelli and Dartnall estimated that the retinal absorption was only 3.5%, in which case the correction would be minute. Other estimates, however, suggests that this figure is too low – about 55% absorption seems more likely (Alpern and Pugh, 1974) – so that the curve in fig. 18.12 should be broadened somewhat.)

The visual cycle

It is evident that light must produce a chemical change in rhodopsin and that eventually the rhodopsin must be restored to its former state so as to be able to respond to light again. The processes that are involved in these changes constitute the visual cycle, a term introduced by Wald in 1934. It soon became clear that the cycle involved a splitting of rhodopsin into its protein portion and the substance detected by Wald, which he called retinene, and that reconstitution probably proceeded via vitamin A. Fig. 18.13*a* shows the cycle as given by Morton in 1944.

Morton suggested that retinene might be the aldehyde of vitamin A, retinal. Evidence that this was so was soon forthcoming (Morton and Goodwin, 1944; Ball *et al.*, 1948). The protein chain in rhodopsin is called opsin. Retinal is attached to opsin via the ε-amino group of a lysine residue, by means of a Schiff's base link (Morton and Pitt, 1955; Bownds, 1967), i.e.

$$C_{19}H_{27}CH{=}O + H_2N.\text{opsin} \rightleftharpoons C_{19}H_{27}CH{=}N.\text{opsin} + H_2O.$$

Retinal can exist in a number of different stereoisomeric forms, some of which are shown in fig. 18.14. The most stable form is all-*trans*, with

Fig. 18.13. The visual cycle of rhodopsin in mammals, as understood in 1944 (*a*; Morton, 1944) and 1982 (*b*). In *b* photochemical reactions are shown by wavy lines, others by straight lines. The λ_{max} of each intermediate is shown as a subscript, and the isomeric form of retinal which it contains is shown in brackets. Rate constants refer to physiological temperatures. (*b* from Fein and Szuts, 1982.)

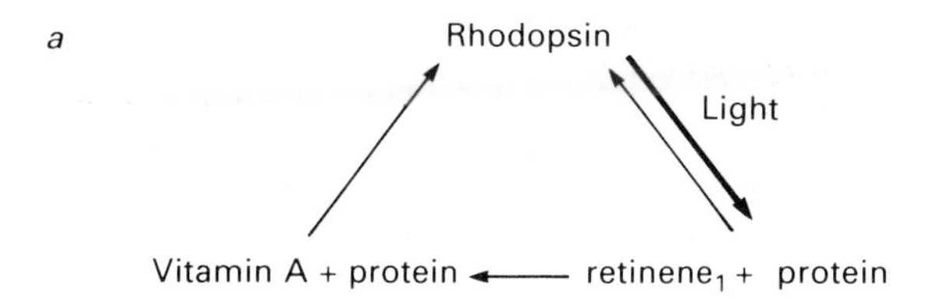

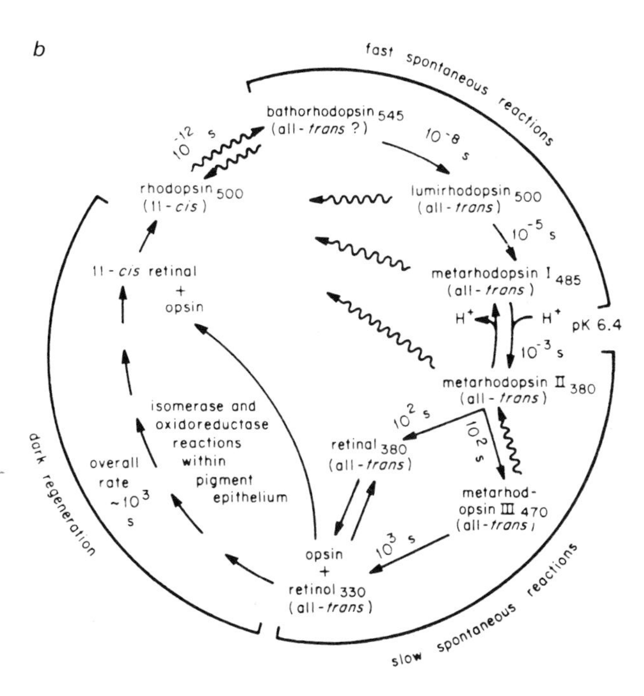

a straight chain. The 9-*cis* and 13-*cis* forms are common, but it was expected that the 11-*cis* form would be less stable, since steric hindrance between the methyl group on C-13 and the hydrogen atom on C-10 should twist the molecule at the *cis* linkage.

Rhodopsin can be reconstituted by mixing opsin with 11-*cis* retinal (Hubbard and Wald, 1952); of the other isomers, only 9-*cis* retinal forms a photosensitive pigment (which has been named 'isorhodopsin'), and its difference spectrum is different from that of natural rhodopsin. The retinal released after exposure of rhodopsin to light, on the other hand, is in the all-*trans* configuration. Rhodopsin can also be broken down in the dark by heating so as to denature the protein part of the molecule, but in this case the retinal liberated is largely in the 11-*cis* form (Hubbard, 1959). From these various results we can conclude that the retinal in rhodopsin is in the 11-*cis* form, but that it is converted into the all-*trans* form by the action of light.

The breakdown of rhodopsin to all-*trans* retinal and opsin after exposure to light takes place in a series of stages (fig. 18.13*b*). The direct action of light appears to be simply to change the retinal from the 11-*cis* to the all-*trans* form (Hubbard and Kropf, 1958); as a result of this a transient intermediate is produced, bathorhodopsin, which can only be observed at temperatures below −140 °C. At higher temperatures there follows a series of changes that are apparently alterations in the internal structure of the opsin molecule. These conformational changes appear to be fairly minor in nature during the lumirhodopsin and metarhodopsin I stages; the evidence for this view is that metarhodopsin I is produced within microseconds of illumination at room temperatures, that totally dry rhodopsin is converted to metarhodopsin I by

Fig. 18.14. Structural formulae of vitamin A_1, shown in full and in shorthand notation, and of three stereoisomers of retinal, shown in shorthand notation.

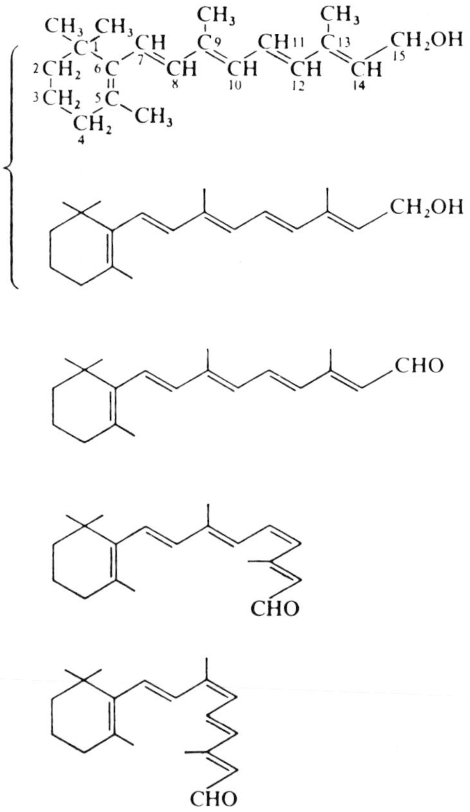

Vitamin A_1 (retinol)

All-*trans* retinal

11-*cis* retinal
(neo-*b*)

9-*cis* retinal
(iso-*a*)

the action of light, and the overall change in the absorption spectrum up to this stage is not very great. The next stage, to metarhodopsin II, appears to involve a more drastic alteration in the structure of the molecule, since it takes longer (the half-time for the change is about 1 ms at room temperature), it requires hydrogen ions and it involves a large change in absorption spectrum which is readily visible as 'bleaching'.

Hubbard and Kropf (1958) discovered that metarhodopsin II and the intermediates preceding it in the visual cycle can all absorb light to produce rhodopsin. This process is called photoregeneration; it seems to be more important in invertebrate visual cycles than in vertebrates. The normal regeneration process in vertebrates is enzymatic, and takes place largely in the pigment epithelium. A number of specialized binding proteins are concerned with the transport of retinol (vitamin A) and retinal within and between the pigment epithelium and the receptor cells (Bridges, 1976; Bridges *et al.*, 1984).

Other visual pigments

We have already seen that the density spectrum of human rhodopsin is slightly different from that of frog rhodopsin, a circumstance which indicates that the two rhodopsins are different pigments. In fact, there are a large number of different pigments, produced by combination of retinal with different opsins. Different authors use different nomenclatures for these pigments. Wald described all retinal pigments extracted from rods as 'rhodopsins' and all those extracted from cones as 'iodopsins'. In Dartnall's system the pigment is named from its λ_{max} value, with a subscript to indicate that it contains retinal; thus frog rhodopsin is 'visual pigment 502_1', and human rhodopsin is 'visual pigment 497_1'.

In addition to these retinal pigments, there is another set of pigments which contains 3-dehydroretinal, which used to be known as retinene$_2$. 3-Dehydroretinal is the aldehyde of vitamin A_2, which differs from vitamin A_1 in possessing a double bond between carbon atoms 3 and 4 (fig. 18.15). Wald described these pigments as 'porphyropsins' when they are extracted from rods; when 3-dehydroretinal is combined with cone opsin (from iodopsin) an artificial pigment, 'cyanopsin', is produced. In Dartnall's nomenclature, 3-dehydroretinal pigments are indicated by the subscript; thus the porphyropsin of tadpoles is 'visual pigment 523_2'.

Retinal densitometry

We shall now consider some of the applications of a very elegant technique, known as retinal densitometry, which is used to measure the concentrations and characteristics of visual pigments in the living eye. The principle of the method is to shine lights of two wavelengths, of which one is absorbed by the pigment and the other not, into the eye and measure the intensity of the light reflected back from the choroid layer behind the retina. Since this light passes through the retina twice, some of it will be absorbed by the visual pigment, and so the absorptive properties of the pigment can be determined from comparison of the intensities of the reflected beams of the two lights.

The retinal densitometry technique was developed by Rushton and his colleagues, among others (see Rushton, 1972). Thus Campbell and Rushton (1955) measured the concentration of rhodopsin in the human eye under various conditions. They

Fig. 18.15. Vitamin A_2 and all-*trans* 3-dehydroretinal.

Vitamin A_2 CH_2OH

All-*trans* 3-dehydroretinal CHO

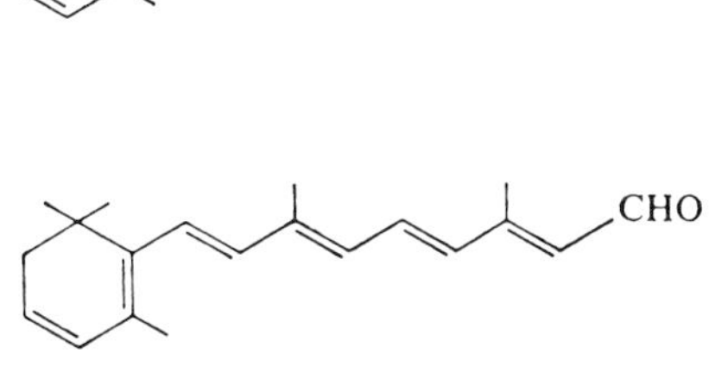

used blue–green light for absorption by rhodopsin, with orange light for the reference beam.

Fig. 18.16 shows the results of one of Campbell and Rushton's experiments. The eye was dark adapted and the reflection was measured from a patch about 15° away from the fovea. The white circles show the effect on the rhodopsin concentration of bleaching with white lights. When the bleaching light intensity was not too high, the rhodopsin concentration reached a steady level at which its rate of bleaching was counter-balanced by regeneration. With increasing bleaching light intensities, this steady level occurred at lower concentrations. The black circles show the recovery of rhodopsin concentration in the dark. The time course of this curve is very similar to that of the scotopic dark adaptation curve when this is plotted in the normal way as log threshold against time. In other words, the *logarithm* of the scotopic sensitivity is proportional to the concentration of rhodopsin in the patch of retina which is illuminated.

Campbell and Rushton performed various other experiments to check that the difference between the reflection of the blue–green and orange beams really was caused by absorption of the blue–green light by rhodopsin. Rhodopsin is a rod pigment, and therefore the differential absorption in different parts of the retina should be proportional to the rod concentration. Fig. 18.17 shows that this is so (notice, particularly, that there is no absorption at the fovea or at the blind spot). The effectiveness of bleaching of lights of different wavelengths should be related to the scotopic sensitivity curve. This, again, was found to be so: lights which looked as though they were equally bright produced equal amounts of bleaching when brightened by the same factor. Rushton (1956) measured the reflection from the back of the eye at different wavelengths. If such measurements were made before and after bleaching, the difference spectrum of the absorbing substance could be obtained; it was very similar to that of rhodopsin in solution.

Fig. 18.16. Measurement of rhodopsin concentration in the human eye during bleaching under three levels of illumination (white circles) and its subsequent regeneration in the dark (black circles). The displacement of the purple wedge is proportional to the optical density of the retinal rhodopsin. Further explanation in the text. (From Campbell and Rushton, 1955.)

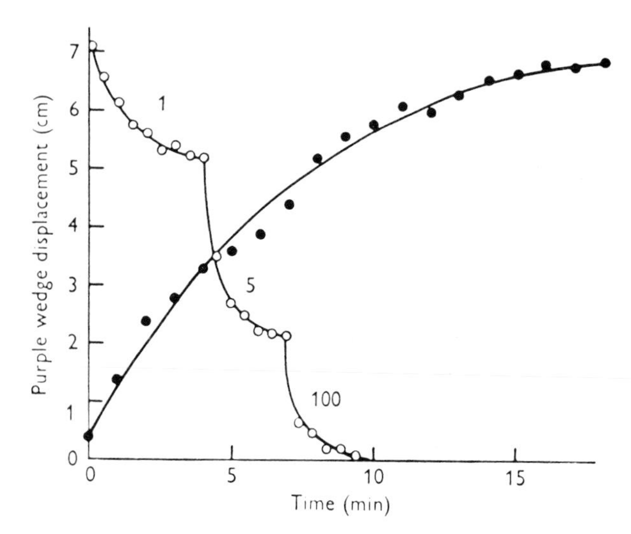

Rushton also applied the retinal densitometry technique to the measurement of cone pigments. To avoid interference from the rhodopsin in the rods, he confined the measurements to the fovea. Colour vision in the central part of the fovea is dichromatic, that is to say any colour seen by the fovea can be matched by a combination of two other colours of suitable brightness. This suggests that there are two different pigments present. In the type of colour blindness known as protanopia, it is impossible to distinguish between red and green, and the fovea is apparently monochromatic, lacking the red-sensitive pigment. Thus the fovea of the protanope offers a useful method of investigating a particular cone pigment in isolation.

Using retinal densiometry on the fovea of a subject with protanopia, Rushton (1963) detected a green-sensitive visual pigment with a λ_{max} near to 540 nm. Recovery after bleaching was much faster than for rod rhodopsin, being complete within 6 min. By comparison with the fovea of a subject with normal colour vision, the properties of a second, red-sensitive, pigment could be inferred.

Microspectrophotometry

It would clearly be of considerable interest to measure the light-absorbing properties of individual rods and cones. Microspectrophotometry is a method for doing this. It is not an easy technique: a very small beam of light has to be focused on the receptor cells of isolated retinas.

Difficulties arise in eliminating light scattering and in measuring the low light intensities that have to be used in order to avoid excessive bleaching of the pigments (see Liebman, 1972; Knowles and Dartnall, 1977).

First results with the technique established a number of crucial facts. Liebman (1962) identified rhodopsin and some of its photoproducts in frog rods. Marks (1965) found that the difference spectra of goldfish cones fell into three separate groups, suggesting that there is one of three different pigments in each cone. Similar results were obtained for human and monkey retinas by Marks *et al.* (1964) and Brown and Wald (1964). These findings were very exciting, since they fitted precisely the requirements of the Young–Helmholtz theory of colour vision.

In recent years more measurements have been made with improved apparatus. Let us look at some results obtained by Bowmaker and Dartnall (1980) on human rods and cones. A retina was obtained from a patient whose eye had to be removed because of a malignant choroid tumour. Pieces of it were teased apart on a microscope slide and the photoreceptor cells were immobilized under a coverslip. The microspectrophotometer compares the intensity of two monochromatic light beams, one passing through the outer segment of a rod or cone, and the other not. The wavelength of the light is scanned from 700 to 350 nm and back. This double scan checks that no appreciable bleaching has taken place during the determination of the absorbance spectrum.

A typical record, giving the absorbance spectrum of a rod, is shown in fig. 18.18. Absorbance spectra from cones fell into three groups, corresponding to blue-sensitive, green-sensitive and red-sensitive cones. The mean absorbance curves for these four types are shown in fig. 18.19.

Similar results were obtained from a further six subjects (Dartnall *et al.*, 1983). The mean λ_{max} values were 496 nm for the rods, and 419, 531 and 558 nm for the cones. For the red- and green-sensitive cones, there was some evidence that different cones may have slightly different pigments, with λ_{max} values separated by up to 9 nm.

Microspectrophotometry has now been applied to a variety of vertebrates, especially among fish, birds and primates (see Bowmaker, 1984). Individual variation in colour pigments, comparable

Fig. 18.17. Relative rhodopsin concentrations in the human eye at different positions in the horizontal plane, as measured by retinal densitometry. White and black circles indicate the change in absorption following bleaching (two different methods) with a constant quantity of light. The curve shows the number of rods per square millimetre, as determinined by Østerberg. (From Campbell and Rushton, 1955.)

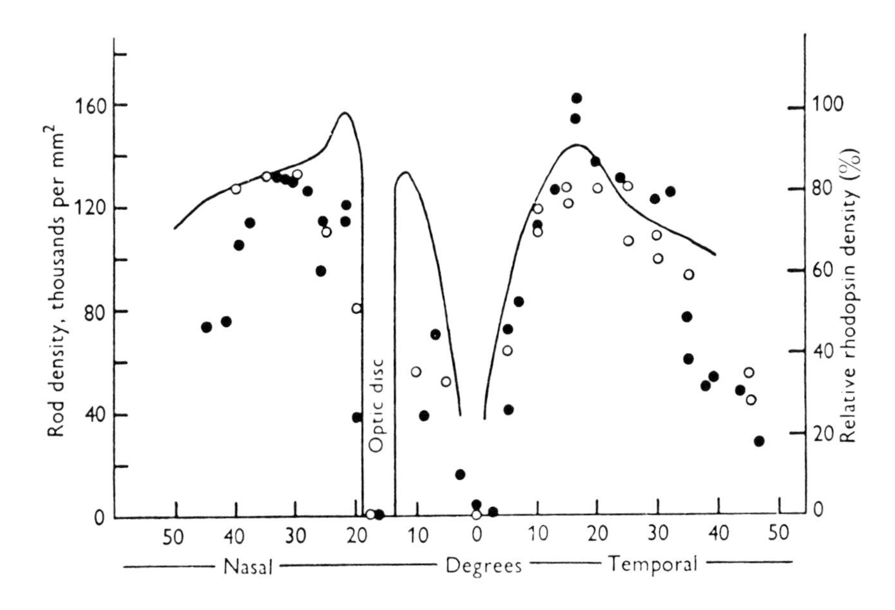

to the situation described by Dartnall and his colleagues in man, occurs in a number of primate species (Jacobs, 1986).

The organization of rhodopsin in rods

If the molecules of a light-absorbing substance are not arranged at random, but are oriented in one particular plane or one particular direction, then light waves vibrating in a particular plane are preferentially absorbed. Thus we can determine whether the visual pigment molecules in a photoreceptor cell are so organized by observing the absorption of light polarized in different directions. This was first done by Schmidt (1938), using the outer segments of frog rods. He found that when light was shone along the axis of the rods, they appeared to be red (indicating that light was absorbed) whatever the plane of polarization. However, with illumination from the side, the rods appeared to be red when the plane of polarization was parallel to the transverse plane of the outer segment, but colourless (indicating that no light was absorbed) when it was parallel to the longitudinal axis (fig. 18.20). These results were confirmed and extended by Denton (1959) and by Wald, Brown and Gibbons (1962). They imply that the rhodopsin molecules (in particular, the retinal moieties) are oriented in transverse planes across the outer segment, but that there is no preferential direction of orientation within these planes. The most obvious way in which this could be done is by incorporation of the rhodopsin molecules in or on the membranes of the sacs in the outer segment.

Such an arrangement must increase the efficiency of light absorption by the rod pigment, since light arrives 'end-on' at the rods in the intact eye; any molecules oriented so as to absorb light vibrating in a plane parallel to the longitudinal axis would be ineffective in vision. The absorption of unidirectional light by rhodopsin solutions (in which the molecules are, of course, oriented at random) is therefore less than that of rhodopsin in the rods, by a factor of ⅔.

The molecular structure of rhodopsin

The amino acid sequence of cattle rhodopsin has been determined by using the techniques we have seen in previous chapters (Ovchinnikov *et al.*, 1983; Hargrave *et al.*, 1983). It consists of a single chain of 348 amino acids, which is shown in fig. 10.13. Structural analysis suggests that there are seven stretches of hydrophilic residues which form α-helices and cross the membrane from one side to the other (figs. 18.21 and 18.22).

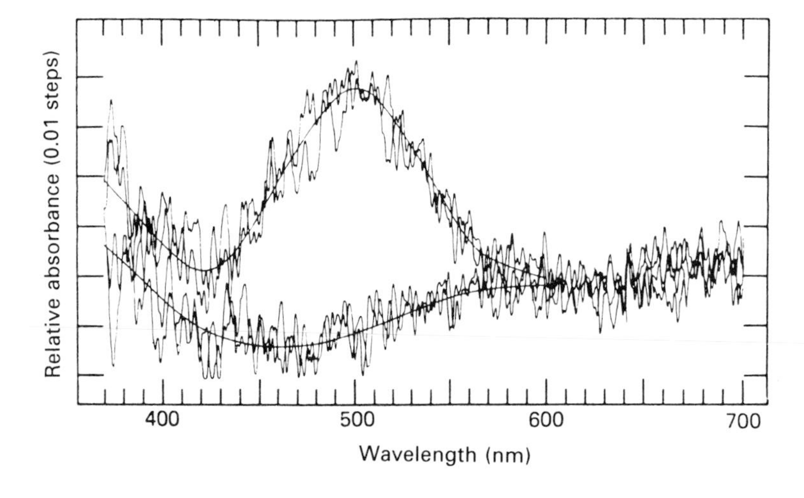

Fig. 18.18. Absorbance spectra of a single human rod outer segment, determined by microspectrophotometry. The microspectrophotometer beam was scanned from blue (short wavelengths) to deep red and back again, giving the noisy traces shown; continuous curves are drawn through the middle of the noise. The upper curve was determined for a dark-adapted rod exposed only to dim deep red light after surgery, and the lower curve shows a similar record after bleaching with white light. (From Bowmaker and Dartnall, 1980.)

Fig. 18.19. Mean absorbance spectra of the four classes of human photoreceptors. The 498 curve is that for rods, the others are for cones: 'blue-sensitive' (420), 'green-sensitive' (534) and 'red-sensitive' (564). Notice that the ratios of the absorbances for the three types of cone are different at different wavelengths; this is the physical basis for colour vision. (From Bowmaker and Dartnall, 1980.)

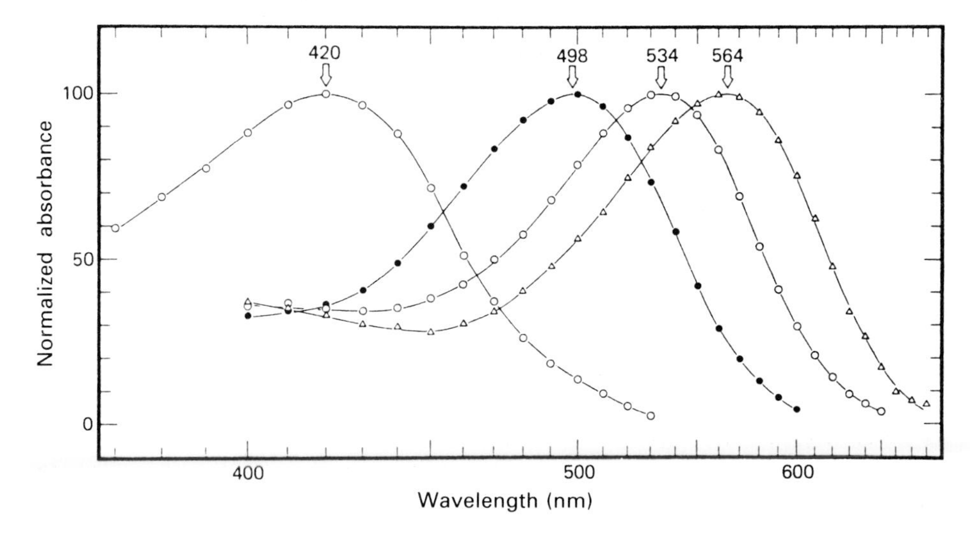

Fig. 18.20. The absorption of polarized light by an intact retina. No absorption occurs (so that the rods appear colourless, instead of red) when the transverse components of the light waves are parallel to the longitudinal axes of the rods.

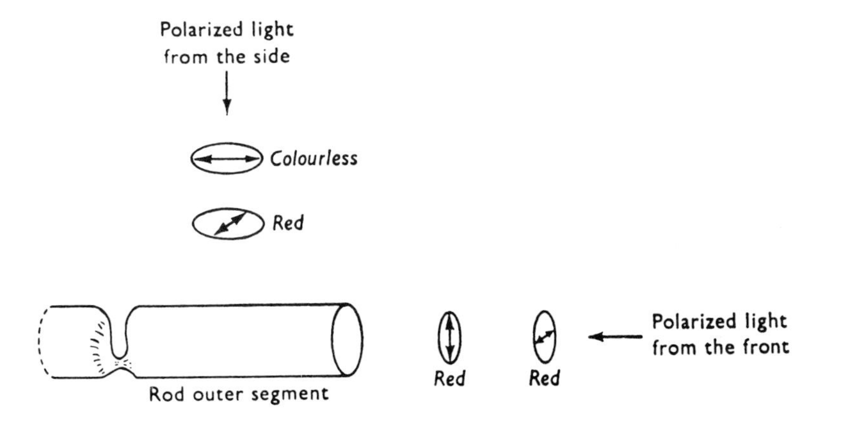

Fig. 18.21. The transmembrane topography of cattle rhodopsin. The graph on the left is a hydrophobicity plot of the amino acid sequence, and the sketch on the right shows how the chain is folded across the rod disc membrane. Squares show sites accessible to hydrophilic probes applied from the luminal side of the membrane, circles show those accessible from the cytoplasmic side. Horizontal bars show the presumed transmembrane segments. Arrows mark the positions of introns (sections of the DNA sequence that are not represented in the protein molecule) in human and cattle rhodopsin genes. (From Nathans, 1987.)

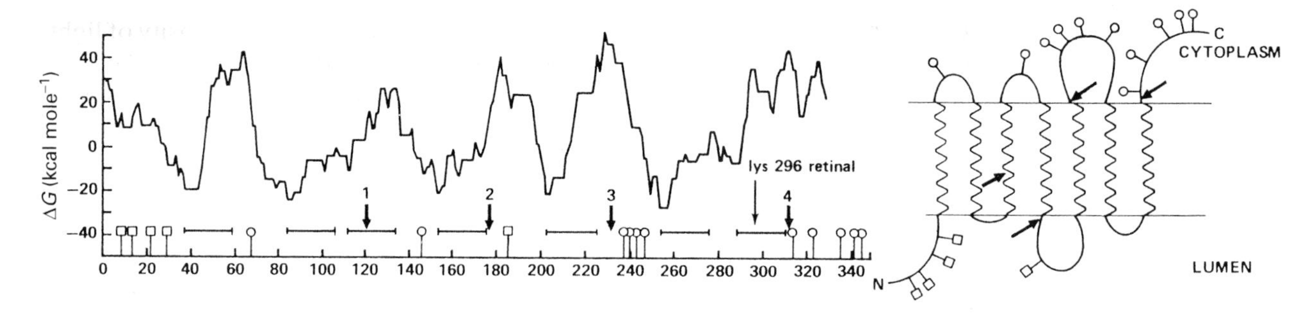

Fig. 18.22. Possible arrangement of the rhodopsin chain as seen from the ctyoplasmic face of the rod disc membrane. The retinal molecule is shown nestling between the transmembrane α-helices. Arrows show possible sites of phosphorylation. Letters show some of the amino acid residues that are also found in the *Drosophila* (fruit fly) visual pigment molecule. (From Baehr and Applebury, 1986.)

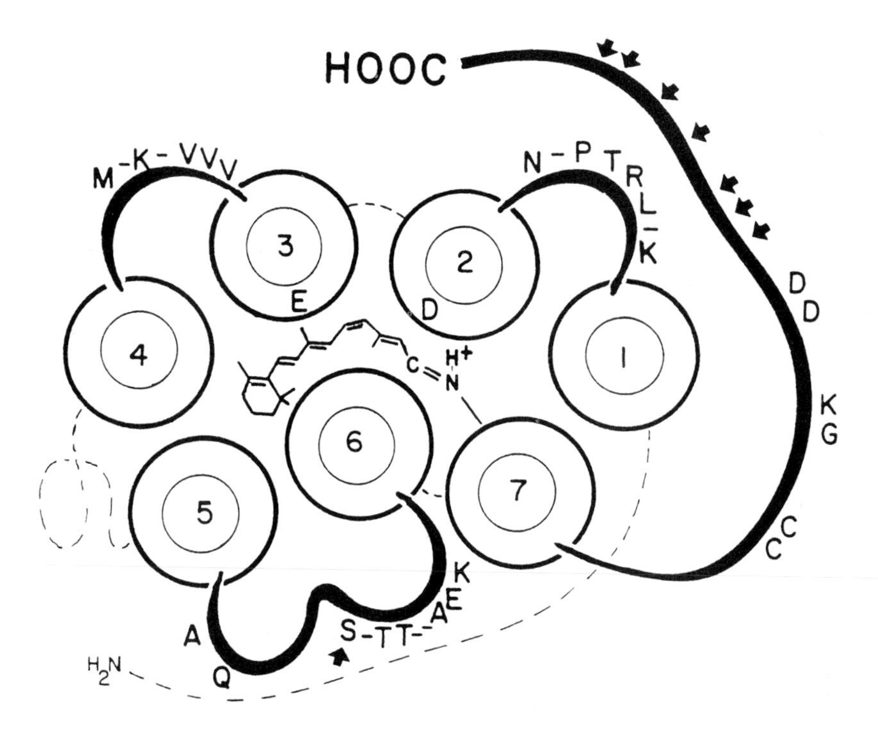

By using radioactive probes that will attach to the hydrophilic sections of the rhodopsin molecules, it is possible to tell which side of the membrane these sections are on. Thus intact outer segment discs will only expose the cytoplasmic side of the disc membrane to such probes, whereas those which have been frozen and thawed will break so as to expose the luminal or intradisc side as well (Ovchinnikov, 1982; Mullen and Akhtar, 1983; Barclay and Findlay, 1984). The results show that the N-terminal end of the molecule is on the luminal side of the membrane (this is equivalent to the extracellular side) and the C-terminal end remains on the cytoplasmic side. They also confirm that there are cytoplasmic sections between helices 1 and 2, 3 and 4, and 5 and 6, and that there is a luminal section between 4 and 5, as is shown by the circles and squares in fig. 18.21.

The retinal chromophore is attached to the lysine residue at position 296, which is in the middle of the seventh transmembrane section. Thus it sits in a largely hydrophobic environment, which is indeed to be expected from its structure and from our knowledge that retinol is one of the fat-soluble vitamins. It may be that it is surrounded by the seven α-helices, as is shown in fig. 18.22.

By using DNA encoding for cattle rhodopsin as a hybridization probe, it has been possible to isolate the genes coding for rhodopsin and the cone pigments from human DNA (Nathans and Hogness, 1984; Nathans *et al.*, 1986*a*). The deduced amino acid sequence for human rhodopsin is identical to the cattle sequence at 94% of its residues. Sequences for the cone pigments show that the red- and green-sensitive pigments are very similar to each other, with 96% of their residues identical. Comparisons between the blue- and green-sensitive pigments, and between each of them and rhodopsin, show greater differences, with 40–45% identity. This suggests that the red- and green-sensitive pigments are derived from a common ancestor relatively recently in evolutionary terms. Extension of this work is leading towards an understanding of the genetic basis of colour vision abnormalities (Nathans *et al.*, 1986*b*; Nathans, 1987).

The absolute threshold

What is the minimum amount of light that can be detected by the fully dark-adapted eye? Because of the quantal nature of light, this question can be rephrased as: what is the minimum number of quanta required for visual excitation? The problem was investigated, in a most elegant set of experiments, by Hecht, Schaer and Pirenne (1942), whose work we shall now examine.

The principle of this work was to ensure that conditions were optimal for the perception of light by human subjects, to measure the intensity of light at the visual threshold, and finally to calculate the number of quanta absorbed by the visual pigment in the retina under these conditions.

In order to secure optimum conditions for perception, a number of precautions were taken. Subjects were fully dark-adapted by remaining in complete darkness for at least half an hour. Using the 'fixation and flash' technique the test flash was positioned so that its image fell 20° to the left (in the left eye) of the fovea, a position which is in the region of maximum sensitivity of the retina. The diameter of the test flash was 10′, which is the angle at which the product of area and threshold intensity is least. The product of threshold intensity times length of flash is constant at times up to about 10 ms, but increases with longer flashes; hence flashes of 1 ms duration were used. The wavelength of the light used was 510 nm, which is near the peak of the scotopic sensitivity curve.

The arrangement of the apparatus is shown in fig. 18.23. The eye looks through the artificial pupil P, fixates the red-light point FP, and sees the test field formed by the lens FL and the diaphragm D. The light for this field comes from the lamp L through the neutral filter F and the wedge W (which together control the intensity), and through the double monochromator M_1M_2 (which selects the wavelength). The shutter S opens for 1 ms, and is controlled by the subject. Careful calibrations of the light intensity at P were made, so that the light energy incident on the cornea during a flash could be calculated for different neutral filters and different settings of the wedge W.

For each subject, a series of flashes of different intensities was presented many times, and the frequency of seeing the flash was determined for each intensity. Graphs of frequency of seeing against intensity were then drawn, and the threshold was regarded as that intensity at which the flash could be seen in 60% of the trials. Seven subjects were tested, some of them more than once. Their thresholds, measured as the light energy incident on the cornea, were in the range $2.1\text{–}5.7 \times 10^{-10}$ erg.

As we have seen, the energy content, $h\nu$, of a quantum of light is dependent upon its wavelength. For light with a wavelength of 510 nm, $h\nu$ is 3.89×10^{-12} erg. Hence the number of quanta incident on the cornea at the absolute threshold was in the range of 54–148.

These values do not represent the number of quanta absorbed by the visual pigment, since some light is lost between the cornea and the retina, and not all of the light incident upon the retina is absorbed by it. Measurements by other workers had shown that about 4% of the light incident upon the cornea is reflected from its surface, and that (at a wavelength of 510 nm) about 50% of the light entering the eye is absorbed by the lens and ocular media before reaching the cornea. The main problem here, then, is the measurement of the proportion of the light incident upon the retina which is absorbed by the visual pigment. Hecht and his colleagues solved this problem in a rather ingenious manner. We have seen that the shape of the absorption spectrum of a pigment depends on its concentration (and therefore on the proportion of light absorbed), being broader at higher concentrations (see fig. 18.9*b*). It follows that the proportion of light absorbed by the rhodopsin in the retina can be obtained by comparing the shape of the scotopic luminosity curve with the shapes of the absorption spectra of different concentrations of rhodopsin in solution. The results suggested that between 5 and 20% of the light incident on the retina is absorbed by the visual pigment; for the purposes of calculation, the 20% value was used. (More recent measurements suggest that this figure was still too low; but we shall return to this point later.)

Putting these figures together, we find that, of the light incident at the cornea at a wavelength of 510 nm, 96% enters the eye, 48% reaches the retina, and about 9.6% is absorbed by the visual pigment. Hence, at the threshold of vision, when 54–148 quanta are incident on the cornea, 5–14 quanta are absorbed by the visual pigment.

This figure is of such importance that it is most desirable to have an independent check on its accuracy. Such a check was obtained by Hecht and his colleagues from measurements of 'frequency-of-seeing' curves and consideration of the statistical properties of light flashes containing small numbers of quanta. Any flash of light of constant *average* quantal content (measured at the retina) will in fact produce fluctuating numbers of quanta; the actual number of quanta absorbed by the retina will also fluctuate, and the relative frequencies of the quantal contents of the flashes will be given by the terms of a Poisson distribution (compare (7.1)). Thus, if n is the number of quanta which it is necessary for the retina to absorb in order to be able to see a flash, and a is the average quantal content per flash in a series of flashes of 'constant' intensity, then the probability P_n that any one flash will yield exactly the necessary number of quanta is given by

$$P_n = \mathrm{e}^{-a} \cdot \frac{a^n}{n!}. \qquad (18.3)$$

From this equation, it is possible to calculate the probabilities of n or more quanta occurring for different values of a and n; the results of this calculation are shown in fig. 18.24.

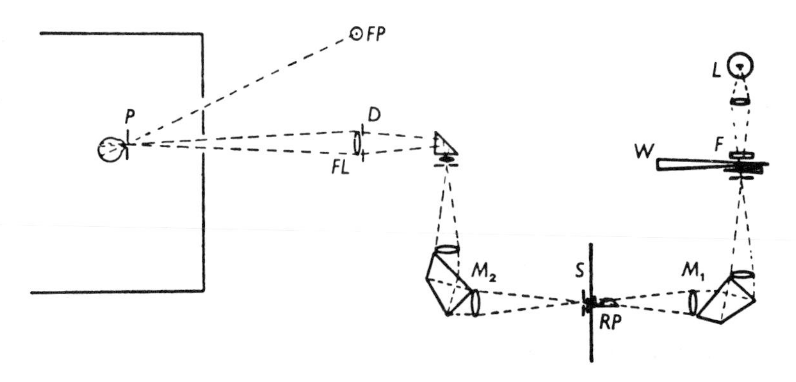

Fig. 18.23. Optical system for determining the absolute threshold of human vision. Explanation in text. (From Hecht *et al.*, 1942.)

The next step was to determine 'frequency-of-seeing' curves at flash intensities in the region of the absolute threshold. The method used was identical with that for determining the absolute threshold in the first instance. Three observers were used, and the results are shown in fig. 18.25. The abscissae in these curves represent the amounts of light incident upon the cornea, whereas that of the theoretical curves in fig. 18.24 is equivalent to the amount of light absorbed by the visual pigment. Hence the theoretical curves must be fitted to the experimental curves by moving them along the x axis. When this was done, it was clear that the value of n must be in the range 5–8. (This argument assumes that the variation in frequency-of-seeing is entirely caused by the fluctuations in the stimulus; if there is also some fluctuation in the stimulus/response ratio, then the frequency-of-seeing curves would be stretched over a greater range of average quantal contents, so that their slopes would be less, and hence the value of n obtained would be too low. These experiments, therefore, show that the true value of n cannot be less than 5–8; it may be a little more.)

Fig. 18.24. Solutions of (18.3) for different values of n. (From Hecht *et al.*, 1942.)

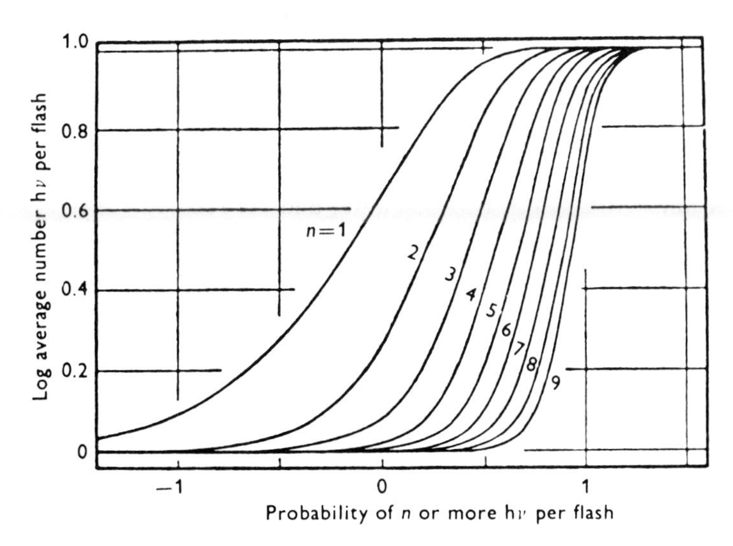

Fig. 18.25. The relation between average quantal content (at the cornea) of a flash of light and the frequency with which it was seen, for three different observers. The curves are drawn as in fig. 18.24 for $n = 5$, 6 and 7, but are moved along the abscissa in order to fit the points (this procedure is permissible, since the abscissa measures the number of quanta incident at the cornea, whereas n is the number absorbed by the retinal rods). (From Hecht *et al.*, 1942.)

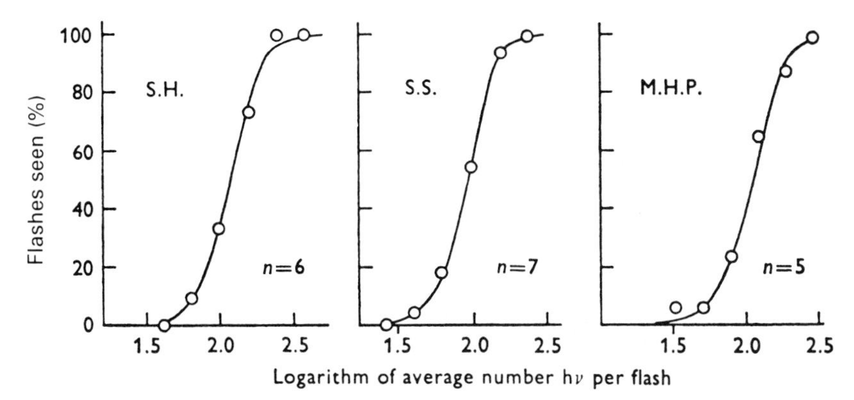

The general conclusion of this investigation is that, under optimal conditions, a minimum of about six quanta must be absorbed by the retina if a light source is to be detected. What does this mean in terms of the excitation of the retinal cells? Let us first consider the requirements for excitation of the rods. We can suppose either (i) that the absorption of one quantum is sufficient to excite a rod, or (ii) that two or more quanta need to be absorbed by a rod before it is excited. We would eliminate the second hypothesis if we could show that the probability of any of the illuminated rods absorbing the requisite number of quanta from a flash of light at the visual threshold is less than the probability of seeing the flash. Our problem can therefore be stated as follows: if n quanta are randomly distributed over N rods, what is the chance of one or more of these rods receiving two or more quanta? The mean number of quanta absorbed per rod is n/N. If the actual numbers of quanta absorbed by each rod are distributed according to a Poisson distribution, then the probability P_x that any particular rod will absorb x quanta is

$$P_x = e^{-n/N} \cdot \frac{(n/N)^x}{x!}.$$

Thus when $x = 0$,

$$P_0 = e^{-n/N}$$

and when $x = 1$,

$$P_1 = e^{-n/N} \cdot \frac{n}{N}.$$

Hence the probability that any particular rod will absorb either 1 or 0 quanta is

$$P_0 + P_1 = e^{-n/N}\left(1 + \frac{n}{N}\right).$$

Therefore the probability that *every* rod in the group of N rods will absorb either 1 or 0 quanta is

$$(P_0 + P_1)^N = e^{-n}\left(1 + \frac{n}{N}\right)^N.$$

Hence the probability, Q, that one or more rods will absorb more than one quantum is given by

$$Q = 1 - e^{-n}\left(1 + \frac{n}{N}\right)^N. \qquad (18.4)$$

We have now to decide what values of n and N are to be inserted in (18.4) for the conditions obtaining at the absolute threshold. The power of the average human eye is about 60 dioptres, which is equivalent to a focal length of 16.67 mm. Thus the image of a 10′ circular field is (16.67 × sin 10′) mm in diameter, i.e. 48.5 μm. Its area is therefore $1.84 \times 10^{-3}\,\text{mm}^2$. According to Østerberg (1935), there are about 150 000 rods per square millimetre in the region of the retina on which the experiments of Hecht *et al.*, were performed, so that a 10′ field will contain about 273 rods. However, the optical image of a 10′ field will be rather larger than this, because of diffraction effects at the pupil; Pirenne (1962) suggests that the actual number of rods in the illuminated field should be about 500. However, let us take a very conservative value of N, 250. Using a similarly conservative value for n, 10, (18.4) becomes

$$Q = 1 - e^{-10}(1.04)^{250}$$
$$= 0.178.$$

Thus, if the two-quanta-per-rod hypothesis were correct, we would expect only 17.8% of the flashes to be seen when the flash intensity is at the visual threshold. But, in fact, 60% of them are seen, since this is how the visual threshold has been defined. We must conclude that a visual sensation arises if a small number of rods absorb a single quantum each.

A complication

Hecht, Schlaer and Pirenne assumed that 20% of the light incident on the retina is absorbed by the photopigment. More direct measurements have since been made which suggest that a much higher figure should be used. Thus Alpern and Pugh (1974), using retinal densitometry conclude that about 55% of the light incident on the retina is absorbed, and Dobelle, Marks and MacNichol (1969) using microspectrophotometry conclude that the figure is at least 50%. How does this affect Hecht, Schlaer and Pirenne's conclusions? Of the light incident at the cornea, about 24% is absorbed by the visual pigment. So at the threshold of vision, when 54–148 quanta are incident on the cornea, 13–35 are absorbed by the visual pigment. This makes the conclusion that one quantum suffices to excite a rod less well founded. Thus, in (18.4), when n is 13, Q is 0.28, when n is 20, Q is 0.53 and when n is 25, Q is 0.69.

However, as Brindley (1970) points out, measurements on thresholds for larger fields are quite conclusive in this respect. Thus Stiles (1939), again using the fixation-and-flash method, measured the threshold for detection of a square of side 1.04° at 5° from the fovea, with a light of 510 nm exposed for 63 ms. His results indicate that such a stimulus had to contain 122 quanta (at the cornea) in order for 50% of the flashes to be seen. So about 59 quanta will reach the retina and about 33 of them will be absorbed by the visual pigment. At 5° from the fovea there are about 90 000 rods per square millimetre, and so the number of rods in the illuminated field is just over 8000. So, applying (18.4) with $n = 33$ and $N = 8000$, we get

$$Q = 1 - e^{-33}\left(1 + \frac{33}{8000}\right)^{8000}$$
$$= 0.066,$$

which is clearly very much less than 0.50, the probability of seeing the flash. Even with $n = 59$ (corresponding to absorption of all the incident quanta by the retina), $Q = 0.195$, which is still a satisfactory figure for the one-quantum-per-rod hypothesis.

An alternative approach

The design of most threshold experiments implies that the subject either sees or does not see a flash of light. He has to make a yes or no decison, and 'false positives' (saying 'yes' when there was no flash) are discouraged. One can argue that this may lead to a cautious approach by the subject, so that an ability to respond to very low numbers of quanta would be missed.

Sakitt (1972) approached the problem by employing signal detection theory (Tanner and Swets, 1954; McNicol, 1972). This assumes that the observer can choose his own response criterion; the lower this is, the higher his frequency of seeing and also the higher his false positive rate. Sakitt's experimental design required the observer to rate the flash of light on a 0–6 scale. A rating of 0 meant that definitely no flash was seen, one of 1 meant that it was very doubtful that a flash was seen, 2 meant slightly doubtful, 3 a dim light, and so on. The experiments were done with three different subjects.

The results of Sakitt's experiments were very interesting. With a rating of 1, the proportion of correct answers was much higher than would be obtained by random guessing, so clearly a rating of 1 must count as detection of the signal. Frequency of seeing curves for ratings of 1 should then give the number of quanta required for detection. The answers were 1, 2 and 3 rod signals for the three subjects. This means that different subjects may choose different criteria for detection, and that these are low and may be as low as a single rod signal.

These results are completely in agreement with the major conclusion of the experiments by Hecht and his colleagues: one quantum suffices to excite a rod. They are largely in agreement over the second conclusion, that a small number of rods needs to be excited in order to produce a visual sensation, but it is a matter for debate as to just what that small number is.

Phototransduction

The quantal sensitivity of the photoreceptor cells has implications for the mechanism of phototransduction, the process whereby the capture of light energy results in an electrical output from the cell. We have seen that the absorption of one photon serves to excite a rod. This means that conformational change in one rhodopsin molecule can produce a response in the whole cell which is sufficient to affect the activity of succeeding cells in the sequence from photoreceptor to brain. What is the mechanism of this remarkably sensitive process? Let us begin by looking at the electrical activity of the photoreceptors.

The electroretinogram

The electroretinogram is a mass electrical response of the retina recorded with fairly large external electrodes, one placed on the cornea and the other at some indifferent point in the body or (when using excised eyes) behind the optic bulb. An enormous amount of work has been done on this response since its discovery by Holmgren in 1865 (see Granit, 1947, 1962), but we shall here just outline the main features of the phenomenon.

Fig. 18.26 shows an electroretinogram recorded by Granit (1933) from a decerebrate cat. The first part of the response is the small, cornea-negative *a*-wave. This followed by a rapid cornea-positive response, the *b*-wave, which is finally followed by a much slower response, the *c*-wave. On the cessation of illumination, there is a small cornea-positive deflection, the *d*-wave. In vertebrates other than mammals, the *d*-wave is usually much larger, and is comparable in size with the *b*-wave.

The components of the electroretinogram can also be detected when records of local activity in the retina are made with a glass micropipette electrode. Using this technique, Brown and Wiesel (1961) determined the amplitude of these various components at different levels in the retina, in an attempt to localize their origin. They found that the *a* and *c*-waves were associated with the receptor layer; there was some evidence that the *a*-wave arises from activity in the rods and cones, whereas the *c*-wave arises from the cells of the pigment epithelium. The *b*-wave was maximal when the electrode was in the inner nuclear region; it appears to be caused by activity in the glial cells (Müller cells, fig. 18.2) which occur at that level (Miller and Dowling, 1970).

Photoreceptor potentials

Since the early stages of the electroretinogram arise in the receptor cells they can be described as receptor potentials. Using intense brief flashes as the light stimulus, Brown and Murakami (1964) found that the early stages of the locally recorded electroretinogram consisted of two parts (fig. 18.27), an initial *early receptor potential* with very brief latency, followed by a larger *late receptor potential.* The rising phase of the late receptor potential corresponds to the *a*-wave of the electroretinogram. Cone (1964) showed that the early receptor potential in the rat could be seen in electroretinograms recorded by conventional methods, that its action spectrum was similar to that of the visual pigment, and that its size was linearly proportional to the amount of pigment bleached by the flash. Many different observations suggest that the early receptor potential is not brought about by changes in membrane conductance (see Pak, 1968; Arden, 1969). It is resistant to anoxia and to treatment with a wide variety of different chemical agents that affect most other bioelectric potentials, such as potassium chloride solutions, acids and alkalis, fixatives and so on.

It seems probable that the early receptor potential reflects charge displacement in the visual pigment molecules during the conformational changes that follow the absorption of photons. In other words, it is a consequence of light absorption, and not part of the causal sequence of events that leads to visual sensation. The late receptor potential, however, appears to be produced by currents flowing through the receptor cell membranes, and is therefore an essential component of retinal signalling. For more information, it was desirable to record from individual photoreceptors with intracellular electrodes.

Rods and cones are rather small cells and it is technically not easy to make intracellular recordings from them. Tomita and his colleagues were

Fig. 18.27. The early and late receptor potentials, as recorded by a tungsten microelectrode in the extrafoveal retina of a monkey. (From Brown and Murakami, 1964.)

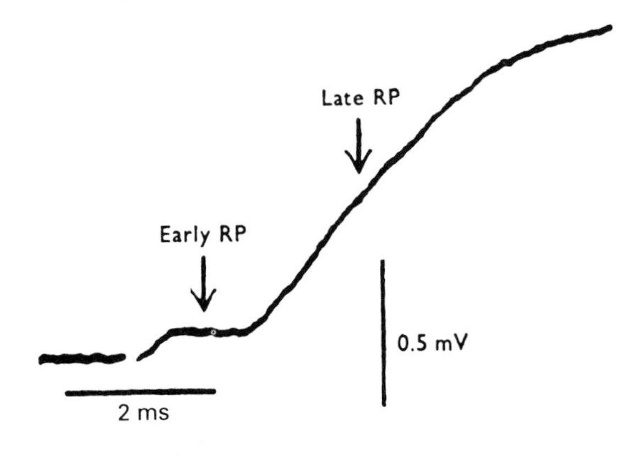

Fig. 18.26. The electroretinogram of a dark-adapted decerebrate cat, to show the *a, b, c* and *d*-waves. The eye was illuminated for the duration of the thick line below the record, on which time intervals of 0.5 s are marked. (From Granit, 1933.)

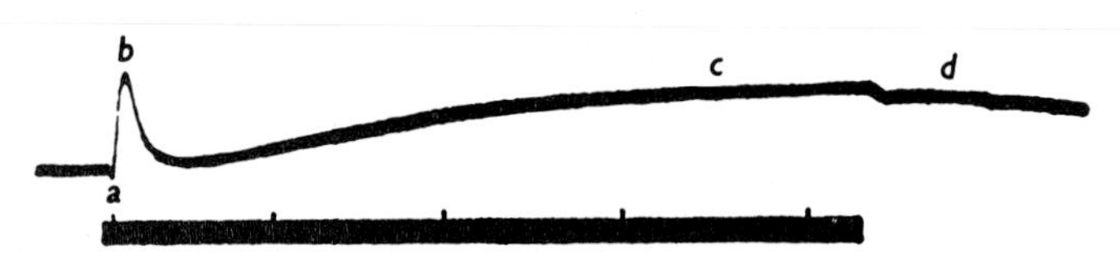

able to insert high-resistance microelectrodes into single cones of the carp by jolting the retina upwards towards the electrode in short steps of about 1 μm (Tomita, 1965, 1970, 1972; Tomita *et al.*, 1967). The microelectrode then records the membrane potential of the inner segment of the cone; outer segments are too small to give satisfactory records, but of course electrical changes in the outer segment cell membrane will be recorded, with some attenuation, in the inner segment. They found that illumination of the cones produced a hyperpolarization, as is shown in fig. 18.28.

Investigations on other preparations (gecko and frog rods, mudpuppy and turtle cones) have produced similar results. These hyperpolarizations produced by vertebrate photoreceptors on illumination are in marked contrast to the response of most other sensory cells (including most invertebrate photoreceptors), in which stimulation causes a depolarization of the cell membrane. Further, it is found that during the hyperpolarization the membrane resistance increases (Toyoda *et al.*, 1969; Baylor and Fuortes 1970), suggesting that illumination reduces the conductance of the cell membrane to some ion or ions. Fig. 18.29 shows some of the evidence for this conclusion: the voltage change produced by brief current pulses increases on illumination.

Fig. 18.28. Intracellular recordings from cones of the carp, showing the hyperpolarizations produced by light. In each case the cone was illuminated with a series of flashes of monochromatic light in steps of 20 nm across the spectrum. The scale at the top gives the light wavelength. These records show the three types of spectral sensitivity found, with the maximal responses in the blue (*a*), green (*b*) and red (*c*) regions of the spectrum. (From Tomita *et al.*, 1967.)

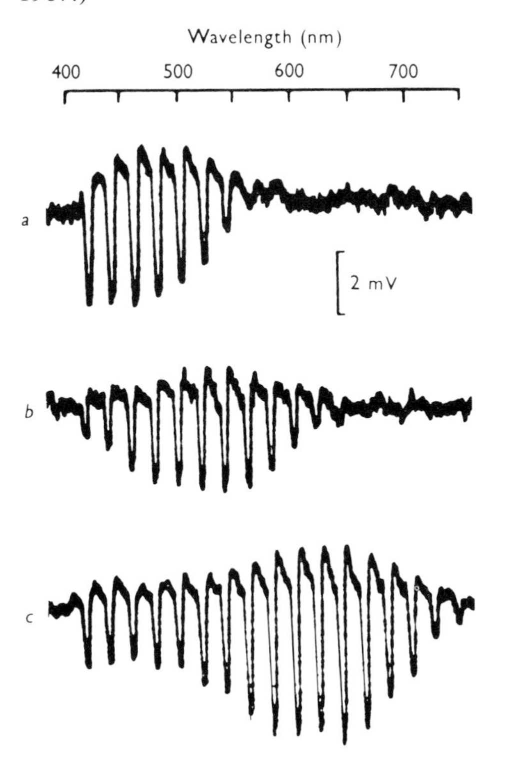

Current flow in rods

The current flow associated with the late receptor potential and also in the dark was investigated by Hagins and his colleagues (Penn and Hagins, 1969; Hagins *et al.*, 1970; Hagins, 1972). They used slices of rat retina which they were able to keep alive mounted on a microscope stage. In order to examine these slices without bleaching their rhodopsin, they were illuminated by infrared light and visualized by means of image converters. Microelectrodes could be accurately placed at different depths in the retina so that the extracellular potentials at these depths could be measured. They found that, in the dark, the potential difference between the measuring electrode and a reference electrode placed at the tips of the rod outer segments increased with increasing penetration into the rod layer (fig. 18.30).

If the resistance of the extracellular space were constant, then the longitudinal current flow would simply be the rate of change of voltage with distance (compare (4.2)) and therefore could be directly obtained from fig. 18.30. However, it is not, and must therefore be measured. To do this, two microelectrodes with tips separated by a depth of 10 μm, were driven through the retina, and the potential (ΔV) between them recorded. At the same time current pulses were passed across the whole retina: the size of ensuing voltage pulses is proportional to the resistance (ΔR) between the two electrodes. The longitudinal flow between the two electrodes (I) is then given by Ohm's law:

$$I = \frac{\Delta V}{\Delta R}.$$

The results (fig. 18.31) show that the longitudinal dark current increases steadily as the electrode traverses the layer of rod outer segments and then begins to fall as they enter the layer beneath them. This means that there is a steady inflow of current into the rod outer segments, supplied by a corresponding outflow from the inner segments and nuclear regions. In similar experiments on frog

Fig. 18.29. Evidence that light increases the membrane resistance in turtle cones. Superimposed tracings of six responses to brief flashes of light at zero time. For each trace a brief pulse of depolarizing current, of the same strength but occurring at different times, was applied through the microelectrode. The voltage changes produced by this pulse are larger during the hyperpolarization following the flash of light. (From Baylor and Fuortes, 1970.)

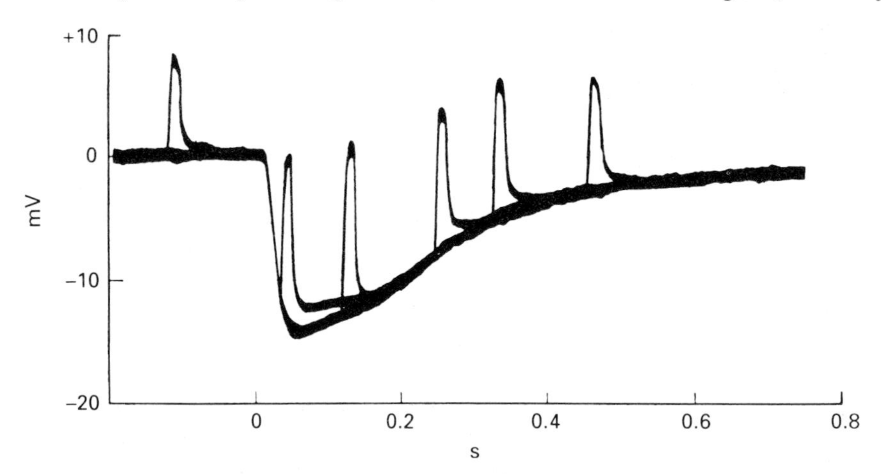

Fig. 18.30. Extracellular potentials recorded from rat retina in the dark. One electrode was positioned at the tips of the rod outer segments, the other had penetrated the retina from the side to a depth of about 100 μm and was being withdrawn while the potential between the two electrodes was recorded. Three separate records are shown; *T* marks the tips of the outer segments. Brief light flashes of constant intensity were delivered at *s*, producing the photovoltages shown as vertical lines. (From Hagins *et al.*, 1970.)

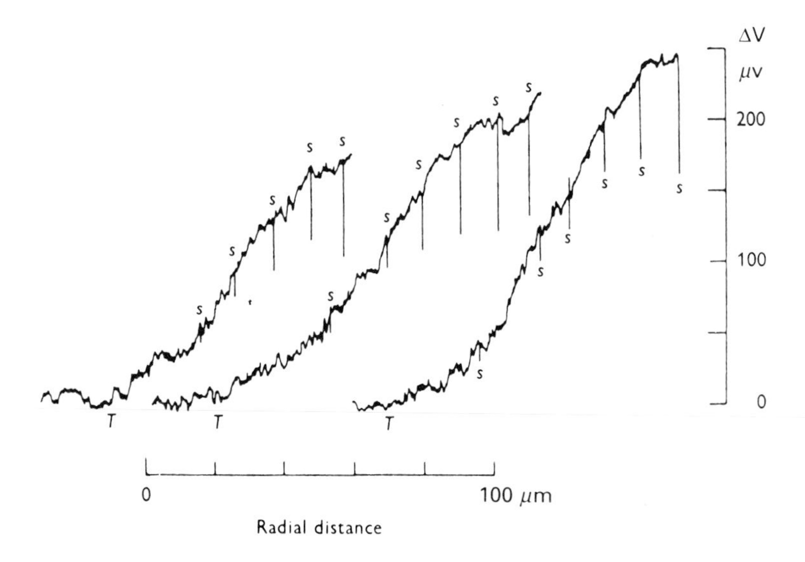

rods, which are rather larger than rat rods, Zuckerman (1973) found that there is also some inward current in the synaptic region, as is shown in fig. 18.32.

Hagins and his colleagues found that illumination of the retina caused a reduction in the potential recorded by the extracellular measuring electrode, by an amount which was roughly proportional to the size of the dark potential measured at that point (as shown at the points marked '*s*' in fig. 18.30). This decrease in potential indicates a reduction in the magnitude of the dark current. (This reduction in the dark current is sometimes known as the 'photocurrent'; but this is a concept which can cause confusion if not handled carefully.) The dark current is completely eliminated by a flash of light which allows 200 or more photons to be absorbed per rod, and is reduced to half by 30–50 photons per rod (Penn and Hagins, 1972). The relation between light absorption and response is linear at levels up to about 30 photons per rod, so we can see that absorption of one photon reduces the dark current by about 0.5 pA per rod. The response to a very brief flash reaches a peak after about 0.1 s and decays to zero within a second, and hence the reduction in dark current corresponds to about 0.1–0.2 pC of electric charge or 1 to 2×10^{-18} moles of univalent ions. Since each mole contains 6×10^{23} ions, we can say that the absorption of one photon by a rat rod causes about a million ions not to flow through the rod membrane.

What is the nature of the dark current? The ERG and the receptor potential with which it begins are dependent upon the presence of sodium ions (Furukawa and Hanawa, 1955; Sillman *et al.*, 1969), and frog rods hyperpolarize and become unresponsive to light in sodium-free solution (Brown and Pinto, 1974). It seems reasonable to suggest, then, that the dark current is caused by an inflow of sodium ions into the outer segments, which is reduced by the action of light. This would account for the hyperpolarization and increase in resistance of rods and cones in the light.

Some nice evidence in favour of this idea was provided by Korenbrot and Cone (1972) from measurements on the osmotic behaviour of rod outer segments. Suspensions of these can be obtained by shaking the retina gently in Ringer

Fig. 18.31. The longitudinal (radial) distribution of the dark current in the rod retina. The three symbols show results from three separate measurements. The curve is calculated on the assumption that the whole of the rod outer segment acts as a sink for the current, whose source is the inner segment and nuclear region. (From Hagins *et al.*, 1970; redrawn.)

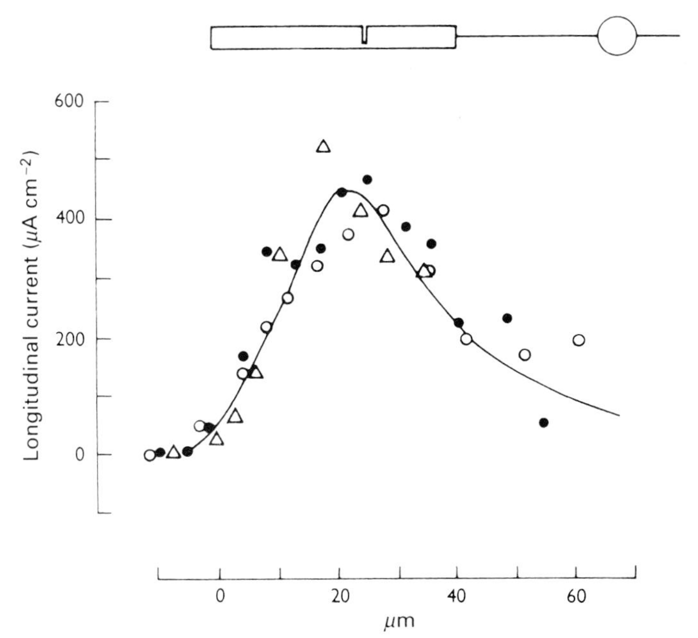

solution, and their volume can easily be measured from photomicrographs. When the outer segments were put into a strong solution of potassium chloride (3.5 times the isotonic concentration) in the dark, their volume shrank to about 70% of its initial value within about 3 s, and then remained constant. This suggests that there is a rapid exit of water from the outer segments in the first 3 s, after which they are in osmotic equilibrium; hence their plasma membranes must be impermeable to potassium ions. Illumination had no effect on this response. When they were put from Ringer into a similar strong solution of sodium chloride in the dark, however, their volume fell equally rapidly to 70% and then began to rise again, reaching 80% after 10 s. This suggests that after the initial rapid exit of water there is a slower entry of sodium ions which allows water to flow back into the outer segments. Illumination prevents the later increase in volume and therefore makes the plasma membrane impermeable to sodium ions. All these results are in delightful agreement with expectation.

If sodium ions are continually flowing into the outer segment in the dark, there must be some region in the rod cell where they are extruded. Ouabain, an inhibitor of the sodium pump in many cells, greatly reduces the dark current within a minute or so of its application to the retina (Yoshikami and Hagins, 1973; Zuckerman, 1973), but Korenbrot and Cone found that it had no effect on the osmotic behaviour of isolated outer segments. So it looks as though there is a metabolically driven sodium extrusion pump in the plasma membrane of the inner segment (where there are many mitochondria) and perhaps in the nuclear region also. This pump serves to maintain the ionic gradient of sodium ions that provides the driving force

Fig. 18.32. Current flow in a frog rod in the dark. The membrane current is calculated from measurements of longitudinal current (similar in general to those of fig. 18.31) by application of (4.3). (From Zuckerman, 1973, with lines of current flow added.)

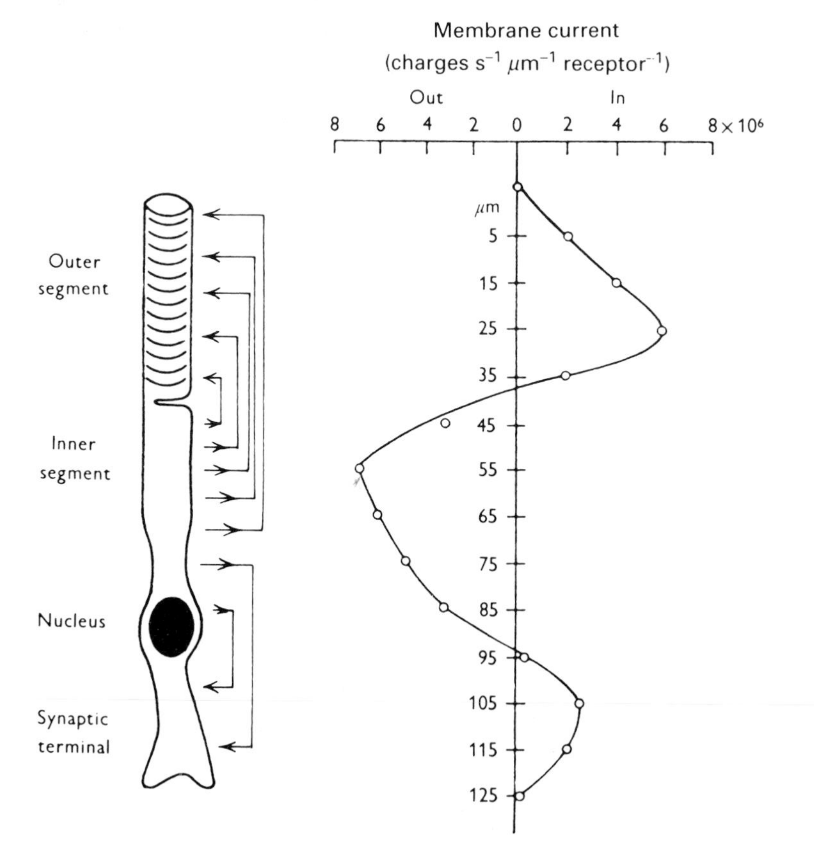

for the flow of current into the outer segment through the open channels in its plasma membrane.

The extracellular currents measured by Hagins and his colleagues were produced by the massed activity of a number of rods. Intracellular measurements also examine responses with some degree of averaging because of the existence of electrical coupling between the cells (p. 449). It would be highly desirable to look at individual outer segments, with the aim of examining the immediate electrical response to light.

A technique for doing this was developed by Baylor and his colleagues (1979*a*). They used a pipette electrode with a rounded tip just large enough for a toad rod outer segment to be sucked into it. The current flowing through the outer segment could then be measured with a feedback circuit similar to that used in patch clamp experiments, as is shown in fig. 18.33. The retina and the electrode system were mounted on a microscope stage, and manipulations were carried out using infrared illumination with an image converter.

Fig. 18.34 shows the response of the outer segment to a brief flash of light. There is a current change in the outward direction of about 20 pA; this is maintained for a few seconds, with recovery after about 10 s or so. For technical reasons it was not possible to determine the baseline of zero current without damaging the rod, so the 20 pA change could represent an increase in an outward current, a decrease in an inward current, or perhaps a mixture of both. However, when the outer segment was broken off from the rest of the rod at the ciliary segment by means of a sharp tap (as at the arrows in fig. 18.34), there was abrupt change in the current baseline by about 22 pA, and the response to illumination was abolished. This means that the baseline represents the dark current of the rod, an inward current of about 22 pA. The 'photocurrent' (the change in current on illumination) is thus a reduction in the dark current, which is almost completely abolished in bright light.

These results agree well with the conclusions drawn by Hagins and his colleagues from extracellular measurements.

The size of the photocurrent is dependent upon the size of the flash (fig. 18.35). Bright flashes caused saturation at currents up to 27 pA.

Fig. 18.33. Measurement of the current through a rod outer segment using the suction pipette electrode technique. (Based on Baylor *et al.*, 1979*a*.)

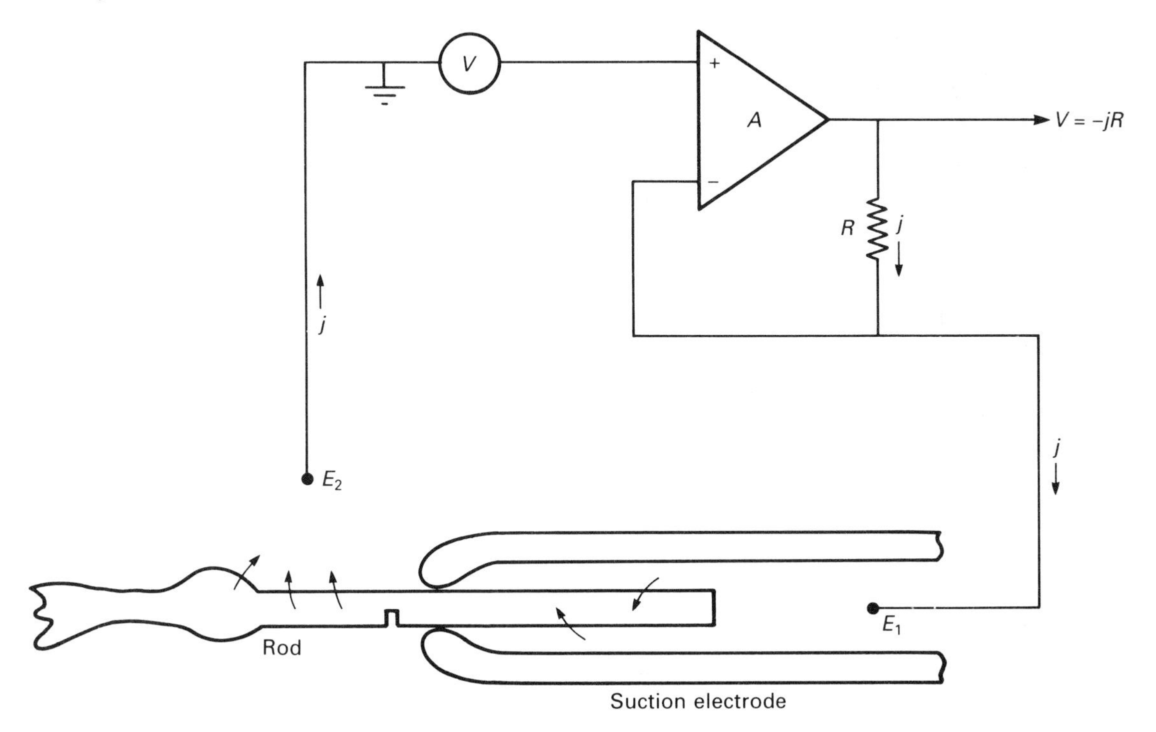

The relation between intensity and response was well fitted by the Michaelis equation:

$$r/r_{max} = i/(i + i_0), \quad (18.5)$$

where r is the amplitude of the response, r_{max} the value of r at saturation, i is the intensity of the flash and i_0 the intensity at half-saturation. The flash intensity at half-saturation was about 1.5 photons per square micrometre on average; this corresponds to a few hundred photons per outer segment. It is clear that, with about 3 billion rhodopsin molecules in the outer segment, saturation occurs when only a tiny fraction of them are photoactivated.

These experiments also provided information about the response of the outer segment to the absorption of single photons (Baylor *et al.*, 1979*b*). Fig. 18.36 shows the current flow through a rod outer segment in response to a series of flashes of light at low intensity. Clearly some flashes fail to produce a response while others are followed by small discrete current changes of about 1 pA, with occasional larger responses. This suggests that the responses reflect the absorption of zero, one or more photons. The overall distribution of response sizes was well fitted by a Poisson distribution with some added variance in the amplitude of the quantal responses.

The suction pipette method has been used by Hodgkin *et al.* (1984) to investigate the ionic basis of the dark current. They used isolated rods with either the outer or the inner segment sucked into the pipette, and looked at the effects of changing the external solution surrounding the part of the rod protruding from it. Fig. 18.37 shows the results of one of their experiments. Reduction of the sodium concentration surrounding the outer segment produced a corresponding reduction in the dark current, but a similar change for the inner segment was without effect. This confirms the view that the dark current is essentially a flow of sodium ions into the outer segment.

Some other interesting features emerged from these experiments. The external calcium concentration had a marked effect on the dark current. A typical rod had a dark current of 18 pA in Ringer solution with a calcium ion concentration of 1 mM. This fell to 3 pA when the concentration was 10 mM and increased to about 300 pA when it was 1 μM

Fig. 18.34. Toad rod outer segment currents measured with the suction electrode method. For technical reasons the dark current could not be measured directly, and the current scale therefore shows the difference between current in the dark at the beginning of the trace and that after subsequent treatments. A light flash (F) at time zero produces a saturating response. After recovery from this the rod was broken at the ciliary segment by giving the apparatus a sharp tap (at the arrows). This caused a change in current to a new level after which a second flash of light had no effect. The difference between old and new baselines gives the magnitude of the dark current. The difference between the original baseline and the response to light is sometimes called the photocurrent. (From Baylor *et al.*, 1979*a*.)

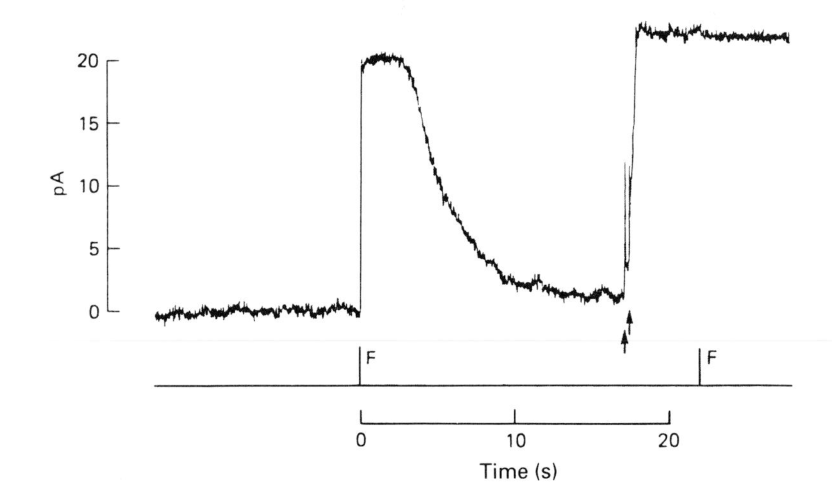

or less. Substitution of lithium for sodium did not affect the dark current, but it made the recovery from a saturating flash much slower. The explanation suggested for this effect is that there is some entry of calcium during a flash, that normally this is removed by the action of a sodium–calcium exchange system (which will not work with lithium in place of sodium), and that the recovery process is affected by the internal calcium concentration. And finally, the selectivity of the light-sensitive channels is not very great: the permeability coefficient for potassium is about 0.8 that for sodium, and calcium ions can also pass through them (Hodgkin *et al.*, 1985).

Fig. 18.35. Rod outer segment responses to flashes of increasing intensity, measured with the suction electrode method. *a* shows averaged time courses of the various responses. *b* shows the peak responses from *a* plotted against intensity on a logarithmic scale. The curve in *b* follows equation (18.5). (From Baylor *et al.*, 1979*a*.)

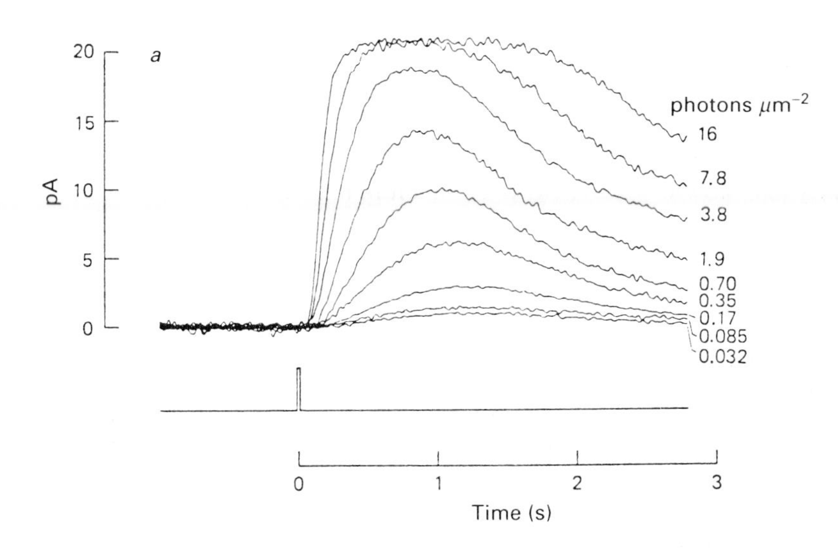

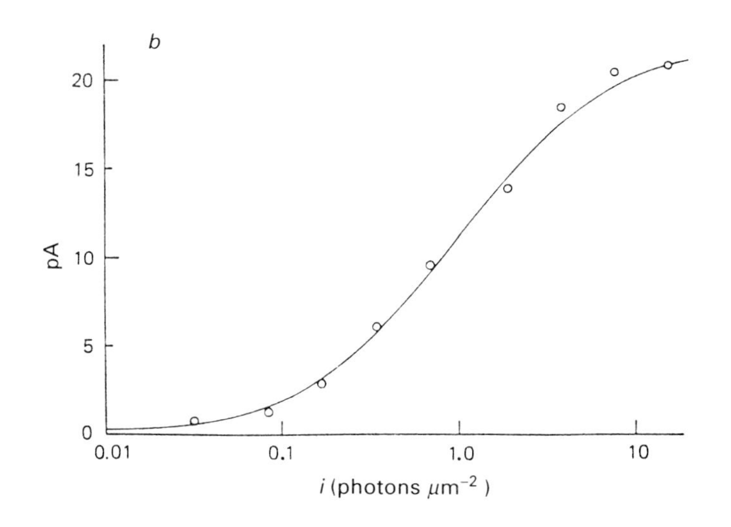

Internal messenger hypotheses

The results in the previous section show that photoisomerization of one rhodopson molecule causes a reduction in the dark current of about 0.5 pA in rat rods or about 1 pA in toad rods. The rhodopsin molecules are embedded in the disc membranes, which are not continuous with the plasma membrane across which the dark current flows. Consideration of this situation led to the suggestion that some internal messenger must carry information between the rhodopsin molecules on the discs and the plasma membrane. Such a system would permit amplification of the signal: one rhodopsin molecule might trigger the release of many messenger molecules, resulting in the closure of numbers of light-sensitive channels.*

The internal messenger hypothesis was stated in rather general terms by Fuortes and Hodgkin (1964), as part of a model system to describe the response of *Limulus* photoreceptors to light. Applications to vertebrate photoreceptors were made by Baylor and Fuortes (1970), Hagins (1972) and Cone (1973), among others.

Two alternative hypotheses about the nature of the internal messenger flourished in the nineteen seventies and early eighties (fig. 18.38). The first of these was the calcium hypothesis, which postulated that light produced an internal release of calcium ions and that these acted to close the light-sensitive channels in the outer segment plasma membrane. The second hypothesis invoked cyclic GMP (cyclic guanosine 3′,5′-monophosphate) as the internal messenger. The conflict between these two alternatives now seems to have been resolved in favour of the cyclic GMP hypothesis (see, for example, Lamb, 1986; Pugh and Cobbs, 1986; Pugh, 1987; Yau *et al.*, 1986). It is instructive to see how this has come about.

The calcium hypothesis was first proposed by Hagins and Yoshikami (Hagins, 1972; Yoshikami and Hagins, 1973; Hagins and Yoshikami, 1974). They found that the dark current is reduced as the external calcium ion concentration is raised, and suggested that this would raise the internal calcium ion concentration and so block the light-sensitive channels. Brown *et al.* (1977) injected calcium ions into the outer segments of toad rods and found that brief hyperpolarizations ensued, similar in time course to the responses to light. Gold and Korenbrot (1980) used calcium sensitive electrodes in the extracellular spaces of the retinal photoreceptor layer to show that light produced a rapid efflux of calcium ions from rods; they attributed this to a light-determined increase in intracellular calcium ion concentration.

Hagins (1972) suggested that calcium ions were released from a rod disc when one of its rhodopsin molecules absorbed a photon; at that time it seemed reasonable to suggest that the rhodopsin molecule might actually act as a calcium channel. In cones, he suggested, the photon-induced calcium entry occurred across the infolded plasma membrane where the visual pigment molecules are located.

More recent experiments have produced results which are not in agreement with the calcium hypothesis. Light appears to cause a decrease, not

* We adopt here the common usage of the term 'light-sensitive channels' to indicate channels in the outer segment plasma membrane which close on illumination. They are not themselves sensitive to light, of course; it is the rhodopsin molecules which are actually light-sensitive.

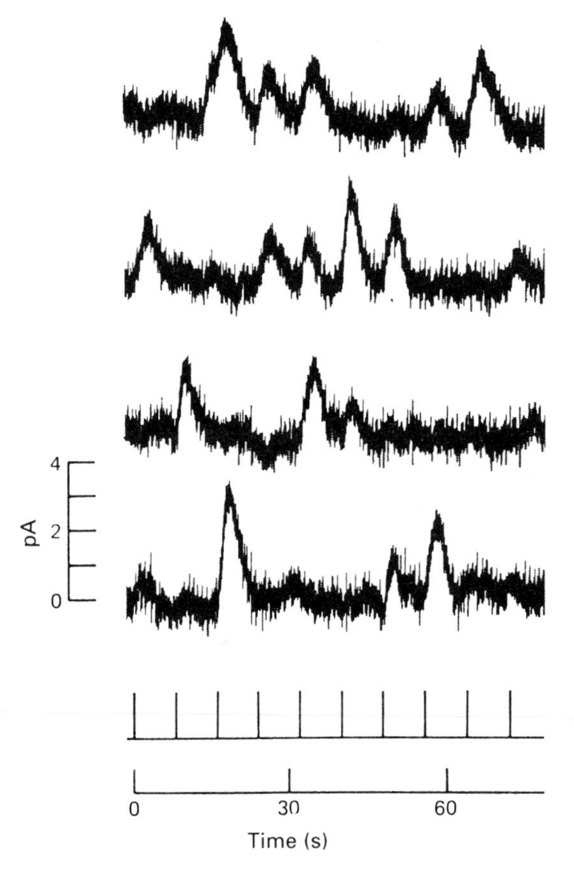

Fig. 18.36. Quantal responses in the toad rod outer segment. The record shows responses to a series of 40 consecutive dim flashes of light at 8 s intervals. (From Baylor *et al.*, 1979*b*.)

an increase, in the calcium ion concentration in the outer segment. Evidence for this comes from the measurement of membrane currents in isolated rods (Yau and Nakatani, 1985) and, more directly, from experiments in which the calcium-sensitive protein aequorin is injected into rods (McNaughton *et al.*, 1986). Furthermore, when the internal calcium ion concentration in rod outer segments was reduced by the incorporation of the calcium buffer compound BAPTA, the dark current and the response to light increased (Matthews *et al.*, 1985). And finally calcium was ineffective in increasing the conductance of inside-out patches of rod outer segment membrane when applied to the cytoplasmic side (Fesenko *et al.*, 1985). It is time for us to look at the alternative hypothesis.

The cyclic GMP cascade

In 1972, Pannbacker and his colleagues found that cattle rods contain a phosphodiesterase (PDE) which will hydrolyse the cyclic nucleotides cyclic AMP and cyclic GMP. Since the enzyme was rather more active with cyclic GMP, they suggested that the latter could be an intermediate in the phototransduction process. The presence of the enzyme for synthesizing cyclic GMP, guanylate cyclase, in the outer segments was demonstrated by Pannbacker (1973). Gorodis *et al.* (1974) showed that light produced a rapid drop in cyclic GMP levels in the retinas of cattle and frogs. In cattle, for example, cyclic GMP was present at the rather high concentration of 64 pmoles per milligram of protein in retinas which had been kept in

Fig. 18.37. Demonstration that there is an inward flow of sodium ions across the outer segment membrane in the dark, using the suction electrode method. *a* shows the effect of reducing the external sodium ion concentration by replacing half of it with choline. The response to saturating flashes of light decreases, indicating that the dark current has decreased. In *b* the rod inner segment is exposed to reduced sodium, but here there is no reduction in dark current. In each record there is some baseline drift for technical reasons. (From Hodgkin *et al.*, 1984.)

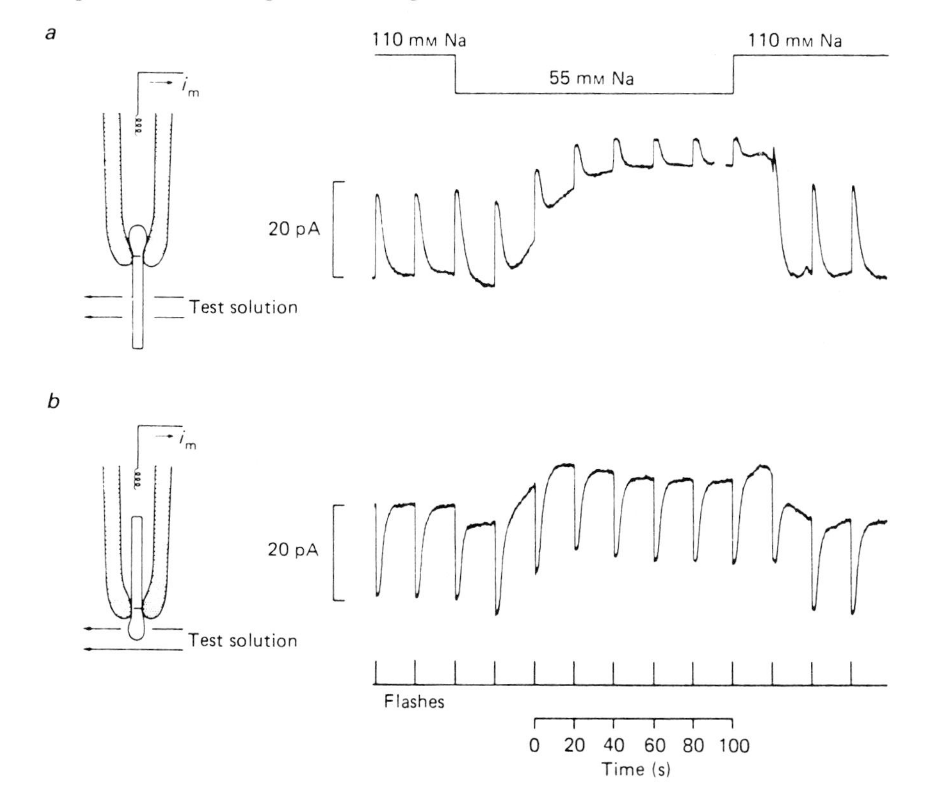

the dark. This level was reduced by 70% after 15 s of illumination.

These observations served to establish cyclic GMP as a worthy candidate for the internal messenger role. There followed a period of biochemical detective work in which the proteins of the outer segment and their roles in controlling cyclic GMP levels were investigated. The essential conclusion of this work is illustrated in fig. 18.39: there is a cascade such that the photoactivation of one rhodopsin molecule leads to the breakdown of perhaps half a million or so cyclic GMP molecules. Let us look at some of the evidence for this conclusion; fuller accounts are given by Chabre (1985), Kühn (1986), Stryer (1986) and Hurley (1987).

Yee and Liebman (1978) used the pH change which accompanies cyclic GMP hydrolysis to measure its rate. They found that each photoactivated rhodopsin molecule (R^*) stimulated the hydrolysis of 4×10^5 cyclic GMP molecules. But since each PDE molecule would only hydrolyse 800 molecules of cyclic GMP per second, they concluded that there was a multiplier mechanism such that photoactivation of one rhodopsin molecule would lead to activation of about 500 PDE molecules. Woodruff and Bownds (1979) and Cote *et al.* (1984) confirmed the rapidity of the reaction by using rapid-quench techniques in which the reaction mixture is plunged into percholoric acid a fraction of a second after illumination. Fig. 18.40 illustrates one of these experiments, and shows that the light-induced fall in cyclic GMP concentration precedes the reduction of the dark current.

The next step was to show that R^* does not activate PDE directly. The presence of a light-activated membrane-dependent GTPase (an enzyme which hydrolyses guanosine triphosphate, GTP, to guanosine diphosphate, GDP) in the rod outer segments was detected by a number of workers (Wheeler *et al.*, 1977; Godchaux and

Fig. 18.38. Two hypotheses for phototransduction in rods. The calcium hypothesis (*a*) suggested that activation of rhodopsin (Rh) releases calcium ions from the rod disc lumina, and that these calcium ions would then close the dark current channels in the outer segment plasma membrane. In the cyclic GMP hypothesis (*b*), now generally accepted, the channels through which the dark current flows are held open by cyclic GMP; activated rhodopsin acts via transducin, a G protein (G), to activate the enzyme phosphodiesterase (PDE), which then reduces the cyclic GMP concentration and so closes the channels. (From Attwell, 1986.)

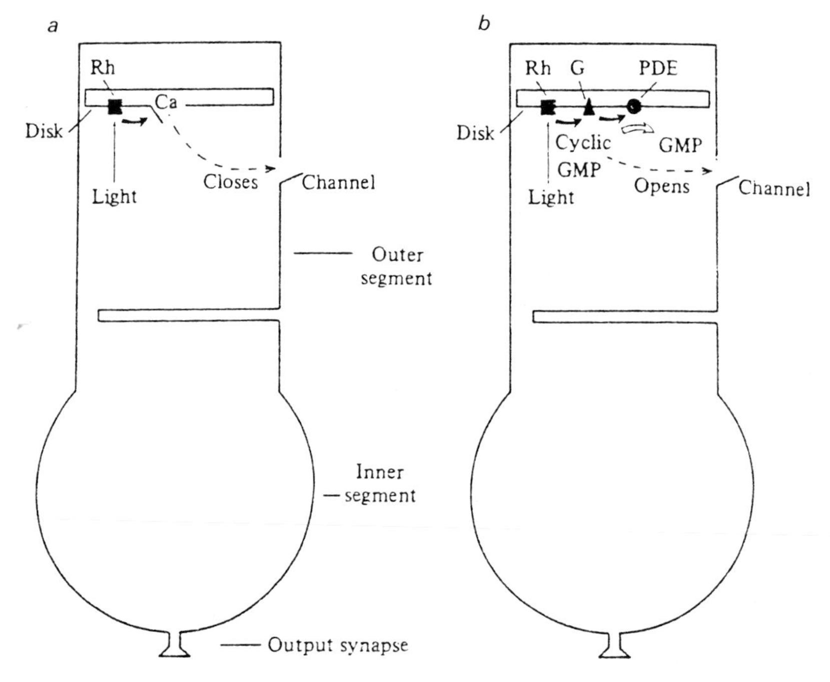

Zimmerman, 1979; Fung and Stryer, 1980). This protein was named transducin by Fung *et al.* (1981); it is one of the family of G proteins which we have met in chapter 10.

There are thus two amplification stages in the cascade shown in fig. 18.39. Each R^* molecule activates about 500 transducin molecules. That is to say, R^* is an enzyme which catalyses the replacement of bound GDP by GTP in transducin. Each activated transducin molecule itself activates a PDE molecule. The second amplification stage is then the breakdown of about 1000 cyclic GMP molecules by each activated PDE molecule.

The activation of PDE by transducin involves a one-to-one combination and so does not produce an enzymic amplification stage. It seems likely that GTP binds to the α-subunit of transducin (T_α), and produces dissociation of T_α from the β- and γ-subunits (Fung *et al.*, 1981). Inactive PDE consists of three subunits; it is activated by removal of its γ-subunit (Hurley and Stryer, 1982). T_α–GTP probably dissociates the inhibitory γ-subunit from PDE by forming a complex with it (Yamazaki *et al.*, 1983).

The idea that rhodopsin might act as an enzyme at the beginning of a multistage amplification process was suggested many years ago by Wald, and he was encouraged in this view by the discovery of the cascade of enzymic actions in blood clotting (Wald, 1965). But a clot of blood is an end product, whereas a visual response is a transient phenomenon: the photoreceptor cell must be reset so that it can respond again to light a few moments later. So it is necessary that the individual steps in the cyclic GMP cascade should be cyclic rather than one-way, so that the various components can be restored to their original states.

How this is done can be outlined as follows. Rhodopsin kinase is a 65 kDa protein which catalyses the phosphorylation of R^* by ATP. The phosphorylated R^* will then combine with another protein, called the 48 kDa protein or arrestin; it is then unable to combine with transducin, so its catalytic activity is brought to an end. T_α–GTP, bound to PDE_γ, shows its GTPase activity in conversion to T_α–GDP, and then dissociates from the PDE_α and recombines with $T_{\beta\gamma}$. The PDE_γ combines with $PDE_{\alpha\beta}$ and inactivates it. Cyclic GMP is resynthesized from GTP by the enzyme guanylate cyclase.

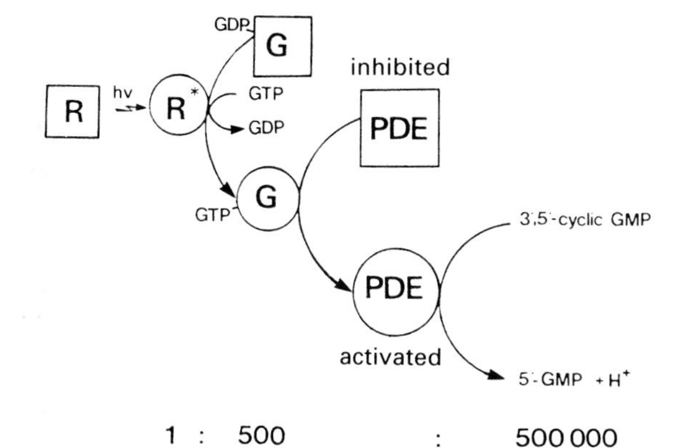

Fig. 18.39. The cyclic GMP cascade activated by light in rods. Each photoactivated rhodopsin molecule (R^*) contacts about 500 transducin molecules (G) and activates them by catalysing the exchange of GTP for bound GDP. Each of the activated transducin molecules then activates a phosphodiesterase (PDE) molecule and each PDE molecule then hydrolyses about 1000 cyclic GMP molecules per second. Thus the absorption of one photon can lead to the breakdown of half a million cyclic GMP molecules. (From Kuhn, 1986.)

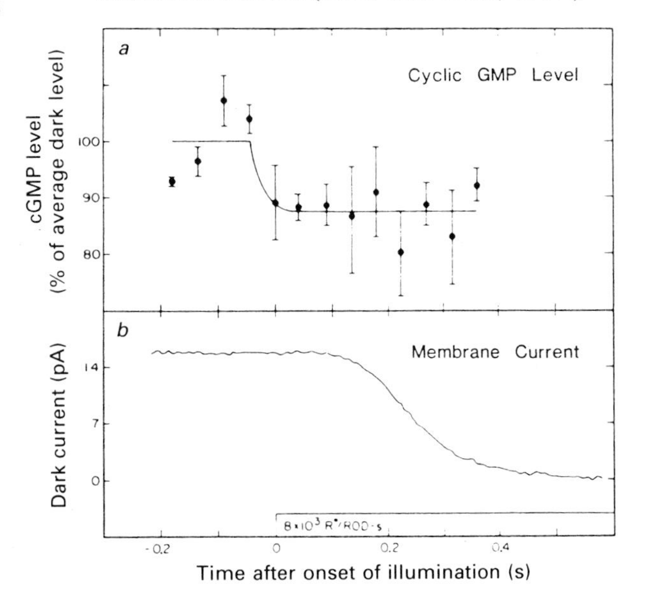

Fig. 18.40. Results of a rapid-quench experiment, showing that the fall in cyclic GMP content in the rods precedes the fall in membrane current. (From Cote *et al.*, 1984.)

Channels opened by cyclic GMP

The existence of the cyclic GMP cascade implies rather strongly that cyclic GMP must play an important role in visual transduction, but it does not tell us just what this role is. Miller and Nicol (1979) were among the first to get to grips with this question; they found that a rapid depolarization followed injection of cyclic GMP into the rod outer segments of intact retinas. Matthews *et al.* (1985) and Cobbs and Pugh (1985) used the suction electrode technique with isolated salamander rods and introduced cyclic GMP via a patch electrode applied to the inner segment. They found that this produced dramatic increases in the dark current and that recovery from light flashes was much delayed (fig. 18.41).

The mode of action of injected cyclic GMP was not evident from these experiments. Cyclic nucleotides usually act by activating a kinase, which is an enzyme that then phosphorylates another protein (a channel protein, a pump or an enzyme, for example) by the transfer of a phosphate group from ATP or GTP.

The situation was greatly clarified following some elegant experiments by Fesenko *et al.* (1985), using the patch-clamp technique. Their recording method is shown in fig. 18.42*a*. A frog retina was treated with trypsin and shaken gently to release isolated rod cells and outer segments. The patch-clamp electrode was brought into contact with the rod, brief suction produced a gigaohm seal, and a sharp tap excised a patch of rod surface membrane. The cytoplasmic side of the excised patch was thus exposed to the bathing solution in the chamber. The conductance of the patch was determined by applying voltage changes and observing the currents that these produced.

Fig. 18.42*b* shows the effect of applying cyclic GMP to the inside of the membrane. The currents produced by 10 mV pulses increased from about 1 pA to nearly 4 pA, indicating a fourfold increase in membrane conductance. Fesenko and his colleagues were surprised to find that this effect was independent of the presence of ATP or GTP; this indicates that the cyclic GMP acts directly on the membrane channels, not indirectly by activat-

Fig. 18.41. The effect of cyclic GMP on the rod dark current. The record shows the outer segment current of an isolated salamander rod held in a suction pipette electrode. Lines on the top trace show light flashes. At the first arrow a patch electrode containing 2 mM cyclic GMP was sealed against the inner segment membrane, and the patch was broken at the second arrow, so that cyclic GMP could enter the rod. Notice how the dark current was greatly increased after a few seconds. Later flashes of light produced the usual reduction in dark current, which recovered as more cyclic GMP diffused into the outer segment. The rod was repositioned slightly at the time shown by the dot. (From Matthews *et al.*, 1985.)

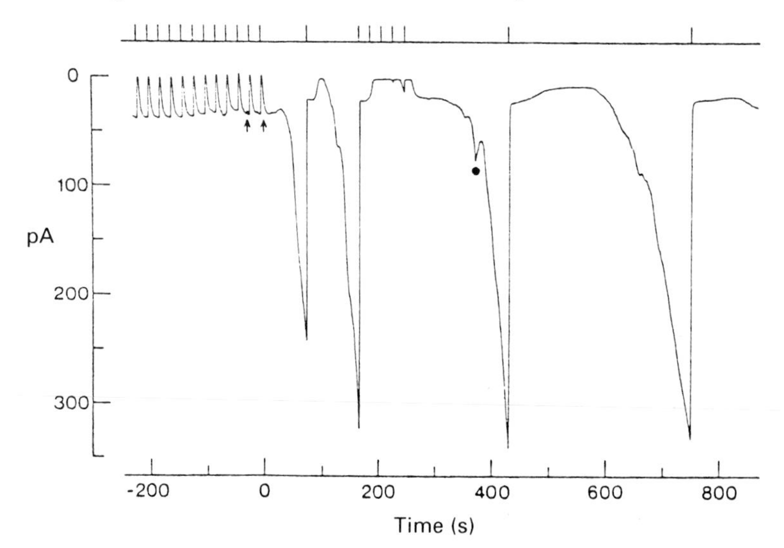

ing a protein kinase. Changing the calcium concentrations on the cytoplasmic side had no effect on the conductance in the absence of cyclic GMP and only a small effect in its presence. This, as we have seen, provides strong evidence against the idea that calcium ions are the internal messenger in phototransduction.

Similar experiments on the effects of cyclic GMP on the conductance of isolated patches were performed by Haynes and Yau (1985) on the outer segment membranes of catfish cones. The results were very similar to those on rod membranes, indicating that cones are also activated by a cyclic GMP cascade. Baylor and Hodgkin (1973) calculated that the mean hyperpolarization produced by a single photon was 25 μV in turtle red- and green-sensitive cones, but 130 μV in the rods. This smaller quantal response of the cones may be due to a smaller gain in the cyclic GMP cascade, i.e. each photoactivated visual pigment molecule may cause the breakdown of fewer cyclic GMP molecules.

Is the membrane conductance which is increased by cyclic GMP in excised patches the same as that which is reduced by light in intact rods? Some pleasing experiments on this question were performed by Yau and Nakatani (1985; see also Nakatani and Yau, 1988). They recorded the currents flowing through a rod outer segment drawn partly into a suction electrode; the response to a flash of light is shown in fig. 18.43*a*. Then the inner segment and the basal part of the outer segment of the rod were knocked off with a probe, leaving an open-ended outer segment whose interior was now accessible to the bath solution. This arrangement allowed the concentrations of ions and small soluble molecules in the outer segment to

Fig. 18.42. The effect of cyclic GMP on a patch of rod outer segment membrane. Constant voltage pulses were applied to the patch (lower trace), and the currents produced by them were measured (upper trace); the currents are thus proportional to the membrane conductance. When cyclic GMP was applied to the cytoplasmic side of the patch the conductance increased fourfold, indicating the presence of channels that are held open by cyclic GMP. (From Fesenko *et al.*, 1985.)

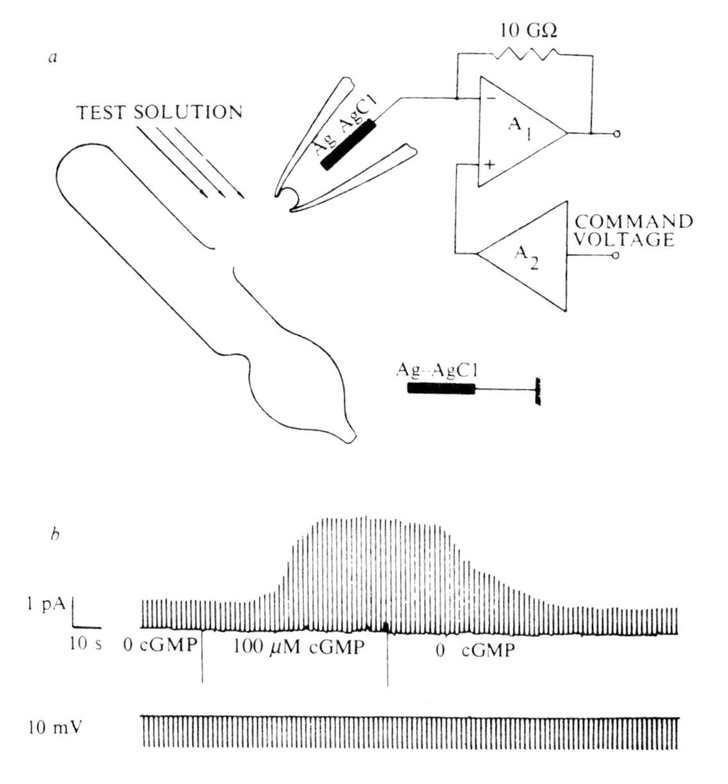

be controlled, while leaving most of its proteins and membrane systems intact.

Fig. 18.43*b* shows that a large inward current could be induced by cyclic GMP. This current was not affected by light unless GTP was also present in the bath solution, when it was greatly reduced (figs. 18.43*c* and *d*). GTP is necessary for the activation of transducin by photoactivated rhodopsin, so there will be no activation of PDE in its absence (see fig. 18.39). Thus the conductance which is reduced by light is the same as that which is increased by cyclic GMP.

The relation between cyclic GMP concentration and the currents produced in Yau and Nakatani's experiments is shown in fig. 18.44. The curve is initially S-shaped, and then saturates above about 200 μm GMP. Half-saturation occurs at about 45 μm. A very similar curve was obtained by Fesenko and his colleagues in their patch-clamp experiments. The sigmoid foot of the curve indicates that two or three cyclic GMP molecules are required to open each channel.

Fesenko and his colleagues did not observe individual channel opening and closings in their patch-clamp records, indicating that the single-channel conductance was much smaller than in many of the other channels which have been investigated. Bodoia and Detwiler (1985) measured the

Fig. 18.44. The relation between cyclic GMP concentration and inward current in a dialysed rod outer segment. Each point on the curve was determined as in fig. 18.43*b*. (From Yau and Nakatani, 1985.)

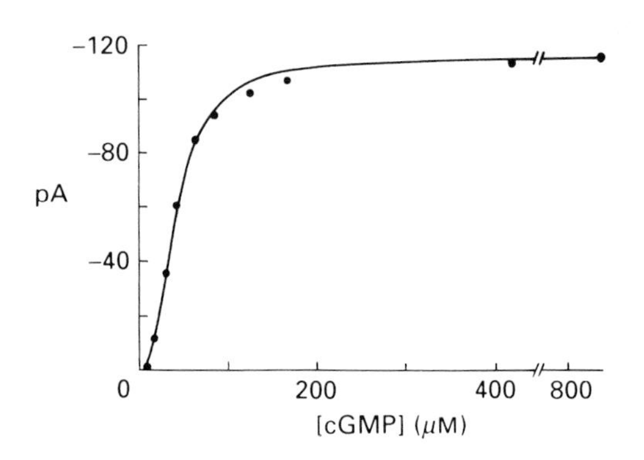

Fig. 18.43. Membrane currents in dialysed rod outer segments. In *a* an intact isolated toad rod is held in a suction pipette electrode; the record shows the dark current and the response to a flash of light. The inner segment and basal part of the outer segment were then knocked off, leaving the inside of the rest of the outer segment membrane open to the bath solution. Raising the cyclic GMP concentration from zero to 0.2 mM then produced a large inward current (*b*). The current induced by cyclic GMP was not sensitive to light (*c*) unless GTP (which is necessary for the activation of transducin) was also present (*d*). (From Yau and Nakatani, 1985.)

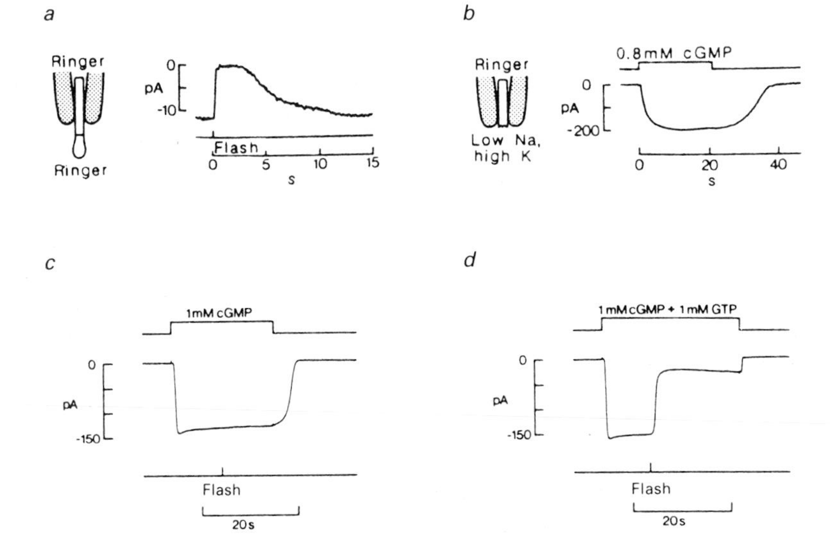

current noise in patch-clamp records from the outer segments of isolated rods. Analysis of the high-frequency component of the noise suggested that it was composed of single channel currents of only 3–5 fA. Thus a response to a single photon, producing a dark current reduction of about 1 pA, would be brought about by the closure of 200–300 channels. A 20 pA dark current in a whole outer segment would require about 5000 open channels. Similar results and conclusions were obtained by Gray and Attwell (1985).

Direct observation of channel opening has been made by using patch-clamp methods in the absence of divalent cations (Haynes *et al.*, 1986; Zimmerman and Baylor, 1986). Under these conditions brief single channel currents in the region of 1.5 pA could be observed, implying a conductance of about 25 pS. These conductance levels are high enough to rule out the possibility that the light-sensitive conductance involves carriers rather than channels. It would be interesting to know how calcium and magnesium ions cause the drop in channel conductance. One possibility is that the channel pore contains binding sites with a high affinity for divalent cations; other ions might not then be able to pass through while these sites are occupied, so the average current will be reduced (Owen, 1987).

The patch technique has been most elegantly employed by Matthews and Watanabe (1987) to provide further evidence that the light-sensitive channels are those which are opened by cyclic GMP. They attached a patch-clamp electrode to the outer segment of a toad rod, recorded single channel activity in the dark and its suppression in the light, and then excised the patch and applied cyclic GMP to it and again recorded single channel currents. The results of these experiments (fig. 18.45) showed that the channel characteristics in the two situations were essentially indistinguishable.

The ionic selectivity of the light-sensitive channels has been investigated by Yau and Nakatani (1984*a*) and Hodgkin *et al.* (1985). Both groups concluded that, in addition to sodium, potassium and other monovalent cations, calcium and magnesium could all pass through the channel. Under normal conditions the dark current consists largely of sodium ion inflow, with calcium ions accounting for about ten percent of the inward current.

What is the role of calcium ions?

There is good evidence for the existence of a sodium–calcium exchange system in the rod plasma membrane, such that three sodium ions are transported into the rod for each calcium ion extruded (Yau and Nakatani, 1984*b*; Hodgkin *et al.*, 1985). This system is shown in action in fig. 18.46. Here a toad rod outer segment is loaded with calcium and then illuminated in the absence of sodium ions; under these conditions there will be no current through the light-sensitive channels since they are all closed and there will be no current through the sodium–calcium exchanger since there are no sodium ions to flow inwards. When the normal external sodium ion concentration is restored in the light, there is an inward current for a time as the exchange system acts to remove the excess internal calcium.

Under normal conditions in the dark, therefore, there must be a balance between the inward leakage of calcium ions through the light-sensitive channels and their extrusion through the sodium–calcium exchange system. On exposure to light, the light-sensitive channels close but the exchange system continues for a time, so there should be a net extrusion of calcium from the rods. This would explain the observation by Gold and Korenbrot (1980) of an increase in extracellular calcium under these conditions, which was interpreted at the time as suggesting an increase in internal calcium ion concentration. It is also in accordance with the later evidence, from experiments with aequorin, that the internal free calcium ion concentration decreases in the light (McNaughton *et al.*, 1986).

Does this reduction in internal calcium ion concentration in the light have any effect upon the phototransduction process? Torre *et al.* (1986) approached this question by introducing a calcium buffer into the cell so as to prevent or reduce the change in calcium concentration. They used the chelating agent 1,2-*bis*(*o*-aminophenoxy)ethane-N,N,N′,N′-tetraacetic acid (BAPTA) as the buffer; a suitable mixture of K_2CaBAPTA and K_4BAPTA will keep the calcium ion concentration steady at the normal level of the cell in darkness, 0.5–1 μM.

To get the buffer into a cell, the outer segment of an isolated salamander rod was drawn into a suction pipette so as to record its membrane currents.

A patch electrode containing a BAPTA buffer solution was then applied to the inner segment and the patch was ruptured by suction. For the next few minutes the buffer would diffuse from the patch electrode into the cell, and then the electrode was carefully withdrawn so that the membrane resealed around the patch. This left a fixed quantity of BAPTA within the cell, so that the cell calcium level would be maintained at a fairly constant level during the rest of the experiment.

Fig. 18.47 shows responses to flashes of light of different intensities. Under normal conditions the time to peak is reduced with increasing light intensity, whereas this does not happen in rods with intracellular BAPTA. Rods with BAPTA produce longer and larger responses, but the initial rate of rise of the response is the same as in normal rods. This suggests that the reduction in internal calcium that normally occurs serves to speed up the time course of the response and reduce its sensitivity. These features are characteristic of light adaptation, the process whereby the visual process responds to increased light intensities.

Torre and his colleagues concluded from these and other experiments that the internal calcium ion concentration plays a central role in phototransduction by contributing to the regulation of light adaptation, so enabling the photoreceptors to function over a wide range of light intensities. It may do this by regulating the activity of the enzyme phosphodiesterase. In the normal response to a flash of light, it is suggested, the reduction in

Fig. 18.45. Patch-clamp recordings of single-channel activity in toad rod outer segment membrane. The patch electrode was filled with a solution containing no divalent cations so as to increase the size of the single channel currents. The traces on the left show channel activity in a patch that was still attached to the cell. The bottom trace was recorded during saturating light, all others were recorded in darkness. Those on the right are from the same membrane patch after it had been pulled away from the cell. The solution bathing the intracellular face of the patch contained 10 μM cyclic GMP, except for the bottom trace, when it did not. (From Matthews and Watanabe, 1987.)

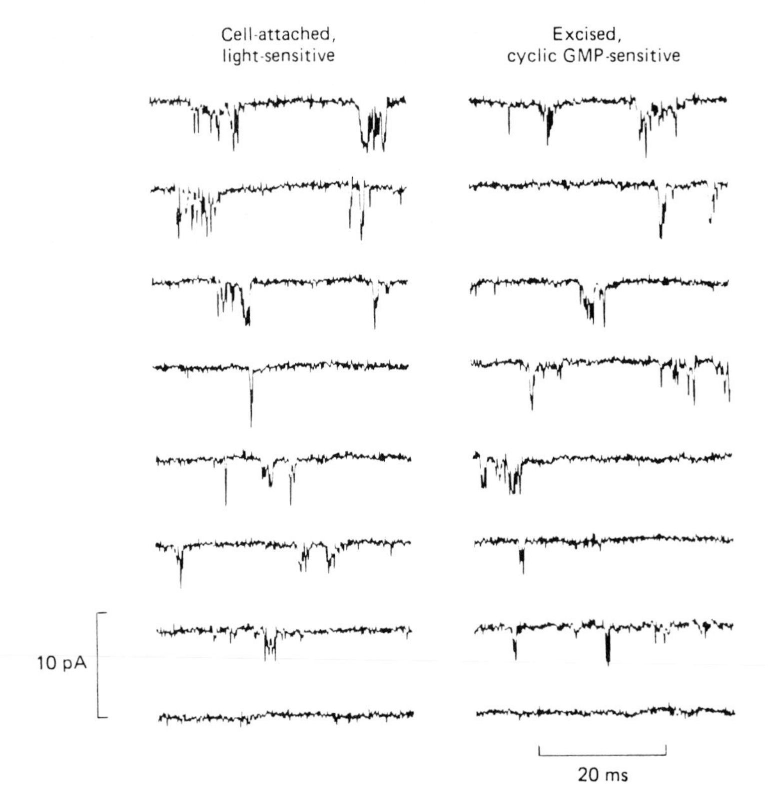

internal calcium that occurs reduces the rate at which cyclic GMP is broken down, so that recovery (i.e. the re-opening of the light-sensitive channels) is more rapid than it would otherwise be. In the presence of background illumination, the reduced internal calcium will reduce PDE activity so that photoactivation of rhodopsin will not break down so many cyclic GMP molecules. The system is thus thought to act as an automatic gain control loop, with calcium ions acting as the feedback link between the output (current flow through the cell membrane) and the gain of the cyclic GMP cascade.

Signal processing in the retina

As mentioned earlier, the retina arises as an outgrowth of the embryonic brain, and contains several different types of cells with much synaptic interaction between them. On structural grounds alone, therefore, we would expect to find some considerable processing of the retinal image before it is transferred to the brain via the optic nerve. Physiological investigations have fully confirmed this expectation. The details on this interaction between the various retinal cells are complicated and fall outside the scope of this book, but it may be useful here to take a brief look at the outlines of the process, with particular reference to the situation in the cat. Let us begin with the first cells to be investigated, the ganglion cells whose axons carry the output from the retina up the optic nerve into the brain.

The ganglion cells

Hartline (1938) found that the ganglion cells of the frog could be divided into three types according to their response to illumination. 'On' units responded to the onset of illumination, 'off' units responsed to the cessation of illumination, and 'on–off' units responded to both onset and cessation.

Fig. 18.46. Electrogenic sodium–calcium exchange in a toad rod. Records show currents recorded from an intact toad rod with the inner segment in the suction pipette electrode and the outer segment exposed to solutions of different ionic concentrations. The record on the left shows the currents involved in loading the rod with calcium ions; there is no current in zero sodium, then an inward current associated with sodium inflow in 55 mM sodium, then a brief large current in isotonic calcium as calcium ions flow into the cell through the light-sensitive channels. Integration of the calcium current shows that about 1.8 billion calcium ions entered the cell at this time, giving an intracellular calcium ion concentration of about 2 mM. The outer segment was then put in zero sodium and zero calcium again, and illuminated to close all the light-sensitive channels. On restoring the external sodium ion concentration there is an appreciable inward current for a few seconds, as shown on the right. This current did not occur if the cell had not previously been loaded with calcium ions. The existence of the current shows that more than twice as many sodium ions are moving in as calcium ions are moving out. (From Yau and Nakatani, 1984*b*.)

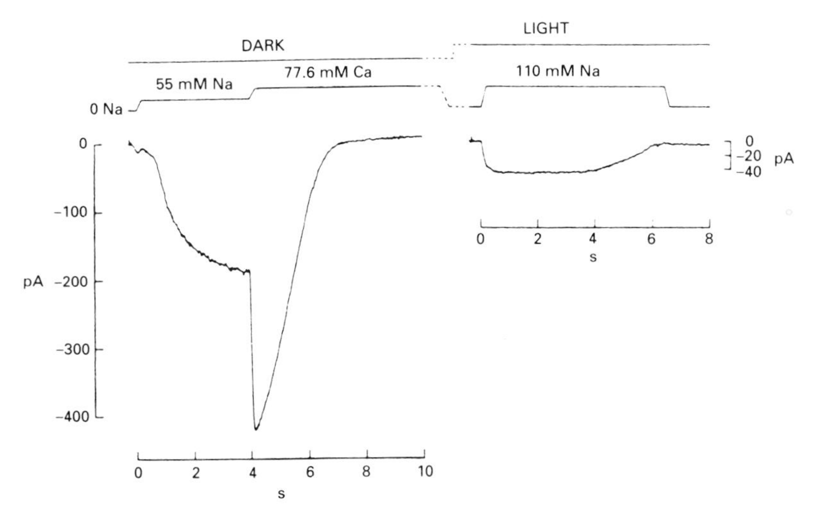

Studies on the structure of the retina show that, as we have seen, a large number of photoreceptor cells may converge, via the bipolars, on any one ganglion cell. Thus a single ganglion cell can respond to illumination anywhere within a certain area of the retina: this area is known as the *receptive field* of the ganglion cell. The receptive fields of cat retinal ganglion cells were investigated by Kuffler (1953), using very small spots of light as stimuli. He found that many cells responded with *on* discharges to illumination of the centre of the field, but with *off* discharges when the periphery was illuminated, and with *on–off* discharges in an intermediate region (fig. 18.48); this type of cell is called an *on*-centre unit. Other cells showed the reverse type of organization (*off*-centre units), giving *off* responses to illumination at the centre of the field and *on* responses to illumination of its periphery.

This type of organization is sometimes called centre-surround antagonism. Since the receptive fields of the ganglion cells show considerable overlap, a spot of light which produces an *on* response in a ganglion cell immediately under it may produce *off* responses in ganglion cells some distance away.

The organization of the receptive fields differs according to the state of adaptation of the eye. Kuffler suggested that they increase in size during dark adaptation, but, in a later investigation (Barlow *et al.*, 1957), it was found that the inhibitory effects in the periphery of the field disappear during dark adaptation.

The nature of ganglion cell receptive fields has been further investigated by a combination of physiological and anatomical methods. It seemed likely that the boundaries of the *on*- or *off*-centre should be defined by the edge of the dendritic field of the ganglion cell and that the surround should be related to the area covered by the amacrine cells which synapse with it (Dowling and Boycott, 1966). Work by Wässle and his colleagues has largely confirmed that this is so (Peichl and Wässle, 1979, 1981; Wässle *et al.*, 1981*a, b*).

Fig. 18.47. The role of internal calcium ions in rod responses to light. The first set of records (*a*) shows a family of responses to flashes of light of increasing intensities; records were made by the suction electrode method from a salamander rod outer segment. Then a patch pipette containing the calcium-buffering agent BAPTA/Ca–BAPTA was brought into contact with the inner segment, and the patch was broken by suction so as to allow the buffering system to enter the cell. Then the cell was subjected to another series of flashes, with the results shown in *b*. (From Torre *et al.*, 1986.)

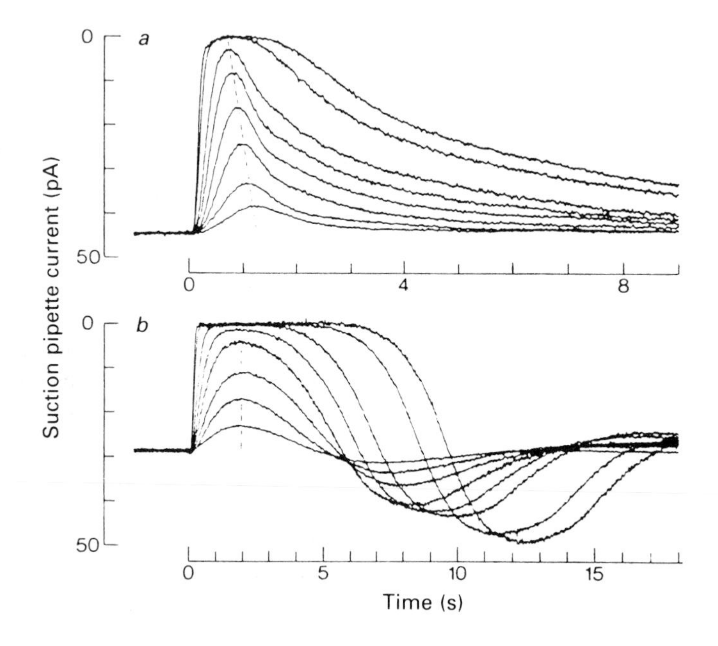

Boycott and Wässle (1974) distinguished three different types of ganglion cells in the cat retina on anatomical grounds. For α and β cells there was a gradient of dendritic field size from small in the centre to larger in the periphery. At any one point the dendritic fields and cell bodies of α cells were much larger than in β cells. A third category, called γ cells, included cells with small cell bodies and large dendritic fields with rather sparse dendrites.

Cleland, Dubin and Levick (1971) and Cleland and Levick (1974) produced a tripartite classification of cat ganglion cells on the basis of the action potentials produced in response to stimuli. Most cells produced 'brisk' responses, with short latencies, but a few were 'sluggish'. The brisk responses were divided into 'transient', with just a few action potentials immediately after a change in illumination, and 'sustained'. The brisk-sustained cells corresponded to the X cells of an earlier classification by Enroth-Cugell and Robson (1966), and the brisk-transient cells to their Y cells. By comparison of the size of receptive fields with Boycott and Wässle's data, Cleland and Levick suggested that the brisk-transient cells are α cells, the brisk-sustained cells are β cells, and the sluggish cells are γ cells.

Famiglietti and Kolb (1976) showed that ganglion cells can also be separated into two classes according to the positions of their dendrites in the inner plexiform layer. They suggested that those with dendrites in the outer part of the inner plexiform layer (sublamina *a*) were *off*-centre cells and those with dendrites in the inner part (sublamina *b*) were *on*-centre cells. They confirmed this idea by recording from particular cells and then filling them with fluorescent dye from the recording electrode (Nelson *et al.*, 1978).

The cat retina is thus covered by four separate arrays of ganglion cells (fig. 18.49). The large α cells are probably concerned more with movement detection, while the β cells are used for finer resolution of static features of the visual field. Each of these groups has an array of *on*-centre cells superimposed on one of *off*-centre cells. The γ cells may perhaps be used for a variety of more complex responses. The next question for us to consider is, how do the various retinal interneurons allow the rods and cones to project into these different ganglion cell arrays?

Fig. 18.48. The distribution of discharge patterns within the receptive field of a ganglion cell in a cat retina. The cell is located at the tip of the electrode. The field was explored with a spot of light 0.2 mm in diameter. 'On' responses (crosses and unshaded zone) occurred in the centre of the field, 'off' responses (circles and diagonal shading) in its periphery. 'On–off' responses (horizontal shading) occurred in an intermediate zone. (From Kuffler, 1953.)

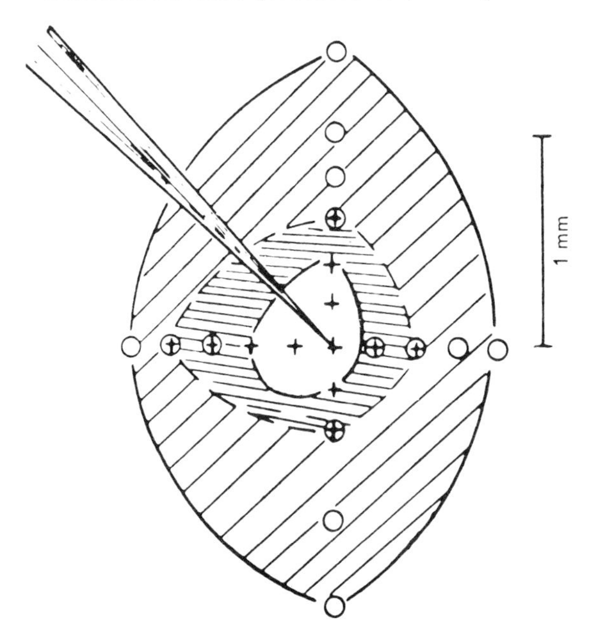

From photoreceptor to ganglion cell

The rods and cones are connected to the ganglion cells via the bipolar cells (and in some cases via amacrine cells as well), with lateral connections via horizontal cells and amacrine cells (figs. 18.2, 18.3). The ganglion cells send their axons into the brain and their activity is readily measurable as a series of action potentials.

Responses of *bipolar cells* were investigated by Werblin and Dowling (1969) in *Necturus* and by Kaneko (1970) in the goldfish. In studies of this type it is necessary to mark the cell from which recordings have been made by afterwards injecting a dye (Niagara Sky Blue or, better, Procion Yellow) by electrophoresis from the electrode. The retina is then fixed, sectioned and examined histologically so as to determine which sort of retinal cell had been impaled and recorded from. The results showed that bipolar cells respond to light by means of graded hyperpolarizations or depolarizations; they do not produce action potentials. Further, they show a feature known as centre-surround antagonism: stimulation of the centre of their receptive fields produces one type of response whereas stimulation of a peripheral area surrounding this produces its opposite. In fig. 18.50, for example, stimulation of the centre of its receptive field by a spot of light produces a hyperpolarization of the bipolar cell, whereas stimulation of the periphery by an annulus produces a depolarization. In about half of the cells, the responses were the opposite of this, in which illumination of the centre caused depolarization and illumination of the periphery caused hyperpolarization.

The size of the central area corresponds approximately to that of the receptive fields of the bipolar cells, and so it would seem that these central responses are caused by activity in photoreceptors to which the bipolars are directly connected. The dimensions of the surround fit reasonably well with the dendritic fields of horizontal cells, and it therefore seems likely that photoreceptors in the peripheral field of a bipolar cell produce their action on it via horizontal cells.

There are a number of different types of bipolar cell in the cat (see Sterling, 1983, and Sterling *et al.*, 1986). Rod bipolars connect rods to a particular type of amacrine cell that itself synapses with ganglion cells. Cone bipolars are of two different types; in invaginating bipolars the dendrite is partly engulfed within the terminal region of the cone, whereas in flat bipolars the contacts are more superficial. Invaginating cone bipolars depolarize when their receptive field centres are illuminated, whereas flat bipolars hyperpolarize. Each of these types can be further subdivided into

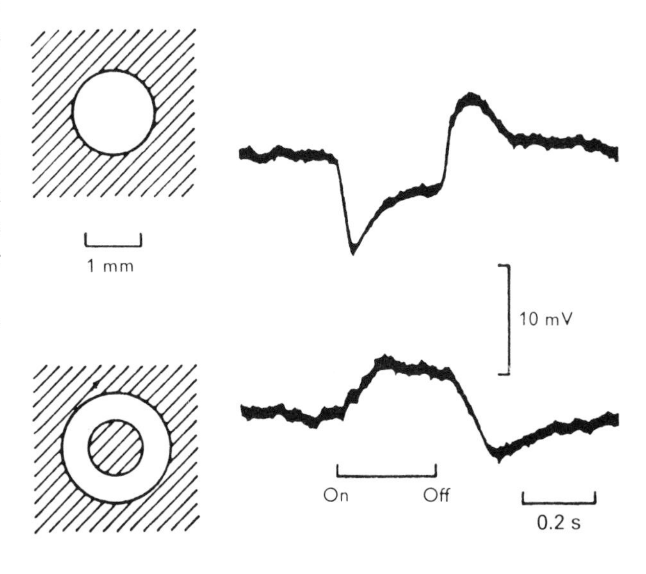

Fig. 18.50. Intracellular records from a bipolar cell which is hyperpolarized by a spot of white light (upper trace) and depolarized by an annulus (lower trace). (From Kaneko, 1970.)

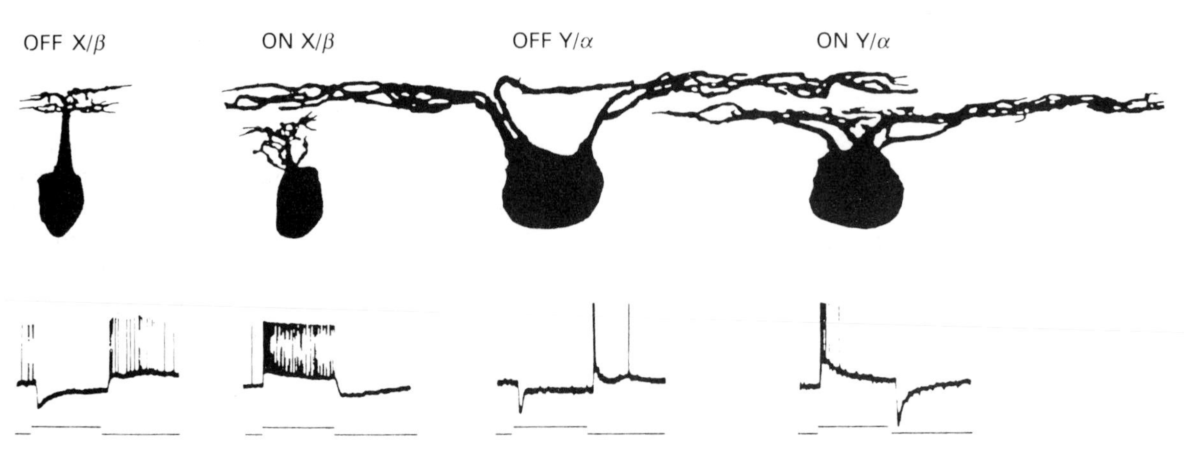

Fig. 18.49. Ganglion cells in the cat retina. β cells give sustained responses to light; α cells have larger receptive fields and give transient responses to light. *Off*-centre cells form synapses in the outer part of the inner plexiform layer; *on*-centre cells form them in the inner part. (From Sterling *et al.*, 1986.)

those which synapse with ganglion cells in sublamina *a* or sublamina *b* of the inner plexiform layer.

The *horizontal cells* themselves respond to illumination with graded hyperpolarizing potentials. Their receptive fields are fairly extensive and do not show centre-surround antagonism. Their responses are relatively easily recorded and were for a long time named '*S*-potentials' (although it is possible that the full range of *S*-potentials, described for example, by MacNichol and Svaetichin, 1958, includes potentials from retinal glial cells as well).

Now let us consider the means whereby these responses in horizontal and bipolar cells are produced by the *photoreceptors*. Light, we have seen, causes hyperpolarization of the rods and cones, and it is pertinent firstly to enquire whether this hyperpolarization is an essential element in the subsequent transmission of information. Baylor and Fettiplace (1976) have performed some useful experiments on turtle retinae which show quite clearly that it is. They found that hyperpolarization of the cone by electric current causes responses in the ganglion cell similar to those produced by light, and that depolarization of the cone can prevent the ganglion cell response to light from occurring.

The synaptic terminals of the rods and cones contain synaptic vesicles and are separated by the usual synaptic cleft from the bipolar and horizontal cells, so it seems very likely that these are chemically transmitting synapses. There are two possible ways in which they might act: either (i) the terminals are inactive in the dark and respond to hyperpolarization (produced by light) by releasing a transmitter or (ii) they are continually releasing transmitter in the dark and respond to hyperpolarization by ceasing to do so. Evidence for the second possibility has been provided, among others, by Dowling and Ripps (1973): they found that magnesium ions, which inhibit transmission at most chemical synapses, cause hyperpolarization of skate horizontal cells in the dark and abolish their responses to light. So it looks as though the transmitter substance causes depolarization of the horizontal cells and some of the bipolars, and hyperpolarization of other bipolars.

The *amacrine cells* show transient depolarizing responses, including what are apparently all-or-nothing action potentials, at the onset or cessation of light. Ganglion cell responses are also depolarizing, with all-or-nothing action potentials, as we have seen.

In 1970, Dowling summarized the electrical activity of the various retinal cells in the diagram shown here as fig. 18.51. Three important generalizations emerge: (1) light produces hyperpolarization in the initial stages of the sequence (i.e. in receptors, horizontals and many of the bipolars), but depolarization in the later stages, (2) lateral connections occur at two levels, via the horizontals and the amacrines, and (3) only the amacrines and the ganglion cells produce action potentials.

Receptor coupling

So far we have assumed that single rods and cones are independent units, each one only responding when light is absorbed in its own outer segment. However, it has become clear that the receptors may be to some extent electrically coupled to each other, so that a change in potential in one cell can spread into neighbouring cells. Baylor, Fuortes and O'Bryan (1971) measured the membrane potentials of turtle cones illuminated with small or large spots of light of the same intensity. The small (4 μm radius) spot would be sufficient in size to illuminate the whole of the cone, the large one (70 μm radius would also illuminate many neighbouring cones. They found that the resulting hyperpolarizations were larger with the large spots than with the small ones. A single cone would produce detectable hyperpolarizations in response to illumination of other cones up to about 40 μm away. They also passed current through one cone and recorded potential changes in a neighbouring one.

Further evidence was provided by some very interesting experiments by Fain (1975; see also Fain, Gold and Dowling, 1976) on the responses of toad rods to diffuse flashes of dim light. He found that the response of an individual rod to a series of flashes was much less variable than would be predicted by the Poisson distribution. For example in one series, when the flashes were of an intensity to bleach an average of 1.4 rhodopsin molecules per receptor, the Poisson distribution (equation (7.1) predicts that about 25% of the flashes would have bleached no rhodopsin molecules in that rod, 35% would have bleached one molecule, 24% two and 16% three or more. And yet the hyperpolarizations recorded from the rods all fell in the narrow range of 440–660 μV. Fain concluded that a rod is able to respond even when no rhodopsin molecules are bleached in its own outer segment, because it receives signals from other rods, and that at least 85% of the response recorded from a single rod is generated by rhodopsin molecules activated in other rods.

The structural basis of coupling lies in the existence of gap junctions which exist between the terminals of neighbouring receptors. The functional reasons for this phenomenon of receptor coupling are discussed by Attwell (1986). Perhaps it serves to increase the sensitivity of the system, but it would seem that this must be at the expense of visual acuity. Receptor coupling is much less evident in mammalian retinas.

Fig. 18.51. Diagram to show the intracellular responses of the various retinal cells and their connections with each other. The receptor on the left is illuminated with a brief flash of light imposed on a dim background which illuminates both receptors. *R*, receptors; *H*, horizontal cell; *B*, bipolar cells; *A*, amacrine cells; *G*, ganglion cells. The responses were actually obtained from the retina of *Necturus*. (From Dowling, 1970.)

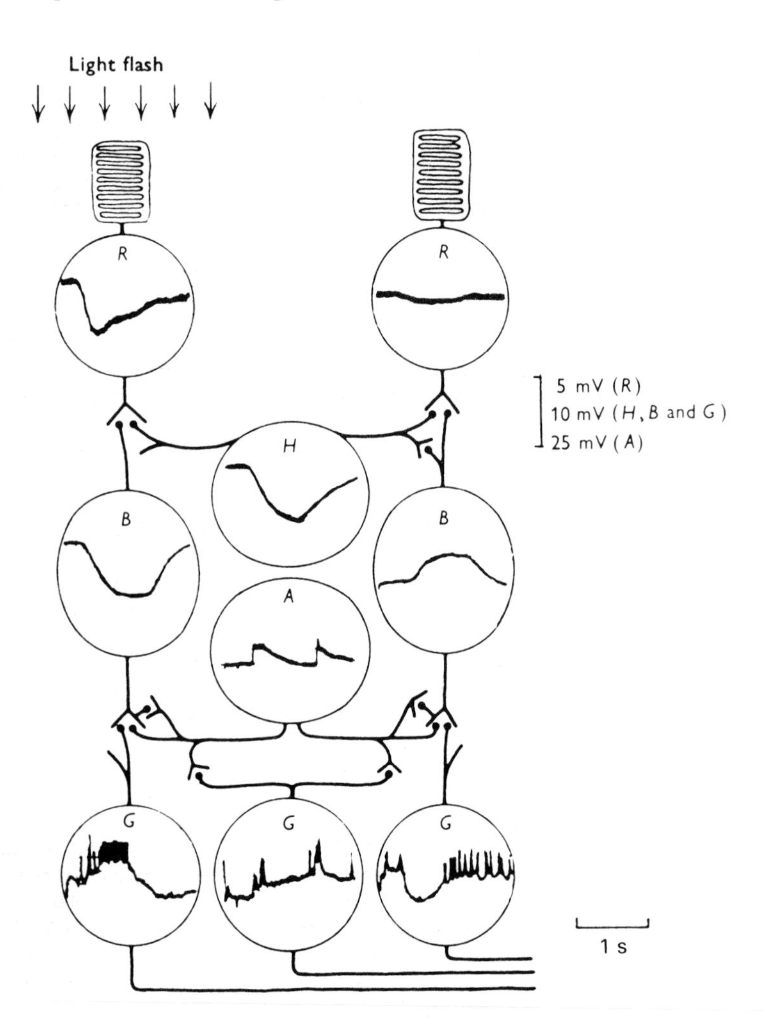

References

Abbott, B. C. and Wilkie, D. R. (1953). The relation between velocity of shortening and the tension-length curve of skeletal muscle. *J. Physiol.* **120**, 214–23.

Achazi, R. K. (1982). Catch muscle. In *Basic Biology of Muscles: a Comparative Approach* (ed. B. M. Twarog, R. J. C. Levine and M. M. Dewey), pp. 291–308. New York: Academic Press.

Adams, D. J., Dwyer, T. M. and Hille, B. (1980). The permeability of endplate channels to monovalent and divalent metal cations. *J. gen. Physiol.* **75**, 493–510.

Adams, P. R. and Brown, D. A. (1982). Synaptic inhibition of the M-current: slow post-synaptic potential mechanism in bullfrog sympathetic neurones. *J. Physiol.* **332**, 263–72.

Adams, P. R., Brown, D. A. and Constanti, A. (1982). M-currents and other potassium currents in bullfrog sympathetic neurones. *J. Physiol.* **330**, 537–72.

Adams, P. R., Jones, S. W., Pennefather, P., Brown, D. A., Koch, C. and Lancaster, B. (1986). Slow synaptic transmission in frog sympathetic ganglia. *J. exp. Biol.* **124**, 259–85.

Adrian, E. D. (1926). The impulses produced by sensory nerve endings. I. *J. Physiol.* **61**, 47–72.

Adrian, E. D. and Zotterman, Y. (1926). The impulses produced by sensory nerve endings. II. The response of a single end-organ. *J. Physiol.* **61**, 151–71.

Adrian, R. H., Chandler, W. K. and Hodgkin, A. L. (1969). The kinetics of mechanical activation in frog muscle. *J. Physiol.* **204**, 207–30.

Adrian, R. H., Chandler, W. K. and Hodgkin, A. L. (1970). Slow changes in potassium permeability in skeletal muscle. *J. Physiol.* **208**, 645–68.

Adrian, R. H., Chandler, W. K. and Rakowski, R. F. (1976). Charge movement and mechanical repriming in skeletal muscle. *J. Physiol.* **254**, 361–88.

Agnew, W. S., Levinson, S. R., Brabson, J. S. and Raftery, M. A. (1978). Purification of the tetrodotoxin-binding component associated with the voltage-sensitive channel from *Electrophorus electricus* electroplax membranes. *Proc. Natl. Acad. Sci. USA* **75**, 2606–11.

Agnew, W. S., Miller, J. A., Ellisman, M. H., Rosenberg, R. L., Tomiko, S. A. and Levinson, S. R. (1983). The voltage-regulated sodium channel from the electroplax of *Electrophorus electricus. Cold Spr. Harb. Symp. Quant. Biol.* **48**, 165–79.

Ahlquist, R. P. (1948). A study of the adrenotropic receptors. *Am. J. Physiol.* **153**, 586–600.

Aidley, D. J. (1965*a*). The effect of calcium ions on potassium contracture in a locust leg muscle. *J. Physiol.* **177**, 94–102.

Aidley, D. J. (1965*b*). Transient changes in isotonic shortening velocity of frog rectus abdominis muscles in potassium contracture. *Proc. R. Soc. Lond.* B **163**, 214–23.

Aidley, D. J. (1985). Muscular contraction. In *Comprehensive Insect Physiology, Biochemistry and Pharmacology* (ed. G. A. Kerkut and L. I. Gilbert), vol. 5, pp. 407–37. Oxford: Pergamon Press.

Aidley, D. J. and White, D. C. S. (1969). Mechanical properties of glycerinated fibres from the tymbal muscles of a Brazilian cicada. *J. Physiol.* **205**, 179–92.

Akert, K., Pfenninger, K., Sandri, C. and Moor, H. (1972). Freeze etching and cytochemistry of vesicles and membrane complexes in synapses of the central nervous system. In *Structure and Function of Synapses* (ed. G. D. Pappas and D. P. Purpura), pp. 67–86. New York: Raven Press.

Alberts, B., Bray, D., Lewis, J., Raff, M., Roberts, K. and Watson, J. D. (1983). *Molecular Biology of the Cell.* New York: Garland.

Aldrich, R. W., Corey, D. P. and Stevens, C. F. (1983). A reinterpretation of mammalian sodium channel gating based on single channel recording. *Nature* **306**, 436–41.

Aldrich, R. W. and Stevens, C. F. (1983). Inactivation of open and closed sodium channels determined separately. *Cold Spr. Harb. Symp. Quant. Biol.* **48**, 147–53.

Allen, D. G. and Blinks, J. R. (1978). Calcium transients in aequorin-injected frog cardiac muscle. *Nature* **273**, 509–13.

Allen, D. G., Jewell, B. R. and Murray, J. W. (1974). The contribution of activation processes to the length-tension relation of cardiac muscle. *Nature* **248**, 606–7.

Allen, D. G. and Kentish, J. C. (1985). The cellular basis of the length-tension relation in cardiac muscle. *J. Mol. Cell. Cardiol.* **17**, 821–40.

Almers, W. and McCleskey, E. W. (1984). Non-selective conductance in calcium channels of frog muscle: calcium selectivity in a single-file pore. *J. Physiol.* **353**, 585–608.

Almers, W., McCleskey, E. W. and Palade, P. T. (1984). A non-selective cation conductance in frog muscle membrane blocked by micromolar external calcium ions. *J. Physiol.* **353**, 565–83.

Almers, W., McCleskey, E. W. and Palade, P. T. (1986). The mechanism of ion selectivity in calcium channels of skeletal muscle membrane. In *Membrane Control of Cellular Activity* (ed. H. C. Lüttgau), pp. 61–73. Stuttgart: G. Fischer Verlag.

Alpern, M. and Pugh, E. N., Jnr (1974). The density and photosensitivity of human rhodopsin in the living retina. *J. Physiol.* **237**, 341–70.

Altamarino, M., Coates, C. W. and Grundfest, H. (1955). Mechanisms of direct and neural excitability in electroplaques of electric eel. *J. gen. Physiol.* **38**, 319–60.

Amara, S. G., Jones, V., Rosenfeld, M. G., Ong, E. S. and Evans, R. M. (1982). Alternative RNA processing in calcitonin gene expression generates mRNAs encoding different polypeptide products. *Nature* **298**, 240–4.

Amos, L. A. (1985). Structure of muscle filaments studied by electron microscopy. *Ann. Rev. Biophys.* **14**, 291–313.

Amos, L. A., Henderson, R. and Unwin, P. N. T. (1982*a*). Three-dimensional structure determination by electron microscopy of two-dimensional crystals. *Prog. Biophys. molec. Biol.* **39**, 183–231.

Amos, L. A., Huxley, H. E., Holmes, K. C., Goody, R. S. and Taylor, K. A. (1982*b*). Structural evidence that myosin heads may interact with two sites on F-actin. *Nature* **299**, 467–9.

Anderson, C. R. and Stevens, C. F. (1973). Voltage clamp analysis of acetylcholine produced end-plate current fluctuations at frog neuromuscular junction. *J. Physiol.* **235**, 655–91.

Aprison, M. H. and Werman, R. (1965). The distribution of glycine in cat spinal cord and roots. *Life Sci.* **4**, 2075–83.

Araki, T., Eccles, J. C. and Ito, M. (1960). Correlation of the inhibitory postsynaptic potential of motoneurones with the latency and time course of inhibition of monosynaptic reflexes. *J. Physiol.* **154**, 354–77.

Araki, T. and Terzuolo, C. A. (1962). Membrane currents in spinal motoneurones associated with the action potential and synaptic activity. *J. Neurophysiol.* **25**, 772–89.

Arden, G. B. (1969). The excitation of photoreceptors. *Progr. Biophys.* **19**, 371–421.

Armett, C. J., Gray, J. A. B., Hunsperger, R. and Lal, S. (1962). The transmission of information in primary receptor neurones and second order neurones of a phasic system. *J. Physiol.* **164**, 395–421.

Armstrong, C., Huxley, A. F. and Julian, F. J. (1966). Oscillatory responses in frog skeletal muscle fibres. *J. Physiol.* **186**, 26–7*P*.

Armstrong, C. M. (1971). Interaction of tetraethylammonium ion derivatives with the potassium channels of giant axons. *J. gen. Physiol.* **58**, 413–37.

Armstrong, C. M. (1975). K^+ pores of nerve and muscle. In *Membranes: a Series of Advances* (ed. G. Eisenman), vol. 3, 325–58. New York: Dekker.

Armstrong, C. M. (1981). Sodium channels and gating currents. *Physiol. Rev.* **61**, 644–83.

Armstrong, C. M. and Bezanilla, F. M. (1973). Current related to the movement of the gating particles of the sodium channels. *Nature.* **242**, 459–61.

Armstrong, C. M. and Bezanilla, F. (1974). Charge movement associated with the opening and closing of the activation gates of the Na channels. *J. gen. Physiol.* **63**, 675–89.

Armstrong, C. M. and Bezanilla, F. (1977). Inactivation of the sodium channel. II. Gating current experiments. *J. gen. Physiol.* **70**, 567–90.

Armstrong, C. M., Bezanilla, F. M. and Rojas, E. (1973). Destruction of sodium conductance inactivation in squid axon perfused with pronase. *J. gen. Physiol.* **62**, 375–91.

Arrang, J.-M., Garbarg, M., Lancelot, J.-C., Lecomte, J.-M., Pollard, H., Robba, M., Schunack, W. and

Schwartz, J.-C. (1987). Highly potent and selective ligands for histamine H_3-receptors. *Nature* **327**, 117–22.

Arrang, J.-M., Garbarg, M. and Schwartz, J.-C. (1983). Auto-inhibition of brain histamine release mediated by a novel class (H_3) of histamine receptor. *Nature* **302**, 832–7.

Ascher, P., and Nowak, L. (1987). Electrophysiological studies of NMDA receptors. *Trends Neurosci.* **10**, 284–8.

Ashcroft, F. M. (1981). Calcium action potentials in the skeletal muscle fibres of the stick insect *Carausius morosus*. *J. exp. Biol.* **93**, 257–67.

Ashley, C. C. and Moisescu, D. G. (1977). Effect of changing the composition of the bathing solutions upon the isometric tension – pCa relationship in bundles of crustacean myofibrils. *J. Physiol.* **270**, 627–52.

Ashley, C. C. and Ridgeway, E. B. (1968). Simultaneous recording of membrane potential, calcium transient and tension in single muscle fibres. *Nature* **219**, 1168–9.

Ashley, C. C. and Ridgeway, E. B. (1970). On the relationships between membrane potential, calcium transient and tension in single barnacle muscle fibres. *J. Physiol.* **209**, 105–30.

Ashmore, J. F. and Brownell, W. E. (1986). Kiloherz movements induced by electrical stimulation in outer hair cells isolated from the guinea-pig cochlea. *J. Physiol.* **377**, 41*P*.

Atwater, I., Bezanilla, F. and Rojas, E. (1969). Sodium influxes in internally perfused squid giant axon during voltage clamp. *J. Physiol.* **201**, 657–64.

Atwell, D. (1986). Ion channels and signal processing in the outer retina. *Quart. J. Exp. Physiol.* **71**, 497–536.

Aubert, X. (1956). La relation entre la force et la vitesse d'allongement et de raccourcissement du muscle strié. *Archs int. Physiol. Biochim.* **64**, 121.

Auerbach, A. and Sachs, F. (1983). Flickering of a nicotinic ion channel to a subconductance state. *Biophys. J.* **42**, 1–10.

Axelsson, J. and Thesleff, S. (1959). A study of supersensitivity in denervated mammalian skeletal muscle. *J. Physiol.* **147**, 178–93.

Babu, Y. S., Sack, J. S., Greenhough, T. J., Bugg, C. E., Means, A. R. and Cook, W. J. (1985). Three dimensional structure of calmodulin. *Nature* **315**, 37–40.

Baehr, W. and Applebury, M. L. (1986). Exploring visual transduction with recombinant DNA techniques. *Trends Neurosci.* **9**, 198–203.

Bagshaw, C. R. (1982). *Muscle Contraction.* London: Chapman and Hall.

Bailey, K. (1948). Tropomyosin: a new asymmetrical protein component of the muscle fibril. *Biochem. J.* **43**, 271–9.

Baker, P. F. (1986). The sodium-calcium exchange system. In *Calcium and the Cell*. pp. 73–86. Ciba Foundation Symposium **122**. Chichester: John Wiley.

Baker, P. F., Blaustein, M. P., Keynes, R. D., Mansil, J., Shaw, T. I. and Steinhardt, R. A. (1969). The ouabain-sensitive fluxes of sodium and potassium in squid giant axons. *J. Physiol.* **200**, 459–96.

Baker, P. F. and Crawford, A. C. (1972). Mobility and transport of magnesium in squid giant axons. *J. Physiol.* **227**, 855–74.

Baker, P. F. and Dipolo, R. (1984). Axonal calcium and magnesium homeostasis. In *The Squid Axon* (ed. P. F. Baker). *Current Topics in Membranes and Transport* vol. **22**, pp. 195–247. Orlando: Academic Press.

Baker, P. F., Hodgkin, A. L. and Ridgeway, E. B. (1971). Depolarization and calcium entry in squid giant axons. *J. Physiol.* **218**, 709–55.

Baker, P. F. Hodgkin, A. L. and Shaw, T. I. (1961). Replacement of the protoplasm of a giant nerve fibre with artificial solutions. *Nature* **190**, 885.

Baker, P. F., Hodgkin, A. L. and Shaw, T. I. (1962*a*). Replacement of the axoplasm of giant nerve fibres with artificial solutions. *J. Physiol.* **164**, 330–54.

Baker, P. F., Hodgkin, A. L. and Shaw, T. I. (1962*b*). The effects of changes in internal ionic concentrations on the electrical properties of perfused giant axons. *J. Physiol.* **164**, 355–74.

Baker, P. F. and Schlaepfer, W. W. (1978). Uptake and binding of calcium by axoplasm isolated from giant axons of *Loligo* and *Myxicola*. *J. Physiol.* **276**, 103–25.

Baker, P. F. and Willis, J. S. (1972). Binding of the cardiac glycoside ouabain to intact cells. *J. Physiol.* **224**, 441–62.

Ball, S., Goodwin, T. W. and Morton, R. A. (1948). Studies on vitamin A. 5. The preparation of retinene$_1$ – vitamin A aldehyde. *Biochem. J.* **42**, 516–23.

Banks, R. W., Barker, D. and Stacey, M. J. (1982). Form and distribution of sensory terminals in cat hindlimb muscle spindles. *Phil. Trans. R. Soc. Lond.* B **299**, 329–64.

Banks, R. W., Barker, D. and Stacey, M. J. (1985). Form and classification of motor endings in mammalian muscle spindles. *Proc. R. Soc. Lond.* B **225**, 195–212.

Banks, R. W., Harker, D. W. and Stacey, M. J. (1977). A study of mammalian intrafusal muscle fibres using a combined histochemical and ultrastructural approach. *J. Anat.* **123**, 783–96.

Bar, R. S., Deamer, D. W. and Corwell, D. G. (1966). Surface area of human erythrocyte lipids: reinvestigation of experiments on plasma membrane. *Science* **153**, 1010–12.

Barclay, P. L. and Findlay, J. B. C. (1984). Labelling of the cytoplasmic domains of ovine rhodopsin with hydrophilic chemical probes. *Biochem. J.* **220**, 75–84.

Bargmann, W. and Scharrer, E. (1951). The site of origin of the hormones of the posterior pituitary. *Am. Sci.* **29**, 255–9.

Barker, D. and Banks, R. W. (1986). The muscle spindle. In *Myology* (ed. A. G. Engel and B. Q. Banker), pp. 309–41. New York: McGraw-Hill.

Barker, D., Emonet-Dénand, F., Harker, D. W., Jami, L. and Laporte, Y. (1976). Distribution of fusimotor axons to intrafusal muscle fibres in cat tenuissimus spindles as determined by the glycogen-depletion method. *J. Physiol.* **261**, 49–69.

Barker, D., Emonet-Dénand, F., Harker, D. W., Jami, L. and Laporte, Y. (1977). Types of intra- and extrafusal muscle fibre innervated by dynamic skeleto-fusimotor axons in cat peroneus brevis and tenuissimus muscles, as determined by the glycogen depletion method. *J. Physiol.* **266**, 713–26.

Barker, J. L. and Ransom, B. R. (1978). Phentobarbitone pharmacology of mammalian central neurones grown in tissue culture. *J. Physiol.* **280**, 355–72.

Barlow, H. B., Fitzhugh, R. and Kuffler, S. W. (1957). Change of organization in the receptive fields of the cat's retina during dark adaptation. *J. Physiol.* **137**, 338–54.

Barlow, H. B. and Mollon, J. D. (eds.) (1982). *The Senses.* Cambridge: Cambridge University Press.

Barnard, E. A., Miledi, R. and Sumikawa, K. (1982). Translation of exogenous messenger RNA coding for nicotinic acetylocholine receptors produces functional receptors in *Xenopus* oocytes. *Proc. R. Soc. Lond.* B **215**, 241–6.

Barnard, E. A., Wieckowski, J. and Chiu, T. H. (1971). Cholinergic receptor molecules and cholinesterase molecules at mouse skeletal muscle junction. *Nature* **234**, 207–9.

Barnett, V. A. and Thomas, D. D. (1984). Saturation transfer electron paramagnetic resonance of spin-labelled muscle fibres. *J. Mol. Biol.* **179**, 83–102.

Barrett, J. N., Magleby, K. L. and Pallotta, B. S. (1982). Properties of single calcium-activated potassium channels in cultured rat muscle. *J. Physiol.* **331**, 211–30.

Barron, D. H. and Matthews, B. H. C. (1938). The interpretation of potential changes in the spinal cord. *J. Physiol.* **92**, 276–321.

Bass, A. H. (1986). Electric organs revisited. In *Electroreception* (ed. T. H. Bullock and W. Heiligenberg), pp. 13–70. New York: John Wiley.

Baylor, D. A. and Fettiplace, R. (1976). Transmission of signals from photoreceptors to ganglion cells in the eye of the turtle. *Cold Spr. Harb. Symp. Quant. Biol.* **40**, 529–36.

Baylor, D. A. and Fuortes, M. G. F. (1970). Electrical responses of single cones in the retina of the turtle. *J. Physiol.* **207**, 77–92.

Baylor, D. A., Fuortes, M. G. F. and O'Bryan, P. M. (1971). Receptive fields of cones in the retina of the turtle. *J. Physiol.* **214**, 265–94.

Baylor, D. A. and Hodgkin, A. L. (1973). Detection and resolution of visual stimuli by turtle photoreceptors. *J. Physiol.* **234**, 163–98.

Baylor, D. A., Lamb, T. D. and Yau, K.-W. (1979*a*). The membrane current of single rod outer segments. *J. Physiol.* **288**, 589–611.

Baylor, D. A., Lamb, T. D. and Yau, K.-W. (1979*b*). Responses of retinal rods to single photons. *J. Physiol.* **288**, 613–34.

Beeler, G. W. and Reuter, H. (1970). Membrane calcium current in ventricular myocardial fibres. *J. Physiol.* **207**, 191–209.

Begenisich, T. and Cahalan, M. D. (1980). Sodium channel permeation in squid axons. I. Reversal potential experiments. *J. Physiol.* **307**, 217–42.

Bekkers, J. M., Greef, N. G. and Neumcke, B. (1983). The conductance of sodium channels in the squid giant axon. *J. Physiol.* **343**, 24–25*P*.

Bendall, J. R. (1953). Further observations on a factor (the 'Marsh' factor) effecting relaxation of ATP-shortened muscle fibre models, and the effect of Ca and Mg ions on it. *J. Physiol.* **121**, 232–54

Benham, C. D. and Tsien, R. W. (1987). A novel receptor-operated Ca^{2+}-permeable channel activated by ATP in smooth muscle. *Nature* **328**, 275–8.

Bennett, M. R. (1972). *Autonomic Neuromuscular Transmission.* Cambridge: Cambridge University Press.

Bennett, M. R., Fisher, C., Florin, T., Quine, M. and Robinson, J. (1977). The effect of calcium ions and temperature on the binomial parameters that control acetylcholine release by a nerve impulse at amphibian neuromuscular synapses. *J. Physiol.* **271**, 641–72.

Bennett, M. V. L. (1960). Electrical connections between supermedullary neurons. *Fed. Proc.* **19**, 282.

Bennett, M. V. L. (1965). Electroreceptors in mormyrids. *Cold Spr. Harb. Symp. Quant. Biol.* **30**, 245–62.

Bennett, M. V. L. (1966). Physiology of electrotonic junctions. *Ann. N.Y. Acad. Sci.* **137**, 509–39.

Bennett, M. V. L. (1968). Neural control of electric organs. In *The Central Nervous System and Fish Behaviour* (ed. D. Ingle), pp. 147–69. Chicago: University of Chicago Press.

Bennett, M. V. L. (1971). Electric Organs. In *Fish Physiology*, (ed. W. S. Hoar and D. J. Randall), **5**, pp. 347–492. New York: Academic Press.

Bennett, M. V. L. (1977). Electrical transmission: a functional analysis and comparison to chemical transmission. In *Handbook of Physiology*, section 1, vol. 1 (ed. E. R. Kandel), pp. 357–416. Bethesda, Maryland; American Physiological Society

Bennett, M. V. L. (1985). Nicked by Occam's razor: unitarianism in the investigation of synaptic transmission. *Biol. Bull.* **168**, (suppl.), 159–67.

Bennett, M. V. L., Aljure, E., Nakajima, Y. and Pappas, G. D. (1963). Electrotonic junctions between teleost spinal neurones: electrophysiology and ultrastructure. *Science* **141**, 262–4.

Bennett, M. V. L. and Grundfest, H. (1959). Electrophysiology of electric organ in *Gymnotus carapo. J. gen. Physiol.* **42**, 1067–104.

Bennett, M. V. L. and Grundfest, H. (1961*a*). The electrophysiology of electric organs of marine electric fishes. II. The electroplaques of main and accessory organs of *Narcine brasiliensis. J. gen. Physiol.* **44**, 805–18.

Bennett, M. V. L. and Grundfest, H. (1961*b*). The electrophysiology of electric organs of marine electric fishes. III. The electroplaques of the stargazer. *Astroscopus y-graceum. J. gen. Physiol.* **44**, 819–42.

Bennett, M. V. L. and Grundfest, H. (1961*c*). Studies on the morphology and electrophysiology of electric organs. III. Electrophysiology of electric organs in mormyrids. In *Bioelectrogenesis* (ed. C. Chagas and A. Paes de Carvalho), pp. 113–14. Amsterdam; Elsevier.

Bennett, M. V. L. and Spray, D. C. (1985). *Gap Junctions*. New York: Cold Spring Harbor Laboratory.

Bennett, M. V. L., Wurzel, M. and Grundfest, H. (1961). The electrophysiology of electric organs of marine electric fishes. I. Properties of electroplaques of *Torpedo nobiliana*. *J. gen. Physiol.* **44**, 757–804.

Bennett, P., Craig, R., Starr, R. and Offer, G. (1986). The ultrastructural location of C-protein, X-protein and H-protein in rabbit muscle. *J. Mus. Res. Cell. Motil.* **7**, 550–67.

Bernard, C. (1857). *Lecons sur les Effects des Substances Toxiques et Medicamenteuses*. Paris.

Bernstein, J. (1902). Untersuchungen zur Thermodynamik der bioelektrischen Ströme. *Pflügers Arch. ges. Physiol.* **92**, 521–62.

Berridge, M. J. (1985). The molecular basis of communication within the cell. *Sci. Amer.* **253** (4), 124–34.

Berridge, M. J. (1986). Regulation of ion channels by inositol trisphosphate and diacylglycerol. *J. exp. Biol.* **124**, 323–35.

Berridge, M. J. and Irvine, R. F. (1984). Inositol trisphosphate, a novel second messenger in cellular signal transduction. *Nature* **312**, 315–21.

Bessou, P. and Laporte, Y. (1962). Responses from primary and secondary endings of the same neuromuscular spindle of the tenuissimus muscle of the cat. In *Symposium on Muscle Receptors* (ed. D. Barker). Hong Kong: University Press.

Betz, H. (1987). Biology and structure of the mammalian glycine receptor. *Trends Neurosci.* **10**, 113–17.

Bezanilla, F. (1986). Voltage-dependent gating: current measurement and interpretation. In *Ionic Channels in Cells and Model Systems* (ed. R. Latorre). New York: Plenum Press.

Bezanilla, F., Rojas, E. and Taylor, R. E. (1970). Sodium and potassium conductance changes during a membrane action potential. *J. Physiol.* **211**, 729–51.

Bialek, W. (1987). Physical limits to sensation and perception. *Ann. Rev. Biophys.* **16**, 455–78.

Bianchi, C. P. and Shanes, A. M. (1959). Calcium influx in skeletal muscle at rest, during activity, and during potassium contracture. *J. gen. Physiol.* **42**, 803–15.

Binstock, L. and Goldman, L. (1969). Current- and voltage-clamped studies on *Myxicola* giant axons. Effect of tetrodotoxin. *J. gen. Physiol.* **54**, 730–40.

Binstock, L. and Lecar, H. (1969). Ammonium ion currents in the squid giant axon. *J. gen. Physiol.* **53**, 342–61.

Birks, R., Huxley, H. E. and Katz, B. (1960). The fine structure of the neuromuscular junction of the frog. *J. Physiol.* **150**, 134–44.

Birks, R. and Macintosh, F. C. (1961). Acetylcholine metabolism of a sympathetic ganglion. *Can. J. Biochem. Physiol.* **39**, 787–827.

Black, J. W., Duncan, W. A. M., Durrant, C. J., Ganellin, C. R. and Parsons, E. M. (1972). Definition and antagonism of histamine H_2-receptors. *Nature* **236**, 385–90.

Blaustein, M. P. and Hodgkin, A. L. (1969). The effect of cyanide on the efflux of calcium from squid axons. *J. Physiol.* **200**, 497–527.

Blinks, J. R., Wier, W. G., Hess, P. and Prendergast, F. G. (1982). Measurement of Ca^{2+} concentrations in living cells. *Prog. Biophys. molec. Biol.* **40**, 1–114.

Bliss, T. V. P. and Lomo, T. (1973). Long-lasting potentiation of synaptic transmission in the dentate area of the anaesthetized rabbit following stimulation of the perforant path. *J. Physiol.* **232**, 331–56.

Bodoia, R. D. and Detwiler, P. B. (1985). Patch-clamp recordings of the light-sensitive dark noise in retinal rods from the lizard and frog. *J. Physiol.* **367**, 183–216.

Boeckh, J., Kaissling, K.-E. and Schneider, D. (1965). Insect olfactory receptors. *Cold Spr. Harb. Symp. Quant. Biol.* **30**, 263–80.

Boistel, J. and Fatt, P. (1958). Membrane permeability change during transmitter action in crustacean muscle. *J. Physiol.* **144**, 176–91.

Bolis, L., Keynes, R. D. and Maddrell, S. H. P. (eds.) (1984). *Comparative Physiology of Sensory Systems*. Cambridge: Cambridge University Press.

Boll, F. (1877). Zur Anatomie und Physiologie der Retina. *Arch. Anat. Physiol.* **1877**, 4–35.

Bolton, T. B. (1979). Mechanisms of action of transmitters and other substances on smooth muscle. *Physiol. Rev.* **59**, 606–718.

Bolton, T. B. (1981). Action of acetylcholine on the smooth muscle membrane. In *Smooth Muscle* (ed. E. Bülbring *et al.*) pp. 199–217. London: Edward Arnold.

Bolton, T. B. and Large, W. A. (1986). Are junction potentials essential? Dual mechanism of smooth muscle cell activation by transmitter released from autonomic nerves. *Q. J. Exp. Physiol.* **71**, 1–28.

Bolton, T. S. (1986). Calcium metabolism in vascular smooth muscle. *Brit. Med. Bull.* **42**, 421–9.

Bone, Q. (1978). Locomotor muscle. In *Fish Physiology* (ed. W. S. Hoar and D. J. Randall), vol. 7, pp. 361–424. New York: Academic Press.

Bormann, J., Hamill, O. P. and Sakmann, B. (1987). Mechanism of anion permeation through channels gated by glycine and γ-aminobutyric acid in mouse cultured spinal neurones. *J. Physiol.* **385**, 243–86.

Boulter, J., Evans, K., Goldman, D., Martin, G., Treco, D., Heinemann, S. and Patrick, J. (1986). Isolation of a cDNA clone for a possible neural nicotinic acetylcholine receptor α-subunit. *Nature* **319**, 368–74.

Bowery, N. G., Hill, D. R., Hudson, A. L., Doble, A., Middlemiss, D. N., Shaw, J. and Turnbull, M. (1980). (−)Baclofen decreases neurotransmitter release in the mammalian CNS by an action at a novel GABA receptor. *Nature* **283**, 92–4.

Bowmaker, J. K. (1984). Microspectrophotometry of vertebrate photoreceptors. *Vision Res.* **24**, 1641–50.

Bowmaker, J. K. and Dartnall, H. J. A. (1980). Visual pigments of rods and cones in a human retina. *J. Physiol.* **298**, 501–11.

Bownds, D. (1967). Site of attachment of retinal in rhodopsin. *Nature* **216**, 1178–81.

Boycott, B. B. and Wässle, H. (1974). The morphological

types of ganglion cells of the domestic cat's retina. *J. Physiol.* **240**, 397–419.
Boyd, I. A. (1962). The structure and innervation of the nuclear bag muscle fibre system and nuclear chain muscle fibre system in mammalian muscle spindles. *Phil. Trans. R. Soc. Lond.* B **245**, 81–136.
Boyd, I. A. (1981). The muscle spindle controversy. *Sci. Prog., Oxf.* **67**, 205–21.
Boyd, I. A. and Gladden, M. (1986). Morphology of mammalian muscle spindles: review. In *The Muscle Spindle* (ed. I. A. Boyd and M. Gladden), pp. 3–22. Basingstoke: Macmillan.
Boyd, I. A., Gladden, M. H., McWilliam, P. N. and Ward, J. (1977). Control of dynamic and static nuclear bag fibres and nuclear chain fibres by gamma and beta axons in isolated cat muscle spindles. *J. Physiol.* **265**, 133–62.
Boyle, P. J. and Conway, E. J. (1941). Potassium accumulation in muscle and associated changes. *J. Physiol.* **100**, 1–63.
Bradford, H. F. (1986). *Chemical Neurobiology*. New York: Freeman.
Bradley, K., Easton, D. M. and Eccles, J. C. (1953). An investigation of primary or direct inhibition. *J. Physiol.* **122**, 474–88.
Bradley, P. B. and Wolstencroft, J. H. (1965). Actions of drugs on single neurones in the brain-stem. *Br. med. Bull.* **21**, 15–18.
Bretscher, M. S. (1973). Membrane structure: some general principles. *Science* **181**, 622–9.
Bridges, C. D. B. (1976). Vitamin A and the role of the pigment epithelium during bleaching and regeneration of rhodopsin in the frog eye. *Exp. Eye Res.* **22**, 435–55.
Bridges, C. D. B., Alvarez, R. A., Fong, S.-L., Gonzalez-Fernanez, F., Lam, D. M. K. and Liou, G. I. (1984). Visual cycle in the mammalian eye: retinoid-binding proteins and the distribution of 11-cis retinoids. *Vision Res.* **24**, 1581–94.
Brindley, G. S. (1970). *Physiology of the Retina and the Visual Pathway*, 2nd edn. London: Arnold.
Brinley, F. J. (1978). Calcium buffering in squid axons. *Ann. Rev. Biophys. Bioeng.* **7**, 363–92.
Brisson, A. and Unwin, P. N. T. (1985). Quaternary structure of the acetylcholine receptor. *Nature* **315**, 474–7.
Brock, L. G., Coombs, J. S. and Eccles, J. C. (1952). The recording of potentials from motoneurones with an intracellular electrode. *J. Physiol.* **117**, 431–60.
Brock, L. G., Eccles, R. M. and Keynes, R. D. (1953). The discharge of individual electroplates in *Raia clavata*. *J. Physiol.* **122**, 4–6*P*.
Brooke, M. H. and Kaiser, K. K. (1970). Muscle fibre types: how many and what kind? *Arch. Neurol.* **23**, 369–79.
Brown, A. M., Camerer, H., Kunze, D. L. and Lux, H. D. (1982). Similarity of unitary Ca^{2+} currents in three different species. *Nature* **299**, 156–8.
Brown, A. M., Lee, K. S. and Powell, T. (1981). Sodium current in single rat heart muscle cells. *J. Physiol.* **318**, 479–500.
Brown, D. A. and Adams, P. R. (1980). Muscarinic suppression of a novel voltage-sensitive K^+-current in a vertebrate neurone. *Nature* **283**, 673–6.
Brown, G. L. (1965). The release and fate of transmitter liberated by adrenergic nerves. *Proc. R. Soc. Lond.* B **162**, 1–19.
Brown, G. L., Dale, H. H. and Feldburg, W. (1936). Reactions of the normal mammalian muscle to acetylcholine and to eserine. *J. Physiol.* **87**, 394–424.
Brown, H. F., Kimura, J., Noble, D., Noble, S. J. and Taupignon, A. (1984). The ionic currents underlying pacemaker activity in rabbit sino-atrial node: experimental results and computer simulations. *Proc. R. Soc. Lond.* B **222**, 329–47.
Brown, J. E. and Pinto, L. H. (1974). Ionic mechanism for the photoreceptor potential of the retina of *Bufo marinus*. *J. Physiol.* **236**, 575–91.
Brown, J. E., Coles, J. A. and Pinto, L. H. (1977). Effects of injections of calcium and EGTA into the outer segments of retinal rods of *Bufo marinus*. *J. Physiol.* **269**, 707–22.
Brown, J. H. and Brown, S. L. (1984). Agonists differentiate muscarinic receptors that inhibit cyclic AMP formation from those that stimulate phosphoinositide metabolism. *J. Biol. Chem.* **254**, 3777–81.
Brown, K. T. and Murakami, M. (1964). A new receptor potential of the monkey retina with no detectable latency. *Nature* **201**, 626–8.
Brown, K. T. and Wiesel, T. N. (1961). Localization of origins of electroretinogram components by intraretinal recording in the intact cat eye. *J. Physiol.* **158**, 257–80.
Brown, M. C. and Butler, R. G. (1973). Studies on the site of termination of static and dynamic fusimotor fibres within spindles of the tenuissimus muscles of the cat. *J. Physiol.* **233**, 553–73.
Brown, M. C., Crowe, A. and Matthews, P. B. C. (1965). Observations on the fusimotor fibres of the tibialis posterior muscle of the cat. *J. Physiol.* **177**, 140–59.
Brown, M. C., Nuttall, A. L. and Masta, R. I. (1983). Intracellular recordings from cochlear inner hair cells: effects of stimulation of the crossed olivocochlear efferents. *Science* **222**, 69–72.
Brown, P. K. and Wald, G. (1964). Visual pigments in single rods and cones of the human retina. *Science* **144**, 45–52.
Brownell, W. E., Bader, C. R., Bertrand, D. and de Ribaupierre, Y. (1985). Evoked mechanical responses of isolated cochlear outer hair cells. *Science* **227**, 194–6.
Bülbring, E. (1955). Correlation between membrane potential, spike discharge and tension in smooth muscle. *J. Physiol.* **128**, 200–21.
Bülbring, E., Brading, A. F., Jones, A. W. and Tomita, T. (eds.) (1981*a*). *Smooth Muscle: an Assessment of Current Knowledge*. London: Edward Arnold.
Bülbring, E., Burnstock, G. and Holman, M. (1958). Excitation and conduction in the smooth muscle of the isolated taenia coli of the guinea pig. *J. Physiol.* **142**, 420–37.
Bülbring, E., Ohashi, H. and Tomita, T. (1981*b*).

Adrenergic mechanisms. In *Smooth Muscle* (ed. E. Bülbring *et al.*), pp. 219–48. London: Edward Arnold.
Bullard, B., Luke, B. and Winkelman, L. (1973). The paramyosin of insect flight muscle. *J. Mol. Biol.* **75**, 359–67.
Buller, A. J. (1975). *The Contractile Behaviour of Mammalian Skeletal Muscle* (Oxford Biology Reader No. 36). London: Oxford University Press.
Bullock, T. H. and Dieke, F. P. J. (1956). Properties of an infra-red receptor. *J. Physiol.* **134**, 47–87.
Bullock, T. H. and Hagiwara, S. (1957). Intracellular recording from the giant synapse of the squid. *J. gen. Physiol.* **40**, 565–77.
Bullock, T. H., Hagiwara, S., Kusand, K. and Negishi, K. (1961). Evidence for a category of electroreceptors in the lateral line of gymnotid fishes. *Science* **134**, 1426–7.
Bullock, T. H. and Heiligenberg, W. (eds.) (1986). *Electroreception.* New York: John Wiley.
Burke, R. E., Levine, D. N., Tsairis, P. and Zasac, F. E. (1973). Physiological types and histochemical profiles in motor units of the cat gastrocnemius. *J. Physiol.* **234**, 723–48.
Burkhardt, D. (1958). Die Sinnersorgane des Skeletmuskels und die nervöse Steuerung der Muskeltätigkeit. *Ergebn. Biol.* **20**, 27–66.
Burnstock, G. (1972). Purinergic nerves. *Pharmacol. Rev.* **24**, 509–81.
Burnstock, G. (1981). Neurotransmitters and trophic factors in the autonomic nervous system. *J. Physiol.* **313**, 1–35.
Burnstock, G., Campbell, G., Satchell, D. G. and Smyth, A. (1970). Evidence that adenosine triphosphate or a related nucleotide is the transmitter substance released by non-adrenergic inhibitory nerves in the gut. *Br. J. Pharmacol.* **40**, 668–88.
Bylund, D. B. and U'Prichard, D. C. (1983). Characterization of α_1-and α_2-adrenergic receptors. *Int. Rev. Neurobiol.* **24**, 343–431.
Cahalan, M. D. (1975). Modification of sodium channel gating in frog myelinated nerve fibres by *Centruroides sculpturatus* scorpion venom. *J. Physiol.* **244**, 511–34.
Cain, D. F., Infante, A. A. and Davies, R. E. (1962). Chemistry of muscle contraction. Adenosine triphosphate and phosphoryl creatine as energy supplies for single contractions of working muscle. *Nature* **196**, 214–17.
Caldwell, P. C., Hodgkin, A. L., Keynes, R. D. and Shaw, T. I. (1960). The effects of injecting 'energy-rich' phosphate compounds on the active transport of ions in the giant axons of *Loligo*. *J. Physiol.* **152**, 561–90.
Caldwell, P. C. and Walster, G. E. (1963). Studies on the micro-injection of various substances into crab muscle fibres. *J. Physiol.* **169**, 353–72.
Callec, J. J. (1974). Synaptic transmission in the central nervous system of insects. In *Insect Neurobiology* (ed. J. E. Treherne), pp. 119–85. Amsterdam: North-Holland.
Campbell, F. W. and Rushton, W. A. H. (1955). Measurement of the scotopic pigment in the living human eye. *J. Physiol.* **130**, 131–47.
Campbell, G. (1987). Cotransmission. *Ann. Rev. Pharmacol. Toxicol.* **27**, 51–70.
Carafoli, E. and Zurini, M. (1982). The Ca^{2+} pumping ATPase of plasma membranes. *Biochim. Biophys. Acta* **683**, 279–301.
Carafoli, E., Zurini, M. and Benaim, G. (1986). The calcium pump of plasma membranes. In *Calcium and the Cell.* Ciba Foundation Symposium 122, pp. 58–65. Chichester: John Wiley.
Carlsson, A. (1987). Perspectives on the discovery of central monoaminergic neurotransmission. *Ann. rev. Neurosci.* **10**, 19–40.
Carlsson, A., Falck, B. and Hillarp, N.-Å. (1962). Cellular localization of brain monoamines. *Acta Physiol. Scand.* **56** (suppl. 196), 1–27.
Cartaud, J., Benedetti, E. L., Cohen, J. B., Meunier, J.-C. and Changeux, J.-P. (1973). Presence of a lattice structure in membrane fragments rich in nicotinic receptor protein from the electronic organ of *Torpedo marmorata*. *FEBS Letters* **33**, 109–13.
Cartaud, J., Benedetti, E. L., Sobel, A. and Changeux, J.-P. (1978). A morphological study of the cholinergic receptor protein from *Torpedo marmorata* in its membrane environment and in its detergent-extracted purified form. *J. Cell Sci.* **29**, 313–37.
Castellani, L. and Cohen, C. (1987). Myosin rod phosphorylation and the catch state of molluscan muscles. *Science* **235**, 334–7.
Catterall, W. A. (1980). Neurotoxins that act on voltage-sensitive sodium channels in excitable membranes. *Ann. Rev. Pharmacol. Toxicol.* **20**, 15–43.
Catterall, W. A. (1986*a*). Voltage-dependent gating of sodium channels: correlating structure and function. *Trends Neurosci.* **9**, 7–10.
Catterall, W. A. (1986*b*). Molecular properties of voltage-sensitive sodium channels. *Ann. Rev. Biochem.* **55**, 953–85.
Chabre, M. (1985). Trigger and amplification mechanisms in visual phototransduction. *Ann. Rev. Biophys. Chem.* **14**, 331–60.
Chabre, M. (1987). The G protein connection: is it in the membrane or the cytoplasm? *Trends Biochem. Sci.* **12**, 213–15.
Chalovich, J. M. and Eisenberg, E. (1982). Inhibition of actomyososin ATPase activity by troponin-tropomyosin without blocking the binding of myosin to actin. *J. Biol. Chem.* **257**, 2432–7.
Chandler, W. K., Hodgkin, A. L. and Meves, H. (1965). The effect of changing the internal solution on sodium inactivation and related phenomena in giant axons. *J. Physiol.* **180**, 821–36.
Chandler, W. K. and Meves, H. (1965). Voltage clamp measurements on internally perfused giant axons. *J. Physiol.* **180**, 788–820.
Chandler, W. K., Rakowski, R. F. and Schneider, M. F. (1976). Effects of glycerol treatment and maintained depolarization on charge movement in skeletal muscle. *J. Physiol.* **254**, 285–316.
Chang, M. M., Leeman, S. E. and Niall, H. D. (1971). Amino-acid sequence of substance P. *Nature New Biol.* **232**, 86–7.

Chantler, P. D. and Szent-Györgyi, A. G. (1980). Regulatory light chains and scallop myosin: full dissociation, reversibility and co-operative effects. *J. Mol. Biol.* **138**, 473–92.

Charest, R., Prpic, V., Exton, J. H. and Blackmore, P. F. (1985). Stimulation of inositol trisphosphate formation in hepatocytes by vasopressin, epinephrine and angiotensin II and its relationship to changes in cytosolic free Ca^{2+}. *Biochem. J.* **227**, 79–90.

Cherubini, E. and North, R. A. (1985). μ and κ opioids inhibit transmitter release by different mechanisms. *Proc. Natl. Acad. Sci. USA* **82**, 1860–3.

Chiu, S. Y. (1980). Asymmetry currents in the mammalian myelinated nerve. *J. Physiol.* **309**, 499–519.

Chiu, S. Y. and Ritchie, J. M. (1981). Evidence for the presence of potassium channels in the paranodal region of acutely demyelinated mammalian single nerve fibres. *J. Physiol.* **313**, 415–37.

Chiu, S. Y. and Ritchie, J. M. (1982). Evidence for the presence of potassium channels in the internode of frog myelinated nerve fibres. *J. Physiol.* **322**, 485–501.

Chiu, S. Y. and Ritchie, J. M. (1984). On the physiological role of internodal potassium channels and the security of conduction in myelinated nerve fibres. *Proc. R. Soc. Lond.* B **220**, 415–22.

Chiu, S. Y., Ritchie, J. M., Rogart, R. B. and Stagg, D. (1979). A quantitative description of membrane currents in rabbit myelinated nerve. *J. Physiol.* **292**, 149–66.

Choi, D. W., Farb, D. H. and Fischbach, G. D. (1977). Chlordiazepoxide selectively augments GABA action in spinal cord cultures. *Nature* **269**, 342–4.

Civan, M. M. and Podolsky, R. J. (1966). Contraction kinetics of striated muscle fibres following quick changes in load. *J. Physiol.* **184**, 511–34.

Claudio, T., Ballivet, M., Patrick, J. and Heinemann, S. (1983). Nucleotide and deduced amino acid sequences of *Torpedo californica* acetylcholine receptor α subunit. *Proc. Natl. Acad. Sci. USA* **80**, 1111–15.

Clay, J. R. and DeFelice, L. J. (1983). Relationship between membrane excitability and single channel open-close kinetics. *Biophys. J.* **42**, 151–7.

Cleland, B. G., Dubin, M. W. and Levick, W. R. (1971). Sustained and transient neurones in the cat's retina and lateral geniculate nucleus. *J. Physiol.* **217**, 473–96.

Cleland, B. G. and Levick, W. R. (1974). Brisk and sluggish concentrically organized ganglion cells in the cat's retina. *J. Physiol.* **240**, 421–56.

Close, R. I. (1972). Dynamic properties of mammalian skeletal muscle. *Physiol. Rev.* **52**, 129–97.

Cobbs, W. H. and Pugh, E. N. (1985). Cyclic GMP can increase rod outer-segment light-sensitive current 10-fold without delay of excitation. *Nature* **313**, 585–7.

Cockroft, S. (1987). Polyphosphoinositide phosphodiesterase: regulation by a novel guanine nucleotide binding protein, G_p. *Trends Biochem. Sci.* **12**, 75–8.

Cockroft, S. and Gomperts, B. D. (1985). Role of guanine nucleotide binding protein in the activation of polyphosphoinositide phosphodiesterase. *Nature* **314**, 534–6.

Cohen, I. and Van Der Kloot, W. (1985). Calcium and transmitter release. *Int. Rev. Neurobiol.* **27**, 299–336.

Cohen, M. J. (1964). The peripheral organization of sensory systems. In *Neural Theory and Modelling* (ed. R. F. Reiss), pp. 273–92. Stanford University Press.

Cohen, M. J., Hagiwara, S. and Zotterman, Y. (1955). The response spectrum of taste fibres in the cat: a single fibre analysis. *Acta physiol. Scand.* **33**, 316–32.

Cohen, P. (1982). The role of protein phosphorylation in neural and hormonal control of cellular activity. *Nature* **296**, 613–20.

Cohen, S. (1982). Matching molecules in the catch mechanism. *Proc. Natl. Acad. Sci. USA* **79**, 3176–8.

Cole, K. S. (1949). Dynamic electrical characteristics of the squid axon membrane. *Arch. Sci. Physiol.* **3**, 253–8.

Cole, K. S. (1968). *Membranes, Ions and Impulses.* Berkeley: University of California Press.

Cole, K. S. and Curtis, H. J. (1939). Electric impedance of the squid giant axon during activity. *J. gen. Physiol.* **22**, 649–70.

Collingridge, G. L. and Bliss, T. V. P. (1987). NMDA receptors – their role in long-term potentiation. *Trends Neurosci.* **10**, 288–93.

Collingridge, G. L., Kehl, S. J. and Maclennan, H. (1983). Excitatory amino acids in the Schaffer collateral-commisural pathway of the rat hippocampus. *J. Physiol.* **334**, 33–46.

Collins, C. A., Rojas, E. and Suarez-Isla, B. A. (1982*a*). Activation and inactivation characteristics of the sodium permeability in muscle fibres from *Rana temporaria. J. Physiol.* **324**, 297–318.

Collins, C. A., Rojas, E. and Suarez-Isla, B. A. (1982*b*). Fast charge movements in skeletal muscle fibres from *Rana temporaria. J. Physiol.* **324**, 319–45.

Collins, J. H., Potter, J. D., Horn, M. J., Wiltshire, G. and Jackman, N. (1973). The amino acid sequence of rabbit skeletal muscle troponin C: gene replication and homology with calcium-binding proteins from carp and hake muscle. *FEBS Lett.* **36**, 268–72.

Colquhoun, D. and Hawkes, A. G. (1977). Relaxation and fluctuations of membrane current that flow through drug-operated ion channels. *Proc. R. Soc. Lond.* B **199**, 231–62.

Colquhoun, D. and Hawkes, A. G. (1981). On the stochastic properties of single ion channels. *Proc. R. Soc. Lond.* B **211**, 205–34.

Colquhoun, D. and Hawkes, A. G. (1982). On the stochastic properties of bursts of single ion channel openings and of clusters of bursts. *Phil. Trans. R. Soc. Lond.* B **300**, 1–59.

Colquhoun, D. and Sakmann, B. (1981). Fluctuations in the microsecond time range of the current through single acetylcholine receptor ion channels. *Nature* **294**, 464–6.

Colquhoun, D. and Sakmann, B. (1983). Bursts of openings in transmitter-activated ion channels. In *Single Channel Recording* (ed. B. Sakmann and E. Neher), pp. 345–64. New York: Plenum Press.

Colquhoun, D. and Sakmann, B. (1985). Fast events in single-channel currents activated by acetylcholine and its analogues at the frog muscle end-plate. *J. Physiol.* **369**, 501–57.

Cone, R. A. (1964). The early receptor potential of the vertebrate retina. *Nature* **204**, 736–9.

Cone, R. A. (1973). The internal transmitter model for visual excitation: some quantitative implications. In *Biochemistry and Physiology of Visual Pigments* (ed. H. Langer), pp. 275–82. Berlin: Springer.

Connor, J. A. and Stevens, C. F. (1971*a*). Voltage clamp studies of a transient outward membrane current in gastropod neural somata. *J. Physiol.* **213**, 21–30.

Connor, J. A. and Stevens, C. F. (1971*b*). Prediction of repetitive firing behaviour from voltage clamp data on an isolated neurone soma. *J. Physiol.* **213**, 31–53.

Conti, F., DeFelice, L. J. and Wanke, E. (1975). Potassium and sodium ion current noise in the membrane of the squid giant axon. *J. Physiol.* **248**, 45–82.

Conti, F., Hille, B., Neumcke, B., Nonner, W. and Stämpfli, R. (1976*a*). Measurement of the conductance of the sodium channel from current fluctuations at the node of Ranvier. *J. Physiol.*, **262**, 699–727.

Conti, F., Hille, B., Neumcke, B., Nonner, W. and Stämpfli, R. (1976*b*). Conductance of the sodium channel in myelinated nerve fibres with modified sodium inactivation. *J. Physiol.* **262**, 729–42.

Conti, F., Hille, B. and Nonner, W. (1984). Nonstationary fluctuations of the potassium conductance at the node of Ranvier of the frog. *J. Physiol.* **353**, 199–230.

Conti, F. and Neher, E. (1980). Single channel recordings of K currents in squid axons. *Nature* **285**, 140–3.

Conti, M. A. and Adelstein, R. S. (1980). Phosphorylation by cyclic adenosine 3′:5′-monophosphate-dependent protein kinase regulates myosin light chain kinase. *Fed. Proc.* **39**, 1569–73.

Conway, E. J. (1957). Nature and significance of concentration relations of potassium and sodium ions in skeletal muscle. *Physiol. Rev.* **37**, 84–132.

Cooke, R., Crowder, M. S., Wendt, C. H., Barnett, V. A. and Thomas, D. D. (1984). Muscle cross-bridges: do they rotate? In *Contractile Mechanisms in Muscle* (ed. G. H. Pollack and H. Sugi), pp. 413–23. Tokyo: University of Tokyo Press.

Cooley, L. B., Johnson, W. H. and Krause, S. (1979). Phosphorylation of paramyosin and its possible role in the catch mechanism. *J. Biol. Chem.* **254**, 2195–8.

Coombs, J. S., Curtis, D. R. and Eccles, J. C. (1957*a*). The interpretation of spike potentials of motoneurones. *J. Physiol.* **139**, 198–231.

Coombs, J. S., Curtis, D. R. and Eccles, J. C. (1957*b*). The generation of impulses in motoneurones. *J. Physiol.* **139**, 232–49.

Coombs, J. S., Eccles, J. C. and Fatt, P. (1955*a*). The electrical properties of the motoneurone membrane. *J. Physiol.* **130**, 291–235.

Coombs, J. S., Eccles, J. C. and Fatt, P. (1955*b*). The specific ionic conductances and the ionic movements across the motoneuronal membrane that produce the inhibitory postsynaptic potential. *J. Physiol.* **130**, 326–73.

Coombs, J. S., Eccles, J. C. and Fatt, P. (1955*c*). Excitatory synaptic action in motoneurones. *J. Physiol.* **130**, 374–95.

Coombs, J. S., Eccles, J. C. and Fatt, P. (1955*d*). The inhibitory suppression of reflex dicharges from motoneurones. *J. Physiol.* **130**, 396–413.

Cooper, S. (1961). The responses of the primary and secondary endings of muscle spindles with intact motor innervation during applied stretch. *Q. J. exp. Physiol.* **46**, 398–98.

Cooper, S. and Daniel, P. M. (1956). Human muscle spindles. *J. Physiol.* **133**, 1–3*P*.

Cornelius, F. (1982). Tonic contraction and the control of relaxation in a chemically skinned molluscan smooth muscle. *J. gen. Physiol.* **79**, 821–34.

Costantin, L. L. (1970). The role of sodium current in the radial spread of contraction in frog muscle Fibres. *J. gen. Physiol.* **55**, 703–15.

Costantin, L. L. (1974). Contractile activation in frog skeletal muscle. *J. gen. Physiol.* **63**, 657–74.

Cote, R. H., Biernbaum, M. S., Nicol, G. D. and Bownds, M. D. (1984). Light-induced decreases in cGMP concentration precede changes in membrane permeability in frog rod photoreceptors. *J. Biol. Chem.* **259**, 9635–41.

Cotman, C. W., Monaghan, D. T., Ottersen, O. P. and Storm-Mathisen, J. (1987). Anatomical organization of excitatory amino acid receptors and their pathways. *Trends Neurosci.* **10**, 273–80.

Couceiro, A. and de Almeida, D. F. (1961). The electrogenic tissue of some Gymnotidae. In *Bioelectrogenesis* (ed. C. Chagas and A. Paes de Carvalho), pp. 3–13. Amsterdam: Elsevier.

Couteaux, R. (1960). Motor end-plate structure. In *The Structure and Function of Muscle*, vol. 1 (ed. G. H. Bourne). New York: Academic Press.

Couteaux, R. and Pecot-Dechavassine, M. (1974). Les zones specialisées des membranes presynaptiques. *C.R. Acad. Sci. Paris* **278**, 291–3.

Craig, R. (1977). Structure of A-segments from frog and rabbit skeletal muscle. *J. Mol. Biol.* **109**, 69–81.

Craig, R. (1986). The structure of the contractile filaments. In *Myology* (ed. A. G. Engel and B. Q. Banker), pp. 73–123. New York: McGraw-Hill.

Craig, R. and Megerman, J. (1977). Assembly of smooth muscle myosin into side-polar filaments. *J. Cell Biol.* **75**, 990–6.

Craig, R. and Offer, G. (1976). Axial arrangement of crossbridges in thick filaments of vertebrate skeletal muscle. *J. Mol. Biol.* **102**, 325–32.

Crawford, B. H. (1949). The scotopic visibility function. *Proc. phys. Soc.* B **62**, 321–34.

Crescitelli, F. and Dartnall, H. J. A. (1953). Human visual purple. *Nature* **172**, 195–6.

Crick, F. H. C. (1953). The packing of α-helices: simple coiled-coils. *Acta Cryst.* **6**, 689–97.

Croucher, M. J., Collins, J. F. and Meldrum, B. S. (1982). Anticonvulsant action of excitatory amino acid antagonists. *Science* **216**, 899–901.

Crowe, A. and Matthews, P. B. C. (1964). The effects of stimulation of static and dynamic fusimotor fibres on the response to stretching of the primary endings of the muscle spindles. *J. Physiol.* **174**, 109–31.

Cuello, A. C. (ed.) (1982). *Co-transmission.* London: Macmillan.

Cuello, A. C. and Sofroniew, M. V. (1984). The anatomy of the CNS cholinergic neurons. *Trends Neurosci.* **7**, 74–8.

Cull-Candy, S. G., Miledi, R. and Parker, I. (1980). Single glutamate-activated channels recorded from locust muscle fibres with perfused patch-clamp electrodes. *J. Physiol.* **321**, 195–210.

Cull-Candy, S. G. and Usowicz, M. M. (1987). Multiple-conductance channels activated by excitatory amino acids in cerebellar neurons. *Nature* **325**, 525–8.

Curtin, N. A. and Woledge, R. C. (1978). Energy changes and muscular contraction. *Physiol. Rev.* **58**, 690–761.

Curtin, N. A. and Woledge, R. C. (1979). Chemical change and energy production during contraction of frog muscle: how are their time-courses related? *J. Physiol.* **288**, 353–66.

Curtin, N. A. and Woledge, R. C. (1981). Effect of muscle length on energy balance in frog skeletal muscle. *J. Physiol.* **316**, 453–68.

Curtis, D. R. (1965). Actions of drugs on single neurons in the spinal cord and thalamus. *Brit. med. Bull.* **21**, 1–9.

Curtis, D. R. (1969). The pharmacology of spinal postsynaptic inhibition. *Progress in Brain Research* **31**, 171–89.

Curtis, D. R., Duggan, A. W., Felix, D. and Johnston, G. A. R. (1970). GABA, bicuculline and central inhibition. *Nature* **226**, 1222–4.

Curtis, D. R. and Eccles, J. C. (1959). The time courses of excitatory and inhibitory synaptic actions. *J. Physiol.* **145**, 529–46.

Curtis, D. R. and Eccles, R. M. (1958*a*). The excitation of Renshaw cells by pharmacological agents applied electrophoretically. *J. Physiol.* **141**, 435–45.

Curtis, D. R. and Eccles, R. M. (1958*b*). The effect of diffusional barriers upon the pharmacology of cells within the central nervous system. *J. Physiol.* **141**, 446–63.

Curtis, D. R., Phillis, J. W. and Watkins, J. C. (1960). The chemical excitation of spinal neurones by certain acidic amino acids. *J. Physiol.* **150**, 656–82.

Curtis, D. R. and Ryall, R. W. (1964). Nicotinic and muscarinic receptors of Renshaw cells. *Nature* **203**, 652.

Curtis, D. R. and Watkins, J. C. (1960). The excitation and depression of spinal neurones by structurally related amino acids. *J. Neurochem.* **6**, 117–41.

Curtis, H. J. and Cole, K. S. (1942). Membrane resting and action potentials in giant fibres of squid nerve. *J. cell. comp. Physiol.* **19**, 135–44.

Dale, H. H. (1914). The action of certain esters and ethers of choline, and their relation to muscarine. *J. Pharmac. exp. Ther.* **6**, 147–90.

Dale, H. H. (1933). Nomenclature of fibres in the autonomic system and their effects. *J. Physiol.* **80**, 10–11.

Dale, H. H. and Dudley, H. W. (1929). The presence of histamine and acetylcholine in the spleen of the ox and the horse. *J. Physiol.* **68**, 97–123.

Dale, H. H., Feldburg, W. and Vogt, M. (1936). Release of acetylcholine at voluntary motor nerve endings. *J. Physiol.* **86**, 353–80.

Darian-Smith, I. (ed.) (1984). *Handbook of Physiology*, section 1, vol. 3, *Sensory Processes.* Bethesda, Maryland: American Physiological Society.

Darnell, J. E., Lodish, H. and Baltimore, D. (1986). *Molecular Cell Biology*. New York: Scientific American Books.

Dartnall, H. J. A. (1952). Visual pigment 467, a photosensitive pigment present in tench retinae. *J. Physiol.* **116**, 259–89.

Dartnall, H. J. A. (1953). The interpretation of spectral sensitivity curves. *Brit. Med. Bull.* **9**, 24–30.

Dartnall, H. J. A. (1957). *The Visual Pigments.* London: Methuen.

Dartnall, H. J. A., Arden, G. B., Ikeda, H., Luck, C. P., Rosenberg, M. E., Pedler, C. M. H. and Tansley, K. (1965). Anatomical electrophysiological and pigmentary aspects of vision in the bush baby: an interpretative study. *Vision Res.* **5**, 399–424.

Dartnall, H. J. A., Bowmaker, J. K. and Mollon, J. D. (1983). Human visual pigments: microspectrophotometric results from the eyes of seven persons. *Proc. R. Soc. Lond.* B **220**, 115–30.

Darwin, C. (1859). *The Origin of Species.* London: John Murray.

Davis, H. (1961). Some principles of sensory receptor action. *Physiol. Rev.* **41**, 391–416.

Davis, H. *et al.* (1953). Acoustic trauma in the guinea pig. *J. acoust. Soc. Am.* **25**, 1180–9.

Davson, H. and Danielli, J. F. (1943). *The Permeability of Natural Membranes.* Cambridge: Cambridge University Press.

Dawson, W. W. and Enoch, J. M. (eds.) (1984). *Foundations of Sensory Science.* Berlin: Springer-Verlag.

Decoursey, T. E., Dempster, J. and Hutter, O. F. (1984). Inward rectifier current noise in frog skeletal muscle. *J. Physiol.* **349**, 299–327.

DeFelice, L. J. (1981). *Introduction to Membrane Noise.* New York: Plenum Press.

Del Castillo, J. and Engbaek, L. (1954). The nature of the neuromuscular block produced by magnesium. *J. Physiol.* **124**, 370–84.

Del Castillo, J. and Katz, B. (1954*a*). The membrane change produced by the neuromuscular transmitter. *J. Physiol.* **125**, 546–65.

Del Castillo, J. and Katz, B. (1954*b*). Quantal components of the end-plate potential. *J. Physiol.* **124**, 560–73.

Del Castillo, J. and Katz, B. (1954*c*). Statistical factors involved in neuromuscular facilitation and depression. *J. Physiol.* **124**, 574–85.

Del Castillo, J. and Katz, B. (1955). On the localization of acetylcholine receptors. *J. Physiol.* **128**, 157–81.

Del Castillo, J. and Katz, B. (1956). Biophysical aspects of neuromuscular transmission. *Progr. Biophys.* **6**, 121–70.

Del Castillo, J. and Katz, B. (1957). Interaction at end-plate receptors between different choline derivatives. *Proc. R. Soc. Lond.* B **146**, 369–81.

Del Castillo, J. and Moore, J. W. (1959). On increasing the velocity of a nerve impulse. *J. Physiol.* **148**, 665.

De Lorenzo, A. J. D. (1963). Studies on the ultrastructure and histophysiology of cell membranes, nerve fibres and synaptic junctions in chemoreceptors. In *Olfaction and Taste* (ed. Y. Zotterman). Oxford: Pergamon Press.

Den Otter, C. J. (1977). Single sensillum responses in the male moth *Adoxophyes orana* (F.v.R.) to female sex pheromone components and their geometrical isomers. *J. Comp. Physiol.* **121**, 205–22.

Denton, E. J. (1959). The contributions of the oriented photosensitive and other molecules to the absorption of whole retina. *Proc. R. Soc. Lond.* B **150**, 78–94.

De Robertis, E. (1960). Some observations on the ultrastructure and morphogenesis of photoreceptors. *J. gen. Physiol.* **43**, suppl. 1–13.

De Robertis, E. and Bennett, H. S. (1954). Submicroscopic vesicular component in the synapse. *Fed. Proc.* **13**, 38.

De Robertis, E. and Franchi, C. M. (1956). Electron microscope observations on synaptic vesicles in synapses of the retinal rods and cones. *J. biophys. biochem. Cytol.* **2**, 307–18.

DeRosier, D. J. and Klug, A. (1968). Reconstruction of three dimensional structures from electron micrographs. *Nature* **217**, 130–4.

Dethier, V. G. (1953). Summation and inhibition following contralateral stimulation of the tarsal chemoreceptors of the blowfly. *Biol. Bull. mar. biol. Lab., Woods Hole* **105**, 257–68.

Dethier, V. G. (1955). The physiology and histology of the contact chemoreceptors of the blowfly. *Quart. rev. Biol.* **30**, 348–71.

Dethier, V. G. (1976). *The Hungry Fly*. Cambridge, Massachusetts: Harvard University Press.

Dethier, V. G. and Hanson, F. E. (1968). Electrophysiological responses of the chemoreceptors of the blowfly to sodium salts of fatty acids. *Proc. Natl. Acad. Sci. USA* **60**, 1296–303.

Devillers-Thiery, A., Giraudat, J., Bentaboulet, M. and Changeux, J.-P. (1983). Complete mRNA coding sequence of the acetylcholine binding α subunit of *Torpedo marmorata* acetylcholine receptor: a model for the transmembrane organization of the peptide chain. *Proc. Natl. Acad. Sci. USA* **80**, 2067–71.

Diamond, J., Gray, J. A. B. and Inman, D. R. (1958). The relation between receptor potentials and the concentration of sodium ions. *J. Physiol.* **142**, 382–94.

DiFrancesco, D. (1981*a*). A new interpretation of the pace-maker current in calf Purkinje fibres. *J. Physiol.* **314**, 359–76.

DiFrancesco, D. (1981*b*). A study of the ionic nature of the pace-maker current in calf Purkinje fibres. *J. Physiol.* **314**, 377–93.

DiFrancesco, D. and Noble, D. (1985). A model of cardiac electrical activity incorporating ionic pumps and concentration changes. *Phil. Trans R. Soc. Lond.* B **307**, 353–98.

DiFrancesco, D. and Ojeda, C. (1980). Properties of the current i_f in the sino-atrial node of the rabbit compared with those of the current i_{K2} in Purkinje fibres. *J. Physiol.* **308**, 353–67.

Dijkgraaf, S. (1934). Untersuchungen über die Funktion der Seitenorgane an Fischen. *Z. vergl. Physiol.* **20**, 162–214.

Dijkgraaf, S. (1952). Bau und Funktionen der Seitenorgane und des Ohrlabyrinths bei Fischen. *Experientia* **8**, 205–16.

Dijkgraaf, S. (1963). The functioning and significance of the lateral line organs. *Biol. Rev.* **38**, 51–105.

Dijkgraaf, S. and Kalmijn, A. J. (1963). Untersuchungen über die Funktion der Lorenzinischen Ampullen an Haifischen. *Z. vergl. Physiol.* **47**, 438–56.

Dipolo, R., Rojas, H., Vergara, J., Lopez, R. and Caputo, C. (1983). Measurements of intracellular ionized calcium in squid giant axons using calcium-selective electrodes. *Biochim. Biophys. Acta* **728**, 311–18.

Dixon, R. A. F., Kobilka, B. K., Strader, D. J., Benovic, J. L., Dohlman, H. G., Frielle, T., Bolanowski, M. A., Bennett, C. D., Rands, E., Diehl, R. E., Mumford, R. A., Slater, E. E., Signal, I. S., Caron, M. G., Lefkowitz, R. J. and Strader, C. D. (1986). Cloning of the gene and cDNA for mammalian β-adrenergic receptor and homology with rhodopsin. *Nature* **321**, 75–9.

Dobelle, W. H., Marks, W. B. and MacNichol, E. F., Jnr. (1969). Visual pigment density in single primate foveal cones. *Science* **166**, 1508–10.

Dodge, F. A. and Frankenhaeuser, B. (1958). Membrane currents in isolated frog nerve fibre under voltage clamp conditions. *J. Physiol.* **143**, 76–90.

Dodge, F. A. and Frankenhaeuser, B. (1959). Sodium currents in the myelinated nerve fibres of *Xenopus laevis* investigated with the voltage clamp technique. *J. Physiol.* **148**, 188–200.

Dodge, F. A. and Rahamimoff, R. (1967). On the relationship between calcium concentration and the amplitude of the end-plate potential. *J. Physiol.* **189**, 90–2*P*.

Dohlman, G. (1935). Some practical and theoretical points in labyrinthology. *Proc. R. Soc. Med.* **28**, 1371–80.

Douglass, J., Civelli, O. and Herbert, E. (1984). Polyprotein gene expression: generation of diversity of neuroendocrine peptides. *Ann. Rev. Biochem.* **53**, 665–715.

Dowdall, M. J., Boyne, A. F. and Whittaker, V. P. (1974). Adenenosine triphosphate: a constitutent of cholinergic synaptic vesicles. *Biochem. J.* **140**, 1–12.

Dowling, J. E. (1968). Synaptic organization of the frog retina: an electron miscroscopic analysis comparing the retinas of frogs and primates. *Proc. R. Soc. Lond.* B **170**, 205–28.

Dowling, J. E. (1970). Organization of vertebrate retinas. *Investigative Ophthalmol.* **9**, 655–80.

Dowling, J. E. (1987). *The Retina: an Approachable part of the Brain*. Cambridge, Massachusetts: Harvard University Press.

Dowling, J. E. and Boycott, B. B. (1966). Organization of the primate retina: electron microscopy. *Proc. R. Soc. Lond.* B **166**, 80–111.

Dowling, J. E. and Dubin, M. W. (1984). The vertebrate

retina. In *Handbook of Physiology*, section 1, vol. 3 *Sensory Processes* (ed. I. Darian-Smith), pp. 317–39. Bethesda, Maryland: American Physiological Society.

Dowling, J. E. and Ripps, H. (1973). Effect of magnesium on horizontal cell activity in skate retina. *Nature* **242**, 101–3.

Draper, M. H. and Weidmann, S. (1951). Cardiac resting and action potentials recorded with an intracellular electrode. *J. Physiol.* **115**, 74–94.

Du Bois Reymond, E. (1877). *Gesammelte Abhandl.d.allgem. Muskel und Nervenphysik* **2**, 700.

Dudel, J. and Kuffler, S. W. (1961). Presynaptic inhibition at the crayfish neuromuscular junction. *J. Physiol.* **155**, 543–62.

Duncan, G. (1989). *Physics in the Life Sciences*. Oxford: Blackwell Scientific.

Du Vieneaud, V. (1956). Hormones of the posterior pituitary gland: oxytocin and vasopressin. *Harvey Lect.* **50**, 1–26.

Dwyer, T. M., Adams, D. J. and Hille, B. (1980). The permeability of the endplate channel to organic cations in frog muscle. *J. gen. Physiol.* **75**, 469–92.

Eastwood, A. B., Wood, D. S., Bock, K. L. and Sorenson, M. M. (1979). Chemically skinned mammalian skeletal muscle. 1. The structure of skinned rabbit psoas. *Tiss. Cell* **11**, 553–66.

Eaton, R. C., Bombardieri, R. A. and Meyer, D. L. (1977). The Mauther-initiated startle response in teleost fish. *J. exp. Biol.* **66**, 65–81.

Ebashi, S., Endo, M. and Ohtsuki, I. (1969). Control of muscle contraction. *Quart. Rev. Biophys.* **2**, 351–84.

Ebashi, S. and Kodama, A. (1966*a*). A new factor promoting aggregation of tropomyosin. *J. Biochem.* **58**, 107–8.

Ebashi, S. and Kodama, A. (1966*b*). Native tropomyosin-like action of troponin on trypsin-treated myosin B. *J. Biochem.* **60**, 733–4.

Ebashi, S., Kodama, A. and Ebashi, F. (1968). Troponin. 1. Preparation and physiological function. *J. Biochem.* **64**, 465–77.

Eccles, J. C. (1957). *The Physiology of Nerve Cells*. London: Oxford University Press.

Eccles, J. C. (1964). *The Physiology of Synapses*. Berlin: Springer-Verlag.

Eccles, J. C. (1976). From electrical to chemical transmission in the central nervous system. *Notes and Records R. Soc. Lond.* **30**, 219–230.

Eccles, J. C., Eccles, R. M. and Fatt, P. (1956). Pharmacological investigations on a central synapse operated by acetylcholine. *J. Physiol.* **131**, 154–69.

Eccles, J. C., Eccles, R. M. and Magni, F. (1961). Central inhibitory action attributable to presynaptic depolarization produced by muscle afferent volleys. *J. Physiol.* **159**, 147–66.

Eccles, J. C., Magni, F. and Willis, W. D. (1962). Depolarization of central terminals of group I afferent fibres from muscle. *J. Physiol.* **160**, 62–93.

Eckert, R. and Chad, J. E. (1984). Inactivation of Ca channels. *Prog. Biophys. molec. Biol.* **44**, 215–67.

Eckert, R. and Tillotson, D. L. (1981). Calcium-mediated inactivation of the calcium conductance in caesium-loaded giant neurons of *Aplysia californica*. *J. Physiol.* **314**, 265–80.

Edman, K. A. P. and Schild, H. O. (1962). The need for calcium in the contractile responses induced by acetylcholine and potassium in the rat uterus. *J. Physiol.* **161**, 424–41.

Egelman, E. H. (1985). The structure of F-actin. *J. Mus. Res. Cell Motil.* **6**, 129–51.

Egelman, E. H. and DeRosier, D. J. (1983). A model for F-actin derived from image analysis of isolated filaments. *J. Mol. Biol.* **166**, 623–9.

Egelman, E. H., Francis, N. and DeRosier, D. J. (1982). F-actin is a helix with a random variable twist. *Nature* **298**, 131–5.

Egelman, E. H. and Padrón, R. (1984). X-ray diffraction evidence that actin is a 100 Å filament. *Nature* **307**, 56–8.

Eggleton, P. and Eggleton, G. P., (1927). The inorganic phosphate and a labile form of organic phosphate in the gastrocnemius of the frog. *Biochem. J.* **21**, 190–5.

Ehrenstein, G. and Gilbert, D. L. (1966). Slow changes of potassium permeability in the squid giant axon. *Biophys. J.* **6**, 553–66.

Einthoven, W. (1924). The string galvanometer and measurement of the action currents of the heart. Nobel Lecture. Republished in 1965 In *Nobel lectures, Physiology or Medicine 1921–41*. Amsterdam: Elsevier.

Eisenberg, E. and Greene, L. E. (1980). The relation of muscle biochemistry to muscle physiology. *Ann. Rev. Physiol.* **42**, 293–309.

Eisenberg, E. and Hill, T. L. (1978). A cross-bridge model of muscle contraction. *Prog. Biophys. Molec. Biol.* **33**, 55–82.

Eisenberg, E. and Hill, T. L. (1985). Muscle contraction and free energy transduction in biological systems. *Science* **227**, 999–1006.

Elliott, A. and Bennett, P. M. (1982). Structure of the thick filaments in molluscan adductor muscle. In *Basic Biology of Muscle: a Comparative Approach* (ed. B. M. Twarog, R. J. C. Levine and M. M. Dewey), pp. 11–28. New York: Raven Press.

Elliott, A. and Offer, G. (1978). Shape and flexibility of the myosin molecule. *J. Mol. Biol.* **123**, 505–19.

Elliott, G. F., Lowy, J. and Millman, B. M. (1965). X-ray diffraction from living striated muscle during contraction. *Nature* **206**, 1357–8.

Elliott, T. R. (1904). On the action of adrenaline. *J. Physiol.* **31**, 20*P*.

Ellisman, M. H. and Levinson, S. R. (1982). Immunocytochemical localization of sodium channel distributions in the excitable membranes of *Electrophorus electricus*. *Proc. Natl. Acad. Sci. USA* **79**, 6707–11.

Elizinga, M., Collins, J. H., Kuehl, W. M. and Adelstein, R. S. (1973). Complete aminoacid sequence of actin of rabbit skeletal muscle. *Proc. Natl. Acad. Sci. USA*. **70**, 2687–91.

Emonet-Dénand, F., Hunt, C. C. and Laporte, Y. (1985).

Effects of stretch on dynamic fusimotor after-effects in cat muscle spindles. *J. Physiol.* **360**, 201–13.

Engberg, I., Flatman, J. A. and Lambert, J. D. C. (1979). The actions of excitatory amino acids on motoneurones in the feline spinal cord. *J. Physiol.* **288**, 227–61.

Engelhardt, W. A. and Ljubimowa, M. N. (1939). Myosine and adenosinetriphosphatase. *Nature* **144**, 668–9.

Engström, H. and Wersäll, J. (1958). The ultrastructural organization of the organ of Corti and the vestibular sensory epithelia. *Exp. Cell Res.* suppl. **5,** 460.

Enroth-Cugell, C. and Robinson, J. G. (1966). The contrast sensitivity of retinal ganglion cells of the cat. *J. Physiol.* **187**, 517–52.

Erlanger, J. and Gasser, H. S. (1937). *Electrical Signs of Nervous Activity.* Philadelphia: University of Pennsylvania Press.

Evans, D. R. and Mellon, De F. (1962). Electrophysiological studies of a water receptor associated with the taste sensilla of the blowfly. *J. gen. Physiol.* **45**, 487–500.

Evans, E. F. (1972). The frequency response and other properties of single fibres in the guinea-pig cochlear nerve. *J. Physiol.* **226**, 263–87.

Evans, E. F. (1974). The effects of hypoxia on the tuning of single cochlear nerve fibres. *J. Physiol.* **338**, 65–67P.

Evans, E. F. (1975). Cochlear nucleus. In *Handbook of Sensory Physiology* (ed. H. Autrum *et al.*) **5**(2), 1–108. Berlin: Springer.

Evans, E. F. and Wilson, J. P. (1975). Cochlear tuning properties: concurrent basilar membrane and single nerve fiber measurements. *Science* **190**, 1218–21.

Evans, P. D. (1985). Octopamine. In *Comprehensive Insect Physiology, Biochemistry and Pharmacology* (ed. G. A. Kerut and L. I. Gibert), vol. 11, pp. 499–530. Oxford: Pergamon Press.

Exton, J. H. (1985). Mechanisms involved in α-adrenergic phenomena. *Am. J. Physiol.* **248**, E633-47.

Eyzaguirre, C. and Kuffler, S. W. (1955*a*). Processes of excitation in the dendrites and in the soma of single isolated sensory nerve cells of the lobster and crayfish. *J. gen. Physiol.* **39**, 87–119.

Eyzaguirre, C. and Kuffler, S. W. (1955*b*). Synaptic inhibition in an isolated nerve cell. *J. gen. Physiol.* **39**, 155–84.

Fabiato, A. (1983). Calcium-induced release of calcium from the cardiac sarcoplasmic reticulum. *Am. J. Physiol.* **245**, C1-C14

Fabiato, A. (1985). Time and calcium dependence of activation and inactivation of calcium-induced release of calcium from the sarcoplasmic reticulum of a skinned canine cardiac Purkinje cell. *J. gen. Physiol.* **85**, 247–89.

Fabiato, A. and Fabiato, F. (1975). Dependence of the contractile activation of skinned cardiac cells on the sarcomere length. *Nature* **256**, 54–6.

Faber, D. S. and Korn, H. (1978). Electrophysiology of the Mauthner cell: basic properties, synaptic mechanisms, and associated networks. In *Neurobiology of the Mauthner Cell* (ed. D. Faber and H. Korn), pp. 47–131. New York: Raven Press.

Fain, G. L. (1975). Quantum sensitivity of rods in the toad retina. *Science* **187**, 838–41.

Fain, G. L., Gold, G. H. and Dowling, J. E. (1976). Receptor coupling in the toad retina. *Cold Spr. Harb. Symp. Quant. Biol.* **40**, 547–61.

Fairclough, R. H., Finer-Moore, J., Love, R. A., Kristofferson, D., Desmeules, P. J. and Stroud, R. M. (1983). Subunit organization and structure of an acetylcholine receptor. *Cold Spr. Harb. Symp. Quant. Biol.* **48**, 9–20.

Famiglietti, E. V. and Kolb, H. (1976). Structural basis of 'ON' and 'OFF'-centre responses in retinal ganglion cells. *Science* **194**, 193–5.

Fatt, P. and Katz, B. (1951). An analysis of the end-plate potential recorded with an intracellular electrode. *J. Physiol.* **115**, 320–69.

Fatt, P. and Katz, B. (1952). Spontaneous subthreshold activity at motor nerve endings. *J. Physiol.* **117**, 109–28.

Fawcett, D. W. and McNutt, N. S. (1969). The ultrastructure of the cat myocardium. I. Ventricular papillary muscle. *J. Cell Biol.* **42**, 1–45.

Fechner, G. T. (1862). *Elemente der Pscyhophysik.* Leipzig: Breitkopf und Härtel.

Fein, A. and Szuts, E. Z. (1982). *Photoreceptors: their Role in Vision.* Cambridge: Cambridge University Press.

Feldburg, W. and Gaddum, J. H. (1934). The chemical transmitter at synapses in a sympathetic ganglion. *J. Physiol.* **81**, 305–19.

Fenn, W. O. (1923). A quantitative comparison between the energy liberated and the work performed by the isolated sartorius of the frog. *J. Physiol.* **58**, 175–203.

Fenn, W. O. (1924). The relation between the work performed and the energy liberated in muscular contraction. *J. Physiol.* **58**, 373–95.

Fenn, W. O. and Marsh, B. S. (1935). Muscular force at different speeds of shortening. *J. Physiol.* **85**, 277–97.

Fenwick, E. M., Marty, A. and Neher, E. (1982). Sodium and calcium channels in bovine chromaffin cells. *J. Physiol.* **331**, 599–635.

Ferguson, D. G., Schwarz, H. W. and Franzini-Armstrong, C. (1984). Subunit structure of junctional feet in triads of skeletal muscle: a freeze-drying, rotary shadowing study. *J. Cell Biol.* **99**, 1735–42.

Fertuck, H. C. and Salpeter, M. M. (1974). Localization of acetylcholine receptor by ^{125}I-labelled α-bungarotoxin binding at mouse motor end-plates. *Proc. Nat. Acad. Sci. USA* **71**, 1376–8.

Fesenko, E. E., Kolesnikov, S. S. and Lyubarsky, A. L. (1985). Induction by cyclic GMP of cationic conductance in plasma membrane of retinal rod outer segment. *Nature* **313**, 310-13.

Fessard, A. (ed.) (1974). Electroreceptors and other specialized receptors in lower vertebrates. *Handbook of Sensory Physiology* (ed. H. Autrum *et al.*) **3**(3). Berlin: Springer.

Fessard, A. and Szabo, T. (1961). Mise en évidence d'un récepteur sensible à l'électricité dans la peau d'un mormyre. *C. R. Acad. Sci.* **253**, 1859–60.

Finer-Moore, J. and Stroud, R. M. (1984). Amphipathic

analysis and possible formation of the ion channel in an acetylcholine receptor. *Proc. Natl. Acad. Sci. USA* **81**, 155–9.
Finkel, A. S. and Redman, S. J. (1983). The synaptic current evoked in cat spinal motoneurones by impulses in single group Ia axons. *J. Physiol.* **342**, 615–32.
Fisher, L. A., Kikkawa, D. O., Rivier, J. E., Amara, S. G., Evans, R. M., Rosenfeld, M. G., Vale, W. W. and Brown, M. R. (1983). Stimulation of noradrenergic sympathetic outflow by calcitonin gene-related peptide. *Nature* **305**, 534–6.
Fiske, C. H. and Subbarow, Y. (1927). The nature of the 'inorganic phosphate' in voluntary muscle. *Science* **65**, 401–3.
Flagg-Newton, J. L., Simpson, I. and Loewenstein, W. R. (1979). Permeability of the cell-to-cell junction. *Science* **205**, 404–7.
Fletcher, P. and Forrester, T. (1975). The effect of curare on the release of acetylcholine from mammalian motor nerve terminals and an estimate of quantum content. *J. Physiol.* **251**, 131–44.
Fletcher, W. M. and Hopkins, F. G. (1907). Lactic acid in amphibian muscle. *J. Physiol.* **35**, 247–309.
Flock, Å. (1964). Structure of the macula utriculi with special reference to directional interplay of sensory responses as revealed by morphological polarization. *J. Cell Biol.* **22**, 413–31.
Flock, Å. (1965). Transducing mechanisms in the lateral line canal organ receptors. *Cold Spr. Harb. Sympt. quant. Biol.* **30**, 133–44.
Flock, Å. (1971). Sensory transduction in hair cells. In *Handbook of Sensory Physiology* (ed. H. Autrum *et al.*) **1**, 396–441. Berlin: Springer.
Flock, Å. (1977). Physiological properties of sensory hairs in the ear. In *Psychophysics and Physiology of Hearing* (ed. E. F. Evans and J. P. Wilson), pp. 15–25. London: Academic Press.
Flock, Å. and Cheung, H. C. (1977). Actin filaments in sensory hairs of inner ear receptor cells. *J. Cell Biol.* **75**, 339–43.
Flood, P. R. (1977). The sarcoplasmic reticulum and associated plasma membrane of trunk muscle lamellae in *Branchiostoma lanceolatum* (Pallas). *Cell Tiss. Res.* **181**, 169–96.
Ford, L. E., Huxley, A. F. and Simmons, R. M. (1977). Tension responses to sudden length change in stimulated frog muscle fibres near slack length. *J. Physiol.* **269**, 441–515.
Ford, L. E., Huxley, A. F. and Simmons, R. M. (1981). The relation between stiffness and filament overlap in stimulated frog muscle fibres. *J. Physiol.* **311**, 219–49.
Ford, L. E., Huxley, A. F. and Simmons, R. M. (1985). Tension transients during steady shortening of frog muscle fibres. *J. Physiol.* **361**, 131–50.
Ford, L. E., Huxley, A. F. and Simmons, R. M. (1986). Tension transients during the rise of tetanic tension in frog muscle fibres. *J. Physiol.* **372**, 595–609.
Frank, G. B. (1960). Effects of changes in extracellular calcium concentration on the potassium-induced contracture of frog's skeletal muscle. *J. Physiol.* **151**, 518–38.
Frank, K. and Fuortes, M. G. F. (1957). Presynaptic and postsynaptic inhibition of monosynaptic reflexes. *Fed. Proc.* **16**, 39–40.
Frank, O. (1895). Zur Dynamik des Herzmuskels. *Z. Biol.* **32**, 370–447.
Frankenhaeuser, B. (1965). Computed action potential in nerve from *Xenopus laevis*. *J. Physiol.* **180**, 780–7.
Frankenhaeuser, B. and Hodgkin, A. L. (1957). The action of calcium on the electrical properties of squid axons. *J. Physiol.* **137**, 218–44.
Franzini-Armstrong, C. (1970a). Studies of the triad. I. Structure of the junction in frog twitch fibres. *J. Cell Biol.* **47**, 488–99.
Franzini-Armstrong, C. (1970b). Natural variability in the length of thin and thick filaments in single fibres from a crab, *Portunus depurator*. *J. Cell. Sci.* **6**, 559–92.
Franzini-Armstrong, C. (1975). Membrane particles and transmission at the triad. *Fed. Proc.* **34**, 1382–9.
Franzini-Armstrong, C. (1980). Structure of the sarcoplasmic reticulum. *Fed. Proc.* **39**, 2403–9.
Franzini-Armstrong, C. (1986). The sarcoplasmic reticulum and the transverse tubules. In *Myology* (ed. A. E. Engel and B. Q. Banker), pp. 125–53. New York: McGraw-Hill.
Franzini-Armstrong, C. and Porter, K. R. (1964). Sarcolemmal invaginations constituting the T system in fish muscle fibres. *J. Cell Biol.* **22**, 675–96.
French, A. S. (1984). Action potential adaptation in the femoral tactile spine of the cockroach, *Periplaneta americana*. J. Comp. Physiol. A **155**, 803–12.
Fukuda, K., Kubo, T., Akiba, I., Maeda, A., Mishina, M. and Numa, S. (1987). Molecular distinction between muscarinic acetylcholine receptor subtypes. *Nature* **327**, 623–5.
Fung, B. K.-K., Hurley, J. B. and Stryer, L. (1981). Flow of information in the light-triggered cyclic nucleotide cascade of vision. *Proc. Natl. Acad. Sci. USA* **78**, 152–6.
Fung, B. K.-K. and Stryer, L. (1980). Photolyzed rhodopsin catalyzes the exchange of GTP for bound GDP in rod outer segments. *Proc. Natl. Acad. Sci. USA* **77**, 2500–4.
Fuortes, M. G. F. (1959). The initiation of impulses in visual cells of *Limulus*. *J. Physiol.* **148**, 14–28.
Fuortes, M. G. F. and Hodgkin, A. L. (1964). Changes in time scale and sensitivity in the ommatidia of *Limulus*. *J. Physiol.* **172**, 239–63.
Fuortes, M. G. F. and Mantegazzini, F. (1962). Interpretation of the repetitive firing of nerve cells. *J. gen. Physiol.* **45**, 1163–79.
Furness, D. N. and Hackney, C. M. (1985). Crosslinks between stereocilia in the guinea pig cochlea. *Hearing Res.* **18**, 177–88.
Furshpan, E. J. and Potter, D. D. (1959). Transmission at the giant synapses of the crayfish. *J. Physiol.* **145**, 289–325.
Furukawa, T. and Furshpan, E. J. (1963). Two inhibitory

mechanisms in the Mauthner neurons of goldfish. *J. Neurophysiol.* **26**, 140–76.

Furukawa, T. and Hanawa, I. (1955). Effects of some common cations on electroretinogram of the toad. *Jap. J. Physiol.* **5**, 289–300.

Furukawa, T., Hayashida, Y. and Matsuura, S. (1978). Quantal analysis of the excitatory post-synaptic potentials at synapses between hair cells and afferent nerve fibres in goldfish. *J. Physiol.* **276**, 211–26.

Furukawa, T. and Matsuura, S. (1978). Adaptive rundown of excitatory post-synaptic potentials at synapses between hair cells and eighth nerve fibres in the goldfish. *J. Physiol.* **276**, 193–209.

Gabella, G. (1981). Structure of smooth muscles. In *Smooth Muscle* (ed. Bülbring *et al.*), pp. 1–46. London: Edward Arnold.

Gabella, G. (1984). Structural apparatus for force transmission in smooth muscles. *Physiol. Rev.* **64**, 455–77.

Galvan, M. and Sedlmeir, C. (1984). Outward currents in voltage-clamped rat sympathetic neurones. *J. Physiol.* **356**, 115–33.

Garrahan, P. J. and Glynn, I. M. (1967). The stoicheiometry of the sodium pump. *J. Physiol.* **192**, 217–35.

Gasser, H. S. and Hill, A. V. (1924). The dynamics of muscular contraction. *Proc. R. Soc. Lond.* B **96**, 398–437.

Gauthier, G. F. (1986). Skeletal muscle fibre types. In *Myology* (ed. A. G. Engel, and B. Q. Banker), pp. 253–83. New York: McGraw-Hill.

Geren, B. B. (1954). The formation from the Schwann cell surface of myelin in the peripheral nerves of chick embryos. *Exp. Cell Res.* **7**, 558.

Gerschenfeld, H. M. (1966). Chemical transmitters in invertebrate nervous systems. *Symp. Soc. exp. Biol.* **20**, 299–324.

Gerschenfeld, H. M. and Chiarandini, D. J. (1965). Ionic mechanism associated with non-cholinergic synaptic inhibition in molluscan neurons. *J. Neurophysiol.* **28**, 710–23.

Gilbert, C., Kretschmar, K. M., Wilkie, D. R. and Woledge, R. C. (1971). Chemical change and energy output during muscular contraction. *J. Physiol.* **218**, 163–93.

Gilbert, D. L. and Ehrenstein, G. (1984). Membrane surface charge. In *The Squid Axon* (ed. P. F. Baker). *Current Topics in Membranes and Transport*, vol. 22. pp. 407–43. Orlando: Academic Press.

Giles, W. and Noble, S. J. (1976). Changes in membrane currents in bullfrog atrium produced by acetylcholine. *J. Physiol.* **261**, 103–23.

Gilly, W. F. and Armstrong, C. M. (1980). Gating current and potassium channels in the giant axon of the squid. *Biophys. J.* **29**, 485–92.

Gilly, W. F. and Hui, C. S. (1980). Mechanical activation in slow and twitch skeletal muscle fibres of the frog. *J. Physiol.* **301**, 137–56.

Gilman, A. G. (1984). G proteins and dual control of adenylate cyclase. *Cell* **36**, 577–9.

Gilman, A. G. (1987). G Proteins: transducers of receptor-generated signals. *Ann. Rev. Biochem.* **56**, 615–49.

Ginsborg, B. L. (1960). Some properties of avian skeletal muscle fibres with multiple neuromuscular junctions. *J. Physiol.* **154**, 581–98.

Gintzler, A. R. and Hyde, D. (1983). A specific substance P antagonist attenuates non-cholinergic electrically induced contractures of the guinea-pig isolated ileum. *Neurosci. Lett.* **40**, 75–9.

Glynn, I. M. (1984). The electrogenic sodium pump. In *Electrogenic Transport* (ed. M. P. Blaustein and M. Lieberman), pp. 33–48. New York: Raven Press.

Go, V. L. W. and Yaksh, T. L., (1987). Release of substance P from the cat spinal cord. *J. Physiol.* **391**, 141–67.

Godchaux, W. and Zimmerman, W. F. (1979). Membrane-dependent guanine nucleotide binding and GTPase activities of soluble proteins from bovine rod outer segments. *J. Biol. Chem.* **254**, 7874–84.

Gold, G. H. and Korenbrot, J. I. (1980). Light-induced calcium release by intact retinal rods. *Proc. Natl. Acad. Sci. USA.* **77**, 5557–61.

Goldman, D., Deneris, E., Luyten, W., Kochhar, A., Patrick, J. and Heinemann, S. (1987). Members of a nicotinic acetylcholine receptor gene family are expressed in different regions of the mammalian central nervous system. *Cell* **48**, 965–73.

Goldman, D. E. (1943). Potential, impedance, and rectification in membranes. *J. gen. Physiol.* **27**, 37–60.

Goldman, Y. E. (1987). Kinetics of the actomyosin ATPase in muscle fibres. *Ann. Rev. Physiol.* **49**, 637–54.

Goldman, Y. E., Hibberd, M. G. and Trentham, D. R. (1984*a*). Relaxation of rabbit psoas muscle fibres from rigor by photochemical generation of adenosine-5′-triphosphate. *J. Physiol.* **354**, 577–604.

Goldman, Y. E., Hibberd, M. G. and Trentham, D. R. (1984*b*). Initiation of active contraction by photogeneration of adenosine-5′-triphosphate in rabbit psoas muscle fibres. *J. Physiol.* **354**, 605–24.

Gonzalez-Serratos, H. (1971). Inward spread of activation in vertebrate muscle fibres. *J. Physiol.* **212**, 777–99.

Göpfert, H. and Schaefer, H. (1938). Über den direkt und indirekt erregeten Aktionsstrom und die Funktion der motorischen Endplatte. *Pflügers Arch. ges Physiol.* **239**, 597–619.

Gordon, A. M., Huxley, A. F. and Julian, F. J. (1966*a*). Tension development in highly stretched vertebrate muscle fibres. *J. Physiol.* **184**, 143–69.

Gordon, A. M., Huxley, A. F. and Julian, F. J. (1966*b*). The variation in isometric tension with sarcomere length in vertebrate muscle fibres. *J. Physiol.* **184**, 170–92.

Gorman, A. L. F. and Thomas, M. V. (1978). Changes in the intracellular concentration of the free calcium ions in a pace-maker neurone, measured with the metallochromic indicator dye arsenazo III. *J. Physiol.* **275**, 357–76.

Gorman, A. L. F. and Thomas, M. V. (1980). Intracellular calcium accumulation during depolarization in a molluscan neurone. *J. Physiol.* **308**, 259–85.

Gorodis, C., Virmaux, N., Cailla, H. L. and Delaage, M.

A. (1974). Rapid, light-induced changes of retinal cyclic GMP levels. *FEBS Lett.* **49**, 167–9.

Gorter, E. and Grendel, F. (1925). On bimolecular layers of lipoids on the chromocytes of blood. *J. exp. Med.* **41**, 439–43.

Granger, B. L. and Lazarides, E. (1978). The existence of an insoluble Z disc scaffold in chicken skeletal muscle. *Cell.* **15**, 1253–68.

Granit, R. (1933). The components of the retinal action potential and their relation to the discharge in the optic nerve. *J. Physiol.* **77**, 207–40.

Granit, R. (1947). *Sensory Mechanisms of the Retina.* London: Oxford University Press.

Granit, R. (1962). The visual pathway. In *The Eye* (ed. H. Davson), vol. 2, 535–763.

Graves, R. (1958). What food the centaurs ate. In *Steps.* London: Cassell.

Gray, E. G. (1957). The spindle and extrafusal innervation of a frog muscle. *Proc. R. Soc. Lond.* B **146**, 416–30.

Gray, E. G. (1959). Axosomatic and axodendritic synapses of the cerebral cortex: an electron microscope study. *J. Anat.* **93**, 420–33.

Gray, E. G. (1962). A morphological basis for presynaptic inhibition? *Nature* **193**, 82–3.

Gray, E. G. (1974). The synapse. In *The Cell in Medical Science*, vol. 2 (ed. F. Beck and J. B. Lloyd), pp. 385–416. London: Academic Press.

Gray, E. G. (1978). Synaptic vesicles and microtubules in frog motor endplates. *Proc. R. Soc. Lond.* B **203**, 219–27.

Gray, E. G. (1983). Neurotransmitter release mechanisms and microtubules. *Proc. R. Soc. Lond.* B **218**, 253–8.

Gray, E. G. and Guillery, R. W. (1966). Synaptic morphology in the normal and degenerating nervous system. *Int. Rev. Cytol.* **19**, 111–82.

Gray, J. A. B. (1962). Coding in systems of primary receptor neurons. *Symp. Soc. exp. Biol.* **16**, 345–54.

Gray, J. A. B. and Matthews, P. B. C. (1951). A comparison of the adaptation of the Pacinian corpuscle with the accommodation of its own axon. *J. Physiol.* **114**, 454–64.

Gray, P. and Atwell, D. (1985). Kinetics of light sensitive channels in vertebrate photoreceptors. *Proc. R. Soc. Lond.* B **223**, 379–88.

Greaser, M. L. and Gergely, J. (1973). Purification and properties of the components from troponin. *J. Biol. Chem.* **248**, 2125–33.

Greaser, M. L., Yamaguchi, M., Brekke, C., Potter, J. and Gergely, J. (1973). Troponin subunits and their interactions. *Cold Spr. Harb. Symp. Quant. Biol.* **37**, 235–44.

Green, H. J. (1986). Muscle power: recruitment, metabolism and fatigue. In *Human Muscle Power* (ed. N. L. Jones, N. McCartney and A. J. McComas), pp. 65–79. Champaign, Illinois: Human Kinetics Publishers.

Greenblatt, R. E., Blatt, Y. and Montal, M. (1985). The structure of the voltage-sensitive sodium channel. *FEBS Letters* **193**, 125–34.

Greengard, P. (1976). Possible role for cyclic nucleotides and phosphorylated membrane proteins in postsynaptic actions of neurotransmitters. *Nature* **260**, 101–8.

Greer Walker, M. and Pull, G. (1973). Skeletal muscle function and sustained swimming speeds in the coalfish *Gadus virens* L. *Comp. Biochem. Physiol.* A **44**, 495–501.

Gregory, R. (1970). *The Intelligent Eye.* London: Weidenfeld.

Gregory, R. (1986). *Odd Perceptions.* London: Methuen.

Grenningloh, G., Rienitz, A., Schmitt, B., Methfessel, C., Zensen, M., Beyreuther, K., Gundelfinger, E. D. and Betz, H. (1987). The strychnine-binding subunit of the glycine receptor shows homology with nicotinic acetylcholine receptors. *Nature* **328**, 215–20.

Griffin, D. R. (1958). *Listening in the Dark.* New Haven: Yale University Press.

Griffin, D. R. and Galambos, R. (1941). The sensory basis of obstacle avoidance by flying bats. *J. exp. Zool.* **86**, 481–506.

Griffith, O. H., Dehlinger, P. J. and Van, S. P. (1974). Shape of the hydrophilic barrier of phospholipid bilayers. *J. Membrane Biol.* **15**, 159–92.

Gunderson, C. B., Miledi, R. and Parker, I. (1983). Voltage-operated channels induced by foreign messenger RNA in *Xenopus* oocytes. *Proc. R. Soc. Lond.* B **220**, 131–40.

Gunderson, C. B., Miledi, R. and Parker, I. (1984). Messenger RNA from human brain induces drug- and voltage-activated channels in *Xenopus* oocytes. *Nature* **308**, 421–4.

Gurdon, J. B. (1974). *The Control of Gene Expression in Animal Development.* Oxford University Press.

Gurdon, J. B., Lane, C. D., Woodland, H. R. and Marbaix, G. (1971). Use of frog eggs and oocytes for the study of messenger RNA and its translation in living cells. *Nature* **233**, 177–82.

Guy, H. R. (1984). A structural model of the acetylcholine receptor channel based on partition energy and helix packing calculations. *Biophys. J.* **45**, 249–61.

Guy, H. R. and Seetharamulu, P. (1986). Molecular model of the action potential sodium channel. *Proc. Natl. Acad. Sci. USA* **83**, 508–12.

Haga, K., Haga, T., Ichiyama, A., Katada, T., Kurose, H. and Ui, M. (1985). Functional reconstitution of purified muscarinic receptors and inhibitory guanine nucleotide regulatory protein. *Nature* **316**, 731–3.

Haga, K., Haga, T. and Ichiyama, A. (1986). Reconstitution of the muscarinic acetylcholine receptor. *J. Biol. Chem.* **261**, 10133–40.

Hagedorn, M. (1986). The ecology, courtship and mating of gymnotiform electric fish. In *Electroreception* (ed. T. H. Bullock and W. Heiligenberg), pp. 497–525. New York: John Wiley.

Hagedorn, M. and Carr, C. (1985). Single electrocytes produce a sexually dimorphic signal in a South American electric fish, *Hypopomus occidentalis* (Gymnotiformes, Hypopomidae). *J. Comp. Physiol.* A **156**, 511–23.

Hagins, W. A. (1972). The visual process: excitatory

mechanisms in the primary receptor cells. *A. Rev. Biophys. Bioeng.* **1,** 131–58.

Hagins, W. A., Penn, R. D. and Yoshikami, S. (1970). Dark current and photocurrent in retinal rods. *Biophys. J.* **10,** 380–412.

Hagins, W. A. and Yoshikami, S. (1974). A role for Ca^{2+} in excitation of retinal rods and cones. *Exp. Eye Res.* **18,** 299–305.

Hagiwara, S. and Byerly, L. (1981). Calcium channel. *Ann. Rev. Neurosci.* **4**, 69–125.

Hagiwara, S. and Morita, H. (1962). Electrotonic transmission between two nerve cells in leech ganglion. *J. Neurophysiol.* **25**, 721–31.

Hagiwara, S. and Naka, K. (1964). The initiation of spike potential in barnacle muscle fibres under low intracellular Ca^{++}. *J. gen. Physiol.* **48**, 141–62.

Hagiwara, S. and Ohmori, H. (1983). Studies of single calcium channel currents in rat clonal pituitary cells. *J. Physiol.* **336**, 649–61.

Hagiwara, S., Szabo, T. and Enger, P. S. (1965). Electroreceptor mechanisms in a high-frequency weakly electric fish, *Sternarchus albifrons. J. Neurophysiol.* **28**, 784–99.

Hagiwara, S. and Tasaki, I. (1958). Study of mechanism of impulse transmission across the giant synapse of the squid. *J. Physiol.* **143**, 114–37.

Hagiwara, S., Watanabe, A. and Saito, N. (1959). Potential changes in syncytial neurons of lobster cardiac ganglion. *J. Neurophysiol.* **22**, 554–72.

Hagopian, M. (1966). Myofilament arrangement in femoral muscle of the cockroach, *Leucophaea maderae* Fabricius. *J. Cell. Biol.* **28**, 545–62.

Hagopian, M. and Spiro, D. (1968). The filament lattice of cockroach thoracic muscle. *J. Cell Biol.* **36**, 433–42.

Hall, Z. W. and Kelly, R. B. (1971). Enzymatic detachment of endplate acetylcholinesterase from muscle. *Nature New Biol.* **232**, 62–3.

Hamill, O. P., Marty, A., Neher, E., Sakmann, B. and Sigworth, F. J. (1981). Improved patch-clamp techniques for high-resolution current recording from cells and cell-free membrane patches. *Pflügers Arch.* **391**, 85–100.

Hamill, O. M. and Sakmann, B. (1981). Multiple conductance states of single acetylcholine receptor channels in embryonic muscle cells. *Nature* **294**, 462–4.

Hammer, R. and Giachetti, A. (1982). Muscarinic receptor subtypes: M1 and M2. Biochemical and functional characterization. *Life Sci.* **31**, 2991–8.

Hansen Bay, C. and Strichartz, G. R. (1980). Saxitoxin binding to sodium channels of rat skeletal muscle. *J. Physiol.* **300**, 89–103.

Hanson, J. and Huxley, H. E. (1953). The structural basis of the cross-striations in muscle. *Nature* **172**, 530–2.

Hanson, J. and Huxley, H. E. (1955). The structural basis of contraction in striated muscle. *Symp. Soc. exp. Biol.* **9**, 228–64.

Hanson, J. and Lowy, J. (1960). Structure and function of the contractile apparatus in the muscles of invertebrate animals. In *Structure and Function of Muscle* (ed. G. Bourne) vol. 1, pp. 263–365. New York: Academic Press.

Hanson, J. and Lowy, J. (1963). The structure of F-actin and of actin filaments isolated from muscle. *J. Mol. Biol.* **6**, 46–60.

Hardwicke, P. M. D. and Szent-Györgyi, A. G. (1985). Proximity of regulatory light chains in scallop myosin. *J. Mol. Biol.* **183**, 203–11.

Hardwicke, P. M. D., Wallimann, T. and Szent-Györgyi, A. G. (1983). Light-chain movement and regulation in scallop myosin. *Nature* **301**, 478–82.

Hargrave, P. A., McDowell, J. H., Curtis, D. R., Wang, J. K., Juszczak, E., Fong, S. L., Mohanna Rao, J. K. and Argos, P. (1983). The structure of bovine rhodopsin. *Biophys. Struct. Mech.* **9**, 235–44.

Harrington, W. F. (1979). On the origin of the contractile force in skeletal muscle. *Proc. Natl. Acad. Sci. USA* **76**, 5066–70.

Harrington, W. F. and Rodgers, M. E. (1984). Myosin. *Ann. Rev. Biochem.* **53**, 35–73.

Hartline, H. K. (1938). The response of single optic nerve fibres of the vertebrate eye to illumination of the retina. *Am. J. Physiol.* **121**, 400–15.

Hartline, H. K. and Ratliff, F. (1957). Inhibitory interaction of receptor units in the eye of *Limulus. J. gen. Physiol.* **40**, 357–76.

Hartline, H. K., Ratliff, F. and Miller, W. H. (1961). Inhibitory interaction in the retina and its significance in vision. In *Nervous Inhibition* (ed. E. Florey). Oxford: Pergamon Press.

Hartline, H. K., Wagner, H. G. and Ratliff, F. (1956). Inhibition in the eye of *Limulus. J. gen. Physiol.* **39**, 651–73.

Hartshorne, R. P. and Catterall, W. A. (1984). The sodium channel from rat brain. Purification and subunit composition *J. Biol. Chem.* **259**, 1667–75.

Hartshorne, R. P., Keller, B. U., Talvenheimo, J. A., Catterall, W. A. and Montal, M. (1985). Functional reconstitution of the purified brain sodium channel in planar lipid bilayers. *Proc. Natl. Acad. Sci. USA* **82**, 240–4.

Hartzell, H. C. (1981). Mechanisms of slow postsynaptic potentials. *Nature* **291**, 539–44.

Haselgrove, J. C. (1973). X-ray evidence for a conformational change in the actin-containing filaments of vertebrate striated muscle. *Cold Spr. Harb. Symp. Quant. Biol.* **37**, 341–52.

Haselgrove, J. C. (1983). Structure of vertebrate striated muscle as determined by X-ray-diffraction studies. In *Handbook of Physiology*, section 10 (ed. L. D. Peachey and R. H. Adrian), pp. 143–71. Bethesda, Maryland: American Physiological society.

Haselgrove, J. C. and Huxley, H. E. (1973). X-ray evidence for radial cross-bridge movement and for the sliding filament model in actively contracting muscle. *J. Mol. Biol.* **77**, 549–68.

Hashimoto, T., Hirata, M., Itoh, T., Kanmura, Y. and Kuriyama, H. (1986). Inositol 1,4,5-trisphosphate activates pharmacomechanical coupling in smooth muscle of the rabbit mesenteric artery. *J. Physiol.* **370**, 605–18.

Hasselbach, W. (1964). Relaxing factor and the relaxation of muscle. *Progr. Biophys.* **14**, 167–222.

Hasselbach, W. and Makinose, M. (1963). Über den mechanismus des Calciumtransportes durch die Membranen des Sarkoplasmatichen Reticulums. *Biochem. Z.* **339**, 94–111.

Hasselbach, W. and Oetliker, H. (1983). Energetics and electrogenicity of the sarcoplasmic reticulum calcium pump. *Ann. Rev. Physiol.* **45**, 325–39.

Haynes, L. W., Kay, A. R. and Yau, K.-W. (1986). Single cyclic GMP-activated channel activity in excised patches of rod outer segment membrane. *Nature* **321**, 66–70.

Hecht, S. (1937). Rods, cones, and the chemical basis of vision. *Physiol. Rev.* **17**, 239–90.

Hecht, S. and Schlaer, S. (1938). An adaptometer for measuring human dark adaptation. *J. opt. Soc. Amer.* **28**, 269–75.

Hecht, S., Schlaer, S. and Pirenne, M. (1942). Energy, quanta and vision. *J. gen. Physiol.* **25**, 819–40.

Heiligenberg, W. (1977). *Principles of Electroreception and Jamming Avoidance in Electric Fish.* Berlin: Springer.

Hecker, E. and Butenandt, A. (1984). In *Techniques in Pheromone Research* (ed. H. E. Hummel), p. 1. New York: Springer.

Hellam, D. C. and Podolsky, R. J. (1969). Force measurements in skinned muscle fibres. *J. Physiol.* **200**, 807–19.

Helmholtz, H. (1866). *Handbuch der Physiologischen Optik.* Leipzig.

Helmholtz, H. L. F. (1885). *On the Sensations of Tone.* Reprinted 1954 (translation of the fourth German edition of 1877, by A. J. Ellis, the first edition being in 1863). New York: Dover.

Hensel, H. and Zotterman, Y. (1951). Quantitative Beziehungen zwischen der Entladung einzelner Kältefasern und der Temperature. *Acta physiol. Scand.* **23**, 291–319.

Hermann, L. (1899). Zur Theorie der Erregungsleitung und der elektrischen Erregung. *Pflüg. Arch.* **75**, 574.

Herron, C. E., Lester, R. A. J., Coan, E. J. and Collingridge, G. L. (1986). Frequency-dependent involvement of NMDA receptors in the hippocampus: a novel synaptic mechanism. *Nature* **322**, 265–8.

Herzberg, O. and James, M. N. G. (1985). Structure of the calcium regulatory muscle protein troponin-C at 2.8 Å resolution. *Nature* **313**, 653–9.

Hescheler, J., Rosenthal, W., Trautwein, W. and Schulltz, G. (1987). The GTP-binding protein, G_0, regulates neuronal calcium channels. *Nature* **325**, 445–7.

Heuser, J. (1976). Morphology of synaptic vesicle discharge and reformation at the frog neuromuscular junction. In *Motor Innervation of Muscle* (ed. S. Thesleff), pp. 51–115. London: Academic Press.

Heuser, J. E. and Reese, T. S. (1973). Evidence for recycling of synaptic vesicle membrane during transmitter release at the frog neuromuscular junction. *J. Cell Biol.* **57**, 315–44.

Heuser, J. E. and Reese, T. S. (1977). Structure of the synapse. In *Handbook of Physiology*, section 1, volume 1 (*Cellular Biology of Neurons* Part 1, ed. E. R. Kandel), pp. 261–94. Bethesda, Maryland: American Physiological Society.

Heuser, J. E., Reese, T. S., Dennis, M. J., Jan, Y., Jan, L. and Evans, L. (1979). Synaptic vesicle exocytosis captured by quick-freezing and correlated with quantal transmitter release. *J. Cell Biol.* **81**, 275–300.

Heuser, J. E. and Salpeter, S. R. (1979). Organization of acetylcholine receptors in quick-frozen deep-etched and rotary-replicated *Torpedo* postsynaptic membrane. *J. Cell Biol.* **82**, 150–73.

Hibberd, M. G. and Jewell, B. R. (1982). Calcium- and length-dependent force production in rat ventricular muscle. *J. Physiol.* **329**, 527–40.

Hill, A. V. (1936). The strength–duration relation for electric excitation of medullated nerve. *Proc. R. Soc. Lond.* B **119**, 440–53.

Hill, A. V. (1938). The heat of shortening and the dynamic constants of muscle. *Proc. R. Soc. Lond.* B **126**, 136–95.

Hill, A. V. (1948). On the time required for diffusion and its relation to processes in muscle. *Proc. R. Soc. Lond.* B **135**, 446–53.

Hill, A. V. (1949). The abrupt transition from rest to activity in muscle. *Proc. R. Soc. Lond.* B **136**, 399–420.

Hill, A. V. (1950*a*). A challenge to biochemists. *Biochim. biophys. Acta* **4**, 4–11.

Hill, A. V. (1950*b*). The dimensions of animals and their muscular dynamics. *Sci. Prog. Lond.* **38**, 209–30.

Hill, A. V. (1953). The 'instantaneous' elasticity of active muscle. *Proc. R. Soc. Lond.* B **141**, 161–78.

Hill, A. V. (1960). *The Ethical Dilemma of Science and Other Writings.* New York: Rockefeller Institute Press.

Hill, A. V. (1964*a*). The effect of load on the heat of shortening of muscle. *Proc. R. Soc. Lond.* B **159**, 297–318.

Hill, A. V. (1964*b*). The effect of tension in prolonging the active state in a twitch. *Proc. R. Soc. Lond.* B **159**, 589–95.

Hill, A. V. (1965). *Trails and Trials in Physiology.* London: Edward Arnold.

Hill, A. V. and Hartree, W. (1920). The four phases of heat production of muscle. *J. Physiol.* **54**, 84–128.

Hill, A. V. and Howarth, J. V. (1957). The effect of potassium on the resting metabolism of the frog's sartorius. *Proc. R. Soc. Lond.* B **147**, 21–43.

Hill, D. K. (1953). The effect of stimulation on the diffraction of light by striated muscle. *J. Physiol.* **119**, 501–12.

Hill, D. R. and Bowery, N. G. (1981). ^{3}H-baclofen and ^{3}H-GABA bind to bicuculine-insensitive $GABA_B$ sites in rat brain. *Nature* **290**, 149–52.

Hill, D. R., Bowery, N. G. and Hudson, A. L. (1984). Inhibition of $GABA_B$ receptor binding by guanyl nucleotides. *J. Neurochem.* **42**, 652–7.

Hill, R. B. and Usherwood, P. N. R. (1961). The action of 5-hydroxytryptamine and related compounds on neuromuscular transmission in the locust *Schistocerca gregaria*. *J. Physiol.* **157**, 393–401.

Hill, T. L. (1974). Theoretical formalism for the sliding filament model of contraction of straited muscle. Part I. *Prog. Biophys. Molec. Biol.* **28**, 267–340.

Hill, T. L. (1976). Theoretical formalism for the sliding filament model of contraction of striated muscle. Part II. *Prog. Biophys. Molec. Biol.* **29**, 105–59.

Hille, B. (1966). Common mode of action of three agents that decrease the transient change in sodium permeability in nerves. *Nature* **210**, 1220–2.

Hille, B. (1967). The selective inhibition of delayed potassium currents in nerve by tetraethylammonium ion. *J. Gen. Physiol.* **50**, 1287–302.

Hille, B. (1968). Pharmacological modifications of the sodium channels of frog nerve. *J. gen. Physiol.* **51**, 199–219.

Hille, B. (1971*a*). Voltage clamp studies on myelinated nerve fibres. In *Biophysics and Physiology of Excitable Membranes* (ed. W. J. Adelman, Jr), pp. 230–46. New York: Van Nostrand Reinhold.

Hille, B. (1971*b*). The permeability of the sodium channel to organic cations in myelinated nerve. *J. gen. Physiol.* **58**, 599–619.

Hille, B. (1973). Potassium channels in myelinated nerve; selective permeability to small cations. *J. gen. Physiol.* **61**, 669–86.

Hille, B. (1984). *Ionic Channels of Excitable Membranes.* Sunderland, Massachusetts: Sinauer Associates.

Hille, B., Woodhull, A. M. and Shapiro, B. I. (1975). Negative surface charge near sodium channels of nerve: divalent ions, monovalent ions, and pH. *Phil. Trans. R. Soc. Lond.* B **270**, 301–18.

Hirokawa, N. and Neuser, J. E. (1982). Internal and external differentiations of the postsynaptic membrane at the neuromuscular junction. *J. Neurocytol.* **11**, 487–510.

Hirano, A. (1981). Structure of normal central myelinated fibres. *Adv. Neurol.* **31**, 51–68.

Hochberg, J. (1984). Perception. In *Handbook of Physiology*, section, 1 vol. 3, *Sensory Processes* (ed. I. Darian-Smith), pp. 75–102. Bethesda, Maryland: American Physiological Society.

Hodgkin, A. L. (1937). Evidence for electrical transmission in nerve. *J. Physiol.* **90**, 183–232.

Hodgkin, A. L. (1938). The subthreshold potentials in a crustacean nerve fibre. *Proc. R. Soc. Lond.* B **126**, 87–121.

Hodgkin, A. L. (1939). The relation between conduction velocity and the electrical resistance outside a nerve. *J. Physiol.* **94**, 560–70.

Hodgkin, A. L. (1948). The local electric changes associated with repetitive action in a non-medulated axon. *J. Physiol.* **107**, 165.

Hodgkin, A. L. (1951). The ionic basis of electrical activity in nerve and muscle. *Biol. Rev.* **26**, 339–409.

Hodgkin, A. L. (1954). A note on conduction velocity. *J. Physiol.* **125**, 221–4.

Hodgkin, A. L. (1958). Ionic movements and electrical activity in giant nerve fibres. *Proc. R. Soc. Lond.* B **148**, 1–37.

Hodgkin, A. L. (1964). *The Conduction of the Nervous Impulse.* Liverpool: University Press.

Hodgkin, A. L. (1975). The optimum density of sodium channels in an unmyelinated nerve. *Phil. Trans. R. Soc. Lond.* B **270**, 297–300.

Hodgkin, A. L. and Horowicz, P. (1957). The differential action of hypertonic solutions on the twitch and action potential of a muscle fibre. *J. Physiol.* **136**, 17P.

Hodgkin, A. L. and Horowicz, P. (1959). The influence of potassium and chloride ions on the membrane potential of single muscle fibres. *J. Physiol.* **148**, 127–60.

Hodgkin, A. L. and Horowicz, P. (1960). Potassium contractures in single muscle fibres. *J. Physiol.* **153**, 386–403.

Hodgkin, A. L. and Huxley, A. F. (1939). Action potentials recorded from inside a nerve fibre. *Nature* **140**, 710–1.

Hodgkin, A. L. and Huxley, A. F. (1945). Resting and action potentials in single nerve fibres. *J. Physiol.* **104**, 176–95.

Hodgkin, A. L. and Huxley, A. F. (1952*a*). Currents carried by sodium and potassium ions through the membrane of the giant axon of *Loligo. J. Physiol.* **116**, 449–72.

Hodgkin, A. L. and Huxley, A. F. (1952*b*). The components of membrane conductance in the giant axon of *Loligo. J. Physiol.* **116**, 473–96.

Hodgkin, A. L. and Huxley, A. F. (1952*c*). The dual effect of membrane potential on sodium conductance in the giant axon of *Loligo. J. Physiol.* **116**, 497–506.

Hodgkin, A. L. and Huxley, A. F. (1952*d*). A quantitative description of membrane current and its application to conduction and excitation in nerve. *J. Physiol.* **117**, 500–44.

Hodgkin, A. L. and Huxley, A. F. (1953). Movements of radioactive potassium and membrane current in a giant axon. *J. Physiol.* **121**, 403–14.

Hodgkin, A. L., Huxley, A. F. and Katz, B. (1952). Measurement of current–voltage relations in the membrane of the giant axon of *Loligo. J. Physiol.* **116**, 424–48.

Hodgkin, A. L. and Katz, B. (1949). The effect of sodium ions on the electrical activity of the giant axon of the squid. *J. Physiol.* **108**, 37–77.

Hodgkin, A. L. and Keynes, R. D. (1953). The mobility and diffusion coefficient of potassium in giant axons from *Sepia. J. Physiol.* **119**, 513–28.

Hodgkin, A. L. and Keynes, R. D. (1955*a*). Active transport of cations in giant axons from *Sepia* and *Loligo. J. Physiol.* **128**, 28–60.

Hodgkin, A. L. and Keynes, R. D. (1955*b*). The potassium permeability of a giant nerve fibre. *J. Physiol.* **128**, 61–88.

Hodgkin, A. L. and Keynes, R. D. (1957). Movements of labelled calcium in squid giant axons. *J. Physiol.* **138**, 253–81.

Hodgkin, A. L., McNaughton, P. A. and Nunn, B. J. (1985). The ionic selectivity and calcium dependence of the light-sensitive pathway in toad rods. *J. Physiol.* **358**, 447–68.

Hodgkin, A. L., McNaughton, P. A., Nunn, B. J. and Yau,

K.-W. (1985). Effect of ions on retinal rods from *Bufo marinus*. *J. Physiol.* **350**, 649–80.

Hodgkin, A. L. and Rushton, W. A. H. (1946). The electrical constants of a crustacean nerve fibre. *Proc. R. Soc. Lond.* B **133**, 444.

Hodgson, E. S., Lettvin, J. Y. and Roeder, K. D. (1955). Physiology of a primary chemoreceptor unit. *Science* **122**, 417–8.

Hodgson, E. S. and Roeder, K. D. (1956). Electrophysiological studies of arthropod chemoreception. I. General properties of the labellar chemoreceptors of Diptera. *J. cell. comp. Physiol.* **48**, 51–76.

Hökfelt, T., Elfvin, L. G., Elde, R., Schultzberg, M., Goldstein, M. and Luft, R. (1977). Occurrence of somatostatin-like immunoreactivity in some peripheral sympathetic noradrenergic neurons. *Proc. Natl. Acad. Sci. USA* **74**, 3587–91.

Hökfelt, T., Everitt, B., Holets, V. R., Meister, B., Melander, T., Schalling, M., Staines, W. and Lundberg, J. M. (1986). Coexistence of peptides and other active molecules in neurons: diversity of chemical signalling potential. In *Fast and Slow Chemical Signalling in the Nervous System* (ed. L. L. Iversen and E. Goodman), pp. 205–31. Oxford: Oxford University Press.

Hökfelt, T., Johansson, O., Ljungdahl, A., Lundberg, J. M. and Schultzberg, M. (1980). Peptidergic neurones. *Nature* **284**, 515–21.

Hökfelt, T., Lundberg, J. M., Skirboll, L., Johannson, O., Schultzberg, M. and Vincent. S. R. (1982). Coexistence of classical transmitters and peptides in neurones. In *Co-transmission* (ed. A. C. Cuello), pp. 77–125. London: Macmillan.

Holman, M. E. and Neild, T. O. (1979). Membrane properties. *Brit. Med. Bull.* **35**, 235–41.

Holmes, W. (1942). The giant myelinated nerve fibres of the prawn. *Phil. Trans. R. Soc. Lond.* B **231**, 293–311.

Holmgren, F. (1865). Method att objectivera effecten av ljusintryck på retina. *Upsala Läkarefförh.* **1**, 177–91.

Holton, T. and Hudspeth, A. J. (1986). The transduction channel of hair cells from the bull-frog characterized by noise analysis. *J. Physiol.* **375**, 195–227.

Holz, G. G., Rane, S. G. and Dunlap, K. (1986). GTP-binding proteins mediate transmitter inhibition of voltage-dependent calcium channels. *Nature* **319**, 670–2.

Homsher, E. (1987). Muscle enthalpy production and its relationship to actomyosin ATPase. *Ann Rev. Physiol.* **49**, 673–90.

Homsher, E., Irving, M. and Wallner, A. (1981). High-energy phosphate metabolism and energy liberation associated with rapid shortening in frog skeletal muscle. *J. Physiol.* **321**, 423–36.

Homsher, E. and Kean, C. J. C. (1982). Unexplained enthalpy production in contracting skeletal muscles. *Fed. Proc.* **41**, 149–54.

Homsher, E., Lacktis, J., Yamada, T. and Zohman, G. (1987). Repriming and reversal of the isometric unexplained enthalpy in frog skeletal muscle. *J. Physiol.* **393**, 157–70.

Hopkins, C. D. (1974). Electric communication in fish. *Am. Sci.* **62**, 426–37.

Hopkins, C. D. and Bass, A. H. (1981). Temporal coding of species recognition signals in electric fish. *Science* **212**, 85–7.

Horn, R. and Vandenberg, C. A. (1984). Statistical properties of single sodium channels. *J. gen. Physiol.* **84**, 505–34.

Hornykiewicz, O. (1973). Parkinsons' disease: from brain homogenate to treatment. *Fed. Proc.* **32**, 183–90.

Horowits, R., Kempner, E. S., Bisher, M. E. and Podolsky, R. J. (1986). A physiological role for titin and nebulin in skeletal muscle. *Nature* **323**, 160–4.

Hoyle, G. (1955*a*). The anatomy and innervation of locust skeletal muscle. *Proc. R. Soc. Lond.* B **143**, 281–92.

Hoyle, G. (1955*b*). Neuromuscular mechanisms of a locust skeletal muscle. *Proc. R. Soc. Lond.* B **143**, 343–67.

Hoyle, G. (1959). The neuromuscular mechanism of an insect spiracular muscle. *J. Insect Physiol.* **3**, 378–94.

Hoyle, G., Dagan, D., Moberly, B. and Colquhoun, W. (1974). Dorsal unpaired median insect neurons make neurosecretory endings on skeletal muscle. *J. Exp. Zool.* **187**, 159–65.

Hoyle, G. and Wiersma, C. A. G. (1958). Excitation at neuromuscular junctions in Crustacea. *J. Physiol.* **143**, 403–25.

Hubbard, J. I. (1963). Repetitive stimulation at the mammalian neuromuscular junction, and the mobilisation of transmitter. *J. Physiol.* **169**, 641–62.

Hubbard, R. (1959). The thermal stability of rhodopsin and opsin. *J. gen. Physiol.* **42**, 259–80.

Hubbard, R. and Kropf, A. (1958). The action of light on rhodopsin. *Proc. Natl. Acad. Sci. USA* **44**, 130–9.

Hubbard, R. and Wald, G. (1952). *Cis-trans* isomers of vitamin A and retinene in the rhodopsin system. *J. gen. Physiol.* **36**, 269–315.

Hubbard, S. J. (1958). A study of rapid mechanical events in a mechanoreceptor. *J. Physiol.* **141**, 198–218.

Hudson, R. C. L. (1973). On the function of the white muscles in teleosts at intermediate swimming speeds. *J. exp. Biol.* **58**, 509–22.

Hudspeth, A. J. (1982). Extracellular current flow and the site of transduction by vertebrate hair cells. *J. Neurosci.* **2**, 1–10.

Hudspeth, A. J. (1983). Mechanoelectrical transduction by cells in the acousticolateralis sensory system. *Ann. Rev. Neurosci.* **6**, 187–21.

Hudspeth, A. J. (1985). The cellular basis of hearing: the biophysics of hair cells, *Science* **230**, 745–52.

Hudspeth, A. J. and Corey, D. P. (1977). Sensitivity, polarity, and conductance change in the response of vertebrate hair cells to controlled mechanical stimuli. *Proc. Natl. Acad. Sci. USA* **74**, 2407–11.

Hudspeth, A. J. and Jacobs, R. (1979). Stereocilia mediate transduction in vertebrate hair cells. *Proc. Natl. Acad. Sci. USA* **76**, 1506–9.

Hughes, J., Smith, T. W., Kosterlitz, H. W. Fothergill, L. A. Morgan, B. A. and Morris, H. R. (1975). Identification of two related pentapeptides from the brain with potent opiate agonist activity. *Nature* **258**, 577–9.

Hunt, C. C. (1954). Relation of function to diameter in afferent fibers of muscle nerves. *J. gen. Physiol.* **38**, 117–31.

Hunt, C. C. and Ottoson, D. (1975). Impulse activity and receptor potential of primary and secondary endings of isolated mammalian muscle spindles. *J. Physiol.* **252**, 259–81.

Hurley, J. B. (1987). Molecular properties of the cGMP cascade of vertebrate photoreceptors. *Ann. Rev. Physiol.* **49**, 793–812.

Hurley, J. B. and Stryer, L. (1982). Purification and characterization of the γ subunit of the cyclic GMP phosphodiesterase from retinal rod outer segments. *J. Biol. Chem.* **257**, 11094–9.

Hursh, J. B. (1939). Conduction velocity and diameter of nerve fibres. *Am. J. Physiol.* **127**, 131–9.

Hutter, O. F. (1961). Ion movements during vagus inhibition of the heart. In *Nervous Inhibition* (ed. E. Florey), pp. 114–23. Oxford: Pergamon Press.

Hutter, O. F. and Noble, D. (1960). Rectifying properties of heart muscle. *Nature* **188**, 495.

Huxley, A. F. (1957). Muscle structure and theories of contraction. *Progr. Biophys.* **7**, 255–318.

Huxley, A. F. (1974). Muscular contraction. *J. Physiol.* **243**, 1–43.

Huxley, A. F. (1980). *Reflections on Muscle.* Liverpool: Liverpool University Press.

Huxley, A. F. and Niedergerke, R. (1954). Structural changes in muscle during contraction. Interference microscopy of living muscle fibres. *Nature* **173**, 971–3.

Huxley, A. F. and Peachey, L. D. (1961). The maximum length for contraction in vertebrate striated muscle. *J. Physiol.* **156**, 150–65.

Huxley, A. F. and Simmons, R. M. (1971). Proposed mechanism of force generation in striated muscle. *Nature* **233**, 533–8.

Huxley, A. F. and Simmons, R. M. (1973). Mechanical transients and the origin of muscular force. *Cold Spr. Harb. Symp. quant. Biol.* **37**, 669–80.

Huxley, A. F. and Stämpfli, R. (1949). Evidence for saltatory conduction in peripheral myelinated nerve fibres. *J. Physiol.* **108**, 315–39.

Huxley, A. F. and Stämpfli, R. (1951). Effect of potassium and sodium on resting and action potentials of single myelinated nerve fibres. *J. Physiol.* **122**, 496–508.

Huxley, A. F. and Straub, R. W. (1958). Local activation and interfibrillar structures in striated muscle. *J. Physiol.* **143**, 40–1*P*.

Huxley, A. F. and Taylor, R. E. (1955). Function of Krause's membrane. *Nature* **176**, 1068.

Huxley, A. F. and Taylor, R. E. (1958). Local activation of striated muscle fibres. *J. Physiol.* **144**, 426–41.

Huxley, H. E. (1953*a*). Electron microscope studies of the organisation of the filaments in striated muscle. *Biochem. biophys. Acta* **12**, 387–94.

Huxley, H. E. (1953*b*). X-ray analysis and the problem of muscle. *Proc. R. Soc. Lond.* B **141**, 59–66.

Huxley, H. E. (1957). The double array of filaments in cross-striated muscle. *J. biophys. biochem. Cytol.* **3**, 631–48.

Huxley, H. E. (1960). Muscle Cells. In *The Cell* (ed. J. Brachet and A. E. Mirsky) vol. 4, pp. 365–481. New York: Academic Press.

Huxley, H. E. (1963). Electron microscope studies on the structure of natural and synthetic protein filaments from striated muscle. *J. Mol. Biol.* **7**, 281–308.

Huxley, H. E. (1964). Evidence for continuity between the central elements of the triads and extracellular space in frog sartorius muscle. *Nature* **202**, 1067–71.

Huxley, H. E. (1969). The mechanism of muscular contraction. *Science* **164**, 1356–66.

Huxley, H. E. (1971). The structural basis of muscular contraction. *Proc. R. Soc. Lond.* B **178**, 131–49.

Huxley, H. E. (1972). Molecular basis of contraction in cross-striated muscles. In *The Structure and Function of Muscle*, 2nd edn (ed. G. H. Bourne), vol. 1, pp. 301–87. New York: Academic Press.

Huxley, H. E. (1973). Structural changes in the actin- and myosin-containing filaments during contraction. *Cold Spr. Harb. Symp. quant. Biol.* **37**, 361–76.

Huxley, H. E. (1979). Time-resolved X-ray diffraction studies on muscle. In *Cross-bridge Mechanism in Muscle Contraction* (ed. H. Sugi and G. H. Pollock), pp. 391–401. Tokyo: University of Tokyo Press.

Huxley, H. E. (1983). Molecular basis of contraction in cross-striated muscles and relevance to motile mechanisms in other cells. In *Muscle and Nonmuscle Motility* (ed. A. Stracher), vol. 1, pp. 1–104. New York: Academic Press.

Huxley, H. E. and Brown, W. (1967). The low-angle X-ray diagram of vertebrate striated muscle and its behaviour during contraction and rigor. *J. Mol. Biol.* **30**, 383–434.

Huxley, H. E., Brown, W. and Holmes, K. C. (1965). Constancy of axial spacings in frog sartorius muscle during contraction. *Nature* **206**, 1358.

Huxley, H. E. and Faruqi, A. R. (1983). Time-resolved X-ray diffraction studies on vertebrate striated muscle. *Ann. Rev. Biophys, Bioeng.* **12**, 381–417.

Huxley, H. E., Faruqi, A. R., Kress, M., Borda, J. and Koch, M. H. J. (1982). Time-resolved X-ray diffraction studies of the myosin layer-line reflections during muscular contraction. *J. Mol. Biol.* **158**, 637–84.

Huxley, H. E. and Hanson, J. (1954). Changes in the cross-striations of muscle during contraction and stretch and their structural interpretation. *Nature* **173**, 973–6.

Huxley, H. E. and Kress, M. (1985). Crossbridge behaviour during muscle contraction. *J. Mus. Res. Cell Motil.* **6**, 153–61.

Hynes, T. R., Block, S. M., White, B. T. and Spudich, J. A. (1987). Movement of myosin fragments in vitro: domains involved in force production. *Cell* **48**, 953–63.

Imoto, K., Methfessel, C., Sakmann, B., Mishina, M., Mori, Y., Konno, T., Fukuda, K., Kurasaki, M., Bujo, H., Fujita, Y. and Numa, S. (1986). Location of a δ-subunit region determining ion transport through the acetylcholine receptor channel. *Nature* **324**, 670–4.

Inesi, G. and Scales, D. (1976). Assembly of ATPase protein in sarcoplasmic reticulum membranes. *Biophys. J.* **16**, 735–51.

Infante, A. A. and Davies, R. E. (1962). Adenosine triphosphate breakdown during a single isotonic twitch of frog sartorius muscle. *Biochem. biophys. Res. Commun.* **9**, 410–15.

Irisawa, H. and Kokobun, S. (1983). Modulation by intracellular ATP and cyclic AMP of the slow inward current in isolated single ventricular cells of the guinea-pig. *J. Physiol.* **338**, 321–37.

Ishikawa, S. and Sperelakis, N. (1987). A novel class (H_3) of histamine receptors on perivascular nerve terminals. *Nature* **327**, 158–60.

Iversen, L. L. (1963). Uptake of noradrenaline by isolated perfused rat heart. *Brit. J. Pharmacol.* **21**, 523–37.

Iversen, L. L. (1967). *The Uptake and Storage of Noradrenaline in Sympathetic Nerves.* Cambridge University Press.

Iversen, L. L. (1971). Role of transmitter uptake mechanisms in synaptic neurotransmission. *Brit. J. Pharmacol.* **41**, 571–91.

Iversen, L. L. (1984). Amino acids and peptides: fast and slow chemical signals in the nervous system? *Proc. R. Soc. Lond.* B **221**, 245–60.

Iversen, L. L. and Bloom, F. E. (1972). Studies of the uptake of ^{3}H-GABA and ^{3}H-glycine in slices and homogenates of rat brain and spinal cord by electron microscopic autoradiography. *Brain. Res.* **41**, 131–43.

Jack, J. J. B., Noble, D. and Tsien, R. W. (1975). *Electric Current Flow in Excitable Cells.* Oxford University Press.

Jack, J. J. B., Redman, S. J. and Wong, K. (1981). The components of synaptic potentials evoked in cat spinal motoneurones by impulses in single group Ia afferents. *J. Physiol.* **321**, 65–96.

Jacobs, B. L. (1987). How hallucinogenic drugs work. *Amer. Sci.* **75**, 386–91.

Jacobs, G. H. (1986). Colour vision variations in non-human primates. *Trends Neurosci* **9**, 320–3.

Jahr, C. E. and Stevens, C. F. (1987). Glutamate activates multiple single channel conductances in hippocampal neurons. *Nature* **325**, 522–5.

Jami, L., Murthy, K. S. K. and Petit, J. (1982). A quantitative study of skeletofusimotor innervation in the rat peroneus tertius muscle. *J. Physiol.* **325**, 125–44.

Jan, L. Y. and Jan, Y. N. (1982). Peptidergic transmission in sympathetic ganglia of the frog. *J. Physiol.* **327**, 219–46.

Jan, Y. N., Jan, L. Y. and Kuffler, S. W. (1979). A peptide as a possible transmitter in sympathetic ganglia of the frog. *Proc. Natl. Acad. Sci. USA* **76**, 1501–5.

Jaslove, S. W. and Brink, P. R. (1986). The mechanism of rectification at the electronic motor giant synapse of the crayfish. *Nature* **323**, 63–5.

Jewell, B. R. (1959). The nature of the phasic and tonic responses of the anterior byssal retractor of *Mytilus*. *J. Physiol.* **149**, 154–77.

Jewell, B. R. and Rüegg, J. C. (1966). Oscillatory contraction of insect fibrillar muscle after glycerol extraction. *Proc. R. Soc. Lond.* B **164**, 428–59.

Jewell, B. R. and Wilkie, D. R. (1958). An analysis of the mechanical components in frog striated muscle. *J. Physiol.* **143**, 515–40.

Jewell, B. R. and Wilkie, D. R. (1960). The mechanical properties of relaxing muscle. *J. Physiol.* **152**, 30–47.

Jöbsis, F. F. and O'Connor, M. J. (1966). Calcium release and reabsorption in the sartorius muscle of the toad. *Biochem. Biophys. Res. Comm.* **25**, 246–52.

Johnston, I. A. (1980). Specialization of fish muscle. In *Development and Specialization of Skeletal Muscle* (ed. D. F. Goldspink), pp. 123–48. Cambridge: Cambridge University Press.

Johnston, I. A., Davison, W. and Goldspink, G. (1977). Energy metabolism of carp swimming muscles. *J. Comp. Physiol.* **114**, 203–16.

Johnston, M. F. and Ramon, F. (1982). Voltage independence of an electrotonic synapse. *Biophys. J.* **39**, 115–17.

Johnstone, B. M. and Boyle, A. J. F. (1967). Basilar membrane vibration examined with the Mössbauer technique. *Science* **158**, 389–90.

Johnstone, B. M., Taylor, J. J. and Boyle, A. J. (1970). Mechanics of guinea-pig cochlea. *J. acoust. Soc. Amer.* **47**, 504–9.

Jones, S. W. (1985). Muscarinic and peptidergic excitation of bull-frog sympathetic neurones. *J. Physiol.* **366**, 63–87.

Jorgensen, P. L. (1975). Purification and characterization of (Na^+, K^+)-ATPase. V. Conformational changes in the enzyme. Transitions between the Na-form and the K-form studied with tryptic digestion as a tool. *Biochem. Biophys. Acta* **401**, 399–415.

Jorgensen, P. L. (1982). Mechanism of the Na^+, K^+ pump. Protein structure and conformations of the pure ($Na^+ + K^+$)-ATPase. *Biochim. Biophys. Acta* **694**, 27–68.

Jorgensen, P. L. (1985). Conformational E_1-E_2 transformations in αβ units related to cation transport by pure Na, K-ATPase. In *The Sodium Pump* (ed. I. Glynn and C. Ellory), 83–96. Cambridge: Company of Biologists.

Julian, F. J. (1969). Activation in a skeletal muscle contraction model with a modification for insect fibrillar muscle. *Biophys. J.* **9**, 547–70.

Julian, F. J. (1971). The effect of calcium on the force-velocity relation of briefly glycerinated frog muscle fibres. *J. Physiol.* **218**, 117–45.

Julian, F. J. and Sollins, M. R. (1973). Regulation of force and speed of shortening in muscle contraction. *Cold Spr. Harb. Symp. quant. Biol.* **37**, 635–46.

Kabsch, W., Mannherz, H. G. and Suck, D. (1985). Three-dimensional structure of the complex of actin and DNase I at 4.5 Å resolution. *EMBO J.* **4**, 2113–18.

Kaczmarek, L. K. (1986). Phorbol esters, protein phosphorylation and the regulation of neuronal ion channels. *J. exp. Biol.* **124**, 375–92.

Kaczmarek, L. K. (1987). The role of protein kinase C in

the regulation of ion channels and neurotransmitter release. *Trends Neurosci.* **10**, 30–4.

Kaissling, K.-E. (1971). Insect olfaction. In *Handbook of Sensory Physiology*, vol. 4, part 1 (ed. L. M. Beidler), pp. 351–431. Berlin: Springer.

Kaissling, K.-E. (1986). Chemo-electrical transduction in insect olfactory receptors. *Ann. Rev. Neurosci.* **9**, 121–45.

Kaissling, K.-E., Kasang, G., Bestmann, H. J., Stransky, W. and Vostrowsky, O. (1978). A new pheromone of the silkworm moth *Bombyx mori*. Sensory pathway and behavioural effects. *Naturwissenschaften* **65**, 382–4.

Kaissling, K.-E. and Priesner, E. (1970). Die Riechschwelle des Seidenspinners. *Naturwissenschaften* **57**, 23–8.

Kalmijn, A. J. (1971). The electric sense of sharks and rays. *J. exp. Biol.* **55**, 371–83.

Kalmijn, A. J. (1982). Electric and magnetic field detection in elasmobranch fishes. *Science* **218**, 916–18.

Kalmijn, A. J. (1984). Theory of electromagnetic orientation: a further analysis. In *Comparative Physiology of Sensory Systems* (ed. l. Bolis *et al.*), pp. 525–60. Cambridge: Cambridge University Press.

Kamm, K. E. and Stull, J. T. (1985). The function of myosin and myosin light chain kinase phosphorylation in smooth muscle. *Ann. Rev. Pharmacol. Toxicol.* **25**, 593–620.

Kanaujia, S. and Kaissling, K. E. (1985). Interactions of pheromone with moth antennae: adsorption, desorption and transport. *J. Insect Physiol.* **31**, 71–81.

Kaneko, A. (1970). Physiological and morphological identification of horizontal, bipolar and amacrine cells in goldfish retina. *J. Physiol.* **207**, 623–33.

Kanno, Y. and Loewenstein, W. R. (1964). Intercellular diffusion. *Science* **143**, 959–60.

Kao, P. N., Dwork, A. J., Kaldawy, R.-R. J., Silver, M. L., Wideman, J., Stein, S. and Karlin, A. (1984). Identification of the α subunit half-cystine specifically labelled by an affinity reagent for the acetylcholine receptor binding site. *J. Biol. Chem.* **259**, 11662–5.

Karbon, E. W., Dunman, R. S. and Enna, S. J. (1984). γ-Aminobutyric acid B receptors and norepinephrine-stimulated cyclic AMP production in rat brain cortex. *Brain Res.* **306**, 327–32.

Karn, J., Brenner, S. and Barnett, L. (1983). Protein structural domains in the *Caenorhabditis elegans unc*-54 myosin heavy chain gene are not separated by introns. *Proc. Natl. Acad. Sci. USA* **80**, 4253–7.

Kasai, H., Kameyama, M., Yamaguchi, K. and Fukuda, J. (1986). Single transient K channels in mammalian sensory neurons. *Biophys. J.* **49**, 1243–7.

Katz, A. M. (1977). *Physiology of the Heart.* New York: Raven Press.

Katz, B. (1937). Experimental evidence for a non-conducted response of nerve to subthreshold stimulation. *Proc. R. Soc. Lond.* B **124**, 244–76.

Katz, B. (1949). Les constants électriques de la membrane du muscle. *Arch. Sci. physiol.* **3**, 285.

Katz, B. (1950). Depolarization of sensory terminals and the initiation of impulses in the muscle spindle. *J. Physiol.* **111**, 261–82.

Katz, B. (1962). The transmission of impulses from nerve to muscle, and the subcellular unit of synaptic action. *Proc. R. Soc. Lond.* B **155**, 455–79.

Katz, B. (1969). *The Release of Neural Transmitter Substances.* Liverpool: Liverpool University Press.

Katz, B. and Miledi, R. (1965*a*). Propagation of electrical activity in motor nerve terminals. *Proc. R. Soc. Lond.* B **161**, 453–82.

Katz, B. and Miledi, R. (1965*b*). The measurement of synaptic delay, and the time course of acetylcholine release at the neuromuscular junction. *Proc. R. Soc. Lond.* B **161**, 483–95.

Katz, B. and Miledi, R. (1966). Input/output relation of a single synapse. *Nature* **212**, 1242–5.

Katz, B. and Miledi, R. (1967*a*). Tetrodotoxin and neuromuscular transmission. *Proc. R. Soc. Lond.* B **167**, 8–22.

Katz, B. and Miledi, R. (1967*b*). The timing of calcium action during neuromuscular transmission. *J. Physiol.* **189**, 535–44.

Katz, B. and Miledi, R. (1968). The role of calcium in neuromuscular facilitation. *J. Physiol.* **195**, 481–92.

Katz, R. and Miledi, R. (1970). Membrane noise produced by acetylcholine. *Nature* **226**, 962–3.

Katz, R. and Miledi, R. (1972). The statistical nature of the acetylcholine potential and its molecular components. *J. Physiol.* **224**, 665–700.

Katz, B. and Miledi, R. (1977). Transmitter leakage from motor nerve endings. *Proc. R. Soc. Lond.* B **196**, 59–72.

Katz, B. and Miledi, R. (1981). Does the motor nerve impulse evoke 'non-quantal' transmitter release? *Proc. R. Soc. Lond.* B **212**, 131–7.

Katz, B. and Thes!eff, S. (1957*a*). A study of the 'desensitization' produced by acetylcholine at the motor endplate. *J. Physiol.* **138**, 63–80.

Katz, B. and Thesleff, S. (1957*b*). On the factors which determine the amplitude of the 'miniature end-plate potential'. *J. Physiol.* **137**, 267–78.

Katzung, B. G. (1987). *Basic and Clinical Pharmacology*, 3rd edn. Norwalk, Connecticut: Appleton and Lange.

Kawakami, K., Noguchi, S., Noda, M., Takahashi, H., Ohta, T., Kawamura, M., Nojima, H., Nagano, K., Hirose, T., Inayama, S., Hayashida, H., Miyata, T. and Numa, S. (1985). Primary structure of the α subunit of *Torpedo californica* ($Na^+ + K^+$) ATPase deduced from cDNA sequence. *Nature* **316**, 733–6.

Kebabian, J. W. and Calne, D. B. (1979). Multiple receptors for dopamine. *Nature* **277**, 93–6.

Keeton, W. T. (1981). The orientation and navigation of birds. In *Animal Migration* (ed. D. J. Aidley), pp. 81–104. Cambridge: Cambridge University Press.

Kelly, J. S. and Rogawski, M. A. (1985). Acetylcholine. In *Neurotransmitter Actions in the Vertebrate Nervous System* (ed. M. A. Rogawski and J. L. Barker), pp. 143–97. New York: Plenum Press.

Kemp, D. T. (1978). Stimulated acoustic emissions from

within the human auditory system. *J. Acoust. Soc. Am.* **64**, 1386–91.

Kendrick-Jones, J., Lehman, W. and Szent-Györgyi, A. G. (1970). Regulation in molluscan muscles. *J. Mol. Biol.* **54**, 313–26.

Kerr, L. M. and Yoshikami, D. (1984). A venom peptide with a novel presynaptic blocking action. *Nature* **308**, 282–4.

Keynes, R. D. (1951). The ionic movements during nervous activity. *J. Physiol.*, **114**, 119.

Keynes, R. D. (1963). Chloride in the squid giant axon. *J. Physiol.* **169**, 690–705.

Keynes, R. D. (1983). Voltage-gated ion channels in the nerve membrane. *Proc. R. Soc. Lond.* B **220**, 1–30.

Keynes, R. D. and Aidley, D. J. (1981). *Nerve and Muscle.* Cambridge: Cambridge University Press.

Keynes, R. D., Bennett, M. V. L. and Grundfest, H. (1961). Studies on the morphology and electrophysiology of electric organs. II. Electrophysiology of the electric organ of *Malapterurus electricus.* In *Bioelectrogenesis* (ed. C. Chagas and A. Paes de Carvalho), pp. 102–12. Amsterdam: Elsevier.

Keynes, R. D. and Lewis, P. R. (1951). The sodium and potassium content of cephalopod nerve fibres. *J. Physiol.* **114**, 151–82.

Keynes, R. D. and Martins-Ferreira, H. (1953). Membrane potentials in the electroplates of the electric eel. *J. Physiol.* **119**, 315–51.

Keynes, R. D. and Ritchie, J. M. (1984). On the binding of labelled saxitoxin to the squid giant axon. *Proc. R. Soc. Lond.* B **222**, 147–53.

Keynes, R. D. and Rojas, E. (1974). Kinetics and steady-state properties of the charged system controlling sodium conductance in the squid giant axon. *J. Physiol.* **239**, 393–434.

Kiang, N. Y.-S. (1965). *Discharge Patterns of Single Fibers in the Cat's Auditory Nerve.* Cambridge, Massachusetts: MIT Press.

Kiang, N. Y.-S., Liberman, M. C., Gage, J. S., Northrop, C. C., Dodds, L. W. and Oliver, M. E. (1984). Afferent innervation of the mammalian cochlea. In *Comparative Physiology of Sensory Systems* (ed. L. Bolis *et al.*), pp. 143–61. Cambridge: Cambridge University Press.

Kiang, N. Y.-S., Moxon, E. C. and Levine, R. A. (1970). Auditory nerve activity in cats with normal and abnormal cochleas. In *Sensorineural Hearing Loss* (ed. G. E. W. Wolstenholme and J. Knight), pp. 241–68. London: Churchill.

Kimura, H., McGeer, P. L., Peng, F. and McGeer, E. G. (1980). Choline acetyl-transferase-containing neurons in rodent brain demonstrated by immunohistochemistry. *Science* **208**, 1057–9.

Klein, R. L. and Lagercrantz, H. (1981). Noradrenergic vesicles: composition and function. In *Chemical Neurotransmission: 75 Years* (ed. L. Stjärne et al.), pp. 69–83.

Knappeis, G. G. and Carlsen, F. (1962). The ultrastructure of the Z disc in skeletal muscle. *J. Cell. Biol.* **13**, 323–35.

Knappeis, G. G.and Carlsen, F. (1968). The ultrastructure of the M line in skeletal muscle. *J. Cell Biol.* **38**, 202–11.

Knight, P. J., Erickson, M. A., Rodgers, M. E., Beer, M. and Wiggins, J. W. (1986). Distribution of mass within native thick filaments of vertebrate skeletal muscle. *J. Mol. Biol.* **189**, 167–77.

Knowles, A. and Dartnall, H. J. A. (1977). *The Photobiology of Vision.* Vol. 2B of *The Eye* (2nd edn, ed. H. Davson). New York: Academic Press.

Kobayashi, H. and Libet, B. (1970). Actions of noradrenaline and acetylcholine on sympathetic ganglion cells. *J. Physiol.* **208**, 353–72.

Kobilka, B. K., Dixon, R. A. F., Frielle, T., Dohlman, H. G., Bolanowski, M. A., Sigal, I. S., Yang-Feng, T. L., Francke, U., Caron, M. E. and Lefkowitz, R. J. (1987). cDNA for the human β_2-adrenergic receptor: a protein with multiple membrane-spanning domains and encoded by a gene whose chromosomal location is shared with that of the receptor for platelet-derived growth-factor. *Proc. Natl. Acad. Sci. USA* **84**, 46–50.

Koelle, G. B. and Friedenwald, J. S. (1949). A histochemical method for localising cholinesterase activity. *Proc. Soc. exp. Biol. Med.* **70**, 617–22.

Kordas, M. (1969). The effect of membrane polarization on the time course of the end-plate current in frog sartorius muscle. *J. Physiol.* **204**, 493–502.

Korenbrot, J. I. and Cone, R. A. (1972). Dark ionic flux and effects of light in isolated rod outer segments. *J. gen. Physiol.* **60**, 20–45.

Korn, E. D. (1982). Actin polymerization and its regulation by proteins from nonmuscle cells. *Physiol. Rev.* **62**, 672–737.

Kravitz, E. A., Kuffler, S. W. and Potter, D. D. (1963). Gamma-aminobutyric acid and other blocking compounds in Crustacea. III. Their relative concentrations in separated motor and inhibitory axons. *J. Neurophysiol.* **26**, 739–51.

Kress, M., Huxley, H. E., Faruqi, A. R. and Hendrix, J. (1986). Structural changes during activation of frog muscle studied by time-resolved X-ray diffraction. *J. Mol. Biol.* **188**, 325–42.

Kretschmar, K. M. and Wilkie, D. R. (1969). A new approach to freezing tissues rapidly. *J. Physiol.* **202**, 66–7*P*.

Kreuger, B. K., Worley, J. F. and French, R. J. (1983). Single sodium channels from rat brain incorporated into planar lipid bilayers. *Nature* **303**, 172–5.

Krnjević, K. (1986). Amino acid transmitters: 30 years' progress in research. In *Fast and Slow Chemical Signalling in the Nervous System* (ed. L. L. Iversen and E. C. Goodman), pp. 3–15, Oxford: University Press.

Krnjević, K. and Mitchell, J. F. (1961). The release of acetylcholine in the isolated rat diaphragm. *J. Physiol.* **155**, 246–62.

Krnjević, K. and Phillis, J. W. (1963*a*). Iontophoretic studies of neurones in the mammalian cerebral cortex. *J. Physiol.* **165**, 274–304.

Krnjević, K. and Phillis, J. W. (1963*b*). Acetylcholine-sensitive cells in the cerebral cortex. *J. Physiol.* **166**, 296–327.

Krnjević, K. and Schwarz, S. (1967). The action of gamma-aminobutyric acid on cortical neurones. *Exp. Brain Res.* **3**, 320–36.

Kubo, T., Fukuda, K., Mikami, A., Maeda, A., Takahashi, H., Mishina, M., Haga, T., Haga, K., Ichiyama, A., Kangawa, K., Kojima, M., Matsuo, H., Hirose, T. and Numa, S. (1986*a*). Cloning, sequencing and expression of complementary DNA encoding the muscarinic acetylcholine receptor. *Nature* **323**, 411–16.

Kubo, T., Maeda, A., Sugimoto, K., Akiba, I., Mikami, A., Takahashi, H., Haga, T., Haga, K., Ichiyama, A., Kangawa, K., Matsuo, H., Hirose, T. and Numa, S. (1986*b*). Primary structure of porcine cardiac muscarinic acetylcholine receptor deduced from the cDNA sequence. *FEBS Letters* **209**, 367–72.

Kuffler, S. W. (1942). Electrical potential changes in an isolated nerve-muscle junction. *J. Neurophysiol.* **5**, 18–26.

Kuffler, S. W. (1946). The relation of electrical potential changes to contracture in skeletal muscle. *J. Neurophysiol.* **9**, 367–77.

Kuffler, S. W. (1953). Discharge patterns and functional organization of mammalian retina. *J. Neurophysiol.* **16**, 37–68.

Kuffler, S. W. (1980). Slow synaptic responses in the autonomic ganglia and the pursuit of a peptidergic transmitter. *J. exp. Biol.* **89**, 257–86.

Kuffler, S. W. and Edwards, C. (1958). Mechanism of gamma-aminobutyric acid (GABA) action and its relation to synaptic inhibition. *J. Neurophysiol.* **21**, 589–610.

Kuffler, S. W., Hunt, C. C. and Quilliam, J. P. (1951). Function of medullated small-nerve fibres in mammalian ventral roots: efferent muscle spindle innervation. *J. Neurophysiol.* **14**, 29–54.

Kuffler, S. W., Nicholls, J. G. and Martin, A. R. (1984). *From Neuron to Brain.* 2nd edn. Sunderland, Massachusetts: Sinauer.

Kuffler, S. W. and Vaughan Williams, E. M. (1953*a*). Small-nerve junctional potentials. The distribution of small motor nerves to frog skeletal muscle, and the membrane characteristics of the fibres they innervate. *J. Physiol.* **121**, 289–317.

Kuffler, S. W. and Vaughan Williams, E. M. (1953*b*). Properties of the 'slow' skeletal muscle fibres of the frog. *J. Physiol.* **121**, 318–40.

Kuffler, S. W. and Yoshikami, D. (1975*a*). The distribution of acetylcholine sensitivity at the post-synaptic membrane of vertebrate skeletal twitch muscles: iontophoretic mapping in the micron range. *J. Physiol.* **244**, 703–30.

Kuffler, S. W. and Yoshikami, D. (1975*b*). The number of transmitter molecules in a quantum: an estimate from iontophoretic application of acetylcholine at the neuromuscular synapse. *J. Physiol.* **251**, 465–82.

Kuhar, M. J., De Souza, E. B. and Unnerstall, J. R. (1986). Neurotransmitter receptor mapping by autoradiography and other methods. *Ann. Rev. Neurosci.* **9**, 27–59.

Kühn, H. (1986). Proteins involved in the control of cyclic GMP phosphodiesterase in retinal rod cells. In *Membrane Control of Cellular Activity* (ed. H. C. Lüttgau), pp. 289–97. Stuttgart: Gustav Fischer Verlag.

Kühne, W. (1864). *Untersuchungen über das Protoplasma und die Contractilität.* Leipzig: W. Engelmann.

Kühne, W. (1878). *On the Photochemistry of the Retina and on Visual Purple.* (Edited with notes by M. Foster). London: Macmillan.

Kuno, M. and Rudomin, P. (1966). The release of acetylcholine from the spinal cord of the cat by antidromic stimulation of motor nerves. *J. Physiol.* **187**, 177–93.

Kushmerick, M. J. (1983). Energetics of muscle contraction. In *Handbook of Physiology*, section 10, *Skeletal Muscle* (ed. L. D. Peachey). Bethesda, Maryland: American Physiological Society.

Kushmerick, M. J. and Davies, R. E. (1969). The chemical energetics of muscular contraction II. The chemistry, efficiency and power of maximally working sartorius muscles. *Proc. R. Soc. Lond.* B **174**, 315–53.

Kyte, J. and Doolittle, R. F. (1982). A simple method for displaying the hydropathic character of a protein. *J. Mol. Biol.* **157**, 105–32.

Lai, F. A., Erickson, H. P., Rousseau, E., Liu, Q.-Y. and Meissner, G. (1988). Purification and reconstitution of the calcium release channel from skeletal muscle. *Nature* **331**, 315–19.

Lamb, T. D. (1986). Transduction in vertebrate photoreceptors: the roles of cyclic GMP and calcium. *Trends Neurosci* **9**, 224–8.

Lamvic, M. K. (1978). Muscle thick filament mass measured by electron scattering. *J. Mol. Biol.* **122**, 55–68.

Langer, S. Z. (1974). Presynaptic regulation of catecholamine release. *Biochem. Pharmacol.* **23**, 1793–800.

Langer, S. Z. (1981). Presynaptic regulation of the release of catecholamines. *Pharmacol. Rev.* **32**, 337–62.

Lännergren, J. (1979). An intermediate type of muscle fibre in *Xenopus laevis. Nature* **279**, 254–6.

Lanzavecchia, G. (1968). Studi sulla muscolatura elicoidale e paramiosinica. II. Meccanismo di contrazione dei muscoli elicoidale. *Atti Accad. Naz. Lincei* **44**, 575–83.

Lapique, L. (1907). *J. Physiol., Path. gén.* **9**, 622.

Latorre, R., Coronado, R. and Vergara, C. (1984). K^+ channels gated by voltage and ions. *Ann. Rev. Physiol.* **46**, 485–95.

Latorre, R. and Miller, C. (1983). Conduction and selectivity in potassium channels. *J. Membrane Biol.* **71**, 11–30.

Lawrence, T. S., Beers, W. H. and Gilula, N. B. (1978). Transmission of hormonal stimulation by cell-to-cell communication. *Nature* **272**, 501–6.

Lazarides, E. (1980). Intermediate filaments as mechanical integrators of cellular space. *Nature* **283**, 249–56.

Lazarides, E. and Hubbard, B. D. (1976). Immunological characterization of the subunit of the 100 Å filaments from muscle cells. *Proc. Natl. Acad. Sci. USA* **73**, 4344–8.

Lee, C. Y. (1970). Elapid neurotoxins and their mode of action. *Clinical Toxicol.* **3**, 457–72.

Lehman, W. (1976). Phylogenetic diversity of the proteins regulating muscular contraction. *Int. Rev. Cytol.* **44**, 55–92.

Leksell, L. (1945). The action potential and excitatory effects of the small ventral root fibres to skeletal muscle. *Acta physiol. scand. Suppl.* **31**, 1–84.

Lennon, V. A., McCormick, D. J., Lambert, E. H., Griesmann, G. E. and Atassi, M. A. (1985). Region of peptide 125–147 of acetylcholine receptor α subunit is exposed at neuromuscular junction and induces experimental autoimmune myasthenia gravis, T-cell immunity, and modulating autoantibodies. *Proc. Natl. Acad. Sci. USA* **82**, 8805–9.

Levin, A. and Wyman, J. (1927). The viscous elastic properties of muscle. *Proc. R. Soc. Lond.* B **101**, 218–43.

Levitzki, A. (1986). β-adrenegic receptors and their mode of coupling to adenylate cyclase. *Physiol. Rev.* **66**, 819–54.

Lewis, C. A. (1979). Ion-concentration dependence of the reversal potential and the single-channel conductance of ion channels at the frog neuromuscular junction. *J. Physiol.* **286**, 417–45.

Liberman, M. C. (1982). Single-neuron labelling in the cat auditory nerve. *Science* **216**, 1239–41.

Liebman, P. A. (1962). *In situ* microspectrophotometric study of the pigments of single retinal rods. *Biophys. J.* **2**, 161–78.

Liebman, P. A. (1972). Microspectrophotometry of photoreceptors. In *Handbook of Sensory Physiology*, vol. VII/1, *Photochemistry of Vision* (ed. H. J. A. Dartnall), pp. 481–528. Berlin: Springer.

Liley, A. W. (1956). The effects of presynaptic polarization on the spontaneous activity at the mammalian neuromuscular junction. *J. Physiol.* **134**, 427–43.

Lindauer, M. and Martin, H. (1972). Magnetic effect in dancing bees. In *Animal Orientation and Navigation* (ed. S. R. Galler, K. Schmidt-Koenig, G. J. Jacobs and R. E. Belleville), pp. 559–67. Washington, DC: National Aeronautics and Space Administration.

Linder, T. M. and Quastel, D. M. J. (1978). A voltage-clamp study of the permeability change induced by quanta of transmitter at the mouse end-plate. *J. Physiol.* **281**, 535–56.

Lindstrom, J., Criado, M., Hoschschwender, S., Fox, J. L. and Sarin, V. (1984). Immunochemical tests of acetylcholine receptor subunit models. *Nature* **311**, 573–5.

Lindstrom, J., Merlie, J. and Yogeeswaran, G. (1979). Biochemical properties of acetylcholine receptor subunits from *Torpedo californica. Biochemistry* **18**, 4465–70.

Ling, G. and Gerard, R. W. (1949). The normal membrane potential of frog sartorius fibers. *J. Cell. Comp. Physiol.* **34**, 383–96.

Lipmann, F. (1941). Metabolic generation and utilization of phosphate bond energy. *Adv. Enzymol.* **1**, 99–162.

Lissmann, H. W. (1951). Continuous electrical signals from the tail of a fish, *Gymnarchus niloticus* Cuv. *Nature* **167**, 201.

Lissmann, H. W. (1958). On the function and evolution of electric organs in fish. *J. exp. Biol.* **35**, 156–91.

Lissmann, H. W. and Machin, K. E. (1958). The mechanism of object location in *Gymnarchus niloticus* and similar fish. *J. exp. Biol.* **35**, 451–86.

Livingston, R. B. (1959). Central control of receptors and sensory transmission systems. In *Handbook of Physiology* section 1, vol. 1, pp. 741–60. Washington, DC: American Physiological Society.

Llinas, R., Steinberg, I. Z. and Walton, K. (1981*a*). Presynaptic calcium currents in squid giant synapse. *Biophys. J.* **33**, 289–322.

Llinas, R., Steinberg, I. Z. and Walton, K. (1981*b*). Relationship between presynaptic calcium current and postsynaptic potential in squid giant synapse. *Biophys. J.* **33**, 323–52.

Lloyd, D. P. C. (1943). Neuron patterns controlling transmission of ipsilateral hind limb reflexes in cat. *J. Neurophysiol.* **6**, 293–315.

Lloyd, D. P. C. (1949). Post-tetanic potentiation of response in monosynaptic reflex pathways of the spinal cord. *J. gen. Physiol.* **33**, 147–70.

Loewenstein, W. R. (1956). Modulation of cutaneous mechanoreceptors by sympathetic stimulation. *J. Physiol.* **132**, 40–60.

Loewenstein, W. R. (1981). Junctional intercellular communication: the cell-to-cell membrane channel. *Physiol. Rev.* **61**, 829–913.

Loewi, O. (1921). Über humorale Übertragbarkeit der Herznervenwirkung. *Pflügers Arch. ges. Physiol.* **189**, 239–42.

Lohmann, K. (1929). Über die Pyrophosphatfraktion im Muskel. *Naturwissensch.* **17**, 624–5.

Lohmann, K. (1934). Über die enzymatische Aufspaltung der Kreatinphosphorsäure: zugleich ein Beitrag zum Chemismus der Muskelkontraktion. *Biochem. Z.* **271**, 264–77.

Longenecker, H. E., Hurlbut, W. P., Mauro, A. and Clark, A. W. (1970). Effects of black widow spider venom on the frog neuromuscular junction. *Nature* **225**, 701–5.

Lowenstein, O., Osborne, P. and Wersäll, J. (1964). Structure and innervation of the sensory epithelia in the labyrinth of the thornback ray (*Raja clavata*). *Proc. R. Soc. Lond.* B **160**, 1–12.

Lowenstein, O. and Roberts, T. D. M. (1950). The equilibrium function of the otolith organs of the thornback ray (*Raja clavata*). *J. Physiol.* **110**, 392–415.

Lowenstein, O. and Roberts, T. D. M. (1951). The localization and analysis of the responses to vibration from the isolated elasmobranch labyrinth. A contribution to the problem of the evolution of hearing in vertebrates. *J. Physiol.* **114**, 471–89.

Lowenstein, O. and Sand, A. (1936). The activity of the horizontal semicircular canal of the dogfish, *Scyllium canicula. J. exp. Biol.* **13**, 416–28.

Lowenstein, O. and Sand, A. (1940*a*). The mechanism of the semicircular canal. A study of the responses of single-fibre preparations to angular accelerations and to rotation at constant speed. *Proc. R. Soc. Lond.* B **129**, 256–75.

Lowenstein, O. and Sand, A. (1940*b*). The individual and integrated activity of the semicircular canals of the elasmobranch labyrinth. *J. Physiol.* **99**, 89–101.

Lowenstein, O. E. and Wersäll, J. (1959). A functional interpretation of the electronmicroscopic structure of the sensory hairs in the cristae of the elasmobranch *Raja clavata* in terms of directional sensitivity. *Nature* **184**, 1807.

Lowey, S. (1986). The structure of vertebrate muscle myosin. In *Myology* (ed. A. G. Engel and B. Q. Banker), pp. 563–87. New York: McGraw-Hill.

Lowey, S. and Risby, D. (1971). Light chains from fast and slow muscle myosins. *Nature* **234**, 81–5.

Lowey, S., Slayter, H. S., Weeds, A. G. and Baker, H. (1969). Substructure of the myosin molecule. I. Subfragments of myosin by enzymic degradation. *J. Mol. Biol.* **42**, 1–29.

Lowy, J. and Hanson, J. (1962). Ultrastructure of invertebrate smooth muscle. *Physiol. Rev.* **42**, suppl. 5, 34–42.

Lowy, J. and Millman, B. M. (1963). The contractile mechanism of the anterior byssus retractor muscle of *Mytilus edulis*. *Phil. Trans. R. Soc. Lond.* B **246**, 105–48.

Lowy, J., Poulson, F. R. and Vibert, P. J. (1970). Myosin filaments in vertebrate smooth muscle. *Nature* **225**, 1053–4.

Ludwigh, E. and McCarthy, E. F. (1938). Absorption of visible light by the refractive media of the human eye. *Archs. Ophthal., NY* **20**, 37–51.

Lummis, S. C. R. and Satelle, D. B. (1985). Binding of N-[propionyl-^{3}H] propionylated α-bungarotoxin and L-[benzilic-4, 4^1-^{3}H] quinuclidinyl benzilate to CNS extracts of the cockroach *Periplaneta americana*. *Comp. Biochem. Physiol.* **80C**, 75–83.

Lundberg, A. and Quilisch, H. (1953). On the effect of calcium on presynaptic potentiation and depression at the neuromuscular junction. *Acta physiol. scand. suppl.* **111**,121–9.

Lundsgaard, E. (1930*a*). Untersuchungen über Muskelkontraktionen ohne Milchsäurebildung. *Biochem. Z.* **217**, 162–77.

Lundsgaard, E. (1930*b*). Weitere Untersuchungen über Muskelkontraktionen ohne Milchsäurebildung. *Biochem. Z.* **227**, 51–83.

Luther, P. K. and Squire, J. M. (1978). Three-dimensional structure of the vertebrate muscle M-region. *J. Mol. Biol.* **125**, 313–24.

Lux, H. D. and Brown, A. M. (1984). Single channel studies on inactivation of calcium currents. *Science* **225**, 432–4.

Lux, H. D., Neher, E. and Marty, A. (1981). Single channel activity associated with the Ca^{2+}-dependent outward current in *H. pomatia*. *Pflügers Archiv* **389**, 293–5.

Lymn, R. W. and Taylor, E. W. (1970). Transient state phosphate production in the hydrolysis of nucleoside triphosphates by myosin. *Biochem.* **9**, 2975–83.

Lymn, R. W. and Taylor, E. W. (1971). Mechanism of adenosine triphosphate hydrolysis by actomyosin. *Biochem.* **10**, 4617–24.

Lynch, D. R. and Snyder, S. H. (1986). Neuropeptides: multiple molecular forms, metabolic pathways, and receptors. *Ann. Rev. Biochem.* **55**, 773–99.

Ma, W.-C. (1972). Dynamics of feeding responses in *Pieris brassicae* (Linn.) as a function of chemosensory input: a behavioural, ultrastructural and electrophysiological study. *Meded. Land. Wageningen* **72**, 1–162.

MacDermott, A. B., Mayer, M. L., Westbrook, M. L., Smith, S. J. and Barker, J. L. (1986). NMDA-receptor activation increases cytoplasmic calcium concentration in cultured spinal cord neurones. *Nature* **321**, 519–22.

MacDonald, J. F., Porietis, A. V. and Wojtowicz, J. M. (1982). L-Aspartic acid induces a region of negative slope conductance in the current-voltage relationship of cultured spinal neurons. *Brain Res.* **237**, 248–53.

MacDonald, R. and Barker, J. L. (1978). Benzodiazepines specifically modulate GABA-mediated postsynaptic inhibition in cultured mammalian neurones. *Nature* **271**, 563–4.

Machin, K. E. (1962). Electric receptors. *Symp. Soc. exp. Biol.* **16**, 227–44.

Machin, K. E. and Pringle, J. W. S. (1959). The physiology of insect fibrillar muscle. II. Mechanical properties of a beetle flight muscle. *Proc. R. Soc. Lond.* B **151**, 204–25.

Machin, K. E. and Pringle, J. W. S. (1960). The physiology of insect fibrillar muscle. III. The effect of sinusoidal changes of length on a beetle flight muscle. *Proc. R. Soc. Lond.* B **152**, 311–30.

Machin, K. E., Pringle, J. W. S. and Tamasige, M. (1962). The physiology of insect fibrillar muscle. IV. The effect of temperature on a beetle flight muscle. *Proc. R. Soc. Lond.* B **155**, 493–9.

Maclennan, D. H., Brandl, C. J., Korczak, B. and Green, N. M. (1985). Amino-acid sequence of a Ca^{2+} + Mg^{2+}-dependent ATPase from rabbit muscle sarcoplasmic reticulum, deduced from its complementary DNA sequence. *Nature* **316**, 696–700.

Maclennan, D. H. and Wong, P. T. S. (1971). Isolation of a calcium-sequestering protein from sarcoplasmic reticulum. *Proc. Natl. Acad. Sci. USA* **68**, 1231–5.

MacNichol, E. F. and Svaetichin, G. (1958). Electric responses from isolated retinas of fishes. *Am. J. Ophthal.* **46**, 26–40.

MacVicar, B. and Dudek, F. E. (1981). Electrotonic coupling between pyramidal cells: a direct demonstration in rat hippocampal slices. *Science* **213**, 782–4.

Magleby, K. L. and Stevens, C. F. (1972). The effect of voltage on the time course of end-plate currents. *J. Physiol.* **223**, 151–71.

Magleby, K. L. and Zengel, J. E. (1982). A quantitative description of stimulation-induced changes in transmitter release at the frog neuromuscular junction. *J. gen. Physiol.* **80**, 613–38.

Mahmood, R. and Yount, R. G. (1984). Photochemical probes of the active site of myosin. *J. Biol. Chem.* **259**, 12 956–9.

Makowski, L. (1985). Structural domains in gap junctions: implications for the control of cellular communication. In *Gap Junctions* (ed. M. V. L. Bennett and D. C. Spray). New York: Cold Spring Harbor Laboratory.
Makowski, L., Caspar, D. L. D., Phillips, W. C., Baker, T. S. and Goodenough, D. A. (1984). Gap junction structures. VI. Variation and conservation in connexon conformation and packing. *Biophys. J.* **45**, 208–18.
Makowski, L., Caspar, D. L. D., Phillips, W. C. and Goodenough, D. A. (1977). Gap junction structures II. Analysis of the X-ray diffraction data. *J. Cell Biol.* **74**, 629–45.
Mamalaki, C., Stephenson, F. A. and Barnard, E. A. (1987). The $GABA_A$/benzodiazepine receptor is a heterotetramer of homologous α and β subunits. *EMBO J.* **6**, 561–5.
Marks, W. B. (1965). Visual pigments of single goldfish cones. *J. Physiol.* **178**, 14–32.
Marks, W. B., Dobelle, W. H. and MacNichol, E. F. (1964). Visual pigments of single primate cones. *Science* **143**, 1181–3.
Marmont, G. (1949). Studies on the axon membrane: I. A new method. *J. Cell Comp. Physiol.* **34**, 351–82.
Marnay, A. and Nachmansohn, D. (1938). Choline esterase in voluntary muscle. *J. Physiol.* **92**, 37–47.
Marsh, B. B. (1952). The effects of ATP on the fibre-volume of a muscle homogenate. *Biochim. biophys. Acta* **9**, 247–60.
Marsh, D. (1975). Spectroscopic studies of membrane structure. *Essays in Biochemistry* **11**, 139–80.
Marston, S. B. (1982). The regulation of smooth muscle contractile proteins. *Prog. Biophys. Molec. Biol.* **41**, 1–41.
Marston, S. B. and Smith, C. W. J. (1985). The thin filaments of smooth muscles. *J. Mus. Res. Cell Motil.* **6**, 669–708.
Martin, A. R. (1955). A further study of the statistical composition of the end-plate potential. *J. Physiol.* **130**, 114–22.
Martonosi, A. N. and Beeler, T. J. (1983). Mechanism of Ca^{2+} transport by sarcoplasmic reticulum. In *Handbook of Physiology*, section 10 *Skeletal Muscle* (ed. L. D. Peachey), pp. 417–85. Bethesda, Maryland: American Physiological Society.
Maruyama, K. (1975). A new protein-factor hindering network formation of F-actin in solution. *Biochim. biophys. Acta* **94**, 208–25.
Maruyama, K. (1986). Connectin, an elastic filamentous protein of striated muscle. *Int. Rev. Cytol.* **104**, 81–114.
Maruyama, K., Kimura, S., Ishii, T., Kuroda, M., Ohashi, K. and Muramatsu, S. (1977*a*). β-actinin, a regulatory protein of muscle. *J. Biochem.* **81**, 215–32.
Maruyama, K., Matsubara, S., Natori, S. R., Nonomura, Y., Kimura, S., Ohashi, K., Murakami, F., Handa, S. and Eguchi, G. (1977*b*), Connectin, an elastic protein in muscle. *J. Biochem.* **82**, 317–37.
Masaki, T., Endo, M. and Ebashi, S. (1967). Localization of 6S component of α-actinin at Z-band. *J. Biochem.* **62**, 630–1.
Mason, R. T., Peterfreund, R. A., Sawchenko, P. E., Corrigan, A. Z., Rivier, J. E. and Vale, W. W. (1984). Release of the predicted calcitonin gene-related peptide from cultured rat trigeminal ganglion cells. *Nature* **308**, 653–5.
Matthews, B. H. C. (1933). Nerve endings in mammalian muscle. *J. Physiol.* **78**, 1–53.
Matthews, G. and Watanabe, S.-I. (1987). Properties of ion channels closed by light and opened by guanosine 3′, 5′-cyclic monophosphate in toad retinal rods. *J. Physiol.* **389**, 691–715.
Matthews, H. R., Torre, V. and Lamb, T. D. (1985). Effects on the photoresponse of calcium buffers and cyclic GMP incorporated into the cytoplasm of retinal rods. *Nature* **313**, 582–5.
Matthews, P. B. C. (1962). The differentiation of two types of fusimotor fibre by their effects on the dynamic response of muscle spindle primary endings. *Q. Jl exp. Physiol.* **47**, 324–33.
Matthews, P. B. C. (1964). Muscle spindles and their motor control. *Physiol. Rev.* **44**, 219–88.
Matthews, P. B. C. (1981). Evolving views on the internal operation and functional role of the muscle spindle. *J. Physiol.* **320**, 1–20.
Maxam, A. M. and Gilbert, W. (1980). Sequencing end-labelled DNA with base-specific chemical cleavages. *Methods in Enzymology* **65**, 499–560.
May, D. C., Ross, E. M., Gilman, A. G. and Smigel, M. D. (1985). Reconstitution of catecholamine-stimulated adenylate cyclase activity using three purified proteins. *J. Biol. Chem.* **260**, 15 829–33.
McAfee, D. A. and Henon, B. K. (1985). Adenosine and ATP. In *Neurotransmitter Actions in the Vertebrate Nervous System* (eds. M. A. Rogawski and J. L. Barker), pp. 481–502. New York: Plenum Press.
McAllister, R. E., Noble, D. and Tsien, R. W. (1975). Reconstruction of the electrical activity of cardiac Purkinje fibres. *J. Physiol.* **251**, 1–59.
McLachlan, A. D. (1984). Structural implications of the myosin amino acid sequence. *Ann. Rev. Biophys. Bioeng.* **13**, 167–89.
McLachlan, A. D. and Karn, J. (1982). Periodic charge distributions in the myosin rod amino acid sequence match cross-bridge spacings in muscle. *Nature* **299**, 226–31.
McLachlan, A. D. and Stewart, M. (1976). The 14-fold periodity in α-tropomyosin and the interaction with actin. *J. Mol. Biol.* **103**, 271–98.
McLennan, H. (1963). *Synaptic Transmission.* Philadelphia; Saunders.
McMahan, U. S., Sanes, J. R. and Marshall, L. M. (1978). Cholinesterase is associated with the basal lamina at the neuromuscular junction. *Nature* **271**, 172–4.
McNaughton, P. A., Cervetto, L. and Nunn, B. J. (1986). Measurement of the intracellular free calcium concentration in salamander rods. *Nature* **322**, 261–3.
McNicol, D. (1972). *A Primer of Signal Detection Theory.* London: Allen and Unwin.
Meech, R. W. (1972). Intracellular calcium injection causes

increased potassium conductance in *Aplysia* nerve cells. *Comp. Biochem. Physiol.* **42A**, 493–9.

Meech, R. W. (1974). The sensitivity of *Helix aspersa* neurones to injected calcium ions. *J. Physiol.* **237**, 259–77.

Meijer, G. M., Ritter, F. J., Persoons, C. J., Minks, A. K. and Voerman, S. (1972). Sex pheromones of summer fruit tortrix moth *Adoxophyes orana*: two synergistic isomers. *Science* **175**, 1469–70.

Melzer, W., Rios, E. and Schneider, M. F. (1986*a*). The removal of myoplasmic free calcium following calcium release in frog skeletal muscle. *J. Physiol.* **372**, 261–92.

Melzer, W., Schneider, M. F., Simon, B. J. and Szucs, G. (1986*b*). Intramembrane charge movement and calcium release in frog skeletal muscle. *J. Physiol.* **373**, 481–511.

Meyerhof, O. and Lohmann, K. (1932). Über energetische Wechselbeziehungen zwischem dem Umstatz der Phosphorsäure ester im Muskelextrakt. *Biochem. Z.* **253**, 431–61.

Mihaly, E. and Szent-Györgyi, A. G. (1953). Trypsin digestion of muscle proteins. III Adenosinetriphosphatase activity and actin-binding capacity of the digested myosin. *J. Biol. Chem.* **201**, 211–19.

Miledi, R. (1973). Transmitter release induced by injection of calcium ions into nerve terminals. *Proc. R. Soc. Lond.* B **183**, 421–5.

Miledi, R., Parker, I. and Zhu, P. H. (1982). Calcium transients evoked by action potentials in frog twitch muscle fibres. *J. Physiol.* **333**, 655–79.

Mill, P. J. and Knapp, M. F. (1970). The fine structure of obliquely striated body wall muscles in the earthworm. *Lumbricus terrestris* Linn. *J. Cell. Sci.* **7**, 233–61.

Miller, C. (1984). Integral membrane channels: studies on model membranes. *Physiol. Rev.* **63**, 1209–42.

Miller, C. (1986). Ion channel reconstitution: why bother? In *Ionic Channels in Cells and Model Systems* (ed. R. Latorre), pp. 257–71. New York: Plenum Press.

Miller, C. (1988). *Shaker* shakes out potassium channels. *Trends Neurosci.* **11**, 185–6.

Miller, J. A., Agnew, W. S. and Levinson, S. R. (1983). Principal glycopeptide of the tetrodotoxin/saxitonin binding protein from *Electrophorus electricus*: isolation and partial chemical and physical characterisation. *Biochemistry* **22**, 462–70.

Miller, R. F. and Dowling, J. E. (1970). Intracellular responses of the Müller (glial) cells of the mudpuppy retina: their relation to the b-wave of the electroretinogram. *J. Neurophysiol.* **33**, 323–41.

Miller, R. J. (1987). Multiple calcium channels and neuronal function. *Science* **235**, 46–52.

Miller, S. G. and Kennedy, M. B. (1986). Regulation of brain type II Ca^{2+}/calmodulin-dependent protein kinase by autophosphorylation: a Ca^{2+}-triggered molecular switch. *Cell* **44**, 861–70.

Miller, T. M. and Heuser, J. E. (1984). Endocytosis of synaptic vesicle membrane at the frog neuromuscular junction. *J. Cell Biol.* **98**, 685–98.

Miller, W. H. and Nicol, G. D. (1979). Evidence that cyclic GMP regulates membrane potential in rod photoreceptors. *Nature* **280**, 64–6.

Mishina, M., Kurosaki, T., Tobimatsu, T., Morimoto, Y., Noda, M., Yamamoto, T., Tereo, M., Lindstrom, J., Takahashi, T., Kuno, M. and Numa, S. (1984). Expression of functional acetylcholine receptor from cloned cDNAs. *Nature.* **307**, 604–8.

Mishina, M., Takai, T., Imoto, K., Noda, M., Takahashi, T., Numa, S., Methfessel, C. and Sakmann, B. (1986). Molecular distinction between fetal and adult forms of muscle acetylcholine receptor. *Nature* **321**, 406–11.

Mishina, M., Tobimatsu, T., Imoto, K., Tanaka, K., Fujita, Y., Fukuda, K., Kurasaki, M., Takahashi, H., Morimoto, Y., Hirose, T., Inayama, S., Takahashi, T., Kuno, M. and Numa, S. (1985). Location of functional regions of acetylcholine receptor α-subunit by site-directed mutagenesis. *Nature* **313**, 364–9.

Mitchell, J. F. and Silver, A. (1963). The spontaneous release of acetylcholine from the denervated hemidiaphragm of the rat. *J. Physiol.* **165**, 117–29.

Mollon, J. D. and Sharpe, L. T. (1983). *Colour Vision. Physiology and Psychophysics.* London: Academic Press.

Moore, B. C. J. (1982). *An Introduction to the Psychology of Hearing*. London: Academic Press.

Moore, P. B., Huxley, H. E. and DeRosier, D. J. (1970). Three-dimensional reconstruction of F-actin, thin filaments and decorated thin filaments. *J. Mol. Biol.* **50**, 279–95.

Morgan, D. L., Prochazka, A. and Proske, U. (1984). The after-effects of stretch and fusimotor stimulation on the responses of the primary endings of cat muscle spindles. *J. Physiol.* **356**, 465–77.

Morgan, J. P. and Blinks, J. R. (1982). Intracellular Ca^{2+} transients in the cat papillary muscle. *Can. J. Physiol. Pharmacol.* **60**, 524–9.

Morgan, J. P. and Morgan, K. P. (1984). Alteration of cytoplasmic ionized calcium levels in smooth muscle by vasodilators in the ferret. *J. Physiol.* **357**, 539–51.

Morita, H. and Shiraishi, A. (1985). Chemoreception physiology. In *Comparative Insect Physiology, Biochemistry and Pharmacology* (ed. G. A. Kerkut and L. I. Gilbert), vol. 6, pp. 133–70. Oxford: Pergamon Press.

Morita, H. and Yamashita, S. (1959). Generator potential of an insect chemoreceptor. *Science* **130**, 922.

Morris, R. G. M., Anderson, E., Lynch, G. S. and Baudry, M. (1986). Selective impairment of learning and blockade of long-term potentiation by an N-methyl-D-aspartate receptor antagonist, AP5. *Nature* **319**, 774–6.

Morton, R. A. (1944). Chemical aspects of the visual process. *Nature* **153**, 69–71.

Morton, R. A. and Goodwin, T. W. (1944) Preparation of retinene in vitro. *Nature* **153**, 405–6.

Morton, R. A. and Pitt, G. A. J. (1955). Studies on rhodopsin 9. pH and the analysis of indicator yellow. *Biochem. J.* **59**, 128–34.

Moskowitz, N. and Puszkin, S. (1985). A unified theory of presyaptic chemical neurotransmission. *J. Theoret. Biol.* **112**, 513–34.

Mullen, E. and Akhtar, M. (1983). Structural studies on membrane-bound bovine rhodopsin. *Biochem. J.* **211**, 45–54.

Müller, J. (1826). *Zur vergleichenden Physiologie des Gesichtssinnes.* Leipzig.

Murray, R. W. (1962). The response of the ampullae of Lorenzini of elasmobranchs to electrical stimulation. *J. exp. Biol.* **39**, 119–28.

Nagai, T., Makinose, M. and Hasselbach, W. (1960). Der physiologische Erschlaffungsfaktor und die Muskelgrana. *Biochim. biophys. Acta* **54**, 338–44.

Nakatani, K. and Yau, K.-W. (1988). Guanosine 3′, 5′-cyclic monophosphate-activated conductance studied in a truncated rod outer segment of the toad. *J. Physiol.* **395**, 731–53.

Narahashi, T. (1963). The properties of insect axons. *Adv. Insect Physiol.* **1**, 176–256.

Narahashi, T. (1974). Chemicals as tools in the study of excitable membranes. *Physiol. Rev.* **54**, 812–89.

Narahashi, T., Moore, J. W. and Scott, W. R. (1964). Tetrodotoxin blockage of sodium conductance increase in lobster giant axons. *J. gen. Physiol.* **47**, 965–74.

Nastuk, W. L. (1953). The electrical activity of the muscle cell membrane at the neuro-muscular junction. *J. cell. comp. Physiol.* **42**, 249–72.

Nastuk, W. L. and Hodgkin, A. L. (1950). The electrical activity of single muscle fibres. *J. cell. comp. Physiol.* **35**, 39.

Nathans, J. (1987). Molecular biology of visual pigments. *Ann. Rev. Neurosci.* **10**, 163–94.

Nathans, J. and Hogness, D. S. (1984). Isolation and nucleotide sequence of the gene encoding human rhodopsin. *Proc. Natl. Acad. Sci. USA* **81**, 4851–5.

Nathans, J., Thomas, D. and Hogness, D. S. (1986*a*). Molecular genetics of human colour vision: the genes encoding blue, green and red pigments. *Science* **232**, 193–202.

Nathans, J., Piantanida, T. P., Eddy, R., Shows, T. B. and Hogness, D. S. (1986*b*). Molecular genetics of inherited variation in human colour vision. *Science* **232**, 203–10.

Nawa, H., Hirose, T., Takashima, H., Inayama, S. and Nakanishi, S. (1983). Nucleotide sequences of cloned cDNAs for two types of bovine brain substance P precursor. *Nature* **306**, 32–6.

Nawa, H., Kotani, H. and Nakanishi, S. (1984). Tissue-specific generation of two preprotachykinin mRNAs from one gene by alternative RNA splicing. *Nature* **312**, 729–34.

Needham, D. M. (1971). *Machina Carnis.* Cambridge: Cambridge University Press.

Neher, E. and Sakmann, B. (1976*a*). Single-channel currents recorded from membrane of denervated frog muscle cells. *Nature* **260**, 799–802.

Neher, E. and Sakmann, B. (1976*b*). Noise analysis of drug induced voltage clamp currents in denervated frog muscle fibres. *J. Physiol.* **258**, 705–29.

Neher, E. and Stevens, C. F. (1977). Conductance fluctuations and ionic pores in membranes. *Ann. Rev. Biophys. Bioeng.* **6**, 345–81.

Nelson, R., Famiglietti, E. V. and Kolb, H. (1978). Intracellular staining reveals different levels of stratification for on- and off-centre ganglian cells in cat retina. *J. Neurophysiol.* **41**, 472–84.

Nemeth, P. M., Pette, D. and Vrbová, G. (1981). Comparison of enzyme activities among single muscle fibres within defined motor units. *J. Physiol.* **311**, 489–95.

Nestler, E. J. and Greengard, P. (1984). *Protein Phosphorylation in the Nervous System.* New York: Wiley.

Neumcke, B. and Stämpfli, R. (1982). Sodium currents and sodium-current fluctuations in rat myelinated nerve fibres. *J. Physiol.* **329**, 163–84.

Neyton, J. and Trautmann, A. (1985). Single-channel currents of an intercellular junction. *Nature* **317**, 331–5.

Neville, A. C. (1963). Motor unit distribution of the dorsal longitudinal flight muscles in locusts. *J. exp. Biol.* **40**, 123–36.

Newberry, N. R. and Nicoll, R. A. (1984). Direct hyperpolarizing action of baclofen on hippocampal pyramidal cells. *Nature* **308**, 450–2.

Nicoll, R. A., Eccles, J. C., Oshima, T. and Rubia, F. (1975). Prolongation of hippocampal inhibitory postsynaptic potentials by barbiturates. *Nature* **258**, 625–7.

Nicolson, G. L. (1976). Transmembrane control of the receptors on normal and tumour cells. I. Cytoplasmic influence over cell surface components. *Biochim. biophys. Acta* **457**, 57-108.

Niedergerke, R. (1956). The potassium chloride contracture of the heart and its modification by calcium. *J. Physiol.* **134**, 584–99.

Nishi, S. and Koketsu, K. (1960). Electrical properties and activities of single sympathetic neurons in frogs. *J. cell. comp. Physiol.* **55**, 15–30.

Nishi, S. and Koketsu, K. (1968). Early and late afterdischarges of amphibian sympathetic ganglion cells. *J. Neurophysiol.* **31**, 109–21.

Nishizuka, Y. (1984). The role of protein kinase C in cell surface signal transduction and tumour promotion. *Nature* **308**, 693–8.

Noble, D. (1962). A modification of the Hodgkin-Huxley equations applicable to Purkinje fibre action and pacemaker potentials. *J. Physiol.* **160**, 317–52.

Noble, D. (1966). Applications of Hodgkin–Huxley equations to excitable tissues. *Physiol. Rev.* **46**, 1–50.

Noble, D. (1984). The surprising heart: a review of recent progress in cardiac electrophysiology. *J. Physiol.* **353**, 1–50.

Noble, D. and Noble, S. J. (1984). A model of sino-atrial node electrical activity based on a modification of the DiFrancesco–Noble (1984) equations. *Proc. R. Soc. Lond.* B **222**, 295–304.

Noble, D. and Powell, T. (eds.) (1987). *Electrophysiology of Single Cardiac Cells.* London: Academic Press.

Noble, D. and Stein, R. B. (1966). The threshold conditions for initiation of action potentials by excitable cells. *J. Physiol.* **187**, 129–62.

Noda, M., Takahashi, H., Tanabe, T., Toyosato, M., Furutani, Y., Hirose, T., Asai, M., Inayama, S., Miyata, T., and Numa, S. (1982*a*). Primary structure of α-subunit precursor of *Torpedo californica* acetylcholine receptor deduced from cDNA sequence. *Nature* **299**, 793–7.

Noda, M., Furutani, Y., Takahashi, H., Toyosato, M., Hirose, T., Inayama, S., Nakanishi, S. and Numa, S. (1982*b*). Cloning and sequence analysis of cDNA for bovine adrenal preproenkephalin. *Nature* **295**, 202–6.

Noda, M., Takahashi, H., Tanabe, T., Toyosato, M., Kikyotani, S., Hirose, T., Asai, M., Takashima, H., Inayama, S., Miyata, T., and Numa, S. (1983*a*). Primary structures of β- and δ-subunit precursors of *Torpedo californica* acetylcholine receptor deduced from cDNA sequences. *Nature* **301**, 251–5.

Noda, M., Takahashi, H., Tanabe, T., Toyosato, M., Kikyotani, S., Furutani, Y., Hirose, T., Takashima, H., Inayama, S., Miyata, T., and Numa, S. (1983*b*). Structural homology of *Torpedo californica* acetylcholine receptor units. *Nature* **302**, 528–32.

Noda, M., Shimizu, S., Tanabe, T., Takai, T., Kayano, T., Ikeda, T., Takahashi, H., Nakayama, H., Kanaoka, Y., Minamino, N., Kangawa, K., Matsuo, H., Raftery, M. A., Hirose, T., Inayama, S., Hayashida, H., Miyati, T. and Numa, S. (1984). Primary structure of *Electrophorus electricus* sodium channel deduced from cDNA sequence. *Nature* **312**, 121–7.

Noda, M., Ikeda, T., Kayano, T., Suzuki, H., Takeshima, H., Kurasaki, M., Takahashi, H. and Numa, S. (1986*a*). Existence of distinct sodium channel messenger RNAs in rat brain. *Nature* **320**, 188–92.

Noda, M., Ikeda, T., Suzuki, H., Takeshima, H., Takahashi, T., Kuno, M. and Numa, S. (1986*b*). Expression of functional sodium channels from cloned cDNA. *Nature* **322**, 826–8.

Nonner, W., Rojas, E. and Stämpfli, R. (1975*a*). Gating currents in the node of Ranvier: voltage and time dependence. *Phil. Trans. R. Soc. Lond.* B **270**, 483–92.

Nonner, W., Rojas, E. and Stämpfli, R. (1975*b*). Displacement currents in the node of Ranvier. *Pflügers Archiv.* **354**, 1–18.

North, R. A. (1986). Opioid receptor types and membrane ion channels. *Trends Neurosci.* **9**, 114–17.

North, R. A. and Williams, J. T. (1985). On the potassium conductance increased by opioids in rat locus coeruleus neurones. *J. Physiol.* **364**, 265–80.

Norton, S. J. and Neely, S. T. (1987). Tone-burst-evoked otoacoustic emissions from normal-hearing subjects. *J. Acoust. Soc. Am.* **81**, 1860–72.

Nowak, L., Bregestovski, P., Ascher, P., Herbert, A. and Prochiantz, A. (1984). Magnesium gates glutamate-activated channels in mouse central neurones. *Nature* **307**, 462–5.

Nowycky, M. C., Fox, A. P. and Tsien, R. W. (1985). Three types of neuronal calcium channel with different calcium agonist sensitivity. *Nature* **316**, 440–3.

Numa, S., Noda, M., Takahashi, H., Tanabe, T., Toyosato, M., Furutani, Y. and Kikyotani, S. (1983). Molecular structure of the nicotinic acetylcholine receptor. *Cold Spr. Harb. Symp. Quant. Biol.* **48**, 57–69.

Offer, G. (1974). The molecular basis of muscular contraction. In *Companion to Biochemistry* (ed. A. T. Bull, J. R. Lagnado, J. O. Thomas and K. F. Tipton), pp. 623–71. London: Longman.

Offer, G., Moos, C. and Starr, R. (1973). A new protein of the thick filaments of vertebrate skeletal myofibrils. *J. Mol. Biol.* **74**, 653–76.

Ohmori, H., Yoshida, S. and Hagiwara, S. (1981). Single K channel currents of anomalous rectification in cultured rat myotubes. *Proc. Natl. Acad. Sci. USA* **78**, 4960–4.

Ohsawa, K., Dowe, G. H. C., Morris, S. J. and Whittaker, V. P. (1979). The lipid and protein content of cholinergic synaptic vesicles from the electric organ of *Torpedo marmorata* purified to constant composition: implications for vesicle structure. *Brain Research* **161**, 447–57.

Ohtsuki, I., Masaki, T., Nonomura, Y. and Ebashi, S. (1967). Periodic distribution of troponin along the thin filament. *J. Biochem.* **61**, 817–19.

Okamoto, Y. and Yount, R. G. (1985). Identification of an active site peptide of skeletal myosin after photo-affinity labelling with N-(4-azido-2-nitrophenol)-2-aninoethyl diphosphate. *Proc. Natl. Acad. Sci. USA* **82**, 1575–9.

Okamoto, Y. and Sekine, T. (1987). A new, smaller actin-activatable myosin subfragment-1 which lacks the 20 kDa SH1 and SH2 peptide. *J. Biol. Chem.* **262**, 7951–4.

Olney, J. W. (1969). Brain lesions, obesity and other disturbances in mice treated with monosodium glutamate. *Science* **164**, 719–21.

Olsen, R. W. (1982). Drug interactions at the GABA receptor-ionophore complex. *Ann. Rev. Pharmacol. Toxicol.* **22**, 245–77.

Oron, Y., Dascal, N., Nadler, E. and Lupu, M. (1985). Inositol 1, 4, 5-trisphosphate mimics muscarinic response in *Xenopus* oocytes. *Nature* **313**, 141–3.

Orrego, F. (1979). Criteria for the identification of central neurotransmitters, and their application to studies with some nerve tissue preparations *in vitro*. *Neuroscience* **4**, 1037–57.

Østerberg, E. (1935). Topography of the layer of rods and cones in the human retina. *Acta ophthal.* suppl **6**.

Otsuka M., Iversen, L. L., Hall, Z. W. and Kravitz, E. A. (1966). Release of gamma-aminobutyric acid from inhibitory nerves of lobster. *Proc. Nat. Acad. Sci. USA* **56**, 1110–15.

Ottoson, D. (1964). The effect of sodium deficiency on the response of the isolated muscle spindle. *J. Physiol.* **171**, 109–18.

Ovalle, W. K. and Smith, R. S. (1972). Histochemical identification of three types of intrafusal muscle fibres in the cat and monkey based on myosin ATPase reaction. *Can. J. Physiol. Pharmacol.* **50**, 195–202.

Ovchinnikov, Y. A. (1982). Rhodopsin and bacteriorhodopsin: structure-function relationships. *FEBS Lett.* **148**, 179–91.

Ovchinnikov, Y. A., Abdulaev, N. G., Feigina, M. Y., Artamonov, I. D., Bogachuk, A. S., Zolotarev, A. S., Eganyan, E. R. and Kostetskii, P. V. (1983). Visual

rhodopsin III. Complete amino acid sequence and topography in the membrane. *Biorg. Khim.* **9**, 1331–40.
Owen, F., Crawley, J., Cross, A. J., Crow, T. J., Oldland, S. R., Poulter, M., Veall, N. and Zanelli, G. D. (1985). Dopamine D_2 receptors and schizophrenia. In *Physchopharmacology: recent advances and future prospects* (ed. S. D. Iversen), pp. 216–27. Oxford University Press.
Owen, W. G. (1987). Ionic conductances in rod photoreceptors. *Ann. Rev. Physiol.* **49**, 743–64.
Pace, U., Hanski, E., Salomon, Y. and Lancet, D. (1985). Odorant-sensitive adenylate cyclase may mediate olfactory reception. *Nature* **316**, 255–8.
Page, S. G. (1964). Filament lengths in resting and excited muscles. *Proc. R. Soc. Lond.* B **160**, 460–6.
Page, S. G. (1965). A comparison of the fine structures of frog slow and twitch muscle fibres. *J. Cell. Biol.* **26**, 477–97.
Page, S. G. (1968). The fine structure of tortoise skeletal muscle. *J. Physiol.* **197**, 709–15.
Page, S. G. and Huxley, H. E. (1963). Filament lengths in striated muscle. *J. Cell. Biol.* **19**, 369–90.
Pak, W. L. (1968). Rapid photoresponses in the retina and their relevance to vision research. *Photochem. Photobiol.* **8**, 495–503.
Palade, G. E. and Palay, S. L. (1954). Electron microscope observations of interneuronal and neuromuscular synapses. *Anat. Rec.* **118**, 335.
Pannbacker, R. G. (1973). Control of guanylate cyclase activity in the rod outer segment. *Science* **182**, 1138–40.
Pannbacker, R. G., Fleischman, D. E. and Reed, D. W. (1972). Cyclic nucleotide phosphodiesterase: high activity in a mammalian photoreceptor. *Science* **175**, 757–8.
Papazian, D. M., Schwarz, T. L., Tempel, B. L., Jan, Y. N. and Jan, L. Y. (1987). Cloning of genomic and complementary DNA from *Shaker*, a putative potassium channel gene from *Drosophila. Science* **237**, 749–53.
Parker, I. and Miledi, R. (1987). Inositol trisphosphate activates a voltage-dependent calcium influx in *Xenopus* oocytes. *Proc. R. Soc. Lond.* B **231**, 27–36.
Parry, D. A. D. (1975). Analysis of the primary sequence of α-tropomyosin from rabbit skeletal muscle. *J. Mol. Biol.* **98**, 519–35.
Parry, D. A. D. and Squire, J. M. (1973). Structural role of tropomysin in muscle regulation: analysis of the X-ray diffraction patterns from relaxed and contracting muscles. *J. Mol. Biol.* **75**, 33–55.
Paton, W. D. M. (1958). Central and synaptic transmission in the nervous system. *Ann. Rev. Physiol.* **20**, 431–70.
Patterson, S. and Starling, E. H. (1914). On the mechanical factors which determine the output of the ventricles. *J. Physiol.* **48**, 357–79.
Paul, R. J. (1983). Physical and biochemical energy balance during an isometric tetanus and steady state recovery in frog sartorius at 0 °C. *J. gen. Physiol.* **81**, 337–54.
Peachey, L. D. (1965). The sarcoplasmic reticulum and transverse tubules of the frog's sartorius. *J. Cell Biol.* **25** (part 2), 209–32.
Peachey, L. D. and Franzini-Armstrong, C. (1983). Structure and function of membrane systems of skeletal muscle cells. In *Handbook of Physiology*, section 10 *Skeletal Muscle*, pp. 23–71. Bethesda, Maryland; America Physiological Society.
Peachey, L. D. and Huxley, A. F. (1962). Structural identification of twitch and slow striated muscle fibres of the frog. *J. Cell Biol.* **13**, 117–80.
Peichl, L. and Wässle, H. (1979). Size, scatter and coverage of ganglian cell receptive field centres in the cat's retina. *J. Physiol.* **291**, 117–41.
Peichl, L. and Wässle, H. (1981). Morphological identification of on- and off-centre brisk transient (Y) cells in the cat retina. *Proc. R. Soc. Lond.* B **212**, 139–56.
Pellegrind, R. G. and Ritchie, J. M. (1984). Sodium channels in the axolemma of normal and degenerating rabbit optic nerve. *Proc. R. Soc. Lond.* B **222**, 155–60.
Penn, R. D. and Hagins, W. A. (1969). Signal transmission along retinal rods and the origin of the electroretinographic *a*-wave. *Nature* **223**, 201–5.
Penn, R. D. and Hagins, W. A. (1972). Kinetics of the photocurrent of retinal rods. *Biophys. J.* **12**, 1073–94.
Pennycuick, C. J. (1964). Frog fast muscle. I. Mechanical power output in isotonic twitches. *J. exp. Biol.* **41**, 91–111.
Peroutka, S. J., Lebowitz, R. M. and Snyder, S. H. (1981). Two distinct central serotonin receptors with different physiological functions. *Science* **212**, 827–8.
Perry, S. V. (1986). Activation of the contractile mechanism by calcium. In *Myology* (ed. A. G. Engel and B. Q. Banker), pp. 613–41. New York: McGraw-Hill.
Pert, C. B. and Snyder, S. H. (1973). Opiate receptor: demonstration in nervous tissue. *Science* **179**, 1011–14.
Peter, J. B., Barnard, R. J., Edgerton, V. R., Gillespie, C. A. and Stempel, K. E. (1972). Metabolic profiles of three fibre types of skeletal muscle in guinea pigs and rabbits. *Biochemistry* **11**, 2627–33.
Peters, A., Palay, S. L. and Webster, H. de F. (1976). *The Fine Structure of the Nervous System.* Philadelphia: Saunders.
Pfaffinger, P. J., Martin, J. M. Hunter, D. D., Nathanson, N. M. and Hille, B. (1985). GPT-binding proteins couple cardiac muscarinic receptors to a K channel. *Nature* **317**, 536–8.
Pickles, J. O. (1985). Recent advances in cochlear physiology. *Prog. Neurobiol.* **24**, 1–42.
Pickles, J. O., Comis, S. D. and Osborne, M. P. (1984). Cross-links between stereocilia in the guinea pig organ of Corti, and their possible relation to sensory transduction. *Hearing Res.* **15**, 103–12.
Pierce, G. W. and Griffin, D. R. (1938). Experimental determination of supersonic notes emitted by bats. *J. Mammal.* **19**, 454–5.
Pirenne, M. H. (1943). Binocular and monocular threshold of vision. *Nature* **152**, 698–9.
Pirenne, M. H. (1944). Rods and cones, and Thomas Young's theory of colour vision. *Nature* **154**, 741–2.
Pirenne, M. H. (1962). Visual functions in man. Chaps 1–11 in vol. 2 of H. Davson (ed.), *The Eye*, 2nd edn. London: Academic Press.

Podolsky, R. J. (1960). Kinetics of muscular contraction: the approach to the steady state. *Nature* **188**, 666–8.

Polyak, S. (1941). *The Retina.* Chicago: University Press.

Pollard, T. D. (1987). The myosin crossbridge problem. *Cell* **48**, 909–10.

Popot, J.-L. and Changeux, J.-P. (1984). Nicotinic receptor of acetylcholine: structure of an oligomeric integral membrane protein. *Physiol. Rev.* **64**, 1162–1239.

Poppele, R. E. and Quick, D. C. (1981). Stretch-induced contraction of intrafusal muscle in cat muscle spindle. *J. Neurosci.* **1**, 1069–74.

Popper, K. R. (1963). *Conjectures and Refutations: the Growth of Scientific Knowlege.* London: Routledge and Kegan Paul.

Porter, K. R. and Palade, G. E. (1957). Studies on the endoplasmic reticulum. III. Its form and distribution in striated muscle cells. *J. biophys. biochem. Cytol.* **3**, 269–300.

Portzehl, H. (1957). Die Bindung des Erschlaffungsfaktors von Marsh an die Muskelgrana. *Biochim. biophys. Acta* **26**, 373–7.

Portzehl, H., Caldwell, P. C. and Rüegg, J. C. (1964). The dependence of contraction and relaxation of muscle fibres from the crab *Maia squinado* on the internal concentration of free calcium ions. *Biochim. biophys. Acta* **79**, 581–99.

Post, R. L. and Jolly, P. C. (1957). The linkage of sodium, potassium and ammonium active transport across the human erythrocyte membrane. *Biochim. biophys. Acta* **25**, 118–28.

Priesner, E. (1979). Specificity studies on pheromone receptors of noctuid and tortricid lepidoptera. In *Chemical Ecology: Odour Communication in Animals* (ed. F. J. Ritter), pp. 57–71.

Pringle, J. W. S. (1949). The excitation and contraction of the flight muscles of insects. *J. Physiol.* **108**, 226–32.

Pringle, J. W. S. (1954). The mechanism of the myogenic rhythm of certain insect striated muscles. *J. Physiol.* **124**, 269–91.

Pringle, J. W. S. (1960). Models of Muscle. *Symp. Soc. exp. Biol.* **14**, 41–68.

Pringle, J. W. S. (1981). The evolution of fibrillar muscle in insects. *J. exp. Biol.* **94**, 1–14.

Pugh, E. N. (1987). The nature and identity of the internal excitational transmitter of vertebrate phototransduction. *Ann. Rev. Physiol.* **49**, 715–41.

Pugh, E. N. and Cobbs, W. H. (1986). Visual transduction in vertebrate rods and cones: a tale of two transmitters, calcium and cyclic GMP. *Vision Res.* **26**, 1613–43.

Pumphrey, R. J. and Young, J. Z. (1938). The rates of conduction of nerve fibres of various diameters of cephalopods. *J. exp. Biol.* **15**, 453–66.

Quilliam, T. A. and Sato, M. (1955). The distribution of myelin on nerve fibres from Pacinian corpuscles. *J. Physiol.* **129**, 167–76.

Quinta-Ferreira, M. E., Rojas, E. and Arispe, N. (1982). Potassium currents in the giant axon of the crab *Carcinus maenas. J. Membrane Biol.* **66**, 171–81.

Raftery, M. A., Vandlen, R. L., Reed, K. L. and Lee, T. (1976). Characterization of *Torpedo californica* acetylcholine receptor: its subunit composition and ligand-binding properties. *Cold Spr. Harb. Symp. Quant. Biol.* **40**, 193–202.

Raftery, M. A., Hunkapiller, M. W., Strader, C. D. and Hood, L. E. (1980). Acetylcholine receptor: complex of homologous subunits. *Science* **208**, 1454–7.

Rall, W. (1962). Electrophysiology of a dendritic neuron model. *Biophys. J.* **2** suppl., 145–67.

Rall, W. (1977). Core conductor theory and cable properties of neurons. In *Handbook of Physiology*, section 1, volume 1 (ed. E. R. Kandel), pp. 39–97. Bethesda, Maryland: American Physiological Society.

Rasmussen, G. L. (1953). Further observations on the efferent cochlear bundle. *J. comp. Neurol* **99**, 61–74.

Redman, S. and Walmsley, B. (1983*a*). The time course of synaptic potentials evoked in cat spinal motoneurones at identified group Ia synapses. *J. Physiol.* **343**, 117–33.

Redman, S. and Walmsley, B. (1983*b*). Amplitude fluctuations in synaptic potentials evoked in cat spinal motoneurones at identified group Ia synapses. *J. Physiol.* **343**, 135–45.

Reedy, M. K. (1964). Contribution to discussion. *Proc. R. Soc. Lond.* B **160**, 458–60.

Reedy, M. K., Holmes, K. C. and Tregear, R. T. (1965). Induced changes in orientation of the cross-bridges of glycerinated insect flight muscle. *Nature* **207**, 1276–80.

Reedy, M., Leonard, K. R., Freeman, R. and Arad, T. (1981). Thick filament mass determination by electron scattering measurements with the scanning transmission electron microscope. *J. Mus. Res. Cell Motil.* **2**, 45–64.

Rees, C. J. C. (1969). Chemoreceptor specificity associated with the choice of feeding site by the beetle *Chrysolina brunsvicensis* on its foodplant *Hypericum hirsutum. Entomologia exp. appl.* **12**, 565–83.

Reichardt, W. (1962). Theoretical aspects of neural inhibition in the lateral eye of *Limulus. Proc. Int. Un. Physiol. Sci.* **3**, 65–84.

Requena, J., Mullins, L. J., Whittembury, J. and Brinley, F. J. (1986). Dependence of ionized and total Ca in squid axons on Na_0-free or high K_0 conditions. *J. gen. Physiol.* **87**, 143–59.

Reuter, H. (1983). Calcium channel modulation by neurotransmitters, enzymes and drugs. *Nature* **301**, 569–74.

Reuter, H. (1984). Ion channels in cardiac cell membranes. *Ann. Rev. Physiol.* **46**, 473–84.

Reuter, H. (1986). Voltage-dependent mechanisms for raising intracellular free calcium concentration: calcium channels. In *Calcium and the Cell.* Ciba Foundation Symposium 122, pp. 5–14. Chichester: Wiley.

Reuter, H. and Scholz. H. (1977). The regulation of the calcium conductance of cardiac muscle by adrenaline. *J. Physiol.* **264**, 49–62.

Revel, J. P. (1962). The sarcoplasmic reticulum of the bat cricothyroid muscle. *J. Cell Biol.* **12**, 571–88.

Revel, J. P. and Karnovsky, M. J. (1967). Hexagonal array of subunits in intercellular junctions of the mouse heart and liver. *J. Cell. Biol.* **33**, C7–12.

Reynolds, J. A. and Karlin, A. (1978). Molecular weight in

detergent solution of acetylcholine receptor from *Torpedo californica. Biochemistry* **17**, 2035–8.
Rhode, W. S. and Robles, L. (1974). Evidence from Mössbauer experiments for nonlinear vibration in the cochlea. *J. Acoust. Soc. Am.* **55**, 588–96.
Richardson, B. P., Engel, G., Donatsch, P. and Stadler, P. A. (1985). Identification of serotonin M receptor subtypes and their specific blockade by a new class of drugs. *Nature* **316**, 126–31.
Ripley, S. H., Bush, B. M. H. and Roberts, A. (1968). Crab muscle receptor which responds without impulses. *Nature* **218**, 1170–1.
Ripps, H. and Weale, R. A. (1976). The visual photoreceptors. In *The Eye* (ed. H. Davson), vol. 2A *Visual Function in Man*, pp. 5–41. New York: Academic Press.
Ritchie, J. M. (1954). The effect of nitrate on the active state of muscle. *J. Physiol.* **126**, 155–68.
Ritchie, J. M. (1971). Electrogenic ion pumping in nervous tissue. In *Current topics in Bioenergetics*, **4**, 327–56. New York: Academic Press.
Ritchie, J. M. (1982). On the relation between fibre diameter and conduction velocity in myelinated nerve fibres. *Proc. Roy. Soc. Lond.* B **217**, 29–35.
Ritchie, J. J., Rogart, R. B. and Strichartz, G. R. (1976). A new method for labelling saxitoxin and its binding to non-myelinated fibres of the rabbit vagus, lobster walking leg, and garfish olfactory nerves. *J. Physiol.* **261**, 477–94.
Ritchie, J. M. and Straub, R. W. (1975). The movement of potassium ions during electrical activity, and the kinetics of the recovery process, in the non-myelinated fibres of the garfish olfactory nerve. *J. Physiol.* **249**, 327–48.
Robertson, J. D. (1960). The molecular structure and contact relationships of cell membranes. *Progr. Biophys.* **10**, 343.
Rodbell, M. (1980). The role of hormone receptors and GTP-regulatory proteins in membrane transduction. *Nature* **284**, 17–22.
Roeder, K. D. (1951). Movements of the thorax and potential changes in the thoracic muscles of insects during flight. *Biol. Bull. mar. biol. Lab., Woods Hole* **100**, 95–106.
Romanes, G. J. (1951). The motor cell columns of the lumbosacral spinal cord of the cat. *J. comp. Neurol.* **94**, 313–63.
Romey, G. and Lazdunski, M. (1982). Lipid-soluble toxins thought to be specific for Na^+ channels block Ca^{2+} channels in neuronal cells. *Nature* **297**, 79–80.
Rosenberg, R. L., Tomiko, S. A. and Agnew, W. S. (1984). Single-channel properties of the reconstituted voltage-regulated Na channel isolated from the electroplax of *Electrophorus electricus. Proc. Natl. Acad. Sci. USA* **81**, 5594–8.
Rosenbluth, J. (1968). Obliquely striated muscle. *J. Cell Biol.* **36**, 245–59.
Rosenbluth, J. (1969). Sarcoplasmic reticulum of an unusually fast-acting crustacean muscle. *J. Cell Biol.* **42**, 534–47.
Rosenbluth, J. (1974). Substructure of the amphibian motor end plate. Evidence for a granular component projecting from the outer surface of the receptive membrane. *J. Cell Biol.* **62**, 755–66.
Rosenfeld, M. G., Mermod, J.-J., Amara, S. G., Swanson, L. W., Sawchenko, P. E., Rivier, J., Vale, W. W. and Evans, R. M. (1983). Production of a novel neuropeptide encoded by the calcitonin gene via tissue-specific RNA processing. *Nature* **304**, 129–35.
Rothman, J. E. and Lenard, J. (1977). Membrane asymmetry. *Science* **195**, 743–53.
Rothman, S. J. and Olney, J. W. (1987). Excitotoxicity and the NMDA receptor. *Trends Neurosci.* **10**, 299–302.
Rowlerson, A., Gorza, L. and Schiaffino, S. (1985). Immunohistochemical identification of spindle fibre types in mammalian muscle using type-specific antibodies to isoforms of myosin. In *The Muscle Spindle* (ed. I. A. Boyd and M. H. Gladden), pp. 29–34.
Rüegg, J. C. (1986). *Calcium in Muscle Activation.* Berlin: Springer.
Rüegg, J. C. and Pfitzer, G. (1985). Modulation of calcium sensitivity in guinea pig taenia coli: skinned fiber studies. *Experientia* **41**, 997–1001.
Rüegg, J. C. and Tregear, R. T. (1966). Mechanical factors affecting the ATPase activity of glycerol-extracted insect fibrillar flight muscle. *Proc. R. Soc. Lond.* B **165**, 497–512.
Ruffini, A. (1898). On the minute anatomy of the neuromuscular spindles of the cat, and on their physiological significance. *J. Physiol.* **23**, 190–208.
Rushton, W. A. H. (1951). A theory of the effects of fibre size in medullated nerve. *J. Physiol.* **115**, 101–22.
Rushton, W. A. H. (1956). The difference spectrum and the photosensitivity of rhodopsin in the living human eye. *J. Physiol.* **134**, 11–29.
Rushton, W. A. H. (1963). A cone pigment in the protanope. *J. Physiol.* **168**, 345–59.
Rushton, W. A. H. (1972). Visual pigments in man. In *Handbook of Sensory Physiology* vol. VIII/1, *Photochemistry of Vision.* (ed. H. J. A. Darnall), pp. 364–94. Berlin: Springer.
Russell, I. J. (1971). The role of the lateral-line efferent system in *Xenopus laevis. J. exp. Biol.* **54**, 621–41.
Russell, I. J., Cody, A. R. and Richardson, G. P. (1986). The responses of inner and outer hair cells in the basal turn of the guinea-pig cochlea and in mouse cochlea grown in vitro. *Hearing Res.* **22**, 199–216.
Russell, I. J. and Sellick, P. M. (1978). Intracellular studies of hair cells in the mammalian cochlea. *J. Physiol.* **284**, 261–90.
Russell, J. M. (1984). Chloride in the squid giant axon. In *The Squid Axon* (ed. P. F. Baker). (*Current Topics in Membranes and Transport* vol. 22). Orlando: Academic Press.
Saito, A., Kimura, S., and Goto, K. (1986). Calcitonin gene-related peptide as potential neurotransmitter in guinea pig right atrium. *Am. J. Physiol.* **250**, H693–8.
Sakitt, B. (1972). Counting every quantum. *J. Physiol.* **223**, 131–50.
Sakmann, B., Methfessel, C., Mishina, M., Takahashi, T.,

Takai, T., Kurasaki, M., Fukuda, K. and Numa, S. (1985). Role of acetylcholine receptor subunits in gating of the channel. *Nature* **318**, 538–43.

Sakmann, B. and Neher, E., (eds.) (1983). *Single-channel Recording*. New York: Plenum Press.

Sakmann, B. and Neher, E. (1984). Patch clamp techniques for studying ionic channels in excitable membranes. *Ann. Rev. Physiol.* **46**, 455–72.

Sakmann, B., Noma, A. and Trautwein, W. (1983). Acetylcholine activation of single muscarinic K^+ channels in isolated pacemaker cells of the mammalian heart. *Nature* **303**, 250–3.

Salkoff, L., Butler, A., Wei, A., Scavarda, N., Giffen, K., Ifune, C., Goodman, R. and Mandel, G. (1987). Genomic organization and deduced amino acid sequence of a putative sodium channel gene in *Drosophila*. *Science* **237**, 744–9.

Salkoff, L. and Wyman, R. (1981). Genetic modification of potassium channels in *Drosophila Shaker* mutants. *Nature* **293**, 228–30.

Salpeter, M. M., Rogers, A. W., Kasprzak, H. and McHenry, F. A. (1978). Acetylcholinesterase in the fast extraocular muscle of the mouse by light and electron microscope autoradiography. *J. Cell Biol.* **78**, 274–85.

Sand, A. (1937). The mechanism of the lateral sense organs of fishes. *Proc. R. Soc. Lond.* B **123**, 472–95.

Sandow, A. (1944). General properties of latency relaxation. *J. cell comp. Physiol.* **24,** 221–56.

Sandow, A. (1965). Excitation-contraction coupling in skeletal muscle. *Pharmacol. Rev.* **17**, 265–320.

Sattelle, D. B. (1985). Acetylcholine receptors. In *Comprehensive Insect Physiology Biochemistry and Pharmacology* (ed. G. A. Kerkut and L. I. Gilbert), vol. 11, pp. 395–434. Oxford; Pergamon Press.

Scarpa, A. (1979). Measurement of calcium ion concentrations with metallochromic indicators. In *Detection and Measurement of Free Ca^{2+} in Cells* (ed. C. C. Ashley and A. K. Campbell), pp. 85–115. Amsterdam: Elsevier/North-Holland.

Scharrer, B. (1987). Neurosecretion: beginnings and new directions in neuropeptide research. *Ann. Rev. Neurosci.* **10**, 1–17.

Scharrer, E. and Sharrer, B. (1945). Neurosecretion. *Physiol. Rev.* **25**, 171–81.

Schaumburg, H. H., Byck, R., Gerstl, R. and Mashman, J. H. (1969). Monosodium L-glutamate: its pharmacology and role in the Chinese restaurant syndrome. *Science* **163**, 826–8.

Scheich, H., Langner, G., Tidemann, C., Coles., R. B. and Gupp, A. (1986). Electroreception and electrolocation in platypus. *Nature* **319**, 401–2.

Scher, A. M. (1965). Mechanical events in the cardiac cycle. In *Physiology and Biophysics* (ed. T. C. Ruch and H. D. Patton). Philadelphia: Saunders.

Scher, A. M. and Spach, M. S. (1979). Cardiac depolarization and repolarization and the electrocardiogram. In *Handbook of Physiology*, section 2, vol. 1 (ed. R. M. Berne and N. Sperelakis), pp. 357–92. Bethesda, Maryland: American Physiological Society.

Schmidt, R. F. (1971). Presynaptic inhibition in the vertebrate central nervous system. *Ergeb. Physiol.* **63**, 21–108.

Schmidt, W. J. (1938). Polarizations-optische Analyse eines Eiweiss-Lipoid-Systems erläutert am Aussenglied der Schzellen. *Kolloidzeitschrift* **85**, 137–48.

Schmidt-Nielsen, K. (1983). *Animal Physiology* 3rd edn. Cambridge: Cambridge University Press.

Schneider, D. (1965). Chemical sense communication in insects. *Symp. Soc. Exp. Biol.* **20**, 273–97.

Schneider, D. (1984). Insect olfaction – our research endeavour. In *Foundations of Sensory Science* (ed. W. W. Dawson and J. M. Enoch), pp. 381–418.

Schneider, D., Lacher, V. and Kaissling, K.-E. (1964). Die Reacktionweise und das Reacktionsspaktrum von Reichzellen bei *Antherea pernyi* (Lepidoptera, Saturniidae). *Z. vergl. Physiol.* **48**, 632–62.

Schneider, M. F. and Chandler, W. K. (1973). Voltage dependent charge movement in skeletal muscle: a possible step in excitation-contraction coupling. *Nature* **242**, 244–6.

Schoffeniels, E. (1959). Ion movements studied with single isolated electroplax. *Ann. N.Y. Acad. Sci.* **81**, 285–306.

Schofield, P. R., Darlison, M. G., Fujita, N., Burt, D. R., Stephenson, F. A., Rodriguez. H., Rhee, L. M., Ramachandran, J., Reale, V., Glencorse, T. A., Seeburg, P. H. and Barnard, E. A. (1987). Sequence and functional expression of the $GABA_A$ receptor shows a ligand-gated receptor super-family. *Nature* **328**, 221–7.

Schultze, M. (1866). Zur Anatomie und Physiologie der Retina. *Arch. mikrosk. Anat. EntwMech.* **2**, 175–286.

Schwartz, J.-C., Arrang, J.-M., Garbarg, M. and Korner, M. (1986). Properties and roles of the three subclasses of histamine receptors in brain. *J. exp. Biol.* **124**, 203–24.

Schwarz, J.-C., Malfroy, B. and De La Baume, S. (1981). Biological inactivation of enkephalins and the role of enkephalin-dipeptidyl-carboxypeptidase -'enkephalinase') as neuropeptidase. *Life Sci.* **29**, 1715–40.

Schwartz, J. P. and Costa, E. (1986). Hybridization approaches to the study of neuropeptides. *Ann. Rev. Neurosci.* **9**, 277–304.

Schwarz, J. R. and Vogel, W. (1971). Potassium inactivation in single myelinated nerve fibres of *Xenopus laevis*. *Pflügers Archiv* **330**, 61–73.

Schwarz, T. L., Tempel, B. L., Papazian, D. M., Jan, Y. N. and Jan, L. Y. (1988). Multiple potassium-channel components are produced by alternative splicing at the *Shaker* locus in *Drosophila*. *Nature* **331**, 137–42.

Schwarz, W., Palade, P. T. and Hille, B. (1977). Local anaesthetics: effect of pH on use-dependent block of sodium channels in frog muscle. *Biophys. J.* **20**, 343–68.

Schwarz, W. and Passow, H. (1983). Ca^{2+}-activated K^+ channels in erythrocytes and excitable cells. *Ann. Rev. Physiol.* **45**, 359–74.

Schwarzmann, G., Wiegandt, H., Rose, B., Zimmerman,

A., Ben-Haim, D. and Loewenstein, W. R. (1981). Diameter of the cell-to-cell junctional channels as probed with neutral molecules. *Science* **213**, 551–3.

Seizinger, B. R., Liebisch, D. C., Gramsch, C., Herz, A., Weber, E., Evans, C. J., Esch, F. S. and Böhlen, P. (1985). Isolation and structure of a novel C-terminally amidated opioid peptide, amidorphin, from bovine adrenal medulla. *Nature* **313**, 57–9.

Selby, C. C. and Bear, R. S. (1956). The structure of the actin-rich filaments of muscles according to X-ray diffraction. *J. biophys. biochem. Cytol.* **2**, 71–85.

Sellick, P. M., Patuzzi, R. and Johnstone, B. M. (1982). Measurement of basilar membrane motion in the guinea pig using the Mössbauer technique. *J. Acoust. Soc. Am.* **72**, 131–41.

Sheridan, J. D. and Atkinson, M. M. (1985). Physiological roles of permeable junctions: some possibilities. *Ann. Rev. Physiol.* **47**, 337–53.

Sheterline, P. (1983). *Mechanisms of Cell Motility*. London: Academic Press.

Shoenberg, C. F. and Needham, D. M. (1976). A study of the mechanism of contraction in vertebrate smooth muscle. *Biol. Rev.* **51**, 53–104.

Shull, G. A., Schwarz, A. and Lingrel (1985). Aminoacid sequence of the catalytic subunit of the $(Na^+ + K^+)$ATPase deduced from a complementary DNA. *Nature* **316**, 691–5.

Siegelbaum, S. A., Camardo, J. S. and Kandel, E. R. (1982). Serotonin and cyclic AMP close single K^+ channels in *Aplysia* sensory neurones. *Nature* **299**, 413–17.

Sigel, E. and Barnard, E. A., (1984). A γ-aminobutyric acid/benzodiazepine receptor complex from bovine cerebral cortex. *J. Biol. Chem.* **259**, 7219–23.

Sigworth, F. J. (1980*a*). The variance of sodium current fluctuations at the node of Ranvier. *J. Physiol.* **307**, 97–129.

Sigworth, F. J. (1980*b*). The conductance of sodium channels under conditions of reduced current at the node of Ranvier. *J. Physiol.* **307**, 131–42.

Sigworth, F. J. and Neher, E. (1980). Single Na^+ channel currents observed in cultured rat muscle cells. *Nature* **287**, 447–9.

Silinsky, E. M. (1975). On the association between transmitter secretion and the release of adenine nucleotides from mammalian motor nerve terminals. *J. Physiol.* **247**, 145–62.

Silinsky, E. M. (1985). The biophysical pharmacology of calcium-dependent acetylcholine secretion. *Pharmacol. Rev.* **37**, 81–132.

Sillman, A. J., Ito, H. and Tomita, T. (1969). Studies on the mass receptor potential of the isolated frog retina. II. On the basis of the ionic mechanism. *Vision Res.* **9**, 1443–51.

Simantov, R. and Snyder, S. H. (1976). Morphine-like peptides in mammalian brain: isolation, structure elucidation, and interactions with the opiate receptor. *Proc. Natl. Acad. Sci. USA* **73**, 2515–19.

Simmons, R. M. and Szent-Györgyi, A. G. (1985). A mechanical study of regulation in the striated adductor muscle of the scallop. *J. Physiol.* **358**, 47–64.

Sine, S. M. and Steinbach, J. H. (1987). Activation of acetylcholine receptors on clonal mammalian BC 3H-1 cells by high concentrations of agonist. *J. Physiol.* **385**, 325–59.

Singer, S. J. and Nicolson, G. L. (1972). The fluid mosaic model of the structure of cell membranes. *Science* **175**, 720–31.

Skoglund, S. (1956). Anatomical and physiological studies on knee joint innervation in the cat. *Acta physiol. Scand. Suppl.* **124**, 1–101.

Skou, J. C. (1957). The influence of some cations on an adenosine triphosphatase from peripheral nerves. *Biochim. Biophys. Acta* **23**, 394–401.

Slayter, H. S. and Lowey, S. (1967). Substructure of the myosin molecule as visualized by electron microscopy. *Proc. Natl. Acad. Sci. USA* **58**, 1611–18.

Small, J. V. (1974). Contractile units in vertebrate smooth muscle cells. *Nature* **249**, 324–7.

Small, J. V. (1977). The contractile apparatus of the smooth muscle cell: structure and composition. In *Biochemistry of Smooth Muscle* (ed. N. L. Stephens), pp. 413–43.

Small, J. V. and Squire, J. M. (1972). Structural basis of contraction in vertebrate smooth muscle. *J. Mol. Biol.* **67**, 117–49.

Smith, D. S., Gupta, B. L. and Smith, U. (1966). The organization and myofilament array of insect visceral muscles. *J. Cell Sci.* **1**, 49–57.

Sneddon, P. and Westfall, D. P. (1984). Pharmacological evidence that adenosine triphosphate and noradrenaline are co-transmitters in the guinea-pig vas deferens. *J. Physiol.* **347**, 561–80.

Sobiescek, A. (1973). The fine structure of the contractile apparatus of the anterior byssus retractor muscle of *Mytilus edulis*. *J. Ultrastruct. Res.* **43**, 313–43.

Sobue, K., Muramoto, Y., Fujita, M. and Kakiuchi, S. (1981). Purification of a calmodulin binding protein from chicken gizzard that interacts with F-actin. *Proc. Natl. Acad. Sci. USA* **78**, 5652–5.

Somlyo, A. V., Shuman, H. and Somlyo, A. P. (1977) Elemental distribution in striated muscle and the effects of hypertonicity. *J. Cell Biol.* **74**, 828–57.

Somlyo, A. V., Gonzalez-Serratos, H., Shuman, G., McClellan, G. and Somlyo, A. P. (1981). Calcium release and ionic changes in the sarcoplasmic reticulum of tetanised muscle: an electron-probe study. *J. Cell Biol.* **90**, 577–94.

Sonnenblick, E. H. and Skelton, C. L. (1974). Reconsideration of the ultrastructural basis of cardiac length–tension relations. *Circ. Res.* **35**, 517–26.

Sosinsky, G., Caspar, D. L. D., Baker, J. S., Makowski, L., Maulik, S., Phillips, W. and Goodenough, D. (1986). Analysis of gap junction electron micrographs based on a three-dimensional model. *Biophys. J.* **49**, 204*a*.

Sotavalta, O. (1953). Recordings of high wing-stroke and thoracic vibration frequency in some midges. *Biol. Bull. mar. biol. Lab. Woods Hole* **104**, 439–44.

Spoendlin, H. (1975). Neuranatomical basis of cochlear coding mechanisms. *Audiology* **14**, 383–407.
Spoendlin, H. (1984). Efferent innervation of the cochlea. In *Comparative Physiology of Sensory Systems* (ed. L. Bolis *et al.*), pp. 163–88. Cambridge: Cambridge University Press.
Spray, D. C., Harris, A. L. and Bennett, M. V. L. (1979). Voltage dependence of junctional conductance in early amphibian embryos. *Science* **204**, 432–4.
Spray, D. C., White, R. L., Verselis, V. and Bennett, M. V. L. (1985). General and comparative physiology of gap junction channels. In *Gap Junctions* (ed. M. V. L. Bennett and D. C. Spray), pp. 139–53. New York: Cold Spring Harbor Laboratory.
Squire, J. M. (1974). Symmetry and three-dimensional arrangement of filaments in vertebrate striated muscle. *J. Mol. Biol.* **90**, 153–60.
Squire, J. (1981). *The Structural Basis of Muscular Contraction.* New York: Plenum Press.
Squire, J. M. (1986). *Muscle; Design, Diversity and Disease.* Menlo park, California: Benjamin/Cummings.
Squire, J. M., Harford, J. J., Edman, C. and Sjöström, M. (1982). Fine structure of the A-band in cryo-sections. III Crossbridge distribution and the axial structure of the human C-zone. *J. Mol. Biol.* **155**, 467–94.
Städler, E. (1984). Contact chemoreception. In *Chemical Ecology of Insects* (ed. W. J. Bell and R. T. Cardé), pp. 3–35. London: Chapman and Hall.
Standen, N. F., Gray, P. T. A. and Whitaker, M. J. (1987). *Microelectrode Techniques. The Plymouth Workshop Handbook.* Cambridge: Company of Biologists.
Stanfield, P. R. (1986). Voltage-dependent calcium channels of excitable membranes. *Brit. Med. Bull.* **42**, 359–67.
Starke, K. (1981). Presynaptic receptors. *Ann. Rev. Pharmacol. Toxicol.* **21**, 7–30.
Starr, R. and Offer, G. (1983). H-protein and X-protein: two new components of the thick filaments of vertebrate skeletal muscle. *J. Mol. Biol.* **170**, 675–98.
Stein, L. A., Schwarz, R. P., Chock, P. B. and Eisenberg, E. (1979). Mechanism of actomyosin adenosine triphospatase. Evidence that adenosine 5′-triphosphate hydrolysis can occur without dissociation of the actomyosin complex. *Biochem.* **18**, 3895–909.
Steinbrecht, R. A. and Schneider, D. (1980). Pheromone communication in moths. In *Insect Biology in the Future* (ed. M. Locke and D. S. Smith), pp. 685–703. New York: Academic Press.
Steinert, P. M. and Parry, D. A. D. (1985). Intermediate filaments. *Ann. Rev. Cell Biol.* **1**, 41–65.
Steinhausen, W. (1931). Über den Nachweis der Bewegung der Cupula in der intakten Bogengangsampulle des Labyrinths bei der natürlichen rotatorischen und calorischen Reizung. *Pflügers Archiv* **228**, 322–8.
Steinhausen, W. (1933). Über die Beobachtung der Cupula in den Bogengangsampullen des Labyrinths des lebenden Hechets. *Pflügers Archiv* **232**, 500–12.
Stephenson, D. G. and Williams, D. A. (1981). Calcium-activated force responses in fast- and slow-twitch skinned muscle fibres of the rat at different temperatures. *J. Physiol.* **317**, 281–302.
Stephenson, D. G. and Williams, D. A. (1982). Effects of sarcomere length on the force-pCa relation in fast- and slow-twitch skinned muscle fibres from the rat. *J. Physiol.* **333**, 637–53.
Sterling, P. (1983). Microcircuitry of the cat retina. *Ann. Rev. Neurosci.* **6**, 149–85.
Sterling, P., Freed, M. and Smith, R. G. (1986). Micro-circuitry and functional architecture of the cat retina. *Trends Neurosci.* **9**, 186–92.
Stevens, C. F. (1972). Inferences about membrane properties from electrical noise measurements. *Biophys. J.* **12**, 1028–47.
Stevens, C. F. (1986). Are there two functional classes of glutamate receptors? *Nature* **322**, 210–11.
Stevens, S. S. (1961*a*). To honor Fechner and repeal his law. *Science* **133**, 80–6.
Stevens, S. S. (1961*b*). The psychophysics of sensory function. In *Sensory Communication* (ed. W. A. Rosenblith), pp. 1–33. Cambridge, Massachusetts: MIT Press.
Stevens, S. S. (1970). Neural events and the psychophysical law. *Science* **170**, 1043–50.
Stiles, W. S. (1939). The directional sensitivity of the retina and the spectral sensitivities of the rods and cones. *Proc. R. Soc. Lond.* B **127**, 64–105.
Stimers, J. R., Bezanilla, F. and Taylor, R. E. (1985). Sodium channel activation in the squid giant axon. *J. gen. Physiol.* **85**, 65–82.
Stone, T. W. (1981). Physiological roles for adenosine and adenosine 5′-triphosphate in the nervous system. *Neuroscience* **6**, 523–55.
Stone, D., Sodek, J., Johnson, P. and Smillie, L. B. (1974). Tropomyosin: correlation of amino acid sequence and structure. *Proc. IX FEBS Meeting (Budapest)* **31**, 125–36.
Storm-Mathisen, J. and Iversen, L. L. (1979). Uptake of (^{3}H)gluatmic acid in excitatory nerve endings. *Neuroscience* **4**, 1237–53.
Straub, F. B. (1943). Actin. *Stud. Inst. Med. Chem. Univ. Szeged* **2**, 3–15.
Streb, H., Irvine, R. F., Berridge, M. J. and Schultz, I. (1983). Release of Ca^{2+} from a nonmitochondrial intracellular store in pancreatic acinar cells by inositol 1, 4, 5-trisphosphate. *Nature* **306**, 67–8.
Strehler, E. E., Carlsson, E., Eppenberger, H. M. and Thornell, L.-E. (1983). Ultrastructural localization of M-band proteins in chicken breast muscle as revealed by combined immunocytochemistry and ultramicroscopy. *J. Mol. Biol.* **166**, 141–58.
Strehler, E. E., Strehler-Page, M.-A., Perriard, J.-C., Periasamy, M. and Nadal-Ginard, B. (1986). Complete nucleotide and encoded amino acid sequence of a mammalian myosin heavy chain gene. *J. Mol. Biol.* **190**, 291–317.
Strichartz, G. R. (1973). The inhibition of sodium currents

in myelinated nerve by quaternary derivatives of lidocaine. *J. gen. Physiol.* **62**, 37–57.

Strumwasser, F. (1962). Postsynaptic inhibition and excitation produced by different branches of a single neuron and the common transmitter involved. *Proc. Int. Un. Physiol. Socs.* **2**, 801.

Stryer, L. (1986). Cyclic GMP cascade of vision. *Ann. Rev. Neurosci.* **9**, 87–119.

Stryer, L. and Bourne, H. R. (1986). G proteins: a family of signal transducers. *Ann. Rev. Cell Biol.* **2**, 391–419.

Study, R. E. and Barker, J. L. (1981). Diazepam and (−)pentobarbitol; fluctuation analysis reveals different mechanisms for potentiation of γ-aminobutyric acid responses in cultured central neurons. *Proc. Natl. Acad. Sci. USA* **78**, 7180–4.

Su, C. (1983). Purinergic neurotransmission and neuromodulation. *Ann. Rev. Pharmacol. Toxicol.* **23**, 397–411.

Suematsu, E., Hirata, M., Hashimoto, T. and Kuriyama, H. (1984). Inositol 1, 4, 5-trisphosphate releases Ca^{2+} from intracellular store sites in skinned single cells of porcine coronary artery. *Biochem. Biophys. Res. Comm.* **120**, 481–5.

Sugiyama, H., Ito, I. and Hirono, C. (1987). A new type of glutamate receptor linked to inositol phospholipid metabolism. *Nature* **325**, 531–3.

Sumikawa, K., Houghton, M., Emtage, J. S., Richards, B. M. and Barnard, E. A. (1981). Active multi-unit Ach receptor assembled by translation of heterologous mRNA in *Xenopus* oocytes. *Nature* **242**, 862–4.

Sumikawa, K., Houghton, M., Smith, J. C., Bell. L., Richards, B. M. and Barnard, E. A. (1982). The molecular cloning and characterisation of cDNA coding for the α subunit of the acetylcholine receptor. *Nucleic Acids Research* **10**, 5802–22.

Sumikawa, K., Parker, I. and Miledi, R. (1984). Partial purification and functional expression of brain mRNAs coding for neurotransmitter receptors and voltage-operated channels. *Proc. Natl. Acad. Sci. USA* **81**, 7994–8.

Summers, R. D., Morgan, R. and Reimann, S. P. (1943). The semicircular canals as a device for vectorial resolution. *Arch. Otolaryng.* **37**, 219–37.

Sutoh, K., Yamamoto, K. and Wakabayashi, T. (1986). Electron microscopic visualization of the ATPase site of myosin by photoaffinity labelling with a biotinylated photoreactive ADP analog. *Proc. Natl. Acad. Sci. USA* **83**, 212–16.

Sutherland, E. W., Oye, I. and Butcher, R. W. (1965). The action of epinephrine and the role of the adenyl cyclase system in hormone action. *Rec. Prog. Hormone Res.* **21**, 623–42.

Sutherland, E. W. and Rall, T. W. (1960). Relation of adenosine 3′, 5′ phosphate and phosphorylase to the action of catecholamines and other hormones. *Pharmacol. Rev.* **12**, 265–99.

Swenson, R. P. (1982). Inactivation of potassium current in a squid axon by a variety of quaternary ammonium ions. *J. gen. Physiol.* **77**, 255–71.

Szabo, T. (1965). Sense organs of the lateral line system in some electric fish of the Gymnotidae, Mormyridae and Gymnarchidae. *J. Morph.* **117**, 229–50.

Szabo, T. (1974). Anatomy of the specialized lateral line organs of electroreception. In *Handbook of Sensory Physiology*, vol. III/3 (ed. A. Fessard), pp. 13–58. Berlin: Springer.

Szent-Györgyi, A. (1949). Free energy relations and contraction of actomyosin. *Biol. Bull.* **96**, 140–61.

Szent-Györgyi, A. G. (1951). The reversible depolymerisation of actin by potassium iodide. *Arch. Biochem. Biophys.* **31**, 97.

Szent-Györgyi, A. G., Cohen, C. and Kendrick-Jones, J. (1971). Paramyosin and the filaments of molluscan 'catch' muscles. II. Native filaments: isolation and characterization. *J. Mol. Biol.* **56**, 239–58.

Szent-Györgyi, A. G., Szentkiralyi, E. M. and Kendrick-Jones, J. (1973). The light chains of scallop myosin as regulatory subunits. *J. Mol. Biol.* **74**, 179–203.

Takai, T., Noda, M., Mishina, M., Shimizu, S., Furutani, Y., Kayano, T., Ikeda, T., Kubo, T., Takahashi, H., Takahashi, T., Kuno, M. and Numa, S. (1985). Cloning, sequencing and expression of cDNA for a novel subunit of acetylcholine receptor from calf muscle. *Nature* **315**, 761–4.

Takeuchi, A. and Takeuchi, N. (1959). Active phase of frog's end-plate potential. *J. Neurophysiol.* **22**, 395–411.

Takeuchi, A. and Takeuchi, N. (1960). On the permeability of the end-plate membrane during the action of the transmitter. *J. Physiol.* **154**, 52–67.

Takeuchi, A. and Takeuchi, N. (1964). The effect on crayfish muscle of iontophorectically applied glutamate. *J. Physiol.* **170**, 296–317.

Takeuchi, A. and Takeuchi, N. (1965). Localised action of gamma-aminobutyric acid on the crayfish muscle. *J. Physiol.* **177**, 225–38.

Takeuchi, N. (1963*a*). Some properties of conductance changes at the end-plate membrane during the action of acetylcholine. *J. Physiol.* **167**, 128–40.

Takeuchi, N. (1963*b*). Effects of calcium on the conductance change of the end-plate membrane during the action of the transmitter. *J. Physiol.* **167**, 141–55.

Tallman, J. F., Paul, S. M., Skolnick, P. and Gallager, D. W. (1980). Receptors for the age of anxiety: pharmacology of the benzodiazepines. *Science* **207**, 274–81.

Tanabe, T., Takeshima, H., Mikami, A., Flockerzi, V., Takahashi, H., Kangawa, K., Kojima, M., Matsuo, H., Hirose, T. and Numa, S. (1987). Primary structure of the receptor for calcium channel blockers from skeletal muscle. *Nature* **328**, 313–18.

Tanaka, J. C., Eccleston, J. F. and Barchi, R. L. (1983). Cation selectivity characteristics of the reconstituted voltage-dependent sodium channel purified from rat skeletal muscle sarcolemma. *J. Biol. Chem.* **258**, 7519–26.

Tanner, W. P. and Swets, J. A. (1954). A decision-making theory of visual detection. *Psychol. Rev.* **61**, 401–9.

Tasaki, I. (1953). *Nervous Transmission.* Springfield, Illinois; Charles C. Thomas.

Tasaki, I. (1954). Nerve impulses in individual auditory nerve fibres of guinea pig. *J. Neurophysiol.* **17**, 97–122.

Tasaki, I., Davis, H. and Legoux, J.-P. (1952). The space-time pattern of the cochlear microphonics (guinea-pig), as recorded by differential electrodes. *J. Acoust Soc. Am.* **24**, 502–19.

Tasaki, I., Davis, H. and Eldredge, D. H. (1954). Exploration of cochlear potentials in guinea pig with a microelectrode. *J. Acoust. Soc. Am.* **26**, 765–73.

Tasaki, I., Ishi, K. and Ito, H. (1943). On the relation between the conduction-rate, the fiber-diameter and the internodal distance of the myelinated nerve fiber. *Jap. J. Med. Sci., Biophys.* **9**, 189–99.

Tasaki, I. and Shimamura, M. (1962). Further observations on resting and action potential of intracellularly perfused squid giant axon. *Proc. Natl. Acad. Sci. USA* **48**, 1571–7.

Tasaki, I., Singer, I. and Takenaka, T. (1965). Effects of internal and external ionic environment on excitability of squid giant axon. *J. gen. Physiol.* **48**, 1095–123.

Tasaki, I. and Takeuchi, T. (1942). Weitere Studien über den Aktionsstrom der markhaltigen Nerfenvaser und über die elektrosaltorische Übertragung des Nervenimpulses. *Pflügers Arch. ges. Physiol.* **245**, 764–82.

Tashiro, T. and Stadler, H. (1978). Chemical composition of cholinergic synaptic vesicles from *Torpedo marmorata* based on improved purification. *Eur. J. Biochem.* **90**, 479–87.

Tauc, L. and Gerschenfeld, H. M. (1961). Cholinergic transmission mechanisms for both excitation and inhibition in molluscan central synapses. *Nature* **192**, 366–7.

Tauc, L. and Gerschenfeld, H. M. (1962). A cholinergic mechanism of inhibitory synaptic transmission in a molluscan nervous system. *J. Neurophysiol.* **25**, 236–62.

Taylor, K. A. and Amos, L. A. (1981). A new model for the geometry of the binding of myosin crossbridges to muscle thin filaments. *J. Mol. Biol.* **147**, 297–324.

Taylor, K. A. T., Reedy, M. C., Cordova, L. and Reedy, M. K. (1984). Three dimensional reconstruction of rigor insect flight muscle from tilted thin sections. *Nature* **310**, 285–91.

Taylor, R. E. (1963). Cable Theory. In *Physical Techniques in Biological Research* (ed. W. L. Nastuk) vol. 5. New York: Academic Press.

Tempel, B. L., Papazian, D. M., Schwarz, T. L., Jan, Y. N. and Jan, L. Y. (1987). Sequence of a probable potassium channel component encoded at *Shaker* locus of *Drosphila. Science* **237**, 770–5.

Ter Keurs, H. E. D. J., Rijnsburger, W. H., Van Heuningen, R. and Naglesmit, M. J. (1980). Tension development and sarcomere length in rat cardiac trabeculae: evidence of length-dependent activation. *Circulation Res.* **46**, 703–14.

Thesleff, S. (1955). The mode of neuromuscular block caused by acetylcholine, nicotine, decamethonium and succinylcholine. *Acta physiol. Scand.* **34**, 218–31.

Thomas, D. D. (1987) Spectroscopic probes of muscle cross-bridge rotation. *Ann. Rev. Physiol.* **49**, 691–709.

Thomas, D. D. and Cooke, R. (1980) Orientation of spin-labelled myosin heads in glycerinated muscle fibres. *Biophys. J.* **32**, 891–906.

Thomas, R. C. (1969). Membrane current and intracellular sodium changes in a snail neurone during extrusion of injected sodium. *J. Physiol.* **201**, 495–514.

Thomas, R. C. (1972). Intracellular sodium activity and the sodium pump in snail neurones. *J. Physiol.* **220**, 55–71.

Thorson, J. and White, D. C. S. (1969). Distributed representations for actin-myosin interaction in the oscillatory contraction of muscle. *Biophys. J.* **9**, 360–90.

Thorson, J. and White, D. C. S. (1983). Role of cross-bridge distortion in the small-signal mechanical dynamics of insect and rabbit skeletal muscle. *J. Physiol.* **343**, 59–84.

Thurm, U. (1965). An insect mechanoreceptor. I. Fine structure and adequate stimulus. *Cold Spr. Harb. Symp. Quant. Biol.* **30**, 75–82.

Tilney, L. G., DeRosier, D. J. and Mulroy, M. J. (1980). The organization of actin filaments in the stereocilia of cochlear hair cells. *J. Cell. Biol.* **86**, 244–59.

Timpe, L. C. Schwarz, T. L., Tempel, B. L., Papazian, D. M., Jan, Y. N. and Jan, L. Y. (1988). Expression of functional potassium channels from *Shaker* cDNA in *Xenopus* oocytes. *Nature* **331**, 143–5.

Tokunaga, M., Sutoh, K., Toyoshima, C. and Wakabayashi, T. (1987) Location of the ATPase site of myosin determined by three-dimensional electron microscopy. *Nature* **329**, 635–8.

Tomita, T. (1965). Electrophysiological study of the mechanisms subserving color coding in the fish retina. *Cold Spr. Harb. Symp. Quant. Biol.* **30**, 559–66.

Tomita, T. (1970). Electrical activity of vertebrate photoreceptors. *Quart. Rev. Biophys.* **3**, 179–222.

Tomita, T. (1972). Light-induced potential and resistance changes in vertebrate photoreceptors. In *Handbook of Sensory Physiology* (ed. H. Autrum *et al.*) **7**(2), 483–511. Berlin: Springer.

Tomita, T. (1981). Electrical activity (spikes and slow waves) in gastrointestinal smooth muscles. In *Smooth Muscle* (ed. E. Bülbring *et al.*), pp. 127–56.

Tomita, T., Kaneko, A., Murakami, M. and Pautter, E. L. (1967). Spectral response curves of single cones in the carp. *Vision Res.* **7**, 519–31.

Torre, V., Matthews, H. R. and Lamb, T. D. (1986). Role of calcium in regulating the cyclic GMP cascade of phototransduction in retinal rods. *Proc. Natl. Acad. Sci. USA* **83**, 7109–13.

Torri-Tarelli, F., Grohovaz, F., Fesce, R. and Ceccarelli, B. (1985). Temporal coincidence between synaptic vesicle fusion and quantal secretion of acetylcholine. *J. Cell Biol.* **101**, 1386–99.

Toyoda, J., Nosaki, H. and Tomita, T. (1969). Light-induced resistance changes in single photoreceptors of *Necturus* and *Gekko. Vision Res.* **9**, 453–63.

Toyoshima, C. and Wakabayashi, T. (1985). Three-dimensional image analysis of the complex of thin filaments and myosin molecules from skeletal muscle. V. Assignment of actin in the actin-tropomyosin-myosin subfragment-1 complex. *J. Biochem.* **97**, 245–63.

Toyoshima, Y. Y., Kron, S. J., McNally, E. M., Niebling, K. R., Toyoshima, C. and Spudich, J. A. (1987). Myosin subfragment-1 is sufficient to move actin filaments *in vitro*. *Nature* **328**, 536–9.

Tregear, R. T. (1983). Insect flight muscle. In *Handbook of Physiology*, section 10, *Skeletal Muscle* (ed. L. D. Peachey and R. H. Adrian), pp. 487–506. Bethesda, Maryland: American Physiological Society.

Tregear, R. T. and Marston, S. B. (1979). The crossbridge theory. *Ann. Rev. Physiol.* **41**, 723–36.

Tregear, R. T. and Miller, A. (1969). Evidence of crossbridge movement during contraction of insect flight muscle. *Nature* **222**, 1184–5.

Trinick, J., Knight, P. and Whiting, A. (1984). Purification and properties of native titin. *J. Mol. Biol.* **180**, 331–56.

Tsien, R. W. (1983). Calcium channels in excitable cell membranes. *Ann. Rev. Physiol.* **45**, 341–358.

Turner, A. J. and Whittle, S. R. (1983). Biochemical dissection of the γ-aminobutyrate synapse. *Biochem. J.* **209**, 29–41.

Turner, D. C., Wallimann, T. and Eppenberger, H. M. (1973). A protein that binds specifically to the M-line of skeletal muscle is identified as the muscle form of creatine kinase. *Proc. Natl. Acad. Sci. USA* **70**, 702–5.

Twarog, B. M. and Page, I. H. (1953). Serotonin content of some mammalian tissues and urine and a method for its determination. *Am. J. Physiol.* **175**, 157–61.

Uchizono, K. (1965). Characteristics of excitatory and inhibitory synapses in the central nervous system of the cat. *Nature* **207**, 642–3.

Ueno, H. and Harrington, W. F. (1986) Local melting in the subfragment - 2 region of myosin in activated muscle and its correlation with contractile force. *J. Mol. Biol.* **190**, 69–82.

Unwin, P. N. T. and Ennis, P. D. (1983). Calcium-mediated changes in gap junction structure: evidence from the low angle X-ray pattern. *J. Cell Biol.* **97**, 1459–66.

Unwin, P. N. T. and Ennis, P. D. (1984). Two configurations of a channel-forming membrane protein. *Nature* **307**, 609–13.

Unwin, P. N. T. and Zampigh, G. (1980). Structure of the junction between communicating cells. *Nature* **283**, 545–9.

Usherwood, P. N. R. (1967). Insect neuromuscular mechanisms. *Amer. Zoologist* **7**, 553–82.

Usherwood, P. N. R. and Cull-Candy, S. G. (1975). Pharmacology of somatic nerve-muscle synapses. In *Insect Muscle* (ed. P. N. R. Usherwood), pp. 207–80. London: Academic Press.

Usherwood, P. N. R. and Grundfest, H. (1965). Peripheral inhibition in skeletal muscle of insects. *J. Neurophysiol.* **28**, 497–518.

Ussing, H. H. (1949). The distinction by means of tracers between active transport and diffusion. *Acta physiol. scand.* **19**, 43–56.

Vandekerckhove, J. and Weber, K. (1978*a*). Actin amino-acid sequences. *Eur. J. Biochem.* **90**, 451–62.

Vandekerckhove, J. and Weber, K. (1978*b*). At least six different actins are expressed in a higher mammal: an analysis based on the amino acid sequence of the amino-terminal tryptic peptide. *J. Mol. Biol.* **126**, 783–802.

Vandenberg, C. A. and Horn, R. (1984). Inactivation viewed through single sodium channels. *J. gen. Physiol.* **84**, 535–64.

Vandermaelen, C. P. (1985). Serotonin. In *Neurotransmitter Actions in the Vertebrate Nervous System.* (ed. M. A. Rogawski and J. L. Barker), pp. 201–40. New York: Plenum.

Van Egmond, A. A. J., Groen, J. J. and Jongkees, L. B. W. (1949). The mechanics of the semicircular canal. *J. Physiol.* **110**, 1–17.

Verkleij, A. J., Zwaal, R. F. A., Roelofsen, B., Cumfurius, P., Kastelijn, D. and Von Deenan, L. L. M. (1973). The asymmetric distribution of phospholipids in the human red cell membrane. *Biochim. biophys. Acta* **323**, 178–93.

Verselis, V., White, R. L., Spray, D. C. and Bennett, M. V. L. (1986). Gap junctional conductance and permeability are linearly related. *Science* **234**, 461–4.

Vibert, P. and Craig, R. (1982). Three-dimensional reconstruction of thin filaments decorated with a Ca^{2+}-regulated myosin. *J. Mol. Biol.* **157**, 299–319.

Vibert, P. J., Haselgrove, J. C., Lowy, J. and Poulsen, F. R. (1972). Structural changes in actin-containing filaments in muscle. *Nature New Biol.* **236**, 182–3.

Vizi, E. S. and Vyskocil, F. (1979). Changes in total and quantal release of acetylcholine in the mouse diaphragm during activation and inhibition of membrane ATPase. *J. Physiol.* **286**, 1–14.

Vogt, R. G. and Riddiford, L. M. (1981). Pheromone binding and inactivation by moth antennae. *Nature* **293**, 161–3.

Vogt, R. G., Riddiford, L. M. and Prestwich, G. D. (1985). Kinetic properties of a sex pheromone-degrading enzyme: the sensillar esterase of *Antheraea polyphemus. Proc. Natl. Acad. Sci. USA* **82**, 8827–31.

Von Békésy, G. (1960). *Experiments in Hearing.* New York: McGraw-Hill.

Von Békésy, G. (1962). The gap between the hearing of internal and external sounds. *Symp. Soc. exp. Biol.* **16**, 267–88.

Von Békésy, G. (1964). Concerning the pleasures of observing, and the mechanics of the inner ear. Nobel Lecture 1961. Reprinted in *Nobel Lectures, Physiology or Medicine 1942–1962*, pp. 722–46. Amsterdam: Elsevier for the Nobel Foundation.

Von Békésy, G. and Rosenblith, W. A. (1951). The mechanical properties of the ear. In *Handbook of Experimental Psychology* (ed. S. S. Stevens), pp. 1075–115. New York: John Wiley.

Von Euler, U. S. (1955). *Noradrenaline.* Springfield: Charles C. Thomas.

Von Euler, U. S. and Gaddum, J. H. (1931). An unidentified depressor substance in certain tissue extracts. *J. Physiol.* **72**, 74–87.

Von Frisch, K. (1936). Über den Gehörsinn der Fische. *Biol. Rev.* **11**, 210–46.

Wakabayashi, T., Huxley, H. E., Amos, L. A. and Klug,

A. (1975). Three dimensional image reconstruction of actin-tropomyosin complex and actin-tropomyosin-troponin T-troponin I complex. *J. Mol. Biol.* **93**, 477–97.

Waksman, G., Hamel, E., Fournié-Saluski, M.-C. and Roques, B. P. (1986). Autoradiographic comparison of the distribution of the neutral endopeptidase 'enkephalinase' and of μ and δ opioid receptors in rat brain *Proc. Natl. Acad. Sci. USA* **83**, 1523–7.

Wald, G. (1933). Vitamin A in the retina. *Nature* **132**, 316–17.

Wald, G. (1934). Carotenoids and the vitamin A cycle in vision. *Nature* **134**, 65.

Wald, G. (1965). Visual excitation and blood clotting. *Science* **150**, 1028–30.

Wald, G. (1968). Molecular basis of visual excitation. *Science* **162**, 230–9.

Wald, G., Brown, P. K. and Gibbons, I. R. (1962). Visual excitation: a chemo-anatomical study. *Symp. Soc. exp. Biol.* **16**, 32–57.

Walls, G. L. (1942). *The Vertebrate Eye and its Adaptive Radiation.* Michigan: Cranbrook Institute of Science.

Wang, K. (1984). Cytoskeletal matrix in striated muscle: the role of titin, nebulin and intermediate filaments. In *Contractile Mechanisms in Muscle* (ed. G. H. Pollack and H. Sugi), pp. 285–302. New York: Plenum Press.

Wang, K., McClure, J. and Tu, A. (1979). Titin: major myofibrillar components of striated muscle. *Proc. Natl. Acad. Sci. USA* **76**, 3698–702.

Wässle, H., Boycott, B. B. and Illing, R.-B. (1981*a*) Morphology and mosaic of on- and off-beta cells in the cat retina and some functional considerations. *Proc. R. Soc. Lond.* B **212**, 177–95.

Wässle, H., Peichl, L. and Boycott, B. B. (1981*b*) Morphology and topography of on- and off-alpha cells in the cat retina. *Proc. R. Soc. Lond.* B **212**, 157–75.

Watanabe, A. and Grundfest, H. (1961). Impulse propagation at the septal and commisural junctions of crayfish lateral giant axons. *J. gen. Physiol.* **45**, 267–308.

Watkins, J. C. and Evans, R. H. (1981). Excitatory amino acid transmitters. *Ann. Rev. Pharmacol. Toxicol.* **21**, 165–204.

Watson, J. D., Tooze, J. and Kurtz, D. T. (1983). *Recombinant DNA: a Short Course.* New York: Scientific American Books.

Waxman, S. G. and Bennett, M. V. L. (1972). Relative conduction velocities of small myelinated and non-myelinated fibres in the central nervous system. *Nature New Biol.* **238**, 217–19.

Waxman, S. G., Pappas, G. D. and Bennett, M. V. L. (1972). Morphological correlates of functional differentiation of nodes of Ranvier along single fibers in the neurogenic electric organ of the knife fish *Sternarchus. J. Cell Biol.* **53**, 210–24.

Webb, M. R. and Trentham, D. R. (1983). Chemical mechanism of myosin-catalysed ATP hydrolysis. In *Handbook of physiology*, section 10, *Skeletal Muscle* (ed. L. D. Peachey), pp. 237–55. Bethesda, Maryland: American Physiological Society.

Weber, A. and Herz, R. (1963). The binding of calcium to actomyosin systems in relation to their biological activity. *J. biol. Chem.* **238**, 599–605.

Weber, E., Esch, F. S., Böhlen, P., Paterson, S., Corbett, A. D., McNight, A. T., Kosterlitz, H. W., Barchas, J. D. and Evans, C. J. (1983). Metorphamide: isolation, structure and biological activity of an amidated opioid octapeptide from mammalian brain. *Proc. Natl. Acad. Sci. USA* **80**, 7362–6.

Weber, E. H. (1846). Der Tastsinn und das Gemeingefühl. *Handwörtenbuch d. Physiologie* **3**, no. 2, 481–588.

Weeds, A. G. and Lowey, S. (1971). Substructure of the myosin molecule II. The light chains of myosin. *J. Mol. Biol.* **61**, 701–25.

Weidmann, S. (1951). Effect of current flow on the membrane potential of cardiac muscle. *J. Physiol.* **115**, 227–36.

Weight, F. F. and Votova, J. (1970). Slow synaptic excitation in sympathetic ganglion cells: evidence for synaptic inactivation of potassium conductance. *Science* **170**, 755–8.

Weill, C. L., McNamee, M. G. and Karlin, A. (1974). Affinity-labelling of purified acetylcholine receptor from *Torpedo californica. Biochem. Biophys. Res. Comm.* **61**, 997–1003.

Wells, C. and Bagshaw, C. R. (1985) Calcium regulation of molluscan myosin ATPase in the absence of actin. *Nature* **313**, 696–7.

Werblin, F. S. and Dowling, J. E. (1969). Organisation of the retina of the mudpuppy, *Necturus maculosus.* II. Intracellular recording. *J. Neurophysiol.* **32**, 339–55.

Werman, R. (1966). Criteria for identification of a central nervous system transmitter. *Comp. Biochem. Physiol.* **18**, 745–66.

Werman, R. and Aprison, M. H. (1968). Glycine: the search for a spinal cord inhibitory transmitter. In *Structure and Function of Inhibitory Neuronal Mechanisms* (ed. C. von Euler, S. Skoglund and U. Söderberg). Oxford: Pergamon.

Wernig, A. (1975). Estimates of statistical release parameters from crayfish and frog neuromuscular junctions. *J. Physiol.* **244**, 207–21.

Wersäll, J., Flock, Å. and Lundquist, P.-G. (1965). Structural basis for directional sensitivity in cochlear and vestibular sensory receptors. *Cold Spr. Harb. Symp. Quant. Biol.* **30**, 115–32.

West, R. E. and Miller, R. J. (1983). Opiates, second messengers and cell response. *Brit. Med. Bull.* **39**, 53–8.

Westfall, D. P., Hogaboom, G. K., Colby, J., O'Donnell, J. P. and Fedan, J. S. (1982). Direct evidence against a role of ATP as the nonadrenergic, noncholinergic inhibitory neurotransmitter in guinea pig taenia coli. *Proc. Natl. Acad. Sci. USA* **79**, 7041–5.

Wheeler, G. L., Matuo, Y. and Bitensky, M. W. (1977). Light-activated GTPase in vertebrate photoreceptors. *Nature* **269**, 822–4.

Whitby, L. G., Axelrod, J. and Weil-Malherbe, H. (1961). The fate of H^3-norepinephrine in animals. *J. Pharmacol. exp. Ther.* **132**, 193–201.

White, D. C. S. (1983). The elasticity of relaxed insect fibrillar flight muscle. *J. Physiol.* **343**, 31–57.

White, M. M. and Bezanilla, F. (1985). Activation of squid axon K channels. *J. gen. Physiol.* **85**, 539–54.

White, M. M., Mayne, K. M., Lester, H. A. and Davidson, N. (1985). Mouse *Torpedo* hybrid acetylcholine receptors: functional homology does not equal sequence homology. *Proc. Natl. Acad. Sci. USA* **82**, 4852–6.

Whittaker, V. P. (1984). The structure and function of cholinergic synaptic vesicles. *Biochem. Soc. Transactions* **12**, 561–76.

Whittaker, V. P. and Gray, E. G. (1962). The synapse: biology and morphology. *Brit. med. Bull.* **18**, 223–8.

Whittaker, V. P., Michaelson, J. A. and Kirkland, R. S. (1964). The separation of synaptic vesicles from nerve-ending particles ('synaptosomes'). *Biochem. J.* **90**, 293.

Whittaker, V. P. and Zimmerman, H. (1974). Biochemical studies on cholinergic synaptic vesicles. In *Synaptic Transmission and Neuronal Interaction* (ed. M. V. L. Bennett), pp. 217–38. New York: Raven Press.

Wier, W. G. (1980). Calcium transients during excitation-contraction coupling in mammalian heart: aequorin signals of canine Purkinje fibres. *Science* **207**, 1085–7.

Wilkie, D. R. (1968). Heat work and phosphorylcreatine breakdown in muscle. *J. Physiol.* **195**, 157–83.

Wilkins, M. H. F., Blaurock, A. E. and Engelman, D. M. (1971). Bilayer structure in membranes. *Nature New Biol.* **230**, 72–6.

Williams, T. L. (1976). The location of inward rectification in skeletal muscle. *J. Physiol.* **256**, 125–6*P*.

Wilson, H. R. (1966). *Diffraction of X-rays by Proteins, Nucleic Acids and Viruses.* London: Arnold.

Winklemann, D. A. and Lowey, S. (1986) Probing myosin head structure with monoclonal antibodies. *J. Mol. Biol.* **188**, 595–612.

Wnuk, W., Cox, J. A. and Stein, E. A. (1982). Parvalbumins and other soluble high-affinity calcium-binding proteins from muscle. In *Calcium and Cell Function* (ed. W. Y. Cheung). vol. 2, pp. 243–78.

Wolbarsht, M. L. (1965). Receptor sites in insect chemoreceptors. *Cold Spr. Harb. Symp. Quant. Biol.* **30**, 281–8.

Wolbarsht, M. L. and Dethier, V. G. (1958). Electrical activity in the chemoreceptors of the blowfly. I. Responses to chemical and mechanical stimuli. *J. gen. Physiol.* **42**, 393–412.

Woledge, R. C. (1961). The thermoelastic effect of change of tension in active muscle. *J. Physiol.* **155**, 187–208.

Woledge, R. C. (1963). Heat production and energy liberation in the early part of a muscular contraction. *J. Physiol.* **166**, 211–24.

Woledge, R. C. (1971). Heat production and chemical change in muscle. *Prog. Biophys.* **21**, 37–74.

Woledge, R. C. (1973). In vitro calorimetric studies relating to the interpretation of muscle heat experiments. *Cold Spr. Harb. Symp. Quant. Biol.* **37**, 629–34.

Woledge, R. C., Curtin, N. A. and Homsher, E. (1985). *Energetic Aspects of Muscular Contraction.* London: Academic Press.

Wood, D. S., Zollman, J., Reuben, J. P. and Brandt, P. W. (1975) Human skeletal muscle: properties of the 'chemically skinned' fibre. *Science* **187**, 1075–6.

Woodhead, J. L. and Lowey, S. (1983) An *in vitro* study of the interactions of skeletal muscle M-protein and creatine kinase with myosin and its subfragments. *J. Mol. Biol.* **168**, 831–46.

Woodhull, A. M. (1973). Ionic blockage of sodium channels in nerve. *J. gen. Physiol.* **61**, 687–708.

Woodruff, M. L. and Bownds, M. D. (1979) Amplitude, kinetics and reversibility of a light-induced decrease in guanosine 3′, 5′-cyclic monophosphate in frog photoreceptor membranes. *J. gen. Physiol.* **73**, 629–53.

Worthington, C. R. (1959). Large axial spacings and striated muscle. *J. Mol. Biol.* **1**, 398–401.

Wray, J. S. (1979). Filament geometry and the activation of insect flight muscles. *Nature* **280**, 325–6.

Wright, W. D. (1967) *The Rays are not Coloured.* London: Adam Hilger.

Wu, C. H. (1984). Electric fish and the discovery of animal electricity. *Am. Sci.* **72**, 598–607.

Yagi, N. and Matsubara, I. (1980) Myosin heads do not move on activation in highly stretched vertebrate striated muscle. *Science* **207**, 307–8.

Yagi, N., O'Brien, E. J. and Matsubara, I. (1981) Changes of thick filament structure during contraction of frog striated muscle. *Biophys. J.* **33**, 121–38.

Yamaguchi, N., De Champlain, J. and Nadeau, R. A. (1977). Regulation of norepinephrine release from cardiac sympathetic fibres in the dog by presynaptic α- and β-receptors *Circ. Res.* **41**, 108–17.

Yamamoto, D., Yeh, J. Z. and Narahashi, T. (1984). Voltage-dependent calcium block of normal and tetramethin-modified single sodium channels. *Biophys. J.* **45**, 337–43.

Yamazaki, A., Stein, P. J., Chernoff, N. and Bitensky, M. W. (1983). Activation mechanism of rod outer segment cyclic GMP phosphodiesterase. Release of inhibitor by the GTP/GDP-binding protein. *J. Biol. Chem.* **258**, 8188–94.

Yanagisawa, M., Hamada, Y., Katsuragawa, Y., Imamura, M., Mikawa, T. and Misaki, T. (1987). Complete primary structure of vertebrate smooth-muscle myosin heavy-chain deduced from its complementary-DNA sequence. Implications on topography and function of myosin. *J. Mol. Biol.* **198**, 143–57.

Yarden, Y., Rodriguez, H., Wong, S. K.-F., Brandt, D. R., May, D. C., Burnier, J., Harkins, R. N., Chen, E. Y., Ramachandran, J., Ullrich, A. and Ross, E. M. (1986). The avian β-adrenergic receptor: primary structure and membrane topology. *Proc. Natl. Acad. Sci. USA* **83**, 6795–9.

Yatani, A., Codina, J., Brown, A. M. and Birnbaumer, L. (1987). Direct activation of mammalian atrial muscarinic potassium channels by GTP regulatory protein G_k. *Science* **235**, 207–11.

Yau, K.-W., Haynes, L. W. and Nakatani, K. (1986).

Roles of calcium and cyclic GMP in visual transduction. In *Membrane Control of Cellular Activity* (ed. H. C. Lüttgau), pp. 343–66. Stuttgart: Gustav Fischer Verlag.

Yau, K.-W. and Nakatani, K. (1984*a*). Cation selectivity of light-sensitive conductance in retinal rods. *Nature* **309**, 352–4.

Yau, K.-W. and Nakatani, K. (1984*b*). Electrogenic Na-Ca exchange in retinal rod outer segment. *Nature* **311**, 661–3.

Yau, K.-W. and Nakatani, K. (1985). Light-suppressible, cyclic GMP-sensitive conductance in the plasma membrane of a truncated rod outer segment. *Nature* **317**, 252–5.

Yee, R. and Liebman, P. A. (1978). Light-activated phosphodiesterase of the rod outer segment. *J. Biol. Chem.* **253**, 8902–9.

Yoshikami, S. and Hagins, W. A. (1973). Control of the dark current in vertebrate rods and cones. In *Biochemistry and Physiology of Visual Pigments* (ed. H. Langer), pp. 245–56. Berlin: Springer.

Young, E. F., Ralston, E., Blake, J., Ramachandran, J., Hall, Z. W. and Stroud, R. M. (1985). Topological mapping of acetylcholine receptor: evidence for a model with five transmembrane segments and a cytoplasmic COOH-terminal peptide. *Proc. Natl. Acad. Sci. USA* **82**, 626–30.

Young, J. Z. (1936). The giant nerve fibres and epistellar body of cephalopods. *Q. J. Microsc. Sci.* **78**, 367.

Young, S. (1973). *Electronics in the Life Sciences.* London: Macmillan.

Young, T. (1802). On the theory of light and colours. *Phil. Trans. R. Soc. Lond.* **92**, 12–48.

Zimmerman, A. L. and Baylor, D. A. (1986) Cyclic GMP-sensitive conductance of retinal rods consists of aqueous pores. *Nature* **321**, 70–2.

Zuckerman, R. (1973). Ionic analysis of photoreceptor membrane currents. *J. Physiol.* **235**, 333–54.

Zukin, R. S. and Zukin, S. R. (1981). Multiple opiate receptors: emerging concepts. *Life Sci.* **29**, 2681–90.

Index

Page numbers in italics refer to diagrams